Design of Alloy Metals for Low-Mass Structures

Design of Alloy Metals for Low-Mass Structures

Special Issue Editors

Laszlo S. Toth

Sabine Denis

MDPI • Basel • Beijing • Wuhan • Barcelona • Belgrade • Manchester • Tokyo • Cluj • Tianjin

Special Issue Editors
Laszlo S. Toth
Lorraine University
France

Sabine Denis
Université de Lorraine Nancy
France

Editorial Office
MDPI
St. Alban-Anlage 66
4052 Basel, Switzerland

This is a reprint of articles from the Special Issue published online in the open access journal *Materials* (ISSN 1996-1944) (available at: https://www.mdpi.com/journal/materials/special_issues/design_of_alloy_metals_for_low_mass_structures).

For citation purposes, cite each article independently as indicated on the article page online and as indicated below:

LastName, A.A.; LastName, B.B.; LastName, C.C. Article Title. *Journal Name* **Year**, *Article Number*, Page Range.

ISBN 978-3-03936-158-8 (Pbk)
ISBN 978-3-03936-159-5 (PDF)

Contents

About the Special Issue Editors

Laszlo S. Toth has been an internationally known figure in the mechanics and materials scientific community since about 1990. He earned his high recognition through his broad knowledge in both mechanics and materials, which permits him to bridge these two disciplines. The recent expertise of Prof. Toth is in the field of experiments and mechanical modeling of textures and microstructures of nano-structured metals obtained by SPD, especially on quantitative modeling of grain refinement. Prof. Toth has published 220 refereed papers; his h factor is 43. He founded the 'LEM3' laboratory in 2010, involving 160 persons, and then the Laboratory of Excellence 'Labex DAMAS' in metallurgy in 2012.

Sabine Denis is internationally recognized for her expertise in couplings between temperature phase transformations and internal stresses in metallic alloys as well as for her work in modeling, numerical simulation, and experimental validations at different scales. She is the director of the SI2M department ("Science and Engineering of Materials and Metallurgy") of the Jean Lamour Institute which cofounded 'Labex DAMAS' in 2012, and she is the deputy director. She is also a member of the board of the National Metallurgy Network of France (RNM).

Preface to "Design of Alloy Metals for Low-Mass Structures"

Metals will always be indispensable in many technical applications due to their high load capacity and, especially, their excellent performance at high temperatures. Today's economical and societal challenges are also demanding lighter but more resistant metals that are needed for sustainable development. The design of alloy metals for low-mass structures is a major metallurgical challenge demanding high-level metallurgical expertise.

Lightening of metal structures can be achieved in three ways:

i. Developing new, lighter alloys;

ii. Improving the yield strength of existing alloys;

iii. Better/new design of processes and microstructures.

The Laboratory of Excellence 'DAMAS' (Labex "Design of Alloy Metals for low-mAss Structures") was created in 2012 at the University of Lorraine for cutting-edge research in metallurgy in its three research domains: physical, chemical, and mechanical metallurgy, in the framework of the 'Investment in the future' action of the French Government. A collection of selected articles of the Labex DAMAS is presented in this book, which addresses most of the questions of modern metallurgy:

Five articles deal with the metallurgy of titanium alloys. They deal with solidification, anisotropic plastic flow, nano-indentation, incompatibility stresses, and the effects of texture and grain size on tensile loading behavior.

Among the three articles on Mg alloys, one examines the effects of deformation twinning in AZ31 on the compression behavior via a new micromechanical model. Another focuses on the effect of strain heterogeneities in H-storage capacity of highly deformed Mg. The third paper is devoted to an Al–Cu–Li alloy used in aerospace applications concerning its fracture mechanics.

The five articles on steel and iron are mostly looking into the microstructures of steels: solidification grain structures, precipitation, and granularization behavior, including the modern quench and partitioning steels. One article models the effects of twinning on the texture anisotropy developing during large strain forming of TWIP steels.

Metallurgy by synchrotron is rapidly developing. There are three articles about the use of synchrotron facilities to obtain relevant information on crystal defects, their dynamic behavior in superalloys, and the internal stress states that accompany the phase transformations in metal–matrix composites.

Material architecturing is a new research field. Two articles are presented in this book on this topic: one on shape memory alloys and another on architecturing by additive manufacturing.

Material processing is a key way to obtain microstructures that present high mechanical performance. Three of the six articles on this topic deal with solidification and welding and a new patented process to obtain ultra-fine-grained metals. One article explores the ductility limits of sheet metals considering the recently discovered deviation from Schmid's Law. Two papers present modeling on direct iron ore reduction processes.

Finally, one paper focuses on the elastic properties of the new high-entropy alloys, and another reveals the occurrence of grain boundary sliding in a nano-polycrystalline metal using crystallographic texture.

As can be seen from the above list of topics, many areas of modern metallurgy are present in the 26 articles published in this Special Issue, so we hope that any reader can find his or her interest within this collection.

Laszlo S. Toth, Sabine Denis
Special Issue Editors

Article

A Dislocation-Scale Characterization of the Evolution of Deformation Microstructures around Nanoindentation Imprints in a TiAl Alloy

Antoine Guitton [1,2,*], Hana Kriaa [1,2], Emmanuel Bouzy [1,2], Julien Guyon [1,2] and Nabila Maloufi [1,2]

1 Université de Lorraine, CNRS, Arts et Métiers ParisTech, LEM3, F-57000 Metz, France; hana.kriaa@univ-lorraine.fr (H.K.); emmanuel.bouzy@univ-lorraine.fr (E.B.); julien.guyon@univ-lorraine.fr (J.G.); nabila.maloufi@univ-lorraine.fr (N.M.)

2 Laboratory of Excellence on Design of Alloy Metals for low-mAss Structures (DAMAS)—Université de Lorraine, 57073 Metz, France

* Correspondence: antoine.guitton@univ-lorraine.fr; Tel.: +33-372-747-787

Received: 19 January 2018; Accepted: 12 February 2018; Published: 20 February 2018

Abstract: In this work, plastic deformation was locally introduced at room temperature by nanoindentation on a γ-TiAl-based alloy. Comprehensive analyses of microstructures were performed before and after deformation. In particular, the Burgers vectors, the line directions, and the mechanical twinning systems were studied via accurate electron channeling contrast imaging. Accommodation of the deformation are reported and a scenario is proposed. All features help to explain the poor ductility of the TiAl-based alloys at room temperature.

Keywords: TiAl alloys; dislocation; twinning; nanoindentation; ECCI

1. Introduction

Titanium aluminide alloys have attracted considerable attention due to their unique combination of properties, such as high specific strength and stiffness, good creep properties, and resistance against oxidation and corrosion [1,2], which make them suitable candidate materials for High-Temperature (HT) applications [3,4].

One of the main weaknesses of TiAl alloys is that they are brittle at Room Temperature (RT)—i.e., below their brittle-to-ductile transition temperature, which lies between 800 and 1000 °C [5]. Despite intense research on the HT behavior of TiAl alloys, literature suffers from a lack of understanding of their RT behavior, particularly regarding the elementary deformation mechanisms and the precise role of microstructures [6–8].

Among the several Ti-Al alloy phases, two of them are ordered at RT [4]: γ as the major phase and α_2 as a minor phase. The α_2 phase is hexagonal ($\frac{c}{a} = 0.8$) with a DO19 structure, while the γ phase is tetragonal with a $L1_0$ structure close to cubic ($\frac{c}{a} = \frac{c}{b} = 1.02$). Therefore, six order variants are possible. They can be visualized as generated by a 120° rotation around the $(1\,1\,1)$ plane normal [9].

The microstructures of γ-TiAl alloys are complex. A good compromise for balancing properties between RT plasticity, high strength, and good creep resistance at HT can be obtained for the duplex microstructure. It is constituted of a mixture of monolithic γ grains and small lamellar colonies of γ and α_2 [10,11].

In dual-phase TiAl alloys, plastic deformation mainly occurs on the $\{1\,1\,1\}$ planes of the γ phase by dislocation glide or twinning. It is strongly related to the ordered $L1_0$ structure [12]: along the $\langle\bar{1}\,1\,0]$-directions, there is only one kind of atom (Ti or Al). In this case, dislocations are called ordinary dislocations, and their Burgers vectors are $\frac{1}{2}\langle 1\,1\,0]$ types. Because Ti-atoms and Al-atoms interchange in $\langle 0\,1\,1]$-directions, the $\langle 1\,1\,\bar{2}]$ and the $\langle 1\,0\,1]$ dislocations are called superdislocations. These two types of superdislocations can undergo various dissociations into superpartials (i.e., partial dislocations with the associated planar faults). In addition, true twinning along $\frac{1}{6}\langle 1\,1\,\bar{2}]\{1\,1\,1\}$ occurs that does not alter the ordered $L1_0$ structure of the γ-TiAl. Because of the specific structure of the γ-TiAl, it is relatively easy to know the direction for either slip of ordinary dislocations or for true twinning when the slip/twin plane is known [12]. Note also that at RT twinning and then glide of ordinary dislocations are the easiest deformation modes [2,7,8]. In this manner, Kauffmann et al. suggested that increasing deformation leads to the nucleation of only a few new mechanical twins, since the dislocation movement becomes more dominant with increasing strain [8].

Although it is accepted that the α_2 phase does not contribute to the deformation [6,12], evidence of prismatic slip $\langle 1\,\bar{2}\,1\,0\rangle\{1\,0\,\bar{1}\,0\}$, basal slip $\langle 1\,\bar{2}\,1\,0\rangle(0\,0\,0\,1)$, and pyramidal slip $\langle 1\,1\,\bar{2}\,\bar{6}\rangle\{1\,\bar{2}\,1\,1\}$ was reported [12].

Among the difficulties encountered for understanding the mechanical behavior of TiAl-based alloys, most of our detailed knowledge of their deformation mechanisms has been deduced from Transmission Electron Microscopy (TEM) observations on an electron transparent lamella [7,13]. The investigation presented in this article focuses on the study of deformation mechanisms at the mesoscopic scale. With an original combination of experiments, we investigate the evolution of deformation microstructures at RT in the γ phase of a dual-phase bulk TiAl alloy. Because of the RT brittleness of this material, plastic deformation is induced by nanoindentation. The solid confinement around the indent maintains the integrity of the sample, while applying the load. Note also that nanoindentation is a surface technique, so the stress state at the specimen surface is different from that in the volume. The evolution of the microstructures was characterized by accurate Electron Channeling Contrast Imaging (aECCI) before and after deformation.

2. Materials and Methods

The fully dense Ti–46.8Al–1.7Cr–1.8Nb (at %) sample was obtained in the form of investment cast-bars (diameter 15 mm, height 230 mm) from Howmet. The as-received bars were hot isostatically pressed at 1250 °C and 125 MPa for 4 h, then subjected to a homogenization treatment in a furnace under vacuum at 1270 °C for 24 h [14]. Then, the sample was ground using silicon carbide paper and then polished with a 1 μm diamond suspension. Finally, in order to produce a very flat surface and avoid any work hardening due to conventional grinding, a chemo-mechanical polishing was performed using a colloidal silica suspension.

Because deformation occurs mainly in the γ-phase [5], plastic deformation was locally introduced in the γ phase by nanoindentation using the Ultra Nanoindentation Tester from Anton Paar (Buchs, Switzerland), equipped with a Berkovich indenter. The indents were organized in a regular array of 500 μN indents. For easier recognition, it was surrounded by 20 mN indents at a distance of a few hundred μm.

Detailed characterizations of microstructures before and after deformation were performed by aECCI using a Zeiss Auriga Scanning Electron Microscope (SEM, Oberkochen, Germany) operating at 10 kV. aECCI is a non-destructive method offering the ability to provide—inside a SEM—TEM-like diffraction contrast imaging of sub-surface defects (at a depth of about one hundred of nanometers) on centimetric bulk specimens. Defects such as dislocations can be characterized by applying the TEM extinction criteria [15,16]. Because the yield of BackScattered Electrons (BSE) depends drastically on the orientation of the crystal relative to the incident electron beam (i.e., optic axis of the SEM), obtaining the crystallographic orientation of the grain of interest with an accuracy of 0.1° is a preliminary step to aECCI [16]. The precise orientation of the crystal in the SEM coordinate system is given through Selected Area Channeling Pattern (SACP). To overcome this challenge, rocking the incident electron beam at a pivot point on the surface of a given grain of the sample provides High-Resolution Selected Area Channeling patterns (HR-SACPs) [17]. HR-SACPs cover an angular range of 4.4° and reach an accuracy for the orientation better than 0.1° with a spatial resolution less than 500 nm. Because of this small angular range, in order to obtain the orientation of the grain of interest, the HR-SACP was superimposed on an Electron BackScattered Diffraction (EBSD) pattern simulated at 0° using "Esprit DynamicS" software from Bruker (Billerica, MA, USA). Note that the reason for using an EBSD pattern (acquired at 70°) simulated at 0° is that the specimen is initially placed at 0° for aECCI.

EBSD experiments were carried out on a Zeiss Supra 40 SEM (Oberkochen, Germany) operating at 20 kV. In order to discriminate the different order variants of γ-TiAl, fine EBSD analyses were performed at a step of 75 nm with Channel 5 as the indexation software.

3. Results

3.1. Characterization of the Microstructure around the Regions of Interest

Figure 1a,b show the microstructure around the Regions Of Interest (ROIs): ROI1 on grain A away from any interfaces and ROI2 over both grains A and B. ROI1 and ROI2 are presented in Figures 2 and 3, respectively. Note that References [18,19] mentioned that interfaces play an important role in TiAl alloys, thus controlling the yield stress.

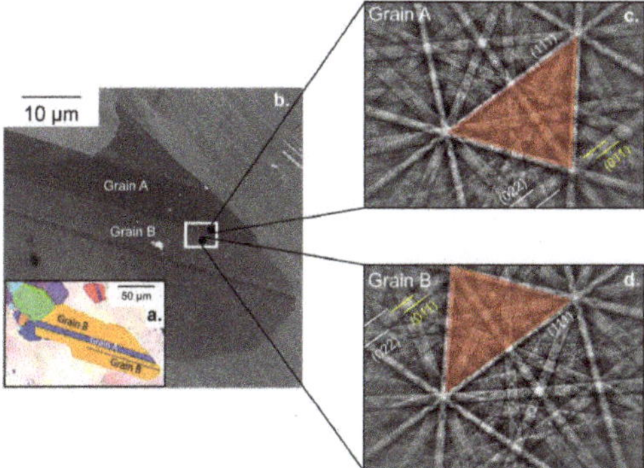

Figure 1. (**a**) Electron backscattered diffraction (EBSD) orientation map of the zone of interest. (**b**) Enhanced BSE image showing the microstructure before deformation. The nanoindentation array is localized in the white rectangle. (**c**,**d**) EBSD patterns corresponding to grains A and B.

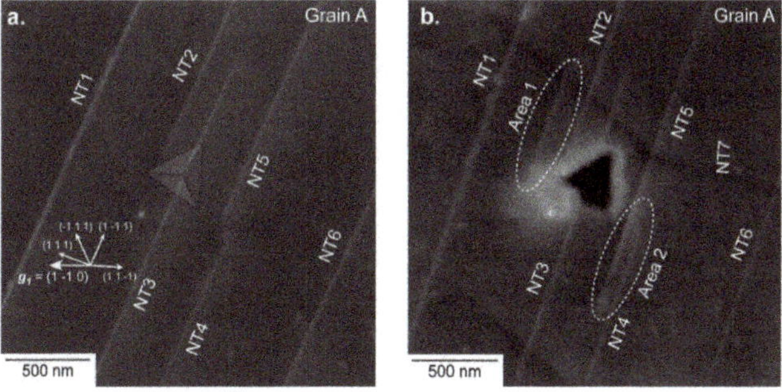

Figure 2. Region of interest 1 (ROI1) for which the surface is close to (4 5 7). (**a**) Accurate electron channeling contrast imaging (aECCI) obtained with $g_1 = (1\,\bar{1}\,0)$ showing six $[1\,1\,\bar{2}](1\,1\,1)$ Nano-Twins (NTs) and the position of the imprint (transparent Berkovich imprint). The white arrows indicate the trace of the $\{111\}$ planes. (**b**) Enhanced BSE image showing the 500 µN indent. Two areas (labelled Areas 1 and 2) have changed. The NT7 slightly visible in (**b**) comes from a neighbor imprint.

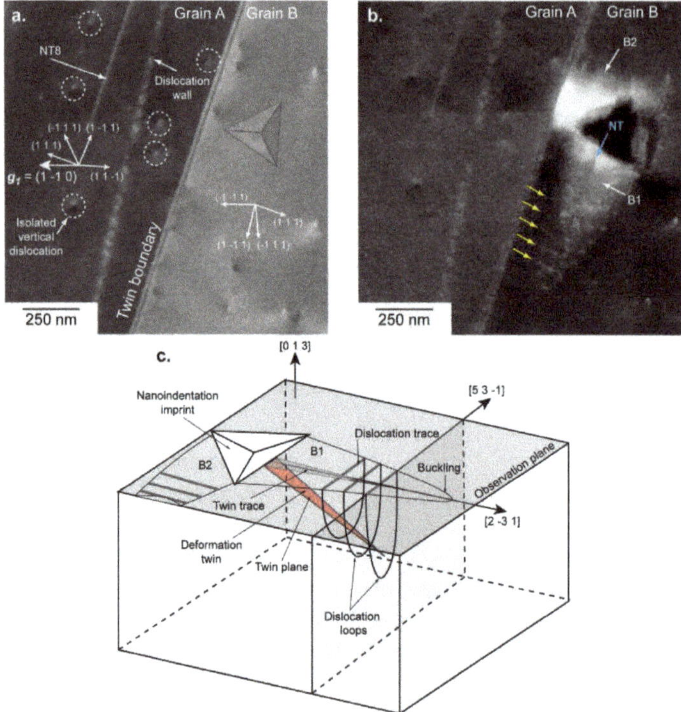

Figure 3. ROI2, where the surface plane is near $(4\,5\,7)$ for twin A (left) and near $(0\,1\,3)$ for grain B. The TB corresponds to $[1\,1\,\overline{2}](1\,1\,1)$ the system. (**a**) aECCI obtained with $g_1 = (1\,\overline{1}\,0)$ with the transparency position of the Berkovich imprint. The white arrows indicate the trace of the $\{1\,1\,1\}$ planes. (**b**) Enhanced BSE image showing two buckling areas (labelled B1 and B2) clearly visible around the 500 μN indent. The blue arrow points to an NT and the yellow to dislocations. (**c**) 3D schematic of B1 and B2.

Experimentally, the twin nature (true or pseudo-twin) was determined using high-resolution spot mode EBSD. Patterns were collected by manually pointing the electron beam at both sides of the Twin Boundary (TB). The corresponding EBSD patterns (Figure 1c,d) clearly indicate that grains A and B are true twin-related: for example, the red triangle formed by the three bands depicted in Figure 1c,d and the $(0\,1\,1)$ superlattice band are in symmetrical position with respect to the unchanged $(1\,1\,1)$ band when going from grain A to grain B.

The evolution of ROI1—before and after deformation—is presented in Figure 2a (ECC image) and Figure 2b (BSE micrograph). Due to a rapid contamination of the sample surface under the electron beam, controlling the channeling conditions after deformation with the required accuracy for aECCI was not possible. However, enhanced BSE images were acquired and gave the necessary information for understanding the evolution of the microstructure already fully characterized before deformation.

EBSD gave $(42\,54\,73) \sim (4\,5\,7)$ as surface plane so that seven channeling conditions or diffracting vectors **g** were accessible by tilting and rotating the specimen: $g_1 = (1\,\overline{1}\,0), g_2 = (1\,1\,\overline{1}), g_3 = (3\,\overline{1}\,\overline{1}), g_4 = (3\,\overline{3}\,1), g_5 = (1\,3\,\overline{3}), g_6 = (1\,\overline{3}\,1), g_7 = (4\,0\,\overline{2})$ (note that only the ECC image taken with g_1 is shown in Figure 2a). In such conditions, all defects are expected to be in contrast. Neither dislocation nor superdislocation are observed before deformation in Figure 2a. Only parallel linear contrasts (labelled NT) are clearly visible. In addition, they are aligned along the $\sim [2\,\overline{3}\,1]$ direction. Such BSE contrast is generally attributed to nano-twins (NTs) and is consistent with $[1\,1\,\overline{2}](1\,1\,1)$ as a true twin system [20,21]. After deformation (see Figure 2b), no dislocation is visible, but clearly identifiable changes are localized in the vicinity of the indent (Area 1 and Area 2 in Figure 2b). In Area 1, near the imprint, a $[1\,1\,\overline{2}](1\,1\,1)$ deformation NT was created. At the other side of the imprint (Area 2) NT5 extends along the $\sim [2\,\overline{3}\,1]$. Note that NT7 visible in Figure 2b comes from a neighbor imprint.

3.2. Microstructure Evolution of ROI2

ROI2 is composed by two twinned grains A (left) and B (right) with their surface plane as (42 54 73)~(4 5 7) and (16 325 946)~(0 1 3) , respectively (see Figure 3a). The common direction on the sample surface for both grains A and B was $[2\,\overline{3}\,1]$. The $\{1\,1\,1\}$-plane, which intercepts both the $(4\,5\,7)$ plane and the $(0\,1\,3)$ plane along $[2\,\overline{3}\,1]$ is the $(1\,1\,1)$. Note also that an NT aligned along $\sim[2\,\overline{3}\,1]$ is visible (labelled NT8 in Figure 3) and consistent with $[1\,1\,\overline{2}](1\,1\,1)$. The vertical dislocations (i.e., almost perpendicular to the sample surface) either isolated or stacked into a wall in grain A (Figure 3a) were analyzed by aECCI in order to determine their Burgers vectors. Using the diffracting conditions g_1 to g_7 previously mentioned with invisibility criteria led to $\pm\frac{1}{2}[1\,1\,0]$ as the Burgers vector.

Unfortunately, good channeling conditions were not reachable in the right $(0\,1\,3)$ grain, resulting in the non-characterization of the isolated vertical dislocations.

Figure 3b and its schematic show ROI2 after deformation. The 500 μN indent was made in the $(0\,1\,3)$ grain near the TB. Around this indent, two similar features (labelled B1 and B2 in Figure 3b) are observed. Parallel to the TB (i.e., in B1), a set of parallel dislocation traces is visible (yellow arrows in Figure 3b). They are localized in an elliptical area forming a buckling (B1) extending far away from the imprint in the $[2\,\overline{3}\,1]$ direction. Such buckling areas were already reported, but not explained for TiAl alloys [18,22].

In addition, an NT contrast (blue arrow in Figure 3b) is observed inside B1, and it is parallel to $[2\,\overline{3}\,1]$, consistent with the $[1\,1\,\overline{2}](1\,1\,1)$ true twinning system.

Perpendicular to the TB (i.e., along $[\overline{5}\,\overline{3}\,1]$), another buckling area B2 is observed, and it could extend because it was blocked by the TB. In the neighbor $(4\,5\,7)$ grain, no change is observed compared to the initial state, even if the TB is distorted locally where B2 was in contact. Outside both buckling areas, no other defect is observed.

4. Discussion

From observations of the evolution of microstructures of ROI1, two assessments can be made:

1. At RT, twinning was observed to be the main deformation mechanism, in agreement with literature [2,7,8]. However, this runs contrary to Zambaldi et al., who prefer to suggest that ordinary dislocation glide is the main deformation mechanism at RT (without totally excluding twinning) from observations by atomic force microscopy around high-load (3000 μN) imprints [18].
2. Deformation was observed to be localized near the indent.

In many materials, buckling areas such as those characterized in ROI2 are associated with a canalization of the deformation, generally taking its origin from the accommodation of twins [23]. Although the accommodation of $\frac{1}{6}\langle1\,1\,\overline{2}]\{1\,1\,1\}$ twin by $\frac{1}{2}\langle1\,1\,0]\{1\,1\,1\}$ ordinary dislocations was already reported by TEM experiments in TiAl alloys [24,25], no mechanism was proposed.

From this knowledge, and taking into account our results, we propose the following scenario (see Figure 3c):

- Under the indent, the $[1\,1\,\overline{2}](1\,1\,1)$ NT was formed.
- The stress concentration at the tip of the $[1\,1\,\overline{2}](1\,1\,1)$ NT nucleated ordinary $\pm\frac{1}{2}[1\,1\,0]$ dislocation loops gliding in the $(1\,\overline{1}\,1)$ planes. The dislocation loops formed an ellipsoid surrounding the NT, thus producing lines after projection on the observation plane.
- The elliptical area or B1 grew by adding successive dislocation loops at its extremity.

- B1 extended until it met an obstacle, such as the TB (for B2 for example).
- At the location where B2 intercepts the TB, a stress concentration appeared. It resulted in a local distortion of the boundary. Therefore, the TB seems to be a strong obstacle to the propagation of the deformation, and at higher load it may cause microcracking at its vicinity, as observed in References [18,25,26].

Furthermore, we can suggest that the low load used (500 μN) was just high enough to generate a complex and non-uniaxial stress field at the tip of the indent. This led to the activation of the main deformation mechanism (i.e., twinning), but it was too low for dislocation glide. For higher loads, both mechanisms were activated subsequently, and led to the formation of a buckling area, according to the previous scenario.

5. Conclusions

In summary, RT nanoindentation tests combined with aECCI observations before and after deformation brought novel insights into the γ-TiAl deformation mechanisms:

1. At RT, twinning was observed to be the main deformation mechanism.
2. Twinning was accommodated by ordinary dislocation mechanism, leading to the canalization of the deformation.
3. TB could play the role of obstacle to the propagation of deformation to neighbor grains, leading to a stress concentration at the vicinity of the boundary. Therefore, the true twin seems to be one of the weak links explaining the poor ductility of γ-TiAl at RT.

Acknowledgments: The authors thank Dr. Nathalie Gey and Dr. Jean-Sébastien Lecomte from the LEM3. This work was supported by the French State through the program "Investment in the future" operated by the National Research Agency (ANR) and referenced by ANR-11-LABX-0008-01 (LabEx DAMAS).

Author Contributions: All experimental observations were performed by Hana Kriaa and Antoine Guitton. Antoine Guitton and Hana Kriaa performed the dislocation analyses. Antoine Guitton wrote the main manuscript. Antoine Guitton, Hana Kriaa, Emmanuel Bouzy, Julien Guyon and Nabila Maloufi participate in the discussion and they reviewed the manuscript.

Conflicts of Interest: The authors declare no conflict of interest.

References

1. Kim, Y.; Dimiduk, D. Progress in the understanding of gamma titanium aluminides. *JOM* **1991**, *43*, 40–47. [CrossRef]
2. Appel, F.; Wagner, R. Microstructure and deformation of two-phase gamma-titanium aluminides. *Mater. Sci. Eng. R Rep.* **1998**, *22*, 187–268. [CrossRef]
3. Loria, E. Quo vadis gamma titanium aluminide. *Intermetallics* **2001**, *9*, 997–1001. [CrossRef]
4. Schuster, J.; Palm, M. Reassessment of the binary aluminum-titanium phase diagram. *J. Phase Equilib. Diff.* **2006**, *27*, 255–277. [CrossRef]
5. Zambaldi, C. *Micromechanical Modeling Gamma-TiAl Based Alloys*; RWTH Aachen University: Aachen, Germany, 2010, ISBN 978-3-8322-9717-6.
6. Appel, F.; Paul, D.; Oehring, M. *Gamma Titanium Aluminide Alloys: Science and Technology*; Wiley-VCH Verlaf GmbJ: Weinheim, Germany, 2011, ISBN 9783527315253.
7. Beran, P.; Heczko, M.; Kruml, T.; Panzner, T.; Van Petegem, S. Complex investigation of deformation twinning in γ-TiAl by TEM and neutron diffraction. *J. Mech. Phys. Sol.* **2016**, *95*, 647–662. [CrossRef]
8. Kauffmann, F.; Bidlingmaier, T.; Dehm, G.; Wanner, A.; Clemens, H. On the origin of acoustic emission during room temperature compressive deformation of a gamma-TiAl based alloy. *Intermetallics* **2000**, *8*, 823–830. [CrossRef]
9. Zambaldi, C.; Zaefferer, C.; Wright, S. Characterization of order domains in γ-TiAl by orientation microscopy based on electron backscatter diffraction. *J. Appl. Crystallogr.* **2009**, *42*, 1092–1101. [CrossRef]

10. Dey, S.; Hazotte, A.; Bouzy, E. Multiscale gamma variant selection in a quaternary near-gamma Ti-Al alloy. *Philos. Mag.* **2006**, *86*, 3089–3112. [CrossRef]
11. Dey, S.; Morawiec, A.; Bouzy, E.; Hazotte, A.; Fundenberger, J.-J. Determination of gamma/gamma interface relationships in a (alpha2 + gamma) TiAl base alloy using TEM Kikuchi patterns obtained by nanoprobe scanning. *Mater. Lett.* **2003**, *60*, 646–650. [CrossRef]
12. Marketz, M.; Fischer, F.; Clemens, H. Deformation mechanisms in TiAl intermetallics—Experiments and modeling. *Int. J. Plast.* **2003**, *19*, 281–321. [CrossRef]
13. Zghal, S.; Coujou, A.; Couret, A. Transmission of the deformation through γ-γ interfaces in a polysynthetically twinned TiAl alloy. *Philos. Mag.* **2001**, *81*, 345–382. [CrossRef]
14. Dey, S.; Hazotte, A.; Bouzy, E.; Naka, S. Development of Widmanstätten laths in a near-gamma TiAl alloy. *Acta Mater.* **2005**, *53*, 3783–3794. [CrossRef]
15. Mansour, H.; Guyon, J.; Crimp, M.; Gey, N.; Beausir, B.; Maloufi, N. Accurate electron channeling contrast analysis of dislocations in fine grained bulk materials. *Scr. Mater.* **2014**, *84*, 11–14. [CrossRef]
16. Kriaa, H.; Guitton, A.; Maloufi, N. Fundamental and experimental aspects of diffraction for characterizing dislocations by electron channeling contrast imaging in scanning electron microscope. *Sci. Rep.* **2017**, *7*, 9742. [CrossRef] [PubMed]
17. Guyon, J.; Mansour, H.; Gey, N.; Crimp, M.; Chalal, S.; Maloufi, N. Sub-micron resolution selected area electron channeling patterns. *Ultromicroscopy* **2015**, *149*, 34–44. [CrossRef] [PubMed]
18. Zambaldi, C.; Raabe, D. Plastic anisotropy of gamma-TiAl revealed by axisymmetric indentation. *Acta Mater.* **2010**, *58*, 3516–3530. [CrossRef]
19. Kad, B.; Asaro, R.J. Apparent Hall-Petch effects in polycrystalline lamellar TiAl. *Philos. Mag. A* **2006**, *75*, 87–104. [CrossRef]
20. Simki, B.; Ng, B.; Crimp, M.; Bieler, T. Crack opening due to deformation twin shear at grain boundaries in near-γ TiAl. *Intermetallics* **2007**, *15*, 55–60. [CrossRef]
21. Ng, B.; Simki, B.; Crimp, M.; Bieler, T. The role of mechanical twinning on microcrack nucleation and crack propagation in a near-γ TiAl alloy. *Intermetallics* **2004**, *12*, 1317–1323. [CrossRef]
22. Gehard, S.; Pyczak, F.; Göken, M. Microstructural and micromechanical characterisation of TiAl alloys using atomic force microscopy and nanoindentation. *Mater. Sci. Eng. A* **2009**, *523*, 235–241. [CrossRef]
23. Hirth, J.P.; Lothe, J. *Theory of Dislocations*, 2nd ed.; Krieger Publishing Company: Malabar, FL, USA, 1982; p. 756, ISBN 0521864364.
24. Gibson, M.; Forwood, C. Slip transfer of deformation twins in duplex γ-based Ti-Al alloys: Part III. Transfer across general large-angle γ-γ grain boundaries. *Philos. Mag. A* **2002**, *82*, 1381–1404. [CrossRef]
25. Simki, B.; Crimp, M.; Bieler, T. A factor to predict microcrack nucleation at gamma-gamma grain boundary in TiAl. *Scr. Mater.* **2003**, *49*, 149–154. [CrossRef]
26. Bieler, T.; Fallahi, A.; Ng, B.; Kumar, D.; Crimp, M.; Simki, B.; Zamiri, A.; Pourboghrat, F.; Mason, D. Fracture initiation/propagation parameters for duplex TiAl grain boundaries based on twinning, slip, crystal orientation and boundary misorientation. *Intermetallics* **2005**, *13*, 979–984. [CrossRef]

Article

Effect of Inoculant Alloy Selection and Particle Size on Efficiency of Isomorphic Inoculation of Ti-Al

Jacob R. Kennedy [1,2,3], Bernard Rouat [1,2], Dominique Daloz [1,2], Emmanuel Bouzy [2,3] and Julien Zollinger [1,2,*]

1 Department of Metallurgy & Materials Science and Engineering, Institut Jean Lamour, Université de Lorraine, Campus ARTEM, Allée André Guinier, F-54011 Nancy, France; jacob-roman.kennedy@univ-lorraine.fr (J.R.K.); bernard.rouat@univ-lorraine.fr (B.R.); dominique.daloz@univ-lorraine.fr (D.D.)
2 Laboratory of Excellence on Design of Alloy Metals for low-mAss Structures (DAMAS), Université de Lorraine, 57073 Metz, France; emmanuel.bouzy@univ-lorraine.fr
3 Université de Lorraine, CNRS, Arts et Métiers ParisTech, LEM3, F-57000 Metz, France
* Correspondence: julien.zollinger@univ-lorraine.fr; Tel.: +33-372-742-669

Received: 29 March 2018; Accepted: 24 April 2018; Published: 25 April 2018

Abstract: The process of isomorphic inoculation relies on precise selection of inoculant alloys for a given system. Three alloys, Ti-10Al-25Nb, Ti-25Al-10Ta, and Ti-47Ta (at %) were selected as potential isomorphic inoculants for a Ti-46Al alloy. The binary Ti-Ta alloy selected was found to be ineffective as an inoculant due to its large density difference with the melt, causing the particles to settle. Both ternary alloys were successfully implemented as isomorphic inoculants that decreased the equiaxed grain size and increased the equiaxed fraction in their ingots. The degree of grain refinement obtained was found to be dependent on the number of particles introduced to the melt. Also, more new grains were formed than particles added to the melt. The grains/particle efficiency varied from greater than one to nearly twenty as the size of the particle increased. This is attributed to the breaking up of particles into smaller particles by dissolution in the melt. For a given particle size, Ti-Al-Ta and Ti-Al-Nb particles were found to have a roughly similar grain/particle efficiency.

Keywords: titanium aluminides; grain refinement; solidification; inoculation

---

1. Introduction

Titanium aluminum alloys are an interesting material for aerospace applications. Their high temperature oxidation resistance, and high specific strength make them particularly promising [1]. Recently, titanium aluminides have replaced nickel based superalloys in the last low pressure stage of the GEnx turbine engine. In order to increase their usefulness in turbines, their properties must be improved for high temperature (>800 °C) applications. Grain refinement is one method to improve the properties of a material, however, the current methods of grain refinement of Ti-Al alloys by inoculation can negatively affect ductility [2]. The Ti-Al alloys of interest for aerospace applications normally fall on the peritectic plateau of the phase diagram at 44–48% Al. The solidification path for alloys in this composition range is complex, and can include multiple phase changes from liquid to room temperature [3]. After solidification the solid state transformations have known orientation relationships, such as: β to α (orientation relationship $(110)_\beta$ // $(0001)_\alpha$ [4]) and α and γ (orientation relationship $(0001)_\alpha$ // $(111)_\gamma$ [5]). Grain refinement can be attempted during any of these phase transformations, however, attempts to grain refine by using the γ massive transformation during heat treatment has been shown to require fast cooling rates, which result in deformations, making the application of the method difficult [6,7]. The location of grain refinement during processing that is most interesting is then the initial solidification of β grains from the liquid.

Inoculation is a method of grain refinement by increasing the number of grains that were formed during solidification [8]. This is achieved by increasing the number of nucleation sites present in the melt for the solid to form from [9]. Powders can be added, which are effective nuclei or alloying additions can be made, which will form precipitates in the melt that act as nuclei [10]. The most common method of inoculating Ti-Al alloys is via boron additions [11]. Relatively small amounts of boron are added to the melt and titanium boride (TiB) or diboride (TiB_2) precipitate and act as nucleants. These precipitates can nucleate α or β phases, depending on alloy composition, and have been found to have orientation relationships with both phases, indicating a good lattice mismatch between the inoculant and nucleated phases [12]. These borides have also been found to be thermodynamically stable in the Ti-Al melt [13]. Boron additions are especially effective at low levels, where they can cause interdendritic nucleation of α phase grains [14]. However, the borides that form may have flake [15] or needle like [11] morphologies in the as-cast state. These morphologies are detrimental to the final mechanical properties of the alloy, and can result in decreased creep resistance and tensile strength, leading to brittle fracture [2]. This illustrates the need for an inoculant in the TiAl system that is more ductile to prevent this embrittlement of the as-cast material.

Nucleation via inoculation is commonly represented by the spherical cap heterogeneous nucleation model [8]. This model is based on a reduction of activation energy when compared to homogenous nucleation that is caused by the presence of solid surfaces that act as nucleation sites. This reduction of activation energy depends on many factors, such as the interfacial energies between the liquid, solid, and nucleus [10], as well as the shape of the solid (i.e., flat, convex, concave) [16]. This model accurately describes less efficient nucleants, however, in the case of inoculants with higher efficiencies, nucleation can be described as atom by atom adsorption onto the solid surface [17] or by particle wetting [18]. Even in these more efficient cases, there is still an energy barrier for nucleation to overcome in order for solidification to progress. A novel solution to this problem was proposed by the authors, called isomorphic self inoculation, where rather than having the inoculants act as centers for nucleation particles are used, which can act instead as centres of growth, bypassing the nucleation step and requisite energy barrier [19]. In this case, as there is no energy barrier to surpass for solidification to progress, each particle that is added can be active in solidification and be the center of a new grain. A similar method of grain refinement was previously investigated by Bermingham et al., for a Ti alloy by pouring the melt over pure Ti powders and into a mold [20].

In order to develop inoculants that function in this manner, four design criteria were proposed: (I) Phase and lattice matching, (II) Thermal and diffusive stability in the melt, (III) Good usability factors, and (IV) No negative effects on final casting. The first two criteria are the phenomenological factors determining if the inoculant can successfully act as an isomorphic inoculant. The final two criteria are related to processing and implementation; if the inoculant is a viable alternative to other inoculants. The critical phenomenological basis for isomorphic self inoculation is phase and lattice matching. By introducing solid particles to the melt which are the same phase as the solidifying alloy with similar lattice parameters the nucleation stage of solidification, and its requisite energy barrier can be circumvented by the direct growth of the particles [19]. Rather than having new solid matrices form on the particles the matrix of the particle itself grow as the melt solidifies epitaxially upon it. The second phenomenological criteria for the design of an isomorphic inoculant is to ensure stability in the melt. Particles that match the phase and lattice of the bulk must survive long enough in the melt to affect solidification in order to be viable isomorphic inoculants. The stability of the particles in the melt does not have to be absolute. Some dissolution or melting is acceptable if the particles can survive past thermal equilibrium with the melt and participate in solidification. The first non-phenomenological design criteria for an isomorphic inoculant is its good usability. This refers to the ease of which the inoculant can be produced and implemented, as well as its behaviour in the melt. These factors may or may not fundamentally affect whether the inoculant can function as designed, but rather affect if it can realistically grain refine a casting. This includes such factors as density of the particles causing them to float or to settle in the melt, the ability of the particles to be easily manufactured, whether the procedure of adding the particles to the melt is difficult and any other factors that affect the actual usage of the particles. The final criteria is that the inoculants do not negatively affect the final properties of the material. It is important to look at the elaboration process as a whole and to ensure that the isomorphic inoculants do not negatively affect either the casting or the final part. This is important for Ti-Al alloys that are normally used in high temperature applications where even if the as cast part may have ideal grain size with no precipitates solid state transformations may occur causing phases, such as σ in the Ti-Al-Ta system [21] or ω in the Ti-Al-Nb system [22], to form which are detrimental to mechanical properties. If any particles do remain heterogeneous in the matrix they must not be too brittle to negatively affect the ductility.

2. Inoculant Selection

In order to determine if the system is suitable and if a suitable composition exists within it for an isomorphic inoculant, each of the four factors above were evaluated. In order to do so, available information on the alloys was required on the thermodynamic description of the system, diffusivity of the alloying elements in β-Ti, the lattice parameter of the alloy, as well as their densities. The thermodynamic description is necessary to ensure that the particles are thermally stable in the melt, as well as exist in the matched phase field with the solidifying bulk phase at the melting temperature of the bulk alloy. The diffusivity of the alloying elements in the inoculant are important for their diffusive stability in the melt. The lattice parameter of the alloy is necessary to check the lattice mismatch with the bulk alloy. The alloy density at high temperature is important for usability issues, as well as at room temperature for calculating the number of inoculants that are introduced during casting trials. In these trials, a base alloy of Ti-46Al (at %) was used. From the phase diagram, the solidification phase for this alloy can be seen to be β-Ti with a melting temperature of ~1540 °C [3]. This means that the inoculant alloys must have melting temperatures that are greater than 1540 °C and be in the β-Ti phase field of their systems at high temperature. In addition to the minimum melting point, the base alloy selection also effects how quickly the inoculants will dissolve in the melt. In order to evaluate the survivability of the inoculants, the diffusion species in the particles at high temperature were considered. Assuming that the particles and the particle/melt interface reach thermal equilibrium with the melt their dissolution should be controlled by the slowest diffusing species out of the particles and into the melt, in order to maintain the conservation of mass across the interface. Slower diffusion rates should correspond with slower diffusion of the inoculant particles and better survivability in the melt.

An equation has been fit to alloys in the Ti-Al-Nb-Ta system to find their lattice parameter at high temperature [23]. The composition of the alloy in atomic percent (X_i) and the temperature in degrees Celsius (T) are taken as inputs and the lattice parameter is the output (α_β) in nm, as can be seen below:

$$\alpha_\beta = \alpha_\beta^{Ti(Pure)} + \sum_{j=1}^{k} e_i X_i + b \quad (1)$$

where $\alpha_\beta{}^{Ti(Pure)}$ is the beta Ti lattice parameter at 0 °C in nm. Using reported lattice parameters for pure Ti at different temperatures, first, the equation could be solved for b, and then each coefficient, respectively, using reported lattice parameters at different compositions. To find the lattice parameter in nm with component compositions in atomic fractions (X_i) at a given temperature in °C (T), the equation was determined to be [23]:

$$\alpha_\beta = 0.328 - 7.4 \times 10^{-5}(X_{Al}) + 1.23 \times 10^{-4}(X_{Nb}) + 4.6 \times 10^{-4}(X_{Ta}) + 7.4 \times 10^{-5}(T) \quad (2)$$

The density of the alloys at high temperature is important as the inoculants should have densities as close as possible to the bulk alloy. If the density of the inoculant is significantly far from that of the bulk alloy problems may arise with segregation of the particles in the melt. A perfect match in densities is not critical in these casting trials as an induction furnace was used, which stirs the melt, helping to avoid segregation. The density of each alloy at high temperature was calculated using the molar mass of the constituent elements and the calculated lattice parameter. Since β-Ti is a Body Centered Cubic (BCC) structure a unit cell contains two atoms. The mass of the unit cell of the alloy (m^a) can then be calculated using the molar mass of each element (M_i) and its amount in the alloy by atomic fraction (X_i) and Avogadro's number (N). Since a BCC structure is cubic the volume of a unit cell is simply the cube of the lattice parameter at any temperature (a_β), and the density of an alloy at a given temperature (ρ_β) can be calculated with:

$$\rho_\beta = \frac{m}{V} = \frac{2\sum M_i X_i}{N(a_\beta)^3} \tag{3}$$

Two refractory metal additions were proposed to form inoculant alloys that would function as isomorphic inoculants, Nb and Ta, their liquidus surface diagrams are shown in Figure 1a,b, respectively. Both of the ternary systems have large β solidifying fields with melting points above the bulk alloy. A plot of the diffusivity of common refractory metals in β-Ti is shown in Figure 1c, where Ta and Nb are shown to diffuse slowly at high temperature when compared to other refractory metals [24].

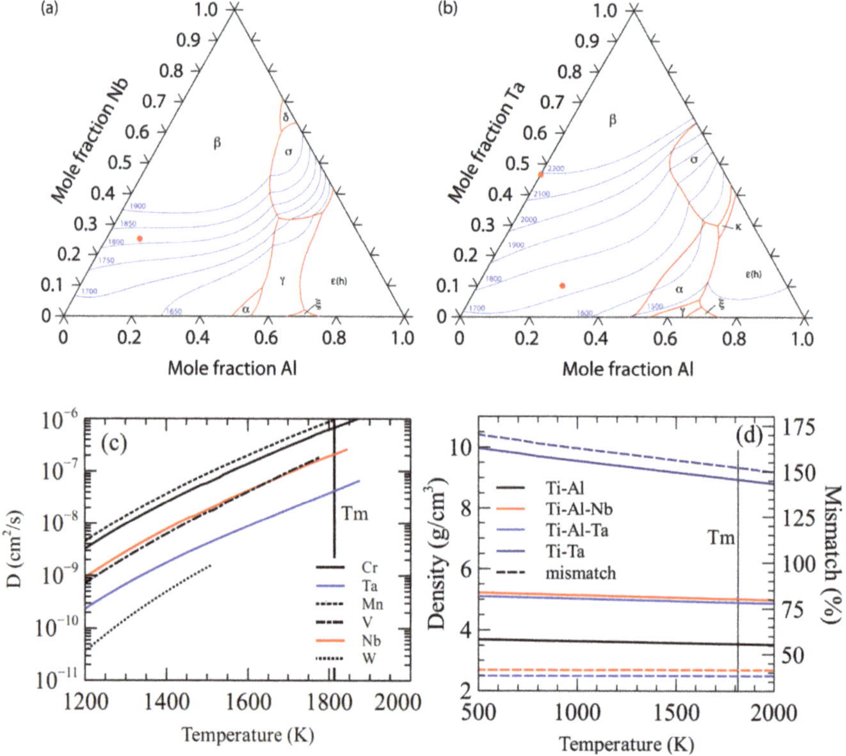

Figure 1. (**a**) Liquidus surface map for the Ti-Al-Nb System (data from ref. [22]), (**b**) Liquidus surface map for the Ti-Al-Ta system (data from ref. [21]) (**c**) Calculated diffusion rates of some refractory metals in high temperature β-Ti (data from ref. [24]), and (**d**) calculated densities (solid lines) and difference from bulk (dashed lines) with temperature.

The first system that was investigated for potential isomorphic self-inoculants was the ternary Ti-Al-Nb. Nb is a beta stabilizing element in the Ti system [25] with a relatively high melting point of 2415 °C [26]. A balance between lattice mismatch, stability, and density was sought to find an ideal inoculant. An inoculant alloy of Ti-10Al-25Nb was selected. It exists as the β-phase at high temperature, has a high melting point (1800 °C [27]), has a small lattice mismatch with the base alloy (<2%), and the Nb content should provide some diffusive stability. The density of the alloy is greater than that of the bulk by roughly 40%, however, the thermal and diffusive stability, along with the low lattice mismatch, make it a good candidate for an isomorphic inoculant.

The second system that was proposed for an isomorphic self-inoculant was Ti-Al-Ta. Ta has a higher melting point (3000 °C [26]), but it is denser than Nb. Again, a compromise was sought between stability, density, and lattice mismatch. The proposed alloy of Ti-25Al-10Ta was selected as a suitable inoculant for Ti-46Al. It has a small lattice mismatch (<2%) with the base alloy, a higher melting point (1725 °C), and it exists as β-Ti at high temperature. The relatively slow diffusion of Ta in β-Ti should provide the inoculant with good diffusive stability.

A third inoculant system was proposed to maximize both the thermal and diffusive stability at the expense of lattice and density mismatch. The alloy that is needed to maintain a relatively small lattice mismatch, but the maximization of stability was given greater priority than minimization of mismatch. To this end, the binary Ti-Ta system was investigated and Ti-47Ta selected, as this was the maximum Ta content possible without risking formation of the σ phase after interaction with the Ti-46Al bulk alloy. Ti-47Ta has a higher lattice mismatch (<5%) that the other inoculant alloys, but it remains low enough to be a suitable inoculant. Its high melting point (2200 °C) and the relatively slow diffusion of Ta should make it the most stable of the inoculants that was proposed for the base metal. The density of the alloy is much greater than the base alloy so the usability of the inoculant may be decreased and its behavior in the melt may be unideal, a comparison of the alloy densities is shown in Figure 1d, and a summary of the inoculant alloy properties is given in Table 1.

Table 1. Summary of inoculant alloy properties.

Alloy	α (nm) (PW)	T_m (°C) [27]	ρ (g/cm³) (PW)		Slowest Diffusing Species	D_β-Tracer (cm²/s) [24]	D_β Interdiffusion (cm²/s)
			25 °C	1540 °C			
Ti-46Al	0.330	1540	3.71	3.55			
Ti-10Al-25Nb	0.335	1800	5.25	5.02	Nb	1.40×10^{-8}	1.02×10^{-8} [28]
Ti-25Al-10Ta	0.336	1725	5.13	4.91	Ta	5.63×10^{-9}	1.22×10^{-8} [29]
Ti-47Ta	0.345	2200	10.14	8.95	Ta	5.63×10^{-9}	2.34×10^{-10} [29]

3. Materials and Methods

Bulk inoculant alloys were fabricated from commercially pure elements in an induction heated cold crucible apparatus. Before melting, the chamber was pumped down to a vacuum of 10^{-3} mbar, and then an Ar flux was applied, allowing for processing to occur at atmospheric pressure without the risk of oxidation. The alloying element with the highest melting point was used as the base for the alloying with other elements added consecutively in order of their melting point from highest to lowest. Once bulk alloys had been fabricated, it was necessary to turn them into powders that could be used for inoculation. This was achieved by drilling into the ingots to produce chips, the majority of which were in the order of mm, too large to be implemented in our laboratory apparatus. In order to reduce their size the powders were cryomilled in a Retsch CryoMill (Retsch–Verder group, Haan, Germany). Cryomilling was chosen as the low temperature would embrittle the particles, as well as prevent them from agglomerating or sticking. Drillings were added to the milling container along with steel balls with a ratio of one steel ball for every gram of powder. The apparatus ran liquid nitrogen around the milling container to maintain a temperature that is near −196 °C. To obtain particles of different sizes drillings of each alloy were milled for different lengths of time, ranging from 1.5 to 11 h. Initially, powders were milled in air, however to minimize the potential effects of oxidation the Ti-Ta powders were milled under Ar, and a further sample of Ti-Al-Nb was milled in Ar as well to compare with those milled in air. Once the powders had been prepared, it was necessary to characterize them. The size of the powders was important to know since it is of critical importance for diffusive stability. Additionally, in order to calculate the number of particles that were added in an inoculation trial, the size distribution must be known. The size distributions were determined by image analysis of the particles taken by Scanning Electronic Microscopy (SEM) (FEI–Thermo Fisher Scientific, Hillsboro, OR, USA). The particle diameters that are reported are from equivalent circular areas from the projected two-dimensional (2D) areas of the particles. The resultant D50 and spans ([D90-D10]/D50) of the distributions are shown in Table 2.

Table 2. Inoculation Trial Conditions.

	Cryo-Milling		Particle Distribution Parameters		
Alloy	**Time (h)**	**Atmosphere**	**D50 (μm)**	**Span**	**Grain Refinement**
Ti-Al-Nb	3	Ar	233	1.56	Yes
	3	Air	191	1.72	Yes
	6	Air	163	1.50	Yes
	9	Air	113	1.67	Yes
	11	Air	48	2.10	Yes
Ti-Al-Ta	3	Air	56	2.03	Yes
	9	Air	54	1.79	Yes
Ti-Ta	1.5	Ar	70	1.62	No
	3	Ar	35	1.10	No
	6	Ar	22	1.17	No
	9	Ar	20	1.26	No

In order to test the ability of the alloys as inoculants, ingots (~40 g) of TiAl were produced both with and without inoculation, and their grain sizes were compared. To inoculate the samples the previously produced powders were formed into pellets by cold pressing the inoculant particles and with an equal amount of aluminum powder (1 g each). These pellets were held by vacuum on a small quartz tube above the Ti-Al alloy. Once the Ti-Al was fully melted the vacuum was released and the pellet fell into the molten alloy. The inoculants remained in the melt for 20 s before the furnace was turned off and the alloy was allowed to solidify. This time period corresponds to the time for the pellet to disperse into the melt and the molten droplet to return to a regular shape within the induction field. The ingots were then bisected, polished, and etched with Krolls reagent. In addition to the inoculation trials, an ingot was produced in the same manner without any inoculants to determine the reference conditions. A further ingot was also produced by increasing the interaction time with a Ti-Al-Nb distribution and remelting after the inoculation three times in order to dissolve all of the inoculants to determine the effect of the Nb alloy addition on the conditions, independently of the inoculation.

4. Results

Example optical images of the ingot cross sections and SEM Back-Scattered Electrons (BSE) images of the equiaxed microstructure for a non-inoculated ingot and ingots that were inoculated with each alloy are shown in Figure 2. Whether or not grain refinement was observed with each size distribution of the alloys is shown in Table 2.

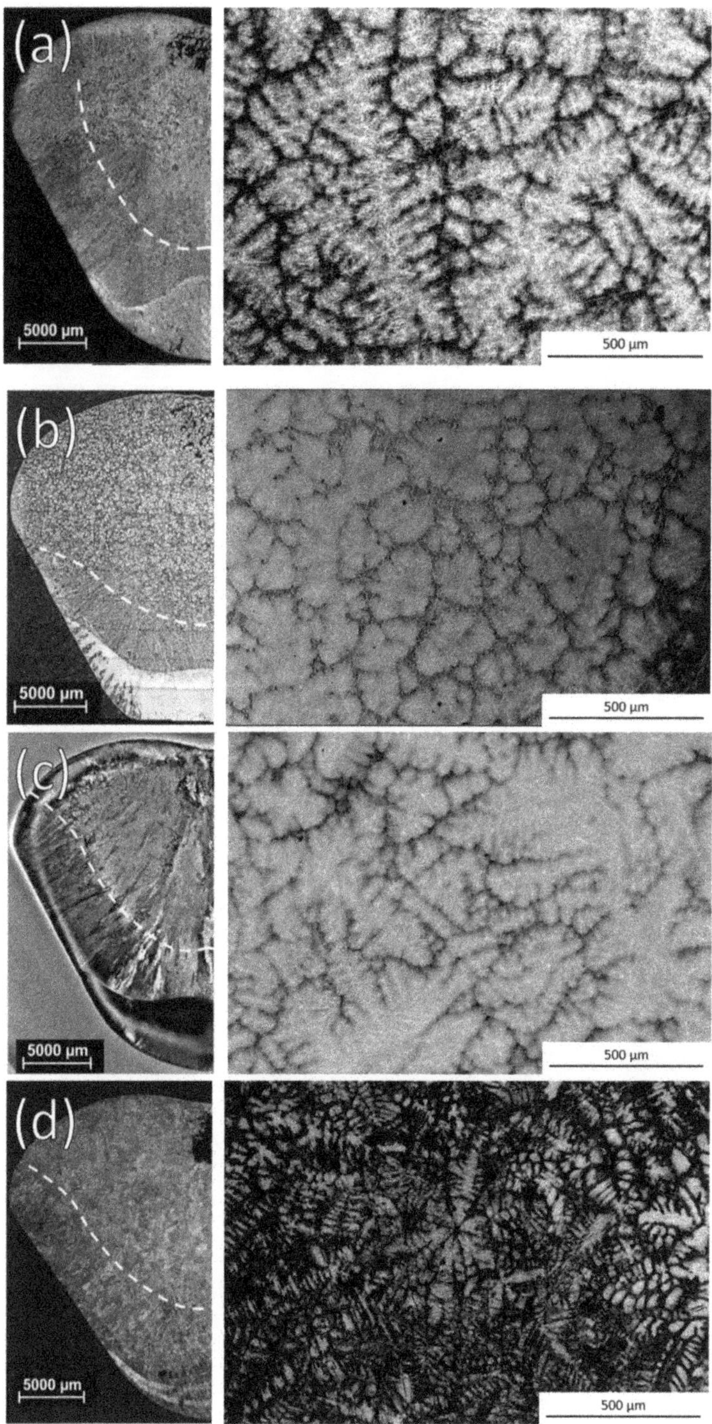

Figure 2. Optical ingot cross sections and SEM BSE images of equiaxed zone after (**a**) No inoculation and inoculation with 3 h cryomilled particles of (**b**) Ti-Al-Nb, (**c**) Ti-Al-Ta, and (**d**) Ti-Ta.

Measuring the size of dendritic grains consistently and accurately is non-trivial. Depending on the plane of the ingot being viewed, dendritic grains may appear very differently, conventional grain size measurement using the line-intercept or a similar method is not possible. In order to ensure that the grain size reported was reflected to be as close to the real grain size as possible, only grains that appeared to be cut in an exact cross section perpendicular to four primary arms were measured. The linear lengths from opposing tips were measured for two pairs of tips as close to perpendicular to one another as possible were used to characterize the size of such grains. These measurements were taken across as many grains as possible in order to obtain an average grain size. Grains were identified and measured manually. This method minimizes the total number of grains measured, since only grains with a near complete cross section are measured, but ensures that each measurement relates precisely to a real dimension of a grain. In order to have comparable results between samples the same method was used in all cases, even if there were globular shaped grains which could have their size evaluated effectively with other methods. The equiaxed fraction of the ingots was also measured. These measured values for the successful inoculation trials along with the final Al contents of the ingots are shown in Table 3.

Table 3. Successful Isomorphic Inoculation Results.

Alloy	Grinding Time (h)	Grain Size (μm)	Equiaxed Fraction	Al Content (at %)
Reference		696	0.3	46.1
Nb Solutal		587	0.31	45.1
Ti-Al-Nb	0	403	0.69	45.6
	3 (Ar)	397	0.68	45.5
	3	336	0.63	45.5
	6	446	0.64	44.4
	9	337	0.64	45.4
	11	188	0.42	45.1
Ti-Al-Ta	3	335	0.52	45.7
	9	308	0.64	45.8

While the Ti-Ta particles were designed to have the greatest stability in the melt none of the inoculation trials that used them were successful at reducing the as-cast equiaxed grain size. While no particles were found in the equiaxed region of any of the inoculated ingots, some Ti-Ta particles were found to have survived processing. These particles were found in the bottom of the ingot, either in the columnar zone or in the semi-solid cap at the bottom of the ingot. Figure 3 shows an SEM BSE image from this semi-solid bottom cap/columnar zone of an ingot inoculated by Ti-Ta. Energy Dispersive Spectrometer X-ray diffraction (EDX) analysis was performed on the particles that were found in the ingots inoculated with Ti-Ta, an EDX map of a representative particle from the 1.5 h inoculation trial is also shown. The particles were confirmed to contain both Ti and Ta with compositions near expected for undissolved inoculant particles (Ti-47Ta).

It can be noted that the mapped particle has a length over 600 μm and a width of 300 μm, this is significantly larger than the both the D50 (70 μm) or D99 (334 μm) for the distribution. The BSE image of the particle shows a somewhat banded structure that can also be seen particularly well in the Ti composition map. This indicates that it exists as a single particle, rather than an agglomeration of smaller particles. This is interesting as such large particles were not detected by SEM image analysis. A single particle of such size has a mass equal to 0.06 wt % of the distribution, when compared to 0.02% for a particle identical to the D99 or 0.0002% for the D50. It then does not take very many large particles existing in the distribution to severely decrease the number of particles that are available for inoculation This could help to explain why the Ti-Al trials did not show any grain refinement, as super large particles unaccounted for in the distributions reduced the number of active particles that are added to the melt. The small number of these particles do not lend well to their detection by imaging methods.

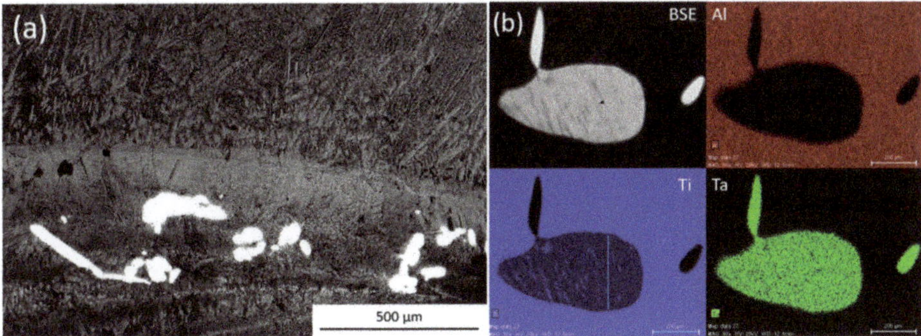

Figure 3. (**a**) BSE images of bottom region of ingots inoculated with Ti-Ta and (**b**) Energy Dispersive Spectrometer X-ray diffraction (EDX) maps of particles found.

In addition to the super large particles shown some smaller particles of Ti-Ta were also found in the solidified ingots that were within the size range of the distributions measured. However, all of these smaller particles were found in the columnar zone of the ingots, or along with the super large particles in the bottom cap. Particles in these regions did not help to form equiaxed grains, as the region that they are in has no equiaxed grains. The question is then how the Ti-Ta particles were deposited in this region when there are no Ti-Al-Ta or Ti-Al-Nb particles that were detected in the same regions of their respective ingots. The fluid flow of the molten Ti-Al alloy should be the same in each inoculation trial, as the processing conditions remain constant, however, the particles may behave differently from one another in the melt as they have different sizes and densities. It is difficult to calculate how the particles move in the melt as the induction field induces complex three-dimensional fluid flows into the molten drop which cannot be easily approximated. The solid bottom cap is humped in the center of the ingot and has a shape matching that of the fluid rolls. Many of the TiTa particles are found in this central humped region, indicating that, rather than being carried by the fluid flow and dispersed throughout the ingot, they have been ejected from the flow and settled in this region where they are not useful for grain refinement of the equiaxed zone.

The ingots inoculated with TiTa also showed dendrite morphologies that matched the α phase (hexagonal) rather than β (cubic). This could be the result of oxygen pickup during the high temperature processing that is required to produce the inoculant alloy. β-TiTa has a large oxygen solubility so that this transition to α was not seen in the bulk inoculant alloy, however, the oxygen solubility is much less in β-TiAl [30]. An increase in oxygen content of the alloy may then also be poisoning its effectiveness at grain refinement as if the solidification phase has changed from β to α the lattice and phase matching for which the inoculant was designed no longer exist.

It is then likely that a combination of these factors lead to the TiTa particles being ineffective. The high density of the particles results in the particles settling to the bottom of the ingot removing the particles from the equiaxed zone where they are useful for decreasing the equiaxed grain size, ensuring that the inoculants are ineffective. The presence of superlarge particles that account for a large portion of the inoculant mass added to the melt reduces the number of inoculants added to the melt. The inoculants that were added to the melt may be poisoning it against them by adding oxygen and inducing a change of the solidification phase from β to α.

All of the inoculation trials with Ti-Al-Nb and Ti-Al-Ta particles were successful at both decreasing the equiaxed grain size and also increasing the fraction of the equiaxed zone in the ingots. The magnitude of the grain refinement and the enlargement of the equiaxed zone depended on the alloy and particle size distribution that was used. It can be seen that the equiaxed grain size tended to decrease as the milling time (Table 3) increased, except in the case of the 6 h milled sample, which has a larger grain size than expected, however the grain size is still reduced when compared to the reference or solutal samples. Milling in Ar resulted in larger equiaxed grains than particles that were equivalently milled in air. The same was true for the resultant equiaxed fractions; milling in Ar was less effective at increasing the equiaxed zone when compared to air. The equiaxed fraction stayed relatively constant between the air milled particles with the exception of the 11 h milled sample that had a large unmelted Ti particle that was inherited from primary ingot elaboration, and larger bottom cap that was present in the cross section. The Ti-Al-Ta particles reduced the grain size roughly the same amount as equivalently milled Ti-Al-Nb particles, however, the three hour Ta containing particles did not increase the equiaxed fraction as much as the others.

It was also important to know the number of inoculant particles added in each inoculation trial. Since the mass of the inoculant particles that were added was measured, it could be used along with the size distributions that were obtained and the calculated densities to determine the number of particles added. This was done assuming the particles were spherical and that all the particles in each size range had a diameter exactly the average of the maximum and minimum values of the range. Using the measured mass of inoculant ($\approx$1 g) and the calculated density of the inoculant alloy, a total volume of inoculant introduced could be found. The particles were also assumed to be spherical when calculating their sized from projected 2D areas, however, if an ellipsoid diverges from a spherical shape (with a constant average radius), its volume will decrease. The calculated number of particles then may be underrepresented if the particles vary significantly from spheres.

The size range of the two Ti-Al-Ta distributions that were tested are between the 9 and 11 h milled Ti-Al-Nb particles. The two particle sizes both reduce the grain size somewhat less than that of the Ti-Al-Nb 11 h milled distribution, which has the closest D50 to the Ti-Al-Ta distributions used. However, when the number of particles introduced is considered the grain sizes are much closer to the trend of the Ti-Al-Nb particles. The Ti-Al-Ta trials introduced more particles than when the 9 h milled Ti-Al-Nb particles were used, but less than with 11 h Ti-Al-Nb particles. As the same mass of inoculant is introduced in each trial, particle size and the number of particles that were introduced are directly correlated. If we compare Ti-Al-Nb (11 h) and Ti-Al-Ta (9 h) with roughly the same median particle size, we can see that the refining power of the Ti-Al-Nb is greater because the number of particles is higher, indicating that the critical parameter is not the size of the particles, but rather the number of particles that are introduced. This is in direct contrast to traditional inoculation, where the size of the particles that are introduced is the most critical as larger particles can nucleate new grains at smaller undercoolings [31].

5. Discussion

The resultant grain size as powder size and number of particles introduced to the melt increase can be seen in Figure 4a,b, respectively, for the Ti-Al-Nb and Ti-Al-Ta inoculation trials. The reported error is the standard deviation of the grain size measurements. Since only equiaxed dendrites with nearly complete cross sections are measured, the number of grains that are measured is limited. Ingots with larger grain sizes will have fewer grains cut in cross section that may be measured, and thus the standard deviation of their measurements will be larger, this is also true of structures that are more dendritic compared to globular. This may obscure some of the trends observed, but it ensures that the reported grain sizes are representative of real structures within the ingots. When evaluated based on the size of the particles introduced the Ar milled sample does not appear to behave significantly differently from those that are milled under air. While milling under Ar for 3 h resulted in larger particles than the same time under air, the resultant grain size and equiaxed fraction are roughly what would be expected for particles that were milled under air with the same size. The Ar milled powders also behaved as the particles milled in air with respect to the resultant equiaxed fraction. The six hour milled sample (D50 = 160 μm) resulted in a larger grain size than expected. Since each bulk alloy was produced individually some variation in the composition is to be expected. This variation may account for the six hour inoculations apparent deviation, as it has the leanest Al composition of all the inoculated ingots. This is important since the grain size has been shown to increase in ingots produced with similar processing as Al content decreases due to the changes in the dendrite fragmentation mechanism [32]. This means that, while the grain size may be larger than expected in the ingot that is inoculated with the six hour milled particles, the reduction in grain size due to the inoculation may be greater than shown, since the bulk alloy would have had a larger grain size than the measured reference sample due to its decreased Al content.

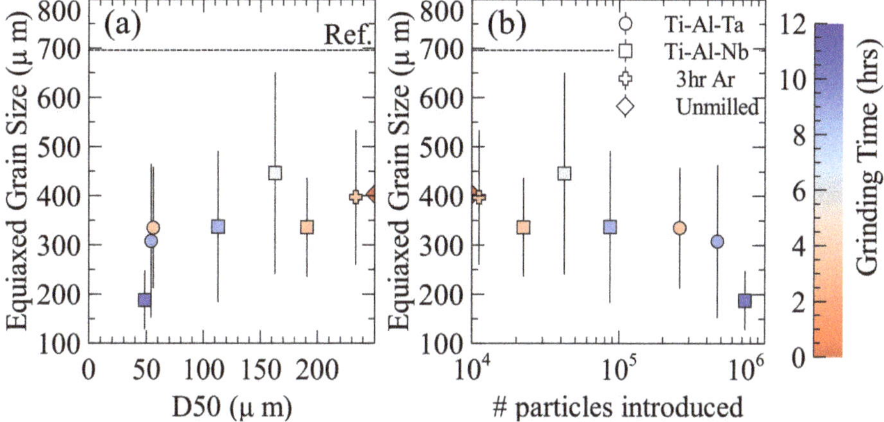

Figure 4. Influence of (**a**) particle size and (**b**) number of particles introduced on equiaxed grain size. The error bars correspond to the standard deviation of grain size measurements.

Using the average equiaxed grain size, the grain density in the equiaxed zone (#/cm^3) can be compared to the density of particles that are added to the melt (#/cm^3), as shown in Figure 5a. This assumes that the all of the particles that were added were evenly distributed in the equiaxed zone. This gives the most pessimistic results of inoculant efficiency, as this is the maximum density the particles could have during solidification, if they were evenly dispersed throughout the ingot their density in the equiaxed zone would be reduced. It is also assumed that no particles completely dissolve, meaning that the same number of particles is present during solidification as was added to the melt. In Figure 5a the dashed line indicated a 1:1 ratio, where each particle would add a single grain to the solidified ingot. In every case, except the 9 h Ti-Al-Ta (which still close to this 1:1 ratio), the grain density is larger than the introduced particle density, which means that more grains were formed than particles that were added to the melt. Moreover, it can be seen that when the number of particles introduced exceeds 5×10^4 particles/cm^3 the efficiency tends towards 1. When fewer particles are added the grain size does not change drastically, rather the efficiency of the particles for isomorphic inoculation increases. This is in strong contradiction to traditional inoculation where the number of grains formed is significantly less than the number of particles that were added to the melt [33,34]. The authors previously proposed that this drastic increase in efficiency was due to the suppression of the nucleation step, allowing for direct particle growth [19] and the increase above a 1:1 ratio of particles to grains due to a process of particle breakup. The effect of particle size on particle break-up is estimated in Figure 5b, where the evolution of the number of equiaxed grains that were formed by single particle is compared with the number of introduced particles, the distribution D50s are indicated. There is a sharp increase in number of grains that were formed per particle with large particles when compared to small particles. The smaller particles (~50 μm) result in one or two grains being formed per particle, with both Ti-Al-Nb and Ti-Al-Ta, while the Ti-Al-Nb particles of intermediate size (100–170 μm) have 2–5 grains formed per particle and the Ti-Al-Nb particles larger than 170 μm result in more than 10 grains formed per particle. It is then possible that the close diffusivity and processing method results in particles that break up in roughly equivalent manners between the two alloys. The difference in effectiveness may then be attributed to the mechanical properties of the particles that behave differently under milling resulting in different distributions, and a greater difference in particle number for roughly equivalent particle D50s. The general efficiency of the inoculation then depends on the number of particles that were introduced, however, the particle size distribution must also be taken into account, since if the particle size is large, significant particle breakup may occur, while when the particle size is low, the breakup factor tends towards one. This means when the introduced particle size is low the correlation between number of particles introduced and the number of grains formed is stronger than with large particles where the distribution may be more affected by interaction with the melt.

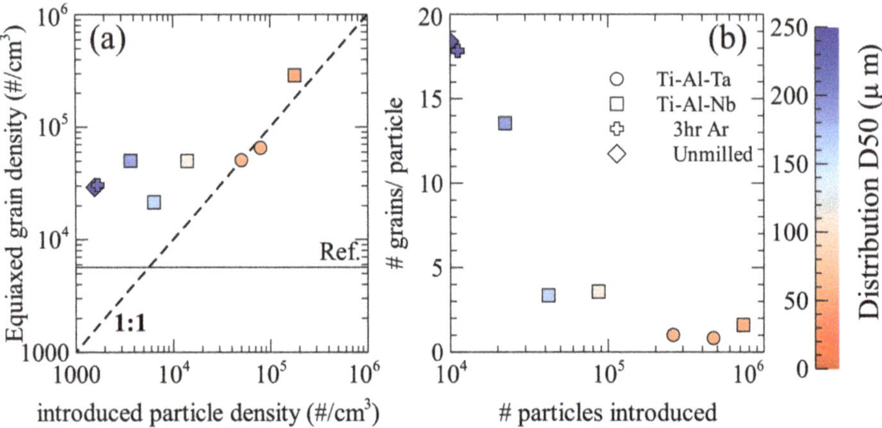

Figure 5. (**a**) Influence inoculant particle density assuming particles only present in the equiaxed zone and (**b**) Relationship between # of grains formed by each inoculant and number of particles introduced if they are only in the equiaxed zone.

The particle break-up mechanism was observed in a Ti-Ta inoculated ingot, and is shown in Figure 6, together with a schematic representation. In Figure 6, the bright Ta rich regions are portions of a particle which are in the process of breaking up while the dark regions are the bulk alloy. By adjusting the brightness and contrast when obtaining the images, the particle shape can be seen to vary from a monolithic single particle to a collection of smaller particles that are separated by regions of lower Ta concentration. This may occur by preferential dissolution [35] and/or impingement of the liquid along the grain boundaries of the particle [36]. Such impingement or wetting of a liquid like phase along grain boundaries has been observed in-situ along Al grain boundaries by liquid Ga [37]. Such wetting may occur if the grain boundary energy is higher than that of two solid liquid interface [38] i.e., $\sigma_{\beta/\beta} > 2\left(\sigma_{\beta/\ell}\right)$, where σ is the surface energy. In this case the liquid will be fully wet along the grain boundary, which would permit the separation of the particles in the melt along the grain boundaries. The dark liquid can be seen to ingress along preferential paths, resulting in the breakup of the particle. As the size of the particles increases, so does the number of grains that are formed by each particle. This could be due to an increased number of cracks or grains present in each large particle when compared to the smaller ones, as they have been milled for less time. These cracks which on further milling result in particle breakup during milling may act as preferential dissolution paths in the melt resulting in a larger breakup effect during inoculation. This would also explain the decrease in the number of grains that are formed by each particle as the number of particles introduced increases. The smaller particles, which were introduced in more plentiful numbers, were also more likely to dissolve completely and less likely to break up into more particles. Again, the breakup and dissolution of particles likely reduces the global maximum and increases the global minimum number of particles that are present on solidification, the number and the size of particles introduced to the melt are indicative of, but not entirely representative of, the distribution present on solidification.

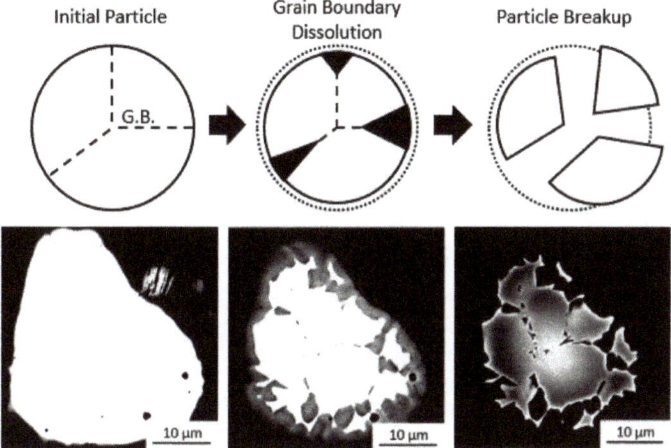

Figure 6. Schematic of particle breakup process: whole polycrystalline particle, preferential grain boundary dissolution, particle breakup, and BSE micrographs of Ti-Ta particle in solidified ingot demonstrating particle breakup by adjusting imaging contrast.

6. Conclusions

In summary, the efficiency of isomorphic inoculation was evaluated based on inoculant alloys selection, accounting for thermal and diffusive stability, particle size distribution, and particle density. The results presented here show that, within this experimental set-up, the most critical parameter is the density difference between the liquid alloy and the inoculant powder. High density differences lead to settling of inoculant particles and prevents grain refinement. In contrast, the diffusive stability of the particles was found to be less critical and only affect the observed as-cast grain size slightly when the other parameters were similar.

Isomorphic inoculation relies on epitaxial growth of the solidifying alloy on inoculant particles. The critical parameter for isomorphic inoculation was proposed to be the number of inoculant particles in [19], as opposed to the particle size distribution as in classical inoculation processing. In this work, it has been shown that particle size distribution also contributes to the efficiency of the isomorphic inoculation process. It was found that larger particle size distributions were more efficient for grain refinement of Ti-Al alloys, this increase is attributed to the trend of the particles to break up into multiple smaller particles by dissolution in the melt. This phenomenon is more apparent if the initial particles are large, showing that there is a critical particle size below which dissolution occurs faster, leading to a decrease in surviving particles, and thus a decrease in inoculation efficiency.

Author Contributions: Jacob R. Kennedy, Julien Zollinger and Emmanuel Bouzy conceived, designed and carried out the experiments; Bernard Rouat assisted with carrying out the experiments. Jacob R. Kennedy wrote the main manuscript, Jacob R. Kennedy, Julien Zollinger, Emmanuel Bouzy and Dominique Daloz participated in the analysis, discussion and reviewed the manuscript.

Acknowledgments: J.R.K. would like to thank Région Lorraine for co-funding his PhD thesis. This work was also supported by the French State through the program "Investment in the future" operated by the National Research Agency (ANR) and referenced by ANR-11-LABX-0008-01 (LabEx DAMAS).

Conflicts of Interest: The authors declare no conflict of interest.

References

1. Kim, Y.-W.; Dimiduk, D.M. Progress in the understanding of gamma titanium aluminides. *JOM* **1991**, *43*, 40–47. [CrossRef]
2. Cowen, C.J.; Boehlert, C.J. Comparison of the microstructure, tensile, and creep behavior for Ti-22Al-26Nb (At. Pct) and Ti-22Al-26Nb-5B (At. Pct). *Metall. Mater. Trans. A* **2007**, *38*, 26–34. [CrossRef]
3. Witusiewicz, V.T.; Bondar, A.A.; Hecht, U.; Rex, S.; Velikanova, T.Y. The Al–B–Nb–Ti system: III. Thermodynamic re-evaluation of the constituent binary system Al–Ti. *J. Alloys Compd.* **2008**, *465*, 64–77. [CrossRef]
4. Burgers, W.G. On the process of transition of the cubic-body-centered modification into the hexagonal-close-packed modification of zirconium. *Physica* **1934**, *1*, 561–586. [CrossRef]
5. Blackburn, M.J. Some Aspects of Phase Transformations in Titanium Alloys. In *The Science, Technology and Application of Titanium*; Jaffee, R.I., Promisel, N.E., Eds.; Pergamon: Oxford, UK, 1970; pp. 633–643. ISBN 978-0-08-006564-9.
6. Sankaran, A.; Bouzy, E.; Fundenberger, J.J.; Hazotte, A. Texture and microstructure evolution during tempering of gamma-massive phase in a TiAl-based alloy. *Intermetallics* **2009**, *17*, 1007–1016. [CrossRef]
7. Wu, X.; Hu, D. Microstructural refinement in cast TiAl alloys by solid state transformations. *Scr. Mater.* **2005**, *52*, 731–734. [CrossRef]
8. Porter, D.A.; Easterling, K.E. *Phase Transformations in Metals and Alloys (Revised Reprint)*, 3rd ed.; CRC Press: Boca Raton, FL, USA, 2009.
9. Greer, A.L. Grain refinement of alloys by inoculation of melts. *Philos. Trans. R. Soc. A Math. Phys. Eng. Sci.* **2003**, *361*, 479–495. [CrossRef]
10. Dantzig, J.A.; Rappaz, M. *Solidification*, 2nd ed.; CRC Press: Boca Raton, FL, USA, 2017.
11. Hyman, M.E.; McCullough, C.; Levi, C.G.; Mehrabian, R. Evolution of boride morphologies in TiAl-B alloys. *Metall. Trans. A* **1991**, *22*, 1647–1662. [CrossRef]
12. Hyman, M.E.; McCullough, C.; Valencia, J.J.; Levi, C.G.; Mehrabian, R. Microstructure evolution in tial alloys with b additions: Conventional solidification. *Metall. Trans. A* **1989**, *20*, 1847–1859. [CrossRef]
13. Witusiewicz, V.T.; Bondar, A.A.; Hecht, U.; Zollinger, J.; Artyukh, L.V.; Velikanova, T.Y. The Al–B–Nb–Ti system: V. Thermodynamic description of the ternary system Al–B–Ti. *J. Alloys Compd.* **2009**, *474*, 86–104. [CrossRef]
14. Hecht, U.; Witusiewicz, V.; Drevermann, A.; Zollinger, J. Grain refinement by low boron additions in niobium-rich TiAl-based alloys. *Intermetallics* **2008**, *16*, 969–978. [CrossRef]
15. Gosslar, D.; Günther, R.; Hecht, U.; Hartig, C.; Bormann, R. Grain refinement of TiAl-based alloys: The role of TiB2 crystallography and growth. *Acta Mater.* **2010**, *58*, 6744–6751. [CrossRef]
16. Qian, M. Heterogeneous nucleation on potent spherical substrates during solidification. *Acta Mater.* **2007**, *55*, 943–953. [CrossRef]
17. Cantor, B. Heterogeneous nucleation and adsorption. *Philos. Trans. R. Soc. A Math. Phys. Eng. Sci.* **2003**, *361*, 409–417. [CrossRef]
18. Kuni, F.M.; Shchekin, A.K.; Rusanov, A.I.; Widom, B. Role of surface forces in heterogeneous nucleation on wettable nuclei. *Adv. Colloid Interface Sci.* **1996**, *65*, 71–124. [CrossRef]
19. Kennedy, J.R.; Daloz, D.; Rouat, B.; Bouzy, E.; Zollinger, J. Grain refinement of TiAl alloys by isomorphic self-inoculation. *Intermetallics* **2018**, *95*, 89–93. [CrossRef]
20. Bermingham, M.J.; McDonald, S.D.; St John, D.H.; Dargusch, M.S. Titanium as an endogenous grain-refining nucleus. *Philos. Mag.* **2010**, *90*, 699–715. [CrossRef]
21. Witusiewicz, V.T.; Bondar, A.A.; Hecht, U.; Voblikov, V.M.; Fomichov, O.S.; Petyukh, V.M.; Rex, S. Experimental study and thermodynamic modelling of the ternary Al–Ta–Ti system. *Intermetallics* **2011**, *19*, 234–259. [CrossRef]
22. Witusiewicz, V.T.; Bondar, A.A.; Hecht, U.; Velikanova, T.Y. The Al–B–Nb–Ti system. *J. Alloys Compd.* **2009**, *472*, 133–161. [CrossRef]
23. Khallouk, S. Developpement D'une Nouvelle Generation D'inoculants Pour les Aluminures de Titane. Master's Thesis, Université de Lorraine, Nancy, France, 2015.
24. Gale, W.F.; Totemeier, T.C. *Smithells Metal Reference Book*, 8th ed.; Elsevier: New York, NY, USA, 2004.

25. Huang, S.C. Alloying considerations in gamma-based alloys. In *Structural Intermetallics*; TMS: Warrendale, PA, USA, 1993; p. 299.
26. Shaffer, P.T.B. *Plenum Press Handbooks of High Temperature Materials: No. 1 Materials Index*; Springer: Berlin, Germany, 1964.
27. Andersson, J.-O.; Helander, T.; Höglund, L.; Shi, P.; Sundman, B. Thermo-Calc & DICTRA, computational tools for materials science. *Calphad* **2002**, *26*, 273–312. [CrossRef]
28. Gibbs, G.B.; Graham, D.; Tomlin, D.H. Diffusion in titanium and titanium-niobium alloys. *Philos. Mag. J. Theor. Exp. Appl. Phys.* **1963**, *8*, 1269–1282. [CrossRef]
29. Ansel, D.; Thibon, I.; Boliveau, M.; Debuigne, J. Interdiffusion in the body cubic centered β-phase of Ta–Ti alloys. *Acta Mater.* **1998**, *46*, 423–430. [CrossRef]
30. Zollinger, J. *Influence de L'oxygene sur le Comportement a la Solidification D'aluminiures de Titane Binares et al. lies au Niobium Bases sur le Compose Intermetallique g-TiAl*; L'institut National Polytechnique de Lorraine: Nancy, France, 2008.
31. Quested, T.; Greer, A. The effect of the size distribution of inoculant particles on as-cast grain size in aluminium alloys. *Acta Mater.* **2004**, *52*, 3859–3868. [CrossRef]
32. Reilly, N.T.; Rouat, B.; Martin, G.; Daloz, D.; Zollinger, J. Enhanced dendrite fragmentation through the peritectic reaction in TiAl-based alloys. *Intermetallics* **2017**, *86*, 126–133. [CrossRef]
33. Gosslar, D.; Gunther, R.; Hartig, C.; Bormann, R.; Zollinger, J.; Steinbach, I. grain refinement of gamma-TiAl alloys by inoculation. In Proceedings of the Materials Research Society Symposium Proceedings, Boston, MA, USA, 1–5 December 2008; Volume 1128, pp. 91–96.
34. Quested, T.E. Solidification of Inoculated Aluminium Alloys. Ph.D. Thesis, University of Cambridge, Cambridge, UK, 2004.
35. Hsieh, T.E.; Balluffi, R.W. Experimental study of grain boundary melting in aluminum. *Acta Metall.* **1989**, *37*, 1637–1644. [CrossRef]
36. Straumal, B.B.; Gornakova, A.S.; Kogtenkova, O.A.; Protasova, S.G.; Sursaeva, V.G.; Baretzky, B. Continuous and discontinuous grain-boundary wetting in Zn_xAl_{1-x}. *Phys. Rev. B* **2008**, *78*. [CrossRef]
37. Pereiro-López, E.; Ludwig, W.; Bellet, D.; Baruchel, J. Grain boundary liquid metal wetting: A synchrotron micro-radiographic investigation. *Nucl. Instrum. Methods Phys. Res. Sect. B Beam Interact. Mater. Atoms* **2003**, *200*, 333–338. [CrossRef]
38. Straumal, B.B.; Mazilkin, A.A.; Kogtenkova, O.A.; Protasova, S.G.; Baretzky, B. Grain boundary phase observed in Al-5 at % Zn alloy by using HREM. *Philos. Mag. Lett.* **2007**, *87*, 423–430. [CrossRef]

Article

Complexity and Anisotropy of Plastic Flow of α-Ti Probed by Acoustic Emission and Local Extensometry

Mikhail Lebyodkin [1,2,*], Kékéli Amouzou [3], Tatiana Lebedkina [4], Thiebaud Richeton [1,2] and Amandine Roth [5]

1 Laboratoire d'Etude des Microstructures et de Mécanique des Matériaux (LEM3), CNRS, Université de Lorraine, Arts & Métiers ParisTech, F-57000 Metz, France; thiebaud.richeton@univ-lorraine.fr
2 Laboratory of Excellence on Design of Alloy Metals for Low-Mass Structures (DAMAS), Université de Lorraine, F-57073 Metz, France
3 Laboratoire de Mécanique de Lille (LML), CNRS UMR 8107, Université de Sciences et Technologies Lille, Cité Scientifique, Boulevard Paul-Langevin, 59655 Villeneuve d'Ascq CEDEX, France; evamouzou@ymail.com
4 Laboratory of Metallic Materials with Spatial Gradient Structure, Togliatti State University, Belorusskaya St. 14, 445020 Tolyatti, Russia; tleb1959@gmail.com
5 Ascometal—CREAS, Avenue de France, 57300 Hagondange, France; amandine.roth@gmail.com
* Correspondence: mikhail.lebedkin@univ-lorraine.fr; Tel.: +33-(0)3-7274-7771

Received: 25 May 2018; Accepted: 21 June 2018; Published: 22 June 2018

Abstract: Current progress in the prediction of mechanical behavior of solids requires understanding of spatiotemporal complexity of plastic flow caused by self-organization of crystal defects. It may be particularly important in hexagonal materials because of their strong anisotropy and combination of different mechanisms of plasticity, such as dislocation glide and twinning. These materials often display complex behavior even on the macroscopic scale of deformation curves, e.g., a peculiar three-stage elastoplastic transition, the origin of which is a matter of debates. The present work is devoted to a multiscale study of plastic flow in α-Ti, based on simultaneous recording of deformation curves, 1D local strain field, and acoustic emission (AE). It is found that the average AE activity also reveals three-stage behavior, but in a qualitatively different way depending on the crystallographic orientation of the sample axis. On the finer scale, the statistical analysis of AE events and local strain rates testifies to an avalanche-like character of dislocation processes, reflected in power-law probability distribution functions. The results are discussed from the viewpoint of collective dislocation dynamics and are confronted to predictions of a recent micromechanical model of Ti strain hardening.

Keywords: titanium; strain hardening; anisotropy; strain heterogeneity; acoustic emission; statistical analysis; collective dislocation dynamics

1. Introduction

The scope of this paper is two-fold. First, despite intensive investigations of plasticity of materials with a hexagonal close-packed (*hcp*) structure during last time, many aspects of their mechanical behavior are still poorly understood. The main reason for this difficulty is a high anisotropy of the crystal structure, limiting the number of easy slip systems for dislocations. As a result, plastic deformation of *hcp* materials generally involves either additional mechanisms, such as twinning, or slip systems with a large difference in the critical resolved shear stress (CRSS). One of the consequences of this complexity is that contrarily to polycrystals with more isotropic cubic structures, which are usually characterized by monotonous strain hardening in a large range of plastic strain (the so-called stages II and III [1,2]), *hcp* materials often display three-stage strain hardening behavior associated with the elastoplastic transition and resulting in a concave shape of the deformation curve at small strains [3–11]. To distinguish such behavior from the generic work-hardening stages observed in materials with various structures [1,2], these specific stages are usually designated as *A*, *B*, and *C*. The initial decrease in the work hardening rate, Θ, during stage *A* is followed by an increase (stage *B*) and a new decrease (stage *C*). The origin of this behavior is still a matter of debate. It is usually rather pronounced in compression, when a large contribution of twinning is observed [3–8]. Accordingly, it is often ascribed to twinning although even in this unique framework, numerous qualitatively different mechanisms, leading to either hardening or softening effects, were suggested: a dynamic Hall-Petch effect implying that twin boundaries play the role of obstacles to the dislocation motion, formation of sessile dislocation configurations within twins, favorable or, inversely, unfavorable reorientation of the crystal lattice for sliding, and so on [3–7]. Nevertheless, recent experiments showed that three-stage behavior can also occur in tension [9–11]. These results could not be ascribed to twinning because of its low contribution to the total deformation. A remarkable feature of the behavior observed in tension concerned the strain-rate effect on the depth of the depression in the work-hardening rate during stage *A* [10]. More specifically, a reduction of the strain rate reinforced this non-monotonicity in the specimens with the tensile axis parallel to the rolling direction (hereinafter designated as RD) and produced an opposite effect in TD (transverse direction) specimens.

To explain these peculiarities, an elasto-viscoplastic self-consistent model of plastic flow of polycrystalline α-Ti was proposed in [12]. This model was uniquely based on the consideration of the dislocation glide which was subject to several hypotheses supported by various experimental observations: (i) a weaker strain-rate sensitivity of stress in prismatic slip systems than in other systems; (ii) an important role of the dislocation multiplication at the elastoplastic transition (cf. [13,14]); (iii) a higher multiplication rate for the prismatic systems; and (iv) a strong anisotropy of the CRSS. This model allowed for reproducing the peculiar features of the work hardening and confirmed their relationship with the evolution of the average activity of various slip systems. In addition, the model allowed for prediction of some elements inaccessible in experiments, e.g., gradual elastoplastic transition in various grains, evolution of the elastic stored energy, or local mechanical response of grains. On the whole, the model provided a link between the macroscopic scale of plastic flow of a polycrystal and mesoscopic scales associated with the dislocated glide. These results imply a further step to verification of the model predictions using experimental methods that provide an explicit information on such mesoscopic scales, e.g., by recording the acoustic emission (AE), produced due to multiplication and motion of dislocations and, therefore, reflecting the slip activity [15], or by measuring local strain fields on the specimen surface [16]. In particular, this approach is justified by the well-known fact that the evolution of the average AE activity often follows that of the macroscopic work-hardening rate [15]. Such studies constitute the main subject of the present paper.

On the other hand, investigations of the last two decades showed that beyond the average characteristics, the analysis of the statistics of the AE and local strain-rate fields uncovers features that escape from the commonly accepted homogenization schemes of plasticity based on the averaging over large representative volumes [17–21]. In particular, the above model, like many others, tacitly suggests that dislocation interactions are random and have a zero average on the scale of a grain. However, the statistical analysis contradicts the randomness hypothesis. Instead, power-law statistics were observed for all materials studied so far, single crystals or polycrystals, with cubic or hexagonal symmetry. The entirety of results led to a hypothesis of an intrinsically collective, avalanche-like nature of the dislocation motion. This kind of analysis would help to identify the limits of the continuous plasticity models and provide a basis for the further development of more realistic multiscale models of plasticity, which is particularly interesting in the case of highly anisotropic materials. Moreover, whereas the statistical approach has already been applied to quite a few materials, such studies have not been carried out for Ti so far. It is thus not known a priori if the above hypothesis is valid in the case of Ti, too.

With these aspects in view, the present work was aimed at coupling mechanical testing with the AE and local extensometry measurements and pursued a double objective: (i) to provide an independent verification of the above-described experiments and theoretical predictions and develop experimental techniques bringing information on both macro- and mesoscopic scales; (ii) to couple this averaging approach with the new issue stemming from the statistical analysis of fluctuations of the plastic flow on a mesoscopic scale.

2. Materials and Methods

Details of the material microstructure and texture, as well as mechanical testing were described elsewhere [10,12] and will be briefly outlined here. The scheme of the experimental setup is presented in Figure 1. Mechanical tests were performed in a Zwick 1476 testing machine (Zwick, Ulm, Germany) controlled by the software package testExper (Zwick/Roell France, Ars-Laquenexy, France). Flat samples of commercially pure polycrystalline Ti with the oxygen content of 1062 ppm (wt.) and average grain size of 9 μm were deformed by tension. These conditions correspond to a low contribution of twinning to plastic deformation of α-Ti (see, e.g., [22]). The initial texture was typical of rolled Ti [9], with basal planes tilted 30 ± 10° from the normal toward the transverse direction. Dog-bone shaped samples with a 30 × 7 × 1.62 mm^2 gage section were cut from sheets along either the rolling or the transverse direction. The tests were performed at room temperature for five nominal (i.e., test. These records are used to calculate the evolution of the local strain rates $\dot{\varepsilon}_a$, selected in a wide range from 5×10^{-5} s^{-1} to 8×10^{-3} s^{-1}. Two to three specimens of each orientation were tested at each strain rate.

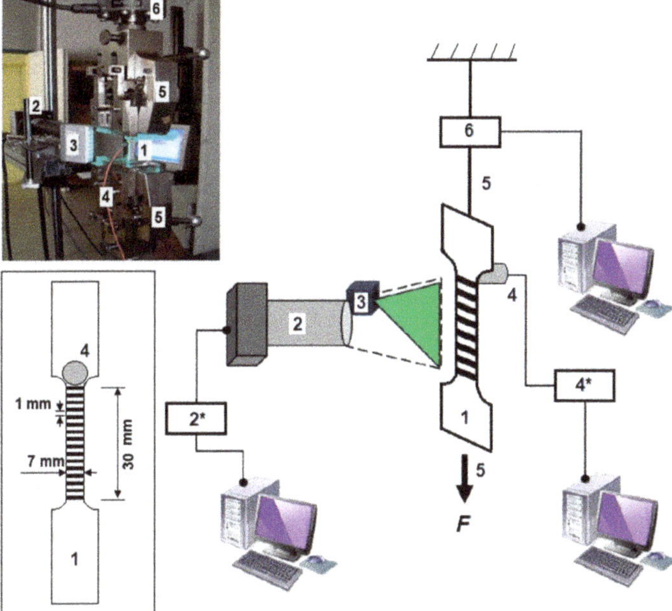

Figure 1. The scheme and a photograph of the experimental setup. Inset represents the scheme of a specimen specifying the relative arrangement of the grid painted on the specimen surface and the piezoelectric transducer. The same designations are used in all parts of the figure: 1—Specimen; 2—Charge-Coupled Device (CCD) camera; 2*—Acquisition block; 3—Laser light; 4—Piezoelectric transducer; 4*—Preamplifier; 5—Fixed and mobile grips; 6—Load cell. The applied force (*F*) indicates the displacement direction of the mobile grip.

AE measurements were performed with the Euro Physical Acoustics system (Mistras Group SA, Division EPA, Sucy en Brie, France). The acoustic signal was captured by a Micro-80 piezoelectric transducer (Mistras Group SA, Division EPA, Sucy en Brie, France) with the operating frequency band 200–900 kHz, clamped to the greased specimen surface just above its gage part. The AE signal was pre-amplified by 40 dB and recorded "continuously" (with a 2 MHz frequency). The continuous data streaming is particularly important for a precise calculation of the average AE count rate (CR) during elastoplastic transition because this deformation stage is characterized by a highly active AE making impossible accurate individualization of AE events ("hits"). As far as the extraction of AE hits for the statistical analysis is concerned, a standard procedure using an amplitude threshold and two temporal parameters was implemented (see, e.g., [21]). In this procedure, an event is considered to begin when the amplitude U of acoustic oscillations exceeds the threshold U_0 corresponding to the background noise and terminate when U remains below U_0 during a Hit Definition Time (HDT). The recording systems is then locked during a Hit Lockout Time (HLT) to filter out echoes. The following set of parameters was used: $\{U_0$; HDT; HLT$\}$ = {27 dB; 300 μs; 40 μs}. More details on the technique and the choice of the parameters can be found, e.g., in [23].

The 1D local extensometry technique was described in detail in [16,24]. About twenty markers are made by painting 1 mm wide and 1 mm distant black stripes across the specimen surface preliminarily painted white. A Charge-Coupled Device (CCD) Line Scan Sensor ZS16D (H.-D. Rudolph GMBH, Reinheim, Germany) seizes the intersections of the black-white transitions with the long axis of the specimen (tensile axis) and records their displacements during the test. These records are used to calculate the evolution of the local strain rates $\dot{\varepsilon}_{loc}(x,t)$ (x is the coordinate along the specimen's centerline). The method allows for surveying the evolution of spatiotemporal maps of the local strain-rate field over a 20 mm length with a 1 mm step and a high spatial (1.3 μm) and time (1 ms) resolution.

In the case of the AE, the statistical distributions were calculated for the square of the peak amplitudes of acoustic events, $I = U^2$. According to [25], it characterizes the energy dissipated in the deformation process giving rise to the acoustic wave. In the case of the local strain-rate bursts, the statistical analysis was applied directly to their amplitudes, Λ, because the mechanical work associated with a strain jump is a product of the strain increment and the stress, the latter being virtually constant during the strain jump duration (cf. [21]). To facilitate the comparison of the distributions of either different variables or the same variable measured in different experimental conditions, probability density functions (PDF) were calculated using data rescaled by the average value of the studied quantity, $X/\langle X \rangle$, where X means either I or Λ.

3. Results

3.1. Three-Stage Work-Hardening Behavior

Figure 2 shows an example of experimental data comparing: (a) the evolution of the stress; (b) the series of amplitudes and durations of AE hits; and (c) the 1D spatiotemporal map of local strain rates for two differently oriented specimens deformed at $\dot{\varepsilon}_a = 5 \times 10^{-4}$ s^{-1}. The following aspects should be noted:

- Both deformation curves (Figure 2a) show a tendency to a concave shape at the elastoplastic transition (it is stronger for the TD orientation at this strain rate). Although the concavity is less pronounced than that reported for compression [3–8], the corresponding $\Theta(\varepsilon)$ curves distinctly display a three-stage character (see Figure 3).
- Figure 2b shows the entire series of AE events recorded for both samples. Each dot in the upper chart represents the logarithmic value U_{log} of the amplitude of an AE hit with the duration τ shown in the bottom chart (For clarity, Figure 2d illustrates one of the diverse waveforms corresponding to a hit. The complexity of such waveforms is one of the signatures of collective nature of the dislocation dynamics [26]). The overall evolution of the series of AE events is typical

of most materials [15]. The AE occurs virtually immediately after the beginning of loading, is very strong during elastic deformation, and decreases after the elastoplastic transition. The strong AE accompanying the elastic stage is usually ascribed to the microplasticity that is characterized by a fast multiplication of dislocations and large flight distances of dislocations between rare obstacles [15,27]. The further exhaustion of AE during stable plastic flow (before necking) is attributed to the accumulation of barriers to the motion of dislocations. It can also be seen that the AE increases again upon the onset of necking leading to strain localization. The last effect goes beyond the scope of the paper and will only be used hereinafter as a benchmark.

- An important feature of the spatiotemporal maps of Figure 2c is that the local strain-rate values attain $\dot{\varepsilon}_a$ during the elastoplastic transition quasi-simultaneously over the entire field of vision [10]. It should be noted that as the sample is stretched during the test, the grid lines located close to the mobile grip gradually leave the field of vision of the CCD camera, so that the overall map only covers about 14 mm of the gage length. However, by restricting the analyzed time interval, it can be recognized that this conclusion is also valid for the other specimen cross-sections. Similar behavior was observed at all strain rates for both orientations. This result allows to invalidate an alternative mechanism of low strain hardening at the stage A [8], based on the suggestion of propagation of a Lüders-like deformation band and requiring conditions favorable for aging of dislocations by solute atoms [28]. Besides the overall pattern, Figure 2c reveals $\dot{\varepsilon}_{loc}$ fluctuations both in space and time. Statistical distributions of such fluctuations will be studied in Section 3.2, alongside with the AE statistics.

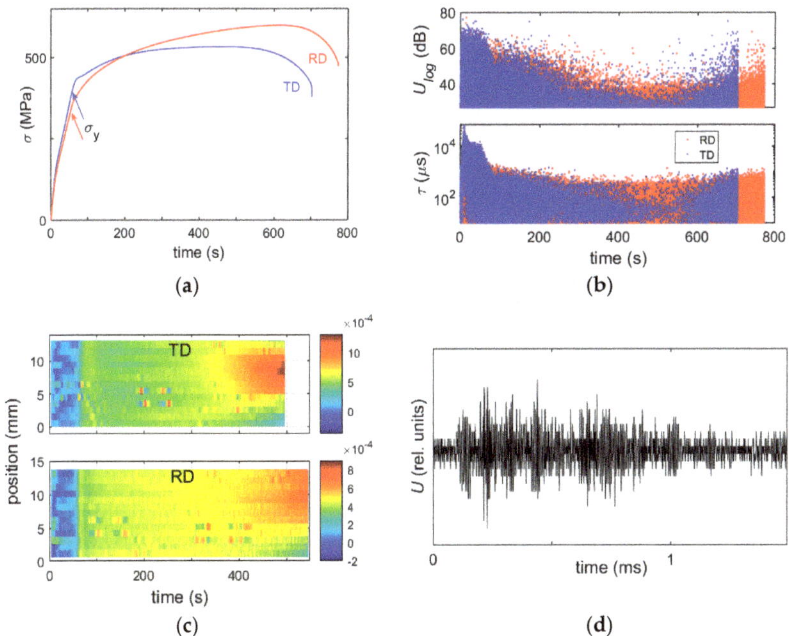

Figure 2. (**a**) Examples of stress-time curves for a transverse direction (TD) and rolling direction (RD) samples deformed at 5×10^{-4} s^{-1}. The offset yield point σ_y corresponds to 0.2% of plastic strain; (**b**) logarithmic amplitude U_{log} and duration τ for the series of AE events detected during the test; (**c**) spatiotemporal pattern displaying the evolution of the local strain rate $\dot{\varepsilon}_{loc}$ along the centerline of the specimen. Color bar is scaled in s^{-1}; and (**d**) example of waveform for an AE event.

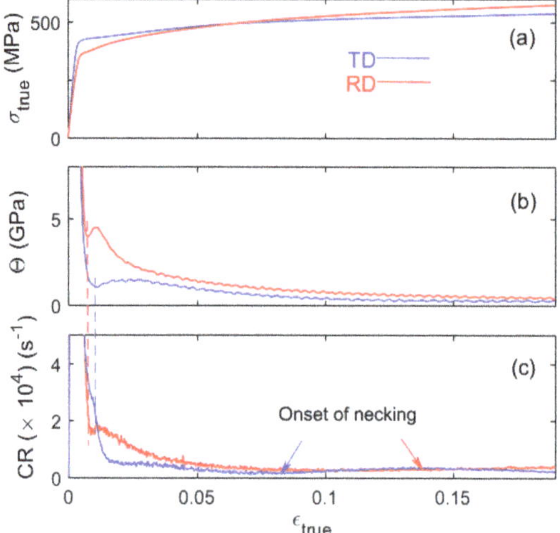

Figure 3. (**a**) True deformation curves for the samples of Figure 2; (**b**) strain dependence of the work-hardening rate Θ; and (**c**) strain dependence of the AE average count rate CR. Vertical dashed lines indicate the *A*/*B* transition.

Figure 3 compares the overall work-hardening behavior and the concomitant AE activity for the same specimen. For this strain rate, the CR was calculated within time intervals of 0.5 s. Θ(ε)-dependences were obtained by numerical differentiation of the true deformation curves (Figure 3a) over strain. They clearly show a non-monotonous character in the strain range corresponding to the elastoplastic transition (Figure 3b). To facilitate the reading of the figures, vertical dashed lines indicate the transition from stage *A* (initial decrease) to stage *B* (transient growth in Θ). All stages are denoted explicitly in the magnification of a part of this plot in Figure 4. These lines also demonstrate that the AE average count rate shows non-monotonous behavior, too, thus confirming the physical origin of the stages in tensile experiments. During stage C, not only the overall trend of CR is similar to that of Θ (before the onset of necking). Also, CR is higher for the RD sample than for its counterpart, in agreement with the similar relationship for Θ. However, the initial CR behavior corresponding to stages *A* and *B* of work hardening shows qualitatively different trends for different specimen orientations, deserving special attention.

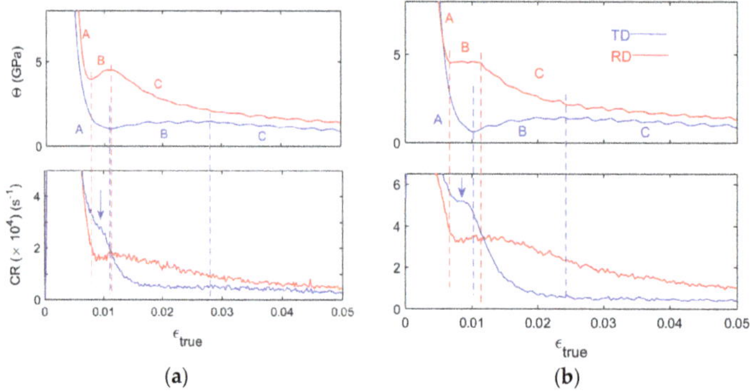

Figure 4. (**a**) Magnification of Figure 3b,c and (**b**) similar plots for two samples deformed at 2×10^{-3} s^{-1}.

These peculiarities are represented in Figure 4 displaying: (a) a magnification of the bottom charts of Figure 3, and (b) similar data for a higher strain rate, $\dot{\varepsilon}_a = 2 \times 10^{-3}\ s^{-1}$. To begin with, the upper charts illustrate the effect of the strain rate on the depth of the well in the Θ dependences. As pointed out in the Introduction, it has an opposite sign for RD and TD samples (see [10,12]). The bottom charts reveal a very good similitude between the Θ stages and their counterparts in the CR in the case of RD specimens. Therewith, the positions of the *B/C* transition closely coincide with the corresponding peaks in the CR curves at all strain rates. In Figure 4a, the preceding bottom of the CR well is reached with some delay regarding the *A/B* transition. Since the wells at the A/B transition are effectively depressed for both CR and Θ when $\dot{\varepsilon}_a$ is increased, this delay becomes indiscernible at the highest strain rate of $8 \times 10^{-3}\ s^{-1}$ (not shown; see [10,12] for illustrations of Θ behavior).

In contrast to RD samples, CR behavior of TD samples qualitatively differs from the respective $\Theta(\varepsilon)$-curves at stages *A* and *B*. It can be seen that C R also shows a peculiarity resembling the stage *B* but it occurs in the strain range corresponding to stage *A* and is quite short regarding the extent of the stage *B*. More exactly, at $2 \times 10^{-3}\ s^{-1}$, the CR curve passes a shallow minimum followed by a slight increase or a plateau (arrow in Figure 4b). This variation is more pronounced at $8 \times 10^{-3}\ s^{-1}$ but only displays a bend in the CR curve at $5 \times 10^{-4}\ s^{-1}$ (arrow in Figure 4a), i.e., the sign of the strain-rate effect on this peculiarity is the same as for the Θ well. This trend is followed by a new decrease in the CR, starting shortly before the *A/B* transition and continuing after it. The decrease is again slowed down during stage *B*. A slight growth towards the *B/C* transition, evoking a certain similitude with Θ behavior, can be seen at slow deformation (Figure 4a) but is almost indiscernible in the faster tests (Figure 4b). On the whole, the overall CR pattern is more complex than that of Θ.

One more consequence of the particular CR pattern in TD samples is that in contrast to the relationship $\Theta_{RD} > \Theta_{TD}$ which is satisfied almost everywhere, except for the very beginning of plastic deformation (beyond the axes limits in Figures 3 and 4 [12]), CR is stronger in TD than in RD samples up to about 1% of true strain.

3.2. Statistics of Deformation Processes on a Mesoscopic Scale

Figure 5 represents PDF for the intensity of the AE collected during deformation of the samples from Figure 2. It should be underlined that the statistical analysis must be applied to a stationary variable. To satisfy the condition of stationarity, the statistics were analyzed within relatively short strain intervals during the stage C characterized by a slow evolution of the AE. The strain intervals were selected as short as possible to deal with a stationary range of AE amplitudes, but large enough to provide representative statistical samples. The length of each selected interval was varied and the PDF calculation was repeated in order to verify the robustness of the calculation result regarding the interval choice. For each specimen, Figure 5 shows the results of analysis within strain intervals corresponding to: (i) the stabilized plastic flow after the elastoplastic transition; (ii) a later deformation stage before necking; (iii) the onset of necking; and (iv) a later phase of necking close to fracture.

The data of Figure 5 allow for a conclusion that like in many materials studied so far, the energy distributions of deformation processes obey power-law statistics. For all strain rates and both orientations (TD and RD), the exponent β of the fitting power-law function, $P(I) \sim I^{-\beta}$, varies during the macroscopically uniform plastic flow within the interval from 1.5 to 1.8, similar to the typical value of 1.5 reported for *hcp* materials [17,20]. It considerably increases close to the onset of necking (more specifically, $\beta \approx 2.3 \pm 0.1$ for the TD sample and 2.6 ± 0.1 for the RD sample in Figure 5) and decreases again when the neck is developed. Since a higher β means a higher probability of smaller events, the β increase indicates that the transition to necking is associated with noisier (uncorrelated) deformation processes, while the development of the strain localization restores the correlation. It can be noted that the numbers in the parenthesis suggest that β takes on a higher value for the RD orientation. Such inequality for the necking region was confirmed at other strain rates, too, except for the highest $\dot{\varepsilon}_a$, where a relatively strong data scatter around the onset of necking prevented from quantitative analysis in this strain range. Some tendency to higher β for the RD orientation could also be remarked at other deformation stages. However, since the span of β is much smaller at these stages, this observation would require additional verification using a large number of samples to reduce data scatter.

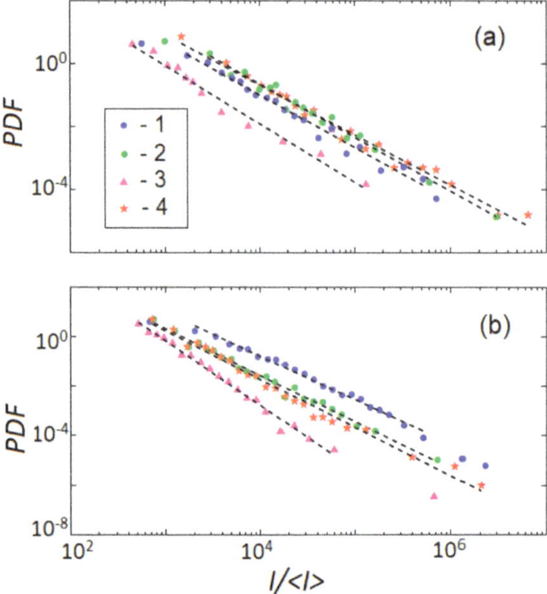

Figure 5. Probability distribution functions (PDF) for the normalized intensity of AE events collected at different stages of deformation of (**a**) TD and (**b**) RD samples of Figure 2: 1—During stabilized plastic flow immediately after the elastoplastic transition; 2—At a later stage before necking; 3—Close to the onset of necking; and 4—During the necking.

The power-law statistics of deformation processes on a mesoscopic scale is also confirmed by the first data of analysis of the local strain-rate bursts. Figure 6a displays a spatiotemporal map $\dot{\varepsilon}_{loc}(x,t)$ for a TD sample deformed at 8×10^{-3} s^{-1} and the corresponding family of $\dot{\varepsilon}_{loc}(t)$-curves served as sources for the map reconstruction. In other words, these curves represent horizontal cross-sections of the map and their fluctuations correspond to the color fluctuations in the above chart. Figure 6b presents the PDF for the amplitudes of fluctuations on one of such curves. Since the time resolution of the extensometry is three orders of magnitude coarser than that of the AE measurements, the statistical sample was collected over the entire range of the uniform plastic flow, starting immediately after the elastoplastic transition ($t \approx 8$ s) and ending before the onset of necking ($t \approx 14$ s). It can be seen that despite the lower resolution of the method, the PDF obeys a power law with β in the same range as for the AE. The feasibility of such analysis was poorer than in the case of AE, most likely because of the insufficient resolution, and the power law was correctly established for a part of data, only. Nevertheless, the entirety of these results corroborate the above conclusions drawn from the AE analysis. Moreover, since the power-law statistics is a mathematical expression of scale-invariant behavior, the very fact that the power law is found using methods with the resolution corresponding to distinct time (amplitude) scales is by itself a strong evidence of scale invariance.

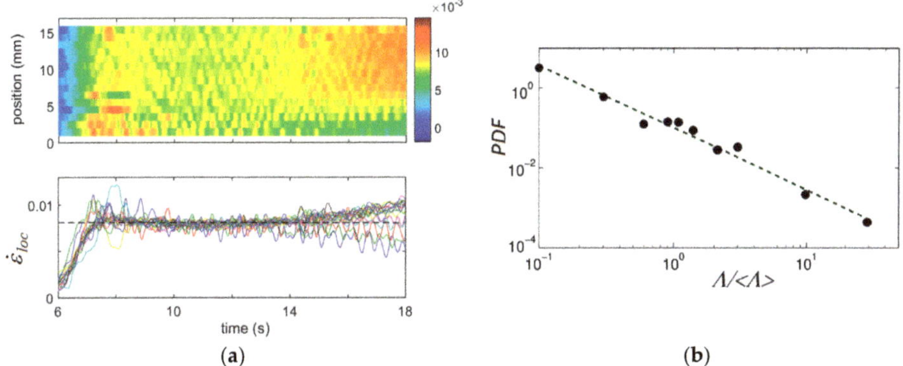

Figure 6. (**a**) Spatiotemporal pattern $\dot{\varepsilon}_{loc}(x,t)$ (top) and the corresponding family of $\dot{\varepsilon}_{loc}(t)$-curves measured at various positions along the centerline of a TD specimen (bottom); (**b**) the corresponding statistical distribution of the normalized amplitudes Λ of fluctuations for one of these curves. The horizontal dashed line indicates the imposed strain rate, $\dot{\varepsilon}_a = 8 \times 10^{-3}\ \mathrm{s}^{-1}$.

4. Discussion

In the present work, high resolution techniques based on the AE and local strain-rate fields were applied, on the one hand, to verify the hypothesis of dislocation mechanism of the non-monotonous work hardening of α-Ti polycrystals deformed by tension [9–12] and, on the other hand, to investigate the deformation processes on finer scales relevant to the so-called collective processes in the dislocation dynamics. The analysis of the AE average count rate allowed, first of all, to corroborate the presence of non-monotonous trends during the elastoplastic transition. Furthermore, the AE revealed specific CR behaviors in the case of TD samples. Taking into account that twinning is a powerful source of AE, so that rare twinning events could be supposed to produce visible AE, it is important to revise whether the AE behavior enters into the same dislocation-based framework and interpret the observed specific behaviors.

As suggested in [10,12], the Θ reduction during stage *A* is caused, in both kinds of samples, by a progressive onset of plastic flow in various grains of the polycrystal. It should be clarified that this statement does not contradict the roughly simultaneous occurrence of plastic activity all over the specimen gage length, revealed in the local strain-rate maps (Figures 2c and 6a). Indeed, the presence of $\dot{\varepsilon}_{loc}$ fluctuations in these figures means that such a conclusion is only valid globally. A magnification of such maps (Figure 7) shows that these fluctuations generate complex structures including not only an apparent disorder but even wavy structures similar to various patterns observed in metals with cubic symmetry (e.g., [16,29,30]).

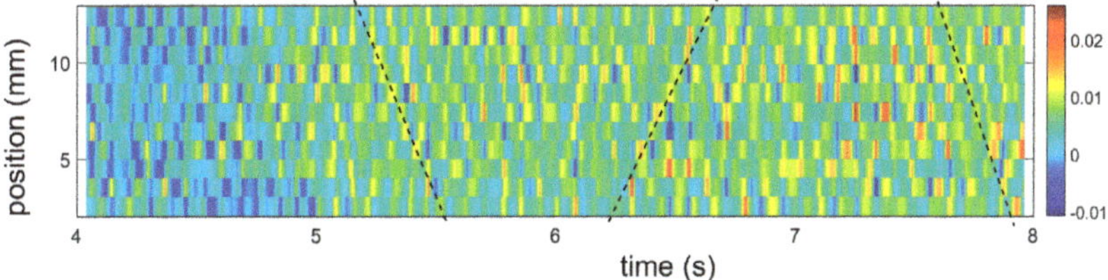

Figure 7. Portion of a spatiotemporal pattern for an RD specimen deformed at $8 \times 10^{-3}\ \mathrm{s}^{-1}$. Inclined dashed lines provide visual guidance indicating an ordering of the intermittent plastic activity (bright spots) into a wave-like propagation along the specimen centerline. A detailed analysis of the local strain-rate behavior will be published elsewhere.

Taking into account that the AE is generally strongly attenuated upon the elastoplastic transition, i.e., when the microplasticity occurring during "elastic" deformation shifts into the macroscopic plastic flow, the conjecture on a correlation between stage *A* and the gradual plastification of grains agrees with the attenuation of the AE during stage *A*. In RD samples, this stage was shown to be mostly governed by prismatic glide, i.e., the motion of dislocations with the Burgers vector <*a*> in prismatic systems. The onset of stage *B*, in these samples, was attributed to the onset of secondary slip of both <*a*> and <*a* + *c*> dislocations in pyramidal systems, while prismatic slip remained the main mechanism of plastic flow during the entire test. Indeed, the forest dislocations are themselves obstacles to the motion of dislocations and, therefore, contribute to an increase in the strain hardening rate. Moreover, they may be effective dislocation sources and lead to multiplication of mobile dislocations. The production of dislocations would result in a strong strain hardening because of dislocation interactions with both forest and collinear dislocations but also lead to an increase in the AE. These reasons can explain why the stage *B* is associated with the increasing branches both in CR and Θ. The following depression in both curves during stage C can also be explained in a conventional way, as a consequence of the reduction in the rate of the dislocations production.

It was remarked in [12] that although the modeling results successfully reproduced the main features of the three-stage work hardening for TD samples, too, the qualitative explanation of the stage *B* is more difficult in this case because the secondary slip starts much earlier than in RD samples. The model predicted that more than 65% of grains are already performing multiple slips at the end of stage *A*. Surprisingly, this difficulty conforms to the peculiarity observed in the AE. Indeed, an analogue of stage *B*, i.e., a transient increase in the CR curves, occurs before the end of stage *A* in TD samples, in consistence with the dislocations multiplication due to the presence of a secondary slip. Moreover, while $\Theta_{TD} < \Theta_{RD}$ in the entire strain range in Figures 3 and 4, CR is initially higher in TD samples.

It is necessary to mention that twinning could be suggested as an alternative explanation of the high CR during stage *A* in TD specimens because they generally manifest a higher fraction of twins than the RD specimens [10]. However, the analysis of the microstructure in the so-called interrupted tests showed that the largest part of twins is formed after the end of stage *B*, when the AE is gradually decreased [12]. Besides, some twins are also found in RD samples the behavior of which directly follows from the predictions of the dislocation-based model.

It can thus be concluded that the models based on the consideration of the dislocation glide in different slip systems are able to explain the non-monotonous nature of both strain hardening and AE accompanying plastic flow of α-Ti. At the same time, it should be underlined that overall the similitude between the CR and Θ dependences is much worse for TD than for RD samples. Not only the stages detected in the AE evolution are shifted regarding their counterparts on the Θ curves, but its overall behavior reveals even more complexity, e.g., a tendency to an additional increase in CR during the stage *B*. It can be suggested that besides the polycrystal plasticity, a realistic model of AE needs to handle the physical mechanisms associated with the generation and propagation of AE in the anisotropic material.

Finally, the measurements using the AE and local extensometry provide a further insight into the heterogeneity of plastic deformation on finer scales. The major conclusion stemming from the first analyses presented in the paper is that the statistics of deformation processes obey a power law. These observations bear evidence to an avalanche-like character of the dislocation dynamics and relate the present results to a more general and intensely studied problem of self-organization of the dislocation dynamics in various materials [31–33]. Adequate modeling of these aspects would require a correct consideration of the spatial coupling associated with local deformations, in particular, internal stresses caused by plastic strain incompatibilities between neighboring grains. Indeed, the observation of a variation of the power-law exponent with deformation, and also between specimens with different orientations, testifies that in order to reproduce the fine-scale heterogeneity, models must be able to consider real grain structures and the role of grain boundaries for the dislocation glide (formation of dislocation pile-ups, absorption or transmission of dislocations through the interface, and so on).

This level of consideration is missing in model [12] and, more generally, in the models based on the homogenization approach, and will present a challenge for future investigations. It is particularly interesting in view of the recent technological developments which show a tendency to fabricate Ti and Ti alloys with complex microstructures containing micro- and nano-size elements, e.g., submicron grains [34–36].

5. Conclusions

In summary, the comparison of several kinds of material response to plastic deformation, including the work hardening, the acoustic emission, and the evolution of the local strain-rate field along the tensile axis, permits the following conclusions to be drawn:

- The average activity of the AE accompanying tensile deformation of α-Ti has a non-monotonous character in the strain range corresponding to the three-stage work-hardening behavior characteristic of *hcp* materials. The totality of data corroborates the recently proposed model predicting that the work-hardening peculiarity can be caused by the dislocation mechanism alone, due to the strong anisotropy of the dislocation glide in different slip systems in *hcp* crystals [12].
- More specifically, the AE activity detected in the samples cut along the rolling direction shows three stages corresponding rather well to the work-hardening stages. Both AE and deformation behaviors allow for a unique interpretation considering the transition from the initially predominant prismatic glide to the multiple slip. The immediate onset of the multiple slip makes more intricate the qualitative interpretation of such results for samples oriented in the transverse direction and would require a further development of the modeling to couple the mechanisms of plasticity and acoustic emission in the anisotropic material.
- On the mesoscopic scales relevant to either AE or local strain rates, the deformation processes in α-Ti obey power-law statistics characteristic of the avalanche dynamics. Such behavior has recently been established for various materials and first and foremost, for *hcp* crystals [31]. The values of the power-law exponents found in the present experiments are also close to the seemingly universal exponent (β ~1.5) reported in the literature. At the same time, β-values determined for different subsets of large AE datasets testify that the power law may depend on the sample orientation and the strain range. In particular, β was found to increase considerably close to the onset of necking and to decrease again during the neck development. These data indicate that the statistical analysis of the data obtained using high-resolution methods of investigation of the plastic flow may be useful for the further progress in the understanding of the plasticity of *hcp* materials.

Author Contributions: Conceptualization, M.L., T.L., and T.R.; Formal analysis, M.L., K.A., T.L., and A.R.; Investigation, M.L., K.A., T.L., T.R., and A.R.; Methodology, M.L. and T.L.; Writing—original draft, M.L.

Funding: This research was funded by the French State through the program "Investment in the future" operated by the National Research Agency (ANR) and referenced by ANR-11-LABX-0008-01 (LabEx DAMAS), and through the ANR project PHIRCILE (ANR 2010 JCJC 0914 01).

Conflicts of Interest: The authors declare no conflict of interest. The founding sponsors had no role in the design of the study; in the collection, analyses, or interpretation of data; in the writing of the manuscript, and in the decision to publish the results.

References

1. Kocks, U.F. Laws for Work-Hardening and Low-Temperature Creep. *J. Eng. Mater. Technol.* **1976**, *98*, 76–85. [CrossRef]
2. Estrin, Y.; Toth, L.S.; Molinari, A.; Bréchet, Y. A dislocation-based model for all hardening stages in large strain deformation. *Acta Mater.* **1998**, *15*, 5509–5522. [CrossRef]

3. Kailas, S.V.; Prasad, Y.V.R.K.; Biswas, S.K. Influence of initial texture on the microstructural instabilities during compression of commercial α-Titanium at 25 °C to 400 °C. *Metall. Mater. Trans. A* **1994**, *25*, 1425–1434. [CrossRef]
4. Wang, B.; Xin, R.; Huang, G.; Liu, Q. Effect of crystal orientation on the mechanical properties and strain hardening behavior of magnesium alloy AZ31 during uniaxial compression. *Mater. Sci. Eng. A* **2012**, *534*, 588–593. [CrossRef]
5. Salem, A.A.; Kalidindi, S.R.; Doherty, R.D. Strain hardening regimes and microstructure evolution during large strain compression of high purity titanium. *Scr. Mater.* **2002**, *46*, 419–423. [CrossRef]
6. Jiang, J.; Godfrey, A.; Liu, W.; Liu, Q. Microtexture evolution via deformation twinning and slip during compression of magnesium alloy AZ31. *Mater. Sci. Eng. A* **2008**, *483–484*, 576–579. [CrossRef]
7. Sarker, D.; Chen, D.L. Detwinning and strain hardening of an extruded magnesium alloy during compression. *Scr. Mater.* **2012**, *67*, 165–168. [CrossRef]
8. Nemat-Nasser, S.; Guo, W.G.; Cheng, J.Y. Mechanical properties and deformation mechanisms of a commercially pure titanium. *Acta Mater.* **1999**, *47*, 3705–3720. [CrossRef]
9. Becker, H.; Pantleon, W. Work-hardening stages and deformation mechanism maps during tensile deformation of commercially pure titanium. *Comp. Mater. Sci.* **2013**, *76*, 52–59. [CrossRef]
10. Roth, A.; Lebyodkin, M.A.; Lebedkina, T.A.; Lecomte, J.S.; Richeton, T.; Amouzou, K.E.K. Mechanisms of anisotropy of mechanical properties of a-titanium in tension conditions. *Mater. Sci. Eng. A* **2014**, *596*, 236–243. [CrossRef]
11. Barkia, B.; Doquet, V.; Couzinié, J.P.; Guillot, I.; Héripré, E. In situ monitoring of the deformation mechanisms in titanium with different oxygen contents. *Mater. Sci. Eng. A* **2015**, *636*, 91–102. [CrossRef]
12. Amouzou, K.E.K.; Richeton, T.; Roth, A.; Lebyodkin, M.A.; Lebedkina, T.A. Micromechanical modeling of hardening mechanisms in commercially pure α-titanium in tensile condition. *Int. J. Plast.* **2016**, *80*, 222–240. [CrossRef]
13. Conrad, H. Effect of interstitial solutes on the strength and ductility of titanium. *Prog. Mater. Sci.* **1981**, *26*, 123–403. [CrossRef]
14. Naka, S. Etude des Mécanismes de Déformation Plastique à Basse Température de Monocristaux de Titane Alpha. Ph.D. Thesis, O.N.E.R.A., Palaiseau, France, 1983.
15. Máthis, K.; Chmelík, F. Exploring plastic deformation of metallic materials by the acoustic emission technique. In *Acoustic Emission*; Sikorski, W., Ed.; InTech: Rijeka, Croatia, 2012; pp. 23–48. ISBN 978-953-51-0056-0.
16. Roth, A.; Lebedkina, T.A.; Lebyodkin, M.A. On the critical strain for the onset of plastic instability in an austenitic FeMnC steel. *Mater. Sci. Eng. A* **2012**, *539*, 280–284. [CrossRef]
17. Richeton, T.; Dobron, P.; Chmelik, F.; Weiss, J.; Louchet, F. On the critical character of plasticity in metallic single crystals. *Mater. Sci. Eng. A* **2006**, *424*, 190–195. [CrossRef]
18. Fressengeas, C.; Beaudoin, A.J.; Entemeyer, D.; Lebedkina, T.; Lebyodkin, M.; Taupin, V. Dislocation transport and intermittency in the plasticity of crystalline solids. *Phys. Rev. B* **2009**, *79*, 014108. [CrossRef]
19. Lebyodkin, M.A.; Kobelev, N.P.; Bougherira, Y.; Entemeyer, D.; Fressengeas, C.; Gornakov, V.S.; Lebedkina, T.A.; Shashkov, I.V. On the similarity of plastic flow processes during smooth and jerky flow: Statistical analysis. *Acta Mater.* **2012**, *60*, 3729–3740. [CrossRef]
20. Weiss, J.; Rhouma, W.B.; Richeton, T.; Dechanel, S.; Louchet, F.; Truskinovsky, L. From mild to wild fluctuations in crystal plasticity. *Phys. Rev. Lett.* **2015**, *114*, 105504. [CrossRef] [PubMed]
21. Lebedkina, T.A.; Zhemchuzhnikova, D.A.; Lebyodkin, M.A. Correlation versus randomization of jerky flow in an AlMgScZr alloy using acoustic emission. *Phys. Rev. E* **2018**, *97*, 013001. [CrossRef] [PubMed]
22. Ghaderi, A.; Barnett, M.R. Sensitivity of deformation twinning to grain size in titanium and magnesium. *Acta Mater.* **2011**, *59*, 7824–7839. [CrossRef]
23. Lebyodkin, M.A.; Shashkov, I.V.; Lebedkina, T.A.; Mathis, K.; Dobron, P.; Chmelik, F. Role of superposition of dislocation avalanches in the statistics of acoustic emission during plastic deformation. *Phys. Rev. E* **2013**, *88*, 042402. [CrossRef] [PubMed]
24. Lebedkina, T.A.; Lebyodkin, M.A.; Chateau, J.-P.; Jacques, A.; Allain, S. On the mechanism of unstable plastic flow in an austenitic FeMnC TWIP steel. *Mater. Sci. Eng. A* **2009**, *519*, 147–154. [CrossRef]
25. Weiss, J.; Grasso, J.R.; Miguel, M.C.; Vespignani, A.; Zapperi, S. Complexity in dislocation dynamics: Experiments. *Mater. Sci. Eng. A* **2001**, *309–310*, 360–364. [CrossRef]

26. Lebyodkin, M.A.; Lebedkina, T.A.; Chmelík, F.; Lamark, T.T.; Estrin, Y.; Fressengeas, C.; Weiss, J. Intrinsic structure of acoustic emission events during jerky flow in an Al alloy. *Phys. Rev. B* **2009**, *79*, 174114. [CrossRef]
27. Maaβ, R.; Derlet, P.M. Micro-plasticity and recent insights from intermittent and small-scale plasticity. *Acta Mater.* **2018**, *143*, 338–363. [CrossRef]
28. Klusemann, B.; Fischer, G.; Böhlke, T.; Svendsen, B. Thermomechanical characterization of Portevin–Le Châtelier bands in AlMg3 (AA5754) and modeling based on a modified Estrin–McCormick approach. *Int. J. Plast.* **2015**, *67*, 192–216. [CrossRef]
29. Zhemchuzhnikova, D.; Lebyodkin, M.; Lebedkina, T.; Mogucheva, A.; Yuzbekova, D.; Kaibyshev, R.O. Peculiar Spatiotemporal Behavior of Unstable Plastic Flow in an AlMgMnScZr Alloy with Coarse and Ultrafine Grains. *Metals* **2017**, *7*, 325. [CrossRef]
30. Mudrock, R.N.; Lebyodkin, M.A.; Kurath, P.; Beaudoin, A.J.; Lebedkina, T.A. Strain-rate fluctuations during macroscopically uniform deformation of a solution-strengthened alloy. *Scr. Mater.* **2011**, *65*, 1093–1096. [CrossRef]
31. Zaiser, M. Scale invariance in plastic flow of crystalline solids. *Adv. Phys.* **2006**, *55*, 185–245. [CrossRef]
32. Zuev, L.B. On the waves of plastic flow localization in pure metals and alloys. *Ann. Phys.* **2007**, *16*, 286–310. [CrossRef]
33. Salje, E.K.H.; Dahmen, K.A. Crackling noise in disordered materials. *Annu. Rev. Condens. Matter Phys.* **2014**, *5*, 233–254. [CrossRef]
34. Gu, Y.; Ma, A.; Jiang, J.; Yuan, Y.; Li, H. Deformation structure and mechanical properties of pure titanium produced by rotary-die equal-channel-angular pressing. *Metals* **2017**, *7*, 297. [CrossRef]
35. Cojocaru, V.D.; Serban, N. Effects of solution treating on microstructural and mechanical properties of a heavily deformed new biocompatible Ti-Nb-Zr-Fe alloy. *Metals* **2018**, *8*, 297. [CrossRef]
36. Ji, Z.; Shen, C.; Wei, F.; Li, H. Dependence of macro- and micro-properties on α plates in Ti-6Al-2Zr-1Mo-1V alloy with tri-modal microstructure. *Metals* **2018**, *8*, 299. [CrossRef]

Article

Micromechanical Modeling of the Elasto-Viscoplastic Behavior and Incompatibility Stresses of β-Ti Alloys

Safaa Lhadi [1,2], Maria-Rita Chini [1], Thiebaud Richeton [1,2], Nathalie Gey [1,2], Lionel Germain [1,2] and Stéphane Berbenni [1,2,*]

1 Laboratoire d'Étude des Microstructures et de Mécanique des Matériaux, Université de Lorraine, CNRS, Arts et Métiers Paris Tech, LEM3, F-57000 Metz, France; safaa.lhadi-khalouche@univ-lorraine.fr (S.L.); rita.chini@univ-lorraine.fr (M.-R.C.); thiebaud.richeton@univ-lorraine.fr (T.R.); nathalie.gey@univ-lorraine.fr (N.G.); lionel.germain@univ-lorraine.fr (L.G.)

2 Laboratory of Excellence on Design of Alloy Metals for low-mAss Structures (DAMAS), Université de Lorraine, F-57000 Metz, France

* Correspondence: stephane.berbenni@univ-lorraine.fr; Tel.: +33-372-747-842

Received: 8 June 2018; Accepted: 15 July 2018; Published: 17 July 2018

Abstract: Near β titanium alloys can now compete with quasi-α or α/β titanium alloys for airframe forging applications. The body-centered cubic β-phase can represent up to 40% of the volume. However, the way that its elastic anisotropy impacts the mechanical behavior remains an open question. In the present work, an advanced elasto-viscoplastic self-consistent model is used to investigate the tensile behavior at different applied strain rates of a fully β-phase Ti alloy taken as a model material. The model considers crystalline anisotropic elasticity and plasticity. It is first shown that two sets of elastic constants taken from the literature can be used to well reproduce the experimental elasto-viscoplastic transition, but lead to scattered mechanical behaviors at the grain scale. Incompatibility stresses and strains are found to increase in magnitude with the elastic anisotropy factor. The highest local stresses are obtained toward the end of the elastic regime for grains oriented with their <111> direction parallel to the tensile axis. Finally, as a major result, it is shown that the elastic anisotropy of the β-phase can affect the distribution of slip activities. In contrast with the isotropic elastic case, it is predicted that {112} <111> slip systems become predominant at the onset of plastic deformation when elastic anisotropy is considered in the micromechanical model.

Keywords: polycrystalline β-Ti; elastic anisotropy; elastic/plastic incompatibilities; elasto-viscoplastic self-consistent scheme (EVPSC); slip activity

---

1. Introduction

Near-β titanium alloys are constituted of hexagonal compact (HCP) α-phase and body-centered cubic (BCC) β-phase, with the latter representing up to 40% of the volume. These alloys can achieve high specific strength thanks to transformation processes, including β and α/β forging before aging. They can now compete with quasi-α or α/β titanium alloys for airframe forging applications such as landing gears, turbine engines, and rotor systems. The microstructure of near-β titanium alloys is complex and multiscale. The α-phase contains primary α nodules and secondary α platelets that are embedded in β grains. It has been recently shown that the α/β forging process does not completely break the prior millimeter-sized β forged grains, but rather fragments them into equiaxed subgrains with close orientations and diameters about 5 μm to 10 μm [1].

The impact of the β-phase on the mechanical properties of near-β titanium alloys has not been completely understood yet. The β-phase (BCC) is elastically anisotropic [2] so that the elastic behavior of β-phase single crystal is strongly dependent on the loading direction. Owing to its cubic symmetry, the β-single crystal elastic stiffness tensor contains three independent elastic constants C_{11}, C_{12}, and C_{44}.

The degree of elastic anisotropy of such a cubic phase can be evaluated by the so-called Zener factor [3], which is defined as:

$$A = \frac{2C_{44}}{C_{11} - C_{12}} \tag{1}$$

The published single-crystal elastic constants (SEC) for the β-phase span very different values, and the reported Zener anisotropy factor A ranges from 1.4 to 8.3 [4–13]. These discrepancies may be due to the specific chemical compositions and thermomechanical treatments of the studied Ti alloys, but also to the difficulty to measure SEC in multiphase polycrystalline specimens.

It is well known that microstructural features such as grain size, crystallographic texture, and grain morphology strongly influence the mechanical properties of polycrystalline materials [14–16]. In addition, the elastic anisotropy may also influence the elastic-viscoplastic transition (microplastic regime, incipient plasticity...) and the development on internal stresses, which might initiate cracks [17]. However, only a few studies have focused on the effect of the β-phase elastic anisotropy in near-β Ti alloys [4,5,18–20]. Therefore, the purpose of the present paper is to perform a comprehensive study of the role of the β-phase elastic anisotropy on the local and overall mechanical behavior of a fully β Ti alloy taken as a model material. Tensile tests performed at different strain rates on Ti-17 alloys with 100% equiaxed β microstructures constitute the experimental basis of the study [4,21,22]. On the other hand, the theoretical investigation is carried out through the use of an advanced elasto-viscoplastic self-consistent model (EVPSC) that considers an affine linearization of the viscoplastic flow rule and one-site self-consistent approximation thanks to the translated field method [23]. It is thought that such a micro–macro scale transition model is well suited to the modeling of disordered materials such as fully β-phase Ti polycrystalline aggregates. Indeed, the EVPSC model can account for crystal elasticity and plasticity, crystallographic texture effects, and strain-rate effects, unlike the elastoplastic self-consistent (EPSC) scheme [5]. Moreover, it is able to consider an extensive number of grains described by their crystallographic orientations, volume fractions, and morphologies. As outputs, it provides the local (at the scale of the grains) and overall effective (at the scale of the polycrystalline aggregate) mechanical behaviors. This study aims to better understand the effects of the elastic anisotropy on the strain-hardening behavior, the local mechanical fields, and the distribution of slip activities.

The paper is organized as follows. Section 2 is devoted to a brief description of the micromechanical model. The governing field equations of the elasto-viscoplastic problem are reviewed, and the single crystal constitutive laws are described. Model parameters and results are reported and discussed in Section 3. Finally, concluding remarks are summarized in Section 4.

2. Micromechanical Model

In this paper, the micro–macro scale transition from single to polycrystal is assessed through the elasto-viscoplastic self-consistent (EVPSC) model based on the translated field method and on an "affine" extension of the viscoplastic strain rate first introduced in [23] for thermo-elasto-viscoplastic heterogeneous materials. Here, thermal effects (i.e., thermal strains) are disregarded. The used EVPSC model has been described in details elsewhere [20,23,24], and therefore, only a brief description is given here.

2.1. Fields Equations

In homogenization theory, the macroscopic stress rate and strain rate tensors ($\dot{\mathbf{\Sigma}}$ and $\dot{\mathbf{E}}$) of a representative volume element (RVE) with volume V are obtained by volume averaging the local (grain) stress rate and strain rate tensors ($\dot{\boldsymbol{\sigma}}$ and $\dot{\boldsymbol{\varepsilon}}$) as follows:

$$\dot{\mathbf{\Sigma}} = \langle \dot{\boldsymbol{\sigma}} \rangle, \quad \dot{\mathbf{E}} = \langle \dot{\boldsymbol{\varepsilon}} \rangle \tag{2}$$

Within the infinitesimal strain framework, the total local strain rate relative to an elasto-viscoplastic behavior (of Maxwell type) is decomposed into:

$$\dot{\varepsilon} = \dot{\varepsilon}^{\mathbf{e}} + \dot{\varepsilon}^{\mathbf{vp}} \tag{3}$$

with $\dot{\varepsilon}^{\mathbf{e}} = \mathbf{s} : \dot{\sigma}$ from the generalized Hooke's law where $\mathbf{s}\ (= \mathbf{c}^{-1})$ is the local elastic compliance tensor, and $\dot{\varepsilon}^{\mathbf{vp}}$ is given by a nonlinear viscoplastic constitutive flow rule that depends on the Cauchy stress tensor $\boldsymbol{\sigma}$. Static equilibrium in the absence of body force and kinematic compatibility complete the field equations of the problem [23].

2.2. Single Crystal's Constitutive Behavior

The behavior of the β single crystal is supposed to be elastic-viscoplastic, where plastic strain and rotation result from crystallographic slips on specific systems "s".

The constitutive viscoplastic model with "pseudo-multiplier" developed by Méric et al. (1991) [25] and Méric and Cailletaud [26] is used to describe the slip rate $\dot{\gamma}^{s}$ for each slip system s:

$$\dot{\gamma}^{s} = \left\langle \frac{|\tau^{s} - x^{s}| - r^{s}}{K} \right\rangle^{n} \operatorname{sign}(\tau^{s} - x^{s}) \tag{4}$$

where $\langle x \rangle = \max(x, 0)$ and τ^{s} is the resolved shear stress, $\tau^{s} = \mathbf{R^{s}} : \boldsymbol{\sigma}$ with $\mathbf{R^{s}}$ is the symmetric Schmid orientation tensor, and x^{s} and r^{s} are associated with kinematic and isotropic hardening on s, respectively:

$$x^{s} = c^{s}\alpha^{s} \quad \text{with} \quad \dot{\alpha}^{s} = \dot{\gamma}^{s} - d^{s}\alpha^{s}|\dot{\gamma}^{s}| \tag{5}$$

and:

$$r^{s} = r_{0}^{s} + \sum_{r=1}^{N} H_{sr} b^{r} Q^{r} q^{r} \quad \text{with} \quad \dot{q}^{r} = (1 - b^{r} q^{r})|\dot{\gamma}^{r}| \tag{6}$$

In the above equations, K and n are two material coefficients characterizing the viscous effect, c, b, d and q are hardening parameters, and $\mathbf{H}$ is an N × N matrix describing the interactions between the different slip systems, where N is the total number of the slip systems (here N = 48, see Section 3.1) of the β single crystal. Therefore, at the grain level, the viscoplastic strain rate is given by:

$$\dot{\varepsilon}^{\mathbf{vp}} = \sum_{s=1}^{N} \mathbf{R^{s}} \left\langle \frac{|\tau^{s} - x^{s}| - r^{s}}{K} \right\rangle^{n} \operatorname{sign}(\tau^{s} - x^{s}) \tag{7}$$

2.3. Self-Consistent Approximation Based on Affine Translated Field Method

Here, the one-site self-consistent approximation is formulated using the translated field method [27–29], which is an internal variable approach. The details of this method with an affine linearization of the viscoplastic flow rule can be found in [20]. Compared to hereditary approaches [30], the numerical implementation is much easier, as no use of Laplace–Carson transform is needed. As a result, the explicit expressions of the tensors of stress rates $\dot{\sigma}$ and strain rates $\dot{\varepsilon}$ at the level of each grain with a given crystallographic orientation can be obtained as a function of applied stress rates and/or strain rates ($\dot{\boldsymbol{\Sigma}}$ and $\dot{\mathbf{E}}$) (mixed boundary conditions) and stress history $\boldsymbol{\sigma}$ in the grain depending on coupled elastic and viscoplastic intergranular accommodations. For details, the reader can find the stress and strain rate concentration equations of the self-consistent model in [20,23,24].

3. Results and Discussions

For all of the simulation results shown in this paper, the considered polycrystalline material has a fully β-phase microstructure. Furthermore, the representative volume element (RVE) always contains 2016 equiaxed grains with the same volume fraction and random crystallographic texture. The texture is first decomposed into 28 fiber components parallel to the loading direction. All of the grain orientations are deduced by a rotation of 5° around the tensile axis (see the inverse pole figure in Figure 1a). Besides, in the whole paper, the tensile loading direction is parallel to the X_3-axis.

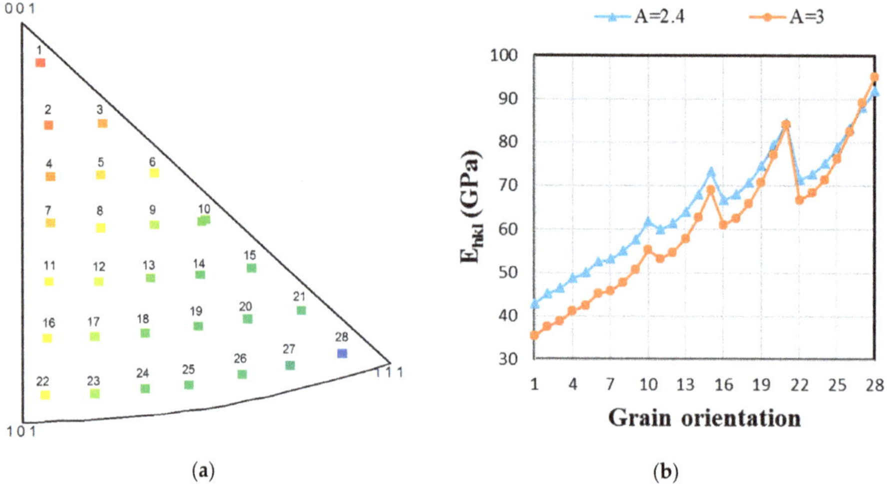

Figure 1. (**a**) Inverse pole figure representing the 2016 grains with random texture decomposed into 28 fiber components with respect to the tensile axis; (**b**) Directional Young's moduli of single crystals with orientations 1 to 28 parallel to the tensile axis obtained from both considered β-phase single-crystal elastic constants (SEC) (A = 2.4 (triangles) and A = 3 (circles)).

3.1. Model Parameters

Two sets of β SEC are considered in this study: The ones considered by Martin (2012) [4], which gives a Zener elastic anisotropy factor of A = 2.4, and the ones of Petry et al. (1991) [6], which gives A = 3. As reported in [20], both SEC give rise to effective Young's modulus close to the measured one for a fully β microstructure Ti-5553 and Ti-17 alloys, which are about 68 GPa and 60 GPa, respectively [21,31]. Table 1 provides the values of the SEC and the corresponding effective Young's modulus computed for a fully β microstructure assuming equiaxed grains and random texture. Unlike the computed effective Young's modulus for the polycrystal with isotropic crystallographic orientation distribution, the single crystalline Young's moduli for the 28 individual orientations are generally different for each SEC (see Figure 1b). As expected, the gap between the directional Young's moduli of single crystals along the directions <100> (i.e., grain orientation 1) and <111> (grain orientation 28) increases with the anisotropy factor A.

Table 1. β-phase single-crystal elastic constants (SEC) considered in this study, Zener elastic anisotropy factor (A) and effective Young's modulus (EYM) computed with the purely elastic self-consistent scheme [32,33] for 100% β-microstructure assuming equiaxed grains and random texture.

Ti-Based Alloy	**SEC (GPa)**			**A**	**EYM (GPa) of 100% β (Calculated)**
	C_{11}	C_{12}	C_{44}		
Ti-17 [4]	100	70	36	2.4	69.4
Pure [6]	134	110	36	3	66.6

Table 2 provides the viscoplastic model parameters that are related to the β single crystal. The Méric-Cailletaud's flow rule parameters [26] were identified by [4] on Ti-5553 and Ti-17 both 100% β equiaxed grains with random texture. Here, the experimental yield strengths and the elastic-viscoplastic transition on the tensile stress–strain curves at both applied strain rates (see Figure 2) were used to fit the initial critical resolved shear stresses (CRSS) for the three slip system families of the BCC β single crystal denoted r_0. The values of r_0 are reported in Table 2. Furthermore, a quasi-linear hardening rate is observed between 2–4% strain (see Figure 2). Therefore, here, a linear kinematic hardening is sufficient ($c \neq 0$, $d = 0$) for the monotonic tensile test simulations, and the isotropic hardening parameters were disregarded ($Q = d = 0$). Both material parameters n and K (see Table 2) were identified to reproduce the low strain rate sensitive behavior between 1–4% strains using the experimental points at both strain rates. The elastic anisotropy factor of A = 3 was used to identify the viscoplastic parameters. For further justifications on the choice of these viscoplastic parameters in the constitutive equations for the present EVPSC scheme, the reader can refer to the recent work by Lhadi et al. [20].

Table 2. Viscoplastic parameters of the β-single crystal for each slip system family.

Slip Family	n	$K(MPa{\cdot}s^{1/n})$	$r_0(MPa)$	c(MPa)
{110} <111> 12 slip systems	20	300	113	200
{112} <111> 12 slip systems	20	300	113	200
{123} <111> 24 slip systems	20	300	123	400

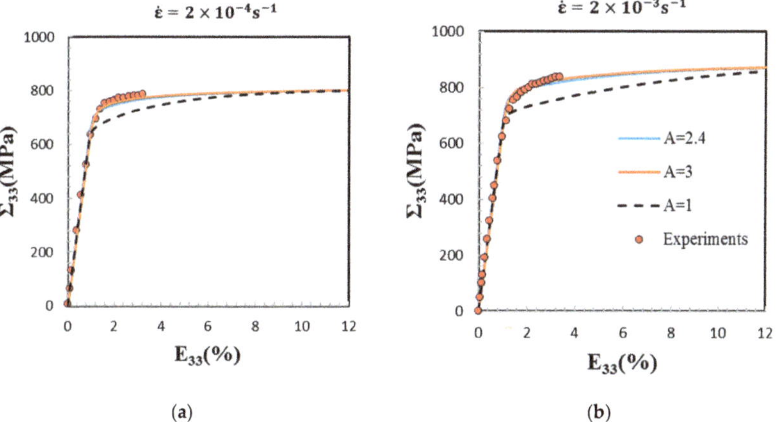

Figure 2. Tensile macroscopic stress (Σ_{33}) vs. macroscopic strain (E_{33}) estimated by the Affine elasto-viscoplastic self-consistent model (EVPSC) model at two strain rates: 2×10^{-4} s^{-1} (**a**) and 2×10^{-3} s^{-1} (**b**) using three anisotropy factors (A). The numerical results are compared to the experimental data provided in [4,21] (red circles). For the calculations, an equiaxed β microstructure with random texture is considered.

3.2. Macroscopic Stress–Strain Responses and Strain-Hardening Behavior

Simulation results show that both sets of SEC (Table 1) predict very well the tensile stress–strain curves obtained on 100% β Ti-17 at two strain rates (2×10^{-4} s^{-1} and 2×10^{-3} s^{-1}) [21]. In all of the figures, we refer to the SEC by their respective anisotropy factor. Figure 2 compares the experimental data and the predictions of the present model. As shown in the figure, the elasto-viscoplastic transition is well captured with both A = 2.4 and 3.

The stress–strain curves obtained for the elastic isotropic case A = 1 is plotted for comparison. In this case, the Young's modulus and Poisson ratio considered at the scale of the β single crystal are 69.4 GPa and 0.35, respectively.

Since the texture is random, the elastic Young's modulus is the same for A = 1, 2.4, and 3. However, the predicted transition from the elastic domain to the viscoplastic one is done earlier for A = 1. As a result, a longer transition between elastic and viscoplastic states is observed. Indeed, as expected by the self-consistent model, the stress–strain curves for A = 1 achieve the same viscoplastic asymptotic state as A = 2.4 and 3 when the elastic strains become negligible (i.e., at large strains ~10% and ~15% for strain rates equal to $2 \times 10^{-4}\ \mathrm{s}^{-1}$ and to $2 \times 10^{-3}\ \mathrm{s}^{-1}$, respectively).

In addition to the macroscopic elasto-viscoplastic responses, the evolution of the overall tensile strain-hardening rate ($\Theta = d\Sigma_{33}/dE^{p}_{33}$) is plotted as a function of the macroscopic tensile plastic strain. From Figure 3, it is clear that both elastic anisotropy and applied strain rate have an influence on the strain-hardening behavior. At $\dot{\varepsilon} = 2 \times 10^{-4}\ \mathrm{s}^{-1}$, the predicted strain-hardening rate exhibits a smooth continuous decrease. At $\dot{\varepsilon} = 2 \times 10^{-3}\ \mathrm{s}^{-1}$, the predicted evolution is less regular, even if there is no increase of the strain-hardening rate. The breaks into the steady decrease of Θ are due to the more important activation of slip systems in the course of plastic deformation, which is given by the threshold type law used for the viscoplastic flow rule (Equation (7)). The differences observed on overall strain-hardening rates between A = 2.4 or A = 3 and A = 1 are due to elastic incompatibilities and their coupling with plastic incompatibilities. Indeed, only plastic incompatibilities are present for A = 1. The coupling between elastic and plastic incompatibilities in the elastic-viscoplastic transition modifies the plastic flow stresses at the scale of slip systems. Hence, most slip systems become active sooner in the anisotropic case compared to the isotropic one, which explains the stronger decrease of Θ in the anisotropic cases A = 2.4 and 3.

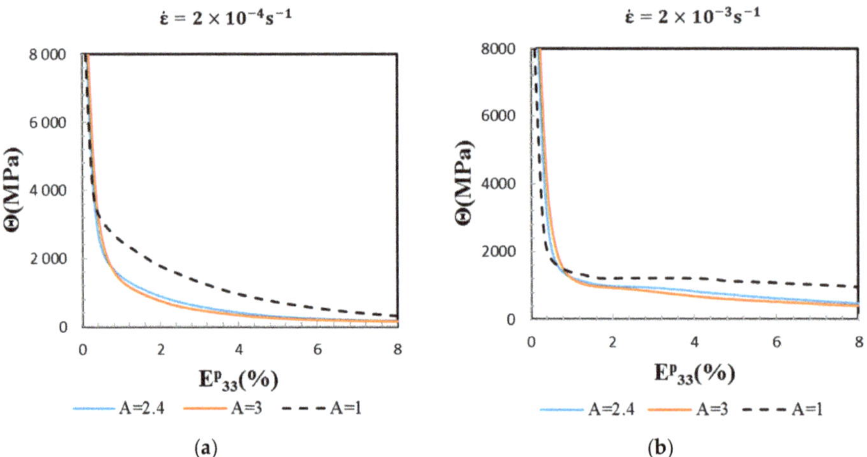

Figure 3. Comparison of overall strain-hardening rate evolutions as a function of macroscopic strain for different anisotropy factors (A = 1, 2.4 and 3) and at two tensile applied strain rates ($2 \times 10^{-4}\ \mathrm{s}^{-1}$ (**a**) and $2 \times 10^{-3}\ \mathrm{s}^{-1}$(**b**)).

3.3. Local Mechanical Fields Analysis

Compared to the macroscopic results, the local behavior depends even more on the elastic anisotropy. Figure 4 shows at the two applied strain rates and for the three anisotropy factors A = 1, 2.4 and 3, the local stresses and the local strains of all of the grains at 1%, 4%, and 10% of macroscopic strain. For the isotropic case A = 1, the elasticity is homogeneous throughout the aggregate, and hence, differences between grains arise only at the onset of plastic deformation in the material. In Figure 4, it is seen that the cloud of dots related to a same simulation forms a line at a given macroscopic strain. The slope of these lines decreases with macroscopic strain, meaning that the range of local strains extends with macroscopic strain. This is in contrast to a simple Taylor's model of uniform strain, which would lead to vertical lines at any strains. For the anisotropic cases, the range of local stresses is at its maximum toward the end of the elastic regime and in the elastic-viscoplastic transition. The cloud of dots obtained with A = 2.4 and 3 are relatively close to each other. The difference between A = 2.4 and 3 is observable when comparing local stress or local strain for a same specific grain orientation, such as <100>-oriented grains parallel to the loading direction (see Figure 5).

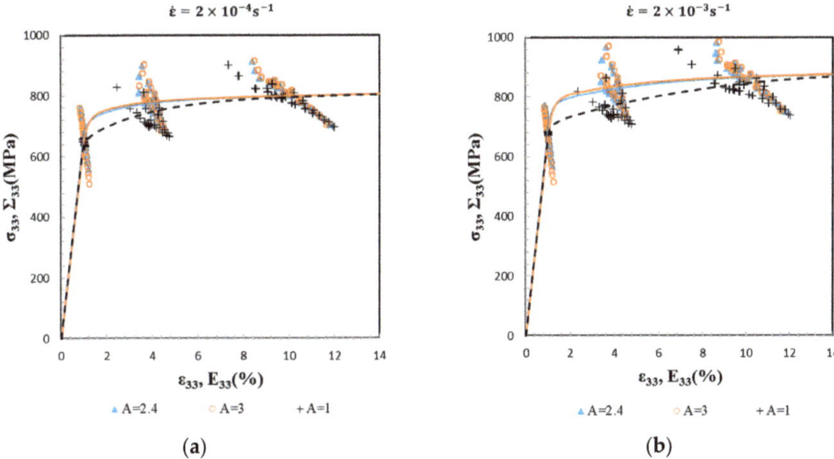

Figure 4. Local stress component (σ_{33}) as a function of local strain component (ε_{33}) at 1%, 4%, and 10% of macroscopic strain for three anisotropy factors (A = 1, 2.4 and 3) (cloud of dots) at two strain rates 2×10^{-4} s^{-1}(**a**) and 2×10^{-3} s^{-1} (**b**). The macroscopic stress–strain responses (Σ_{33} vs. E_{33}) are plotted as references (solid lines).

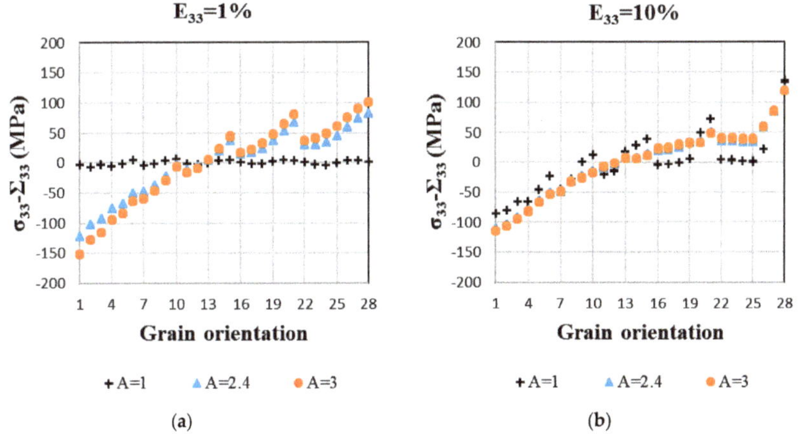

Figure 5. *Cont.*

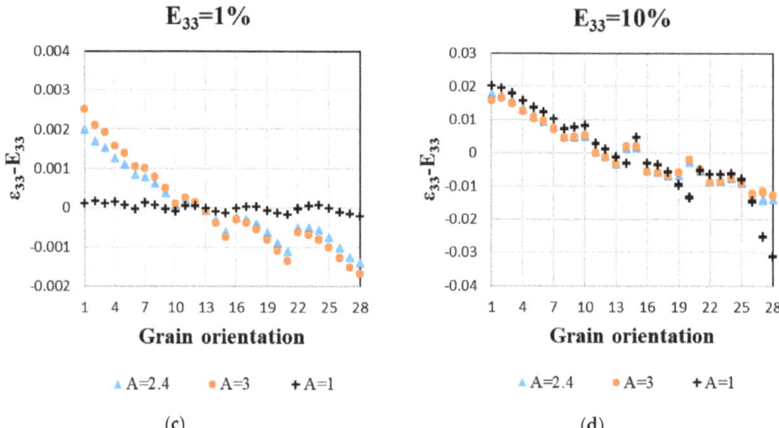

Figure 5. Incompatibility stresses (**a**,**b**) and incompatibility strains (**c**,**d**) (dimensionless) for grain orientations 1 to 28 at 1% and 10% macroscopic strains. The applied strain rate is $\dot{\varepsilon} = 2 \times 10^{-3}\ \mathrm{s}^{-1}$. Inverse pole figure associated with the tensile axis, which defines the studied 28 grain orientations, were reported in Figure 1a.

In Figure 5, incompatibility stresses, which are defined as the difference between the mean local stress (in the grain) and the macroscopic one $\sigma_{33} - \Sigma_{33}$, and incompatibility strains, which are defined as the difference between the local strain and the macroscopic applied one $\varepsilon_{33} - E_{33}$, are plotted as a function of the grain orientations. At 1% macroscopic strain, incompatibility stresses and incompatibility strains obtained for A = 3 are larger in magnitude than the values obtained for A = 2.4, whatever the grain orientation. Incompatibility stresses and incompatibility strains are maximal for grain orientation 1, which is close to a <100> direction, and are minimal for grain orientation 13, which is close to a <529> direction. However, the maximal local stress occurs for grain orientation 28, which is close to a <111> direction. Indeed, in this case, incompatibility stresses have a positive contribution, whereas the contribution is negative in <100> grains (i.e., the local stress is lower than the applied stress). These incompatibilities are due to elastic anisotropy and depend on the grain orientation.

With increasing deformation, plastic strain incompatibilities arise primarily because plastic strains become much larger than elastic strains. Therefore, the incompatibility stresses and incompatibility strains of different anisotropy factor A will all converge to the same set of values. In Figure 5b,d, it can be seen that this convergence is not completely achieved at 10% of macroscopic strain. The full convergence is actually obtained at a macroscopic strain of 15% (not shown).

3.4. *Effect of Anisotropic Elasticity on Relative Slip Activities in 100% β-Ti*

Thanks to the micromechanical model, the contribution of the different slip families to the overall plastic deformation can be deeply analyzed. The relative activity of a slip family can be studied from the definition of the α parameter, which is defined as:

$$\alpha = \frac{\sum_{g=1}^{ng} \sum_{s=p}^{q} f_g \left|\dot{\gamma}_g^s\right|}{\sum_{g=1}^{ng} \sum_{s=1}^{ns} f_g \left|\dot{\gamma}_g^s\right|} \tag{8}$$

In Equation (8), f_g is the grain volume fraction, n_g is the total number of grains, p and q are the first and the last slip system number of the slip family, and n_s is the total number of slip systems (see Table 2).

The evolution of the parameter α related to slip families {110}, {112}, and {123} is plotted on Figure 6 for A = 1, 2.4, and 3. The value of α depends mostly on the texture, the viscoplastic parameters that are specific to slip families, and the intergranular stresses due to elastic/plastic incompatibilities predicted by the EVPSC model. For the present simulations, the initial texture is random, and thus, there is almost no difference in the cumulative distributions considering a classic Schmid factor analysis between the three slip families (Figure 7). From the Schmid factor analysis, it can be noticed that the initial texture should be slightly favorable for {123} systems, and that the distributions of families {110} and {112} are almost equal (Figure 7). However, it is reported from Figure 6 that the evolution of {112}<111> systems activity predicted by the EVPSC model is largely predominant at the onset of plastic deformation. Then, it decreases rapidly while {123} systems are activated first, and are followed closely by {110} systems. The contribution of the {112} systems remains predominant throughout the loading, although the relative contributions of the three families are very close at high strains.

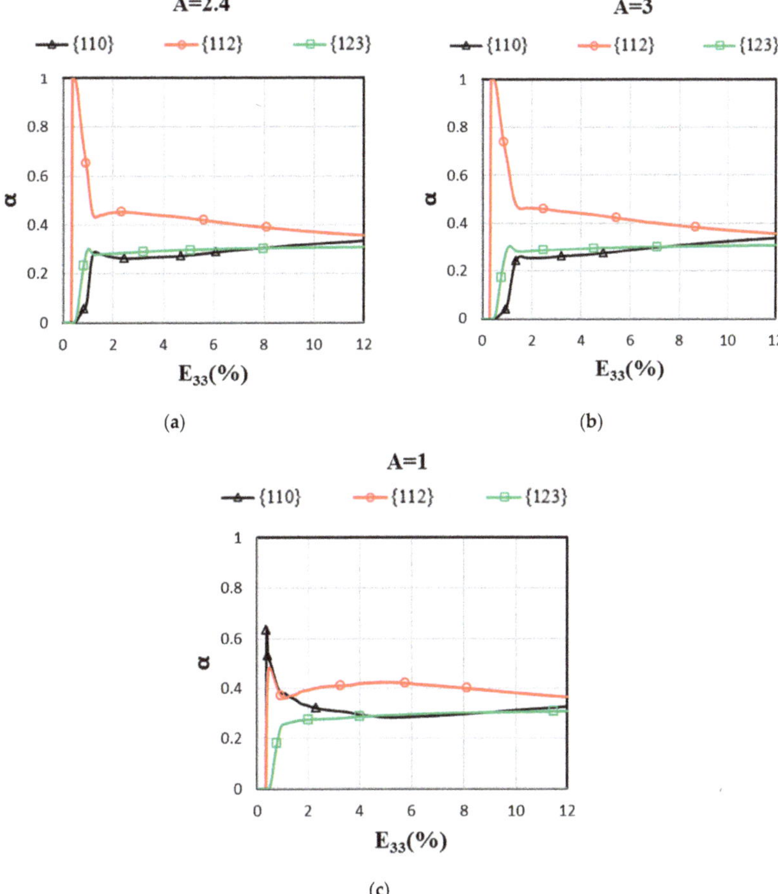

Figure 6. Relative slip plane family activities (dimensionless) as a function of the macroscopic strain predicted by the EVPSC model for A = 2.4 (**a**), 3 (**b**), and 1 (**c**). The applied tensile strain rate in the X_3-direction is $\dot{\varepsilon} = 2 \times 10^{-4}\ s^{-1}$.

Following some trends reported in the literature on body-centered cubic metals [34], the viscoplastic parameters are chosen to be identical for {110} and {112} families, and the {123} family has higher values of r_0 and c (see Table 2). Together with Figure 7, this explains why {123} slip systems are activated after the quasi-simultaneous activation of {110} and {112} systems in the isotropic elasticity case (A = 1) where intergranular stresses are negligible below 2% (see Figure 4). On the contrary, in the anisotropic cases (A = 2.4 and 3), the intergranular stresses that depend on the elastic anisotropy of the β-phase give clear advantage to the activation of the {112} slip family in comparison with the {110} slip family. While the literature shows that there is a strong competition between the activations of the {110} and {112} slip planes as a function of temperature in body-centered cubic metals [34], the present findings need to be checked by slip analysis in a future experimental study. In addition, it was proved from slip analysis and analytical/numerical calculations in face-centered cubic metals that elastic anisotropy and induced incompatibility stresses may drastically change slip activity in the microplastic regime that is observable in polycrystals and bicrystals [35,36].

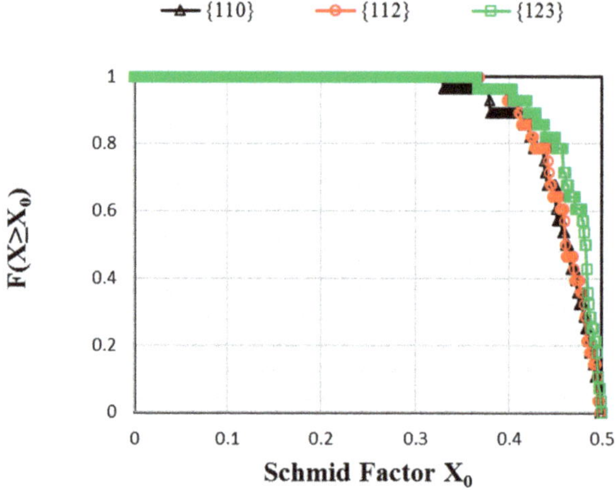

Figure 7. Cumulative distributions $F(X \geq X_0)$ of maximum Schmid factors in grains distributed with an isotropic texture. The representation of the distributions is chosen in such a way that the ordinate of a point X_0 gives the fraction of grains with the Schmid factor X_0 or higher.

4. Conclusions

The impact of the elastically anisotropic β-phase on the tensile behavior of a fully β-phase Ti-17 was studied with an affine extension of the self-consistent scheme for elasto-viscoplastic polycrystals (EVPSC) based on the translated fields (TF) method. Due to the uncertainties on the β-phase single-crystal elastic constants (SEC), two sets of SEC with different Zener anisotropy factors (A = 2.4 and 3) were considered in the simulations, as they both give effective Young's moduli close to the measured one for a 100% β-Ti-17 polycrystal. The numerical results are in good agreement with the experimental data. The main conclusions of this work are summarized as follows:

1. At the macroscopic scale, the two sets of β-phase SEC well predict the tensile elasto-viscoplastic behavior of a randomly oriented and equiaxed β Ti-17 at two strain rates by the EVPSC model. This shows the relevancy of the model for the prediction of the material's hardening and strain rate sensitivity;
2. In contrast, at the grain scale, the two sets of SEC lead to the scattered incompatibility of stresses and strains. These incompatibilities increase in magnitude with the Zener factor A and depend on the grain orientation. The grains oriented with their <100> direction parallel to the tensile axis are more sensitive to the anisotropy factor A (Figure 5). The local mechanical fields analysis

demonstrates that considering only the macroscopic behavior is not sufficient to evaluate the effect of the elastic anisotropy of the β-phase;

3. Lastly, it is shown that the elastic anisotropy of the β-phase can affect the slip activities of a fully β-phase microstructure. Contrary to the isotropic elastic case, {112} <111> slip systems are clearly predominant at the onset of plasticity in the simulations with A = 2.4 and 3 (Figure 6).

Author Contributions: Conceptualization, S.L., T.R. and S.B.; Methodology, S.L., M.-R.C., T.R., N.G., L.G. and S.B.; Investigation, S.L. Writing-Original Draft Preparation, S.L.; Writing-Review & Editing, S.L., M.-R.C., T.R., N.G., L.G. and S.B.; Visualization, S.L., M.-R.C., T.R., N.G., L.G. and S.B.

Funding: This research was funded by the French State through the program "Investment in the future" operated by the National Research Agency (ANR) and referenced by ANR-11-LABX-0008-01 (Laboratory of Excellence 'DAMAS': Design of Alloy Metals for low-mAss Structures).

Conflicts of Interest: The authors declare no conflict of interest.

References

1. Chini, M.R.; Germain, L.; Gey, N.; Andrieu, S.; Duval, T. Advanced Microtexture Analysis of A Ti 10-2-3 Product for Better Understanding of Local Variations in Mechanical Behavior. In Proceedings of the 13th World Conference on Titanium, San Diego, CA, USA, 16–20 August 2015; John Wiley & Sons, Inc.: Hoboken, NJ, USA, 2016; pp. 1943–1948.
2. Chung, D.H.; Buessem, W.R. The Elastic Anisotropy of Crystals. *J. Appl. Phys.* **1967**, *38*, 2010–2012. [CrossRef]
3. Zener, C.M.; Siegel, S. Elasticity and Anelasticity of Metals. *J. Phys. Colloid Chem.* **1949**, *53*, 1468. [CrossRef]
4. Martin, G. Simulation Numérique Multi-Échelles du Comportement Mécanique des Alliages de Titane Bêta-Métastable Ti5553 et Ti17. Ph.D. Thesis, Ecole Nationale Supérieure des Mines de Paris, Paris, France, 2012.
5. Raghunathan, S.L.; Stapleton, A.M.; Dashwood, R.J.; Jackson, M.; Dye, D. Micromechanics of Ti–10V–2Fe–3Al: In situ synchrotron characterisation and modelling. *Acta Mater.* **2007**, *55*, 6861–6872. [CrossRef]
6. Petry, W.; Heiming, A.; Trampenau, J.; Alba, M.; Herzig, C.; Schober, H.R.; Vogl, G. Phonon dispersion of the bcc phase of group-IV metals. I. bcc titanium. *Phys. Rev. B* **1991**, *43*, 10933–10947. [CrossRef]
7. Ledbetter, H.; Ogi, H.; Kai, S.; Kim, S.; Hirao, M. Elastic constants of body-centered-cubic titanium monocrystals. *J. Appl. Phys.* **2004**, *95*, 4642–4644. [CrossRef]
8. Fréour, S.; Lacoste, E.; François, M.; Guillén, R. Determining Ti-17 β-phase single-crystal elasticity constants through X-ray diffraction and inverse scale transition model. *Mater. Sci. Forum* **2011**, *681*, 97–102. [CrossRef]
9. Fréour, S.; Gloaguen, D.; François, M.; Perronnet, A.; Guillén, R. Determination of single-crystal elasticity constants in a cubic phase within a multiphase alloy: X-ray diffraction measurements and inverse-scale transition modelling. *J. Appl. Crystallogr.* **2005**, *38*, 30–37. [CrossRef]
10. Fisher, E.S.; Dever, D. Relation of the c′ elastic modulus to stability of b.c.c. transition metals. *Acta Metall.* **1970**, *18*, 265–269. [CrossRef]
11. Reid, C.N.; Routbort, J.L.; Maynard, R.A. Elastic constants of Ti–40 at.% Nb at 298 °K. *J. Appl. Phys.* **1973**, *44*, 1398–1399. [CrossRef]
12. Kim, J.-Y.; Rokhlin, S.I. Determination of elastic constants of generally anisotropic inclined lamellar structure using line-focus acoustic microscopy. *J. Acoust. Soc. Am.* **2009**, *126*, 2998–3007. [CrossRef] [PubMed]
13. Nejezchlebová, J.; Janovská, M.; Seiner, H.; Sedlák, P.; Landa, M.; Šmilauerová, J.; Stráský, J.; Harcuba, P.; Janeček, M. The effect of athermal and isothermal ω phase particles on elasticity of β-Ti single crystals. *Acta Mater.* **2016**, *110*, 185–191. [CrossRef]
14. Richeton, T.; Wagner, F.; Chen, C.; Toth, L.S. Combined Effects of Texture and Grain Size Distribution on the Tensile Behavior of α-Titanium. *Materials* **2018**, *11*, 1088. [CrossRef] [PubMed]
15. Ji, Z.; Shen, C.; Wei, F.; Li, H. Dependence of Macro- and Micro-Properties on α Plates in Ti-6Al-2Zr-1Mo-1V Alloy with Tri-Modal Microstructure. *Metals* **2018**, *8*, 299. [CrossRef]
16. Cojocaru, V.D.; Şerban, N. Effects of Solution Treating on Microstructural and Mechanical Properties of a Heavily Deformed New Biocompatible Ti–Nb–Zr–Fe Alloy. *Metals* **2018**, *8*, 297. [CrossRef]
17. Li, J.; van Vliet, K.J.; Zhu, T.; Yip, S.; Suresh, S. Atomistic mechanisms governing elastic limit and incipient plasticity in crystals. *Nature* **2002**, *418*, 307–310. [CrossRef] [PubMed]

18. Martin, G.; Naźe, L.; Cailletaud, G. Numerical multi-scale simulations of the mechanical behavior of β-metastable titanium alloys Ti5553 and Ti17. *Procedia Eng.* **2011**, *10*, 1803–1808. [CrossRef]
19. Hounkpati, V.; Fréour, S.; Gloaguen, D.; Legrand, V.; Kelleher, J.; Kockelmann, W.; Kabra, S. In situ neutron measurements and modelling of the intergranular strains in the near-β titanium alloy Ti-β21S. *Acta Mater.* **2016**, *109*, 341–352. [CrossRef]
20. Lhadi, S.; Berbenni, S.; Gey, N.; Richeton, T.; Germain, L. Micromechanical modeling of the effect of elastic and plastic anisotropies on the mechanical behavior of beta-Ti alloys. *Int. J. Plast.* **2018**. [CrossRef]
21. Settefrati, A. Étude Expérimentale et Modélisation par Champ de Phase de la Formation de Alpha dans les Alliages de Titane Bêta-Métastable. Ph.D. Thesis, Université de Lorraine, Lorraine, France, 2012.
22. Duval, T. Analyse Multi-Échelles des Relations Microstructure/Propriétés Mécaniques sous Sollicitation Monotone et Cyclique des Alliages de Titane β-Métastable. Ph.D. Thesis, Ecole Nationale Supérieure de Mécanique et d'Aérotechnique, Chasseneuil-du-Poitou, France, 2013.
23. Mareau, C.; Berbenni, S. An affine formulation for the self-consistent modeling of elasto-viscoplastic heterogeneous materials based on the translated field method. *Int. J. Plast.* **2015**, *64*, 134–150. [CrossRef]
24. Amouzou, K.E.K.; Richeton, T.; Roth, A.; Lebyodkin, M.A.; Lebedkina, T.A. Micromechanical modeling of hardening mechanisms in commercially pure α-titanium in tensile condition. *Int. J. Plast.* **2016**, *80*, 222–240. [CrossRef]
25. Méric, L.; Poubanne, P.; Cailletaud, G. Single Crystal Modeling for Structural Calculations: Part 1—Model Presentation. *J. Eng. Mater. Technol.* **1991**, *113*, 162–170. [CrossRef]
26. Méric, L.; Cailletaud, G. Single Crystal Modeling for Structural Calculations: Part 2—Finite Element Implementation. *J. Eng. Mater. Technol.* **1991**, *113*, 171–182. [CrossRef]
27. Paquin, A.; Sabar, H.; Berveiller, M. Integral formulation and self-consistent modelling of elastoviscoplastic behavior of heterogeneous materials. *Arch. Appl. Mech.* **1999**, *69*, 14–35. [CrossRef]
28. Sabar, H.; Berveiller, M.; Favier, V.; Berbenni, S. A new class of micro–macro models for elastic–viscoplastic heterogeneous materials. *Int. J. Solids Struct.* **2002**, *39*, 3257–3276. [CrossRef]
29. Berbenni, S.; Favier, V.; Lemoine, X.; Berveiller, M. Micromechanical modeling of the elastic-viscoplastic behavior of polycrystalline steels having different microstructures. *Mater. Sci. Eng. A* **2004**, *372*, 128–136. [CrossRef]
30. Masson, R.; Zaoui, A. Self-consistent estimates for the rate-dependentelastoplastic behaviour of polycrystalline materials. *J. Mech. Phys. Solids* **1999**, *47*, 1543–1568. [CrossRef]
31. Clément, N. Phase Transformations and Mechanical Properties of the Ti-5553 Beta-Metastable Titanium Alloy. Ph.D. Thesis, Université Catholique de Louvain (UCL), Louvain-la-Neuve, Belgium, 2010.
32. Hershey, A.V. The elasticity of an isotropic aggregate of anisotropic cubic crystals. *J. Appl. Mech.-Trans. ASME* **1954**, *21*, 236–240.
33. Kröner, E. Computation of the elastic constants of polycrystals from constants of single crystals. *Z. Phys.* **1958**, *151*, 504–518. [CrossRef]
34. Weinberger, C.R.; Boyce, B.L.; Battaile, C.C. Slip planes in bcc transition metals. *Int. Mater. Rev.* **2013**, *58*, 296–314. [CrossRef]
35. Sauzay, M. Cubic elasticity and stress distribution at the free surface of polycrystals. *Acta Mater.* **2007**, *55*, 1193–1202. [CrossRef]
36. Tiba, I.; Richeton, T.; Motz, C.; Vehoff, H.; Berbenni, S. Incompatibility stresses at grain boundaries in Ni bicrystalline micropillars analyzed by an anisotropic model and slip activity. *Acta Mater.* **2015**, *83*, 227–238. [CrossRef]

Article

Combined Effects of Texture and Grain Size Distribution on the Tensile Behavior of α-Titanium

Thiebaud Richeton [1,2,*], Francis Wagner [1,2], Cai Chen [2,3] and Laszlo S. Toth [1,2]

1 Université de Lorraine, CNRS, Arts et Métiers ParisTech, LEM3, F-57000 Metz, France; francis.wagner@univ-lorraine.fr (F.W.); laszlo.toth@univ-lorraine.fr (L.S.T.)
2 Laboratory of Excellence on Design of Alloy Metals for Low-Mass Structures (DAMAS), Université de Lorraine, 57073 Metz, France; cai.chen@njust.edu.cn
3 Nanjing University of Science and Technology, 210094 Nanjing, China
* Correspondence: thiebaud.richeton@univ-lorraine.fr; Tel.: +33-3-72-74-78-02

Received: 19 May 2018; Accepted: 21 June 2018; Published: 26 June 2018

Abstract: This work analyzes the role of both the grain size distribution and the crystallographic texture on the tensile behavior of commercially pure titanium. Specimens with different microstructures, especially with several mean grain sizes, were specifically prepared for that purpose. It is observed that the yield stress depends on the grain size following a Hall–Petch relationship, that the stress–strain curves have a tendency to form a plateau that becomes more and more pronounced with decreasing mean grain size and that the hardening capacity increases with the grain size. All these observations are well reproduced by an elasto-visco-plastic self-consistent model that incorporates grain size effects within a crystal plasticity framework where dislocations' densities are the state variables. First, the critical resolved shear stresses are made dependent on the individual grain size through the addition of a Hall–Petch type term. Then, the main originality of the model comes from the fact that the multiplication of mobile dislocation densities is also made grain size dependent. The underlying assumption is that grain boundaries act mainly as barriers or sinks for dislocations. Hence, the smaller the grain size, the smaller the expansion of dislocation loops and thus the smaller the increase rate of mobile dislocation density is. As a consequence of this hypothesis, both mobile and forest dislocation densities increase with the grain size and provide an explanation for the grain size dependence of the transient low work hardening rate and hardening capacity.

Keywords: grain size; texture; crystal plasticity; elasto-visco-plastic self-consistent (EVPSC) scheme; hardening; dislocation density; titanium

---

1. Introduction

Both crystallographic texture and grain size distribution are recognized to be of prime importance to understand the mechanical behavior of polycrystals [1–4]. In particular, since the works of Hall [5] and Petch [6], the grain size is known to strongly influence the yield stress of polycrystals. For grain sizes larger than 1 μm, the work hardening behavior is, however, generally reported to become almost insensitive to the grain size after a few percent of strain, particularly in face centered cubic metals [7–9]. Texture effects arise because of elastic and plastic anisotropies at the crystal level and are well captured by homogenization schemes like self-consistent approaches which can handle an extensive number of crystallographic orientations within reasonable CPU time simulations. However, only a few approaches do account explicitly for grain size distribution in addition to crystallographic texture [2,4,10–13]. In these mean-field approaches, the individual grain sizes get into the constitutive laws of the model as ad hoc parameters. Otherwise, no grain size effect could be predicted since these classical homogenization schemes do not include any internal length scale. Predicting direct grain size effects is, however, possible from strain gradient theories or generalized continuum plasticity models (e.g., [14]). Though not the grain size, these models consider nevertheless an internal length scale parameter whose value is generally also defined in an ad hoc way. Since the equivalent grain diameter, similarly to the grain orientation Euler angles, can now be readily determined by EBSD analyses, considering the individual grain size in self-consistent models still remains an attractive approach due to its potential good compromise between relative model simplicity, low CPU time and reliability of the predictions. The question is, however, to know how to modify the constitutive laws to include relevant effects of the individual grain size.

The present work focuses on the tensile behavior of commercially pure (cp) α-titanium which exhibits a hexagonal close-packed (hcp) crystallographic structure. In particular, α-Ti is characterized by a high anisotropy of glide resistance [15] and also by non-negligible anisotropy of slip family rate dependence [16–22]. Tensile tests were performed on Ti specimens which were specifically prepared to exhibit various microstructures in order to study the combined role of the texture and the grain size distribution. On the other hand, the constitutive setting of an elasto-visco-plastic self-consistent (EVPSC) scheme [19,23,24] is modified in order to include grain size effects. The constitutive framework is based on crystal plasticity and considers dislocations' densities as state variables. The first goal of the present paper is to present an original way to account for the grain size effect in the dislocation evolution laws. Then, by comparison of the experimental and simulated results, its other purpose is to clarify the respective role of the texture and the grain size distribution and to contribute to a better understanding of the tensile hardening behavior of cp Ti.

The paper is organized as follows. Section 2.1 presents experimental details. Section 2.2 describes the micromechanical model and the way grain size effects are considered. Section 3 shows and discusses the simulations results in comparison with the experimental characterizations. Concluding remarks follow.

2. Materials and Methods

2.1. Preparation, Microstructural and Mechanical Characterization of the Specimens

The as-received material was a 2 mm-thick plate of commercially pure titanium (Ti grade 2) in a fully recrystallized state with a mean grain size D_m of about 10 µm. Specimens cut off from this plate were cold rolled, either along the previous rolling direction or along the previous transverse direction. After rolling, the specimens were annealed under various conditions in a furnace working under secondary vacuum. The samples were small dog-bone-shaped specimens with total length $L_0 = 43$ mm, gauge length $L_g = 11.0$ mm, width $w_0 = 3.0$ mm, thickness $t_0 = 0.5$ mm and could be put entirely in an FEG-SEM. Tensile tests were performed at ambient temperature on a Deben Microtest MT10331-1kN tensile machine (Woolpit, UK) at a constant displacement speed of $0.5\ \mathrm{mm} \cdot \mathrm{mn}^{-1}$, which corresponds to an initial strain rate of $7.5 \cdot 10^{-4}\ \mathrm{s}^{-1}$. The extension direction was parallel either to the last rolling direction or to the last transverse direction. Throughout the paper, the nomenclature to name the specimens is the following: the first letter corresponds to the last rolling direction (R means rolling along the previous rolling direction and T along the previous transverse direction) whereas the second letter refers to the direction of tension (R means tension along the last rolling direction and T along the last transverse direction) and the number to the mean grain size in microns. The names of the seven selected specimens used in this study and their conditions of preparation are reported in Table 1. Figure 1 exhibits examples of experimental tensile curves obtained for three specimens of RR type (i.e., rolling and tension along the previous rolling direction) with different mean grain sizes.

Table 1. Preparation conditions and some metallurgical data of the seven specimens.

Specimen	Cold Rolling Reduction	Annealing Conditions	Number of Grains in the Data Set
RR 2.8	75%	500 °C - 40 mn	6254
TR 2.8	75%	500 °C - 40 mn	7328
RR 4.8	75%	650 °C - 1 h	8793
RR 9.8	75%	730 °C - 2 h	3262
RT 9.8	75%	730 °C - 2 h	3262
RR 11.7	75%	740 °C - 2 h	4075
RR 21.3	30%	840 °C - 4 h	1273

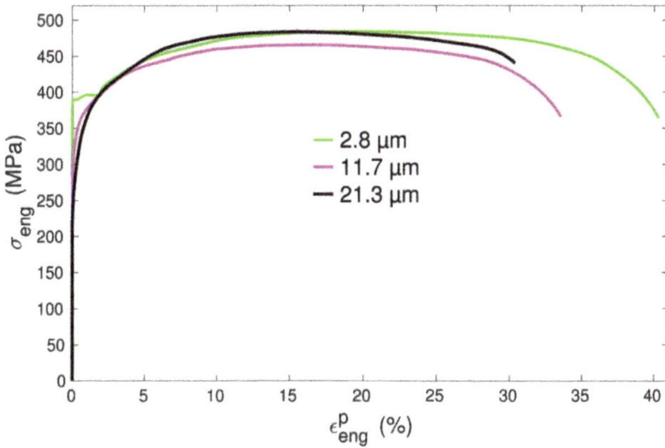

Figure 1. Engineering stress versus engineering plastic strain for three specimens of RR type with different mean grain sizes.

The microstructural characterization was made by EBSD in an FEG-SEM. From each EBSD map, the equivalent diameter D_g and the mean orientation of each grain were collected using a post-processing software (Channel 5 (v5.11, Oxford Instrument, High Wycombe, UK) or ATEX [25]). In all the cases, the specimens were fully recrystallized. The specimens with the smallest mean grain size were close to the end of the primary recrystallization. Figure 2 shows examples of obtained grain size distributions for the two samples with the minimum and maximum mean grain size. In the top part of Figure 2, the scales and the bin sizes used to plot the distributions were exactly the same in order to highlight the differences between the obtained microstructures. When normalized by the mean grain size, all the distributions look, however, pretty similar and follow approximately log-normal probability density functions (see the bottom part of Figure 2).

For each specimen, the data set of the grains (volume fractions and orientations) was used to plot the texture. An example is given in Figure 3 in terms of pole figures. The pole figures correspond to well known textures for this material [17,26–30] where basal planes are tilted 30° ± 10° from the normal toward the transverse direction. When the grains are small, which means that the annealing was stopped close to the end of the primary recrystallization, the texture resembles the deformation one whereas the grain growth texture is obtained after heat treatments made at higher temperatures (see [31] for further information). The data set of the grains (grain diameters in addition to volume fractions and orientations) are the microstructural information used further in the simulations.

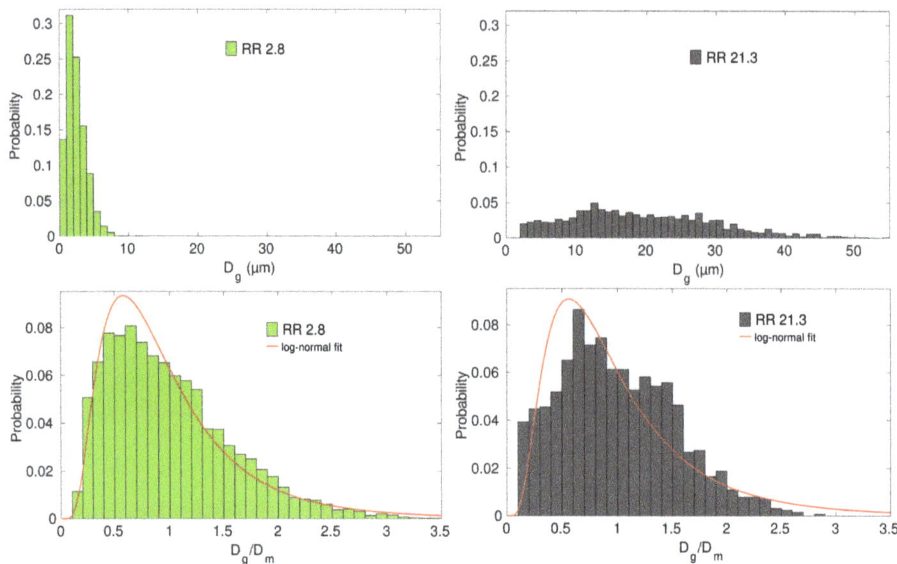

Figure 2. Probability distributions of grain sizes for specimens RR 2.8 and RR 21.3. In the bottom part, the distributions are normalized by the mean grain size D_m and the maximum likelihood estimate for a log-normal distribution computed thanks to the MATLAB software (R2015a, MathWorks, Natick, Massachusetts, US) being superimposed.

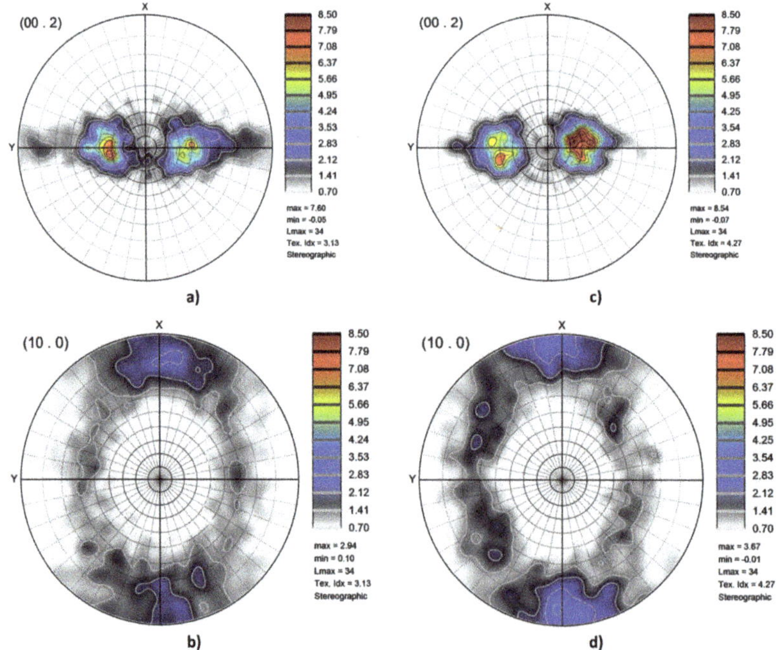

Figure 3. Pole figures for the specimens RR 2.8 (**a**,**b**) and RR 11.7 (**c**,**d**) plotted thanks to the ATEX software [25]. x represents the last rolling direction and y the transverse direction.

2.2. Micromechanical Modeling Including Grain Size Effects

The model used to simulate the tensile behavior is a modified version of the one that was developed in [19] to study hardening mechanisms in cp Ti. The purpose of the modification is to account for grain size effects. The model is still based on an advanced elasto-viscoplastic self-consistent (EVPSC) homogenization scheme which considers a small strain setting and an affine linearization of the viscoplastic flow rule. The 1-site self-consistent approximation is formulated thanks to the translated field method [19,23,24]. Compared to hereditary approaches [32], the numerical implementation of such internal variable approach is much easier as no use of Laplace–Carson transform is needed. In the present model, each grain is represented by a sphere to which is associated a mean crystallographic orientation, a volume fraction and a diameter D_g which hence allows for considering grain size distribution effects. The EVPSC model has been described in detail in several references [19,23,24], and therefore only the main equations are recalled here.

2.2.1. Micro-Macro Scale Transition

In homogenization theory, the macroscopic stress rate and strain rate tensors, $\dot{\sigma}$ and $\dot{\varepsilon}$, of a representative volume element V are obtained by volume averaging the local stress rate and strain rate tensors, $\dot{\sigma}_l$ and $\dot{\varepsilon}_l$ (scale of the grains), as follows:

$$\dot{\sigma} = \frac{1}{V}\int_V \dot{\sigma}_l(\boldsymbol{x})d\boldsymbol{x}\,,\, \dot{\varepsilon} = \frac{1}{V}\int_V \dot{\varepsilon}_l(\boldsymbol{x})d\boldsymbol{x}. \tag{1}$$

Within the small strain framework, the local total strain rate relative to an elasto-viscoplastic behavior splits into an elastic part and a viscoplastic one:

$$\dot{\varepsilon}_l = \dot{\varepsilon}_l^e + \dot{\varepsilon}_l^p. \tag{2}$$

The linear Hooke's law relates $\dot{\varepsilon}_l^e$ to the local Cauchy stress rate $\dot{\sigma}_l$:

$$\dot{\varepsilon}_l^e = \boldsymbol{s} : \dot{\sigma}_l, \tag{3}$$

where the colon : denotes the contracted product of two tensors, and $\boldsymbol{s}$ is the local elastic compliance tensor. The viscoplastic strain rate $\dot{\varepsilon}_l^p$ is a nonlinear tensorial function of the Cauchy stress σ_l. Using a first order affine linearization, it can be written:

$$\dot{\varepsilon}_l^p = \boldsymbol{m} : \sigma_l + \dot{\eta}, \tag{4}$$

where $\boldsymbol{m}$ is the tangent viscoplastic compliance tensor defined by $\boldsymbol{m} = \partial \dot{\varepsilon}_l^p / \partial \sigma_l$, and $\dot{\eta}$ corresponds to a back-extrapolated strain rate. In the absence of volume forces, static equilibrium conditions impose:

$$\text{div}\ \sigma_l = 0\,,\, \text{div}\ \dot{\sigma}_l = 0, \tag{5}$$

whereas kinematic compatibility gives:

$$\dot{\varepsilon} = \frac{1}{2}(\nabla \dot{\boldsymbol{u}} + {}^t\nabla \dot{\boldsymbol{u}}), \tag{6}$$

where $\dot{\boldsymbol{u}}$ is the material velocity field. For uniaxial tensile tests along the x_1-axis, mixed boundary conditions are prescribed as follows: $\dot{\varepsilon}_{11}$ = experimental applied strain rate and $\dot{\sigma}_{22} = \dot{\sigma}_{33} = \dot{\sigma}_{23} = \dot{\sigma}_{31} = \dot{\sigma}_{12} = 0$.

As a result of the translated field method and self-consistent approximations, the explicit expressions of $\dot{\sigma}_l$ and $\dot{\varepsilon}_l$ can be obtained in each grain as a function of $\dot{\sigma}$, $\dot{\varepsilon}$, crystallographic orientation and stress history σ_l in the grain [19,23,24]. These expressions account for coupled elastic and viscoplastic inter-granular accommodations.

2.2.2. Single Crystal Constitutive Laws

Crystal elasticity and plasticity were considered at the single crystal level. Plastic strain and plastic rotation ω_l^p can occur by dislocation-based crystallographic slip only:

$$\dot{\varepsilon}_l^p = \sum_s \boldsymbol{R}^{(s)} \dot{\gamma}^{(s)}, \dot{\omega}_l^p = \sum_s \boldsymbol{S}^{(s)} \dot{\gamma}^{(s)}. \tag{7}$$

$\dot{\gamma}^{(s)}$ denotes the slip rates on systems (s) while $\boldsymbol{R}^{(s)}$ and $\boldsymbol{S}^{(s)}$ are, respectively, the symmetric and the skew-symmetric Schmid orientation tensors associated to system (s). Glide on prismatic, basal, pyramidal <a>, 1st order pyramidal <c+a> and 2nd order pyramidal <c+a> slip systems were taken into account. Twinning was not considered since the twin volume fraction in cp Ti was observed to be very small during tensile tests, especially along the rolling direction [17,19]. The slip rates result from the Orowan relation:

$$\dot{\gamma}^{(s)} = \rho_m^{(s)} b^{(s)} v^{(s)}, \tag{8}$$

where $b^{(s)}$ is the Burgers vector magnitude. From relation 8, it is seen that the evolution of mobile dislocation densities $\rho_m^{(s)}$ and dislocation velocities $v^{(s)}$ are treated separately on each slip system. This allows for accounting for an initial fast multiplication of mobile dislocations, as suggested by Conrad [33] and Naka [34] to explain the presence of a yield plateau in Ti. When dislocations multiply rapidly, the actual strain rate may overcome the imposed strain rate. As a consequence, the average dislocation velocity has to be adjusted so as to retrieve the imposed strain rate, which can cause a significant slow-down of the flow stress increase. The average dislocation velocity on system (s) was assumed to follow a classic power law relationship:

$$v^{(s)} = v_0^{(s)} \left| \frac{\tau^{(s)}}{\tau_c^{(s)}} \right|^{n^{(s)}} \text{sgn}\left(\tau^{(s)}\right), \tag{9}$$

where $v_0^{(s)}$ is a reference velocity, $n^{(s)}$ the inverse of the strain rate sensitivity and $\tau^{(s)}$ the resolved shear stress. It is noteworthy that the components of the tangent viscoplastic compliance tensor $\boldsymbol{m}$ and of the back-extrapolated strain rate tensor $\dot{\boldsymbol{\eta}}$ (Equation (4)) are deduced from the combination of Equations (7)–(9) [19]. Although in elasto-viscoplasticity all non-zero shear stress levels produce some slip, $\tau_c^{(s)}$ can be described as the critical resolved shear stress (CRSS) given the values considered for $n^{(s)}$ (see further Table 2). It writes as the sum of the lattice friction stress $\tau_0^{(s)}$ and a generalized dislocation strengthening relation that accounts for dislocation interactions between systems [35,36]. Moreover, following references [2,4,10–13], $\tau_c^{(s)}$ is also assumed to depend on the individual grain size D_g in a Hall–Petch type relationship:

$$\tau_c^{(s)} = \tau_0^{(s)} + \mu^{(s)} b^{(s)} \sqrt{\sum_l a^{(sl)} \rho_f^{(l)}} + \frac{k_{HP}}{\sqrt{D_g}}. \tag{10}$$

$a^{(sl)}$ denotes the interaction coefficient which is related to the strength of the interaction between system (s) and (l). $\mu^{(s)}$ is the directional shear modulus of system (s). $\rho_f^{(l)}$ depicts forest (or sessile) dislocation density on system (l). k_{HP} is a Hall–Petch type slope resolved on slip system (s). It is noteworthy that, with this supplementary term, the CRSSs are different from grain to grain according to both their orientation and their size. For the same orientation, the smallest grains are harder than the largest ones, which is in agreement with recent experimental characterizations of CRSSs in grade 1 cp Ti by high energy X-ray diffraction microscopy [37]. Moreover, it must be underlined that, through Equation (10), <c+a> systems will display stronger hardening by virtue of the values of Burgers vector magnitude and directional shear modulus (see Table 2), which is in accordance with experimental observations [38,39]. Finally, both mobile and sessile dislocation densities evolve with plastic deformation [19,40]:

$$\dot{\rho}_m^{(s)} = \frac{1}{b^{(s)}}\left[\frac{C_1^{(s)} D_g}{b^{(s)}} - \frac{1}{L^{(s)}}\right]\left|\dot{\gamma}^{(s)}\right|, \tag{11}$$

$$\dot{\rho}_f^{(s)} = \frac{1}{b^{(s)}}\left[\frac{1}{L^{(s)}} - 2k_c^{(s)} b^{(s)} \rho_f^{(s)}\right]\left|\dot{\gamma}^{(s)}\right|. \tag{12}$$

The negative terms in Equation (11), which correspond to mobile dislocation immobilization, appear as positive terms in Equation (12) and stand for storage of forest dislocations. The mean free path of mobile dislocations $L^{(s)}$ is assumed to depend on reactions with other mobile dislocations, which are considered through the parameter $C_2^{(s)}$, and interactions with forest dislocations, which are taken into account through the interaction coefficients $a^{(sl)}$ and the constant $K^{(s)}$:

$$\frac{1}{L^{(s)}} = b^{(s)} C_2^{(s)} \rho_m^{(s)} + \frac{\sqrt{\sum_l a^{(sl)} \rho_f^{(l)}}}{K^{(s)}}. \tag{13}$$

$k_c^{(s)} b^{(s)}$ represents the annihilation distance between dislocations and the term associated to it thus accounts for dynamic recovery.

In Equation (11), the mobile dislocation density production is related to the term $C_1^{(s)} D_g$ and thus is set proportional to the grain diameter. The idea underlying this formalism is that the expansion of dislocation loops (e.g., from a Frank–Read source) is less constrained in larger grains. In smaller grains, dislocation loops will pile-up (as observed in cp Ti [41]) or will be absorbed [42–44] at the grain boundary sooner, i.e., for a smaller increase of dislocation length (= smaller increase of dislocation density). The mechanism of loops expansion should be preserved in cp Ti down to grains of 1 μm or lower (the obstacles spacing d for line tension strengthening may be estimated as $d = 2\mu^{(s)} b^{(s)} / \tau^{(s)}$, which gives $d \approx 0.3$ μm for prismatic systems with $\tau^{(s)} = 150$ MPa). Furthermore, with decreasing grain size, the ratio of grain boundary area over grain volume increases. Hence, the probability of dislocation absorption by grain boundaries should increase when the grain size decreases. Through Equations (11)–(13), it is observed that the grain size, by influencing the increase rate of mobile dislocation densities, will also affect the evolution of forest dislocation densities and hence the hardening behavior. Finally, one may argue that the mean free path of dislocations (Equation (13)), in addition to being dependent on dislocation density, should also depend on the grain size. Accordingly, a term that scales as $1/D_g$ could have been added in Equation (13) as well, as was done for instance in [4]. However, by doing so, one would get higher forest dislocation densities, and thus stronger work hardening in small grain samples, which is the opposite of what is observed (see Figure 1 and further). Hence, our modelling assumption is that the main effect of grain boundaries in cp Ti is to act as barriers or sinks for dislocations [41–44] which, as a consequence, restricts the expansion of dislocation length, rather than having a significant contribution to the creation of forest dislocations, which is mainly accounted for by dislocations' interactions.

2.2.3. Model Parameters

The model is coded in Fortran and solved in an explicit way using very small time steps. It was checked that convergences of the solutions were indeed reached. The values of the model's parameters are given in Table 2 (parameters specific to slip families) and Table 3 (parameters not specific to slip families). Except for $\tau_0^{(s)}$, ρ_{f0}, C_1 and k_c, these values are identical as in [19]. It was shown in [19] that the above described constitutive setting (without the incorporation of grain size effects) was able to explain the opposite effect of strain rate with regard to the orientation of the tensile axis on the three-stage hardening behavior of cp Ti thanks to the consideration of lower strain-rate sensitivity for prismatic systems. Because of the presence of the Hall–Petch type term in the expression of $\tau_c^{(s)}$ (Equation (10)), the values of $\tau_0^{(s)}$ and ρ_{f_0} had to be modified to retrieve CRSSs with about the same level as in [19]. The value of the resolved Hall–Petch type slope k_{HP} corresponds approximately to the macroscopic Hall–Petch slope determined in [31] divided by the Taylor factor. The value of C_1 was adjusted so that the product $C_1 D_g$ is about the former value of C_1 in [19] for a mean grain size of 10 μm.

Table 2. Model parameters that are specific to slip families (Prismatic: P, Basal: B, Pyramidal <a>: $\Pi_1^{<a>}$, 1st order Pyramidal <c+a>: $\Pi_1^{<c+a>}$, 2nd order Pyramidal <c+a>: $\Pi_2^{<c+a>}$).

	P	$\Pi_1^{<a>}$	B	$\Pi_1^{<c+a>}$	$\Pi_2^{<c+a>}$
b (Å)	2.95	2.95	2.95	5.53	5.53
μ (GPa)	35.0	37.1	46.5	47.7	49.2
τ_0 (MPa)	50	90	120	75	150
C_1 (m^{-1})	80	15	15	15	15
n	65	32	32	32	32

Table 3. Model parameters that are not specific to slip families. C_{ij} are the stiffness constants. a_{coli} denotes the interaction coefficient related to collinear interactions, i.e., reactions between dislocations with parallel Burgers vectors gliding in different planes, whereas $a_{\neq coli}$ refers to non-collinear interactions (see details in [19]).

c/a	C_{11} (GPa)	C_{33} (GPa)	C_{44} (GPa)	C_{12} (GPa)	C_{13} (GPa)	k_{HP} (MPa.m$^{0.5}$)	
1.587	**160**	**181**	**46.5**	**90**	**66**	**0.1**	
v_0 (ms^{-1})	ρ_{m_0} (m^{-2})	ρ_{f_0} (m^{-2})	C_2	K	a_{coli}	$a_{\neq coli}$	k_c
3×10^{-5}	1×10^{10}	2×10^{12}	75	80	0.7	0.1	10

3. Results and Discussion

The tensile curves of the seven specimens were simulated using the same set of model parameters. The specific set of grain orientations, volume fractions and equivalent diameters was used for each specimen. The numbers of grains considered in the simulations were the numbers of grains of the EBSD maps indicated in Table 1. Figures 4 and 5 show the simulated curves and their equivalent experimental counterparts. In order to characterize further the mechanical behavior, several quantities were collected from the tensile curves:

- the 0.2% yield stress: σ_e,
- the maximal engineering stress: σ_{max},
- the σ_{max} corresponding engineering strain: ε^*,
- the hardening amplitude: $\Delta\sigma = \sigma_{max} - \sigma_e$.

Both the experimental and simulated values of these quantities are reported in Figure 6 for the seven specimen.

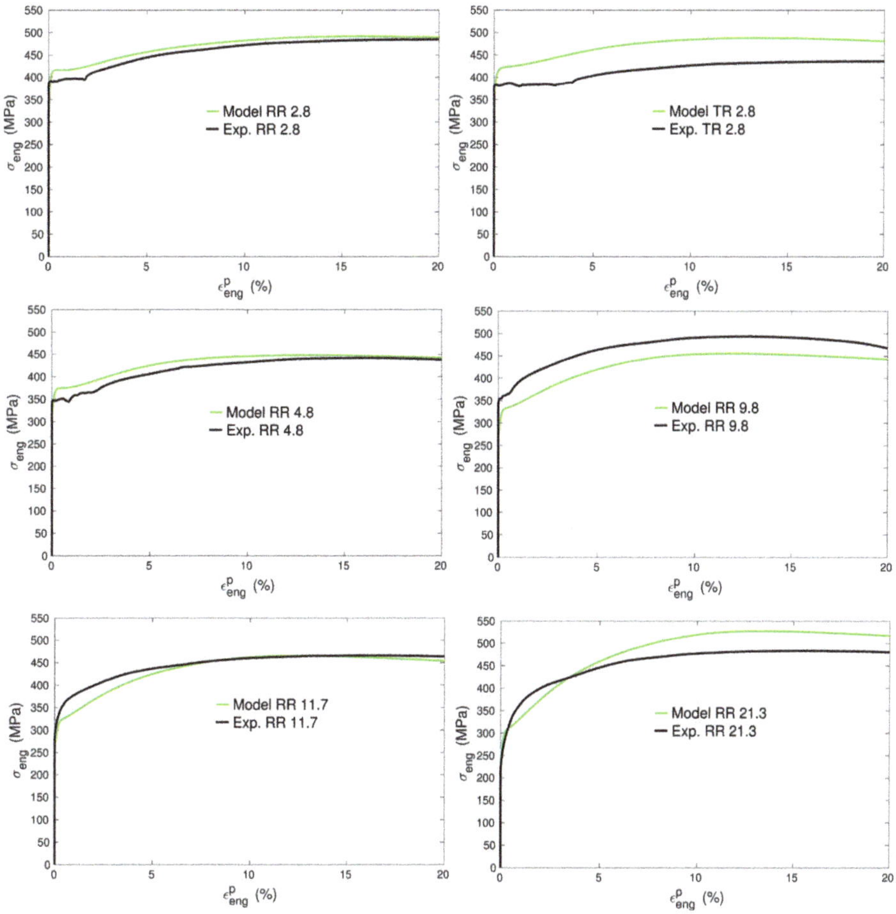

Figure 4. Tensile curves (engineering stress and strain) for model predictions and experimental measurements for the six specimens deformed along the last rolling direction.

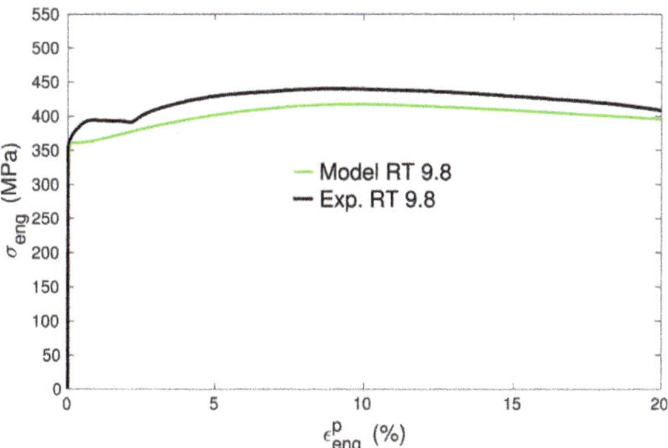

Figure 5. Tensile curves (engineering stress and strain) for model predictions and experimental measurements for the specimen deformed along the last transverse direction.

3.1. Yield Stress

The experimental yield stress increases with decreasing mean grain size D_m and follows rather well a Hall–Petch relationship (Figure 7), i.e., a scaling $\sigma_e \sim 1/\sqrt{D_m}$, as it was already shown on a larger data set of cp Ti specimens that also included partially recrystallized microstructures [31]. The calculated yield stresses follow a pretty correct trend considering the unlikely ranking of measured yield stresses between samples RR 4.8 and RR 9.8 (Figures 6 and 7). The assumption made for the dependence of the CRSS on the grain size (Equation (10)) thus appears relevant. As also found in earlier works [2,4,10–13], it is confirmed that the consideration of a Hall–Petch type dependence at the scale of slip systems enables for retrieving a Hall–Petch law at the macroscopic scale (Figure 7). This transition may, however, not be as direct as it seems since the macroscopic Hall–Petch law deals with the mean grain size, whereas the individual grain size is considered at the scale of slip systems. Hence, the smallest grains of the distribution might act as limiting factors for the complete diffusion of plasticity in the sample.

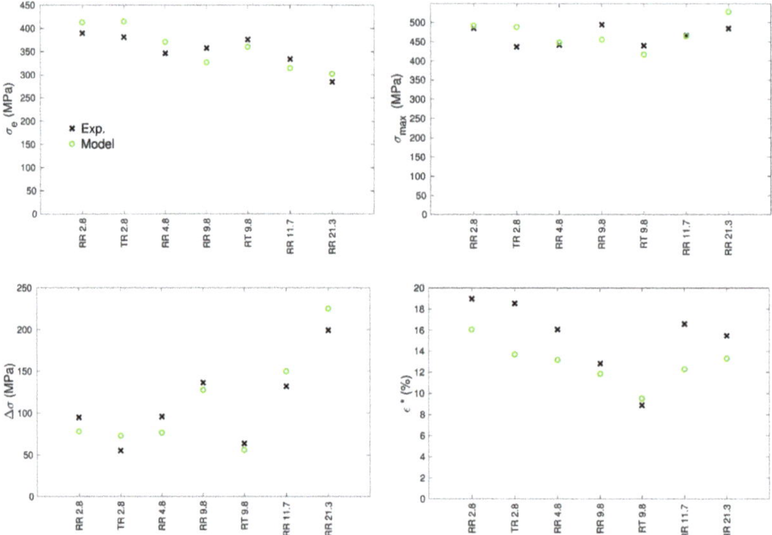

Figure 6. Comparisons between experimental and calculated values of σ_e, σ_{max}, $\Delta\sigma$ and ε^* (see text for definitions).

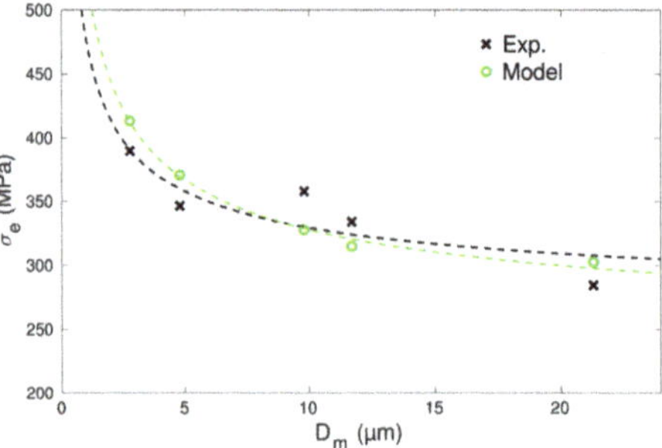

Figure 7. Comparisons of experimental and calculated values of the yield stress σ_e with respect to the mean grain size D_m for the specimens of RR type. Linear fit estimates obtained by MATLAB between σ_e and $1/\sqrt{D_m}$ are shown in green (Model) and black (Exp.) dotted lines.

As already observed [17,19,45], for the same mean grain size, the yield stress is higher when the tensile axis is parallel to the transverse direction (comparisons of samples RT 9.8 and RR 9.8) due to less-well oriented easy prismatic systems.

3.2. Local Mechanical Fields

Figure 8 shows the evolution of the tensile component of local stresses and local strains in the grains for specimens with different average grain sizes. Each dot represents the local stress and local strain values in a given grain. As can be seen, the clouds formed by these dots become larger and larger with increasing macroscopic strain, while their shapes evolve from vertically elongated towards more and more horizontally spread. This result is radically different from the predictions of a simple Taylor model, which would give dots aligned along vertical lines at any strain (uniform strain). The results shown in Figure 8 are in agreement with strain measurements obtained locally in polycrystalline materials [46–48], especially for hexagonal structure [49]. The stress and strain fluctuations arise in the present modeling because of the consideration of non-uniform distributions of grain orientations and sizes coupled with elastic and plastic anisotropies at the single crystal level, as well as grain size dependent CRSSs (Equation (10)) and grain size dependent dislocation density evolution laws (Equation (11)). At 1% macroscopic strain, the fluctuations are mostly due to elastic incompatibilities. With increasing deformation, plastic strains become rapidly larger than elastic strains and plastic incompatibilities become predominant. Moreover, as noticed in [2,4], the grain size dispersion strongly influences the local mechanical fields. Actually, for a same grain size dispersion, $(D_{max} - D_{min})/D_m$, the relative dispersion of the Hall–Petch type term, $k_{HP}/\sqrt{D_g}$ in Equation (10), is invariant whatever the mean grain size is. Since the contribution of the Hall–Petch type term increases with decreasing grain size, the absolute dispersion of the CRSSs becomes larger and larger in samples having smaller and smaller mean grain sizes (considering that the normalized grain size distributions of our samples are quite similar (Figure 2)). It was checked that this effect alone cannot explain totally the impressive increase in the scattering of the local strains and stresses when decreasing the mean grain size in Figure 8. Indeed, there is an additional effect that is related to the grain size dependence of the mobile dislocation density evolution (Equation (11)). The latter impedes significantly the multiplication of mobile dislocations in the smallest grains (see further Figure 11) and thus restricts the value of the slip rates in those grains (Equation (8)). For instance, the smallest grain diameter considered in our simulations, 0.45 μm in sample RR 2.8, is relatively close to the value of 0.14 μm below which $\dot{\rho}_m^{(s)}$ is initially negative in Equation (11). As a result, local stresses almost as high as 1200 MPa are reached in some grains of sample RR 2.8 at a macroscopic strain of 15% (Figure 8). These grains are very small and act like hard particles as they exhibit very little deformation (the corresponding local strains are barely greater than 1%). They represent, however, a very small volume fraction of the whole sample.

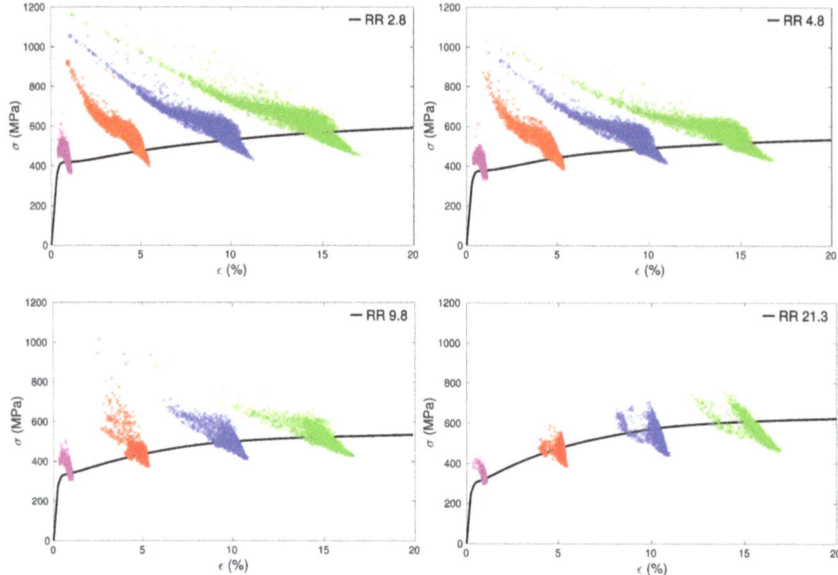

Figure 8. Distribution of predicted tensile stress–strain relationships for each grain at 1% (magenta), 5% (red), 10% (blue) and 15% (green) of macroscopic strain for specimens RR 2.8, RR 4.8, RR 9.8 and RR 21.3. The corresponding model tensile curve is superimposed on each plot (black thick line).

3.3. Slip Activity

From the simulations, it is possible to estimate the relative activity α of a slip family as:

$$\alpha = \frac{\sum_{g=1}^{ng} \sum_{s=p}^{q} f_g |\dot{\gamma}_g^{(s)}|}{\sum_{g=1}^{ng} \sum_{s=1}^{ns} f_g |\dot{\gamma}_g^{(s)}|}, \tag{14}$$

where the slip system numbers of a specific family goes from p to q. ns is the total number of slip systems, ng the total number of grains and f_g the grain volume fraction. The distribution of the slip activity is mainly influenced by the texture, the initial values of the CRSSs and the intergranular stresses that arise from strain incompatibilities. For tensile tests along the rolling direction, it is known that, as a result of the texture, a vast majority of the grains have large macroscopic Schmid factors for prismatic, pyramidal <a> and 1st order pyramidal <c+a> glide and low ones for basal slip [17,31]. On the contrary, along the transverse direction, there is no big difference of Schmid factors among the four main slip families, 1st order pyramidal <a> and <c+a> being nevertheless the most favorably oriented ones [17,31]. As a consequence, for deformation simulation along the last rolling direction, prismatic slip strongly dominates at the very beginning of deformation because of its low CRSS value. Then, it decreases to reach a quasi steady-state of about 50% of the total slip activity whatever the mean grain size (Figure 9). The second most important contribution to plastic deformation arises from 1st order pyramidal <c+a> systems, which exhibit a maximum around 0.5% plastic strain with a rapid initial increase before and a very slight steadily decrease after. This decrease is probably due to the stronger hardening of <c+a> systems. Pyramidal <a> and basal systems have also non negligible contributions, while the one of 2nd order pyramidal systems is completely unimportant due to the much higher value of its CRSS. During the simulations, dislocations of 2nd order pyramidal systems act thus only as a static forest. For deformation along the transverse direction, the fall of the prismatic activity is much stronger and occurs sooner. In steady-state, the contribution of the four main slip families are between 20% and 35%.

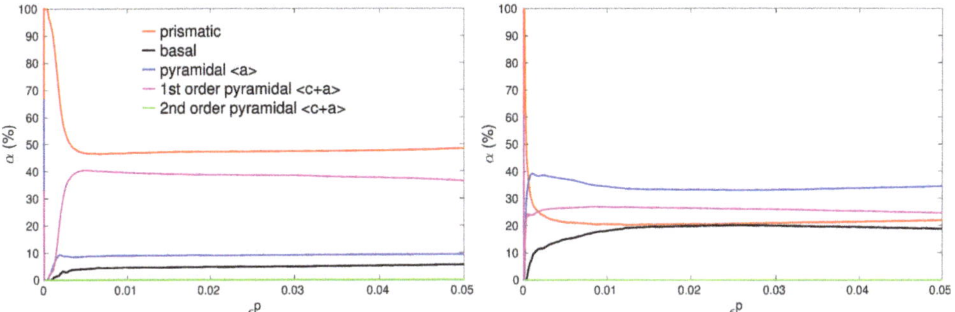

Figure 9. Relative activities of slip families (α) predicted by the model for specimen RR 9.8 (**left**) and RT 9.8 (**right**).

3.4. Hardening Behavior

It can be noticed from the tensile experimental curves (Figures 1 and 4) that the tendency to form an initial plateau becomes more and more pronounced with decreasing mean grain size. This general trend is in agreement with other experiments in titanium [50] and was also characterized as such in copper [9] for instance. The model reproduces very well this trend (Figure 4). Moreover, the evolution of the work hardening rate $\theta = d\sigma/d\varepsilon^P$ for all the simulations of RR type is plotted in Figure 10. In agreement with the present experiments and previous tensile tests on cp Ti [17,19,29,30], a three-stage work hardening behavior is obtained, where an initial fall of θ is followed by an increase and a final progressive decrease. More importantly, the model predicts a growing well depth with decreasing grain size, or, in other words, a transient low work hardening rate with smaller and smaller value as the grain size decreases. In Figure 10, it is also noteworthy that the predicted work hardening rates of all the specimens converge at high strains, meaning that the model correctly predicts an insensitivity to grain size at high strains [9]. The reason is that the mobile dislocation density reaches a saturation value (see further Figure 11).

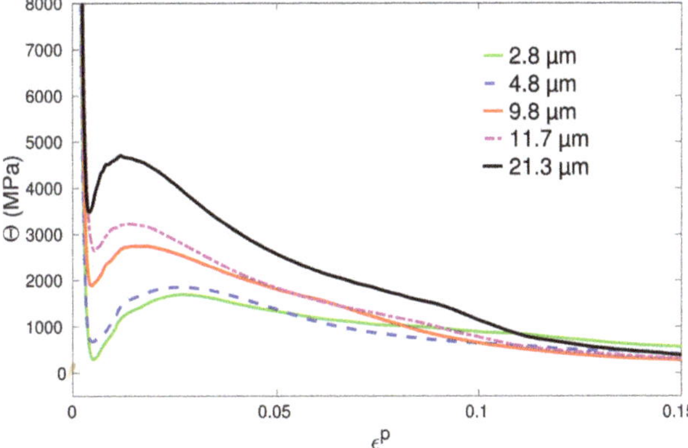

Figure 10. Model predictions for the work hardening rate evolution for the specimens of RR type.

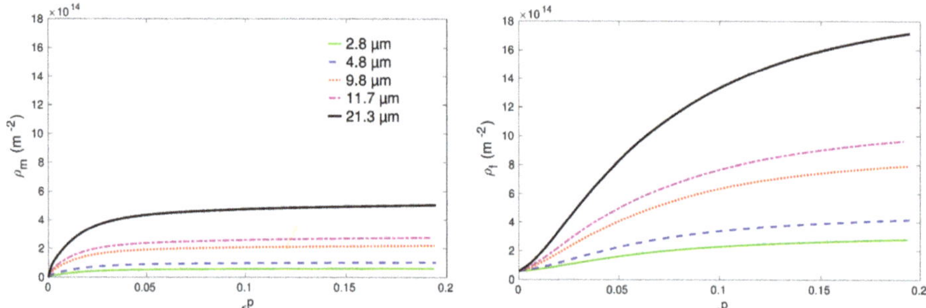

Figure 11. Model predictions for the evolution of the total forest (ρ_f) and mobile (ρ_m) dislocation densities for the specimens of RR type. Densities are cumulated over the 30 slip systems and averaged over the grains' population.

The low point of the work hardening rate precisely corresponds to the moment when the activity of the 1st order pyramidal <c+a> systems approaches its maximum value, around 0.5% of plastic strain (Figure 9). Hence, the appearance of a low point in the work hardening rate evolution seems to be related to the start of multiple slip within grains, or to a kind of balance between <a> and <c+a> slip activities. Under such conditions, plastic strain incompatibilities and the induced intergranular stresses are strongly decreased. Along with the fast multiplication of mobile dislocations (Equation (11)), this gives rise to an initial state with very weak hardening. In agreement with previous experiments, when the tensile axis is parallel to the transverse direction (comparisons of samples RT 9.8 and RR 9.8), the plateau effect is even more pronounced due to an even more balanced slip activity distribution (Figure 9). This texture effect is very well reproduced by the model (Figure 5).

The grain size effect on the work hardening rate can be understood by analyzing the predicted evolutions of dislocation densities in Figure 11. As expected from Equations (11) and (12), the assumption of a mobile dislocation density production that grows with grain size makes both mobile and forest dislocation densities increase with the grain size. This implies that the rate of CRSS increase is more important in large grains since $\tau_0^{(s)}$ and the Hall–Petch type term do not evolve with deformation (Equation (10)). The increase of CRSSs counterbalances the previously mentioned effects, which lead to a low point in the work hardening rate evolution. As a result, the model indeed predicts a less and less pronounced transient low work hardening rate as the grain size increases.

The same explanation holds to interpret the effect of the grain size on the hardening capacity. Both the observations and the model predictions indicate a hardening amplitude $\Delta\sigma$ that tends to increase with the grain size (Figure 6). The CRSSs of the small grains are initially higher than those of the large grains while the latter can display larger increases. By consequence, large grains exhibit a higher hardening capacity.

In addition, the correct prediction of a smaller hardening amplitude obtained for sample RT 9.8 (Figures 5 and 6) results from the choice of a greater C_1 value for prismatic systems compared to other slip families (Table 2). Indeed, since the contribution of prismatic slip is less important for deformation along the transverse direction (Figure 9), this assumption leads to much lower dislocation densities in sample RT 9.8 compared to RR 9.8 and thus to much weaker work hardening (see [19] for further information).

Finally, it can be observed that the model reproduces the Considere point (σ_{max}, ε^*), i.e., a limit for the homogeneous deformation, which matches satisfactorily the experimental results. The maximal engineering stress, σ_{max}, compares rather well with experiments, whereas the corresponding engineering strain, ε^*, is slightly underestimated except for the case RT 9.8 (extension along the transverse direction) (Figure 6). The model, with its assumptions and with the used parameters, is thus able to account for the balance of appearing and disappearing dislocations. The value of ε^* is mainly controlled by the dynamic recovery parameter k_c (Equation (12)). k_c is set to one unique value all along the deformation, which obviously does not allow for capturing totally the complexity of the dynamics of the recovery process (e.g., effect of thermally-activated cross-slip).

4. Conclusions

This study considers tensile curves of a set of cp Ti specimens with different microstructures, especially with several mean grain sizes. It is observed that the yield stress depends on the grain size following a Hall–Petch relationship. The hardening behavior shows also some dependence with the grain size. The tendency of stress–strain curves to form a plateau becomes more and more pronounced with decreasing mean grain size while the hardening amplitude $\Delta\sigma = \sigma_{max} - \sigma_e$ increases with the grain size. All these observations are well reproduced by an EVPSC model that incorporates grain size effects in a crystal plasticity framework where dislocation densities are the state variables. While the CRSSs are made dependent on the individual grain size through the addition of a Hall–Petch type term (Equation (10)), the originality of the model comes from the fact that the multiplication of mobile dislocation densities is also made grain size dependent (Equation (11)). Our assumption is that, due to grain boundaries acting mainly as barriers or sinks for dislocations, the smaller the grain size, the smaller the expansion of dislocation loops and thus the smaller the increase rate of mobile dislocation density. As a consequence of this assumption, both mobile and forest dislocation densities increase with the grain size, which provides an explanation for the grain size dependence of the transient low work hardening rate and hardening amplitude. It is noteworthy, however, that, in agreement with observations, the predicted work hardening rates of different mean grain size samples eventually converge at high strains because the mobile dislocation density reaches a saturation value. The fact that the model provides correct estimates of the homogeneous deformations is also quite positive and shows that the formalism, as well as the choice of the parameters, reflects rather well the complex reality.

Author Contributions: F.W. and C.C conceived and designed the experiments; C.C. performed the experiments; T.R. and F.W. conceived the model and performed the simulations; T.R., F.W., C.C. and L.S.T. analyzed the data; T.R. and F.W. wrote the paper. All the authors participated in the discussion and reviewed the paper.

Funding: This research was funded by by the French State through the program "Investment in the Future" operated by the National Research Agency (ANR) and referenced by grant number [ANR-11-LABX-0008-01] (LabEx DAMAS).

Acknowledgments: The authors thank Yvon Millet from the company Timet for providing cp Ti sheets and Laurent Peltier for performing various heat treatments. This work was supported by the French State through the program "Investment in the Future" operated by the National Research Agency (ANR) and referenced by ANR-11-LABX-0008-01 (LabEx DAMAS).

Conflicts of Interest: The authors declare no conflict of interest. The founding sponsors had no role in the design of the study; in the collection, analyses, or interpretation of data; in the writing of the manuscript, and in the decision to publish the results.

Abbreviations

The following abbreviations are used in this manuscript:

Ti	titanium
cp	commercially pure
RD	rolling direction
TD	transverse direction
EBSD	electron backscatter diffraction
SEM	scanning electron microscopy
FEG	field emission gun
EVPSC	elasto-visco-plastic self-consistent
CRSS	critical resolved shear stress

References

1. Kocks, U.F.; Tomé, C.N.; Wenk, H.R. *Texture and Anisotropy*; Cambridge University Press: Cambridge, UK, 1998.
2. Berbenni, S.; Favier, V.; Berveiller, M. Impact of the grain size distribution on the behaviour of heterogeneous materials. *Int. J. Plast.* **2007**, *23*, 114–142. [CrossRef]
3. Roters, F.; Eisenlohr, P.; Hantcherli, L.; Tjahjanto, D.D.; Bieler, T.R.; Raabe, D. Overview of constitutive laws, kinematics, homogenization and multiscale methods in crystal plasticity finite-element modeling: Theory, experiments, applications. *Acta Mater.* **2010**, *58*, 1152–1211. [CrossRef]
4. Nicaise, N.; Berbenni, S.; Wagner, F.; Berveiller, M.; Lemoine, X. Coupled effects of grain size distributions and crystallographic textures on the plastic behaviour of IF steels. *Int. J. Plast.* **2011**, *27*, 232–249. [CrossRef]
5. Hall, E.O. The deformation and aging of mild steel. Part III: discussion and results. *Proc. Phys. Soc. Lond.* **1951**, *64*, 747–753. [CrossRef]
6. Petch, N.J. The cleavage strength of polycrystals. *J. Iron Steel Inst.* **1953**, *174*, 25–28.
7. Hansen, N.; Ralph, B. The strain and grain size dependence of the flow stress of copper. *Acta Metall.* **1982**, *30*, 411–417. [CrossRef]
8. Hansen, N. Polycrystalline strengthening. *Metall. Trans. A* **1985**, *16A*, 2167–2190. [CrossRef]
9. Sinclair, C.W.; Poole, W.J.; Y.Bréchet. A model for the grain size dependent work hardening of copper. *Scr. Mater.* **2006**, *55*, 732–742. [CrossRef]
10. Bunge, H.J.; Wagner, F.; Houtte, P.V. A new way to include the grain shape in texture simulations with the Taylor model. *J. Phys. Lett.* **1985**, *46*, L1109–L1113. [CrossRef]
11. Raeisinia, B.; Sinclair, C.W.; Poole, W.J.; Tome, C.N. On the impact of grain size distribution on the plastic behaviour of polycrystalline metals. *Model. Simul. Mater. Sci. Eng.* **2008**, *16*, 025001. [CrossRef]
12. Jain, A.; Duygulu, O.; Brown, D.W.; Tomé, C.N.; Agnew, S.R. Grain size effects on the tensile properties and deformation mechanisms of a magnesium alloy, AZ31B, sheet. *Mater. Sci. Eng. A* **2008**, *486*, 545–555. [CrossRef]
13. Fromm, B.S.; Adams, B.L.; S.Ahmadi.; Knezevic, M. Grain size and orientation distributions: Application to yielding of α-titanium. *Acta Mater.* **2009**, *57*, 2339–2348. [CrossRef]
14. Cordero, N.M.; Forest, S.; Busso, E.P.; Berbenni, S.; Cherkaoui, M. Grain size effects on plastic strain and dislocation density tensor fields in metal polycrystals. *Comput. Mater. Sci.* **2012**, *52*, 7–13. [CrossRef]
15. Churchman, A.T. The slip modes of titanium and the effect of purity on their occurence during tensile deformation of single crystals. *Proc. R. Soc. A* **1954**, *226*, 216–226. [CrossRef]
16. Levine, E.D. Deformation mechanisms in Titanium at low temperatures. *Trans. Metall. Soc. AIME* **1966**, *236*, 1558–1565.
17. Roth, A.; Lebyodkin, M.A.; Lebedkina, T.A.; Lecomte, J.S.; Richeton, T.; Amouzou, K.E.K. Mechanisms of anisotropy of mechanical properties of α-titanium in tension conditions. *Mater. Sci. Eng. A* **2014**, *596*, 236–243. [CrossRef]
18. Doquet, V.; Barkia, B. A micromechanical model of the viscoplastic behaviour of titanium accounting for its anisotropic and strain-rate dependent viscosity. *Mech. Time-Depend. Mater.* **2015**, *19*, 153–166. [CrossRef]
19. Amouzou, K.E.K.; Richeton, T.; Roth, A.; Lebyodkin, M.A.; Lebedkina, T.A. Micromechanical modeling of hardening mechanisms in commercially pure α-titanium in tensile condition. *Int. J. Plast.* **2016**, *80*, 222–240. [CrossRef]
20. Zhang, Z.; Jun, T.S.; Britton, T.B.; Dunne, F.P.E. Determination of Ti-6242 α and β slip properties using micro-pillar test and computational crystal plasticity. *J. Mech. Phys. Solids* **2016**, *95*, 393–410. [CrossRef]
21. Zhang, Z.; Jun, T.S.; Britton, T.B.; Dunne, F.P.E. Intrinsic anisotropy of strain rate sensitivity in single crystal alpha titanium. *Acta Mater.* **2016**, *118*, 317–330. [CrossRef]
22. Zhang, Z.; Dunne, F.P.E. Microstructure, Slip Accumulation and Load Shedding in Multiphase Alloys. *Int. J. Fatigue* **2018**, *113*, 324–334. [CrossRef]

23. Mareau, C.; Berbenni, S. An affine formulation for the self-consistent modeling of elasto-viscoplastic heterogeneous materials based on the translated fields method. *Int. J. Plast.* **2015**, *64*, 134–150. [CrossRef]
24. Lhadi, S.; Berbenni, S.; Gey, N.; Richeton, T.; Germain, L. Micromechanical modeling of the effect of elastic and plastic anisotropies on the mechanical behavior of β-Ti alloys. *Int. J. Plast.* **2018**. [CrossRef]
25. Beausir, B.; Fundenberger, J.J. ATEX-Software, Analysis Tools for Electron and X-ray Diffraction, Université de Lorraine. 2018. Available online: http://atex-software.eu/ (accessed on 17 May 2018).
26. Wagner, F.; Bozzolo, N.; Landuyt, O.V.; Grosdidier, T. Evolution of recrystallisation texture and microstructure in low alloyed titanium sheets. *Acta Mater.* **2002**, *50*, 1245–1259. [CrossRef]
27. Wang, Y.N.; Huang, J.C. Texture analysis in hexagonal materials. *Mater. Chem. Phys.* **2003**, *81*, 11–26. [CrossRef]
28. Singh, A.K.; Schwarzer, R.A. Evolution of texture during thermomechanical processing of titanium and its alloys. *Trans. Indian Inst. Met.* **2008**, *61*, 371–387. [CrossRef]
29. Becker, H.; Pantleon, W. Work-hardening stages and deformation mechanism maps during tensile deformation of commercially pure titanium. *Comput. Mater. Sci.* **2013**, *76*, 52–59. [CrossRef]
30. Barkia, B.; Doquet, V.; Couzinié, J.P.; Guillot, I.; Héripré, E. *In situ* monitoring of the deformation mechanisms in titanium with different oxygen contents. *Mater. Sci. Eng. A* **2015**, *636*, 91–102. [CrossRef]
31. Wagner, F.; Ouarem, A.; Richeton, T.; Toth, L.S. Improving Mechanical Properties of cp Titanium by Heat Treatment Optimization. *Adv. Eng. Mater.* **2018**, *20*, 1700237. [CrossRef]
32. Masson, R.; Zaoui, A. Self-consistent estimates for the rate-dependentelastoplastic behaviour of polycrystalline materials. *J. Mech. Phys. Solids* **1999**, *47*, 1543–1568. [CrossRef]
33. Conrad, H. Effect of interstitial solutes on the strength and ductility of titanium. *Prog. Mater. Sci.* **1981**, *26*, 123–403. [CrossRef]
34. Naka, S. Etude des Mécanismes de Déformation Plastique à Basse Température de Monocristaux de Titane α. Ph.D. Thesis, Université de Paris Sud (Orsay), Orsay, France, 1983.
35. Franciosi, P.; Berveiller, M.; Zaoui, A. Latent hardening in copper and aluminium single crystals. *Acta Metall.* **1980**, *28*, 273–283. [CrossRef]
36. Messner, M.C.; M.Rhee.; Arsenlis, A.; Barton, N.R. A crystal plasticity model for slip in hexagonal close packed metals based on discrete dislocation simulations. *Model. Simul. Mater. Sci. Eng.* **2017**, *25*, 044001. [CrossRef]
37. Wang, L.; Zheng, Z.; Phuka, H.; Kenesei, P.; Park, J.S.; Lind, J.; Suter, R.M.; Bieler, T.R. Direct measurement of critical resolved shear stress of prismatic and basal slip in polycrystalline Ti using high energy X-ray diffraction microscopy. *Acta Mater.* **2017**, *132*, 598–610. [CrossRef]
38. Pagan, D.C.; Shade, P.A.; Barton, N.R.; Park, J.S.; Kenesei, P.; Menasche, D.B.; Bernier, J.V. Modeling slip system strength evolution in Ti-7Al informed by in-situ grain stress measurements. *Acta Mater.* **2017**, *128*, 406–417. [CrossRef]
39. Pagan, D.C.; Bernier, J.V.; Dale, D.D.; Ko, J.Y.P.; Turner, T.J.; Blank, B.; Shade, P.A. Measuring Ti-7Al slip system strengths at elevated temperature using high-energy X-ray diffraction. *Scr. Mater.* **2018**, *142*, 96–100. [CrossRef]
40. Estrin, Y.; Kubin, L.P. Local strain hardening and nonuniformity of plastic deformation. *Acta Metall.* **1986**, *34*, 2455–2466. [CrossRef]
41. Britton, T.B.; Wilkinson, A.J. Stress fields and geometrically necessarry dislocation density distributions near the head of a blocked slip band. *Acta Mater.* **2012**, *60*, 5773–5782. [CrossRef]
42. Grabski, M.W.; Korski, R. Grain boundaries as sinks for dislocations. *Philos. Mag.* **1970**, *22*, 707–715. [CrossRef]
43. Pond, R.C.; Smith, D.A. On the absorption of dislocations by grain boundaries. *Philos. Mag.* **1977**, *36*, 353–366. [CrossRef]
44. Smith, D.A. Interaction of dislocations with grain boundaries. *J. Phys. Colloq.* **1982**, *43*, C6 225–C6 237. [CrossRef]
45. Won, J.W.; Park, K.T.; Hong, S.G.; Lee, C.S. Anisotropic yielding behavior of rolling textured High purity titanium. *Mater. Sci. Eng. A* **2015**, *637*, 215–221. [CrossRef]
46. Raabe, D.; Sachtleber, M.; Zhao, Z.; Roters, F.; Zaefferer, S. Micromechanical and macromechanical effects in grain scale polycrystal plasticity experimentation and simulation. *Acta Mater.* **2001**, *49*, 3433–3441. [CrossRef]

47. Allain-Bonasso, N.; Wagner, F.; Berbenni, S.; Field, D.P. A study of plastic heterogeneity in IF steel by EBSD. *Mater. Sci. Eng. A* **2012**, *548*, 56–63. [CrossRef]
48. Wagner, F.; Ouarem, A.; C. F. Gu, N.A.B.; Toth, L.S. A new method to determine plastic deformation at the grain scale. *Mater. Charact.* **2014**, *92*, 106–117. [CrossRef]
49. Grennerat, F.; Montagnat, M.; Castelnau, O.; Vacher, P.; Moulinec, H.; Suquet, P.; Duval, P. Experimental characterization of the intragranular strain field in columnar ice during transient creep. *Acta Mater.* **2012**, *60*, 3655–3666. [CrossRef]
50. Ghaderi, A.; Barnett, M.R. Sensitivity of deformation twinning to grain size in titanium and magnesium. *Acta Mater.* **2011**, *59*, 7824–7839. [CrossRef]

Article

Crystal Plasticity Modeling of Anisotropic Hardening and Texture Due to Dislocation Transmutation in Twinning

Robert M. Allen [1,2,3], Laszlo S. Toth [3,4], Andrew L. Oppedal [1,2] and Haitham El Kadiri [1,2,*]

1 Department of Mechanical Engineering, Mississippi State University, Mississippi State, MS 39762, USA; rma140@msstate.edu (R.M.A.); aoppedal@cavs.msstate.edu (A.L.O.)
2 Center for Advanced Vehicular Systems, Mississippi State University, Mississippi State, MS 39762, USA
3 Laboratory of Excellence on Design of Alloy Metals for Low-Mass Structures (DAMAS), Université de Lorraine, 57045 Metz, France; laszlo.toth@univ-lorraine.fr
4 Laboratoire d'Etude des Microstructures et de Mécanique des Matériaux (LEM3), CNRS UMR 7239, Université de Lorraine, 57045 Metz, France
* Correspondence: elkadiri@me.msstate.edu; Tel.: +1-(662)-325-4777

Received: 10 September 2018; Accepted: 15 September 2018; Published: 28 September 2018

Abstract: In crystalline materials, dislocations are three-dimensional lattice distortions that systematically distort twin interfaces that they encounter. This results in dislocation dissociation events and changes in the atomic structure of the interface. The manner in which the interface distorts drive the product of the dissociation event, and consequently, the incident dislocation core and the magnitude and relative direction of the Burgers vector govern these slip-twin interaction phenomena. Recent characterization studies using transmission electron microscopy as well as advanced molecular dynamic simulations have shown that slip dislocations, whether striking or struck by a $\{10\bar{1}2\}$ twin boundary, dissociate into a combination of twinning disconnections, interfacial disclinations (facets), jogs, and other types of dislocations engulfed inside the twin domains, called transmuted dislocations. While twinning disconnections were found to promote twin propagation, the dislocations incorporated inside the twin are of considerable importance to hardening and damage initiation as they more significantly obstruct slip dislocations accommodating plasticity of the twins. In this work, the dislocation transmutation event and its effect on hardening is captured using a dislocation density based hardening model contained in a visco-plastic self-consistent mean-field model. This is done by allowing the twins to increase their dislocation densities, not only by virtue of slip inside the twin, but also through dislocations that transmute from the parents as the twin volume fraction increases. A correspondence matrix rule is used to determine the type of converted dislocations while tracking and parameterizing their evolution. This hypothesis provides a modeling framework for capturing slip-twin interactions. The model is used to simulate the mechanical response of pure Mg and provides a more physically based approach for modeling stress-strain behavior.

Keywords: magnesium; twinning; hardening; dislocations; crystal plasticity; self consistent methods; modeling; simulation

1. Introduction

The utilization of hexagonal-close-packed (HCP) magnesium (Mg) and magnesium based alloys is of interest to industry because of their low density and high specific strength. Their low crystal symmetry leads to mechanically anisotropic behavior, especially in highly textured polycrystals, in which significant nucleation and growth of $\{10\bar{1}2\}$ tensile twins may occur. $\{10\bar{1}1\}$ compression twinning also occurs in Mg when the $\langle c \rangle$-axis is compressed or contracted, and has been correlated with a high propensity of damage when formed inside $\{10\bar{1}2\}$ tensile twins or after significant pyramidal $\langle c + a \rangle$ slip has occurred [1,2]. Several design approaches have been developed in an effort to produce Mg alloys with higher ductility, mainly relying on the idea that twinning should be reduced [3]. Additions of rare earth elements proved to be useful for improving formability at low temperature, primarily by means of texture weakening. However, energy absorption levels remained shockingly low under high rate regimes relevant to car crash deformation [4,5]. In order to expand the range of industrial applications of Mg alloys the way twinning affects hardening must be better understood and predicted.

Studies of single crystal magnesium revealed that slip along the basal and prismatic planes are often active during deformation, both possessing a relatively low CRSS when compared with other deformation modes [6–8]. However, in order to accommodate deformation along the $\langle c \rangle$-axis, pyramidal-$\langle c + a \rangle$ slip and tensile $\{10\bar{1}2\}$ twinning modes are often activated as well, leading to heavily anisotropic behavior, both observed experimentally and predicted in polycrystal modeling [9–13]. In some cases, highly profuse twinning in which 80% of the parent grain volume is overtaken by twin volume fraction can occur. In other cases, scarcely no twinning at all may take place [6,8]. The reason for this anisotropic behavior is well understood, and is rooted in the polarity of twinning combined with its marked low CRSS compared to any other deformation mode able to accommodate $\langle c \rangle$-axis strain. As $\{10\bar{1}2\}$ twinning cannot facilitate $\langle c \rangle$-axis compression, the crystal or grain is very strong when compressed along the $\langle c \rangle$-axis by virtue of the domination of pyramidal-$\langle c + a \rangle$ slip. However, compression perpendicular to $\langle c \rangle$ forces $\langle c \rangle$-axis extension. In this case the crystal is initially weak when compared to compression along along the $\langle c \rangle$-axis, but quickly hardens as the reorientation by twinning rotates the parent lattice, forcing it back into $\langle c \rangle$-axis compression. Similar anisotropy occurs under tension.

As highlighted in the work of Oppedal et al. [14], if the anisotropic behavior of Mg is informed exclusively by the process described above, then the saturation stress should, in principal, be roughly the same for both compression paths. But this is not the case, and consequently, additional hardening mechanisms are needed to explain this discrepancy. Previously reported crystal plasticity simulation work generally introduced Hall–Petch effects in order to obtain acceptable stress-strain simulation results [15–18]. If the differences in saturation stress associated with the incidence of twinning in Mg were motivated by Hall–Petch effects, then it is logical to expect the observation of dislocation pile-up at twin boundaries. However, pre-strained Mg polycrystals at various levels of strain showed that $\{10\bar{1}2\}$ twins readily pass through any dislocation substructure that they encounter during their propagation, suggesting that twin boundaries are able to either absorb or incorporate those dislocations [2].

Oppedal et al. [14] go on to demonstrate that, rather than being caused by Hall–Petch effects, this anisotropy could be the result of increased dislocation density inside the twin grains. Implementing a modified version of the dislocation density based hardening model of Beyerlein and Tomé [19] in the visco-plastic self-consistent (VPSC) code (VPSC-7d from Los Alamos National Laboratory) the compression of rolled Mg with a highly basal texture along multiple compression loading paths were simulated. By setting the contributions of Hall–Petch mechanisms to zero, Oppedal et al. [14] showed that increasing the amount of dislocation density stored inside twin volumes by a Twin Storage Factor (TSF) recreated the characteristic hardening behavior across multiple load paths. This approach will be referred to as the TSF method in the following. This demonstrated that the assumptions about how slip dislocations interact with twin grain boundaries may require additional investigation in order to be completely understood. More specifically, it was shown that increasing the dislocation density of twin grains could reproduce hardening reactions previously thought to have been caused solely by Hall–Petch effects. However, this empirical approach was limited in that the specific physical process by which dislocation density was increased inside of twin grains was not modeled.

The findings of Oppedal et al. [14] were bolstered by the more recent studies of slip-twin interactions in the works of El Kadiri et al. [20], Barrett et al. [21], and Wang et al. [22], in which the behavior of basal dislocations encountering $\{10\bar{1}2\}$ twin boundaries were investigated by means of molecular dynamic simulations and transmission electron microscopy. These studies showed that such dislocations readily disassociate into twinning disconnections bolstering twin propagation, interfacial disclination dipoles (facets), and other dislocations that transmute into the interior of the twin volume, called transmuted dislocations. Though the complete dislocation reaction at the twin interface was only recently identified by means of atomistic simulations and advanced interfacial defect theory, the part of the dissociation leading to dislocation transmutation has been hypothesized as early as the nineteen sixties by Bilby and Saxl and was later known as the Basinski mechanism [23–25]. Similar to the Basinski mechanism [25], the capacity of dislocations to transmute from parent grains across twin boundaries represents a potential vehicle by which the dislocation density of twin grains might be increased in excess of statistically and geometrically necessary dislocations typically formed to accommodate slip within the twin. Transmutation could then account for the increased dislocation density enforced by the twin storage factor of Oppedal et al. [14]. While previous studies have conceded that a Basinski type transmutation effect may be occuring in twinning Mg, experimental evidence has been sparse thus far [2,15,16,18]. The work of El Kadiri and Oppedal [26] provides a crystal plasticity framework by which such transmutation effects might be modeled in VPSC simulations. They developed a transmutation matrix α where each element α_{ij} represents the proportion of dislocation density that is transmuted to the ith slip mode inside the twin volume from the jth slip mode in the parent volume.

In order to investigate the potential role of dislocation transmutation effects in the hardening of Mg, the work presented herein presents a modified version of the crystal plasticity based dislocation transmutation theory of El Kadiri and Oppedal [26], further adapted to capture the system-wise nature slip dislocation interaction with twin boundaries and allow for the inclusion of both transmutation and dissociation behavior of these dislocations. In this way, a physically based model for the complex interactions of slip and twinning in polycrystal Mg is developed. This model is implemented in the LANL code VPSC-7d, and is calibrated against data from rolled Mg under multiple compression load paths. Finally, it is used to simulate the anisotropic hardening behavior of the material.

2. Model

The hardening model for this work is based on the approach of Beyerlein and Tomé [19] in which the hardening behavior of HCP polycrystals is simulated by the evolution of the critical resolved shear stress (CRSS) used in Hutchinson's strain rate sensitive constitutive equation for slip. In this approach, the CRSS for slip for a given slip system s as part of the slip mode, i, is defined as a function of the dislocation density on that system. The CRSS of each system is given by the additive decomposition

$$\tau_c^s\left(\rho^i\right) = \tau_0^i + \tau_{forest}^i\left(\rho^i\right) + \tau_{deb}^i\left(\rho^i\right) + \tau_{HP}^s, \ \forall s \in i. \tag{1}$$

Here, τ_0^i, τ_{forest}^i, τ_{deb}^i, and τ_{HP}^s are the initial critical stress values, hardening from forest dislocations, hardening from debris formation, and hardening from Hall–Petch effects, respectively. They can be defined as

$$\tau_{forest}^i\left(\rho^i\right) = b^i \chi \mu \sqrt{\rho^i}, \tag{2a}$$

$$\tau_{deb}^i\left(\rho^i\right) = k_{deb}\mu b^i \sqrt{\rho_{deb}} \log\left(\frac{1}{b^i\sqrt{\rho_{deb}}}\right), \text{ and} \tag{2b}$$

$$\tau_{HP}^s = \begin{cases} \mu \mathrm{HP}^i \sqrt{\frac{b^i}{d_g}} \text{ for grains without twins,} \\ \mu \mathrm{HP}^{ij} \sqrt{\frac{b^i}{d_{mfp}^{s,PTS}}} \text{ for grains with twins.} \end{cases} \tag{2c}$$

In Equation (2a), b^i and ρ^i are the Burgers vector and dislocation density on the ith slip mode. χ and μ are the dislocation interaction coefficient and shear modulus, respectively. Equation (2b) also contains the debris formation parameter k_{deb} and debris density ρ_{deb}. The Hall–Petch contributions in Equation (2c) depend on the presence of twins. In grains where there is no active twinning, these contributions are the same for all $s \in i$, with a bulk interaction term HP^i and grain size d_g. For grains containing twins, the values of τ^s_{HP} cannot be assumed to be the same for all $s \in i$. In this case, the interaction term HP^{ij} describes the interaction between twin mode j, which contains the predominate twin system, and slip mode i. Here, $d^{s,PTS}_{mfp}$ is the mean free path between twin lamella for dislocations on slip system s as described in Proust et al. [17].

In the simulations presented in this work, the forest and debris contributions remain unchanged from the forms presented in Beyerlein and Tomé [19]. However, the constitutive model of Beyerlein and Tomé [19] is based on the Composite Grain (CG) model of Proust et al. [17] and treats twin boundaries as full-stop barriers to dislocation glide. As twin boundaries are not considered to act as barriers in this work by hypothesis, the Hall–Petch contributions to CRSS are treated as being identically zero, hence

$$\tau^s_{HP} = 0, \quad \forall s \in i. \tag{3}$$

In the model of Beyerlein and Tomé [19], the change in dislocation density for a given slip mode α with change in shear on the same system is calculated by a Mecking-Kocks [27] type equation and adopted in this work to calculate the change in dislocation density with shear as

$$\frac{\partial \rho^i}{\partial \gamma^i} = k^i_1 \sqrt{\rho^i} - k^i_2 \left(\dot{\epsilon}, T\right) \rho^i. \tag{4}$$

Equation (4) introduces the phenomenological generation and rate and temperature dependent parameters k^i_1 and k^i_2. In Beyerlein and Tomé [19], this leads to a relationship between these two parameters,

$$\frac{k^i_2\left(\dot{\epsilon}, T\right)}{k^i_1} = \frac{\chi b^i}{g^i} \left(1 - \frac{kT}{D^i (b^i)^3} log\left(\frac{\dot{\epsilon}}{\dot{\epsilon}_0}\right)\right). \tag{5}$$

Above, g^i, D^i are the normalized activation energy and drag stress, respectively, and $\dot{\epsilon}$ and $\dot{\epsilon}_0$ are the strain rate and reference strain rate.

In the present model, the dislocation densities were modified based on changes in twin volume fraction of the composite grain, which were added:

$$\rho^{iTF} = \begin{cases} \rho^i, & \forall s \in \text{parent volume fraction} \\ \rho^{iTF}\left(V\right), & \forall s \in \text{parent volume fraction.} \end{cases} \tag{6}$$

In this way, the hardening contributions from dislocation density to the CRSS of slip systems inside the twin volume fraction of each grain are differentiated from those systems inside the parent volume fraction.

The function $\rho^{iTF}\left(V\right)$ from the work of El Kadiri and Oppedal [26] defining the increase in dislocation density of the ith slip mode type inside the jth twin volume fraction is expressed as

$$d\rho^{iTF}\left(V\right) = \frac{V^T \rho^{iT} + dV \sum_j \alpha_{ij} \rho^{jP}}{V^T + dV}. \tag{7}$$

In the present work, the matrix α includes an additional row in order to account for dislocation transmutation into sessile, higher order dislocations and Equation (7) is modified in order to account for the dissociation of parent dislocations at twin grain boundaries. A normalized dissociation parameter η is introduced, representing the proportion of dislocation density that fails to transmute from the jth slip system in the parent volume to the twin volume and does not contribute to dislocation density in the twin. This leads to the final, modified constitutive law

$$d\rho^{iTF}(V) = \frac{V^T \rho^{iT} + dV \sum_j (1 - \eta_j)\, \alpha_{ij} \rho^{jP}}{V^T + dV}. \tag{8}$$

The model of Beyerlein and Tomé [19] also contains constitutive laws for twin activation, propagation, and interaction. These remained unchanged in the current work and are summarized in Appendix A.

3. Implementation and Calibration

The above described changes to the dislocation density model were implemented in the LANL code VPSC-7b by making changes directly to the subroutines UPDATE_CRSS_CG_disl and DATA_CRYS and were compiled using the GFortran GNU compiler. The model was calibrated using a three stage process. First, the values for α were assigned based on the correspondence method introduced by Bilby and Crocker [23]. After this, values for the elements η_j were assigned, and finally, the slip and twinning parameters respectively associated with dislocation generation, twin nucleation, and twin propagation were adopted from Oppedal et al. [14].

3.1. Correspondence Method for Transmutation for the Construction of the α matrix

A generalized correspondence method for mapping a vector on a slip plane in a parent grain on to a corresponding slip plane inside a twin grain was first presented in the work of Bilby and Crocker [23]. This method was applied to twin modes in FCC and HCP crystals in the works of Niewczas [28,29]. Summarized in Appendix B, the precise mathematical expression of the general theory of the correspondence method shown in Niewczas [29] in the case of Mg was adapted for the purpose of assigning values to the transmutation matrix α.

The onto mapping of slip systems in the parent volume to corresponding slip systems in the twin volume implemented by Niewczas [29] provides explicit calculations for transmuting the plane normal and directional vectors for each slip system in a parent grain to its corresponding slip system inside a twin. Rather than using the method to calculate values of α at each deformation step, the transmutation matrix for $\{10\bar{1}2\}$ twins was constructed in pre-processing. $\{10\bar{1}1\}$ twins were not assumed to contribute to transmutation in this work as the onset of compression twinning in Mg is taking place at higher strains, and very closely associated with the nucleation of damage in the material, leading quickly to brittle failure. As such, the contributions of compression twins to dislocation transmutation were assumed to be negligible. Strictly for the purposes of the element wise construction of α, the following assumptions were made:

1. The entirety of the dislocation density from a slip system in the volume fraction of the parent grain overtaken by a twin mode is considered to transmute to its corresponding slip system inside the twin grain. This pairing of slip systems is defined by the mappings defined by the correspondence method used by Niewczas [29] for HCP materials.
2. The dislocation density of slip mode in the volume fraction of the parent grain overtaken by a twin volume is considered to be evenly distributed across the systems of that mode.
3. In this simulation, the dislocation density of any parent slip system corresponding to a slip system inside the twin grain that was not part of the prismatic, basal, or 2nd order pyramidal $< c+a >$ slip modes was assumed to contribute to debris formation inside the twin as part of ρ_{deb}.

Working from these assumptions, the value of each element α_{ij} can be calculated as the proportion of systems from the jth slip mode in the parent volume fraction which contribute to dislocation density in the ith slip mode in the twinned volume. This can be expressed as

$$\alpha_{ij} = \frac{n_j^C}{n_j^{tot}}, \tag{9}$$

where n_j^C is the number of slip systems in the jth slip mode of parent volume mapped onto slip systems in the ith slip mode inside the twin volume and n_j^{tot} is the total number of slip systems in the jth slip mode of the parent volume. The values of α_{ij} for Mg used in this work are summarized in Appendix C.

3.2. Parameters for Dissociation

The molecular dynamics simulations in the work of El Kadiri et al. [20] suggest that the type of dislocation and its orientation relative to the advancing twin boundary play a significant role in determining whether a given dislocation will transmute across the boundary or dissociate upon contact. In the simulations presented in this work, screw type basal dislocations were allowed to transmute perfectly across tensile twin boundaries. Edge type dislocations would either transmute across these twin boundaries or dissociate into twinning disconnections according to their orientation relative to the twin boundary, with positively oriented dislocations transmuting across the boundary and negatively oriented ones dissociating.

The simulations presented in this work were conducted at relatively low strain regimes, with $\epsilon \leq 0.25$ for all simulations. As such, the following assumptions were made:

1. Dislocations in the prismatic, basal, and 2nd order pyramidal $< c + a >$ modes are assumed to transmute.
2. The incidence of screw type dislocations at low strain regimes is quite low. As such, for the purposes of this work, it was assumed that the contributions to dislocation density inside the twin volume fractions made by these types of dislocations were negligible.
3. Relative to twin boundaries, it was assumed that dislocations with positive and negative Burgers vector occur in equal measure.

The proportion of dislocation density dissociated at twin boundaries rather than transmuted is set by the η parameter at

$$\eta_{prismatic} = \eta_{basal} = \eta_{pyramidal} = \eta_{higherorder} = 0.5, \tag{10}$$

and summarized again in Appendix C.

Parameters for Dislocation Generation and Twin Nucleation and Propagation

Simulations from the work of Oppedal et al. [14] were taken as a baseline for the presented work, and parameters for the constitutive laws governing dislocation generation, drag stress, normalized activation energy, critical stress values, and deactivated Hall–Petch effects were taken from this work. Transmutation effects were used in place of twin storage factor, reducing the latter value to zero. The values of these parameters are summarized in Appendix A.

3.3. Simulation

Experimental data from the work of Oppedal et al. [14] was utilized to establish a control data set, where possible. In the case where such data were not available, comparisons were made to simulated results from the same work. In this work, rolled, pure magnesium with a highly basal texture in the normal direction was subjected to uniaxial compression along the normal and transverse directions, hereafter referred to as through thickness compression (TTC) and in-plane compression

(IPC), respectively. This was done in order to capture the anisotropic mechanical behavior of textured magnesium, and, in particular, to motivate both scant and profuse twin volume growth. The chemical composition of the pure magnesium used to generate the experimental data in Oppedal et al. [14] is summarized in Table 1. The initial texture data was obtained via neutron diffraction at LANL and pole figures generated from this data are shown in Figure 1.

Table 1. Composition of pure magnesium in ppm.

Al	Ca	Mn	Zr	Zn	Sn	Si	Pb	Mg
30	10	40	10	130	10	40	10	Balance

Shown in Figure 2, the load paths for the simulations presented in this work were performed along both TTC and IPC directions on polycrystals consisting of 1944 grains with a highly basal texture in order to recreate the conditions under which the literature data were collected. The loading was conducted at room temperature conditions, with a strain increment of $\Delta\epsilon = 0.001$, up to a total compressive true strain of $\epsilon = 0.25$. The initial texture of the compressed Mg specimen is shown in Figure 1.

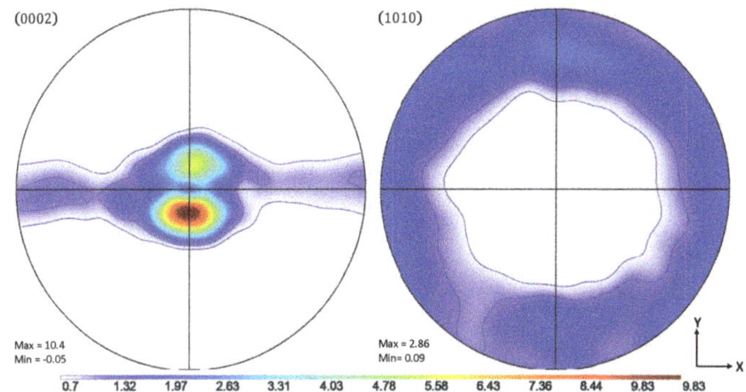

Figure 1. Pole figures showing the initial rolled texture of the Mg specimen.

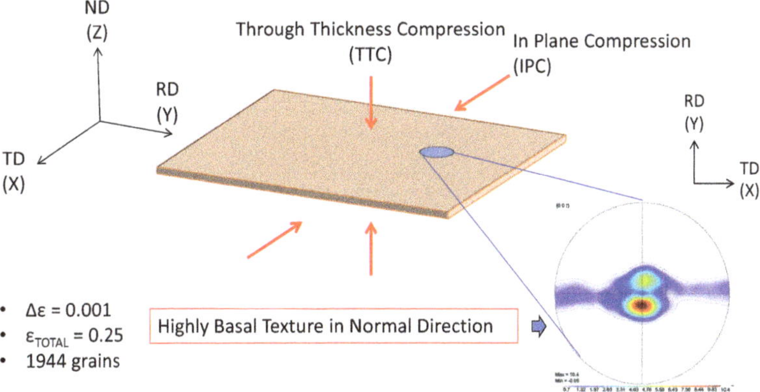

Figure 2. Simulated load paths and initial texture for simulations of rolled pure magnesium.

4. Results

4.1. TTC Load Path

As highlighted in Figure 3, the TTC simulations reproduced both the texture and hardening curves from the experimental literature data with good accuracy. Although the strain softening observed in the experimental data at approximately $\epsilon = 0.1$ was not reproduced in the simulation, capturing such behavior lies beyond the scope of the current work. The simulated modal activity, being defined as the proportional contribution to strain from each mode at each deformation increment, and evolution of twin volume fraction shown in Figure 4, both followed trends seen in simulations utilizing twin storage factor effects. Good agreement between the two simulation approaches was observed, however, tension twin volume was somewhat under predicted relative to simulations utilizing the twin storage factor approach.

From the literature, the texture in TTC simulations was not expected to change much over the course of loading. This was evident in Figure 5 showing results of the simulations utilizing the transmutation scheme, with the basal texture remaining consistent throughout the compression loading.

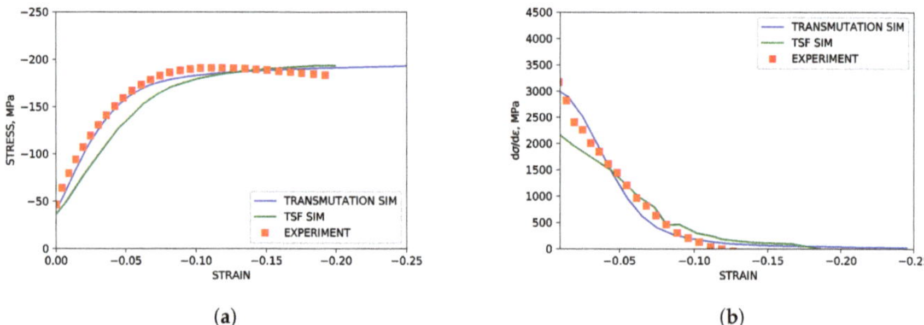

Figure 3. Simulated and experimental mechanical response under TTC compression. (**a**) Stress-strain. (**b**) Hardening rate vs. strain.

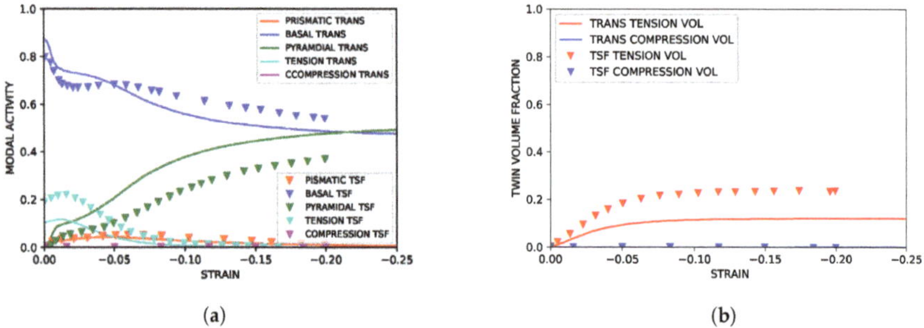

Figure 4. Simulated slip and twinning activity under TTC compression compared to simulated results from Oppedal et al. [14]. (**a**) Simulated modal contributions vs. strain. (**b**) Simulated twin volume growth vs. strain.

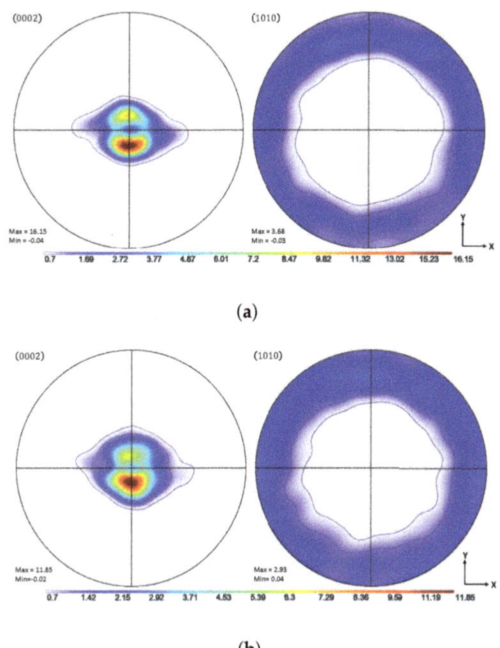

(a)

(b)

Figure 5. Comparison of simulated and experimental textures under TTC load conditions at $\epsilon = 0.09$. (**a**) Simulated $\{0\,0\,0\,2\}$ and $\{1\,0\,\bar{1}\,0\}$ pole figures. (**b**) Experimental $\{0\,0\,0\,2\}$ and $\{1\,0\,\bar{1}\,0\}$ pole figures.

4.2. IPC Load Path

The sigmoidal stress curve associated with a high incidence of twinning was successfully captured in the IPC simulations. The associated simulated hardening rate was reasonably similar to experimental data from the literature. These results are shown in Figure 6. Shown in Figures 7 and 8, modal contributions to strain were nearly identical to simulation results from previous simulation literature as well. The growth of secondary compression twin volume fraction did exceed that seen in simulations utilizing the twin storage factor approach, but it is assumed that this can be neglected as the onset of compression twinning is, at higher strains, associated with void nucleation along compression twin grain boundaries leading to brittle failure. This fact was reinforced by the failure of the experimental control specimen coinciding with the formation of significant (>10%) volume fraction of secondary compressive twins within the primary twin volume.

In Figure 9, the textural evolution is shown to be consistent with experimental data from the literature, with high concentrations of orientations in the neighborhood of 90° from the normal specimen axis. This type of reorientation is consistent with the high degree of tensile twinning.

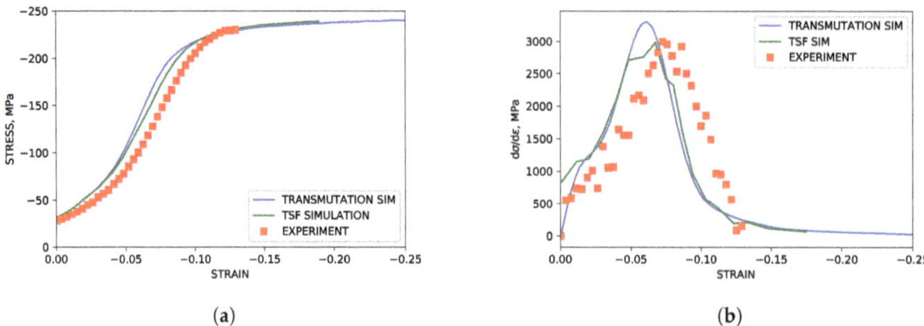

(a) (b)

Figure 6. Simulated and experimental mechanical response under IPC compression. (**a**) Stress-strain. (**b**) Hardening rate vs. strain.

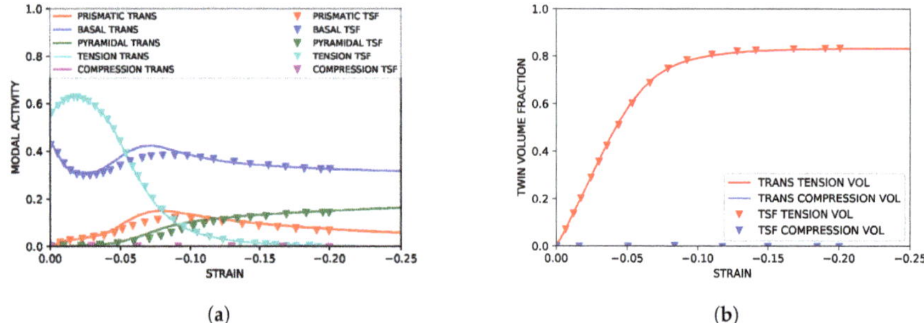

(a) (b)

Figure 7. Simulated parent slip and primary twinning activity under IPC compression compared to simulated results from Oppedal et al. [14]. (**a**) Simulated modal contributions vs. strain. (**b**) Simulated twin volume growth vs. strain.

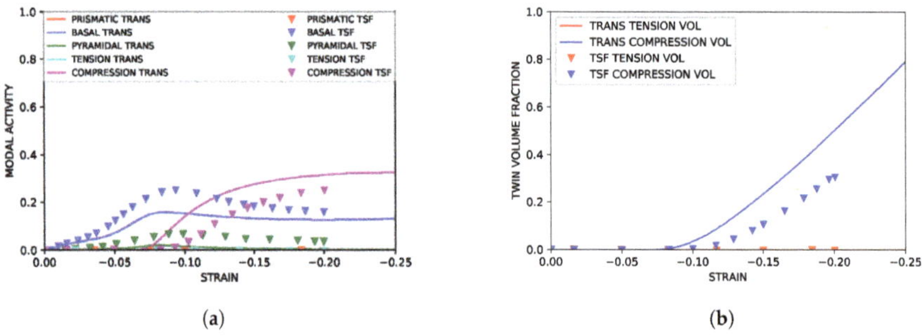

(a) (b)

Figure 8. Simulated twin slip and secondary twinning activity under IPC compression compared to simulated results from Oppedal et al. [14]. (**a**) Simulated modal contributions to strain vs. strain. (**b**) Simulated twin volume growth vs. strain.

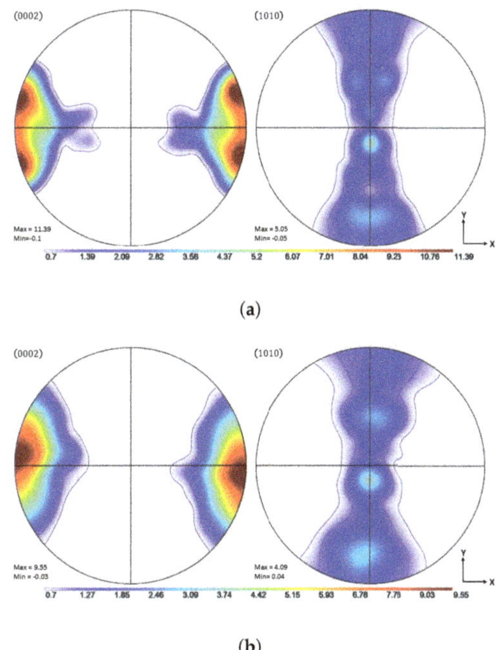

Figure 9. Texture comparison of simulated and experimental textures under IPC load conditions at $\epsilon = 0.12$. (**a**) Simulated $\{0002\}$ and $\{10\bar{1}0\}$ pole figures. (**b**) Experimental $\{0002\}$ and $\{10\bar{1}0\}$ pole figures.

5. Discussion

As the evolution of yield stress and twin volume with strain of pure magnesium is heavily anisotropic, it becomes necessary to consider a measure for "goodness of fit" of simulated data along multiple load paths when evaluating the degree to which the transmutation model and the method for calibrating its parameters are validated. This measure must also be extended to simulated data obtained from simulations utilizing the TSF model in order to make truly meaningful conclusions regarding connections between the mechanical behavior resulting from artificially increased dislocation density in twin grains and the increased dislocation density resulting from transmutation effects. To this end the normalized mean squared error (NMSE) was calculated across the simulated strain sub-domains for which experimental data was available ($\epsilon = 0$ to 0.2 for TTC and $\epsilon = 0$ to 0.125 for IPC). Summarized in Table 2, these values were calculated for both load paths and for both modeling approaches.

Table 2. Normalized mean squared error for transmutation and TSF modeling approach.

	NMSE
Transmutation (TTC)	5.7613%
TSF (TTC)	16.0268%
Transmutation (IPC)	21.0105%
TSF (IPC)	16.7439%

For simulations of the TTC load path, the normalized mean squared error of the transmutation model is approximately 10% lower than that of the TSF model. In the IPC load path simulations, the normalized mean squared error is roughly 5% higher. Taking both of these comparisons into consideration, it can be concluded that while the transmutation model does suffer from some loss of accuracy in IPC simulations compared to the TSF approach, this loss is marginal at worst and more than offset by gains in accuracy in TTC simulations. This conclusion is further reinforced when one considers that the TSF model is a more empirical approach and, therefore, more limited than transmutation model by the availability of sound experimental data to insure its predictive capabilities.

η Sensitivity

Although stress-strain curves, texture evolution, and modal activities were consistent both with previous experimental and simulation work, further validation of the modeling approach is necessary to conclude that predicted results are indeed the product of increased dislocation density inside the twinned volume fraction of the simulated polycrystal.

In experiments and in simulations using both the TSF method and the method from the current work, the saturation stress observed in the TTC and IPC load paths differs significantly by approximately 60 MPa. In order to confirm that the "gulf" observed between the saturation stress levels of TTC and IPC simulations was the result of increased hardness due to higher dislocation density brought on by dislocation transmutation, additional simulations in which transmutation effects were deactivated were performed for both load paths. This corresponds to simulations in which all of the dislocations dissociate upon reaching the twin boundaries, leading to dissociation parameter values of $\eta = 1.0$. In these cases, the saturation stress levels for both TTC and IPC simulations were shown to be nearly equal, at roughly $\sigma_{saturation} = 180$ MPa. These results are summarized in Figure 10.

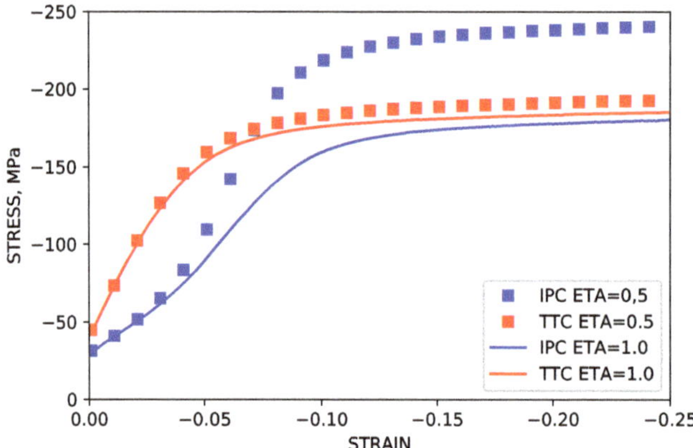

Figure 10. Simulated stress-strain curves with a dissociation parameter of η = 0.5 and dissociation parameter η = 1.0. These values correspond to a state of active dislocation transmutation and a state of no transmutation, respectively.

Figure 11a shows the evolution of the dislocation density of each slip as a function of IPC compression. Dislocation activity correlates well with the modal activity trends shown in Figure 7a, showing increasing dislocation density across strain sub-domains in modes with more modal activity. As shown in Figure 11b, the dislocation density evolution in IPC simulations is consistent with the TSF parameters of Oppedal et al. [14]. At the point in which twin volume fraction was fully realized (roughly 80% total volume fraction) the dislocation density of the twin volume fraction lies in the range of twice that of the the dislocation density of the parent volume fraction.

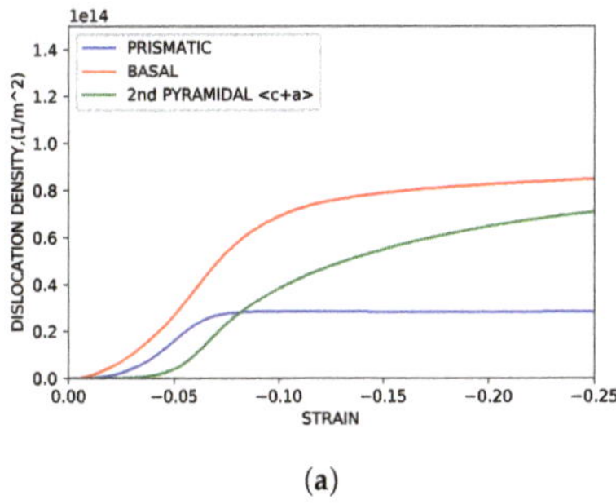

(a)

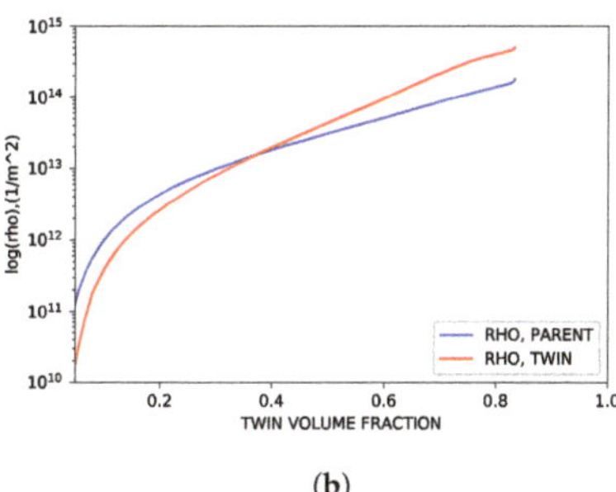

(b)

Figure 11. Simulated dislocation density evolution. (**a**) Modal dislocation density vs. strain. (**b**) Dislocation density for parent and primary twin volume fractions vs. primary twin volume fraction.

The assumption that dislocations exist on the prismatic $\langle a \rangle$ and 2nd order pyramidal $\langle c + a \rangle$ was also put to the test. IPC compression was simulated under the assumption that only basal dislocations would be allowed to transmute across twin grain boundaries. This corresponds to dissociation input parameters of $\eta_{basal} = 0.5$ and $\eta_{prismatic} = \eta_{2^{nd}\,order\,pyramidal} = 1.0$. Shown in Figure 12, the stress levels after the onset of twinning were noticeably under predicted. While stress levels for simulations with only basal dislocations actively transmuting were closer to approximating experimental stress levels than simulations in which no transmutation was activated, the relative similarities can be accounted for by acknowledging that deformation by basal slip, and therefore basal type dislocation density, is significantly higher than both prismatic $\langle a \rangle$ and 2nd order pyramidal $\langle c + a \rangle$ in general. This is supported by observations of simulated dislocation density by mode, summarized in Figure 11a.

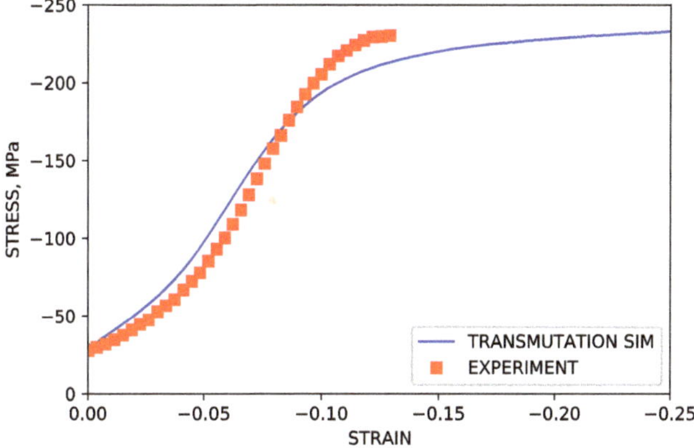

Figure 12. Simulated IPC stress curve with transmutation active in the basal mode only.

6. Conclusions

In summary, methods for simulating the effects of twin transmutation and dissociation at twin grain boundaries were implemented in a visco-plastic self-consistent framework in place of more conventional approaches that utilize Hall–Petch effects, and used to simulate uniaxial compression along multiple load paths in order to capture the behavior of magnesium under conditions which twin volume growth was both profuse and sparse. The results of these simulations were compared to both experimental data and simulated data from previous approaches. The results permit highlighting the following points:

1. In place of Twin Storage Factor and Hall–Petch effects, a model for dislocation and twin boundary interactions was implemented. These methods and parameter selections were used to simulate the behavior of pure, basal textured, rolled magnesium subjected to uniaxial compression along (TTC) and perpendicular to (IPC) the dominant $\langle c \rangle$-axis of the texture. The simulation results for stress, hardening, and texture development were consistent with observed experimental results. Modal activity and twin volume growth were largely similar to simulation work performed using the TSF approach.
2. The large difference between the saturation stress of the simulated IPC and TTC load paths was confirmed to be the result of transmutation effects included in the model by disabling the contributions of transmuted dislocations to the dislocation density of twin volume fractions.
3. It can be stated that Hall–Petch effects cannot be assumed to be the sole or even primary source of twinning induced anisotropy in the mechanical behavior of pure magnesium.

Author Contributions: The research work presented in this manuscript was planned and carried out as a collaboration between all authors listed above. Initial conceptualization was developed by H.E.K., A.L.O. and R.M.A. Methodologies and software were developed by and the investigation was carried out by R.M.A. Validation, and formal analysis was carried out by H.E.K., A.L.O., L.S.T. and R.M.A. Writing and draft preparation was done by R.M.A. This project was supervised by H.E.K. and L.S.T.

Funding: The research presented herein was sponsored by the French State through the program "Investment in the future" operated by the National Research Agency (ANR) and referenced by ANR-11-LABX-0008-01 (LabEX DAMAS), by the Chateaubriand Fellowship of the Office for Science & Technology of the Embassy of France in the United States, and by the Army Research Laboratory under Cooperative Agreement Number W911NF-15-2-0025. The views and conclusions contained in this document are those of the authors and should not be interpreted as representing the official policies, either expressed or implied, of the Army Research Laboratory or the U.S. Government. The U.S. Government is authorized to reproduce and distribute reprints for Government purposes notwithstanding any copyright notation herein.

Conflicts of Interest: The authors declare no conflict of interest.

Abbreviations

The following abbreviations are used in this manuscript:

VPSC	Visco-Plastic Self-Consistent
HCP	Hexagonal Close Packed
Mg	Magnesium
CRSS	Critical Resolved Shear Stress
HP	Hall–Petch
LANL	Los Alamos National Laboratory
TSF	Twin Storage Factor
MD	Molecular Dynamics
CG	Composite Grain
TTC	Through Thickness Compression
IPC	In-Plane Compression

Appendix A. Parameters for Dislocation Density Based Hardening

Constitutive equations for twin hardening evolution in Beyerlein and Tomé [19]:

$$\tau_c^t = \tau_0^k + \tau_{HP}^t + \tau_{slip}^k \tag{A1}$$

$$\tau_{HP}^t = \begin{cases} \dfrac{HP^k}{\sqrt{d_g}} \text{ no twinning or when } t = \text{predominate twin system,} \\ \dfrac{HP^{kk'}}{\sqrt{d_{mfp}^s}}, \text{ when } t \neq \text{predominate twin system.} \end{cases} \tag{A2}$$

$$\tau_{slip}^k = \mu \sum_i C^{ik} b^k b^i \rho^i \tag{A3}$$

Hardening and dislocation generation parameters adopted from Oppedal et al. [14]:

Table A1. Parameters for hardening evolution of slip.

	Prismatic	**Basal**	**Pyramidal** $\langle c+a \rangle$
$k_1\ ^1/_m$	2.0×10^9	0.25×10^9	2.0×10^9
$\dot{\epsilon}\ s^{-1}$	1.0×10^7	1.0×10^7	1.0×10^7
g^α	0.0035	0.0035	0.003
D_0^α MPa	3.4×10^3	10.0×10^3	0.08×10^3
τ_0^α MPa	30	11	50
χ	0.9	0.9	0.9
HP^α	0	0	0
$HP^{\alpha\beta}$	0	0	0

Table A2. Parameters for hardening evolution of twinning.

	Tensile Twin	**Compression Twin**
τ_0^β MPa	$\tau_{crit} = 15, \tau_{prop} = 10$	$\tau_{crit} = 185, \tau_{prop} = 185$
HP^β	0	0
$HP_{TW}^{\beta\beta}$	0	0
$C^{\beta 1}$	0	0
$C^{\beta 2}$	0	0
$C^{\beta 3}$	0	0

Appendix B. Correspondence Method; Summary of Equations

Presented here is a summary of the general Correspondence Method equations from Niewczas [28], Niewczas [29] and Bilby and Crocker [23]. In this, a vector $\mathbf{u}_M$ on in the parent lattice is mapped onto the vector $\mathbf{v}_T$ in the twin vector using the matrices **S**, **X**, and **U**. These matrices are constructed from elements of **m** and **n** describing the twinning direction and normal to the twin plane.

$$\mathbf{S} = \begin{bmatrix} 1+sm_1n_1 & sm_1n_2 & sm_1n_3 \\ sm_2n_1 & 1+sm_2n_2 & sm_2n_3 \\ sm_3n_1 & sm_3n_2 & 1+sm_3n_3 \end{bmatrix} \tag{A4}$$

$$\mathbf{R} = \begin{bmatrix} -1 & 0 & 0 \\ 0 & -1 & 0 \\ 0 & 0 & 1 \end{bmatrix} \tag{A5}$$

$$\mathbf{X} = \begin{bmatrix} m_1 & m_2 & m_3 \\ m_3n_2 - m_2n_3 & m_1n_3 - m_3n_1 & m_2n_1 - m_1n_2 \\ n_1 & n_2 & n_3 \end{bmatrix} \tag{A6}$$

$$\mathbf{v}_M = \mathbf{S}\mathbf{u}_M \tag{A7}$$

$$\mathbf{U} = \mathbf{X}^{-1}\mathbf{R}\mathbf{X} \tag{A8}$$

$$\mathbf{v}_T = \mathbf{U}\mathbf{S}\mathbf{u}_M. \tag{A9}$$

Appendix C. Parameters for Transmutation and Dissociation: Alpha Matrix Composition

Final values of transmutation matrix α used in simulations of Mg, with prismatic, basal, 2nd order pyramidal $< c + a >$ modes and debris corresponding to indices 1, 2, 3, and 4, respectively:

$$\alpha = \begin{bmatrix} 0.0 & 0.333\bar{3} & 0.333\bar{3} \\ 0.333\bar{3} & 0.0 & 0.0 \\ 0.666\bar{6} & 0.0 & 0.0 \\ 0.0 & 0.666\bar{6} & 0.666\bar{6} \end{bmatrix} \tag{A10}$$

References

1. Ma, Q.; El Kadiri, H.; Oppedal, A.L.; Baird, J.C.; Li, B.; Horstemeyer, M.F.; Vogel, S.C. Twinning effects in a rod-textured AM30 Magnesium alloy. *Int. J. Plast.* **2012**, *29*, 60–76. [CrossRef]
2. El Kadiri, H.; Kapil, J.; Oppedal, A.L.; Hector, L.G.; Agnew, S.R.; Cherkaoui, M.; Vogel, S.C. The effect of twin–twin interactions on the nucleation and propagation of $\{10\,1\,\bar{2}\}$ twinning in magnesium. *Acta Mater.* **2013**, *61*, 3549–3563. [CrossRef]
3. Imandoust, A.; Barrett, C.D.; Al-Samman, T.; Inal, K.A.; Kadiri, H.E. A review on the effect of rare-earth elements on texture evolution during processing of magnesium alloys. *J. Mater. Sci.* **2017**, *52*, 1–29. [CrossRef]
4. Imandoust, A.; Barrett, C.D.; Oppedal, A.L.; Whittington, W.R.; Paudel, Y.; El Kadiri, H. Nucleation and preferential growth mechanism of recrystallization texture in high purity binary magnesium-rare earth alloys. *Acta Mater.* **2017**, *138*, 27–41. [CrossRef]
5. Imandoust, A.; Barrett, C.D.; Al-Samman, T.; Tschopp, M.A.; Essadiqi, E.; Hort, N.; El Kadiri, H. Unraveling Recrystallization Mechanisms Governing Texture Development from Rare-Earth Element Additions to Magnesium. *Metall. Mater. Trans. A* **2018**, *49*, 1809–1829. [CrossRef]
6. Bhattacharya, B.; Niewczas, M. Work-hardening behavior of Mg single crystals oriented for basal slip. *Philos. Mag.* **2011**, *91*, 2227–2247. [CrossRef]
7. Burke, E.C.; Hibbard, W.R. Plastic Deformation of Magnesium Single Crystals. *J. Met.* **1952**, *4*, 295–303. [CrossRef]
8. Hirsch, P.B.; Lally, J.S. The Deformation of Magnesium Single Crystals. *Philos. Mag.* **1965**, *12*, 595–648. [CrossRef]
9. Chun, Y.B.; Davies, C.H.J. Twinning-induced anomaly in the yield surface of highly textured Mg-3Al-1Zn plate. *Scr. Mater.* **2011**, *64*, 958–961. [CrossRef]
10. Oppedal, A.; El Kadiri, H.; Tomé, C.; Vogel, S.C.; Horstemeyer, M. Anisotropy in hexagonal close-packed structures: improvements to crystal plasticity approaches applied to magnesium alloy. *Philos. Mag.* **2013**, *93*, 4311–4330. [CrossRef]
11. Li, X.; Al-Samman, T.; Mu, S.; Gottstein, G. Texture and microstructure development during hot deformation of ME20 magnesium alloy: Experiments and simulations. *Mater. Sci. Eng. A* **2011**, *528*, 7915–7925. [CrossRef]
12. Jiang, L.; Jonas, J.; Mishra, R.; Luo, A.; Sachdev, A.; Godet, S. Twinning and texture development in two Mg alloys subjected to loading along three different strain paths. *Acta Mater.* **2007**, *55*, 1359–6454. [CrossRef]
13. Jonas, J.J.; Mu, S.; Al-Samman, T.; Gottstein, G.; Jiang, L.; Martin, E. The role of strain accommodation during the variant selection of primary twins in magnesium. *Acta Mater.* **2011**, *59*, 2046–2056. [CrossRef]
14. Oppedal, A.L.; El Kadiri, H.; Tomé, C.N.; Kaschner, G.C.; Vogel, S.C.; Baird, J.C. Effect of dislocation transmutation on modeling hardening mechanisms by twinning in magnesium. *Int. J. Plast.* **2012**, *30–31*, 41–61. [CrossRef]
15. Barnett, M.R.; Davies, C.H.J.; Ma, X. An analytical constitutive law for twinning dominated flow in magnesium. *Scr. Mater.* **2005**, *52*, 627–632. [CrossRef]
16. Barnett, M.; Keshavarz, Z.; Beer, A.; Atwell, D. Influence of grain size on the compressive deformation of wrought Mg–3Al–1Zn. *Acta Mater.* **2004**, *52*, 5093–5103. [CrossRef]
17. Proust, G.; Tomé, C.; Kaschner, G. Modeling texture, twinning and hardening evolution during deformation of hexagonal materials. *Acta Mater.* **2007**, *55*, 2137–2148. [CrossRef]
18. Cáceres, C.; Lukáč, P.; Blake, A. Strain hardening due to $\{10\,12\}$ twinning in pure magnesium. *Philos. Mag.* **2008**, *88*, 991–1003. [CrossRef]

19. Beyerlein, I.; Tomé, C. A dislocation-based constitutive law for pure Zr including temperature effects. *Int. J. Plast.* **2008**, *24*, 867–895. [CrossRef]
20. El Kadiri, H.; Barrett, C.D.; Wang, J.; Tomé, C.N. Why are twins profuse in magnesium? *Acta Mater.* **2015**, *85*, 354–361. [CrossRef]
21. Barrett, C.; Wang, F.; Agnew, S.; Kadiri, H.E. Transmutation of Basal Dislocations by $\{10\bar{1}2\}$ Twinning in Magnesium. In *Magnesium Technology 2017*; The Minerals, Metals & Materials Series; Springer: Cham, Switzerland, 2017; pp. 147–152.
22. Wang, F.; Hazeli, K.; Molodov, K.D.; Barrett, C.D.; Al-Samman, T.; Molodov, D.A.; Kontsos, A.; Ramesh, K.T.; El Kadiri, H.; Agnew, S.R. Characteristic dislocation substructure in $\{10\bar{1}2\}$ twins in hexagonal metals. *Scr. Mater.* **2018**, *143*, 81–85. [CrossRef]
23. Bilby, B.A.; Crocker, A.G. The theory of the crystallography of deformation twinning. *Proc. R. Soc. Lond. A Math. Phys. Eng. Sci.* **1965**, *288*, 240–255. [CrossRef]
24. Saxl, I. The incorporation of slip dislocations in twins. *Czechoslov. J. Phys. B* **1968**, *18*, 39–49. [CrossRef]
25. Basinski, Z.; Szczerba, M.; Niewczas, M.; Embury, J.; Basinski, S. The transformation of slip dislocations during twinning of copper-aluminium alloy crystals. *Rev. Metall. Cah. d'Inform. Tech. (Fr.)* **1997**, *94*, 1037–1043.
26. El Kadiri, H.; Oppedal, A. A crystal plasticity theory for latent hardening by glide twinning through dislocation transmutation and twin accommodation effects. *J. Mech. Phys. Solids* **2010**, *58*, 613–624. [CrossRef]
27. Mecking, H.; Kocks, U. Kinetics of flow and strain-hardening. *Acta Mater.* **1981**, *29*, 1865–1875. [CrossRef]
28. Niewczas, M. Dislocations and twinning in face centred cubic crystals. *Dislocat. Solids* **2007**, *13*, 263–364.
29. Niewczas, M. Lattice correspondence during twinning in hexagonal close-packed crystals. *Acta Mater.* **2010**, *58*, 5848–5857. [CrossRef]

Article

Effect of Strain Heterogeneities on Microstructure, Texture, Hardness, and H-Activation of High-Pressure Torsion Mg Consolidated from Different Powders

Subrata Panda [1,2], Laszlo S. Toth [1,2], Jianxin Zou [2,3] and Thierry Grosdidier [1,2,*]

1 Laboratoire d'Etude des Microstructures et de Mécanique des Matériaux (LEM3), Université de Lorraine, CNRS, Arts et Métiers ParisTech, LEM3, 57000 Metz, France; laszlo.toth@univ-lorraine.fr
2 Laboratory of Excellence on Design of Alloy Metals for low-mAss Structures (DAMAS), Université de Lorraine, 57073 Metz, France; subrata.panda@univ-lorraine.fr
3 National Engineering Research Center of Light Alloy Net Forming & State Key Laboratory of Metal Matrix Composites, School of Materials Science and Engineering, Shanghai Jiao Tong University, Shanghai 200240, China; zoujx@sjtu.edu.cn
* Correspondence: thierry.grosdidier@univ-lorraine.fr; Tel.: +33-03-72-74-78-36

Received: 29 June 2018; Accepted: 24 July 2018; Published: 1 August 2018

Abstract: Severe plastic deformation techniques, such as high-pressure torsion (HPT), have been increasingly applied on powder materials to consolidate bulk nanostructured materials. In this context, the aim of the present study is to compare the plastic deformation characteristics during HPT of two distinct Mg-based powder precursors: (i) atomized micro-sized powder and (ii) condensed and passivated nanopowder. Dynamic recrystallization could take place during HPT consolidation of the atomized powder particles while the oxide pinning of the grain boundaries restricted it for the condensed powder. Consequently, there have been substantial differences in the development of the microstructure, texture, local strain heterogeneities, and hardness in the two types of consolidated products. Different types of local strain heterogeneities were also revealed in the consolidated products. The associated diversity in microstructure within the same consolidated product has been demonstrated to have an effect on the hydrogen activation kinetics to form hydrides for these Mg-based materials that could be suitable for solid state H-storage applications.

Keywords: magnesium powders; HPT consolidation; microstructure; hardness; H-activation

1. Introduction

Severe plastic deformation (SPD) techniques are well known for their effective potential to produce ultrafine-grained/nanocrystalline bulk materials with enhanced mechanical and functional properties for a large variety of metal systems [1,2]. It is only recently that they have been employed for powder consolidation, not only because of their ability to fabricate bulk nanostructured materials but also for their capability of low-temperature solid-state densification under large shear strains [3–5]. Among the existing SPD processes, high-pressure torsion (HPT) [6] is an effective process for powder consolidation due to its unique working principle: simple shear deformation of individual particles by introducing extremely large amount of shear strains under high hydrostatic pressures. In addition, compared to other SPD routes, for example, equal-channel angular pressing (ECAP), in HPT it is simpler to control the processing parameters and is also easier to apply on powder materials. Therefore, a significant amount of research involving HPT assisted powder consolidation has been conducted on a wide range of material systems such as metallic materials [7–10], metal-matrix composites [11–15], etc.

Magnesium-based materials are regarded as potential structural elements in the automotive, aerospace, and biomedical applications mainly due to their lightweight, low cost and excellent specific strength [16]. The design and manufacture of high quality structural components made of Mg remain challenging. In general, Mg alloys are produced by casting processes, which generate coarse and segregated microstructures with residual porosities; factors that are limiting the mechanical strength [17,18]. Also, the significant grain growth occurring during high-temperature post-treatments may alter the materials properties. In contrast, the powder metallurgy (PM) processes represent an effective way for mechanical improvement through reduction in segregation scale, improvement in densification and microstructural refinement [19–21]. One of the major drawbacks of the PM process in Mg is the lack of ductility inherited from the presence of oxides at the surface of the powder particles which are difficult to remove during "conventional" sintering [20,22,23]. Therefore, an additional plastic deformation step is generally required [24,25]. Among the plastic deformation processes, solid-state sintering through SPD techniques are the most desirable, offering real alternatives to the conventional press/sinter routes. Indeed, the particle-particle interactions during severe plastic deformation are mainly governed by shearing rather than diffusion to achieve the required bonding and, thus, the bonding can be obtained at much lower temperatures.

To date, there have been limited studies [14,15,26–29] involving Mg based powder consolidation through SPD processing routes. Moss et al. [26] employed the ECAP technique to produce bulk materials by consolidating Mg and its alloy powders. Back pressure equal channel angular consolidation (BP-ECAC) of micrometer size Mg powder at 200 °C using route C produced a dense consolidated Mg sample with enhanced mechanical properties [27]. Also, it has been observed that ECAP processing of pure Mg and Mg-Ti powders led to fully dense consolidated samples with efficient structural refinement comparable to those obtained by mechanical milling [28]. An $Mg_{95}Zn_{4.3}Y_{0.7}$ alloy was processed by HPT consolidation at room temperature as well as at 100 °C, and the superior mechanical properties of this alloy were associated with the ultrafine-grained microstructures accompanied with a high density of dislocations and also due to the significant equilibrium grain boundaries attained at elevated temperature [29].

Although the feasibility of HPT processing is, in the vast majority of cases, appreciated for fabricating thin samples with thickness of around 1 mm [5,7–9,11–13], some recent studies have revealed that there are significant structural heterogeneities across the disk-thickness when relatively thick samples are subjected to HPT [10,30–32]. In particular, Zhao et al. [10] reported the development of gradient microstructures in consolidated bcc-iron powder while using a constrained HPT facility. Substantial shear strain localization at the middle thickness of the HPT disks was noticed while processing bulk Mg sample through quasi-constrained HPT conditions [32].

In this context, motivated by the potential advantages of SPD powder consolidation, and also by the difficulties associated with the conventional processing of Mg-based materials, the present study focuses on HPT consolidation of two kinds of Mg powder precursors: atomized micro-sized powder produced by gas atomization and condensed ultrafine powder obtained by arc-plasma condensation. Since the initial powder particles were very different [33], it could be expected that the nature of the initial powder precursor influenced the powder consolidation and the associated developments in microstructures and mechanical properties in the resultant consolidated products. Following the pioneering works of Skripnyuk et al. using ECAP and high-energy ball milling (HEBM) [34,35], several routes of SPD processing have been employed for enhancing the hydrogen sorption properties of Mg and its alloys [33,36–38]. Here, the effect of HPT on the Mg powders and the associated structural heterogeneities on the kinetics of hydrogen activation was analyzed.

2. Materials and Methods

2.1. Materials and Consolidation by HPT

Two kinds of Mg powder precursors—atomized micro-sized powder produced by gas atomization and condensed ultrafine powder obtained by arc-plasma condensation—were employed for HPT-assisted powder consolidation at room temperature. Commercial purity Mg (99.8%) was gas atomized by SFM SA (Martigny, Switzerland) to produce micro-sized Mg powder with a particle size distribution in the range of 10–70 μm. Figure 1a shows a scanning electron microscopy (SEM) image illustrating the morphological aspects of the micro-sized powder particles. The atomization process led to fairly rounded particles with an average particle size of about 15 μm. The surfaces of the particles observed at higher magnification are shown in the inset of Figure 1a. Marks present at the powder surface, which were created by shrinkage during the solidification process, witness grains having a maximum size of about 10 μm. The nano-sized Mg powder was obtained by an arc-plasma method at NERC-LAF, SJTU (Shanghai, China) by arc evaporation of commercial purity Mg (~99.9%) to generate particle sizes in the sub-micrometer range. In order to preclude these Mg nano-particles from burning in open atmosphere, they were passivated in a mixture of Ar and normal air before collecting them from the evaporation/condensation chamber. The details of this procedure can be found in Ref. [39]. Figure 1b displays representative SEM and tunneling electron microscopy (TEM) images of the condensed powder illustrating the morphological features of the Mg particles. It was estimated from several TEM images that the major particle size distribution was in the range of 50–800 nm with an average particle size below 300 nm, and that the fine powder particles were single crystals with hexagonal symmetry (see the inset of Figure 1b).

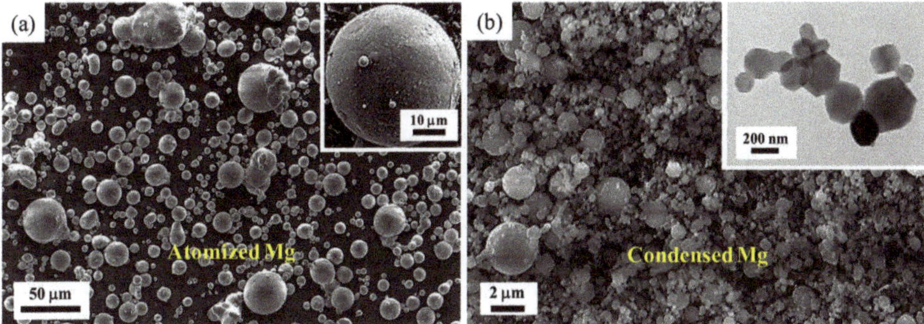

Figure 1. Morphological features of the initial powder precursors: (**a**) scanning electron microscopy (SEM) image of the atomized Mg particles accompanied with a magnified image of a particle showing grain structures (inset); (**b**) SEM image and a typical bright field tunneling electron microscopy (TEM) image (inset) of the condensed Mg particles.

Both Mg powder precursors were separately processed by HPT processing at room temperature. A two-step HPT procedure was followed: step 1—a pre-compacted body in disk shape was produced by uniaxial compression of the powder, and step 2—torsional straining by HPT was imposed on the intermediate disk. Both steps were carried out at room temperature. Details can be found in our recent publication [33] where a sketch of the two-step HPT procedure is also given. The powder was first compressed into an intermediate disk with 20 mm in diameter and 3 mm in thickness by uniaxial compression under a pressure of 1.5 GPa and a holding time of 10 min. Subsequent torsional straining was carried out by HPT on the pre-compacted disks under a pressure of 1.2 GPa. This final step was conducted under quasi-constrained conditions [6], where the materials lateral flow is only partially restricted. A total maximum shear strain (γ) of about 42 at the periphery of the disk was achieved by rotating the lower anvil at 0.125 rpm up to 2 revolutions. In the subsequent sections, these HPT-consolidated samples are designated as micro-HPT and nano-HPT product according to their original powder precursors: the atomized micro-sized powder and the condensed nanopowder, respectively.

2.2. Characterization

The phases in the initial powders as well as in the HPT-consolidated products were identified by X-ray diffraction (XRD) using a powder X-ray diffraction apparatus (Rigaku, D/max 2550VL/PCX, Tokyo, Japan) using a diffracted beam monochromator and a conventional copper target x-ray tube set to 40 KV and 30 mA. The XRD data were obtained under the following conditions: step size 0.02°, scan speed 10°/min. The JADE software (Jade Software Corp., Christchurch, Nouvelle-Zélande) [40] was utilized to analyze the phase components and the associated crystallographic parameters of the structural elements.

The morphology and the microstructure of the initial powders and their consolidated products were examined by scanning electron microscopy (SEM) using a field emission gun scanning electron microscope (SEM, Zeiss Supra40, Oberkochen, Germany). Mechanical polishing was preliminary carried out on these specimens and was found sufficient to observe the secondary phases, while a subsequent etching with an acetic glycol solution was required to reveal the grain boundaries. The grain orientations and the related microtexture were materialized by using regular electron backscatter diffraction (EBSD) technique for the micro-HPT product, and a SEM-based transmission Kikuchi diffraction (SEM-TKD) facility for the nano-HPT product. For the EBSD preparation, the samples were mechanically polished down to 0.05 μm diamond using a mixture of glycerol and ethanol (Ratio 1 for 3) lubricant and a final rising using pure ethanol. The details of this TKD facility and its acquisition techniques can be found in Ref. [41]. Accordingly, very small size samples were cut from the periphery and from the middle-section of the HPT-disk, then mechanically ground followed by diamond paste cloth polishing. Finally, thin-foils were prepared from the polished samples using a focused ion beam (FIB) technique for the TKD measurements. EBSD maps were treated using the HKL Channel 5 (Oxford Instruments, Abingdon-on-Thames, UK), and the Atom software (Laboratoire d'Etude des Microstructures et de Mécanique des Matériaux, LEM3, Metz, France) [42], while the microtexture based on EBSD measurements was analyzed with the Jtex software (Laboratoire d'Etude des Microstructures et de Mécanique des Matériaux, LEM3, Metz, France) [43].

For hardness measurements, the HPT-disks were cut diagonally, and the cross-sections were polished mechanically with abrasive papers and with cloths using diamond paste to a mirror-like finishing. Vickers microhardness was measured using a Leitz Micro Hardness tester, for a dwell time of 15 s with a load of 50 g. In order to obtain the radial as well as axial distributions of the microhardness, the measurements were conducted at 0.1R, 0.3R, 0.5R, 0.7R, and 0.9R radial positions in the middle plane and also on some additional cross-sectional planes across the thickness-height (R is the radius of the HPT-disk, R = 10 mm).

The hydrogen activation characteristics of the two HPT-consolidated products were investigated using a Sievert-type pressure-composition-temperature (PCT) volumetric apparatus (Type PCT-2, Shanghai Institute of Microsystem and Information Technology, Shanghai, China). Prior to hydrogenation, the bulk HPT-products were broken into micrometer size particles (using a ceramic mortar) to fit into the testing vessel. The hydrogen activation was carried out at 400 °C for 8 h under a hydrogen pressure of 3.5 MPa. In order to get insights into the effects of severe plastic deformation (SPD) processing routes on the hydrogen sorption properties of Mg, the initial powder precursors were also tested using identical experimental conditions.

3. Results

3.1. Structural Characterizations of the Consolidated Products

3.1.1. X-ray Diffraction (XRD)

Figure 2 compares the XRD patterns of the initial powder precursors to their HPT-consolidated products. The phase compositions and crystallographic parameters were calculated from the XRD profile refinement using the JADE software [40]; the results are given in Table 1. Besides the major contribution from hexagonal Mg in all XRD profiles, a minor peak from MgO is also detected for the condensed powder and its nano-HPT product (Figure 2b). The amount of oxides in the condensed powder and its nano-HPT product was estimated to be about 5 wt.% (Table 1). In addition, a trace amount of $Mg(OH)_2$ was also identified through its (101) reflexion in the condensed powder and its nano-HPT product. This phase along with MgO oxide was formed during the controlled passivation process conducted in the mixture of Ar + normal air. In contrast, the atomized powder and its micro-HPT product did not show such kinds of additional peaks (see Figure 2a), which is due to the lower surface area available for such "contamination."

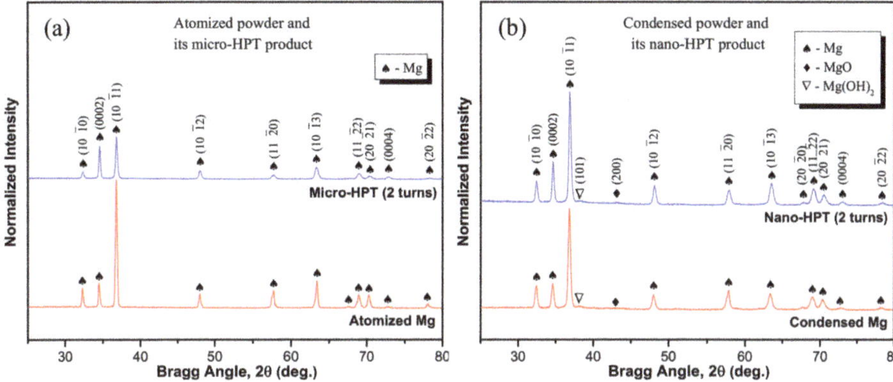

Figure 2. Normalized X-ray diffraction (XRD) patterns recorded on the initial powder precursors and their high-pressure torsion (HPT) products: (**a**) atomized Mg powder and its micro-HPT product; (**b**) condensed Mg powder and its nano-HPT product.

Table 1. Phase composition and crystallographic data for the two types of HPT products obtained from the XRD profile refinement using the JADE software [40].

Samples	Mg (wt.%)	MgO (wt.%)	$Mg(OH)_2$ (wt.%)	Lattice Constants (nm)	Intensity Ratio ($I_{0002}/I_{10\overline{1}0}$)
Atomized Mg	100	-	-	a = b = 0.3208, c = 0.5209	1.24
2 turns micro-HPT	100	-	-	a = b = 0.3200, c = 0.5192	4.11
Condensed Mg	93	5	2	a = b = 0.3207, c = 0.5206	1.08
2 turns nano-HPT	92	5	3	a = b = 0.3194, c = 0.5185	1.73

The lattice parameters of the Mg phase determined for the initial powders and their HPT-products are listed in Table 1. The results reveal that the lattice constants of the Mg phase in the condensed powder were: a = b = 0.3207 nm and c = 0.5206 nm, while the parameters for its nano-HPT product were a = b = 0.3194 nm and c = 0.5185 nm. The differences in the lattice parameters indicate that there might have been some lattice distortions in the Mg crystal structure during the intense shear straining by HPT. However, these changes in the lattice parameters seem to be less pronounced in case of the micro-HPT product obtained from the less contaminated atomized powder (atomized powder: a = b = 0.3208 nm and c = 0.5209 nm and its micro-HPT product: a = b = 0.3200 nm and c = 0.5192 nm). Furthermore, it is interesting to notice from Table 1 that the HPT consolidation of the powder precursors led to significant differences in their peak intensities. For example, there are considerable differences in the peak intensity ratio of (0002) to ($10\overline{1}0$) planes before and after HPT processing. This ratio substantially increased after HPT treatments of the atomized powder precursor while the increment is not significant for the nano-HPT counterpart. The discrepancy in the peak ratios imply the development of different kinds of textures after shear straining by HPT when different types of powders were used.

3.1.2. SEM Observations

Keeping in mind that the products sintered by HPT consolidation for relatively large thicknesses, such as used in the present study, can be characterized by significant microstructural heterogeneities within their thickness-height [10,33], the SEM study was conducted at the periphery and across the disk-thickness. Typical SEM micrographs obtained under backscattered electron (BSE) imaging conditions are displayed in Figure 3. These images essentially compare the developments of microstructure at the periphery and within the thickness-height for the micro-HPT product, see Figure 3a–c for the nano-HPT product and Figure 3d–f after 2 HPT turns.

In the micro-HPT product consolidated from the atomized powder precursor, the microstructure at the middle-section of the HPT-disk displays a recrystallized equiaxed grain structure, see Figure 3b. This observation is consistent with the results obtained for such a thick sample in the bulk [32], where the shear strain was extremely localized in the middle thickness. This localization of the shear strain induced a dynamic recrystallization process and resulted in an average grain size of 0.5 to 1 μm. In contrast, the top and bottom parts of the micro-HPT disk exhibited fairly heterogeneous microstructures with some deformation-unaffected local domains (Figure 3a,c). This is clearly visible, for example, by the presence of inter-particle oxide layers (as shown by the arrows) that witness the presence of poorly deformed atomized powder particles. Since the presence of the MgO phase was not detected by the XRD measurements for the atomized power (see Figure 2a), it is likely that the oxide content was so small that it did not give rise to any XRD signal under the applied experimental conditions.

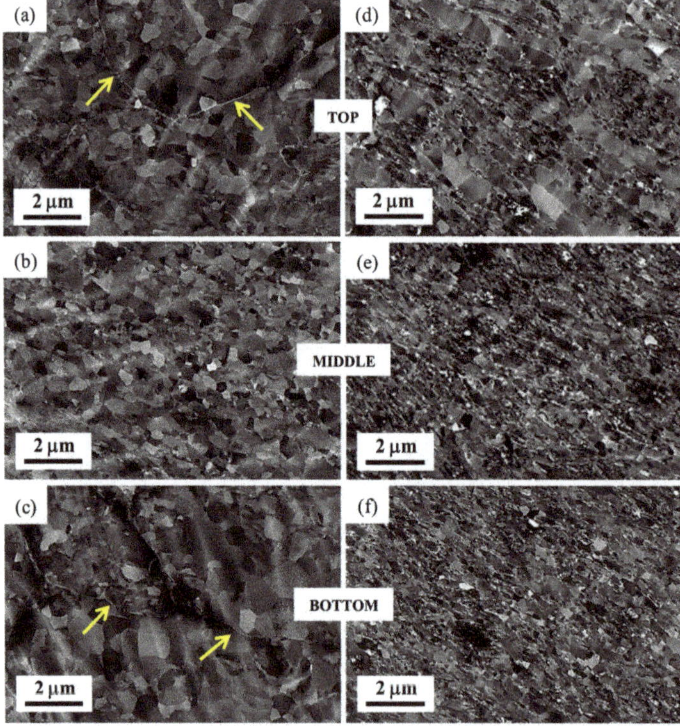

Figure 3. SEM backscattered electron micrographs acquired at the outer-edge and across the through-thickness of the 2 turns HPT disks for (**a**–**c**) the micro-HPT product and (**d**–**f**) the nano-HPT product.

In the case of the nano-HPT product, it is readily apparent from the right-side images in Figure 3 that the whole sample was rather more homogeneously deformed, leading to the formation of slightly elongated domains across the disk-thickness (Figure 3d–f). However, the microstructure at the bottom part seems to be less elongated while the width of the Mg domains is thicker at the top; so, a gradient-microstructure developed across the thickness-height. Since the ultrafine powder particles were coated with a thin layer of Mg oxides, it is thereby reasonable to anticipate that these oxide layers were sheared and fragmented into fine particles (appeared as white spots under BSE imaging conditions). Also, it is interesting to notice that most of these oxides were aligned along the grain boundaries that were essentially inclined at about 45°. Figure 4 displays a FIB micrograph obtained from the middle-section of the nano-HPT disks sample prepared for TKD measurements. The microstructural features in the FIB image were consistent with the SEM observations. It is interesting to note that this micrograph clearly shows some residual porosity (as shown by the arrows) even after 2 HPT turns of the nanopowder precursor. It may imply that the densification of the nano-HPT product was not completely achieved.

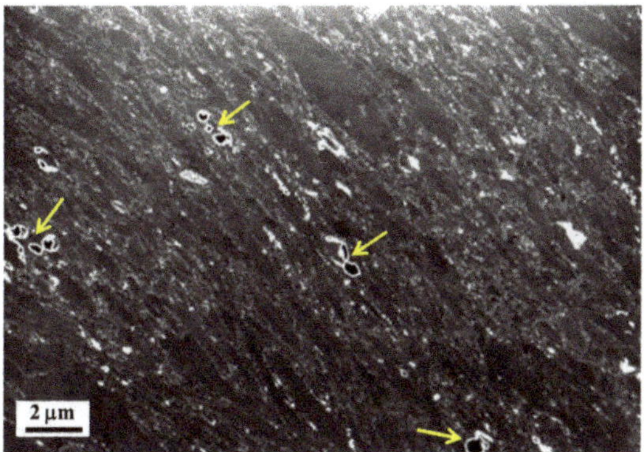

Figure 4. In-situ focused ion beam (FIB) micrograph of the nano-HPT product recorded during the foil preparation for TKD measurements. Arrows indicate the vacancies left after 2 turns HPT consolidation.

3.2. EBSD and TKD Characterizations

In order to gain some insights into the HPT consolidation deformation behavior of the Mg powder particles, some additional samples were processed at lower level of shear strains (i.e., $N = 1/2$ and 3/4) for both powder systems under identical experimental conditions. Grain orientation mapping for the deformed samples was carefully acquired by using a normal EBSD technique for the micro-HPT product, while a SEM-based TKD technique was used for the nano-HPT products due to their ultra-fine structures. The local EBSD and TKD measurements were carried out at the periphery and on the middle-thickness of the HPT-disk, and the corresponding results are described in the following subsections for the two types of consolidated products.

3.2.1. Micro-HPT Product Consolidated from Atomized Powder

The EBSD orientation maps accompanied with grain misorientation distribution and microtexture for the micro-HPT product deformed up to 1/2, 3/4 and 2 turns are plotted in Figure 5. A schematic diagram of an HPT-disk showing the location and the acquisition direction of the EBSD measurements for all samples is provided in the inset of Figure 5b. The color code in the inset of Figure 5e refers to the crystallographic directions parallel to the normal direction of the sample surface. Furthermore, it should be noted that low-angle grain boundaries (LAGBs, 3–15°) and high-angle grain boundaries (HAGBs, >15°) are denoted by white and black lines, respectively, in all EBSD maps.

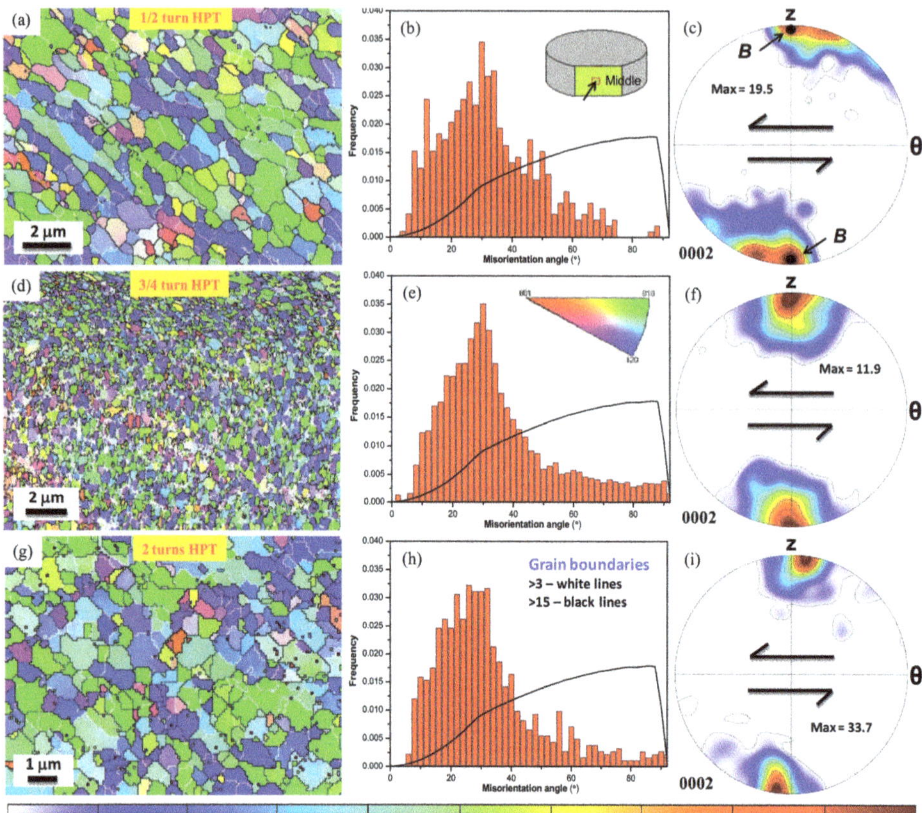

Figure 5. Electron backscatter diffraction (EBSD) orientation images accompanied with grain boundary misorientations and associated microtexture in (0002) pole figure for the 1/2 turn (**a–c**), 3/4 turn (**d–f**), and 2 turns (**g–i**) micro-HPT products acquired at the outer edge and middle-section of the HPT-disk (see the inset of Figure 5b). The Mackenzie distribution of misorientation angles (uncorrelated) is plotted by solid black line. Note the differences in scale bar.

The HPT consolidation of atomized powder only up to 1/2 turn has led to the formation of recovered/recrystallized microstructures (see Figure 5a). The grains were elongated at an inclination angle of about 45° from the direction of the imposed shear. Figure 5b displays the grain misorientation angle distribution. A high frequency of medium-angle boundaries can be noticed in the early stages of HPT processing, which is pronounced at around 30°, and can be attributed to the development of a basal fibre texture in Mg materials [44]. Moreover, it appears that the LAGBs are basically limited within the elongated grain domains indicating a close orientation relationship between the neighboring grains, which can be further appreciated from the color gradients within the original grain boundaries. This clearly suggests that, at this stage of deformation, the grain-structure was only partially recovered from its highly deformed state, a phenomenon occurring in materials such as pure Zn, Al, and Mg metal systems deformed at low homologous temperatures [32,36,45]. Figure 5c presents the evolution of the microtexture in (0002) pole figure after 1/2 turn. It is readily apparent that there is a strong *B* fibre, which is the main ideal fibre of hexagonal simple shear textures [46]. It has been observed previously that upon free-end torsion [46], or in HPT processing [47] of pure Mg, the *c*-axis—initially aligned with the shear plane—basically rotates by 90° to become parallel with the torsion axis. In the present study, the *c*-axes of the grains in the powder sample are initially randomly oriented and similarly become significantly aligned with the shear plane normal due to the large shear strain (of about 10 for 1/2 revolution).

Further torsional straining by 3/4 and 2 turns led to recrystallized microstructures accompanied with equiaxed grain-structures (see Figure 5d,g). The elongated shape of the grain-domains after 1/2 HPT turn was transformed into equiaxed grain-structures in the 3/4 turn HPT-product, while the well-defined equiaxed grain-structures in the 2 HPT turns product were inclined at 45° from the shear plane, Figure 5g. These microstructure features are suggesting that the generation of equiaxed microstructures—particularly at the middle-portion—can be associated with the occurrence of complete dynamic recrystallization [32]. The average grain size for these microstructures was in the range of 1 to 1.5 μm for the 3/4 and 0.5 to 1 μm for 2 HPT turns product, respectively. The frequency distribution of grain misorientation shown in Figure 5e and h for the 3/4 and 2 turns infers that most of the grain boundaries were concentrated at around 30°, which also supports the occurrence of a dynamic recovery/recrystallization process. In the case of the 2 HPT turns product, the shear texture component appears tilted from its ideal position. The sense of this tilt is opposite to the direction of the applied shear direction by about 5° (see Figure 5i).

3.2.2. Nano-HPT Product Consolidated from Condensed Powder

The microstructure of the HPT-compacted nanopowder was examined by transmission Kikuchi diffraction (TKD) because of the much smaller grain size. The TKD technique used in our laboratory is unique in the sense that it is applied in a direct on-axis configuration permitting rapid and accurate measurements [41]. At the same time, the maximum size of the examined surface is limited; much smaller than in EBSD. For this reason, the statistical data (disorientation distribution, crystallographic texture) are less representative compared to the EBSD results shown in Figure 5 for the micro-product. Nevertheless, important morphological features and in-grain disorientations can be characterized by these TKD measurements. The evolution of grain orientations and the corresponding microtexture for the nano-HPT product deformed up to 1/2, 3/4, and 2 turns are given in Figure 6.

At the early stages of deformation (i.e., after 1/2 turn), the TKD orientation map in Figure 6a reveals a rather heterogeneous grain-structure consisting of a wide range of grain/particle size distribution. The larger particles likely to be inclined at around 45° to the shear direction while the smaller ones did not respond much to the applied shear strain (i.e., $\gamma = 10$ for 1/2 revolution). As mentioned earlier, since the surfaces of ultrafine Mg particles have been originally coated with a thin layer of Mg oxides, these oxide layers were not indexed in the TKD measurements and appeared as non-indexed zones along the particle boundaries. The frequency distribution of grain misorientation shown in Figure 6b clearly indicates mixed type of grain boundaries for this microstructure. As expected from the microstructural heterogeneities, the texture in this sample was not developed as expected from the applied strain (see Figure 6c). The *c*-axes should be positioned at the top and bottom parts of the pole figure in Figure 6c, which is visible. However, large population of grains were very far from these positions.

For the subsequent deformation through 3/4 and 2 turns, a rather homogeneously deformed microstructure was developed in the HPT-products (see Figure 6d,g). In both cases, the grains/particles were significantly elongated with an inclination angle of about 45° from the applied shear. It is readily apparent that some parts of the microstructure seem extremely fragmented to grain sizes less than 100 nm, while the coarser grain-structures were probably generated from initially larger powder particles. Also, the grain/powder particle boundaries, which essentially consisted of oxide layers, were not indexed during the TKD measurements. Since the given TKD map for the 2 HPT turns sample contained a limited number of grains, Figure 6i was constructed from several small TKD maps in order to obtain statistically representative texture using the Jtex software [43].

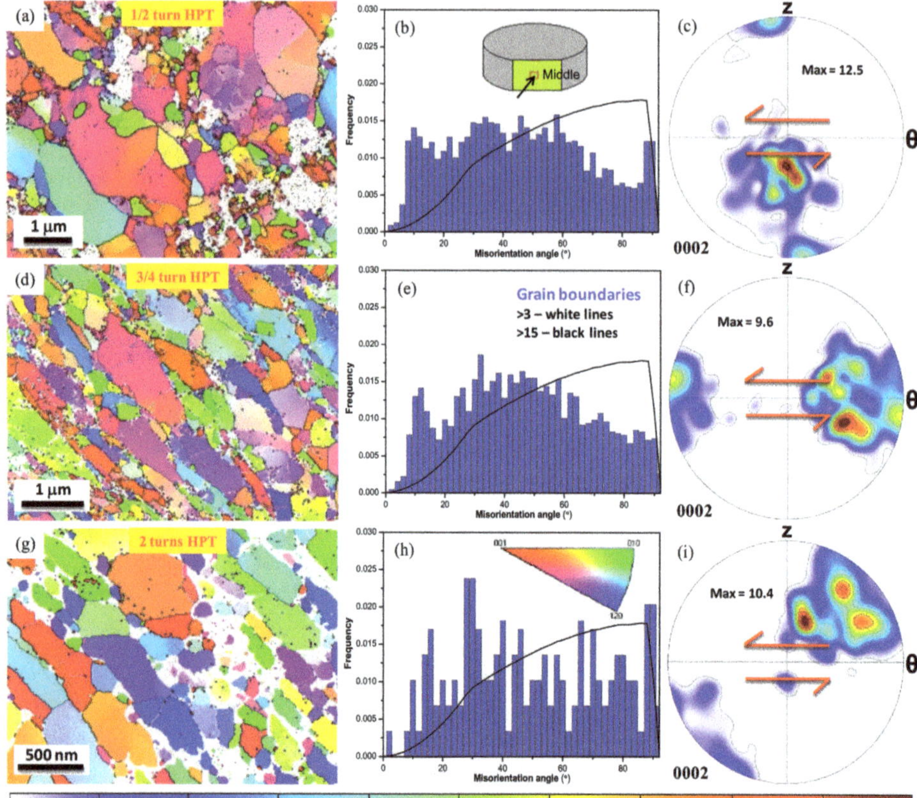

Figure 6. Transmission Kikuchi diffraction (TKD)-based EBSD orientation maps accompanied with grain boundary misorientations and associated microtexture in (0002) pole figure for the 1/2 turn (**a**–**c**), 3/4 turn (**d**–**f**), and 2 turns (**g**–**i**) nano-HPT products, obtained at the outer edge and middle-section of the HPT-disk (see the inset of Figure 6b). Note the differences in scale bar.

3.3. Microhardness Evolution

The mechanical strengths of the 2 turns HPT products were characterized by Vickers microhardness across the radial as well as the axial directions; the results are shown in Figure 7. Since it was demonstrated in our recent work [32] that for relatively thick HPT specimens the evolution of microhardness can vary significantly across the through thickness, it was carefully examined on several cross-sectional planes across the thickness height. For comparison, the hardness values obtained for HPT bulk Mg [32,36] are also given in Figure 7.

Because of the presence of finer grains and oxide particles, the hardness of the nano-HPT product is always twice higher than that of the micro-HPT product. It is apparent from Figure 7 for the micro-HPT product that there was a radical drop in the microhardness in the middle section displaying the lowest value at the outer edge of the HPT disk. This decrease in hardness in the mid-thickness is consistent with the observation of microstructural evolutions in the middle-section of the micro-HPT disk (see Figure 3b) in which a recrystallized microstructure developed during the intense shear straining by HPT. Comparatively, there is an increase in hardness values towards the outer-edge for the bottom and top locations. This is due to the fact that while the strain increases towards the periphery of the disk, it does not reach sufficient amount (contrary to what happens at the middle section) to trigger dynamic recrystallization. Such increase in hardness for the HPT-processed sample with increasing strain followed by a hardness drop due to recrystallization was also depicted for pure bulk Mg in Refs. [32,36]. Finally, having an intermediate behavior, the hardness values at the disk center that are associated with comparatively lower strains were almost the same across the disk-thickness.

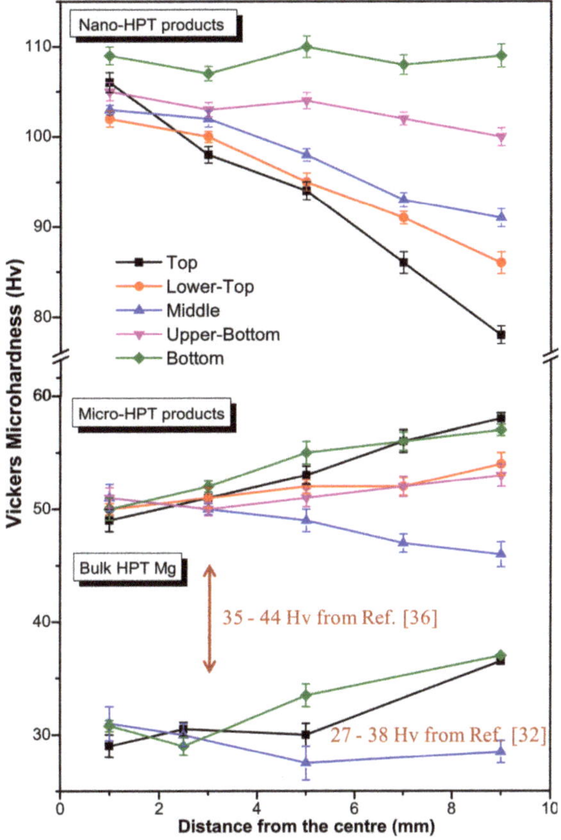

Figure 7. Variations of Vickers microhardness (Hv) along the radial direction as well as across the disk-thickness for the 2 turns nano-HPT and micro-HPT products. The reference hardness values for the HPT deformed bulk Mg samples are also included from Refs. [32,36].

As recalled by the data given in Figure 7, it is also interesting to notice that the hardness values in our micro-HPT product obtained from atomized powder precursor are always higher than the ones reported on HPT pure Mg for which the precursor was a bulk sample [32,36]. This must be due to the presence of a small fraction of oxide always present at the periphery of the Mg powder particles that provided with some hardening and pinned the grain boundaries to reach finer grain size after deformation. Recently, evaluating the different contributions on strengthening of spark plasma sintered AZ91 Mg alloy, Mondet et al. [20] have reported that the Hall-Petch effect was the major contributor to reach significant improvements in hardness and yield strength.

In the case of the nano-HPT product (see Figure 7), the hardness evolution demonstrates a completely different hardening response than that of its micro-counterpart. Because of the small powder particle size and the presence of oxides, the hardness values were always higher at the disk-center, with an average value of more than 105 Hv for all cross-sectional planes. While the bottom surface exhibited consistently higher hardness values across the radial direction, the other cross-sectional planes showed a monotonic drop in the microhardness values from the disk-centre to the periphery. It is also clearly evident that at the outer-edge of the HPT-disk, the hardness values continuously decreased from the bottom to the top surface. These features are in fact consistent with the microstructural development examined by SEM, where the bottom part revealed a more significant grain refinement than the other parts (see Figure 3d–f).

3.4. Hydrogen Activation

The 2 turns HPT products were analyzed for H-sorption properties during the first hydrogenation. Figure 8 compares the activation characteristics of the two HPT products with that of their initial powder precursors. The hydrogen activation experiments were carried out at 400 °C for 8 h under a hydrogen pressure of 3.5 MPa. Since the micro-HPT disk were characterized by a heterogeneous microstructure across the disk-thickness (see Figure 3a–c), a heterogeneity also materialized by different hardness modifications (see Figure 7), the activation characteristics were tested for the overall sample as well as for the middle section that was recrystallized (Figure 8a). For the nano-HPT product that was more uniformly deformed across the thickness height, the overall sample was tested as a whole (Figure 8b).

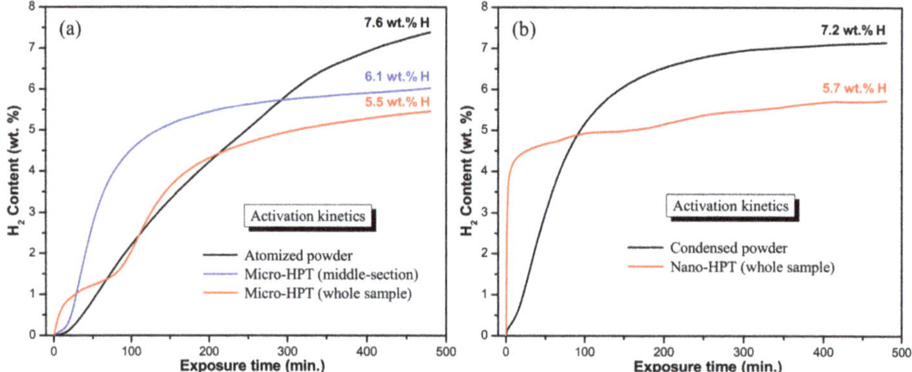

Figure 8. First hydrogenation profiles obtained by dynamic hydrogen absorption of the as-prepared samples at 400 °C for 8 h under a hydrogen pressure of 3.5 MPa: (**a**) atomized powder and its 2 turns micro-HPT product; (**b**) condensed powder and its 2 turns nano-HPT product.

Figure 8a shows that the atomized powder particles could store up to the theoretical H-storage capacity of Mg (i.e., 7.6 wt.%) within 8 h (480 min). This is substantially higher than the amount of H stored for the different zones of the micro-HPT product within the same amount of time. While the middle-part of the sample could uptake 6.1 wt.% H, under the identical experimental conditions, the whole sample could reach only 5.5 wt.%. However, the interest of the HPT deformation is that the kinetics of H pick up was improved compared to the loose powder. Within the different bulk micro-HPT parts, the sample obtained from the middle section that was heavily deformed and recrystallized to a fine grain size showed the fastest overall absorption kinetics. Interestingly, the whole sample exhibited a kind of transient behavior with the kinetics slowing down before reaching again a more enhanced absorption rate. Hence, the fine recrystallized microstructures developed in the middle-section enhanced the hydrogenation kinetics compared to the overall performances of the whole sample that consisted of heterogeneous microstructure with a combination of finer recrystallized and coarser deformed grain structures.

For the ultrafine condensed powder and its nano-HPT product (see Figure 8b), the effect of the severe plastic deformation imparted by the HPT processing gave the same trends: an acceleration of the H-storage kinetics but, unfortunately, accompanied with a lower H-storage capacity. While the condensed powder particles took 68 min to reach 4 wt.% of H-uptake, the effect of HPT was that the nano-HPT product required only 5 min for the same 4 wt.% uptake. Concomitantly, the HPT processing has impaired the H-storage capacity of the condensed powder precursor (5.7 wt.% vs. 7.2 wt.% H).

4. Discussion

It was revealed from the present study that the nature of the initial powder precursors had significant impacts on the powder consolidation behavior. The evolutions of the microstructure, the texture and the microhardness for the two HPT-consolidated products were significantly dissimilar in nature. They also showed distinct hydrogen sorption characteristics for the first hydrogenation cycle. These aspects will be discussed hereafter.

4.1. Differences in Consolidation Mechanisms, Grain Sizes, and Texture Developments

It has been a great concern for long time that processing of Mg based materials through powder metallurgy routes inherently induces significant difficulties due to the high reactivity of the powder particles towards atmosphere that rapidly generates impervious oxide layer around the powder particles that prevent subsequent sintering [20,22,23]. During HPT processing, huge amount of strain can be introduced into the material assisted by the large applied hydrostatic pressures which can prevent crack formation. As a consequence, the HPT consolidation of powders is mainly governed by particle-particle shearing (rather than diffusion when other thermally assisted sintering processes are used) under which the massive shear straining can fragment the oxide layers and lead to the required bonding, even at low processing temperatures. Concomitantly, while the oxide layer is fractured into nano-sized particles, these oxide particles contributed to the grain size refinement that is known to be one of the major contributors to hardening in Mg alloy [20]. Thus, as illustrated in Figures 3d–f and 4, our nano-HPT composite consolidated from ultrafine condensed and passivated powder particles that contained about 5 wt.% of MgO particles which prevented dynamic recrystallization developed an elongated recovered structure characterized by thin elongated Mg domains pinned by oxide particles. When the HPT process was carried out on the atomized powder that contained a much lower amount of oxide, the material was free to recrystallize if a sufficient amount of plastic deformation was reached. This was the case in the middle portion of the disk, as illustrated in Figure 3b, where a rather refined equiaxed microstructure has formed.

The nature of the powder precursors had significant influences on the development of the microstructure. This is particularly true for the final attainable grain size which depended essentially on the amount of oxide particles present in the deformed products. While a saturation grain size of about 2 to 3 μm is generally obtained for dynamically recrystallized HPT Mg disk deformed from bulk precursors that did not contain oxides [32,36], the HPT consolidation of the atomized Mg powder particles that contained some native oxide at their periphery led to a further refine recrystallized structure with a grain size of 1 μm (see Figure 3b). These differences in grain sizes between the recrystallized products obtained from bulk [32,36] and atomized precursors explain the differences in hardness between these two classes of products, as recalled in Figure 7. As the refinement in grain size is associated with the pinning effect of the oxide particles on the grain boundaries, a finer microstructure developed for the Mg/MgO based nano-HPT product where the grain growth was restricted to elongated Mg domains having an average thickness in the range 0.1 μm (Figure 3f) to 0.5 μm (Figure 3d) depending on the local imparted amount of strain.

As the presence of Mg oxides has affected the mechanisms of deformation and microstructure evolution, it has also affected the texture evolution in the two types of HPT products. A strong shear texture was developed for the micro-HPT product at large shear strains. It clearly implies that the HPT consolidation of the atomized powder particles was mainly governed by simple shear plastic deformation of the powder particles under very high hydrostatic pressure [3–5,10]. However, it was noticed, that the location of the *B* fibre was somewhat deviated from its ideal position with increasing shear strain; it was tilted against the shear sense (see Figure 5i). This is consistent with the findings of Bonarski et al. [47] who reported that upon HPT deformation of pure bulk Mg, the *c*-fibre axis deviates from its ideal position and these angular deviation increases while the texture strength decreases by increasing the hydrostatic pressure. According to Skrotzki et al. [48], the occurrence of dynamic recrystallization processes at relatively large shear-strains, as the one taking place in our micro-HPT product, can prevent the rotation of the *c*-axes from reaching the ideal position. Furthermore, the appearance of a large fraction of LAGBs along with the deformed microstructural features after 2 turns suggests that, due to the continuous shear deformation in HPT processing, the grain fragmentation and the associated texture evolution is the consequence of a steady state regime of repeated deformation/recrystallization phenomenon [49–51]. In contrast to the micro-HPT product, the texture evolution in the nano-HPT Mg/MgO composite obtained from the condensed and passivated powder particles was more complicated probably due to the influence of the MgO layer that surrounded the Mg core. It was observed from the sequential torsional straining of the condensed powder precursor that the presence of Mg oxide particles somewhat hindered the evolution toward the ideal shear texture in the nano-HPT composite, although a significant densification was achieved through the particle fragmentation followed by adequate bonding between the metallic phases after 2 turns HPT deformation. In spite of the high HPT rotation, the texture depicted in Figure 6 showed many orientations out of the ideal positions. This must imply that many nano-particles did not deform plastically in a proper manner but were probably only rotating against each other under the applied shear.

4.2. Effectiveness of Powder Consolidation: Strain Heterogeneities, Associated Microstructure, and Hardness Evolutions

The effectiveness of the powder consolidation was assessed by measuring the relative density of the bulk HPT products using the Archimedes principle; the results are given in Table 2. A relative density of about 92% and 96% was achieved for the micro-HPT and nano-HPT product, respectively, after 2 turns cold consolidation. The lower bulk density for the micro-HPT product can be associated with the relatively poor consolidation at the top and bottom parts of the HPT disk as witnessed from the SEM micrographs (see Figure 3a–c). Indeed, it is well established that products of relatively large thickness (say 3 mm) and deformed by HPT can be characterized by significant microstructure variation due to the development of strain heterogeneities [10,32]. This is reflected here by variations in microstructure and hardness evolutions at different locations of the consolidated product; the hardness being essentially affected by the local grain size and local amount of oxide particles. Thus, the hardness of the nano-HPT product obtained from the condensed and passivated powder exhibited significantly higher microhardness; almost twice higher than the micro-HPT product (nano-HPT: 108 Hv and micro-HPT: 55 Hv). The evolution of the microhardness across the sample volume however demonstrates distinct hardening mechanisms for the two types of HPT products.

Table 2. Relative density of the HPT products consolidated from the atomized Mg and condensed Mg powder particles.

HPT products	Calculated Density (d_c, g cm^{-3})	Theoretical Density (d_t, g cm^{-3})	Relative Density (d_c/d_t, %)
2 turns micro-HPT	1.596	1.738	91.8 ± 0.5
2 turns nano-HPT	1.715	1.784 (5 wt.% MgO)	96.1 ± 0.8

The hardening in the micro-HPT product, that contained limited amount of oxide particles, was affected by the occurrence of dynamic recrystallization. Consistently with the results of HPT bulk deformed Mg [32], the maximum amount of plastic strain was obtained in the middle section of the HPT disk where recrystallization (Figure 3b) and an associated softening took place (Figure 7). As the amount of strain increases towards the periphery of the disk, the hardness logically decreases slightly in Figure 7. Comparatively, as confirmed by the presence of some poorly sheared powder particles at the top (Figure 3a) and bottom (Figure 3c) parts of the disk, the amount of strain imparted locally did allow to reach a sufficient amount of plastic deformation to trigger recrystallization. Consequently, the amount of structural defects present in the disk at the bottom and top sections must increase toward the periphery of the disk, and, concomitantly, the hardness increases as well (Figure 7). This hardening followed by a softening behavior at large shear strains is a common feature for pure magnesium [36] as well as for other metal systems having low homologous temperatures that facilitate grain boundary migrations through diffusion processes [45].

The strengthening for the nano-HPT composite is correlated to the presence of the extremely small oxide particles that also pin the grain boundary to prevent recrystallization. Thus, because of the ultra fine size of the condensed powder particles and their core-shell structure coated by a fine layer of oxide, the maximum hardness is always recorded in the center of the disk that was poorly deformed, whatever the height was. In this type of HPT product, the microstructure was less heterogeneous, and the overall disk sustained a fair amount of deformation leading to elongated recovered grains with oxides aligned in the flow direction and pinned their boundaries. However, as for the case depicted in Ref. [10], a strain gradient has developed along the height of the disk from top to bottom. At the bottom of the disk that sustained less strain, the size of the elongated Mg domains remained rather fine, at about 0.1 mm (Figure 3f), and the hardness was rather constant from the center to the periphery of the disk. Comparatively, when going toward the top of the disk, the increasing amount of strain has generated some grain boundary migration that generated recovered elongated Mg domains having larger thicknesses, even up to 0.5 μm (Figure 3d). Consequently, the local average thickness of the elongated Mg domains (Figure 3d–f) and the local hardness developed in the nano-HPT product are directly related to the local amount of strain that was sustained. This is nicely depicted in Figure 7 by the continuous and monotonous decreases in hardness that are visible from the center toward the periphery of the disk, the slope of which increases from the bottom to the top of the disk.

4.3. H-Sorption Properties

The first hydrogenation of Mg based material is generally considered to be the most difficult step due to the formation of an impervious oxide layer around the surface of Mg material [37,38]. The activation of the type of powders used in this investigation has already been the subject of previous publications. It was claimed that because of their much finer size and their core-shell structure with the presence of MgO catalysts, the condensed powder particles react much faster than the atomized ones but can store a lower amount of hydrogen. This is confirmed here by comparing the curves of the non-consolidated precursors given in Figure 8a and b for the atomized and condensed powder, respectively.

When properly deformed (i.e., the middle section of the micro-HPT product and the whole nano-HPT one) the HPT-consolidated products always showed faster absorption kinetics in comparison with their original powder precursors. Indeed, the torsional straining of the powder particles introduced numerous structural defects in the Mg matrix (nanostructuring, dislocations, vacancies, etc.) that are known to promote the hydrogen activation [36,37,52,53]. However, this improvement in kinetics is always accompanied by a reduction in the storage capacity. A possible explanation for such behavior lies in the fact that the severe deformation has generated a much higher density of MgH_2 nucleus at the surface of the components that subsequently led to the formation of a rather continuous and compact layer of hydride at their periphery. The high density of nucleus has improved the initial kinetics but, as was previously demonstrated that hydrogen capacity is drastically influenced by the nucleation rate of the hydride [54,55], must have reduced the capacity because of the low diffusivity of hydrogen through the dense and continuous MgH_2 layer [56]. When the overall micro-HPT sample that contained also poorly deformed domains was tested, a change in the kinetics was observed during storage so that neither the kinetics nor the amount of H-uptake was improved (red curve in Figure 8a). Concerning the positive effect of HPT, as revealed in a previous study [33], an interesting feature is that it always substantially reduced the hysteresis between the absorption and desorption plateau pressures when these HPT products were subjected to hydrogen uptake/release thermodynamic testing [33–35].

5. Conclusions

The aim of the present study was to compare the plastic deformation characteristics of 3 mm thick disc obtained by high-pressure torsion (HPT) of two distinct powder precursors: (i) atomized micro-sized Mg (10–70 μm) powder particles containing 2–5 μm grains to consolidate into the so-called micro-HPT disk and (ii) condensed nano-sized Mg (50–800 nm) single grain powder particles that were subsequently passivated to consolidate into the so-called nano-HPT Mg/MgO disk.

The effect of the powder type was examined in terms of (i) microstructure and texture evolution mechanisms, (ii) development of strain heterogeneities and related local mechanical strength (by microhardness), and (iii) H-activation ability by hydrogenation measurements. The following main conclusions are listed hereafter.

1. The nature of the initial powder precursors had significant influences on the microstructure and texture evolutions in the two types of HPT products as well as on the development of local strain heterogeneities.
2. The HPT consolidation of the atomized powder particles led to heterogeneous microstructures across the through thickness: recrystallized equiaxed 1 μm grains at the middle section that sustained the highest strain and rather heterogeneously deformed microstructures at the top and bottom regions of the disk.
3. The HPT consolidation of the condensed powder particles led to a Mg + 5%MgO composite having a rather more homogeneously deformed microstructure consisting of elongated recovered nanometric Mg domains pinned by the nano-sized Mg oxide particles. Some strain heterogeneities have, however, developed from bottom to top of the disk where the average thickness of the elongated recovered grains varied from about 100 to 500 nm.
4. Due to the plastic shear deformation and subsequent dynamic recrystallization, a strong basal texture has developed in the micro-HPT product. Comparatively, the presence of oxides that promoted dynamic recovery and possible powder particle sliding, have hindered the evolution towards the ideal shear texture in the nano-HPT product.
5. The hardness being essentially affected by the local grain size and local amount of oxide particles, the hardness of the nano-HPT product obtained from the fine condensed and passivated powder exhibited twice higher hardness than the micro-HPT product: 108 Hv and 55 Hv, respectively. Also, the hardness values in our micro-HPT product obtained from the atomized powder precursor are always higher than the ones reported on HPT deformed disk of pure Mg for which the precursor was a bulk sample [32,36].
6. The local hardness varied in the consolidated products with the local amount of strain and depended on the mechanism of deformation accommodation. For the micro-HPT product obtained from the atomized powder, the increasing amount of local strain hardened the materials until dynamic recrystallization occurred. Comparatively, for the nano-HPT product obtained from the nano-sized Mg/MgO core-shell powder, the dynamic recovery process led a continuous decrease in local hardness with the amount of strain.
7. After HPT deformation, likely due to an increase in nucleation sites for the hydride formation, the kinetics of the H pick up was significantly improved. For example, the nano-HPT product took only 5 min to reach 4 wt.% H-uptake compared to 68 min for its initial condensed powder precursor. For the heterogeneous micro-HPT product, the fastest kinetics was reported for the central part of the product that was properly deformed and developed a finer recrystallized microstructure. Unfortunately, this faster kinetics was always accompanied with a lower H-storage capacity. For example, in the case of the condensed nano-sized powder, the capacity was reduced from 7.2 wt.% to 5.7 wt.% after HPT.

Author Contributions: The research work presented in this manuscript was designed and conducted as a collaboration between all affiliated authors listed above. Conceptualization, S.P., J.Z., and T.G.; Methodology, S.P. and L.S.T.; Formal Analysis, S.P., L.S.T., and T.G.; Investigation, S.P. and J.Z.; Resources, J.Z.; Writing-Original Draft Preparation, S.P. and T.G.; Writing-Review & Editing, all authors contributed to the scientific design of the study, the thorough discussion of the results and have checked and approved the final manuscript.

Funding: This research received no external funding.

Acknowledgments: This work was supported by the French State through the program "Investment in the Future" operated by the National Research Agency (ANR) and referenced by ANR-11-LABX-0008-01 (Labex DAMAS). This work was also partly supported by the National Key Research and Development Program of China (2016YBF0701203) and the "111" project from China's Ministry of Education (B16032).

Conflicts of Interest: The authors declare no conflict of interest.

References

1. Valiev, R.Z.; Estrin, Y.; Horita, Z.; Langdon, T.G.; Zehetbauer, M.J.; Zhu, Y.T. Producing bulk ultrafine-grained materials by severe plastic deformation. *JOM* **2006**, *58*, 33–39. [CrossRef]
2. Zehetbauer, M.; Grössinger, R.; Krenn, H.; Krystian, M.; Pippan, R.; Rogl, P.; Waitz, T.; Würschum, R. Bulk nanostructured functional materials by severe plastic deformation. *Adv. Eng. Mater.* **2010**, *12*, 692–700. [CrossRef]
3. Alexandrov, I.V.; Zhu, Y.T.; Lowe, T.C.; Valiev, R.Z. Severe plastic deformation: New technique for powder consolidation and grain size refinement. *Powder Metall.* **1998**, *41*, 11–13. [CrossRef]
4. Xia, K. Consolidation of particles by severe plastic deformation: Mechanism and applications in processing bulk ultrafine and nanostructured alloys and composites. *Adv. Eng. Mater.* **2010**, *12*, 724–729. [CrossRef]
5. Bachmaier, A.; Pippan, R. Generation of metallic nanocomposites by severe plastic deformation. *Int. Mater. Rev.* **2013**, *58*, 41–62. [CrossRef]
6. Zhilyaev, A.P.; Langdon, T.G. Using high-pressure torsion for metal processing: Fundamentals and applications. *Prog. Mater. Sci.* **2008**, *53*, 893–979. [CrossRef]
7. Valiev, R.Z.; Mishral, R.S.; Grozal, J.; Mukherjee, A.K. Processing of nanostructured nickel by severe plastic deformation consolidation of ball-milled powder. *Scr. Mater.* **1996**, *34*, 1443–1448. [CrossRef]
8. Sort, J.; Zhilyaev, A.; Zielinska, M.; Nogués, J.; Surinach, S.; Thibault, J.; Baró, M.D. Microstructural effects and large microhardness in cobalt processed by high pressure torsion consolidation of ball milled powders. *Acta Mater.* **2003**, *51*, 6385–6393. [CrossRef]
9. Edalati, K.; Horita, Z.; Fujiwara, H.; Ameyama, K. Cold consolidation of ball-milled titanium powders using high-pressure torsion. *Metall. Mater. Trans. A* **2010**, *41*, 3308–3317. [CrossRef]
10. Zhao, Y.; Massion, R.; Grosdidier, T.; Toth, L.S. Gradient structure in high pressure torsion compacted iron powder. *Adv. Eng. Mater.* **2015**, *17*, 1748–1753. [CrossRef]
11. Stolyarov, V.V.; Zhu, Y.T.; Lowe, T.C.; Islamgaliev, R.K.; Valiev, R.Z. Processing nanocrystalline Ti and its nanocomposites from micrometer-sized Ti powder using high pressure torsion. *Mater. Sci. Eng. A* **2000**, *282*, 78–85. [CrossRef]
12. Bachmaier, A.; Hohenwarter, A.; Pippan, R. New procedure to generate stable nanocrystallites by severe plastic deformation. *Scr. Mater.* **2009**, *61*, 1016–1019. [CrossRef]
13. Bachmaier, A.; Pippan, R. Effect of oxide particles on the stabilization and final microstructure in aluminium. *Mater. Sci. Eng. A* **2011**, *528*, 7589–7595. [CrossRef] [PubMed]
14. Zou, J.X.; Perez-Brokate, C.F.; Arruffat, R.; Bolle, B.; Fundenberger, J.J.; Zeng, X.Q.; Grosdidier, T.; Ding, W.J. Nanostructured bulk Mg+MgO composite synthesized through arc plasma evaporation and high pressure torsion for H-storage application. *Mater. Sci. Eng. B* **2014**, *183*, 1–5. [CrossRef]
15. Grosdidier, T.; Fundenberger, J.J.; Zou, J.X.; Pan, Y.C.; Zeng, X.Q. Nanostructured Mg based hydrogen storage bulk materials prepared by high pressure torsion consolidation of arc plasma evaporated ultrafine powders. *Int. J. Hydrogen Energy* **2015**, *40*, 16985–16991. [CrossRef]
16. Kulekci, M.K. Magnesium and its alloys applications in automotive industry. *Int. J. Adv. Manuf. Tech.* **2008**, *39*, 851–865. [CrossRef]
17. Avedesian, M.M.; Baker, H. *ASM Speciality Handbook: Magnesium and Magnesium Alloys*; ASM International: Ohio, OH, USA, 1999; pp. 78–84.

18. Dobrzański, L.A.; Tański, T.; Čížek, L.; Brytan, Z. Structure and properties of magnesium cast alloys. *J. Mater. Process. Technol.* **2007**, *192*, 567–574. [CrossRef]
19. Tun, K.S.; Gupta, M. Improving mechanical properties of magnesium using nano-yttria reinforcement and microwave assisted powder metallurgy method. *Compos. Sci.* **2007**, *67*, 2657–2664. [CrossRef]
20. Mondet, M.; Barraud, E.; Lemonnier, S.; Guyon, J.; Allain, N.; Grosdidier, T. Microstructure and mechanical properties of AZ91 magnesium alloy developed by Spark Plasma Sintering. *Acta Mater.* **2016**, *119*, 55–67. [CrossRef]
21. Shen, J.; Imai, H.; Chen, B.; Ye, X.; Umeda, J.; Kondoh, K. Deformation mechanisms of pure Mg materials fabricated by using pre-rolled powders. *Mater. Sci. Eng. A* **2016**, *658*, 309–320. [CrossRef]
22. Bettles, C.J. Magnesium powder metallurgy: Process and materials opportunities. *J. Mater. Eng. Perform.* **2008**, *17*, 297–301. [CrossRef]
23. Umeda, J.; Kawakami, M.; Kondoh, K.; Ayman, E.S.; Imai, H. Microstructural and mechanical properties of titanium particulate reinforced magnesium composite materials. *Mater. Chem. Phys.* **2010**, *123*, 649–657. [CrossRef]
24. Straffelini, G.; Dione Da Costa, L.; Menapace, C.; Zanella, C.; Torralba, J.M. Properties of AZ91 alloy produced by spark plasma sintering and extrusion. *Powder Metall.* **2013**, *56*, 405–410. [CrossRef]
25. Burke, P. Investigation of the Sintering Fundamentals of Magnesium Powders. Ph.D. Thesis, Dalhousie University, Nova Scotia, NS, Canada, 2011.
26. Moss, M.; Lapovok, R.; Bettles, C.J. The equal channel angular pressing of magnesium and magnesium alloy powders. *JOM* **2007**, *59*, 54–57. [CrossRef]
27. Wu, X.; Xu, W.; Kubota, M.; Xia, K. Bulk Mg produced by back pressure equal channel angular consolidation (BP-ECAC). *Mater. Sci. Forum* **2008**, *584*, 114–118. [CrossRef]
28. Cakmak, G.; Ozturk, T. ECAP processing and mechanical milling of Mg and Mg-Ti powders: A comparative study. *J. Mater. Sci.* **2011**, *46*, 5559–5567. [CrossRef]
29. Yoon, E.Y.; Lee, D.J.; Kim, T.S.; Chae, H.J.; Jenei, P.; Gubicza, J.; Ungár, T.; Janecek, M.; Vratna, J.; Lee, S.; Kim, H.S. Microstructures and mechanical properties of Mg-Zn-Y alloy consolidated from gas-atomized powders using high-pressure torsion. *J. Mater. Sci.* **2012**, *47*, 7117–7123. [CrossRef]
30. Sakai, G.; Nakamura, K.; Horita, Z.; Langdon, T.G. Developing high-pressure torsion for use with bulk samples. *Mater. Sci. Eng. A* **2005**, *406*, 268–273. [CrossRef]
31. Figueiredo, R.B.; Langdon, T.G. Development of structural heterogeneities in a magnesium alloy processed by high-pressure torsion. *Mater. Sci. Eng. A* **2011**, *528*, 4500–4506. [CrossRef]
32. Panda, S.; Toth, L.S.; Fundenberger, J.J.; Perroud, O.; Guyon, J.; Zou, J.; Grosdidier, T. Analysis of heterogeneities in strain and microstructure in aluminum alloy and magnesium processed by high-pressure torsion. *Mater. Charact.* **2017**, *123*, 159–165. [CrossRef]
33. Panda, S.; Fundenberger, J.J.; Zhao, Y.; Zou, J.; Toth, L.S.; Grosdidier, T. Effect of initial powder type on the hydrogen storage properties of high-pressure torsion consolidated Mg. *Int. J. Hydrogen Energy* **2017**, *42*, 22438–22448. [CrossRef]
34. Skripnyuk, V.M.; Rabkin, E.; Estrin, Y.; Lapovok, R. The effect of ball milling and equal channel angular pressing on the hydrogen absorption/desorption properties of Mg-4.95 wt.% Zn-0.71 wt.% Zr (ZK60) alloy. *Acta Mater.* **2004**, *52*, 405–414. [CrossRef]
35. Skripnyuk, V.M.; Rabkin, E.; Estrin, Y.; Lapovok, R. Improving hydrogen storage properties of magnesium based alloys by equal channel angular pressing. *Int. J. Hydrogen Energy* **2009**, *34*, 6320–6324. [CrossRef]
36. Edalati, K.; Yamamoto, A.; Horita, Z.; Ishihara, T. High-pressure torsion of pure magnesium: Evolution of mechanical properties, microstructures and hydrogen storage capacity with equivalent strain. *Scr. Mater.* **2011**, *64*, 880–883. [CrossRef]
37. Huot, J.; Skryabina, N.Y.; Fruchart, D. Application of severe plastic deformation techniques to magnesium for enhanced hydrogen sorption properties. *Metals* **2012**, *31*, 329–343. [CrossRef]
38. Skryabina, N.; Medvedeva, N.; Gabov, A.; Fruchart, D.; Nachev, S.; de Rango, P. Impact of Severe Plastic Deformation on the stability of MgH_2. *J. Alloys Compd.* **2015**, *645*, S14–S17. [CrossRef]
39. Shao, H.; Wang, Y.; Xu, H.; Li, X. Hydrogen storage properties of magnesium ultrafine particles prepared by hydrogen plasma-metal reaction. *Mater. Sci. Eng. B* **2004**, *110*, 221–226. [CrossRef]
40. MDI JADE 7 Materials Data XRD Pattern Processing, Identification, and Quantification. Available online: http://www.materialsdata.com/ (accessed on 30 May 2018).

41. Fundenberger, J.J.; Bouzy, E.; Goran, D.; Guyon, J.; Yuan, H.; Morawiec, A. Orientation mapping by transmission-SEM with an on-axis detector. *Ultramicroscopy* **2016**, *161*, 17–22. [CrossRef] [PubMed]
42. Beausir, B.; Fundenberger, J.J. ATOM—Analysis Tools for Orientation Maps. Université de Lorraine: Metz, France, 2015. Available online: http://www.atex-software.eu/ (accessed on 31 July 2018).
43. Fundenberger, J.J.; Beausir, B. JTEX—Software for Texture Analysis. Université de Lorraine: Metz, France, 2015. Available online: http://www.atex-software.eu/ (accessed on 31 July 2018).
44. Del Valle, J.A.; Perez-Prado, M.T.; Ruano, O.A. Deformation mechanisms responsible for the high ductility in a Mg AZ31 alloy analyzed by electron backscattered diffraction. *Metall. Mater. Trans. A* **2005**, *36*, 1427–1438. [CrossRef]
45. Edalati, K.; Horita, Z. Significance of homologous temperature in softening behavior and grain size of pure metals processed by high-pressure torsion. *Mater. Sci. Eng. A* **2011**, *528*, 7514–7523. [CrossRef]
46. Beausir, B.; Toth, L.S.; Neale, K.W. Ideal orientations and persistence characteristics of hexagonal close packed crystals in simple shear. *Acta Mater.* **2007**, *55*, 2695–2705. [CrossRef]
47. Bonarski, B.J.; Schafler, E.; Mingler, B.; Skrotzki, W.; Mikulowski, B.; Zehetbauer, M.J. Texture evolution of Mg during high-pressure torsion. *J. Mater. Sci.* **2008**, *43*, 7513–7518. [CrossRef]
48. Skrotzki, W.; Scheerbaum, N.; Oertel, C.G.; Brokmeier, H.G.; Suwas, S.; Toth, L.S. Recrystallization of high-purity aluminium during equal channel angular pressing. *Acta Mater.* **2007**, *55*, 2211–2218. [CrossRef]
49. Zhilyaev, A.P.; Nurislamova, G.V.; Kim, B.K.; Baro, M.D.; Szpunar, J.A.; Langdon, T.G. Experimental parameters influencing grain refinement and microstructural evolution during high-pressure torsion. *Acta Mater.* **2003**, *51*, 753–765. [CrossRef]
50. Sakai, T.; Belyakov, A.; Kaibyshev, R.; Miura, H.; Jonas, J.J. Dynamic and post-dynamic recrystallization under hot, cold and severe plastic deformation conditions. *Prog. Mater. Sci.* **2014**, *60*, 130–207. [CrossRef]
51. Toth, L.S.; Gu, C.F.; Beausir, B.; Fundenberger, J.J.; Hoffman, M. Geometrically necessary dislocations favor the Taylor uniform deformation mode in ultra-fine-grained polycrystals. *Acta Mater.* **2016**, *117*, 35–42. [CrossRef]
52. Leiva, D.R.; Huot, J.; Ishikawa, T.T.; Bolfarini, C.; Kiminami, C.S.; Jorge, A.M.; Botta, W.J. Hydrogen activation behavior of commercial magnesium processed by different severe plastic deformation routes. *Mater. Sci. Forum* **2011**, *667*, 1047–1051. [CrossRef]
53. Edalati, K.; Novelli, M.; Itano, S.; Li, H.W.; Akiba, E.; Horita, Z.; Grosdidier, T. Effect of gradient-structure versus uniform nanostructure on hydrogen storage of Ti-V-Cr alloys: Investigation using ultrasonic SMAT and HPT processes. *J. Alloys Compd.* **2018**, *737*, 337–346. [CrossRef]
54. Gerard, N. Role of nucleation in some hydrogen formation kinetics. *J. Less Comm. Met.* **1987**, *131*, 13–23. [CrossRef]
55. Tien, H.Y.; Tanniru, M.; Wu, C.Y.; Ebrahimi, F. Effect of hydride nucleation rate on the hydrogen capacity of Mg. *Int. J. Hydrogen Energy* **2009**, *34*, 6343–6349. [CrossRef]
56. Töpler, J.; Buchner, H.; Säufferer, H.; Knorr, K.; Prandl, W. Measurements of the diffusion of hydrogen atoms in magnesium and Mg2Ni by neutron scattering. *J. Less Comm. Met.* **1982**, *88*, 397–404. [CrossRef]

Article

Geometrically Nonlinear Field Fracture Mechanics and Crack Nucleation, Application to Strain Localization Fields in Al-Cu-Li Aerospace Alloys

Satyapriya Gupta [1,2], Vincent Taupin [1,2,*], Claude Fressengeas [1,2] and Mohamad Jrad [1]

[1] Université de Lorraine, CNRS, Arts et Métiers ParisTech, LEM3, F-57000 Metz, France; satyapriya.gupta@univ-lorraine.fr (S.G.); claude.fressengeas@univ-lorraine.fr (C.F.); mohamad.jrad@univ-lorraine.fr (M.J.)

[2] Laboratory of Excellence on Design of Alloy Metals for low-mAss Structures (DAMAS), Université de Lorraine, Nancy-Metz, France

* Correspondence: vincent.taupin@univ-lorraine.fr; Tel.: +33-(0)3-72-74-78-43

Received: 7 March 2018; Accepted: 23 March 2018; Published: 27 March 2018

Abstract: The displacement discontinuity arising between crack surfaces is assigned to smooth densities of crystal defects referred to as disconnections, through the incompatibility of the distortion tensor. In a dual way, the disconnections are defined as line defects terminating surfaces where the displacement encounters a discontinuity. A conservation statement for the crack opening displacement provides a framework for disconnection dynamics in the form of transport laws. A similar methodology applied to the discontinuity of the plastic displacement due to dislocations results in the concurrent involvement of dislocation densities in the analysis. Non-linearity of the geometrical setting is assumed for defining the elastic distortion incompatibility in the presence of both dislocations and disconnections, as well as for their transport. Crack nucleation in the presence of thermally-activated fluctuations of the atomic order is shown to derive from this nonlinearity in elastic brittle materials, without any algorithmic rule or ad hoc material parameter. Digital image correlation techniques applied to the analysis of tensile tests on ductile Al-Cu-Li samples further demonstrate the ability of the disconnection density concept to capture crack nucleation and relate strain localization bands to consistent disconnection fields and to the eventual occurrence of complex and combined crack modes in these alloys.

Keywords: disconnection density; displacement discontinuity; crack nucleation; crack opening displacement; digital image correlation; Al-Cu-Li alloys

---

1. Introduction

Earlier work devoted to the applications of differential geometry to fracture processes have suggested that cracks in solids could be modeled by dislocations [1–6]. Indeed, several authors used fictitious dislocation fields as mathematical tools for generating the stress-field around cracks in elastic and elasto-plastic materials [7–11]. However, it is well known that dislocations reflect the discontinuity of the elastic/plastic displacement field across some bounded surface in a continuum [12], whereas fracture results in the discontinuity of the total displacement field. Thus, the dislocations involved in the calculations do not actually describe the disruption of matter inherent to fracture. Line defects distinct from dislocations, and referred to as disconnections, have been introduced to consistently account for such discontinuity [13]. Typically, the topological feature associated with the disconnection field is the crack opening displacement (COD), a quantity clearly distinct from the Burgers vector associated with dislocation fields. The approach is grounded on the assignment of the displacement discontinuity across the crack surfaces to a smooth field of incompatible distortion tensors by using the

Stokes theorem. A similar procedure applied to the plastic displacement discontinuity in the presence of dislocations results in concurrently involving the incompatibility of the plastic distortion and dislocation densities in the analysis. As a consequence, the elastic distortion field has to accommodate the plastic and total incompatibilities concurrently arising from both dislocations and disconnections. Due to smoothness of this field, the stress field is continuous at all points, including in the dislocation core and crack tip areas.

From a dynamic perspective, large stresses almost invariably result in some dislocation motion in the crack tip area in crystalline materials. Thus, plastic relaxation of the elastic stress field should accompany the elastic unloading due to crack growth, and plastic dissipation in the body should result concurrently from crack growth and dislocation motion. Therefore, crack growth is bound to occur if sufficient energy is available for sustaining both mechanisms, and identifying the driving force for crack growth requires a global approach to dissipation, also involving dislocation motion. After other approaches [10,14–16], such a framework was provided in [13] by using a transport scheme for the dislocation and disconnection density tensors: crack growth occurs through disconnection transport, just as plasticity occurs through dislocation transport. The driving force for disconnection motion (and hence for crack growth) is dual to the disconnection velocity in the volumetric dissipation density, in analogy with the Peach–Koehler force on dislocations. Hence, the thermodynamic requirement of non-negative dissipation can be used to formulate guidelines for the constitutive relationship between the driving force for crack growth and the disconnection velocity. As such, the theory provides a dynamic framework for crack growth, and it appears to be well suited for investigating the interplay between crack propagation and plasticity. The role of rotational incompatibility in the stress field of cracks and the influence of rotational defects on plasticity and crack growth was further detailed in [17] for a more accurate investigation of nano-sized neighborhoods of the crack tips. In the present paper, a geometrically nonlinear setting is proposed for the determination of the stress field in the presence of disconnections and dislocations, as well as for their transport scheme, which extends the work of [13] to finite transformations. The implications of this nonlinearity in the modeling of crack nucleation and growth are investigated in elastic-brittle materials. Further insights into the Griffith-based thermally-activated model of crack nucleation [18–20] will result from this approach.

Finally, because the present framework is able to relate the discontinuity of the material displacement in forming cracks to the incompatibility of the total elastic plus plastic distortion field, we think that it can be of practical use to understand complex links between strain localization and fracture in ductile materials. In particular, it can be applied to experimental strain fields, for instance to those obtained as per well-used digital image correlation (DIC) methods [21]. In this paper, we indeed consider DIC methods to investigate particular Al-Cu-Li aerospace alloys where complex interactions between localized plasticity and damage evolution leads to failure, and thus, they are interesting for our proposed framework. Indeed, as shown by recent studies on such alloys [22,23], the latter are prone to early and intermittent strain localization followed by fracture within bands. In an Al-Cu-Mg alloy for instance [23], tearing experiments combined with in situ synchrotron laminography and digital volume correlation showed the occurrence of early and intermittent localization bands, certainly associated with dynamic strain aging. Fracture eventually takes place along these bands from damage induced by strain localization. Therefore, we anticipate that, by integrating in time and space the evolution of disconnection densities during deformation, the eventual occurrence of fracture can be related to the accumulation of strain heterogeneity and incompatibility. The Al-Cu-Li alloys can thus be seen as a model benchmark material for validating our disconnection framework. As will be shown indeed, the disconnection density concept is able to capture crack nucleation and to relate complex strain localization features to the crack configurations, which validates the present framework.

The paper is organized as follows. Notations are settled in Section 2. Section 3 includes the fundamental basis for a field theory of the incompatibility of the lattice distortion and its relations, through elasticity, with the plastic incompatibility arising from the presence of dislocations, in a geometrically nonlinear setting. In Section 4, the solution of the boundary value problem for an elasto-static body containing a distribution of disconnection and dislocation densities is provided. Transport of the disconnection and dislocation densities is presented in Section 5, also in a geometrically nonlinear setting. Several algorithms for the solution of boundary value problems are presented in Section 6. The implications of the theory on the interpretation of crack nucleation phenomena are shown in Section 7 by modeling nucleation in elastic-brittle materials. The application of the framework to DIC strain fields and to the description of failure in ductile Al-Cu-Li alloys follows.

2. Notations

A bold symbol denotes a tensor. When there may be ambiguity, an arrow is superposed to represent a vector: $\vec{\mathbf{V}}$. The symmetric part of the second-order tensor $\mathbf{A}$ is denoted $\mathbf{A}^{sym}$, and its transpose is $\mathbf{A}^t$. The tensor $\mathbf{A}.\mathbf{B}$, with rectangular Cartesian components $A_{ik}B_{kj}$, results from the dot product of tensors $\mathbf{A}$ and $\mathbf{B}$, and $\mathbf{A} \otimes \mathbf{B}$ is their tensorial product, with components $A_{ij}B_{kl}$. A :represents the inner product of the two second-order tensors $\mathbf{A} : \mathbf{B} = A_{ij}B_{ij}$, in rectangular Cartesian components, or the product of a higher order tensor with a second-order tensor, e.g., $(\mathbf{A} : \mathbf{B})_{ij} = A_{ijkl}B_{kl}$. The cross product of a second-order tensor $\mathbf{A}$ and a vector $\mathbf{V}$, the **div** and **curl** operations for second-order tensors are defined row by row, in analogy with the vectorial case. For any base vector $\mathbf{e}_i$ of the reference frame:

$$(\mathbf{A} \times \mathbf{V})^t.\mathbf{e}_i = (\mathbf{A}^t.\mathbf{e}_i) \times \mathbf{V} \tag{1}$$

$$(\mathbf{div}\ \mathbf{A})^t.\mathbf{e}_i = \mathbf{div}(\mathbf{A}^t.\mathbf{e}_i) \tag{2}$$

$$(\mathbf{curl}\ \mathbf{A})^t.\mathbf{e}_i = \mathbf{curl}(\mathbf{A}^t.\mathbf{e}_i). \tag{3}$$

In rectangular Cartesian components:

$$(\mathbf{A} \times \mathbf{V})_{ij} = e_{jkl}A_{ik}V_l \tag{4}$$

$$(\mathbf{div}\mathbf{A})_i = A_{ij,j} \tag{5}$$

$$(\mathbf{curl}\mathbf{A})_{ij} = e_{jkl}A_{il,k}. \tag{6}$$

where e_{jkl} is a component of the third-order alternating Levi–Civita tensor $\mathbf{X}$. A vector $\vec{\mathbf{A}}$ is associated with tensor $\mathbf{A}$ by using its inner product with tensor $\mathbf{X}$:

$$(\vec{\mathbf{A}})_k = -\frac{1}{2}(\mathbf{A} : \mathbf{X})_k = -\frac{1}{2}e_{ijk}A_{ij}. \tag{7}$$

A superposed dot represents a material time derivative. In the component representation, the spatial derivative with respect to a Cartesian coordinate is indicated by a comma followed by the component index.

3. Incompatible Elasto-Plastic Continuum Theory

3.1. Disconnections and the Incompatibility of the Total Distortion

In the framework of field models, the displacement vector $\mathbf{u}$ is usually defined continuously at any point of a body $\mathcal{B}$ undergoing elasto-plastic transformation. If $\mathbf{X}$ and $\mathbf{x}$ are respectively the reference (Lagrange) and current (Euler) position vectors of a material element, the transformation tensor $\mathbf{F}$ is defined as the Jacobian tensor $\partial\mathbf{x}/\partial\mathbf{X}$. The gradient of the displacement is:

$$\mathbf{U} = \mathbf{F} - \mathbf{I} = \mathbf{grad}\ \mathbf{u}, \tag{8}$$

where $\mathbf{I}$ is the identity tensor. As such, $\mathbf{U}$ is curl-free:

$$\mathbf{curl}\ \mathbf{U} = 0, \tag{9}$$

a condition also verified by **F**: **curl F** $= 0$. Equation (9) is a necessary condition for the integrability of the distortion **U**, or equivalently for finding a single-valued displacement field **u** from Equation (8). It is usually referred to as a compatibility condition for **U**. The distortion tensor may be decomposed into its symmetric part, i.e., the strain tensor $\boldsymbol{\epsilon} = \mathbf{U}^{sym}$, and its skew-symmetric part, the rotation tensor $\boldsymbol{\omega} = \mathbf{U}^{skew}$, or equivalently the rotation vector:

$$\vec{\omega} = \frac{1}{2}\,\mathbf{curl}\,\mathbf{u}. \tag{10}$$

From Equation (10), it is seen that a necessary condition for the integrability of the rotation is:

$$\mathbf{div}\,\vec{\omega} = 0, \tag{11}$$

and that the compatibility condition (9) also reads:

$$\mathbf{curl}\boldsymbol{\epsilon} - \mathbf{grad}^t\vec{\omega} = 0. \tag{12}$$

Transposing Equation (12) and taking the curl of the result provides a necessary condition on the integrability of the strain tensor known as Saint-Venant's compatibility relation:

$$\mathbf{curl}\,\mathbf{curl}^t\,\boldsymbol{\epsilon} = 0. \tag{13}$$

Conversely, taken with Equation (11), Equation (13) is sufficient to insure the existence of a single-valued continuous solution **u** to Equation (8), up to a constant translation. More directly, Equation (9) is also a sufficient condition for the existence of this displacement field. However, displacement discontinuities may occur across bounded surfaces Σ, and the displacement field may cease to be single-valued if cracks nucleate and develop along such surfaces. A closed circuit C drawn on the body in the reference configuration and threading once the surface Σ may therefore present a closure defect **f** in the current configuration (see Figure 1). A definition for **f** is:

$$\mathbf{f} \;=\; -\int_C \mathbf{F}.\mathbf{dX} = -\int_C \mathbf{U}.\mathbf{dX} = -\int_S \mathbf{curl}\,\mathbf{U}.\mathbf{n}dS \tag{14}$$

if the following convention is adopted: the circuit C is oriented clockwise; the starting point S and finishing point F coincide in the reference configuration; and the closure defect is $\mathbf{f} = \mathbf{F}'\mathbf{S}'$ in the current configuration, where S' and F' denote the transforms of points S and F. In this relation, the surface S, with unit normal **n** in the reference configuration, is bounded by curve C. **f** actually characterizes a line defect: it is the displacement discontinuity occurring across the surface Σ. Let us denote by $\mathcal{L}$ the closed line bounding Σ and **l** its tangent vector. $\mathcal{L}$ threads surface S and is the front line of the disconnected crack surfaces in the current configuration. The displacement discontinuity **f** is usually referred to as the crack opening displacement (COD). Note that the possibility of a discontinuity of the rotation field is disregarded in the present paper, which limits the accuracy of the analysis in the close vicinity of the front line $\mathcal{L}$. This issue was further investigated in [17] in relation to the development of cracks in nanograined materials. Using the Stokes theorem (The Stokes theorem applies in simply connected bodies, and more generally in all simply connected parts of a multiply-connected body. It is simply required that the complete boundary curve be involved in the line integral, with adequate orientation.), the above displacement incompatibility can be reflected in an average manner over the surface S by the continuous tensorial density $\boldsymbol{\beta}$:

$$\boldsymbol{\beta} \;=\; -\mathbf{curl}\,\mathbf{U}. \tag{15}$$

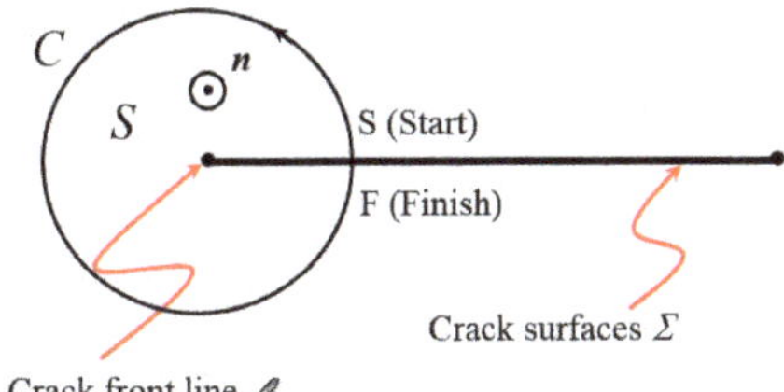

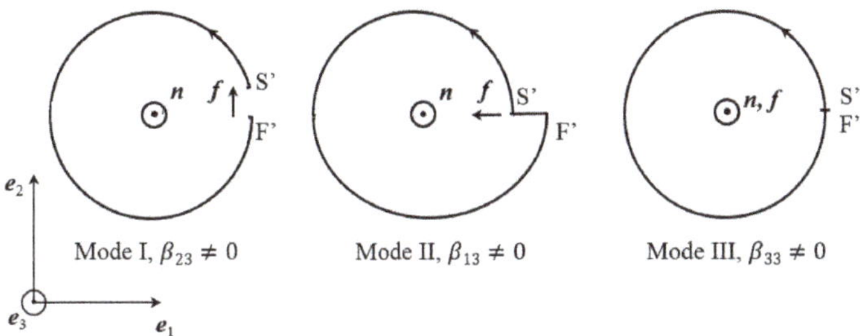

Figure 1. Crack surfaces Σ and front line $\mathcal{L}$ edge-on, test surface S with unit normal $\mathbf{n}$ and oriented bounding circuit C; (**a**) reference configuration, points (S,F) coincide; (**b**) current configuration with crack opening displacement (COD): $\mathbf{f} = \mathbf{F'S'}$ and fracture Modes I, II and III.

Of course, this equation still holds if $\mathbf{U}$ is replaced with $\mathbf{F}$. We shall refer to $\boldsymbol{\beta}$ as the disconnection density tensor (After J.P.Hirth [3], the word "disconnection" has been associated with interfacial defects of dislocation and/or step character in relation to grain boundary migration or twinning/phase transformation mechanisms through interface propagation. Of course, it carries a different meaning in the present context. "Disconnection" presently reflects broken atomic bonds and a locally disconnected material. Beyond being consistent with physical intuition, this usage is in agreement with mathematical terminology. Indeed, if the incipient crack tunnels from side to side through the body, the topological status of the latter shifts from simply connected to multiply connected.). Clearly, the Saint-Venant's compatibility condition (13) is not satisfied if $\boldsymbol{\beta}$ is a non-vanishing tensor. Taking the same steps as above, it can indeed be shown that the tensor $\boldsymbol{\eta} = \mathbf{curl}\,\mathbf{curl}^t\,\boldsymbol{\epsilon}$ is now:

$$\boldsymbol{\eta} = -\mathbf{curl}\boldsymbol{\beta}^t. \tag{16}$$

$\boldsymbol{\eta}$ may be referred to as an incompatibility tensor. The geometrical meaning of Equation (16) is that the continuity of matter is disrupted if the incompatibility tensor $\boldsymbol{\eta}$ is non-vanishing in the presence of disconnections $\boldsymbol{\beta}$. According to Equation (14), $\boldsymbol{\beta}$ is also the dyadic product:

$$\boldsymbol{\beta} = \mathbf{f} \otimes \mathbf{n} \tag{17}$$

for a unit surface of normal $\mathbf{n}$. If a disconnection ensemble of line vector $\mathbf{l}$ and crack opening displacement $\tilde{\mathbf{f}}$ threads the unit surface of normal $\mathbf{l}$, then only the part $\mathbf{l}.\mathbf{n}$ of this ensemble threads the unit surface of normal $\mathbf{n}$. The corresponding crack opening displacement is $\mathbf{f} = (\mathbf{l}.\mathbf{n})\tilde{\mathbf{f}}$, and from Equation (17):

$$\boldsymbol{\beta} = (\mathbf{l}.\mathbf{n})\tilde{\mathbf{f}} \otimes \mathbf{n} = \tilde{\mathbf{f}} \otimes \mathbf{l}. \tag{18}$$

The dimension of the disconnection tensor components is m^{-1}, i.e., m of displacement discontinuity per m^2. Equation (18) represents a Mode I or Mode II fracture when $\tilde{\mathbf{f}}.\mathbf{l} = 0$, whereas a Mode III crack is reflected if $\tilde{\mathbf{f}} \times \mathbf{l} = 0$. For instance, if a planar crack is considered as in Figure 1, with $\mathbf{e}_2$ and $\mathbf{e}_3$ as the unit normal to the fracture plane and the rectilinear fracture line direction respectively in the orthogonal frame $(\mathbf{e}_1, \mathbf{e}_2, \mathbf{e}_3)$, we may refer to the components β_{23} and β_{13} as edge-disconnection densities associated with Mode I and Mode II, respectively, and to β_{33} as a screw-disconnection density associated with Mode III fracture. The emergence of an edge-disconnection density field in relation to crack nucleation is illustrated below by using DIC records of tensile tests on Al-Cu-Li samples in Section 7. From Equation (15), it is seen that $\boldsymbol{\beta}$ necessarily satisfies the relation:

$$\mathbf{div}\, \boldsymbol{\beta} = 0, \tag{19}$$

which has the meaning that disconnection lines cannot end within the body. If they do not exit the body, these lines are indeed loops terminating surfaces of discontinuity of the displacement field.

Further insights into the distortion incompatibility deriving from the presence of disconnections are provided by its Stokes–Helmholtz orthogonal decomposition. Invoking the latter in the space of square-integrable tensor fields with square-integrable first order derivatives (see for example [24], Theorem 5.8), there exists indeed a unique tensor field $\boldsymbol{\chi}$ and a unique (up to a constant) vector field $\mathbf{z}$ (both square-integrable, as well as their derivatives to second-order) such that the distortion field $\mathbf{U}$ reads as the sum:

$$\mathbf{U} = \mathbf{curl}\, \boldsymbol{\chi} + \mathbf{grad}\, \mathbf{z}. \tag{20}$$

with the orthogonality condition $\int_D \mathbf{curl}\, \boldsymbol{\chi} : \mathbf{grad}\, \mathbf{z}\, dv = 0$. Thus, taking the **curl** of $\mathbf{U}$ in Equation (20) extracts $\mathbf{curl}\, \boldsymbol{\chi}$ and discards $\mathbf{grad}\, \mathbf{z}$, whereas $\mathbf{div}\, \mathbf{U}$ extracts $\mathbf{grad}\, \mathbf{z}$ and eliminates $\mathbf{curl}\, \boldsymbol{\chi}$. Therefore, Equation (15) involves only $\mathbf{curl}\, \boldsymbol{\chi}$, which will be identified below as the incompatible part $\mathbf{U}^{\perp}$ of $\mathbf{U}$:

$$\mathbf{curl}\, \mathbf{U}^{\perp} \;=\; \mathbf{curl}\, \mathbf{curl}\, \boldsymbol{\chi} = -\boldsymbol{\beta}. \tag{21}$$

Similarly, $\mathbf{grad}\, \mathbf{z}$ will be the compatible part $\mathbf{U}^{\parallel}$ of the distortion $\mathbf{U}$. Up to a constant, $\mathbf{z}$ will be the compatible part of the displacement, $\mathbf{u}^{\parallel}$. To ensure correctness of this identification, $\mathbf{U}^{\perp}$ must vanish identically throughout the body when $\boldsymbol{\beta} = 0$. Thus, following [24], Equation (21) is augmented with the side conditions $\mathbf{div}\, \mathbf{U}^{\perp} = \mathbf{0}$ and $\mathbf{U}^{\perp}.\mathbf{n} = \mathbf{0}$ on the external boundary with unit normal $\mathbf{n}$. Then, any residual gradient part $\mathbf{grad}\, \mathbf{w}$ of $\mathbf{U}^{\perp}$ satisfies a Poisson equation $\mathbf{div}\, \mathbf{grad}\, \mathbf{w} = 0$, and the boundary condition $\mathbf{grad}\, \mathbf{w}.\mathbf{n} = 0$ guarantees a vanishing $\mathbf{w}$. As a consequence, the incompatible distortion field $\mathbf{U}^{\perp}$ is uniquely determined once the disconnection density field is known. The incompatible part of the displacement, $\mathbf{u}^{\perp}$, which is to be identified with the crack opening displacement, follows from Equations (14), (15) and (21) through an integral on a patch S:

$$\mathbf{u}^{\perp} = \mathbf{f} = \int_S \boldsymbol{\beta}.\mathbf{n} dS. \tag{22}$$

The compatible displacement, $\mathbf{u}^{\parallel}$, and the compatible part $\mathbf{U}^{\parallel}$ of the distortion tensor are used for calculating the Jacobian J, i.e., the determinant of the tensor $\partial \mathbf{x}/\partial \mathbf{X}$, in the statement of mass conservation $\rho J = \rho_0$, where ρ_0 and ρ are respectively the initial and current mass densities.

3.2. Dislocations and the Incompatibility of the Plastic Distortion

When dislocations are present in the body $\mathcal{B}$, a closed circuit C drawn in the reference configuration may also present a closure defect reflecting a discontinuity of the plastic displacement, referred to as the Burgers vector $\mathbf{b}$, in the intermediate local stress-free configuration wherein the body is not elastically distorted. Consistent with earlier assumptions in the present paper, the possibility of a discontinuity of the plastic rotation is disregarded in the present paper. Acknowledging this possibility would amount to considering the presence of disclinations [17,25]. Using the same sign convention as in Equation (14), a definition for $\mathbf{b}$ is:

$$\mathbf{b} \;=\; -\int_C \mathbf{F}_p.\mathbf{dX} = -\int_S \mathbf{curl}\ \mathbf{F}_p.\mathbf{n}dS = -\int_S \mathbf{curl}\ \mathbf{U}_p.\mathbf{n}dS \tag{23}$$

where $\mathbf{F}_p$ is the plastic transformation tensor in the elastic/plastic multiplicative decomposition $\mathbf{F} = \mathbf{F}_e.\mathbf{F}_p$ of the transformation tensor [26] and $\mathbf{U}_p = \mathbf{F}_p - \mathbf{I}$ is the corresponding plastic distortion. As argued above for the disconnection and total distortion tensors, Equation (23) can be used to define Nye's dislocation density tensor $\boldsymbol{\alpha}$ by exchanging the incompatibility of the smooth plastic distortion tensor for the discontinuity $\mathbf{b}$ of the plastic displacement:

$$\mathbf{curl}\ \mathbf{U}_p \;=\; -\boldsymbol{\alpha}. \tag{24}$$

Invoking again the Stokes–Helmholtz decomposition of $\mathbf{U}_p$, there exist a gradient-free incompatible part of the plastic distortion, $\mathbf{U}_p^{\perp}$, such that:

$$\mathbf{curl}\ \mathbf{U}_p^{\perp} \;=\; -\boldsymbol{\alpha}. \tag{25}$$

Again, the same equation holds if $\mathbf{U}_p^{\perp}$ is replaced with $\mathbf{F}_p^{\perp}$. $\mathbf{U}_p^{\perp}$ is uniquely determined from the dislocation density field $\boldsymbol{\alpha}$ if it is demanded that it also satisfies the side conditions $\mathbf{div}\,\mathbf{U}_p^{\perp} = 0$ and $\mathbf{U}_p^{\perp}.\mathbf{n} = 0$ on the external boundary with unit normal $\mathbf{n}$. The compatible part $\mathbf{U}_p^{\parallel}$ of the plastic distortion contributes to the accumulated plastic strain, but does not play a role in the mechanical state of the material or in the definition of the dislocation density $\boldsymbol{\alpha}$. From Equation (25), it is additionally seen that:

$$\mathbf{div}\,\boldsymbol{\alpha} = 0, \tag{26}$$

which means that dislocation lines cannot end within the body. Although $\boldsymbol{\alpha}$ is carrying a topological content $\mathbf{b}$ reflecting the distortion of the material in the intermediate configuration, it operates on lines $\mathbf{t}$ in the reference configuration. In its standard definition, Nye's dislocation density tensor operates on lines in the current configuration, and produces Burgers vectors in the local intermediate configuration [12]. Thus, in the present work, the components α_{ij} of Nye's tensor are dislocation line densities with respect to surface units in the reference configuration, whereas they are referred to surface units in the current configuration in the standard definition. These two dislocation density measures may differ significantly at finite strains, as detailed below in Equations (30) and (31). However, the Burgers vectors obtained from both definitions are strictly identical. Further, the present definition of the dislocation density tensor is natural from a physical perspective, because dislocations are the cause of the incompatible plastic distortion, whereas the incompatible elastic distortion derives from the presence of both dislocations and disconnections. In the presence of dislocations, it is such that either it maintains the continuity of matter in the absence of disconnections, or it also accommodates the total distortion incompatibility in their presence. The derivation of the elastic distortion incompatibility will be the subject of the next section. The formal similarity in the treatment of dislocation and disconnection densities shown above should not hide their fundamental topological difference. Dislocations reflect a discontinuity of the plastic displacement field across the surface they sweep, but continuity of matter is maintained along this surface. Conversely, disconnections disrupt the continuity of matter along their paths, whose evidence therefore remain tangible in the body.

3.3. *Composition of Incompatibilities and Elastic Distortion*

Because the incompatibilities of the total and plastic distortion fields both impact on the elastic state of the body, they must compose and the elastic transformation tensor $\mathbf{F}_e$ of the lattice must adjust to the dislocation and disconnection density fields. Taking the curl of the multiplicative decomposition $\mathbf{F} = \mathbf{F}_e.\mathbf{F}_p$, we obtain successively:

$$\begin{aligned}
(\mathbf{curl}\,\mathbf{F})_{ij} &= e_{jkl}(F^e_{im}F^p_{ml})_{,k} = F^e_{im}e_{jkl}F^p_{ml,k} + e_{jkl}F^e_{im,k}F^p_{ml} \\
(\mathbf{curl}\,\mathbf{F})_{ij} &= F^e_{im}(\mathbf{curl}\,\mathbf{F}_p)_{mj} + e_{jkl}G^e_{imk}F^p_{ml} \\
(\mathbf{curl}\,\mathbf{F})_{ij} &= (\mathbf{F}_e.\mathbf{curl}\,\mathbf{F}_p)_{ij} + e_{jkl}G^{e,T}_{ikm}F^p_{ml},
\end{aligned}$$

which translates into the intrinsic relation:

$$\mathbf{curl}\,\mathbf{F} = \mathbf{F}_e.\mathbf{curl}\,\mathbf{F}_p + \mathbf{G}^T_e.\mathbf{F}_p : \mathbf{X}, \tag{27}$$

where $\mathbf{G}^T_e$ is defined from the second-order elastic distortion $\mathbf{G}_e$ with components $G^e_{imk} = F^e_{im,k}$ by transposing the last two subscripts. Substituting Equations (15) and (24) in (27) then yields the differential equation:

$$\boldsymbol{\beta} = \mathbf{F}_e.\boldsymbol{\alpha} - \mathbf{G}^T_e.\mathbf{F}_p : \mathbf{X} \tag{28}$$

for the elastic distortion $\mathbf{F}_e$, where the defect densities $(\boldsymbol{\alpha}, \boldsymbol{\beta})$ and plastic transformation tensor $\mathbf{F}_p$ are prescribed quantities. In this form, the solution $\mathbf{F}_e$ is not unambiguously defined because the compatible part $\mathbf{F}^{\parallel}_p$ of $\mathbf{F}_p$ arbitrarily depends on the choice of the reference configuration. Hence, we choose $\mathbf{F}^{\parallel}_p = 0$ with no loss of generality, and transform Equation (28) into:

$$\boldsymbol{\beta} = \mathbf{F}_e.\boldsymbol{\alpha} - \mathbf{G}^T_e.\mathbf{F}^{\perp}_p : \mathbf{X} \tag{29}$$

where $\mathbf{F}^{\perp}_p$ derives from the disconnection density field $\boldsymbol{\beta}$ through the solution of Equation (25). Therefore, $\mathbf{F}_e$ can be found uniquely from Equation (29) once the fields $(\boldsymbol{\alpha}, \boldsymbol{\beta})$ are known and boundary conditions are prescribed. This issue will be dealt with in the next section. In the absence of disconnections ($\boldsymbol{\beta} = 0$), $\mathbf{F}^{\perp}_p = \mathbf{F}^{-1}_e$, and Equation (29) leads to the relation:

$$\boldsymbol{\alpha} = \mathbf{F}^{-1}_e.\mathbf{G}^T_e.\mathbf{F}^{-1}_e : \mathbf{X}. \tag{30}$$

The dislocation density tensor found in Equation (30) differs from the conventional definition of the dislocation density tensor in the absence of disconnections:

$$\boldsymbol{\alpha} = -\mathbf{curl}\,\mathbf{F}^{-1}_e, \tag{31}$$

because, as already mentioned, it operates on surface elements in the reference configuration, in place of the current configuration. Of course, by restricting the purpose to small transformations, it is seen that Equation (29) becomes successively:

$$\beta_{ij} \cong \alpha_{ij} - U^e_{im,k}\delta_{ml}e_{klj} \tag{32}$$

$$\beta_{ij} \cong \alpha_{ij} - e_{jkl}U^e_{il,k} \tag{33}$$

$$\mathbf{curl}\,\mathbf{U}^{\perp}_e \cong \boldsymbol{\alpha} - \boldsymbol{\beta}, \tag{34}$$

which coincides with the composition relation provided in [13] and, when $\boldsymbol{\beta} = 0$, with the approximation of Equation (31) at small distortions:

$$\boldsymbol{\alpha} = \mathbf{curl}\,\mathbf{U}^{\perp}_e. \tag{35}$$

Equations (29) and (34) suggest that the $(\boldsymbol{\alpha}, \boldsymbol{\beta})$ tensor fields may be simultaneously non-zero at a given material point, implying that the present description intrinsically belongs to the cohesive approaches of fracture. Indeed, $(\boldsymbol{\alpha}, \boldsymbol{\beta})$ would be mutually exclusive in a non-cohesive description of fracture: $(\boldsymbol{\alpha} \neq 0, \boldsymbol{\beta} = 0)$ or $(\boldsymbol{\alpha} = 0, \boldsymbol{\beta} \neq 0)$. In the geometrically linear context, Equation (34) suggests that a given incompatible elastic distortion field can be obtained from either a distribution of dislocations or the same distribution of disconnections (with changed sign). However, such an identification does not imply equality of the associated stress fields, as zero-traction boundary conditions need to be fulfilled on the crack surfaces in the case of disconnections. In the geometrically nonlinear case, it appears rather that the dislocation and disconnection density fields do not play such a symmetric role in the stress field. Changing the sign of the dislocation distribution in Equation (29) is not equivalent to introducing a disconnection density field. Further details on calculating the stress field associated with a distribution of dislocations and disconnections in this cohesive modeling scheme are provided in the following Section.

4. Elasto-Static Incompatible Media

We assume that the body $\mathcal{B}$ contains a distribution of dislocations $\boldsymbol{\alpha}$ and disconnections $\boldsymbol{\beta}$. Provisionally, we are not interested in the dynamic plastic and fracture processes leading to this distribution, and the history-dependent compatible part of the plastic transformation, $\mathbf{F}_p^{\parallel}$, and plastic displacement, $\mathbf{u}_p^{\parallel}$, are arbitrarily set to zero without loss of generality. The body is in equilibrium under traction loads $\mathbf{F}_{\partial\mathcal{B}_t}(P)$ applied at points P on a part $\partial\mathcal{B}_t$ of its external surface $\partial\mathcal{B}$, including in particular the non-cohesive parts of the crack surfaces $\partial\mathcal{B}_c$ where $\mathbf{F}_{\partial\mathcal{B}_t}(P) = 0$, and other loads leading to prescribed displacements $\mathbf{u}(P)$ on the remaining part $\partial\mathcal{B}_u$ of the external surface. The following equations are therefore satisfied:

$$\mathbf{G}_e^T.\mathbf{F}_p^{\perp} : \mathbf{X} = \mathbf{F}_e.\boldsymbol{\alpha} - \boldsymbol{\beta} \tag{36}$$

$$\mathbf{T} = \mathbf{T}(\mathbf{F}_e) \tag{37}$$

$$\mathbf{div}\,\mathbf{T} = 0. \tag{38}$$

In this set of equations, geometric nonlinearity is assumed, and Equation (36) is reproduced after Equation (29). The stress tensor $\mathbf{T}$ is provided as a function of the elastic transformation tensor $\mathbf{F}_e$, for instance through the right Cauchy–Green tensor $\mathbf{F}_e^t.\mathbf{F}_e$. In the balance of momentum (38), the inertial terms and volumetric load density are neglected as being unessential in the present context. As a first step of the solution process, an incompatible solution $\mathbf{F}_e^{\perp}$ of Equation (36) is searched for by appending to the latter the side conditions:

$$\mathbf{div}\,\mathbf{F}_e = 0 \tag{39}$$

$$\mathbf{F}_e.\mathbf{n} = 0 \text{ on } \partial\mathcal{B}. \tag{40}$$

As already explained in Section 3.1, the conditions (39),(40) guarantee that $\mathbf{F}_e$ does not have a gradient component and is therefore incompatible. Generally, the field $\mathbf{F}_e^{\perp}$ does not satisfy the balance of momentum equation, and we consequently look for the complementary compatible transformation $\mathbf{F}_e^{\parallel}$ such that the total elastic transformation $\mathbf{F}_e$ satisfies Equations (37) and (38) under the boundary conditions detailed above. In the present nonlinear context, it is not possible to prove the existence and uniqueness of such a solution, but in a simplified version where the material response is assumed to be linear, it can be shown that such a solution does exist. Let indeed the elastic response be specified as:

$$\mathbf{T} = \mathbf{C} : \mathbf{U}_e^{sym} = \mathbf{C} : \boldsymbol{\epsilon}_e \tag{41}$$

where $\mathbf{C}$ is the fourth-order tensor of elastic moduli and $\boldsymbol{\epsilon}_e$ is the symmetric part of the elastic distortion tensor, i.e., the elastic strain tensor. Separating the compatible and incompatible parts of the latter:

$$\boldsymbol{\epsilon}_e = \boldsymbol{\epsilon}_e^{\perp} + \boldsymbol{\epsilon}_e^{\parallel}, \tag{42}$$

and combining Equations (38), (41) and (42), we find:

$$\mathbf{div}\,(\mathbf{C}:\boldsymbol{\epsilon}_e^{\parallel}) + \mathbf{f}^{\perp} = 0, \qquad \mathbf{f}^{\perp} = \mathbf{div}\,(\mathbf{C}:\boldsymbol{\epsilon}_e^{\perp}) \tag{43}$$

where $\mathbf{f}^{\perp}$ is a volumetric density of loads reflecting the stress field $\mathbf{C}:\boldsymbol{\epsilon}_e^{\perp}$ that arises in the body from the presence of the dislocation and disconnection fields $(\boldsymbol{\alpha},\boldsymbol{\beta})$. In addition to Equation (43), the compatible elastic strain $\boldsymbol{\epsilon}_e^{\parallel}$ and corresponding elastic displacement $\mathbf{u}_e^{\parallel}$ must also satisfy the boundary conditions:

$$\mathbf{u}_e^{\parallel}(P) = \mathbf{u}(P) - \mathbf{u}_e^{\perp}(P) - \mathbf{u}^{\perp}(P), \quad \forall P \in \partial\mathcal{B}_u \tag{44}$$

$$\mathbf{C}:\boldsymbol{\epsilon}_e^{\parallel}.\mathbf{n} = \mathbf{F}_{\partial D_t}(P) - \mathbf{C}:\boldsymbol{\epsilon}_e^{\perp}.\mathbf{n}, \quad \forall P \in \partial\mathcal{B}_t \tag{45}$$

where $\mathbf{u}_e^{\perp}(P)$ results from the incompatible elastic distortion $\mathbf{U}_e^{\perp}$ and $\mathbf{u}^{\perp}(P)$ is found from Equation (22) through an integral on a patch S around point P:

$$\forall P \in \mathcal{B}, \mathbf{u}^{\perp}(P) = \int_S \boldsymbol{\beta}.\mathbf{n}dS. \tag{46}$$

For given dislocation and disconnection densities $(\boldsymbol{\alpha},\boldsymbol{\beta})$, this is a conventional Navier-type elasticity problem, whose solution, i.e., the elastic displacement field $\mathbf{u}_e^{\parallel}$, is unique. Once the displacement is arbitrarily chosen at some point $P_0 \in \partial\mathcal{B}_u$ where it is uniquely defined, the compatible displacement field at any point $P \in D$ is computed from the path-integral:

$$\forall P \in \mathcal{B}, \mathbf{u}^{\parallel}(P) = \int_{P_0}^{P} \mathbf{U}_e^{\parallel}.\mathbf{dX} + \mathbf{u}(P_0) \tag{47}$$

Clearly, the integral (47) is path-dependent in the presence of dislocations and disconnections.

5. Transport

Because the dislocation and disconnection density tensors are areal densities of lines carrying a topological content (the Burgers and COD vectors), their evolution is governed by statements for the conservation of this content. We assume that the dislocation and the disconnection densities have associated velocities with respect to the material, $\mathbf{V}_\alpha$ and $\mathbf{V}_\beta$ respectively. When the resolution length scale is small enough, these density fields represent individual defects, and the corresponding values of the velocity fields represent the motion and deformation of the core regions of the defects. When, instead, the resolution length scale is sufficiently large, the surface S bounded by circuit C in Equations (14) and (23) may be threaded by several defect lines. Then, the velocity fields reflect the collective motion of these lines with respect to the material. Their advance represents the spreading of one topologically-charged piece of the body at the expense of the other. Let us consider an element of a disconnection ensemble, of line vector $\mathbf{t}_0$ in the reference configuration $\mathcal{B}_0$ of the body $\mathcal{B}$ and COD $\mathbf{f}$ per unit surface of the current configuration $\mathcal{B}_t$, having a velocity $\mathbf{V}_\beta$ with respect to the lattice in this configuration. The disconnection density tensor associated with this disconnection element is $\boldsymbol{\beta} = \mathbf{f} \otimes \mathbf{t}_0$. In its motion, this disconnection element crosses an oriented closed curve C_0 drawn in the initial configuration $\mathcal{B}_0$ and enters (or exits) the surface S_0 bounded by C_0. Thus, some COD flux $\mathcal{F}$ is generated through the differential element $\mathbf{dx}_0$ tangent to curve C_0, and the rate of CODs through the surface $\mathbf{dS}_0 = \mathbf{t}_0 \times \mathbf{dx}_0$ is:

$$\mathcal{F}.\mathbf{dx}_0 = \mathbf{f}(\mathbf{V}_\beta^0.\mathbf{dS}_0) = \mathbf{f}(\mathbf{V}_\beta^0.(\mathbf{t}_0 \times \mathbf{dx}_0)), \tag{48}$$

where $\mathbf{V}_\beta^0$ is the disconnection velocity with respect to the reference frame. Composing with the material velocity $\mathbf{v}$ with respect to the reference frame, the latter is also $\mathbf{V}_\beta^0 = \mathbf{v} + \mathbf{V}_\beta$. Using a permutation of the mixed product allows extracting: $\mathbf{dx}_0$

$$\mathcal{F}.\mathbf{dx}_0 = -\mathbf{f}((\mathbf{t}_0 \times \mathbf{V}^0_\beta).\mathbf{dx}_0) = -\mathbf{f} \otimes (\mathbf{t}_0 \times \mathbf{V}^0_\beta).\mathbf{dx}_0, \tag{49}$$

which leads to the identification of the COD flux tensor as:

$$\mathcal{F} = -(\mathbf{f} \otimes \mathbf{t}_0) \times \mathbf{V}^0_\beta = -\boldsymbol{\beta} \times \mathbf{V}^0_\beta = -\dot{\mathbf{U}}. \tag{50}$$

Hence, $\mathcal{F}$ appears as being opposed to the distortion rate tensor $\dot{\mathbf{U}}$. The evolution of the fracture line is typically characterized by $\dot{\mathbf{U}}$. Indeed, assume that the resolution length scale is small enough to envision a single defect, and consider a planar crack in the orthonormal basis $(\mathbf{e}_1, \mathbf{e}_2, \mathbf{e}_3)$ with $\mathbf{e}_2$ and $\mathbf{e}_3$ as the unit normal to the fracture plane and the rectilinear crack line direction respectively. If $\mathbf{V}^0_\beta = V_1\mathbf{e}_1 + V_2\mathbf{e}_2$ is the crack line velocity, then it is found from Equation (50) that $\dot{\mathbf{U}}.\mathbf{e}_2 = \beta_{23}V_1\mathbf{e}_2$ in Mode I for the disconnection tensor $\boldsymbol{\beta} = \beta_{23}\mathbf{e}_2 \otimes \mathbf{e}_3$, $\dot{\mathbf{U}}.\mathbf{e}_2 = \beta_{13}V_1\mathbf{e}_1$ in Mode II for $\boldsymbol{\beta} = \beta_{13}\mathbf{e}_1 \otimes \mathbf{e}_3$, and in Mode III: $\dot{\mathbf{U}}.\mathbf{e}_2 = \beta_{33}V_1\mathbf{e}_3$ for $\boldsymbol{\beta} = \beta_{33}\mathbf{e}_3 \otimes \mathbf{e}_3$, as could be expected. Indeed, the distortion rate induces crack opening by tension, shear perpendicular to the crack line and shear parallel to the crack line in Modes I, II and III, respectively.

It seems reasonable to postulate the following statement of balance: the rate of change of the COD content of the disconnections threading S_0 is the difference between the incoming and outgoing COD fluxes if no COD is independently generated or absorbed inside S_0. Employing the COD flux defined in Equation (48), the corresponding balance equation is:

$$\frac{d}{dt}\int_{S_0} \boldsymbol{\beta}.\mathbf{n}_0 dS_0 = \int_{C_0} \mathcal{F}.\mathbf{dx}_0. \tag{51}$$

This statement is required to hold for all such patches in the body $\mathcal{B}$. Applying the Stokes theorem to the line integral on the right-hand side and pushing the time derivative inside the integral on the left-hand side, we obtain:

$$\int_{S_0} (\dot{\boldsymbol{\beta}} - \mathbf{curl}\,\mathcal{F}).\mathbf{n}_0 dS_0 = 0. \tag{52}$$

For reasons of continuity of the integrand, the equivalent point-wise statement can be extracted:

$$\dot{\boldsymbol{\beta}} - \mathbf{curl}\,\mathcal{F} = 0 \tag{53}$$

at all points in $\mathcal{B}$. Here, $\dot{\boldsymbol{\beta}}$ represents the time derivative of $\boldsymbol{\beta}$ in the reference frame. Since S_0 is invariant, Equation (53) is valid for finite transformations. By substituting the relation (50) in Equation (53), this balance statement takes the form:

$$\dot{\boldsymbol{\beta}} + \mathbf{curl}\,\dot{\mathbf{U}} = \dot{\boldsymbol{\beta}} + \mathbf{curl}\,(\boldsymbol{\beta} \times \mathbf{V}^0_\beta) = \dot{\boldsymbol{\beta}} + \mathbf{curl}\,(\boldsymbol{\beta} \times (\mathbf{v} + \mathbf{V}_\beta)) = 0. \tag{54}$$

Equation (54) may be read as:

$$\dot{\boldsymbol{\beta}} + \mathbf{curl}\,(\boldsymbol{\beta} \times \mathbf{V}_\beta) = \mathbf{s}_v, \tag{55}$$

where the term $\mathbf{s}_v = -\mathbf{curl}\,(\boldsymbol{\beta} \times \mathbf{v})$ appears as a source for disconnections in the material frame. This source derives from the incompatibility of the distortion rate arising in the motion of the material points supporting the disconnection density. Further analysis of the crack nucleation arising from this source is provided below in Section 7 in the context of elastic-brittle materials. If small transformations can be assumed, the transport Equation (54) simplifies into:

$$\dot{\boldsymbol{\beta}} + \mathbf{curl}\,(\boldsymbol{\beta} \times \mathbf{V}_\beta) = 0. \tag{56}$$

As for the evolution of the dislocation density field $\boldsymbol{\alpha}$, it has been shown from the kinematics of flux of lines moving in and out of the area patch S through its bounding curve C in the current configuration that similar transport equations are obtained for the dislocation densities [27], resulting in:

$$\dot{\boldsymbol{\alpha}} + \mathbf{curl}\,(\boldsymbol{\alpha} \times (\mathbf{v} + \mathbf{V}_{\alpha})) = 0 \tag{57}$$

for finite transformations, and:

$$\dot{\boldsymbol{\alpha}} + \mathbf{curl}\,(\boldsymbol{\alpha} \times \mathbf{V}_{\alpha}) = 0 \tag{58}$$

for small transformations. Similar to Equation (55), it can be stated here that the term $\mathbf{s}_v = -\mathbf{curl}\,(\boldsymbol{\alpha} \times \mathbf{v})$ appears as a source for dislocations in the material frame [28]. If suitable constitutive assumptions are made for the motion of dislocations and disconnections, i.e., if the velocities $(\mathbf{V}_{\alpha}, \mathbf{V}_{\beta})$ for the transport of dislocation and disconnection densities are specified in terms of appropriate driving forces, then Equations (54), (56)–(58) may be used as governing equations for the dynamics of dislocations and disconnections. Such constitutive assumptions should in particular confer to the transport scheme the ability to reflect a Griffith-type behavior for fracture growth. In addition, if annihilation is pervasive in the dynamics of dislocations and is conveniently treated by Equations (57) and (58) [29], disconnection annihilation may not be as customary because healing of pre-existing cracks can be difficult in certain materials, such as metals below melting temperature in usual pressure conditions. Such material behavior also needs to be reflected by the constitutive assumptions made for the disconnection velocity $\mathbf{V}_{\beta}$ in Equations (54) and (56). These assumptions were discussed in [13], based on thermodynamical arguments originally introduced in [30].

6. Elasto-Plastic Incompatible Media

Gathering the above analyses together with results from [13] regarding the driving forces on dislocations and disconnections, the equations of the elasto-plastic boundary value problem with crack propagation can be summarized as follows:

$$
\begin{aligned}
\mathbf{curl}\,\mathbf{U}_p^{\perp} &= -\boldsymbol{\alpha} && (59)\\
\mathbf{curl}\,\mathbf{U}^{\perp} &= -\boldsymbol{\beta} && (60)\\
\mathbf{G}_e^{\perp,T}.\mathbf{F}_p^{\perp} : \mathbf{X} &= \mathbf{F}_e^{\perp}.\boldsymbol{\alpha} - \boldsymbol{\beta} && (61)\\
\mathbf{div}\,\mathbf{F}_e^{\perp} &= 0 && (62)\\
\mathbf{T} &= \mathbf{C} : (\boldsymbol{\epsilon}_e^{\parallel} + \boldsymbol{\epsilon}_e^{\perp}) && (63)\\
\mathbf{div}\,\mathbf{T} &= 0 && (64)\\
\dot{\boldsymbol{\alpha}} + \mathbf{curl}\,(\boldsymbol{\alpha} \times (\mathbf{v} + \mathbf{V}_{\alpha})) &= 0 && (65)\\
\dot{\boldsymbol{\beta}} + \mathbf{curl}\,(\boldsymbol{\beta} \times (\mathbf{v} + \mathbf{V}_{\beta})) &= 0 && (66)\\
\mathbf{V}_{\alpha} &= \frac{1}{B_{\alpha}}\mathbf{T}^t.\boldsymbol{\alpha} : \mathbf{X} && (67)\\
\mathbf{V}_{\beta} &= -\frac{1}{B_{\beta}}\mathbf{T}^t.\boldsymbol{\beta} : \mathbf{X} && (68)\\
\dot{\mathbf{U}}_p^{\parallel} &= \boldsymbol{\alpha} \times \mathbf{V}_{\alpha}. && (69)
\end{aligned}
$$

Here, nonlinearity of the geometrical setting is imposed but linearity of elasticity is assumed in Equation (63) for the sake of simplicity. The relations (67) and (68) reflect the dependence of the crystal defect velocities $(\mathbf{V}_{\alpha}, \mathbf{V}_{\beta})$ on the respective driving forces [13]. (B_{α}, B_{β}) are positive drag parameters through which the physics of dislocation and disconnection motion can be prescribed. In such a framework, climb of dislocations is allowed, as well as pressure-dependence of the plastic distortion rate.

As in Section 4, the boundary conditions are prescribed on the compatible displacement field, $\mathbf{u}^{\parallel}$, and the traction vector at the external surfaces of the body $\mathcal{B}$:

$$\mathbf{u}^{\parallel} = \mathbf{u}(P) - \mathbf{u}^{\perp}(P), \quad \forall P \in \partial\mathcal{B}_u \tag{70}$$

$$\mathbf{C} : \epsilon_e^{\parallel}.\mathbf{n} = \mathbf{F}_{\partial\mathcal{B}_t}(P) - C : \epsilon_e^{\perp}.\mathbf{n}, \quad \forall P \in \partial\mathcal{B}_t. \tag{71}$$

However, Equations (70) and (71) differ substantially from the boundary conditions (44) and (45) in the elasto-static problem, because $\mathbf{u}^{\parallel}$ and $\mathbf{U}^{\parallel}$ now include an evolving plastic part. Boundary conditions on the dislocation density tensor $\boldsymbol{\alpha}$ may be additionally prescribed, particularly on the crack surfaces where $\boldsymbol{\alpha}$ may also vanish in non-cohesive regions, but no boundary condition is imposed on the disconnection density tensor $\boldsymbol{\beta}$. The unknowns are the compatible displacement field and the dislocation and disconnection fields.

7. Disconnection Nucleation

Thermally-Activated Crack Nucleation in Elastic-Brittle Materials

The current understanding of crack nucleation and growth is based on ideas first initiated by [18]. Following this original presentation, let us recall that creating a circular crack of radius l in a perfect elastic solid under a tensile stress σ perpendicular to the crack costs an amount of energy of the order of $2\gamma\pi l^2$, if γ is the energy needed to break atomic bonds and generate new surfaces over a unit area. The stored elastic energy density in the solid being $\sigma^2/2E$ with E as Young's modulus, this crack allows releasing the stored elastic energy $-(4\pi l^3/3)(\sigma^2/2E)$ from the volume $4\pi l^3/3$. The total energy variation in nucleating a crack of length l is therefore:

$$W = 2\gamma\pi l^2 - \frac{2\pi l^3\sigma^2}{3E}. \tag{72}$$

Crack initiation shall therefore result from a competition between elastic energy relaxation and atomic bond stretching. When l is varied, W goes through the maximum:

$$W_m = \frac{8\pi\gamma^3E^2}{3\sigma^4} \tag{73}$$

for $l = l_c = 2\gamma E/\sigma^2$. Increasing the crack length costs a positive amount of energy when $l < l_c$, but releases a negative amount if $l > l_c$, which fuels crack growth. The critical condition that the crack may extend therefore occurs for $l = l_c$ at the maximum of energy, because crack growth becomes sustained beyond this point. Yet, crack growth may occur when $l < l_c$ if thermal fluctuations of the lattice activate the phenomenon. Then, according to the Arrhenius law, the lifetime t_N of the undamaged solid is such that:

$$\frac{1}{t_N} = \nu_N = \nu_0 exp(-\frac{W_m}{kT}) = \nu_0 exp(-\frac{8\pi\gamma^3E^2}{3kT\sigma^4}) \tag{74}$$

where ν_N is a nucleation rate, ν_0 a reference rate, k the Boltzmann constant, T the temperature and W_m, as given by Equation (73), appears as an energy barrier for the nucleation process [19]. Careful experiments on wood and fiberglass samples confirmed the lifetime vs. applied stress dependence $ln\, t_N \sim \sigma^{-4}$ implied by Equation (74) [31], suggesting that crack growth is indeed a thermally-activated phenomenon. However, under realistic stress and temperature conditions, Equation (74) predicts exceedingly large nucleation times t_N, which points rather to irrelevance of the mechanism. The conundrum can be solved by assuming the presence of pre-existing defects, whose existence eases crack growth by lowering the energy barrier. Indeed, thermal fluctuations need only to augment a pre-existing defect of radius l_0 to the critical value l_c, rather than create a crack of radius l_c from scratch. The energy barrier is consequently $W_m - W_0$, where:

$$W_0 = 2\gamma\pi l_0^2 - \frac{2\pi l_0^3\sigma^2}{3E} \tag{75}$$

is positive for $0 < l_0 < l_c$. The nucleation time t_N therefore becomes:

$$t_N = \nu_0^{-1} exp(\frac{2\pi}{kT}(\frac{4\gamma^3 E^2}{3\sigma^4} + \frac{\sigma^2 l_0^3}{3E} - \gamma l_0^2)), \tag{76}$$

which can be shown to take realistic values at room temperature, while not altering the correct lifetime vs. stress dependence [20]. Both ab initio and molecular dynamics simulations further confirm that microcracks may be created from thermally activated fluctuations of the atomic order in diamond or silicon crystals otherwise perfect [32–34]. Such crystal defects were shown to be at the origin of the brown coloration of diamonds [33]. In silicon, the nucleation of microcracks could be responsible for the strong chemical reactivity and electrical activity in the wake of moving dislocations [34]. Molecular dynamics simulations further suggest that microcracks may be formed in a perfect two-dimensional elastic crystal in tension at high strain rate, in a direction normal to the tensile axis, as a result of a competition between stress relaxation and atomic bond stretching [32]. Note that a common feature of microcrack nucleation in diamond and silicon is the concomitant nucleation of dislocation loops.

Despite successful predictions regarding the sample lifetime under stress, the thermal activation model for crack initiation does not provide the dynamics of the microcrack size l. Assuming a modified Langevin-type dynamics $\dot{l} = -M\partial W/\partial l + \eta(t)$ where M is a phenomenological mobility constant and $\eta(t)$ a thermal noise contribution, [35] arrives at a lifetime vs. applied stress dependence at variance with the experimental data [31], which tends to rule out diffusive dynamics. Alternatively, by describing microcracks as disconnection loops, the present continuous modeling scheme provides a natural dynamics for crack growth, governed by the transport law (66) appended with the mobility relationship (68) between the driving force and the disconnection velocity. In order to investigate microcrack nucleation and growth in a simple setting by using this framework and in analogy with the calculations of [32], let us consider a sample loaded in simple tension at constant driving strain rate $\dot{\epsilon} > 0$ along the loading direction $\mathbf{e}_2$, with half-width L in the transverse direction $\mathbf{e}_1$. Plane strain parallel to the plane $(\mathbf{e}_1, \mathbf{e}_2)$ is assumed. Again, the material response is taken to be isotropic linear elastic, characterized by Young's modulus E and Poisson's ratio ν. The sample is initially free of cracks, but we look for conditions possibly allowing the nucleation of edge disconnections $\beta_{23}(x_1, t)$ with transverse velocity $\mathbf{V} = V_1\mathbf{e}_1$ with respect to the material, i.e., for conditions possibly allowing Mode II crack nucleation and growth. For such disconnections, the transport Equation (54) reduces to:

$$\frac{\partial\beta}{\partial t} + \frac{\partial}{\partial x}(\beta(v + V)) = 0, \tag{77}$$

when the material velocity $\mathbf{v} = v_1\mathbf{e}_1 + v_2\mathbf{e}_2$ with respect to the reference frame is accounted for, and where the indices in α, v, V and x have been omitted. For the sake of simplicity of the model, it is provisionally assumed that the edge disconnection velocity V with respect to the material is:

$$V = V_0\frac{\beta}{|\beta|} = -V_0\, sgn(\beta),\ \beta \neq 0 \tag{78}$$

$$V = 0,\ \beta = 0, \tag{79}$$

where V_0 is a constant velocity. Thus, no account is made of effects of the stress state on the disconnection velocity. It is then straightforward to show that the homogeneous distribution: $\forall(x,t)$, $\beta_0(x,t) = \beta_0(t) = 0$ with $\epsilon = \dot{\epsilon}t$ is solution to Equation (77). Clearly, this fundamental solution reflects the continuous deformation of a perfect elastic crystal without cracks. To analyze the stability of this solution, we look for inhomogeneous solutions $\beta(x,t)$ to Equation (77) built by adding an inhomogeneous perturbation $\delta\beta$ to $\beta_0(t)$: $\beta(x,t) = \beta_0(t) + \delta\beta(x,t)$. With the above Griffith–Pomeau analysis in mind, such perturbations may be thought of as arising from thermal fluctuations of the atomic order in the lattice. In the homogeneous state: $\beta_0(t) = 0$, and the associated velocity is $V = 0$ according to relation (79). The disconnection perturbation $\delta\beta$ is necessarily non-zero and the sign of the associated velocity $\pm V_0$ is opposed to its sign, according to Equation (78). Substituting the inhomogeneous distribution $\beta(x,t) = \beta_0(t) + \delta\beta(x,t)$ into Equation (77) then leads to the evolution equation for $\delta\beta$:

$$\frac{\partial\delta\beta}{\partial t} + \frac{\partial}{\partial x}((v+V)\delta\beta) = 0. \tag{80}$$

The inhomogeneous perturbations $\delta\beta$ are looked for in the spectral form:

$$\delta\beta_k = \delta\hat{\beta}_k e^{\lambda_k t} e^{i\xi_k x} \tag{81}$$

where $(\delta\hat{\beta}_k, \lambda_k, \xi_k)$ are constants. $\xi_k = 2\pi k/L$ is to be interpreted as the perturbation wave number per unit length. λ_k is a (possibly complex) perturbation growth rate and $\delta\hat{\beta}_k$ an initial lattice defect. Using this development in Equation (80), it is found that:

$$(\lambda_k + i\xi_k(v+V) - \nu\dot{\epsilon})\delta\beta = 0. \tag{82}$$

Thus, non-vanishing perturbations $\delta\beta$ are allowed when the eigenvalue λ_k is:

$$\lambda_k = \nu\dot{\epsilon} - i\xi_k(v+V). \tag{83}$$

The real part of λ_k: $Re(\lambda_k) = \nu\dot{\epsilon}$ is the growth-rate of the perturbation. It is positive in tension ($\dot{\epsilon} > 0$), irrespective of the wave number k, implying that all perturbation modes grow and that the fundamental state $\beta_0(t)$ is unstable. In addition, the perturbation growth rate increases with the strain rate $\dot{\epsilon}$, and this result holds true for all initial amplitudes $\delta\hat{\beta}_k$. The resulting disconnection density modes are in the form:

$$\beta_k(x,t) = \delta\hat{\beta}_k e^{\nu\epsilon} e^{-i\xi_k(-\nu\dot{\epsilon}x+V)t} e^{i\xi_k x} \tag{84}$$

where $\epsilon = \dot{\epsilon}t$ denotes the compatible tensile strain achieved since the initial time and $v = -\nu\dot{\epsilon}x$ is the transverse material velocity. Re-arranging and extracting the real part, it is found that:

$$\beta_k(x,t) = \delta\hat{\beta}_k e^{\nu\epsilon} cos\, 2\pi k \frac{X - Vt}{L}, \tag{85}$$

where $X = (1+\nu\epsilon)x$. The meaning of Equation (85) is that a disconnection density spectral component emerges from arbitrarily small thermal fluctuations in the tension of the sample and that the nucleated disconnections travel away from the material point X at velocity $\pm V_0$ with respect to the material. Since positive and negative values of $\delta\hat{\beta}_k$ are equally admissible, dipoles of edge disconnections are actually nucleated. The latter are the traces along the x-axis of disconnection loops representing slit cracks in a three-dimensional setting, and the outward motion of the edge dipoles at velocity $\pm V_0$ reflects the propagation of these cracks. Further, nucleation may be localized in a small area patch by linearly combining all the spectral modes, because all wave numbers k are equally admissible:

$$\beta(x,t) = e^{\nu\epsilon} \sum_{k=1}^{\infty} \delta\hat{\beta}_k cos\, 2\pi k \frac{X - Vt}{L}. \tag{86}$$

The origin for disconnection growth can be traced to shrinkage of the circuit C of Equation (14) due to Poisson's effect in tension and to conservation of the COD **f** in this motion, in the absence of any incoming/outgoing disconnection flux. Thus, thermal fluctuations of the atomic arrangement may be converted into expanding disconnection loops if the applied loads provide adequate conditions.

In order to account for the relaxation of the stress field accompanying crack nucleation, we now evaluate the incompatible strain field arising from the disconnection distribution (86) in a unidimensional approximation. The incompatible distortion satisfies Equation (60), which reduces here to $U^{\perp}_{22,1} = \epsilon^{\perp}_{22,1} = -\beta_{23}$ or, using Equation (86):

$$\frac{\partial \epsilon^{\perp}}{\partial x} = -e^{\nu\epsilon} \sum_{k=1}^{+\infty} \delta\hat{\beta}_k cos\, 2\pi k \frac{X - Vt}{L}. \tag{87}$$

Assuming all initial defects to be identical: $\forall k,\ \delta\hat{\beta}_k = \delta\hat{\beta}$, we find:

$$\epsilon^{\perp} = -\frac{L\delta\hat{\beta}}{2\pi} \frac{e^{\nu\epsilon}}{1+\nu\epsilon} \sum_{k=1}^{+\infty} \frac{sin\, 2\pi k \frac{X-Vt}{L}}{k} = -L\delta\hat{\beta} \frac{e^{\nu\epsilon}}{2(1+\nu\epsilon)} (\frac{1}{2} - \frac{X - Vt}{L}), \tag{88}$$

provided $X - Vt < L$ (see Item 1.441.1 in [36]). A first-order approximation of $\epsilon^{\perp}$ valid in the close vicinity of the nucleation point is therefore:

$$\epsilon^{\perp} \cong -\delta\beta_0 \frac{e^{\nu\epsilon}}{1+\nu\epsilon} \tag{89}$$

where $\delta\beta_0 = L\delta\hat{\beta}/4$ is an initial nondimensional disconnection density. For linear elasticity, the tensile stress accounting for both the compatible and incompatible strain fields is therefore (see Equation (63)):

$$\sigma = E(\epsilon - \delta\beta_0 \frac{e^{\nu\epsilon}}{1+\nu\epsilon}). \tag{90}$$

Clearly, σ first increases with strain up to a maximum and eventually drops down to zero. The elastic energy density $(\epsilon^{\perp})^2/2E$ resulting from the incompatible elastic strain $\epsilon^{\perp}$ appears as the translation in the present continuous analysis of the surface energy of Griffith's model, while the elastic energy density $\epsilon^2/2E$ in the latter is actually the energy arising from the compatible elastic strain energy in the former. Thus, the condition for the maximum tensile stress in the present model corresponds to Griffith's criterion. In addition, the present analysis yields a dynamics of crack growth, which develops by transport from the less-than-critical thermally-activated fluctuations of the atomic order. If, however, it is now demanded that the dislocation velocity V_0 decrease with the tensile stress and vanish below some positive threshold stress according to some Griffith-type criterion, then crack growth eventually stops and the tensile stress σ remains positive. However, the present unidimensional analysis cannot capture this situation, for which numerical simulations are necessary. The present conclusions are very similar to the results of the atomistic simulations [32–34], although the methods used are clearly different. It must be noticed further that similar conclusions may be obtained from a parallel treatment of the nonlinear dislocation transport Equation (57), implying that there is an equal chance to nucleate edge dislocations loops in similar numbers in the process. This remark might help to explain the concomitant presence of microcracks and dislocation loops mentioned above.

8. Application to DIC Methods and Strain Localization-Induced Fracture in Al-Cu-Li Alloys

In this section, we propose that our theoretical framework can be of practical use to complement experimental analysis of links between strain localization and fracture initiation in materials, in terms of strain incompatibility and associated disconnection densities. In particular, targeted applications are DIC-based methods, including standard techniques that use painting speckles [21], as well as more complex three-dimensional or high-resolution techniques that can use small grids or precipitates [22,23,37,38]. The idea is that the evolution of spatial gradients of the material distortion field, which are involved in the estimation of the disconnection densities, can help in predicting the occurrence of cracks in relation to strain localization. We expect in particular that our framework can help in establishing links between strain/strain rate distributions and corresponding crack modes and profiles. Alternatively, it may also be used to complement experimentally validated damage/fracture models. In this work, we propose to apply our disconnection framework to standard two-dimensional DIC data obtained during tensile quasi-static loading of Al-Cu-Li alloys. As aforementioned, these alloys are prone to complex interactions between localized plasticity and damage evolution leading to failure. This study can be considered as a benchmark for our modeling framework. Before describing the association of crack nucleation/characteristics with different components of disconnection densities, we will first explain below the experiments performed, the standard DIC technique used, and the procedure involved in the derivation of disconnection density maps from DIC data.

8.1. DIC Setup and Estimation of Disconnection Densities from DIC Data

Digital image correlation (DIC) is a non-contacting optical technique for the full-field deformation measurements [39,40]. Standard DIC methods, such as the one used in the present work, allow deriving the strain field on a sample surface as per following the evolution of a painting speckle pattern deposited on the sample surface by spray. Such methods have several advantages, such as no particular surface preparation needed, mostly implemented in situ and minimum setup requirement. In the present case, DIC has been applied with the objective to measure the local strain fields, not only all along the deformation test, but also and more importantly just before material failure. DIC methods devoted to fracture already exist in the literature, where special care is to be taken to treat the discontinuity of displacement field at crack faces [41–44]. The method we propose here rather consists of integrating the evolution with strain of some skewed disconnection density components, as detailed below, up to failure, but not at the time of failure. Eventually, we can analyze if the skewed disconnection densities correlate with the fracture of the sample and if they can help in interpreting/anticipating fracture profiles and crack modes in relation to strain/strain rate distributions and evolutions. The 2D DIC setup used for the purpose consists of an imaging source CCD camera (AVT Pike F-421B) with a resolution equal to 2048 × 2048 pixels, a Myutron lens with a 50-mm focal length to control focus, zoom and f-stop and an artificial speckle pattern on the specimen to be examined. Speckle patterns employed can be seen in the Figures 2 and 3. Digital images were captured at a frame rate of 5 to 10 Hz (depending on the strain rate) during the tension tests, and thereafter, post-processing of the images was done using VIC2D commercial software from correlated solutions in order to get the pixel by pixel strain, strain rate and displacement maps at the specimen surface.

As an extended use of DIC, strain rate maps are used as an input to evaluate the distortion rate (symmetric part is strain rate) and thereby some skewed disconnection density component rates, during loading and just before crack nucleation in ductile materials. The compatibility condition (Equation (9)) holds until the onset of a displacement discontinuity; this is the point where a non-vanishing disconnection density field is expected to emerge in accordance with Equation (15). Our intent is to demonstrate the ability of such an approach to characterize the nucleation of a crack. To this end, we consider a flat surface of a dog bone-shaped tensile specimen loaded along direction $\mathbf{e}_2$, in the plane $(\mathbf{e}_1, \mathbf{e}_2)$ of the orthonormal frame $(\mathbf{e}_1, \mathbf{e}_2, \mathbf{e}_3)$. From the DIC recorded data, we can use the in-plane distortion rate fields $(\dot{U}_{11}, \dot{U}_{12}, \dot{U}_{21}, \dot{U}_{22})$ of the surface on a grid (if shear distortion rates are not available they can be evaluated from spatial derivatives of displacement rates), from which the rate of disconnection densities $\dot{\beta}_{13} = \dot{U}_{11,2} - \dot{U}_{12,1}$ and $\dot{\beta}_{23} = \dot{U}_{21,2} - \dot{U}_{22,1}$ may be numerically evaluated. We did it with a very simple MATLAB script that uses DIC mat files generated by VIC2D software. In the absence of a displacement rate discontinuity, $\dot{\beta}_{13} = \dot{\beta}_{23} = 0$, according to the rate form of Equations (8) and (9). However, to anticipate and detect crack nucleation, i.e., the advent of a non-zero disconnection density distribution, we slightly skew the curl operator in Equation (15) and define instead the modified rate of disconnection densities $\dot{\beta}'_{13} = (1+\epsilon)\dot{U}_{11,2} - (1-\epsilon)\dot{U}_{12,1}$ and $\dot{\beta}'_{23} = (1+\epsilon)\dot{U}_{21,2} - (1-\epsilon)\dot{U}_{22,1}$, where $\epsilon << 1$. As a result, the skewed-disconnection densities $(\beta'_{13}, \beta'_{23})$, integrated over time during the deformation of the material, are non-vanishing even if $\beta_{13} = \beta_{23} = 0$. The amplitude of disconnection densities (to avoid the confusion, skewed disconnection densities β'_{13} and β'_{23} are presented as disconnection densities β_{13} and β_{23} itself), as well as their sign both depend on the artificial parameter ϵ, and no quantitative conclusion will be drawn in this paper (strong uncertainty in displacement and strain fields as obtained from the DIC analysis also prevents from providing a reliable quantitative description of fracture). However, we can still expect that monitoring the variations of the disconnection densities will provide very valuable qualitative information on the emergence of a crack as the sample deformation proceeds. Furthermore, we can expect that comparison of relative weights and distributions of available disconnection density components can provide information on emerging crack modes (only Mode I and Mode II in 2D).

To test our extended DIC post-treatment analysis, we used tensile specimens extracted from Al-Cu-Li AA-2198-T8 alloys. Experimental data on such alloys can be found in the literature [22,45]. This material was industrially processed into sheets of thickness 2.5 mm. The processing routes involved rolling and several thermo-mechanical heat-treatments leading to a multi-layered morphological texture with grains strongly elongated in the rolling direction L. Machining may have been additionally used to reduce further the sheet thickness to 1.1 mm from the initial thickness, while keeping the grain thickness unchanged. Thus, the thickness, width and gauge-length of the tested specimens were (2.5 mm, 20 mm, 120 mm) and (1.1 mm, 20 mm, 120 mm). In each case, samples have been loaded in both the rolling L and transverse T direction. Failure of the samples occurred while the load was decreasing after the nominal ultimate tensile strength (UTS), and moderate necking was observed, particularly when loading in the T direction.

8.2. Analysis of Sample Fracture Using Disconnection Densities

Predominantly, the specimens fractured in a mixed mode configuration (both Mode I and Mode II are present). Mixed mode resulted in a crack along slanted bands at an angle with the loading direction. In a first case shown in this paper, the fracture band is rotated around both directions normal to the loading direction (Figure 3). In a second case also shown in the paper, the fracture band is only inclined in the thickness direction (Figure 4). Although it is a mixed mode, we refer to that last case as a Mode I dominated fracture, as compared to the first case. Indeed, the interpretation of this inclination in the thickness direction, in terms of disconnection densities, would require 3D DIC data, which is not available in the present work. As such, with the 2D analysis presented hereafter, we can only describe the Mode I fracture component in the second sample. For this particular alloy and the tensile tests performed here, the crack initiation mechanism was not investigated. Damage evolution and eventual fracture might be the result of strain accumulation induced void growth [22]. Void nucleation may occur at micrometric intermetallic particles after sufficient strain [23]. Grain boundary decohesion might also be involved because of the strongly rolled microstructure [45]. Before starting to explore the connection between fracture nucleation/characteristics and disconnection theory, it is necessary to understand how different fracture modes can be represented in terms of disconnection density dipoles and loading direction [13]. Following the dislocation-based fracture mechanics framework [11], one can characterize cracks in the DIC observation domain in terms of disconnection density dipoles, which manifest opening and closing of a full crack. When the crack face is perpendicular to the tension axis, it is known as a Mode I crack and the latter can be represented a dipole of β_{23} components with the disconnection densities of opposite sign being located at the crack tips and delimiting the crack. This case is well described in a recent paper [13]. This case is shown in Figure 2a. However, if such a dipole was vertical in the figure and the sample was loaded in shear (in plane shear), it would correspond to a Mode II crack (we are leaving out the Mode III for now as we have only access to 2D data). In the present tension tests, Mode II cracking can still be observed as a resultant of combined β_{23} and β_{13} disconnection dipoles, as shown in Figure 2b. We have analyzed several specimens, which were fractured fortunately in the DIC observation window, and among them, two representative cases are discussed here.

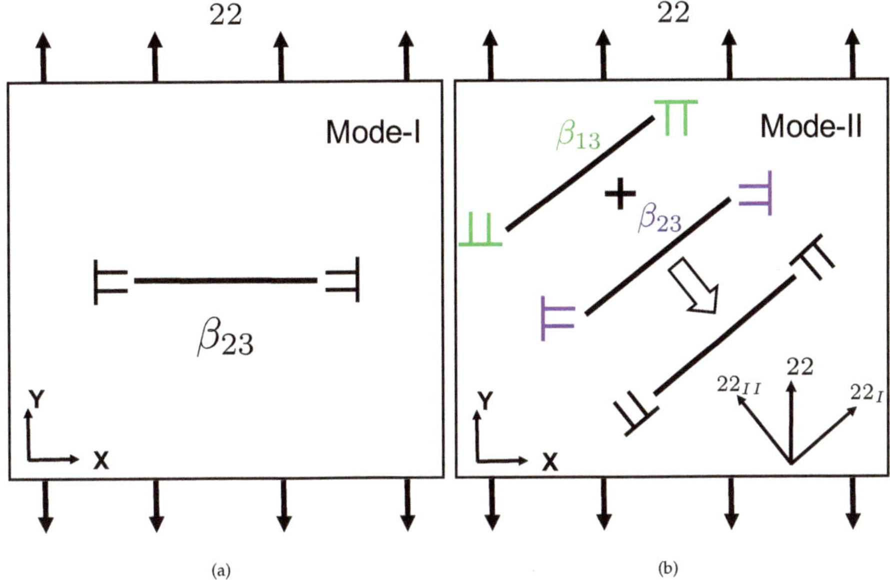

Figure 2. Representation of different fracture modes in terms of disconnection dipoles and loading direction: (**a**) Fracture Mode I; (**b**) Fracture Mode II.

8.3. Mixed Mode Fractured Specimen

This specimen (B2TL5) was tested in the uniaxial tension along transverse direction at a constant strain rate 1×10^{-3}/sec, the global stress strain curve being presented in Figure 3b. We are only interested in the characteristics of the fracture and therefore, DIC maps are presented only for the time frame just before the fracture took place. Figure 3 shows the fracture pattern of the specimen and associated DIC maps for strain rate, strain and disconnection components fetched from the DIC post-treatment analysis. As shown in Figure 3a, the specimen was broken in a slanted manner which can be referred to as mixed mode fracture and can be explained on the basis of maps presented in Figure 3c. Both fracture modes described in Figure 2 (regarding disconnection distribution resulting in Mode I and Mode II) fit well for the β distribution obtained in the present fracture case. Indeed one can see first an horizontal dipole of β_{23}, which corresponds to Mode I, and also an inclined dipole made of both β_{13} and β_{23}, which corresponds to Mode II fracture (dipoles are represented by blue and red boxes on β distribution maps). Looking at the norm of disconnection density tensor $||\beta||$, one can clearly anticipate that the upcoming slanted mixed mode fracture of the sample. Although this slanted profile is hardly visible in the accumulated strain distribution, it can be clearly identified in the disconnection density maps, which correlate also well with the strain rate map.

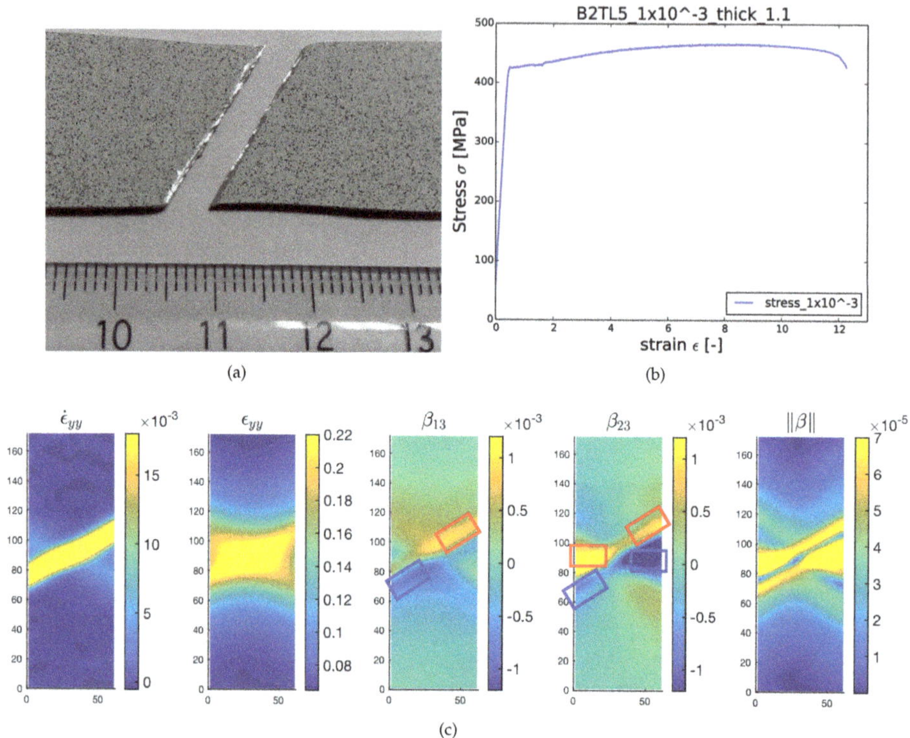

Figure 3. Fracture characteristics of the specimen B2TL5 broken in mixed mode (under uniaxial tension): (**a**) slanted fracture of the sample B2TL5; (**b**) stress/strain curve; (c) DIC maps for strain (no unit), strain rate (/s) and disconnection density fields (arbitrary units) evaluated just before the fracture.

8.4. Mode I Dominated Fractured Specimen

The specimen (B1L6) was tested in uniaxial tension along rolling direction at a constant strain rate 1×10^{-3}/sec, the global stress strain curve being presented in Figure 4b. As mentioned earlier, due to the sole interest in fracture, DIC maps presented here are only associated with the time frame just before the actual fracture took place. Figure 4 demonstrates fracture characteristics of the broken specimen, including fracture pattern and associated DIC maps for strain rate, strain and disconnection density calculated from the DIC analysis. Figure 4a shows that the specimen was broken perpendicular to the loading direction, which can be referred to as Mode I-dominated fracture. As a general observation in all our tests not shown here, it is found that tension tests in transverse direction resulted primarily in mixed mode fracture whereas rolling direction favored the fracture dominated by Mode I with few exceptions. The fracture observed for the B1L6 sample can be explained well with the help of β distribution maps shown in Figure 4c. Analogous to B2TL5, dipoles of β_{13} and β_{23} are also visible in this case (they are again marked with red and blue boxes on β distribution maps). However, the horizontal dipole of β_{23} reflecting Mode I is observed to be much stronger in magnitude than the dipoles manifesting Mode II. In this sample, two inclined dipoles of lower magnitude and corresponding to Mode II fracture can be seen in the maps (blue and red boxes with thinner lines than boxes corresponding to Mode I horizontal dipole). They are disposed in a symmetric manner with respect to the sample and loading direction, as can be seen in the norm of the disconnection density tensor. The latter norm clearly suggests that Mode I fracture will be predominant for this sample. It also suggests that the inclined dipoles corresponding to Mode II are canceling out because of symmetry. Similarly to the B2TL5 sample, the fracture pattern is in good agreement with the distributions of disconnection densities. Here however, while the fracture profile shows also good correlation to the strain rate map for the B2TL5 sample, it does not really fit for the B1L6 sample, and the disconnection density analysis seems to be the most predictive quantity.

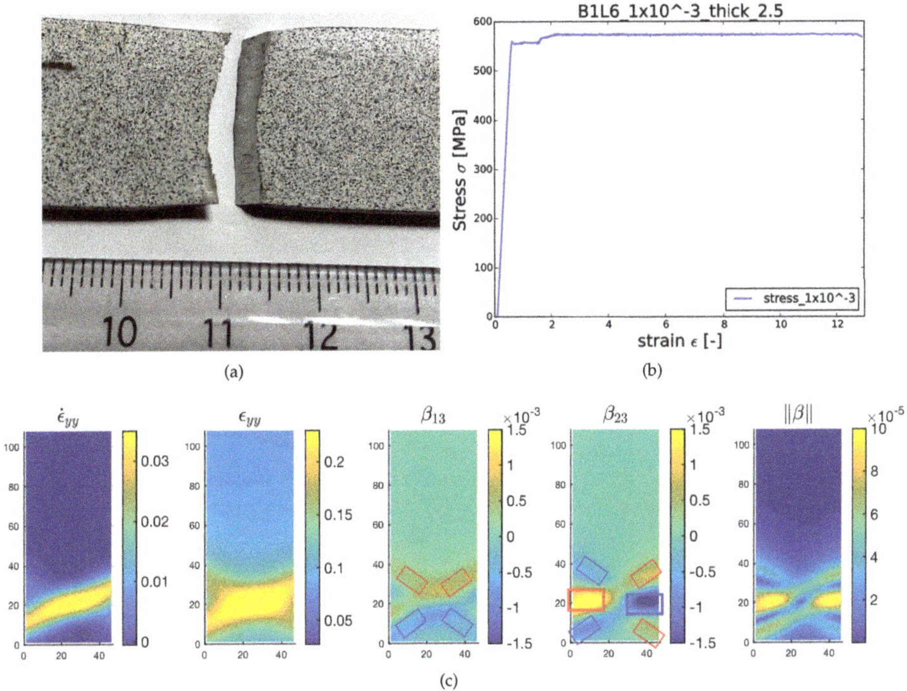

Figure 4. Fracture characterization of the specimen B1L6 predominantly failed in Mode I: (**a**) fracture pattern of broken specimen; (**b**) stress/strain curve; (**c**) digital image correlation (DIC) maps for strain (no unit), strain rate (/s) and disconnection density fields (arbitrary units) evaluated just before the fracture.

9. Conclusions

We proposed a geometrically nonlinear theory of coupled plasticity and fracture in crystalline materials dealing with the statics and dynamics of smooth crystal defect density fields, namely dislocation and disconnection densities, as introduced in [13]. While dislocations are line defects terminating surfaces across which the plastic displacement field encounters a discontinuity, disconnections are defined as line defects terminating surfaces across which the total displacement faces a discontinuity, thus reflecting fracture of the body. The paper builds on the duality existing between the discontinuity of the displacement field and the incompatibility (non-integrability) of the associated smooth distortion field to introduce a field theory of fracture. While the incompatibility of the plastic distortion lends its meaning to the tensorial density of dislocations $\boldsymbol{\alpha}$, the disconnection density tensor $\boldsymbol{\beta}$ acquires its topological significance from the incompatibility of the total distortion. When the latter remains compatible, i.e., when it reduces to a gradient tensor, continuity of the body is maintained, and the present theory reduces to the mechanics of dislocation fields [46]. The achievements of the Peierls model in elucidating basic dislocation physics testifies to the appeal of field descriptions for the analysis of line defects. Such attractiveness lies in that, when viewed at a sufficiently small scale, line defects are better described by a suitably-localized smooth density field than by a singularity. Smoothness of the description is not only desirable from the point of view of mathematical analysis and numerical computation, but also because it allows coping with core properties. Beyond regularizing the description of individual cracks, disconnections may also be useful, if a larger resolution length scale is chosen, in describing damage distributed throughout the body.

Further interest in considering disconnection density fields for fracture modeling arises from their conservation laws, which are unquestionable from a kinematic point of view and, hence, provide a natural framework for their dynamics and consequently for crack growth. This feature is used in the present paper to introduce a set of partial differential equations (PDEs) for the dynamics of the compatible displacement and dislocation/disconnection density fields, reflecting respectively dislocation motion and crack growth. A prominent property of the theory is therefore that energy dissipation may occur concurrently through crack growth and dislocation motion, both in the bulk of the body and along the crack surfaces. The Peach–Koehler force on dislocations, as well as the driving forces for disconnection motion and crack surface separation, are found as dual quantities to the respective defect velocities in the dissipation. Eventual constitutive choices include a Griffith-type criterion for disconnection transport (fracture growth), as well as a cohesive-type behavior for the crack opening displacement. While being based on PDEs in the set of variables involved, i.e., the compatible components of the total displacement, the dislocation and disconnection densities, the present theory is nonlocal in space and time in the standard variables of conventional continuum mechanics. Thus, it provides nonlocal generalization of the latter, leading to well-posed problems in dislocation and fracture dynamics. Generating approximate solutions to the present theory, while being non-trivial, leads to finite-element or spectral FFT-based implementations similar to those already employed in the pure dislocation dynamics case in [29,47] and [48,49]. However, a fundamental difference between disconnections and dislocations lies in that the stress field of disconnections is dependent on the actual location of the surface of displacement discontinuity, because the traction vector vanishes along this surface, whereas the stress field of dislocations does not depend on this position, but only on that of the dislocation line. Hence, the history of the surface of discontinuity must be kept track of in implementing disconnection dynamics.

As an illustration of the effects of nonlinearity, thermally-activated crack nucleation and growth were evidenced in this paper by using the disconnection density concept in elastic-brittle materials. It is revealing that the elastic energy density arising from the incompatible elastic strain field in the present theory plays the role of the surface energy density associated with atomic bond breaking in Griffith's theory of crack nucleation, while the elastic energy density in the latter is actually the energy density arising from the compatible elastic strain field in the former. Here, crack nucleation appears as a natural consequence of the geometrical nonlinearity of transport: an initial whiff of disconnection density due to thermal agitation of the atomic lattice may grow if the deformation process tends to shrink the surrounding material, because the COD content of elementary patches must be conserved. Eventually, the spatio-temporal dynamics of the disconnection density fluctuation is also governed by transport, which results in further propagation of the nascent crack.

Finally, the DIC disconnection density-based analysis of tensile tests performed on ductile Al-Cu-Li samples not only validates the disconnection density framework, but it also shows that disconnection densities can help to anticipate and to describe crack nucleation and morphology in relation to strain and strain rate distribution and evolution. For some samples tested in this work, we observed fracture profiles that were in good agreement with disconnection density maps, but less with strain and strain rate maps. Further work on Al-Cu-Li alloys, consisting of using disconnection densities for modeling strain localization induced fracture, is under progress. We also think that our DIC disconnection extension can provide a valuable and complementary analysis of fracture. Importantly, it is very easy to implement in a DIC analysis code, and the methodology could be applied to recent high-resolution DIC techniques [37,38] and to 3D DIC data [22,23] for a full 3D description of cracks and a fine resolution scale analysis of fracture in relation to strain localization.

Acknowledgments: The authors greatly acknowledge financial support from Constellium under Research contract Frac-Size, from the French State through the National Research Agency under the program Investment in the future (Labex DAMAS referenced as ANR-11-LABX-0008-01) and the Region Grand-Est. The authors express their gratitude to Dr. Juliette Chevy and Prof. Emeritus Armand J. Beaudoin for valuable discussions.

Author Contributions: C.F. developed and wrote the first theoretical part of the paper. S.G. and M.J. performed the tensile tests with DIC. V.T. and S.G. wrote the second experimental part of the manuscript, with feedback from M.J.

References

1. Eshelby, J.D. The determination of the elastic field in an ellipsoidal inclusion, and related problems. *Proc. R. Soc. Lond. A* **1957**, *241*, 376–396.
2. Friedel, J. *Dislocations*; Pergamon Press: Oxford, UK, 1964.
3. Hirth, J.P. Dislocations, steps and disconnections at interfaces. *J. Phys. Chem. Solids* **1994**, *55*, 985–989.
4. Kondo, K. *RAAG Memoirs of the Unifying Study of the Basic Problems in Engineering Sciences by Means of Geometry*; Unifying Study Group: New York, NY, USA, 1955.
5. Marcinkowski, M.J. The differential geometry of fracture. *Acta Mech.* **1978**, *30*, 175–195.
6. Nabarro, F.R.N. *Dislocations in Solids*; Oxford University Press: Oxford, UK, 1967.
7. Barnett, D.M.; Asaro, R.J. The fracture mechanics of slit-like cracks in anisotropic elastic media. *J. Mech. Phys. Solids* **1972**, *20*, 353–366.
8. Bilby, B.A.; Cottrell, B.H.; Swinden, K.H. The spread of plastic yield from a notch. *Proc. R. Soc. Lond. A* **1965**, *285*, 22–23.
9. Hills, D.A.; Kelly, P.A.; Dai, D.N.; Korsunsky, A.M. *Solution of Crack Problems: The Distributed Dislocation Technique*; Springer: Dordrecht, The Netherlands, 1996.
10. Mura, T. *Micromechanics of Defects in Solids*; Martinus Nijhoff Publishers: Dordrecht, The Netherlands, 1987.
11. Weertman, J. *Dislocation Based Fracture Mechanics*; World Publishing Co.: Singapore, 1996.
12. Kröner, E. *Kontinuumstheorie der Versetzungen und Eigenspannungen*; Ergeb. Agnew. Math. 5; Springer: Berlin, Germany, 1958.
13. Fressengeas, C.; Taupin, V. A field theory of distortion incompatibility for coupled fracture and plasticity. *J. Mech. Phys. Solids* **2014**, *68*, 45–65.
14. Maugin, G.A. *Configurational Forces: Thermodynamics, Physics, Mathematics, and Numerics*; Chapman & Hall/CRC: Boca Raton, FL, USA, 2011.
15. Miyamoto, H.; Kageyama, K. Extension of J-integral to the general elasto-plastic problem and suggestion of a new method for its evaluation. In Proceedings of the 1st International Conference on Numerical Methods in Fracture Mechanics, West Glamorgan, UK, 9–13 January 1978; Luxmoore, A.R., Owen, D.R.J., Eds.; pp. 261–275.
16. Simha, N.K.; Fischer, F.D.; Shan, G.X.; Chen, C.R.; Kolednik, O. J-integral and crack driving force in elastic-plastic materials. *J. Mech. Phys. Solids* **2008**, *56*, 2876–2895.
17. Fressengeas, C.; Taupin, V. A field theory of strain/curvature incompatibility for coupled fracture and plasticity. *Int. J. Solids Struct.* **2016**, *82*, 16–38.
18. Griffith, A.A. The phenomenon of rupture and flow in solids. *Phil. Trans. R. Soc. Lond. A* **1921**, *221*, 163–198.
19. Pomeau, Y. Brisure spontanée de cristaux bidimensionnels courbés *C. R. Acad. Sci. Paris* **1992**, *314*, 553–556.
20. Rabinovitch, A.; Friedman, M.; Bahat, D. Failure time in heterogeneous materials—Non-homogeneous nucleation. *Europhys. Lett.* **2004**, *67*, 969–975.
21. Gupta, S.; Beaudoin, A.J.; Chevy, J. Strain rate jump induced negative strain rate sensitivity (NSRS) in aluminum alloy 2024: Experiments and constitutive modeling. *Mater. Sci. Eng. A* **2017**, *683*, 143–152.
22. Morgeneyer, T.F.; Taillandier-Thomas, T.; Helfen, L.; Baumbach, T.; Sinclair, I.; Roux, S.; Hild, F. In situ 3-D observation of early strain localization during failure of thin Al alloy (2198) sheet. *Acta Mater.* **2014**, *69*, 78.
23. Morgeneyer, T.F.; Taillandier-Thomas, T.; Buljac, A.; Helfen, L.; Hild, F. On strain and damage interactions during tearing: 3D in situ measurements and simulations for a ductile alloy (AA2139-T3). *J. Mech. Phys. Solids* **2016**, *96*, 550–571.
24. Jiang, B. The least-squares finite element method. In *Theory and Computation in Fluid Dynamics and Electromagnetics*; Springer Series in Scientific Computation; Springer: Berlin, Germany, 1998.

25. Fressengeas, C.; Taupin, V.; Capolungo, L. An elasto-plastic theory of dislocation and disclination fields. *Int. J. Solids Struct.* **2011**, *48*, 3499–3509.
26. Lee, E.H. Elastic-plastic deformation at finite strains. *J. Appl. Phys.* **1969**, *36*, 1–6.
27. Acharya, A. Jump condition for GND evolution as a constraint on slip transmission at grain boundaries. *Phil. Mag.* **2007**, *87*, 1349–1359.
28. Garg, A.; Acharya, A.; Maloney, C.E. A study of conditions for dislocation nucleation in coarser-than-atomistic scale models. *J. Mech. Phys. Solids* **2015**, *75*, 76–92.
29. Varadhan, S.; Beaudoin, A.J.; Acharya, A.; Fressengeas, C. Dislocation transport using an explicit Galerkin/least-squares formulation. *Modell. Simul. Mater. Sci. Eng.* **2006**, *14*, 1245–1270.
30. Coleman, B.D.; Gurtin, M.E. Thermodynamics with internal state variables. *J. Chem. Phys.* **1967**, *47*, 597–613.
31. Guarino, A.; Ciliberto, S.; Garcimartín, A. Failure time and microcrack nucleation. *Europhys. Lett.* **1999**, *47*, 456–461.
32. Dias, C.L.; Kröger, J.; Vernon, D.; Grant, M. Nucleation of cracks in a brittle sheet. *Phys. Rev. E* **2009**, *80*, 066109.
33. Hounsome, L.S.; Jones, R.; Martineau, P.M.; Fisher, D.; Shaw, M.J.; Briddon, P.R.; Öberg, S. Origin of brown coloration in diamond. *Phys. Rev. B* **2006**, *73*, 125203.
34. Rabier, J.; Pizzagalli, L. Dislocation dipole annihilation in diamond and silicon. *J. Phys. Conf. Ser.* **2011**, *281*, 012025.
35. Golubovic, L.; Feng, S. Rate of microcrack nucleation. *Phys. Rev. A* **1991**, *43*, 5223–5227.
36. Gradshteyn, I.S.; Ryzhik, I.M. *Table of Integrals, Series and Products*; Academic Press: New York, NY, USA, 1965.
37. Di Gioacchino, F.; da Fonseca, J.Q. Plastic Strain Mapping with Sub-micron Resolution Using Digital Image Correlation. *Exp. Mech.* **2013**, *53*, 743–754.
38. Di Gioacchino, F.; da Fonseca, J.Q. An experimental study of the polycrystalline plasticity of austenitic stainless steel. *Int. J. Plast.* **2015**, *74*, 92–109.
39. Chu, T.C.; Ranson, W.F.; Sutton, M.A.; Peters, W.H. Applications of digital-image-correlation techniques to experimental mechanics. *Exp. Mech.* **1985**, *25*, 232–244.
40. Sutton, M.A.; Cheng, M.Q.; Peters, W.H.; Chao, Y.J.; McNeill, S.R. Application of an optimized digital image correlation method to planar deformation analysis. *Image Vis. Comp.* **1986**, *4*, 143–150.
41. Fagerholt, E.; Rvik, T.B.; Hopperstad, O.S. Measuring discontinuous displacement fields in cracked specimens using digital image correlation to mesh adaptation and crack-path optimization. *Opt. Lasers Eng.* **2013**, *51*, 299–310.
42. Liu, J.; Lyons, J.; Sutton, M.A.; Reynolds, A. Experimental characterization of crack tip deformation fields in alloy 718 at high temperatures. *J. Eng. Mat. Tech. Trans. ASME* **1998**, *120*, 71–78.
43. McNeill, S.R.; Peters, W.H.; Sutton, M.A. Estimation of stress intensity factor by digital image correlation. *Eng. Fract. Mech.* **1987**, *28*, 101–112.
44. Valle, V.; Hedan, S.; Cosenza, P.; Fauchille, A.L.; Berdjane, M. Digital Image Correlation Development for the Study of Materials Including Multiple Crossing Cracks. *Exp. Mech.* **2015**, *55*, 379–391.
45. Chen, J.; Madi, Y.; Morgeneyer, T.F.; Besson, J. Plastic flow and ductile rupture of a 2198 Al-Cu-Li aluminum alloy. *Comput. Mater. Sci.* **2011**, *50*, 1365–1371.
46. Acharya, A. A model of crystal plasticity based on the theory of continuously distributed dislocations. *J. Mech. Phys. Solids* **2001**, *49*, 761–784.
47. Roy, A.; Acharya, A. Finite element approximation of Field Dislocation Mechanics. *J. Mech. Phys. Solids* **2005**, *53*, 143.
48. Berbenni, S.; Taupin, V.; Djaka, K.S.; Fressengeas, C. A numerical spectral approach for solving elasto-static field dislocation and g-disclination mechanics. *Int. J. Solids Struct.* **2014**, *51*, 4157–4175.
49. Djaka, K.S.; Taupin, V.; Berbenni, S.; Fressengeas, C. A numerical spectral approach to solve the dislocation density transport equation. *Modell. Simul. Mater. Sci. Eng.* **2015**, *23*, 065008.

Article

Modeling the Effect of Primary and Secondary Twinning on Texture Evolution during Severe Plastic Deformation of a Twinning-Induced Plasticity Steel

Laszlo S. Toth [1,2], Christian Haase [3,*], Robert Allen [4,2], Rimma Lapovok [5], Dmitri A. Molodov [6], Mohammed Cherkaoui [4] and Haitham El Kadiri [4]

1 Laboratoire d'Etude des Microstructures et de Mécanique des Matériaux (LEM3), Université de Lorraine, CNRS, Arts et Métiers ParisTech, LEM3, 57000 Metz, France; laszlo.toth@univ-lorraine.fr

2 Laboratory of Excellence on Design of Alloy Metals for Low-mAss Structures (DAMAS), Université de Lorraine, 57045 Metz, France

3 Steel Institute, RWTH Aachen University, 52072 Aachen, Germany

4 Department of Mechanical Engineering, Mississippi State University, Starkville, MS 39762, USA; rma140@msstate.edu (R.A.); cherkaou@cavs.msstate.edu (M.C.); elkadiri@me.msstate.edu (H.E.K.)

5 Institute for Frontier Materials, Deakin University, Geelong, Victoria 3217, Australia; r.lapovok@deakin.edu.au

6 Institute of Physical Metallurgy and Metal Physics, RWTH Aachen University, 52074 Aachen, Germany; molodov@imm.rwth-aachen.de

* Correspondence: christian.haase@iehk.rwth-aachen.de; Tel.: +49-241-809-5821

Received: 8 May 2018; Accepted: 18 May 2018; Published: 22 May 2018

Abstract: Modeling the effect of deformation twinning and the ensuing twin-twin- and slip-twin-induced hardening is a long-standing problem in computational mechanical metallurgy of materials that deform by both slip and twinning. In this work, we address this effect using the twin volume transfer method, which obviates the need of any cumbersome criterion for twin variant selection. Additionally, this method is capable of capturing, at the same time, secondary or double twinning, which is particularly important for modeling in large strain regimes. We validate our modeling methodology by simulating the behavior of an Fe-23Mn-1.5Al-0.3C twinning-induced plasticity (TWIP) steel under large strain conditions, experimentally achieved in this work through equal-channel angular pressing (ECAP) for up to two passes in a 90° die following route B_C at 300 °C. Each possible twin variant, whether nucleating inside the parent grain or inside a potential primary twin variant was predefined in the initial list of orientations as possible grain of the polycrystal with zero initial volume fraction. A novelty of our approach is to take into account the loss of coherency of the twins with their parent matrix under large strains, obstructing progressively their further growth. This effect has been captured by attenuating growth rates of twins as a function of their rotation away from their perfect twin orientation, dubbed here as "disorientation" with respect to the mother grain's lattice. The simulated textures and the hardening under tensile strain showed very good agreement with experimental characterization and mechanical testing results. Furthermore, upper-bound Taylor deformation was found to be operational for the TWIP steel deformation when all the above ingredients of twinning are captured, indicating that self-consistent schemes can be bypassed.

Keywords: TWIP steel; ECAP; deformation twinning; texture; VPSC; simulation

---

1. Introduction

High-manganese twinning-induced plasticity (TWIP) steels are promising candidates for application in crash-relevant automobile components due to their outstanding mechanical properties [1,2]. These properties, i.e., high strength, ductility, and work-hardening capacity, originate from their low stacking fault energy (SFE), which is typically in the range between ~20 mJ/m^2 and ~50 mJ/m^2 and enables the activation of deformation-induced twinning in addition to planar dislocation slip, while strongly impeding cross-slip [3]. The high work-hardening rates that are responsible for the excellent deformation behavior are attributed to the combination of strong planarity of dislocation motion and the dynamic Hall-Petch effect that facilitates drastic reduction of the dislocations' mean free paths due to the formation of nano-scale deformation twins [4,5].

Besides their effect on the mechanical properties, the deformation mechanisms active in high-manganese steels have a strong influence on texture evolution. So far, this has been analyzed mainly during cold rolling [6–13], recrystallization [7,14–17], and tensile testing [18–21]. In contrast to the conventional processing of TWIP steels, severe plastic deformation of these alloys has only been sparsely investigated, even though it was found to have a significant effect on their microstructural evolution and mechanical properties [22–26]. In particular, the understanding of the correlation between deformation mechanisms and texture evolution is limited. In our recent study [26], the equal-channel angular pressing (ECAP) method was used to deform an Fe-23Mn-1.5Al-0.3C TWIP steel up to four passes at 300 °C following route B_C. Experimental analysis revealed that the microstructure during ECAP was continuously refined, owing to both grain subdivision via deformation twinning and by dislocation-driven grain fragmentation. Due to the increased SFE at 300 °C, the latter mechanism dominated the microstructure evolution. In accordance with the correspondence introduced by Suwas et al. [27,28], a transition texture is formed that contains texture components characteristic of materials with both high and low SFE. Shear bands, if they occur, appear parallel to the applied shear in shear deformation [29]. As they were hardly observed during ECAP of TWIP steel [26], they were not considered in the present modeling work.

Previously developed methods for modeling the evolution of twin volume fractions and their effects on texture and mechanical hardening begin with the pioneering work of Van Houtte [30], which utilized a statistically based criterion for the selection of parent grains to be reoriented into twin orientations. Later works by Tomé et al. [31] and Lebensohn and Tomé [32] introduced the predominant twin reorientation (PTR) scheme, in which the mother grain is replaced completely by its most active twinning variant once its activity surpasses a critical value. Implemented as part of the viscoplastic self-consistent code, VPSC-7d, the PTR scheme was further developed in order to explore the relationship between twinning and hardening evolution [32–34]. However, both methods were limited in that neither allowed for the direct representation of the effects of multiple twin systems on the evolution of texture in simulated polycrystals. Other methods have been developed in order to address this limitation. The volume fraction transfer (VFT) scheme of Tomé et al. [31] calculated changes in texture by discretizing Euler space into separate orientation cells whose centers are occupied by the orientations of physical grains. As the grains are reoriented either continuously through rotations encountered by slip, or discontinuously in the case of twinning, the cells in the orientation space associated with these grains are shifted via translation vectors or by specific vector transformations for twinning, respectively, in Euler space. The volume fraction of each grain is then modified by an amount proportional to the overlap of the newly positioned Euler cells. The volume transfer approach of Kalidindi [33] introduces each twinning variant as a physical grain from the beginning, and the grain-weights are modified according to the twin activity: volume fractions are transferred from the mother grain to each of the twin-variant as a function of the pseudo-slip that a twin contributes to the deformation of the mother grain. This technique was adopted for large deformation simulations in recent works of Toth and co-workers to reveal the effect of nano-twins on texture evolution for rolling of ultra-fine-grained copper [34,35] and for twinning-detwinning activity in the fatigue of Mg AZ31 [36]. Twinning in TWIP steel was modelled by Prakash et al. [37], who created only one twin-variant as a new grain, which was selected by the PTR scheme among the 12 possible ones. The full volume transfer scheme for twinning in TWIP steel is applied for the first time in the present work, and without the use of the PTR scheme.

In this work, we aim to shed more light on the influence of deformation twinning on the texture evolution of face-centered cubic (fcc) alloys with low SFE, such as the high-manganese steel investigated here. To model the texture evolution during ECAP, an extended version of the viscoplastic self-consistent (VPSC) polycrystal code, developed in Metz, was used [38,39]. Nevertheless, the Taylor version of the VPSC model was identified to be applicable for the case of heavy large strain deformation twinning in TWIP steel. In addition to dislocation slip, primary and secondary twinning were implemented in the simulation code by predefining all possible twin variants. The volume fractions of twins were increased according to the activity of the twin systems, which were considered pseudo-slips in their parent grains, without variant selection. One of the main aims of the study is to examine the question as to whether the texture evolution can be simulated without applying any twin variant selection criteria. Another main novelty of our approach is to take into account the loss of twin orientation relationship between the twin and the mother grain in the growth process of the twins. This loss occurs due to the different lattice rotations of the twin and the matrix during the very large strains that are imposed. Finally, the different hardening characteristics of the twins with respect to the matrix are taken into account by a sophisticated self and latent hardening model. The obtained results show that our modeling provides a genuine approach for a quantitative simulation of the deformation behavior of TWIP steels.

2. Material and Methods

The chemical composition of the investigated TWIP steel is given in Table 1. The SFEs at room temperature and at 300 °C were calculated to be ~25 mJ/m^2 and ~75 mJ/m^2, respectively, using a subregular solution thermodynamic model [40].

Table 1. Chemical composition of the investigated alloy.

Element	Fraction (wt.%)
Fe	bal.
Mn	22.46
Al	1.21
C	0.325
Si	0.041
N	0.015
P	0.01

After casting, homogenization, annealing, forging, and hot rolling, rods with a diameter of 10 mm and a length of 35 mm were cut out perpendicular to the rolling direction for further deformation by ECAP. The ECAP rig used for the experiments is described in detail in [41]. The ECAP experiments were performed at 300 °C following route B_C up to two passes using a 90° ECAP die at a pressing rate of 1 mm/s. The chosen value of the pressing speed is based on previous studies [42]. The die geometry and the reference system for the testing is given in Figure 1. In the following, the ED, ND, and TD directions mean extrusion, normal, and transverse directions, respectively. Specimens for microstructure and texture characterization were cut from the deformed rods with dimensions of 8 mm (ED) × 10 mm (ND) × 1 mm (TD). Sample preparation on the ED-ND sections consisted of mechanical grinding and polishing, followed by electro-polishing at room temperature and 22 V using an electrolyte containing 700 mL ethanol (C_2H_5OH), 100 mL butyl glycol ($C_6H_{14}O_2$), and 78 mL perchloric acid (60%) ($HClO_4$). Transmission electron microscopy (TEM) samples (~100 µm initial thickness, 3 mm in diameter) were prepared using the same electrolyte in a double-jet Tenupol-5 electrolytic polisher with a voltage of 25 V at room temperature.

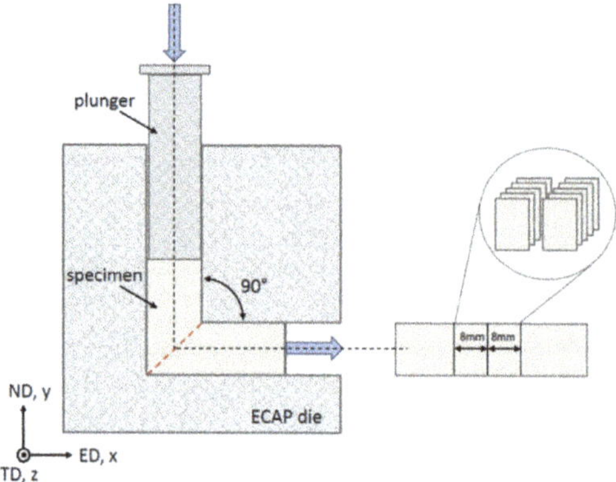

Figure 1. Schematic diagram of the reference coordinate system and sample taking according to the die geometry. ED, ND, and TD denote extrusion, normal, and transverse direction, respectively.

Electron backscatter diffraction (EBSD) measurements were performed using a LEO 1530 field emission gun scanning electron microscope (FEG-SEM) (Carl Zeiss AG, Oberkochen, Germany) operated at a 20 kV accelerating voltage and a working distance of 10 mm. EBSD maps were acquired applying a step size of 0.15 µm and were post-processed utilizing the HKL Channel 5 software (Oxford Instruments plc, Abington, Great Britain). The TEM specimens were analyzed using a JEOL JEM 2000 FX II analytical TEM (JEOL Ltd., Akishima, Japan) operated at 200 kV.

In order to characterize the crystallographic textures, X-ray pole figure measurements were performed by acquiring three incomplete (0–85°) pole figures {111}, {200}, and {220}. A Bruker D8 Advance diffractometer (Bruker Corporation, Billerica, USA), equipped with a HI-STAR area detector, operating at 30 kV and 25 mA, using filtered iron radiation and polycapillary focusing optics was used. The corresponding orientation distribution functions (ODFs) were calculated using the JTEX software (Universite de Lorraine, Metz, France) [43].

3. Experimental Results

The main experimental findings were presented in a previous study [26]. It has been found that ECAP of the investigated TWIP steel at 300 °C led to the formation of a shear texture, with texture components that are characteristic for both low- and high-SFE materials. Therefore, the texture was described as a transition texture, based on the correspondence suggested by Suwas et al. [27,28]. The accommodation of plastic strain by both dislocation driven grain fragmentation and deformation twinning were found to be the reason for the formation of this transition texture.

In the following, the experimental results that serve as basis for the present modeling approach are described briefly.

The initial texture was relatively weak (Figure 2). The strongest component was the {001}<100> cube texture component, with an intensity of about 2.0 in the {100} pole figure. After one ECAP pass, a shear texture appeared, with relatively strong A1, C, and B/Bb components. The A2 and A/Ab components were relatively weak. For the characteristics of these ideal components, see Table 2 and Ref. [44,45]. (Note that Bb means the $\overline{\mathrm{B}}$ component; similarly, Ab is $\overline{\mathrm{A}}$ in previous publications). The texture was comparatively weak with a texture index of 2.14 and the ideal components were generally surrounded by other, weaker texture components.

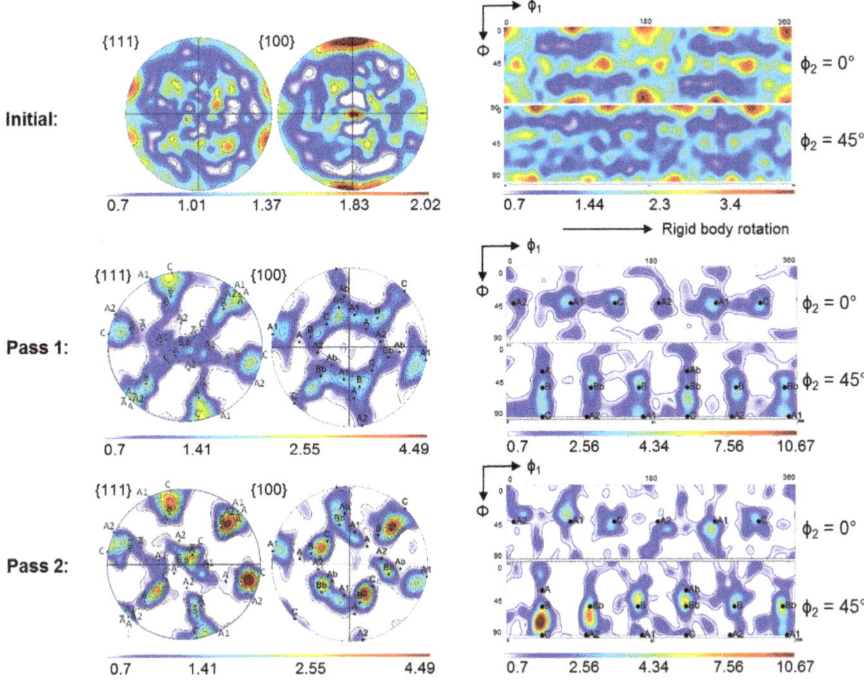

Figure 2. Experimental textures in {111} and {100} pole figures as well as in the $\phi_2 = 0°$ (**top**) and 45° (**bottom**) ODF sections. Reference system for the pole figures: vertical axis (x) is the outgoing channel axis direction, horizontal axis (y) is opposite to the pressing direction, center (z) is the transverse direction. The Euler angles of the ODFs refer to the same (x, y, z) reference system. The range of ϕ_1 angle (horizontal axis) is from 0° to 180°, the Φ is from 0° to 90°, in top to bottom direction.

Table 2. Definition of the ideal texture components for an ECAP die with 90° angle.

Texture Component	ϕ_1 (°)	Φ (°)	ϕ_2 (°)
A_1	80.26/260.26	45	0
	170.26/350.26	90	45
A_2	9.74/189.74	45	0
	99.74/279.74	90	45
C	135/315	45	0
	45/225	90	45
A	45	35.26	45
Ab	225	35.26	45
B	45/165/285	54.74	45
Bb	105/225/345	54.74	45
{111}-/A-fiber	{111}<uvw>		
<110>-/B-fiber	{hkl}<110>		

After two passes (Figure 2), the ODF maximum was located between the C and B components, with higher intensity than the peaks after the first pass (texture index: 2.8). The total surface area of low intensities in the two ODF sections was reduced compared to that after the first pass.

The true stress-true strain curve obtained from tensile testing of the investigated steel in its initial hot-rolled condition before ECAP is displayed in Figure 3. As can be seen, the strain hardening was very substantial and the curve increases nearly linear.

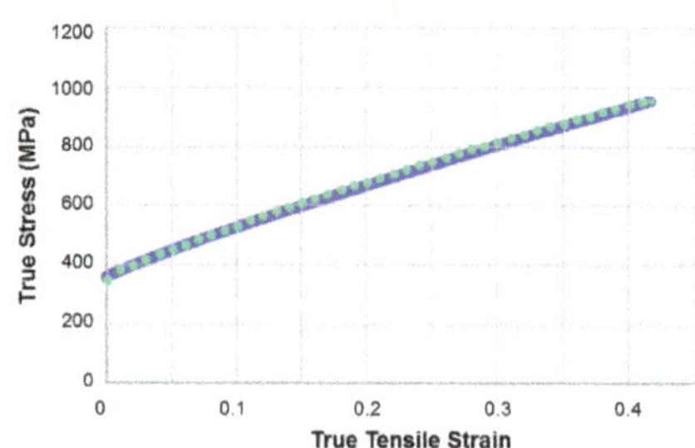

Figure 3. Tensile stress-strain curve of the TWIP steel in initial condition before ECAP, tested in RD of hot rolling. Continuous line: experiment, dotted line: simulation. The experiment was performed at a strain rate of 10^{-3} s^{-1} at room temperature.

As expected for TWIP steel, deformation twinning was pronounced in the studied material during ECAP. Figure 4 shows that the elongated twin lamellae were abundant in the deformation microstructure. Finer details of the microstructure could be seen by TEM, see Figure 5. Very fine twin lamellae were found with thicknesses in the nanometer range, between 40 nm and 100 nm (Figure 5). While, at their creation, the twins must appear in perfect twin-orientation relationship with respect to the parent grain, it is not guaranteed that the ideal orientation relationship will be maintained during plastic deformation up to large strains. In order to verify the loss of orientation relationship between parent grain and its twin, an analysis was carried out on the EBSD-observed microstructure. The inset in Figure 4 shows the analyzed region in the EBSD map after one ECAP pass. Some selected adjacent grains labeled in Figure 4 with 1 to 13 were examined in terms of their misorientation. The mean Euler angles of these 13 grains are given in Table 3.

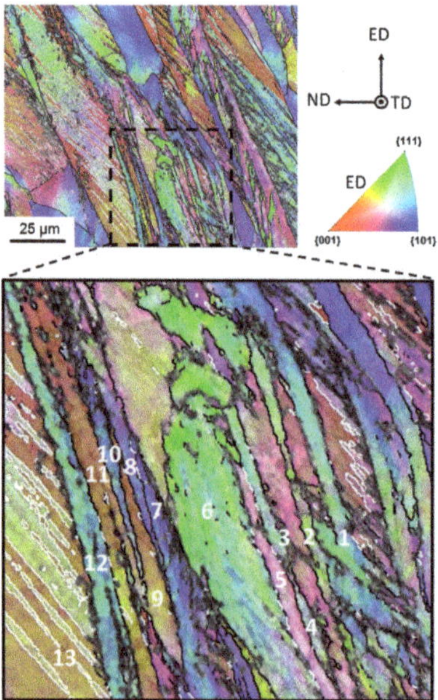

Figure 4. EBSD map with color-coding according to the inverse pole figure after one-pass ECAP. 13 grains were analyzed in the selected zone as illustrated in the inset. Black and white lines denote high-angle grain boundaries ($\theta \geq 15°$) and $\sum 3$ (60°<111>) grain boundaries, respectively.

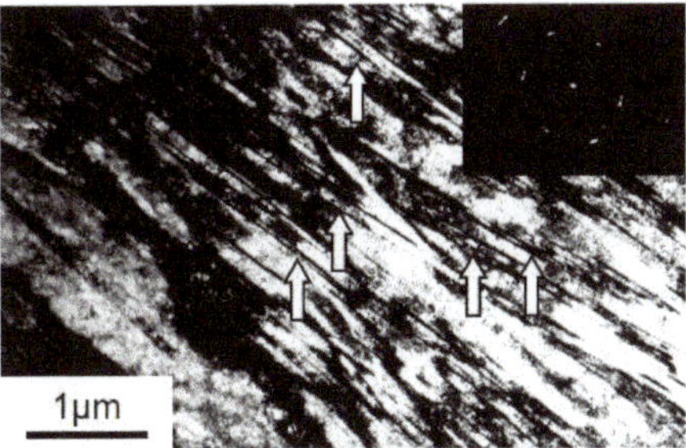

Figure 5. Bright field TEM micrograph and the corresponding selected area diffraction pattern after one ECAP pass showing a high fraction of deformation twins. Selected twins/twin-matrix lamellae are indicated by arrows.

Table 3. The mean orientations of the grains marked in Figure 4.

Grain	Mean ϕ_1 (°)	Mean Φ (°)	Mean ϕ_2 (°)
1	286.45	51.28	45.76
2	44.5	45.35	46.73
3	341.6	48.14	33.77
4	43.81	49.17	33.91
5	215.52	50.33	48.11
6	285.1	50.8	52.95
7	98.82	41.62	39.14
8	279.32	45.41	38.32
9	212.38	46.61	60
10	280.54	50.03	48.95
11	203.3	44.94	68.5
12	285.8	43.34	41.47
13	352.36	43.8	25.28

Table 4 contains the results of the twin-orientation analysis, i.e., the disorientation angle of each pair of neighboring grains together with their nearest disorientation axis, the offset from the nearest rotation axis, and the deviation angle from the ideal 60°<111> twinning orientation relationship. The deviations were typically around 10°, but there were also several larger disorientations, up to 50°. This analysis provided evidence that the twins drifted away from their initial ideal orientation relation with respect to their parent grains during deformation, i.e., they did not co-rotate with the parent grain after large strains.

Table 4. Orientation relationships between twin and matrix for the 13 grains shown in Figure 4.

Combination	Disorientation and Nearest Rotation Axis	Offset from Nearest Rotation Axis (°)	Deviation from Twinning Position (°) (from 60°<111>)
1–2	19.07°<3-20>	6.21	50
2–3	53.13°<-14-3>	2.47	42.45
3–4	55.07°<423>	2.57	35.54
4–5	56.77°<-2-3-3>	2.39	10.8
5–6	48.29°<443>	1.87	15.6
6–7	50.7°<-1-4-4>	0.47	22.5
7–8	55.98°<24-3>	1.98	18.3
8–9	48.29°<433>	3.42	8.87
9–10	58.03°<-3-2-3>	3.87	10.72
10–11	58.13°<-101>	4.14	28.14
11–12	55.84°<343>	1.76	9.94
12–13	56.81°<-3-4-4>	2.52	9.7

4. Modeling Approach

4.1. The Polycrystal Model

In this work, simulation of the texture evolution was performed using the VPSC polycrystal code developed in Metz, in its finite-element tuned version [38,39]. The code itself is considered an 'isotropic' code, so-called because the interaction between a given grain and the homogeneous medium is calculated using the assumption that the medium itself is isotropic. Although not appropriate for hexagonal materials, this assumption can be considered valid for cubic materials without strong textures. As TWIP steels possess an fcc crystal structure and usually show weak textures [46], such conditions are expected to be met. The volume transfer scheme recently introduced in publications by Toth et al. to account for twinning [35,36] was implemented in the VPSC code; simultaneously for both primary and secondary twinning.

4.2. Twinning Approach

In the volume transfer scheme, each twin variant is considered as a separate grain created at the beginning of the simulation with zero initial volume fraction. As deformation progresses during the simulation, volume fraction $(\Delta v)^{twin}$ is transferred from the parent grain to its corresponding twins. This approach can only be applied if the twin-parent orientation relation does not deviate too much from the ideal one. A certain deviation angle can be permitted because the local dislocation mechanisms that make the twin grow can accommodate a small difference. In the present work, the volume attributed to the twin from the parent grain was reduced progressively as a function of the deviation of the orientation of the twin from its exact twinning position. After each strain increment Δt, the disorientation angle (θ) between the ideal twin position and the actual orientation of the twin was calculated. Then the <112> slip system activity was examined in order to transfer some part of the volume of the parent grain to the twins. The volume was calculated in proportion to the <112> slip increment normalized by the shear necessary to form a twin ($\gamma_{twin} = 1/\sqrt{2}$). The deviation from the exact twin position was also taken into account in the amount of volume transfer:

$$(\Delta v)^{twin} = \frac{(10^\circ - \theta)}{10^\circ} \frac{\left(\Delta t \times \dot{\gamma}^{twin}\right)}{\gamma_{twin}} V_{parent}, \text{ for } \theta \leq 10^\circ, \ (\Delta v)^{twin} = 0, \text{ for } \theta > 10^\circ. \quad (1)$$

When the deviation angle exceeded 10°, the volume transfer was stopped. Once formed, the twins continued deforming using their available slip systems. The <112> type slip activity was used for growing the twin; however, they did not grow any more once they rotated out of their ideal twinning position with respect to their parent grains.

In order to avoid artifacts related to twins with extremely small volume fractions, the volume transfer from the mother grain to a twin variant was only initiated if the mother grain was slipping on the corresponding <112> slip at least by 1% of the total slip of the mother grain. (Here, the "corresponding <112> slip" means the twin variant that corresponds to the given <112> slip system of the matrix.) Until this initiation point, the twin orientation was co-rotated with the mother grain. The co-rotation was stopped as soon as the twin had at least 1% volume of the mother grain. Starting from that point, the lattice rotation of the twin is dictated by its own slip system activity, which is different from its mother grain, due to the different orientation of the twin.

Except for a direct reorientation of the twinned volume, twinning also influences the texture evolution due to a strong latent hardening effect [12,47,48]. The twins are assumed to be harder than the parent grains [49–51], which was considered in the present modeling by assigning higher hardening rate for the twins with respect to the matrix (see the hardening approach in Section 4.3). Furthermore, twins observed in TWIP steels appear in stacks of thin, nanoscale lamellae with small thickness (e.g., Figure 5). Therefore, enhancement of the twin volume fraction is dominated by the nucleation of new lamellar structures rather than by the expansion of previously existing ones [52].

Due to their large volume fraction—even if they are hard—the twins must deform plastically due to the large imposed strain. This induces the formation of new twins, i.e., secondary twins are formed inside primary twins. This behavior has been observed frequently in TWIP steels subjected to large strains and/or complex deformation conditions, e.g., during tension [53], cold rolling [13,54], asymmetric rolling [55], and ECAP [25,56]. The present modeling approach also includes secondary twining, treating it in the same manner as primary twining. However, secondary twinning was considered only from the second pass on, as the hardening due to the first pass and the rotation during pass one and two are likely to result in activation of additional twinning systems during the second pass. All secondary variants were assigned as grains at zero strain (of the second pass), initially with zero volume fraction. The VPSC code requires to assign also elliptical shape tensors for considering their interaction with the equivalent homogeneous medium. Initially, spherical shapes were assigned to the parent grains while the twins were approximated by flat disks with an aspect ratio of 5, and were oriented parallel to the twinning plane. The shape was allowed to change with strain according to the operation mode of the self-consistent code.

4.3. Strain Hardening Model

Strain hardening was simulated using the self and latent hardening approach proposed by Kalidindi et al. [57] and Zhou et al. [58]. In that approach, the reference resolved shear strength of a slip system τ_0^α is controlled by the following relation:

$$\dot{\tau}_0^\alpha = \sum_\beta \mathrm{H}^{\alpha\beta}\left|\dot{\gamma}^\beta\right|, \alpha, \beta = 1\ldots 12 \tag{2}$$

Here, α and β are the indices of the slip systems, $\dot{\gamma}^\beta$ is the slip rate and $\mathrm{H}^{\alpha\beta}$ is the hardening matrix. The slip rate is calculated from the constitutive law proposed by Hutchinson [49] for strain rate sensitive slip:

$$\dot{\gamma}^\beta = \dot{\gamma}_0^\beta \mathrm{sign}(\tau^\beta)\left|\frac{\tau^\beta}{\tau_0^\beta}\right|^{1/\mathrm{m}}. \tag{3}$$

Here, τ^β is the resolved shear stress and m is the strain rate sensitivity index. The $\mathrm{H}^{\alpha\beta}$ hardening matrix in Equation (2) is constructed from a slip system interaction matrix $\mathrm{q}^{\alpha\beta}$:

$$\mathrm{H}^{\alpha\beta} = \mathrm{q}^{\alpha\beta}\mathrm{h}_0\{1-(\tau_0^\alpha/\tau_{\mathrm{sat}})\}^{\mathrm{a}}. \tag{4}$$

Here, τ_0^α is the actual slip strength, τ_{sat} is the saturation value, h_0 controls the hardening rate, and 'a' is a parameter. In the $\mathrm{q}^{\alpha\beta}$ interaction matrix only four cases are distinguished [58]; collinear (q_1), coplanar (q_2), perpendicular (q_3), and other slip orientations (q_4) between two slip systems α and β. They are represented by four parameters; $\mathrm{q}_1 = 1$, $\mathrm{q}_2 = 1.2$, $\mathrm{q}_3 = 2$, and $\mathrm{q}_4 = 1.5$, respectively. Based on their definition, there is a physically justified order for three latent hardening parameters: $\mathrm{q}_1 < \mathrm{q}_2 < \mathrm{q}_3$. The fourth parameter is not defined by a specific physical argument, so it was approximated by the average value of the other three ('all other cases'). The 12 {111}<110> slip systems and the 12 {111}<112> pseudo-slip systems for twinning were used. Suitable hardening parameter values were found by fitting to the experimental tensile hardening curve (Figure 3). An iteration procedure led to the following common values for the matrix and the twins: $\tau_0^{\langle 110\rangle} = \tau_0^{\langle 112\rangle} = 167$ MPa, $\tau_{\mathrm{sat}} = 1650$ MPa, and $\alpha = 0.5$. The parameter for the hardening rate was different: $\mathrm{h}_0^{matrix} = 163.5$ MPa, $\mathrm{h}_0^{primary\ twin} = 327$ MPa, and $\mathrm{h}_0^{secondary\ twin} = 489$ MPa. After 0.4 tensile strain in RD, the twinned volume fraction was 0.2, which agrees well with the estimated experimental value. The experimental stress-strain curve was well reproduced, see Figure 3.

4.4. ECAP Texture Modeling Conditions

The simple shear model of ECAP was used to define the deformation gradient at each deformation step, with a 90° die, up to a total shear of $\gamma = 2$ on the 45° plane. The simulation was advanced in 40 deformation steps for one ECAP pass with increments of $\Delta\gamma = 0.05$, applying a macroscopic shear rate of $\dot{\gamma} = 1/s$. 500 initial parent grain orientations were generated from the initial texture; each of them was completed with the 12 primary twin variant grain orientations with zero volume fractions. In this way, the total number of grain orientations was 500 + (500 × 12) = 6500. The strain rate sensitivity exponent was taken to 0.166 in the viscoplastic constitutive law in Equation (3). This value may seem to be relatively high; however, it can be justified when the material is heavily deformed. Namely, it is expected that the yield surface becomes more and more rounded when the dislocation density increases because individual dislocations have different critical resolved shear stresses due to their different curvatures and interactions with other dislocations. While the plastic response of a complicated dislocation structure cannot be modelled, the rounded shape of the yield surface can be considered by using higher m values. The rounding effect of the m parameter on the yield surface was shown first in Ref. [44]. It was also found in the present simulations that higher slip viscosity led to better relative strengths of the texture components. The modeling parameters are collected in Table 5.

Table 5. List of the polycrystal modeling parameters used.

Parameter	Value
Slip systems and initial strength, τ_0	{111}<110>, 167 MPa
Twinning systems and initial strength, τ_0	{112}<110>, 167 MPa
Hardening rate for mother grain, h_0	163.5 MPa
Hardening rate for primary twin, h_0	327 MPa
Hardening rate for secondary twin, h_0	489 MPa
Saturation stress, τ_{sat}	1650 MPa
Strain hardening parameter, a	0.5
Latent hardening parameters	$q_1 = 1$, $q_2 = 1.2$, $q_3 = 2$, $q_4 = 1.5$
Strain rate sensitivity parameter for slip and twinning, m	0.166
Interaction coefficient for Tangent VPSC model, α	0.166
Initial number of grains (mothers + 12 primary twins)	6500
Initial grain shape axis ratios for twins	1:5:5

5. Modeling Results and Discussion

In order to find the best agreement with the experimental textures, the ECAP texture simulation was carried out by iteration. The parameters to fit were the α parameter in the localization equation of the VPSC model [38,39] and the strain rate sensitivity index. The hardening parameters were not varied again, because they were fitted to the tensile strain hardening curve (Figure 3). By varying the α parameter in the VPSC model, one can continuously change the nature of the interaction between a grain and the homogeneous equivalent medium around the grain, which represents the polycrystal. $\alpha = 0$ corresponds to the Static model, where the stress state is the same for all grains, while $\alpha = \infty$ is the Taylor model with uniform strain. Best texture results were obtained by taking the α parameter to 100, which is equivalent to the Taylor model in practice. Concerning the strain rate sensitivity of slip, the m = 0.166 value was adopted.

The simulation results for the crystallographic textures after the first and second pass are presented in Figures 6–9, respectively. In general, a very good agreement between experimental and simulated textures was achieved, when using the Taylor model and taking mechanical twinning into account. Detailed discussion of the texture evolution is presented in the following Section 5.1. For a better comparison with the experimental textures (Figure 2), the same intensity levels were used in the simulated textures. For the construction of the ODFs, the Gaussian spread around the grain orientations was 7° and the harmonic coefficients were calculated up to the rank of 32.

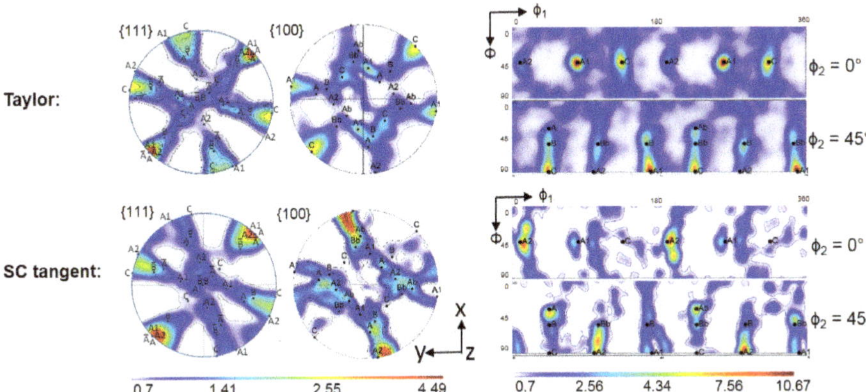

Figure 6. Simulated textures after one-pass ECAP obtained by the Taylor and the self-consistent (SC) (tangent) model in {111} and {100} pole figures, as well as in the $\phi_2 = 0°$ and 45° ODF sections.

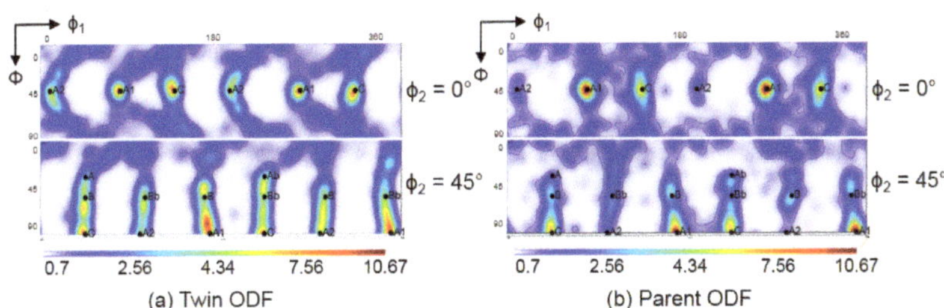

Figure 7. Decomposition of the full, simulated texture obtained by the Taylor model in Figure 6 into (**a**) primary twins only and (**b**) parent grains only. The textures are presented by $\phi_2 = 0°$ and 45° ODF sections.

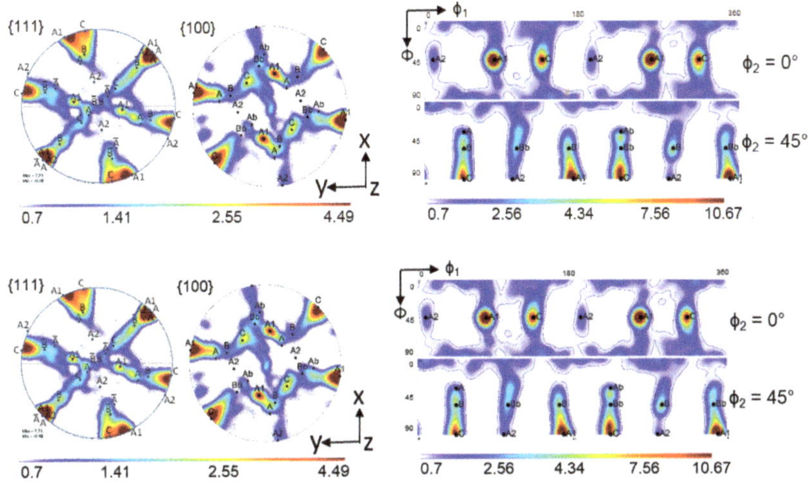

Figure 8. Simulated texture after one-pass ECAP obtained by the Taylor model without twinning in {111} and {100} pole figures as well as in the $\phi_2 = 0°$ and 45° ODF sections.

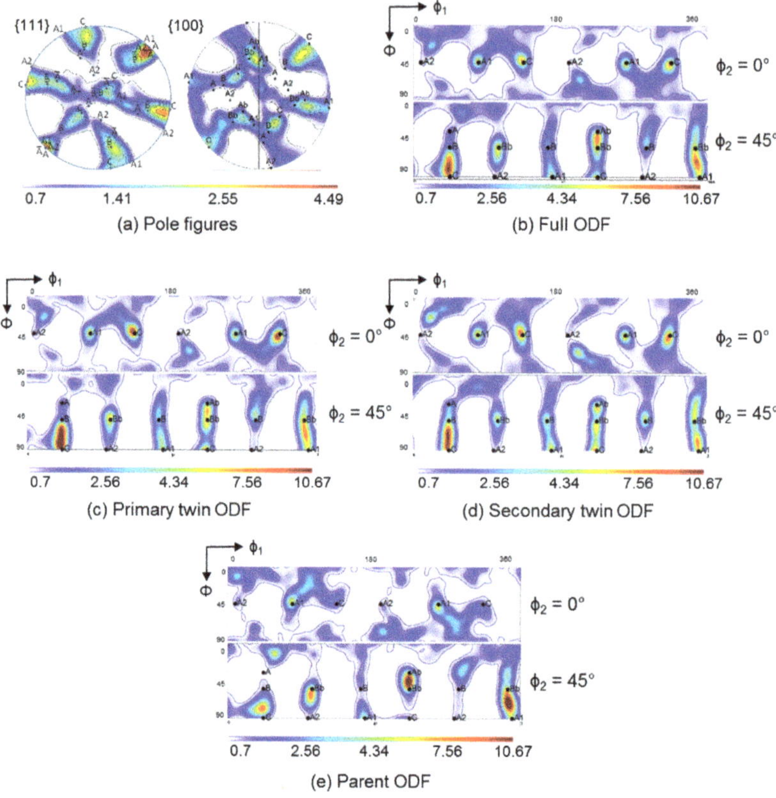

Figure 9. Simulated texture obtained by the Taylor model after two-pass ECAP. (**a**) {111} and {100} pole figures; (**b**) ODF of all grains; (**c**) ODF of primary twins only; (**d**) ODF of secondary twins only; and (**e**) ODF of parent grains only. ODFs are given as $\phi_2 = 0°$ (top) and 45° (bottom) sections.

5.1. Texture Evolution

As can be seen from a comparison between the experimental (Figure 2) and simulated results (Figure 6 (Taylor) and Figure 9), the texture intensities are very similar and the relative strengths of the main texture components was well predicted. The first point of discussion is the meaning of the Taylor deformation mode that was identified by the simulation. This is quite unusual, because previous modeling of fcc polycrystals subjected to large deformation showed low α values [35]. In order to verify the validity of the Taylor model, a calculation of the first ECAP pass using the same parameters, except for the α parameter, was performed with the tangent approach of the self-consistent (SC) model, i.e., with α = m = 0.166. The obtained results are shown in Figure 6. The Taylor model correctly predicted weaker A, Ab and A2, intermediate B and Bb, and stronger A1 and C texture components. In contrast, the SC model overestimated the A and A2 texture components, whereas the main experimental components A1, B, and C were underestimated. The relative strengths of the texture components were also calculated incorrectly by the SC model. The applicability of the Taylor model must come from the high twinning activity. Indeed, the predicted volume fraction of primary twins was 46% after the first pass, and increased up to 74% during the second one. During the second pass, there was also secondary twinning, occupying 26% of the total volume, see Figure 10. They were, however, included into the primary twins, so secondary twinning did not decrease the parent grain volume fraction, which is then also 100% − 74% = 26% after the second pass.

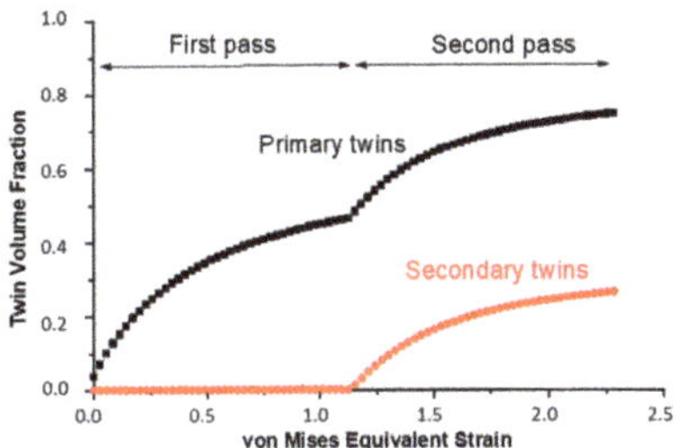

Figure 10. Simulated evolution of the twin volume fraction as a function of the von Mises equivalent strain during the first and second pass. Secondary twins were activated only from the start of the second pass.

The obtained large twinning volume fractions at the end of the first (46%) and second pass (74%) seem to be comparatively high. Indeed, observations by EBSD indicate a lower volume fraction. This difference needs explanation. First of all, it is emphasized that the capability of the EBSD technique to resolve deformation twins with nanoscale width and separation distance quantitatively is highly limited, which leads to an underestimation of the volume fraction, and so does not permit exact quantitative analysis. In addition, EBSD measurements can only capture twins that are near twinning positions and are unable to detect twins that were created before but rotated away from their twin orientation. Those twins can only be recognized from their morphology, that is, by their lamella-type shape. This is why we conducted the analysis presented in Figure 4 and in Tables 3 and 4. Therefore, the simulation result can be reasonable, meaning that only one third of the total twins remained near their twin orientation.

The twins take the form of lamellae and occupy 74% of the volume. They are also very hard (see Section 4.2) and interconnected. According to the microstructure observations, the twins constitute a skeleton that fills up the whole volume of the parent grain. When such a structure is subjected to large plastic deformation, the hard twin skeleton, with its large volume fraction, enforces uniformity of plastic strain in the whole grain. A consequence of the strain homogeneity within individual grains is that all grains deform in the same way. This conclusion comes from the fact that strain heterogeneity between adjacent grains of the polycrystal must be accommodated near the grain boundary regions of individual grains, which is different for each neighbor. This difference is not possible because the strain is the same everywhere in a given grain. Therefore, the homogeneity of strain within each grain of the polycrystal induces a homogeneous deformation of the whole polycrystal, that is, the Taylor deformation mode.

Our investigation led to the fundamental result that the polycrystal is deforming according to the Taylor model, which is a uniform deformation. Therefore, the twins deform the same way as the imposed macroscopic deformation, with the same strain mode as the mother grains. In this way, compatibility is maintained between the twins and the mother grains. Normal slip was deforming the twins when they were formed. Secondary twinning was allowed only at large strains, starting from the second ECAP pass, which was also taking place according to the Taylor model.

5.2. Effect of Twinning on Texture Evolution

It is important to emphasize that the present modeling is for large strains, up to a shear of $\gamma = 4$, during which the initial twins can rotate significantly and differently with respect to their parent grains, according to their own slip system activity. This effect was considered in the present simulation according to the scheme described in Section 4 above and proven experimentally, as shown in Figure 4 and Table 4. Indeed, by the end of the first ECAP pass, the average disorientation from the exact twinning position had increased by up to 8.2°, as shown by the distribution of the disorientations in Figure 11a,b. The disorientation achieves values of up to 48°, which is similar to the maximum value observed experimentally (50°). Therefore, it was important to consider these deviations for controlling the transfer of volume between parent grains and twin variants. Indeed, to account for the increase in deviation, the volume transfer was progressively reduced, and finally stopped for disorientations larger than 10°. After 10° deviation, a twin is no longer a twin, but a normal grain, which follows its own orientation change without further growth. This is why the average rate of volume transfer is progressively reduced with the strain, and only a fraction of twins will continue growing after the first pass (Figure 11b).

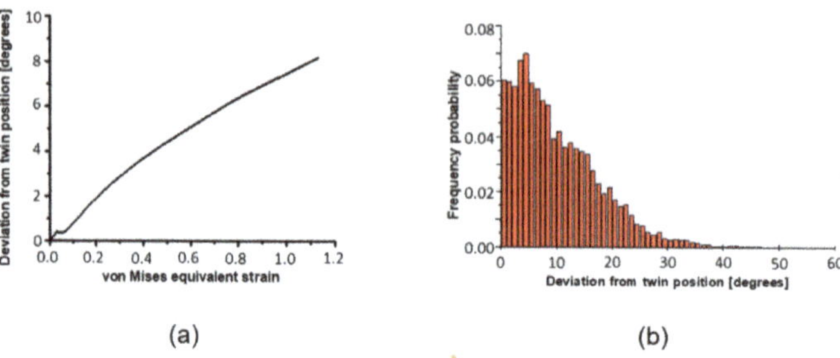

Figure 11. (**a**) Average deviation from exact twin position during the first ECAP pass; (**b**) Histogram showing the distribution of the simulated deviation. (Bin size: 1°, largest deviation: 48°.).

In order to determine the influence of deformation twinning on texture evolution, the simulation of the first ECAP pass was also performed using the Taylor model, with the same simulation parameters, but without taking twinning into account (Figure 8). Comparison of the texture with and without twinning (Figures 6 and 8) evidences that when deformation twinning is not considered as an additional deformation mechanism, the simulated texture deviates strongly from the experimental one. For instance, the A1 and C texture components are strongly overestimated, whereas the B and Bb texture components were not predicted as separate components at the orientation positions (165°/54.74°/45°) and (345°/54.74°/45°), respectively, without twinning. This comparison also shows that deformation twinning leads to strong decrease in the general intensity of the texture, i.e., twinning has a randomization effect. Twins are in different positions with respect to their parent grains, so if the parent grain is near an ideal orientation, the twin cannot approach the same one. This is one of the reasons why the intensity of the simulated texture (Taylor model including twinning) approaches the experimental one. This effect was also observed in quantitative grain fragmentation simulations [50], where the new grains developed orientations that differed from those of their parents.

In order to identify the effect of twinning on the texture evolution, the textures that correspond only to the twin and parent orientations are also plotted separately in Figures 7 and 9. At first, they appear quite similar to each other. Close inspection, however, shows the following major differences: (i) the texture intensity is significantly lower for the twins compared to that in the parent grains; (ii) systematically, the A2 component is very weak for the parent grains; (iii) the A, B, Ab, and Bb texture components are more pronounced in the twin ODF; (iv) for the second pass, where secondary twinning was activated, the A2 and the B component of the ideal texture components of the parent grains are also absent. At the same time, the strong intensity between the B and C components observed experimentally was also well simulated by the primary and secondary twin ODFs.

These texture features can be understood more easily if we study the twinning behavior of the ideal orientations. Figure 12 (The two-fold symmetry operation was applied for the generation of the ODFs in Figure 12, which is applicable for simple shear if the initial texture also displays that symmetry. This symmetry can be applied for the first ECAP pass, but not for the second, because the sample was rotated around its long axis by 90° for the second pass following the Route B_C of ECAP testing. Consequently, the two-fold symmetry was lost. And Figure 13 show the results of an analysis carried out for the twinning behavior of ideal orientations of simple shear textures (here ECAP). For the construction of these figures, the same volume was allocated to the ideal orientations. Afterwards, the twin variants were calculated with the same approach as for the whole polycrystal. The Cube orientation was added to the analysis as it is frequently a strong component in initial textures of fcc polycrystals, as in the present case. Figure 12 displays the generated twin orientations from the ideal ones for ECAP deformation; that is, for simple shear along the 45° oriented intersection plane of the two channels. Except for the Cube orientation, all generated twin orientations lie in the $\phi_2 = 0°$ and 45° sections (for the Cube, there are also twins in the $\phi_2 = 27°$ section of Euler space). The result of the theoretical analysis shows that the A2 texture component twins into A1. Also, there are no twins generated into the A2, C, A and Ab positions. The B/Bb orientations twin into each other, and also into a new component, as indicated by the red color in Figure 12. The Cube texture component twins into stable shear-texture components lying on the so-called B-fiber of shear textures; positioned between the C and B, as well between the C and Bb orientations in the $\phi_2 = 45°$ ODF section (Figure 12). The other twin variants of the Cube orientation do not appear in the ideal positions of shear. The relative intensities of the twin orientations are also indicated with the intensity levels around them in Figure 12. It is clear that the A1 component is the strongest twin orientation, followed by the B/Bb twins.

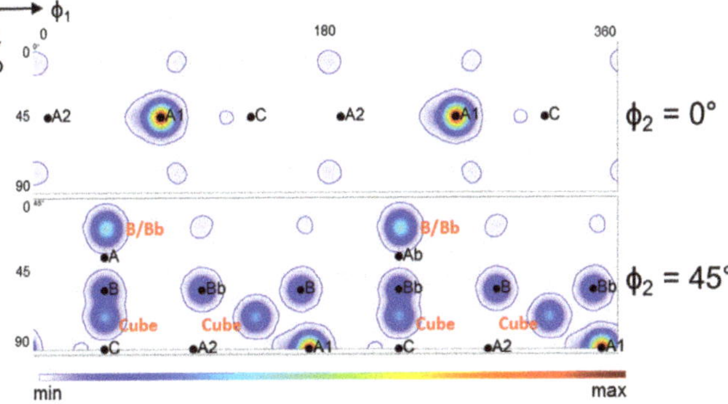

Figure 12. The twin positions and their relative strength formed by the ideal texture components in ECAP, taking all ideal texture components with the same intensity. Ideal orientations twinning into each other are indicated with black letters: the A1 is originating from A2, the B is from Bb, inversely, the Bb from B. The Cube-originated twins are indicated by red letters, also two twin components of B/Bb that are not in ideal positions for simple shear.

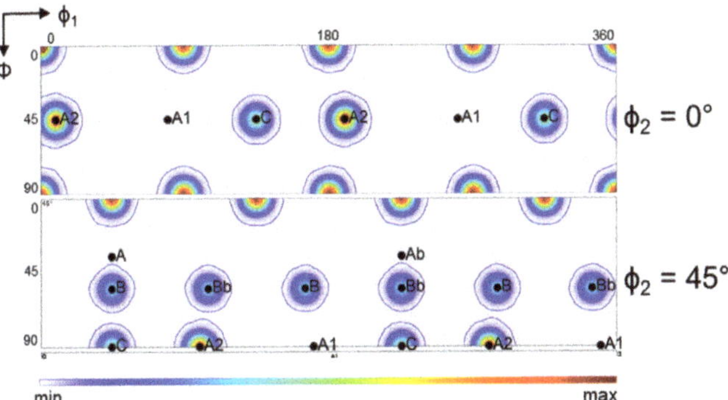

Figure 13. The relative twinning capacity of the ideal ECAP texture components and of the Cube orientation. The Cube orientation is the most prone to twin, followed by the A2, C and the B/Bb components. A1 and A/Ab do not twin.

Another way to study the twinning behavior of the ideal orientations is to plot the twinning capacity of the ideal orientations. This is the inverse of the analysis shown in Figure 12. As shown in Figure 13, the Cube texture component is the most prone to twin, followed by the A2, the C and the B/Bb texture components. The A1 and A/Ab orientations do not twin.

On the basis of the twinning behavior analyzed theoretically in Figures 12 and 13, the following interpretation of the experimentally observed texture evolution can be made:

i. The A2 texture component is very weak, because it twins into A1, which becomes the strongest texture component after the first ECAP pass and remains strong after the second pass. Due to the high fraction of twinned volume, weakening of the A2 texture component is further facilitated with increasing strain, i.e., the intensity of A2 decreases after the second pass. These findings are in agreement with observations made in a previous study on silver, which has a low SFE and shows severe deformation twinning, deformed by ECAP (four passes, route A) [51].
ii. The B/Bb texture components are relatively strong because they are ideal shear texture components and because they twin into each other. Two new components can be formed by the B/Bb twins (indicated in red in Figure 12), which are not ideal components of simple shear textures. They are present in the experimental texture after the first pass, though they are quite weak. The existence of these components can only be explained with a co-rotation that may take place between the ideal B/Bb parent grain and its embedded twin variant in the new orientation. Such lattice co-rotation would require the same slip system activity in the twin as in the parent grain, which is unlikely, as there are more slip systems active in those new positions than in the ideal B/Bb orientations (see in Ref. [59]). Therefore, the two observed slight components are probably only temporary variations of the texture intensity.
iii. Primary twins can increase the strengths of all ideal shear-texture components in the first pass (see Figure 7a), because they rotate out of their twinning positions during large strain (Figure 11). The twins can finally reach an ideal position by their own slip activity. This effect can be seen as the reason for the strengthening of the A, B, Ab, and Bb texture components, which were comparatively weak in the simulated parent ODF, but present in the twin-ODF and in the experimental ODF.
iv. Cube-oriented grains, which dominated the initial hot-rolled texture, twin heavily and can be responsible for the high ODF intensities between the C-B and C-Bb orientations in the first pass. In the second pass, only the orientations between C and B remain strong. The latter is due to the sample rotation between the first and second ECAP pass in Route B_C, which destroys the two-fold symmetry of the texture and impedes the formation of the texture component between C and Bb.

Secondary twinning must be operational during the second ECAP pass, because the experimental texture can only be reproduced if secondary twinning is introduced into the simulation. While secondary twinning might also take place during the first pass, this was not observed experimentally and therefore not modelled. There is no dramatic difference between the ODFs formed by the primary and the secondary twins (Figure 9c,d), while the parent grains display a very different texture (Figure 9e). The reason for the latter is that 74% of the parent grain volumes are converted into twins, so the heavily twinning orientations disappear. More specifically, the experimental ODF maximum between the C and B texture components after two passes was well simulated by the primary and secondary twin ODFs, and the weaker intensity between the C and Bb texture components was also captured. Moreover, not all components can form in the second pass because of the loss of two-fold symmetry.

6. Conclusions

The texture evolution during ECAP of a high-manganese Fe-23Mn-1.5Al-0.3C TWIP steel subjected to one and two ECAP passes was analyzed. The VPSC model was updated to simulate the ECAP texture using a sophisticated volume transfer scheme to account for the effect of twinning. The results of the study provide the basis for the following main conclusions:

1. The new volume transfer scheme was proven to be a robust approach for quantitative modeling of the effect of mechanical twinning on the evolution of the crystallographic texture without employing any variant selection criterion;
2. The twin relation is progressively lost between the twins and their parent grain during large strain, for which it is necessary to control the twin growth and to finally stop after a given disorientation. This mechanism was modeled within the VPSC scheme and led to very good agreement between experimental and simulated textures;
3. Both primary and secondary twinning were accounted for in the present simulations and were found to be essential for correct calculation of the crystallographic textures. The A1, B, and Bb texture components after one pass and the orientations between C and B after two passes are directly related to the occurrence of deformation twinning in fcc alloys with low stacking fault energies;
4. The plastic deformation of the TWIP steel polycrystal was found to take place under the upper-bound Taylor conditions, i.e., under homogeneous deformation, which is due to the composite nature of the parent-twin structure imposing uniform strain within a grain and is subsequently applied to the whole polycrystal. Therefore, it was proven that self-consistent schemes can be bypassed under the conditions investigated.

Author Contributions: The research work presented in this manuscript was planned and carried out as a collaboration between all authors listed above. Conceptualization, L.S.T. and C.H.; Methodology, L.S.T.; Formal Analysis, L.S.T.; Investigation, C.H. and R.A.; Writing-Original Draft Preparation, L.S.T., C.H. and R.A.; Writing-Review & Editing, all authors contributed to the scientific design of the study, the discussion of the results, and have seen and approved the final manuscript.

Funding: This research received no external funding.

Acknowledgments: Christian Haase and Dmitri A. Molodov acknowledge gratefully the financial support of the Deutsche Forschungsgemeinschaft (DFG) within the Collaborative Research Center (SFB) 761 "Steel—ab initio; quantum mechanics guided design of new Fe based materials". Laszlo S. Toth and Robert Allen would like to acknowledge the support from the French State through the program "Investment in the future" operated by the National Research Agency (ANR) and referenced by ANR-11-LABX-0008-01 (LabEx DAMAS).

Conflicts of Interest: The authors declare no conflict of interest.

References

1. Grässel, O.; Frommeyer, G.; Derder, C.; Hofmann, H. Phase Transformations and Mechanical Properties of Fe-Mn-Si-Al TRIP-Steels. *J. Phys. IV* **1997**, *7*, 383–388. [CrossRef]
2. Grässel, O.; Krüger, L.; Frommeyer, G.; Meyer, L.W. High Strength Fe–Mn–(Al, Si) TRIP/TWIP Steels Development—Properties—Application. *Int. J. Plast.* **2000**, *16*, 1391–1409. [CrossRef]
3. Allain, S.; Chateau, J.P.; Bouaziz, O.; Migot, S.; Guelton, N. Correlations Between the Calculated Stacking Fault Energy and the Plasticity Mechanisms in Fe–Mn–C Alloys. *Mater. Sci. Eng. A* **2004**, *387–389*, 158–162. [CrossRef]
4. Bouaziz, O.; Allain, S.; Scott, C. Effect of Grain and Twin Boundaries on the Hardening Mechanisms of Twinning-Induced Plasticity Steels. *Scr. Mater.* **2008**, *58*, 484–487. [CrossRef]
5. Bouaziz, O.; Allain, S.; Scott, C.P.; Cugy, P.; Barbier, D. High Manganese Austenitic Twinning Induced Plasticity Steels: A Review of the Microstructure Properties Relationships. *Curr. Opin. Solid State Mater. Sci.* **2011**, *15*, 141–168. [CrossRef]

6. Vercammen, S.; Blanpain, B.; De Cooman, B.C.; Wollants, P. Cold Rolling Behaviour of an Austenitic Fe–30Mn–3Al–3Si TWIP-Steel: The Importance of Deformation Twinning. *Acta Mater.* **2004**, *52*, 2005–2012. [CrossRef]
7. Bracke, L.; Verbeken, K.; Kestens, L.; Penning, J. Microstructure and Texture Evolution During Cold Rolling and Annealing of a High Mn TWIP Steel. *Acta Mater.* **2009**, *57*, 1512–1524. [CrossRef]
8. Lü, Y.; Molodov, D.A.; Gottstein, G. Correlation Between Microstructure and Texture Development in a Cold-Rolled TWIP Steel. *ISIJ Int.* **2011**, *51*, 812–817. [CrossRef]
9. Saleh, A.A.; Pereloma, E.V.; Gazder, A.A. Texture Evolution of Cold Rolled and Annealed Fe–24Mn–3Al–2Si–1Ni–0.06C TWIP Steel. *Mater. Sci. Eng. A* **2011**, *528*, 4537–4549. [CrossRef]
10. Haase, C.; Chowdhury, S.G.; Barrales-Mora, L.A.; Molodov, D.A.; Gottstein, G. On the Relation of Microstructure and Texture Evolution in an Austenitic Fe-28Mn-0.28C TWIP Steel During Cold Rolling. *Metall. Mater. Trans. A* **2013**, *44*, 911–922. [CrossRef]
11. Haase, C.; Barrales-Mora, L.A.; Roters, F.; Molodov, D.A.; Gottstein, G. Applying the Texture Analysis for Optimizing Thermomechanical Treatment of High Manganese Twinning-Induced Plasticity Steel. *Acta Mater.* **2014**, *80*, 327–340. [CrossRef]
12. Saleh, A.A.; Haase, C.; Pereloma, E.V.; Molodov, D.A.; Gazder, A.A. On the Evolution and Modelling of Brass-Type Texture in Cold-Rolled Twinning-Induced Plasticity Steel. *Acta Mater.* **2014**, *70*, 259–271. [CrossRef]
13. Yanushkevich, Z.; Belyakov, A.; Haase, C.; Molodov, D.A.; Kaibyshev, R. Structural/textural changes and strengthening of an advanced high-Mn steel subjected to cold rolling. *Mater. Sci. Eng. A* **2016**, *651*, 763–773. [CrossRef]
14. Lü, Y.; Hutchinson, B.; Molodov, D.A.; Gottstein, G. Effect of Deformation and Annealing on the Formation and Reversion of ε-Martensite in an Fe–Mn–C Alloy. *Acta Mater.* **2010**, *58*, 3079–3090. [CrossRef]
15. Lü, Y.; Molodov, D.A.; Gottstein, G. Recrystallization Kinetics and Microstructure Evolution During Annealing of a Cold-Rolled Fe–Mn–C Alloy. *Acta Mater.* **2011**, *59*, 3229–3243. [CrossRef]
16. Haase, C.; Barrales-Mora, L.A.; Molodov, D.A.; Gottstein, G. Texture Evolution of a Cold-Rolled Fe-28Mn-0.28C TWIP Steel During Recrystallization. *Mater. Sci. Forum* **2013**, *753*, 213–216. [CrossRef]
17. Haase, C.; Kühbach, M.; Barrales Mora, L.A.; Wong, S.L.; Roters, F.; Molodov, D.A.; Gottstein, G. Recrystallization Behavior of a High-Manganese Steel: Experiments and Simulations. *Acta Mater.* **2015**, *100*, 155–168. [CrossRef]
18. Barbier, D.; Gey, N.; Allain, S.; Bozzolo, N.; Humbert, M. Analysis of the tensile behavior of a TWIP steel based on the texture and microstructure evolutions. *Mater. Sci. Eng. A* **2009**, *500*, 196–206. [CrossRef]
19. Saleh, A.A.; Pereloma, E.V.; Gazder, A.A. Microstructure and Texture Evolution in a Twinning-Induced-Plasticity Steel During Uniaxial Tension. *Acta Mater.* **2013**, *61*, 2671–2691. [CrossRef]
20. Daamen, M.; Haase, C.; Dierdorf, J.; Molodov, D.A.; Hirt, G. Twin-roll strip casting: A competitive alternative for the production of high-manganese steels with advanced mechanical properties. *Mater. Sci. Eng. A* **2015**, *627*, 72–81. [CrossRef]
21. Haase, C.; Zehnder, C.; Ingendahl, T.; Bikar, A.; Tang, F.; Hallstedt, B.; Hu, W.; Bleck, W.; Molodov, D.A. On the deformation behavior of κ-carbide-free and κ-carbide-containing high-Mn light-weight steel. *Acta Mater.* **2017**, *122*, 332–343. [CrossRef]
22. Bagherpour, E.; Reihanian, M.; Ebrahimi, R. On the Capability of Severe Plastic Deformation of Twining Induced Plasticity (TWIP) Steel. *Mater. Des.* **2012**, *36*, 391–395. [CrossRef]
23. Matoso, M.S.; Figueiredo, R.B.; Kawasaki, M.; Santos, D.B.; Langdon, T.G. Processing a Twinning-Induced Plasticity Steel by High-Pressure Torsion. *Scr. Mater.* **2012**, *67*, 649–652. [CrossRef]
24. Wang, H.T.; Tao, N.R.; Lu, K. Strengthening an Austenitic Fe–Mn Steel Using Nanotwinned Austenitic Grains. *Acta Mater.* **2012**, *60*, 4027–4040. [CrossRef]
25. Timokhina, I.B.; Medvedev, A.; Lapovok, R. Severe Plastic Deformation of a TWIP Steel. *Mater. Sci. Eng. A* **2014**, *593*, 163–169. [CrossRef]
26. Haase, C.; Kremer, O.; Hu, W.; Ingendahl, T.; Lapovok, R.; Molodov, D.A. Equal-channel angular pressing and annealing of a twinning-induced plasticity steel: Microstructure, texture, and mechanical properties. *Acta Mater.* **2016**, *107*, 239–253. [CrossRef]
27. Suwas, S.; Tóth, L.S.; Fundenberger, J.-J.; Eberhardt, A.; Skrotzki, W. Evolution of Crystallographic Texture During Equal Channel Angular Extrusion of Silver. *Scr. Mater.* **2003**, *49*, 1203–1208. [CrossRef]

28. Suwas, S.; Tóth, L.S.; Fundenberger, J.-J.; Grosdidier, T.; Skrotzki, W. Texture Evolution in FCC Metals during Equal Channel Angular Extrusion (ECAE) as a Function of Stacking Fault Energy. *Soild State Phenom.* **2005**, *105*, 345–350. [CrossRef]
29. Lapovok, R.; Tóth, L.S.; Molinari, A.; Estrin, Y. Strain localisation patterns under equal-channel angular pressing. *J. Mech. Phys. Solids* **2009**, *57*, 122–136. [CrossRef]
30. Houtte, P.V. Simulation of the rolling and shear texture of brass by the Taylor theory adapted for mechanical twinning. *Acta Metall.* **1978**, *26*, 591–604. [CrossRef]
31. Tomé, C.N.; Lebensohn, R.A.; Kocks, U.F. A Model for Texture Development Dominated by Deformation Twinning: Application to Zirconium Alloys. *Acta Metall. Mater.* **1991**, *39*, 2667–2680. [CrossRef]
32. Lebensohn, R.A.; Tomé, C.N. A Self-Consistent Viscoplastic Model: Prediction of Rolling Textures of Anisotropic Polycrystals. *Mater. Sci. Eng. A* **1994**, *175*, 71–82. [CrossRef]
33. Kalidindi, S.R. Incorporation of Deformation Twinning in Crystal Plasticity Models. *J. Mech. Phys. Solids* **1998**, *46*, 267–290. [CrossRef]
34. Gu, C.F.; Toth, L.S.; Zhang, Y.D.; Hoffman, M. Unexpected brass-type texture in rolling of ultrafine-grained copper. *Scr. Mater.* **2014**, *92*, 51–54. [CrossRef]
35. Gu, C.F.; Hoffman, M.; Toth, L.S.; Zhang, Y.D. Grain size dependent texture evolution in severely rolled pure copper. *Mater. Charact.* **2015**, *101*, 180–188. [CrossRef]
36. Gu, C.F.; Toth, L.S.; Hoffman, M. Twinning effects in a polycrystalline magnesium alloy under cyclic deformation. *Acta Mater.* **2014**, *62*, 212–224. [CrossRef]
37. Prakash, A.; Hochrainer, T.; Reisacher, E.; Riedel, H. Twinning Models in Self-Consistent Texture Simulations of TWIP Steels. *Steel Res. Int.* **2008**, *79*, 645–652. [CrossRef]
38. Molinari, A.; Tóth, L.S. Tuning a self consistent viscoplastic model by finite element results—I. Modeling. *Acta Metall. Mater.* **1994**, *42*, 2453–2458. [CrossRef]
39. Tóth, L.S.; Molinari, A. Tuning a self consistent viscoplastic model by finite element results—II. Application to torsion textures. *Acta Metall. Mater.* **1994**, *42*, 2459–2466. [CrossRef]
40. Saeed-Akbari, A.; Mosecker, L.; Schwedt, A.; Bleck, W. Characterization and Prediction of Flow Behavior in High-Manganese Twinning Induced Plasticity Steels: Part I. Mechanism Maps and Work-Hardening Behavior. *Metall. Mater. Trans. A* **2011**, *43*, 1688–1704. [CrossRef]
41. Lapovok, R.; Tomus, D.; Muddle, B. Low-Temperature Compaction of Ti–6Al–4V Powder Using Equal Channel Angular Extrusion with Back Pressure. *Mater. Sci. Eng. A* **2008**, *490*, 171–180. [CrossRef]
42. Haase, C.; Lapovok, R.; Ng, H.P.; Estrin, Y. Production of Ti–6Al–4V billet through compaction of blended elemental powders by equal-channel angular pressing. *Mater. Sci. Eng. A* **2012**, *550*, 263–272. [CrossRef]
43. Fundenberger, J.-J.; Beausir, B. Université de Lorraine—Metz, 2015, JTEX—Software for Texture Analysis. Available online: http://jtex-software.eu/ (accessed on 8 May 2018).
44. Toth, L.S.; Gilormini, P.; Jonas, J.J. Effect of rate sensitivity on the stability of torsion textures. *Acta Metall.* **1988**, *36*, 3077–3091. [CrossRef]
45. Tóth, L.S.; Arruffat Massion, R.; Germain, L.; Baik, S.C.; Suwas, S. Analysis of Texture Evolution in Equal Channel Angular Extrusion of Copper Using a New Flow Field. *Acta Mater.* **2004**, *52*, 1885–1898. [CrossRef]
46. Haase, C. Texture and Microstructure Evolution during Deformation and Annealing of High-Manganese TWIP Steels. Ph.D. Thesis, RWTH Aachen University, Aachen, Germany, 2016.
47. Leffers, T. The Brass-Type Texture Once Again. In *Texture of Materials. Proceedings of the 11th International Conference on Textures of Materials*; Chu, Y., Ed.; International Academic Publishers: Beijing, China, 1996; Volume 1, pp. 299–306.
48. Leffers, T.; Bilde-Sørensen, J.B. Intra- and Intergranular Heterogeneities in the Plastic Deformation of Brass During Rolling. *Acta Metall. Mater.* **1990**, *38*, 1917–1926. [CrossRef]
49. Hutchinson, J.W. Bounds and Self-Consistent Estimates for Creep of Polycrystalline Materials. *Proc. R. Soc.* **1976**, *348*, 101–127. [CrossRef]
50. Tóth, L.S.; Estrin, Y.; Lapovok, R.; Gu, C. A model of grain fragmentation based on lattice curvature. *Acta Mater.* **2010**, *58*, 1782–1794. [CrossRef]
51. Beyerlein, I.J.; Tóth, L.S.; Tomé, C.N.; Suwas, S. Role of Twinning on Texture Evolution of Silver During Equal Channel Angular Extrusion. *Philos. Mag.* **2007**, *87*, 885–906. [CrossRef]

52. Steinmetz, D.R.; Jäpel, T.; Wietbrock, B.; Eisenlohr, P.; Gutierrez-Urrutia, I.; Saeed–Akbari, A.; Hickel, T.; Roters, F.; Raabe, D. Revealing the Strain-Hardening Behavior of Twinning-Induced Plasticity Steels: Theory, Simulations, Experiments. *Acta Mater.* **2013**, *61*, 494–510. [CrossRef]
53. Gutierrez-Urrutia, I.; Raabe, D. Dislocation and twin substructure evolution during strain hardening of an Fe–22wt.% Mn–0.6wt.% C TWIP steel observed by electron channeling contrast imaging. *Acta Mater.* **2011**, *59*, 6449–6462. [CrossRef]
54. Kusakin, P.; Belyakov, A.; Haase, C.; Kaibyshev, R.; Molodov, D.A. Microstructure Evolution and Strengthening Mechanisms of Fe–23Mn–0.3C–1.5Al TWIP Steel During Cold Rolling. *Mater. Sci. Eng. A* **2014**, *617*, 52–60. [CrossRef]
55. Berrenberg, F.; Wang, J.; Timokhina, I.; Haase, C.; Lapovok, R.; Molodov, D.A. Influence of rolling asymmetry on the microstructure, texture and mechanical behavior of high-manganese twinning-induced plasticity steel. *Mater. Sci. Eng. A* **2018**, *709*, 172–180. [CrossRef]
56. Wang, L.; Benito, J.A.; Calvo, J.; Cabrera, J.M. Twin-Induced Plasticity of an ECAP-Processed TWIP Steel. *J. Mater. Eng. Perform.* **2017**, *26*, 554–562. [CrossRef]
57. Kalidindi, S.R.; Bronkhorst, C.A.; Anand, L. Crystallographic texture evolution in bulk deformation processing of FCC metals. *J. Mech. Phys. Solids* **1992**, *40*, 537–569. [CrossRef]
58. Zhou, Y.; Neale, K.W.; Tóth, L.S. A modified model for simulating latent hardening during the plastic deformation of rate-dependent FCC polycrystals. *Int. J. Plast.* **1993**, *9*, 961–978. [CrossRef]
59. Tóth, L.S.; Neale, K.W.; Jonas, J.J. Stress response and persistence characteristics of the ideal orientations of shear textures. *Acta Metall.* **1989**, *37*, 2197–2210. [CrossRef]

Article

Contribution of Local Analysis Techniques for the Characterization of Iron and Alloying Elements in Nitrides: Consequences on the Precipitation Process in Fe–Si and Fe–Cr Nitrided Alloys

Hugo P. Van Landeghem [1,2,3], Raphaële Danoix [4], Mohamed Gouné [5], Sylvie Bordère [5], Andrius Martinavičius [1,2,4], Peter Jessner [4,5], Thierry Epicier [6], Béatrice Hannoyer [4], Frédéric Danoix [4] and Abdelkrim Redjaïmia [1,2,*]

1 Institut Jean Lamour, UMR 7198, CNRS, Université de Lorraine, Campus ARTEM, 2 allée André Guinier, F-54011 Nancy CEDEX, France; hugo.van-landeghem@grenoble-inp.fr (H.P.V.L.); andriusmart@gmail.com (A.M.)

2 LABoratory of EXcellence Design of Alloy Metals for low-mAss Structure (LabEx DAMAS), Université de Lorraine, Nancy 54011, France

3 SIMAP, CNRS, Grenoble INP, University of Grenoble Alpes, F-38000 Grenoble, France

4 UNIROUEN, INSA Rouen, Normandie Université, CNRS, Groupe de Physique des Matériaux F-76000 Rouen, France; raphaele.danoix@univ-rouen.fr (R.D.); peter.jessner@constellium.com (P.J.); beatrice.hannoyer@univ-rouen.fr (B.H.); frederic.danoix@univ-rouen.fr (F.D.)

5 Institut de Chimie de la matière condensée de Bordeaux, University of Bordeaux, UMR5026, F-33600 Pessac, France; mm.goune@gmail.com (M.G.); sylvie.bordere@icmcb.cnrs.fr (S.B.)

6 Matériaux Ingénierie et Science, UMR CNRS 5510, INSA Lyon, University of Lyon, Bât. Saint-Exupéry, 25 Avenue J. Capelle, F-69621 Villeurbanne CEDEX, France; thierry.epicier@insa-lyon.fr

* Correspondence: abdelkrim.redjaimia@univ-lorraine.fr; Tel.: +33-037-274-2722

Received: 29 July 2018; Accepted: 7 August 2018; Published: 11 August 2018

Abstract: Atom Probe Tomography (APT), Transmission Electron Microscopy (TEM), and 3D mechanical calculations in complex geometry and anisotropic strain fields were employed to study the role of minor elements in the precipitation process of silicon and chromium nitrides in nitrided Fe–Si and Fe–Cr alloys, respectively. In nitrided Fe–Si alloys, an original sequence of Si_3N_4 precipitation was highlighted. Al–N clusters form first and act as nucleation sites for amorphous Si_3N_4 nitrides. This novel example of particle-simulated nucleation opens a new way to control Si_3N_4 precipitation in Fe–Si alloys. In nitrided Fe–Cr alloys, both the presence of iron in chromium nitrides and excess nitrogen in the ferritic matrix are unquestionably proved. Only a certain part of the so-called excess nitrogen is shown to be explained by the elastic accommodation of the misfit between nitride and the ferritic matrix. The presence of immobile excess nitrogen trapped at interfaces can be highly suspected.

Keywords: excess nitrogen; clusters precipitation; Fe–Si and Fe–Cr nitrided alloys; APT and TEM characterization

---

1. Introduction

The automotive industry is constantly looking for ways to reduce the weight of vehicles without sacrificing safety, performance, or durability. The development of both high-strength and low-density steels is a timely and appropriate answer to this challenge. In this context, the substitution of carbon by nitrogen in steel metallurgy could provide a path to such high-performance steels with a reduced manufacturing footprint. First, the eutectoid temperature in the binary Fe–N is much lower than in the Fe–C system (592 °C vs. 723 °C) leading to lower heat treatment energy consumption [1]. Second, the strengthening from nitrogen both in solid solution and precipitated in nitrides leads to high mechanical properties [2,3]. Finally, the presence of nitride forming elements can induce a very fine and intense precipitation of nitrides in α-Fe having a density significantly lower than iron, decreasing the density of the final material through composite effect [4]. However, the low solubility of nitrogen in liquid iron (0.046 wt.% at 1600 °C) constitutes however the main barrier to the development of low-alloy nitrogen steel through regular steelmaking [5]. Nitriding treatments offer an interesting alternative in order to mitigate this difficulty. They consist in making nitrogen atoms diffuse from the surface to the bulk into very thick samples and at a given temperature. The diffusing nitrogen then interacts with iron and/or alloying elements to form nitrides. The presence of these nanometre scale nitrides will affect the structural properties such as density, strength, as well as corrosion and wear resistance [6]. These properties depend on precipitation characteristics (nature, volume fraction, density, and mean size of particles), which are themselves influenced by process parameters and both the nature and the amount of alloying elements. For a better control of the nitride precipitation processes, it is important to understand the role of elements present in the alloy. It is expected that iron, which is in even greater quantity, and trace nitride forming elements such as Al, Ti, or V, will strongly affect the precipitation process. This is the main topic of this paper, in which we will focus on two important aspects. On the one hand, the role of aluminium as a minor element on the nucleation of silicon nitrides during nitriding of Fe–Si alloys. On the other, the study of both the iron content of nanometre nitrides and the so-called "excess nitrogen" in nitrided Fe–Cr alloys.

Regarding the precipitation of silicon nitride during nitriding of Fe–Si alloys, this topic has recently received growing attention [7]. Recent works show that the precipitation of silicon nitrides is taking place in the form of cubic and amorphous particles in Fe–Si alloys. The nanometre size and high density of these particles significantly increase the hardness of the alloy, and the low density of silicon nitrides decreases the mass of a structure [6,8]. A transition to crystalline precipitation occurs when excess nitrogen is removed during annealing [4,9]. The origin of such behaviour is still controversial as evidenced by some recent lively discussions in [4,6–9]. The stability of the amorphous nitride remains an open question all the more so because there are few examples in the literature reporting the formation of an amorphous phase in a crystalline matrix [10–13]. The purpose of this study is thus to investigate the role of minor alloying elements in the precipitation process of amorphous silicon nitrides in nitrided Fe–Si alloys. To this end, atom probe tomography (APT) was employed. The usual overlap in mass spectra between Si^{++} and N^{+} ions has been overcome by using commercially pure ^{15}N.

As for chromium nitride in nitrided Fe–Cr alloys, earlier studies based on atom probe tomography and transmission electron microscopy (TEM) have evidenced that significant levels of iron are present in nitrides [14,15]. However, these results are frequently disregarded because the observed Fe enrichments in CrN can result from experimental artefacts, at least partly. Nevertheless, as will be shown in this paper, recent experimental observations have confirmed the presence of Fe in CrN, even if careful quantitative measurements have shown lower Fe level than initially stated. The consequences of these observations are many. First, if the assumption of stoichiometric MN nitrides (where M stands for metal) is to be ruled out, the volume fraction of particles cannot be simply deduced from a simple lever rule. Unless the amount of iron in the particles is carefully measured, the volume fraction of precipitates remains undetermined. The presence of iron in the precipitates would also modify the thermodynamics of the system and transformation tie-lines. As a result, the kinetics of precipitation, the resulting precipitation state, and mechanical properties of the nitrided layer must be affected and numerical predictions remain inaccurate as long as they do not account for this. Another important potential consequence of the presence of iron in nitrides is the extent of the so-called "nitrogen excess". In the classical view of MN nitride precipitation, the uptake of nitrogen should be limited to the sum of the amount necessary for precipitation of all alloying elements such as stoichiometric nitrides and the equilibrium nitrogen content in ferrite [16]. However, Fe–Cr has demonstrated a considerable capacity for the uptake of larger amounts of nitrogen than expected [17–19]. The additional amount of nitrogen is called excess nitrogen. Keeping the assumption that each metal atom in nitrides binds to one nitrogen atom, that is, nitrides can be considered as (Fe, M)N compounds, the presence of iron in the precipitates leads to an increase in the maximal fraction of nitrides, proportional to the amount of iron they contain. The presence of iron in nitrides could therefore provide a simple explanation to the observation of excess nitrogen as suggested by Ginter et al. [15]. Consequently, it appears essential to confirm the presence of iron in nitrides, and to estimate, as finely as possible, the amount of iron in nitrides, as well as the amount of nitrogen and chromium dissolved in the ferritic matrix (α-matrix). Because of the very fine scale involved, sub-nanometre resolution combined to sensitivity to nitrogen is required. As it fulfils these requirements, APT is the ideal technique for such investigations. Some complementary analyses by analytical TEM (electron energy loss spectrometry, EELS) on two-step extraction replicas are also proposed and reinforce conclusions obtained by APT. The nature of the excess nitrogen uptake, which is a key point in improving our understanding of nitriding treatment, is still matter of debate. In order to address this topic, 3D calculations are performed for determining both the stress field and elastic strain energy induced by the presence of CrN in the α-matrix in complex situations of geometry and anisotropy of elastic properties. The effect of mechanical fields on the nitrogen dissolved in the strained iron lattice is calculated. Comparisons between the measurements and the calculations allow for discussing both the presence and the nature of the so-called excess nitrogen.

2. Experimental Procedure

The materials investigated in the present study are ferritic Fe–5at.% Cr and Fe–3at.% Si alloys. Fe–5at.% Cr samples were nitrided in gaseous atmosphere at 590 °C for 12 h (see [20] for details). The internal structure of the diffusion layer is described in details elsewhere [21]. In order to investigate the stability of iron in the chromium nitrides and the nitrogen content in the matrix, a second set of specimens was prepared. It received an additional thermal treatment of 24 h at 590 °C in vacuum.

Fe–3at.% Si samples were rolled into sheets 1 mm in thickness and plasma nitrided at 570 °C for 1 h using commercially pure ^{15}N in order to avoid the usual overlap in mass spectra between Si^{++} and N^{+} ions at 14Da during atom probe analyses. Details of the method can be found in [4]. Both sides of the sheets were nitrided. Afterwards, samples were annealed at 570 °C in vacuum (10^{-6} mbar) for 20 h.

Depending on the material, and the thickness of the nitrided layers, the specimens for atom probe tomography analyses were prepared combining mechanical grinding and electrochemical etching for the Fe–Cr alloy [20], and focussed ion beam (FIB) lift out and annular milling for the Fe–Si one [22]. Each sample was finished to a tip radius smaller than 50 nm. Field ion microscopy (FIM) (Home built, University of Rouen, Rouen, France) was conducted at a temperature of 80 K and voltages between 10 and 15 kV, using neon as imaging gas. APT measurements on the Fe–Cr alloy were performed at 80 K on a CAMECA energy compensated tomographic atom probe (ECOTAP), equipped with an advanced delay line detector [23], with a pulse repetition rate of 2 kHz and a pulse fraction of 20%. Three-dimensional reconstructions were obtained using the software (Home built, University of Rouen) developed at the University of Rouen, based on the algorithm described in [24]. Concerning the Fe–Si alloy, analyses were conducted on a Cameca LEAP® 4000HR (Cameca Inc.,Madison, WI, USA), using the same experimental conditions, except for the pulse repetition rate, which was increased to 200 kHz. Three-dimensional reconstructions were obtained using Cameca IVAS® 3.6.8 (Cameca Inc., Madison, WI, USA).

The determination of the composition of the precipitates was conducted using a cluster identification method. Threshold concentrations in one or more elements for the phase under consideration are defined. For each atom, the local concentrations in a surrounding sphere of radius 1 nm are calculated, and compared to the threshold values. If the measured concentrations lie within the given limits, the tested atom is considered as being part of a so-called cluster [25,26].

Specimens for TEM investigations were prepared using the two-step extraction replica method [27]. As the iron concentration in nitrides is of interest, any contribution to the determination of the nitride composition arising from the ferritic matrix needs to be eliminated. It is therefore not possible to measure the nitride concentration directly from thin foils. Therefore, extraction replicas on carbon films have to be fabricated from the nitrided Fe–5at.% Cr alloy to meet these requirements. As the region of interest is the diffusion layer, the surface area of a bulk specimen, with a section of 5×5 mm^2, has been removed by grinding and polishing the sample to a depth of 100 µm from the original surface. For extraction replicas, the sample surface is further attacked with Nital 2%. A cellulose-acetate polymer film was glued on the surface and pulled off. Nitride precipitates stick on the polymer-film on which a carbon film, approximately 50 nm thick, is deposited. Subsequently, the polymer is dissolved in a solution of methanol in dichloromethane and the extracted nitrides embedded in the carbon film are recovered on TEM grids. By applying this two-step technique, a possible contamination of iron during the sample preparation process is minimized, as will be discussed below. Electron energy-loss spectrometry (EELS) was performed on a field emission gun JEOL 2010F electron microscope operated at 200 kV. The microscope is equipped with a Gatan DigiPEELS spectrometer with a standard photodiode array detector. EELS spectra were recorded with the specimen at room temperature using the following parameters: Convergent half-angle $\alpha = 10°$, collection half-angle $\beta = 7°$, acquisition time equal to 10 s, probe size focused to about 3 nm in TEM mode, and a nominal energy dispersion of 0.5 eV. The experimental spectra were classically corrected from the dark current and channel-to-channel gain variations of the detector, and then de-convoluted from the multiple scattering using the Fourier-ratio technique. The background under each edge '*i*' (where *i* stands for N, Cr, Fe) of interest is subtracted according to its extrapolation from a power-law, and fitted over an optimized energy window ΔE_i preceding each edge. According to the elements present (Fe, Cr, N), the chemical analysis can be conducted using the classical method proposed by Egerton [28]. To evaluate the accuracy of the measurements, the energy windows used have been optimized on standard spectra CrN and Cr_2N used by Mitterbauer [29]. An accuracy of at least about 5% can be obtained for the atomic ratio [Cr]/[N]. This accuracy is of the same order of magnitude as demonstrated by Courtois in the case of NbCN precipitates examined on extraction replicas [30].

TEM thin foils were prepared from disks of 1.5 mm in radius machined at a distance of about 100 µm from the surface. These disks were electrochemically thinned in a Struers Tenupol at 27 V around 10 °C and cleaned by using a plasma system. They were mainly observed in a cold FEG JEOL ARM 200F and additional observations were carried out in a Philips CM12 (Philips-FEI, Hillsboro, OR, USA) a Philips CM200 (Philips-FEI, Hillsboro, OR, USA) and a JEOL 2010F.

3. Calculation of the Elastic Stresses and Strain Energies

The mechanical calculations were performed using the THETIS code [31]. The simulation system is a cube of representing the ferritic matrix. Periodic conditions and an external pressure $p = 0$ were used to allow the system to relax the anisotropically deformed particle. The pressure p corresponding to the spherical part of the stress tensor $\boldsymbol{\sigma}$ and the shear stress tensor $\boldsymbol{\tau}$ were considered. The relationship between $\boldsymbol{\sigma}$ and $\boldsymbol{\tau}$ is given by:

$$\boldsymbol{\sigma} = -p\boldsymbol{I} + \boldsymbol{\tau} \tag{1}$$

Both the stress and strain fields induced by the formation of the nitride particles into the ferritic matrix were then calculated. The total elastic strain energy density stored into the system was determined from:

$$G^{el,\Phi} = \frac{1}{V_t}\int_{V_t} \frac{1}{2}\boldsymbol{\epsilon} : \boldsymbol{\sigma} dV \tag{2}$$

where $\boldsymbol{\epsilon}$ is the strain tensor and V_t is the total volume of the system.

4. Results and Discussion

4.1. Mechanism of Amorphous Si_3N_4 Precipitation in Nitrided Fe–Si Alloys

Atom probe analyses were carried out on four specimens sampled from the as-nitrided Fe–Si alloy, the typical volume of which was $50 \times 50 \times 200$ nm^3. One such volume shown in Figure 1a was found to be representative of all other volumes. The chemical composition of the alloy based on the four analyses is reported in Table 1. The concentration in Si, which is the major alloying element, was found to be very close to the nominal one (3.1 at.%). The concentration in residual Al, Ti, and Mn are about 0.06, 0.03, and 0.01 at.%, respectively. Traces of C were also detected.

On the one hand, frequency distribution analysis confirmed that Si is homogeneously distributed. Added to the fact that Si concentration in the matrix is close to the bulk concentration, it suggests that both the fraction and number density of silicon nitride are very low, if non-zero, in the as-nitrided condition. On the other hand, residual Al was found to form Al–N clusters that are visible with the naked eye on tomograms, as illustrated in Figure 1a, where they are highlighted with arrows. These clusters were also systematically identified and delineated using 7 at.% Al isosurfaces. Local magnification, a technique-related artefact that causes incorrect positioning of atoms in the immediate vicinity of phase interfaces, prevents any accurate calculation of the exact composition of such small clusters [32,33]. However, their Al/N ratio proved to be roughly 1.1, close to the theoretical ratio of Al/N.

Figure 1b shows the tomogram of a typical atom probe specimen in the annealed condition, in which silicon nitride precipitates are visible. It exhibited lower matrix concentrations in Si, Al, and N. This drop in concentration can be ascribed to the precipitation of the observed silicon nitride. After correction for the overlap of $^{30}Si/^{15}N$ and $^{28}Si/^{14}N$ (residual), Si and N contents in the precipitates was determined to be ≈42 at.% and ≈55.5 at.%, respectively, leading to a Si/N ratio of 0.757. This ratio is in excellent agreement with the theoretical ratio of Si_3N_4, suggesting that the precipitates deviate little from stoichiometry, which confirms similar reports based on different techniques [4,7,9]. Al, Ti, and Mn were found to be present in the precipitates at contents of 1.75, 0.27, and 0.14 at.%, respectively. The concentrations of the precipitates in these elements were found to vary considerably, up to about 40 at.% in the analysed particles, probably due to locally different precipitation states. The residual Fe found in the precipitates is likely to be marginally overestimated due the possible $^{28}Si^{+}/^{56}Fe^{2+}$ and $^{14}(N_2)^{+}/^{56}Fe^{2+}$ peak overlaps and/or trajectory overlap of ions [34].

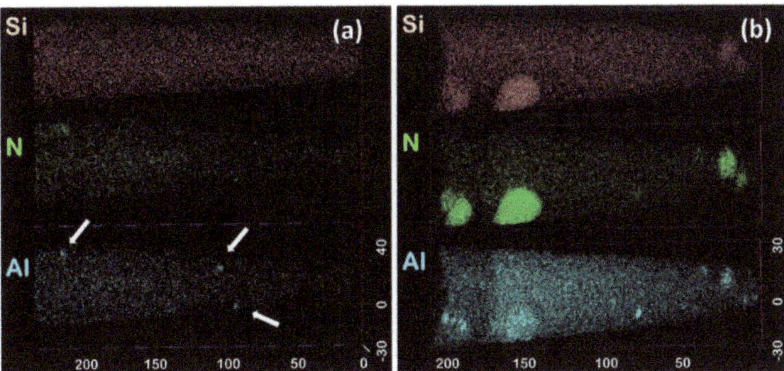

Figure 1. Tomograms of Si, ^{15}N, and Al in (**a**) the as-nitrided sample and (**b**) after 20 h of annealing. The arrows indicate Al–N clusters. Only 50% of detected Si atoms are shown for clarity (Scale in nm).

Table 1. Atomic compositions of the Fe–Si alloy, after cold rolling, nitriding, and annealing. Compositions in the annealed sample are given separately for the ferritic matrix and Si_3N_4 precipitates. Al–N clusters are also excluded for the calculation of matrix composition. The error is calculated from counting statistics, according to [35].

Alloying Elements	Cold Rolled	As-Nitrided	Annealed	
			Matrix	Si_3N_4
Fe	96.77 ± 0.01	96.82 ± 0.01	97.98 ± 0.01	0.10 ± 0.02
Si	3.10 ± 0.01	2.91 ± 0.01	1.79 ± 0.01	42.03 ± 0.35
^{15}N	-	0.18 ± 0.01	0.13 ± 0.01	55.71 ± 0.35
Al	0.06 ± 0.01	0.06 ± 0.01	0.04 ± 0.01	1.75 ± 0.09
Ti	0.04 ± 0.01	0.03 ± 0.01	0.03 ± 0.01	0.27 ± 0.04
Mn	0.01 ± 0.01	0.01 ± 0.01	-	0.14 ± 0.03
C	0.02 ± 0.01	0.01 ± 0.01	0.03 ± 0.01	-

Based on silicon nitride precipitates intercepted in six atom probe specimens, their number density could be estimated to about 10^{21} m^{-3}. Clustering of Al and N atoms is obviously more pronounced here. It is noteworthy that all Si_3N_4 are associated with an Al rich region, but not reciprocally. Al–N clusters were found to be more numerous than Si_3N_4 precipitates, at about 4×10^{21} m^{-3}, which is still the same order of magnitude. Since they are also anterior to the silicon nitride precipitates, it suggests that they act as nucleation sites for Si_3N_4. The proximity histogram (proxigram) shown in Figure 2a, is constructed based on a 25 at.% N isosurface, by calculating concentration as a function of the distance to this interface, from the tomogram shown in Figure 1b [36]. It indicates that silicon nitride precipitates are enriched in Al (mostly at the interface) as well as in Ti and Mn (not shown here). However, since the concentrations are averaged over multiple precipitates, it does not represent accurately the distribution of elements in each individual precipitate. In particular, it has been observed in Figure 2b that Al–N clusters can lie at the interface between the Si_3N_4 precipitates and the matrix.

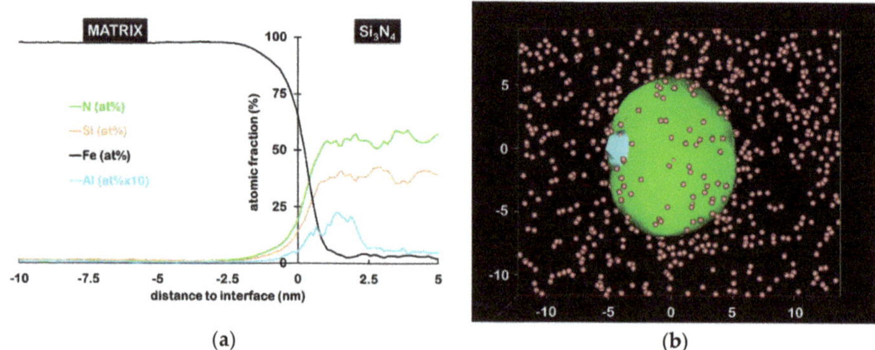

(a) (b)

Figure 2. (**a**) Proxigram across Si_3N_4/matrix interfaces shown in (**b**). Al–N clusters, selected by 7 at.% isosurface (blue), can be seen at the interface between the Si_3N_4 precipitate (isosurface N = 25 at.%, shown in green) and matrix (**b**). For clarity, only Si ions in the matrix are shown.

These results indicate first that the formation of Si_3N_4 precipitates occurs after the formation of Al–N clusters. This is consistent with the high affinity of Al with N and the large volume misfit reported for Si_3N_4 that results in its high nucleation energy barrier [8]. Further, the fact that every observed Si_3N_4 precipitate contain, or is directly adjacent to, Al–N clusters suggests they nucleated heterogeneously at the interface of these clusters. This is once again consistent with too high a homogeneous nucleation barrier leading to heterogeneous precipitation on low energy sites. This has already been observed on dislocations [8], and for other nitrides [37] on Al–N clusters as well. Consequently, the number density of Si_3N_4 nucleation sites is directly related to the number density of those clusters, a point that is often disregarded for simplicity's sake in classical precipitation models, where each substitutional is assumed to be a potential nucleation site [9,38]. The present result shows that this approximation does not hold in the case of silicon nitride and opens new strategies in controlling silicon nitride precipitation.

4.2. Experimental Evidence of Iron in CrN in Nitrided Fe–Cr

4.2.1. Field Ion Microscopy & Transmission Electron Microscopy

The typical chromium nitride precipitates, formed in the diffusion layer of nitrided Fe–Cr alloys are thin disc-shaped platelets with nanometric dimensions [21,39–41]. They exhibit a face centred cubic structure, with the metal atom on the fcc sites and nitrogen occupying the octahedral sites. The [100] direction of the nitride is then parallel to the [110] direction of the bcc matrix, with a misfit of about 2.34%. In the perpendicular direction, the misfit is much larger, close to 45% [42]. Therefore, coherent interfaces along the (001) planes and incoherent interfaces along the (100) and (010) planes develop. The growth rate in the direction normal to the incoherent interface is much faster than in the direction of the coherent interfaces [43]. The latter are forced to migrate by a ledge mechanism, which explains the disk-shaped form of the nitrides. Due to symmetries in the cubic system, three variants of platelets develop, with habit planes lying in the bcc {001} planes.

These platelets can be observed by high-resolution transmission electron microscopy (HR-TEM), provided the thin foil is observed along <001> zone axis. On the HR-TEM, recorded along [001] axis, only two visible variants are developed along [100] and [010] directions, respectively. The third variant is revealed edge-on in the HR-TEM micrograph recorded along [001] zone axis, in Figure 3b.

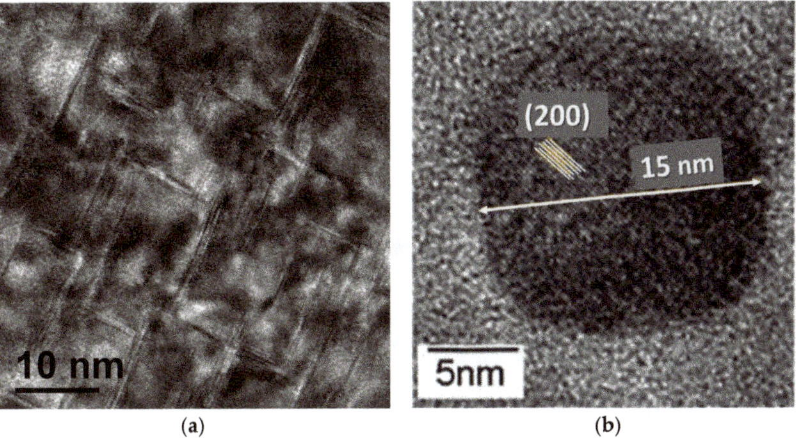

Figure 3. (**a**) High-resolution transmission electron microscopy (HR-TEM) micrograph recorded along the [001] zone axis. In this configuration, only two variants are visible, the third, lying parallel to the surface, is hidden; (**b**) High-resolution view on a single nitride obtained from a replica, with (200) fcc planes in the precipitate indicated.

The same microstructure was investigated using three-dimensional field ion microscopy (3D-FIM) to reveal the three variants simultaneously. Details related to 3D-FIM can be found in [44,45]. Figure 4 shows three sections of the analysed volume: Two along the analysis direction (in orange) and one perpendicular (in black and white). The three variants of CrN appear in dark contrast for all sections.

Combining HR TEM and 3D-FIM, the average diameter of CrN platelets is found to be close to 20 nm, and their thickness about 1 to 2 nm.

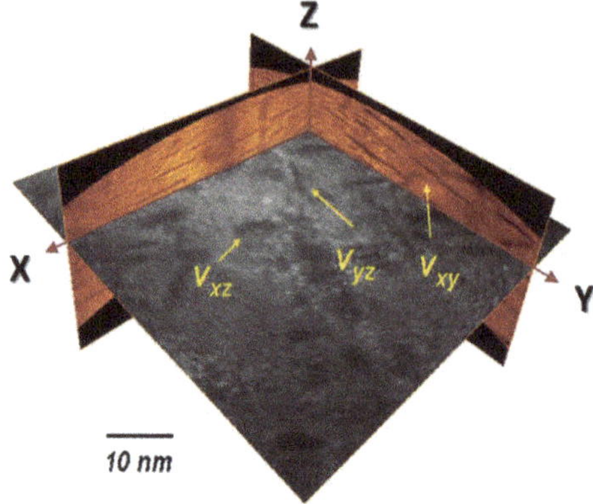

Figure 4. 3D-FIM reconstruction of analysed volume in the nitrided layer, showing the three variants of CrN platelets. CrN platelets appear in dark contrast. The different variants, referred as *Vxy*, *Vyz* and *Vxz* are indicated with yellow arrows.

4.2.2. Atom Probe Tomography

A typical APT reconstruction is shown in Figure 5a. Due to the high density of precipitates, standard atom maps are difficult to interpret, and the nanostructure is better revealed using isoconcentration surfaces. The green surfaces in Figure 3a are surrounding regions where the local Cr + N content is higher than 40 at.%. They clearly correspond to CrN particles. In this reconstruction, the three variants can be observed, and the average dimensions are coherent with those previously derived, as shown in the atom map in Figure 5b.

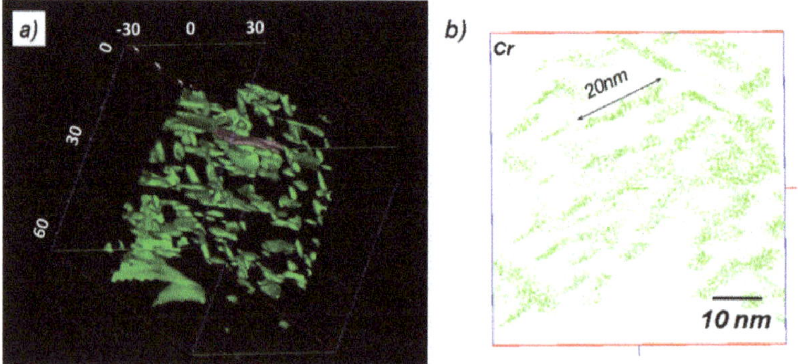

Figure 5. APT reconstruction in the nitrided layer containing CrN: (**a**) isoconcentration Cr + N > 40 at.%, showing the morphology and distribution of CrN platelets; (**b**) atom map showing the location of individual Cr atoms. For clarity, the represented volume is restricted to a 2 nm thick slice.

Concentration profiles are plotted by moving sampling sub-volumes along a given direction. The main advantage of this technique is that the spatial resolution of the profile is defined by the operator (sampling volume thickness), and that the profile direction can be chosen perpendicular to the nitride habit planes, whatever the orientation of the specimen. Together with a lateral extension of the sampling volume smaller than the nitride diameter, concentration profiles with minimized spatial overlaps between matrix and precipitates are obtained, and nitride composition obtained with the highest accuracy. This procedure is illustrated in Figure 6.

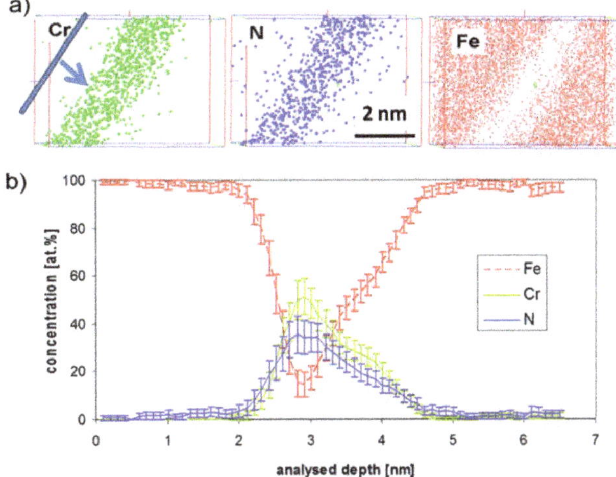

Figure 6. (**a**) Elemental atom maps around a selected nitride and (**b**) corresponding composition profile obtained perpendicular to the nitride habit plane, with a resolution of 0.1 nm. The sampling volume (0.1 nm thick) and profile direction are indicated on the top left of the Figure 6a.

The concentration profile shown in Figure 6b indicates that there is 15 at.% of residual iron, still present in the core region of the nitride.

Because of the experimental artefact of local magnification, this value should be used with caution. Indeed, it is known that the local composition in thin layers (such as platelets in this case) determined by APT depends on the angle between the layer and analysis direction [46]. This is due to a potential overlap of ion trajectories in the close vicinity of the matrix-platelet interfaces, caused by a difference in evaporation field between the two phases, known as local magnification. When the evaporation field is higher in the matrix, ions initially in this phase can erroneously be positioned in the platelets in 3D reconstructions. This effect is absent when the platelet is perpendicular to the analysis direction (inclination angle equal to zero), and increases with the inclination angle. As previously mentioned, nitrides appear in dark contrast in FIM micrographs, indicating that they have a lower evaporation field (or cohesive energy) than the matrix. Consequently, it is expected that the amount of residual iron in the nitride will depend on their orientation with respect to the analysis direction. In order to investigate this effect, the apparent iron content in nitrides has been determined as a function of the inclination angle. Figure 7 reports this evolution, as obtained from different samples having various orientations.

The iron content apparently increases with the inclination angle from 0 to 90°, which is consistent with the local magnification effect. For geometrical reasons [21], the experimental data are fitted by a sine-like function, and the best fit is obtained by a least squares method. The extrapolation of the fit to the zero-inclination angle (where artificial mixing is supposed to be absent) leads to a value of 6.6 at.% Fe, considered as the actual residual amount of Fe in the CrN platelets.

It is worth noting that a nitrogen deficit is always observed during the APT analysis of CrN, leading to a Cr/N ratio smaller than 1. Based on the hypothesis that this deficit is mostly due to evaporation artefacts, the real nitrogen concentration in the nitride is higher than measured by APT, and thus the iron concentration measured has to be seen as an upper limit.

In order to confirm this result based on sine-like fit of the experimental data, the use of a complementary technique is necessary. Therefore, analytical TEM investigations were conducted to independently measure the iron content in nitrides.

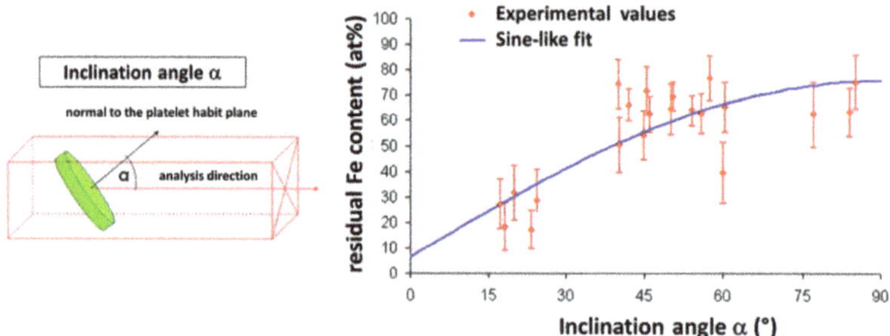

Figure 7. Measured Fe-concentration against the inclination angle between nitride and analysis direction.

4.2.3. Analytical Transmission Electron Microscopy

Previous TEM investigations on nitrides in nitrided systems have been performed on one-step extraction replicas [15,22,47,48]. The preparation technique of these replicas can result in an incomplete dissolution of the ferritic matrix surrounding the nitride particles. Such matrix residue would later produce an unwanted iron signal during TEM analyses. Two-step replica extraction was therefore preferred here, being less likely to produce this artefact. To the knowledge of the authors, this is the first attempt at quantifying iron in chromium nitride using this preparation technique.

The micrograph in Figure 8 shows a group of nitrides on the supporting carbon layer with the corresponding EELS spectrum. Two nitrides, one upright and one horizontal, are marked by dashed lines, indicating their disk-shaped form. They show a diameter of approximately 20 to 30 nm, but their thickness is difficult to access, because their exact orientation on the replica is not known. The EELS spectrum exhibits the 'saw tooth' edge of nitrogen (N-K) and the white lines of the chromium L_{2-3} edge. The iron peak Fe-$L_{2,3}$ is low, but still emerges from the background after the Cr peak. Applying the analytical method proposed by Egerton [28], the ratios of Fe/Cr and Cr/N have been determined to be Cr/N = 0.9 and Fe/Cr = 0.14. This leads to a ratio of (Cr + Fe)/N = 1.03. Both the platelet shaped form and the stoichiometry are consistent with fcc-NaCl-type CrN nitrides. The accuracy of the measurements is estimated to be approximately 5%.

The key point here is to identify the location of detected Fe-atoms: Are they incorporated in the nitrides, or concentrated at their surfaces, due to the replica extraction technique. If iron residues were mostly deposited on the particle surfaces, this would lead to a more pronounced iron concentration along the edges of the particles than in their centres, as the surface over volume ratio would be larger. Figure 9b shows the EELS spectra obtained in both regions of the same particle. The Cr and Fe contents are, within experimental errors, identical in the edge and the core of the particle, strongly indicating that surface contamination can be ruled out. It is concluded that Fe atoms are homogeneously incorporated in the nitrides, partially substituting to chromium, in agreement with previously reported results [49].

This procedure has been reproduced for a total of 44 measurements. The mean chemical composition of the nitride precipitates obtained is $(Cr_{0.89\pm0.02}Fe_{0.11\pm0.01})N_{0.88\pm0.02}$, leading to an average residual amount of iron equal to 5.85 at.%.

This result is fully coherent with the value obtained by APT, where 6.6 at.% was considered as the upper estimate. It can therefore be concluded that two independent measurement lead to the same conclusion, that there is about 6 at.% Fe in the chromium nitrides after nitriding.

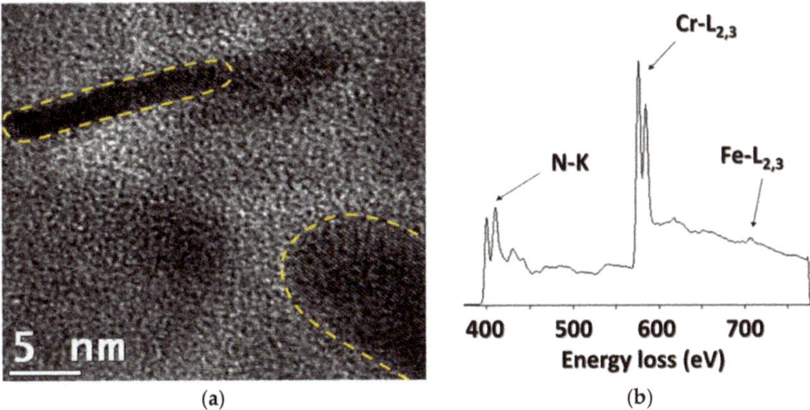

Figure 8. (**a**) TEM micrograph of nitride replicas on carbon film and (**b**) corresponding electron energy loss spectrometry (EELS) spectrum indicating the presence of Fe in Cr-rich nitrides.

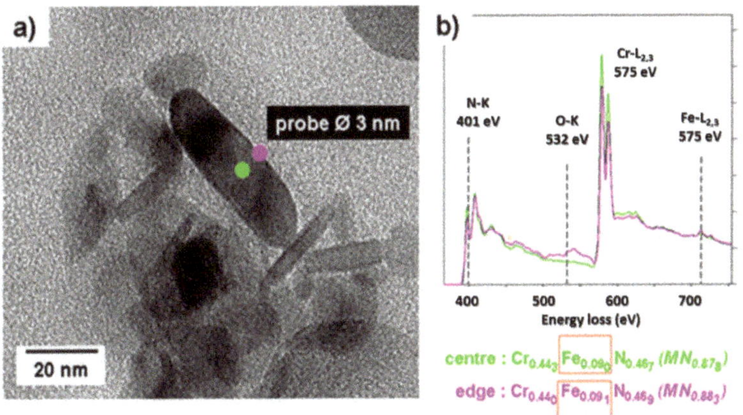

Figure 9. EELS spectra (**b**) of a nitride lying on a carbon replica (**a**) and corresponding Cr and Fe contents. (The Fe-concentration does not differ significantly between the core of the nitride and its edge).

4.2.4. Equilibrium Concentration of Nitrides

If the previous results prove the presence of iron in CrN right after nitriding, its origin is not clear: Is this presence related to a kinetic effect (slow redistribution of iron) or due to the thermodynamics of this ternary system. In order to elucidate this point, an additional heat treatment, consisting in a simple thermal holding was applied to the as-nitrided specimens. The samples were thus aged for 24 h at 590 °C (the nitriding temperature).

Figure 10a shows tomograms obtained after this further thermal treatment. They clearly show that the nanostructure has coarsened, that is, precipitates have grown (mostly in their habit planes), and their number density has significantly dropped. Figure 10b shows the concentration profiles obtained with the same protocol as previously described on the first nitride on the left of Figure 10a. These concentration profiles clearly indicate that iron is still present in the core of the nitrides. This presence cannot be solely explained by the local magnification effect, as the amount of Fe is similar to the one observed in the as nitrided precipitates, for the same inclination angle (here 15°).

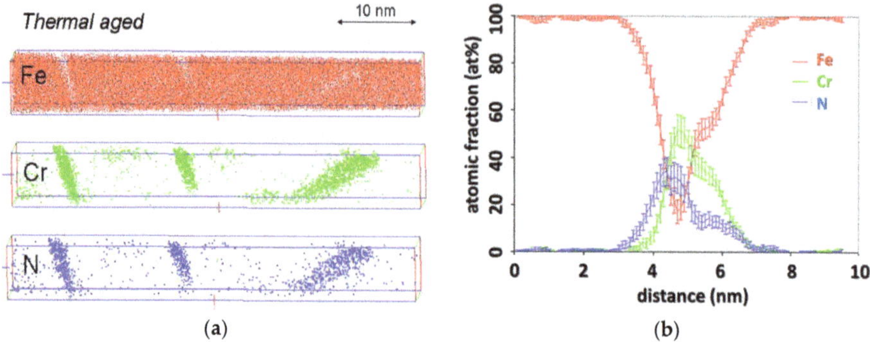

(**a**) (**b**)

Figure 10. (**a**) Atom probe tomography (APT) tomogram and (**b**) corresponding composition profile (**b**) of nitride platelet in nitrided Fe-5at.% Cr alloy followed by thermal treatment in vacuum for 24 h at 590 °C.

This observation leads to the conclusion that the presence of iron in CrN cannot be regarded as a kinetic effect, but rather as a thermodynamic one. It must be accepted that the equilibrium composition of CrN nitrides includes iron, to a level of approximately 6 at.%.

The classical assumption is that only stoichiometric CrN nitrides are formed, excluding any solvent content in the particles. The reason for such a simplifying assumption is obvious, as it reduces the thermodynamic quantities to be assessed to well-known compounds. Driving forces (equilibrium and kinetic) may be calculated from thermodynamic data related to CrN, as implemented in THERMOCALC®. In the ternary Fe–Cr–N system, these data come mainly from the thermodynamic evaluation of the Cr–N and the Fe–N systems [50]. Classically, the formula used for the CrN nitride is $Cr_1(N, Va)_1$ (where Va is vacancies in the second sublattice). It is supposed that only a very small amount of vacancies is allowed on the second sublattice.

The assessment proposed by Frisk [50] supposes that the CrN phase has a very narrow existence range and implies that CrN is considered as a stoichiometric phase.

Therefore, the thermodynamic databases presuppose that the compound formed is stoichiometric and contains no iron. The elastic effects are also neglected. The consequences are twofold. On the one hand, it leads to a distorted description of the system thermodynamics, because the precipitate equilibrium volume fraction is skewed. On the other hand, since thermodynamics and kinetics are coupled, it also affects the description of precipitation kinetics through the evaluation of the driving force for both nucleation and growth. More precisely, the tie lines, which depend on thermodynamics and kinetics, will be affected. A consequence of the present finding is that it may potentially question the classical thermodynamic treatment of nitriding, not only in Fe–Cr, but also in all the systems where nitrides potentially contain iron.

Recently, thermodynamic calculations performed with Thermocalc software show that chromium nitrides can incorporate up to 22 at.% of iron during nitriding of Fe–3at.% Cr alloy at 520 °C [15]. Regarding the nature of the presence of iron in chromium nitrides, it may be addressed in two ways. First, iron is the solvent, and is therefore present in very high quantities in the base material. Second, iron nitride FeN has a lower solubility product constant than chromium nitride CrN. A simplistic description of a mixed nitride in the framework of the ideal solid solution as proposed by Hillert and Staffansson [51] gives additional clarifications. Indeed, the mixed nitride (Fe, Cr)N can be regarded as a combination of the two nitrides CrN and FeN. With the regular solid solution model, the amount of iron in the nitride particles depends on the amount of iron in the matrix, and the relative stability of FeN with respect to CrN. Thermodynamic data show that iron nitride is 10 times more stable than CrN at the nitriding temperature [52]. The calculations performed by this approach explain the presence of iron in particles (from 5% to 95% of iron), with the content of iron depending on the solubility product of FeN with respect to CrN [21]. Consequently, the iron content in CrN precipitate that was measured experimentally is consistent with this modelling approach.

4.3. Excess Nitrogen in Fe–Cr–N System

4.3.1. Effects of Iron

Some systems (Fe–V, Fe–Cr) have a considerable capacity for the uptake of so-called excess nitrogen. Indeed, it was observed that the nitrogen uptake was larger than necessary for simultaneously precipitating the entire alloying element content as stoichiometric nitride and reaching the equilibrium nitrogen content in the ferritic matrix. The origin of that behaviour is not fully understood but it can be suspected that the presence of iron in the corresponding nitride (CrN in our case) could explain or influence the observed excess nitrogen. Indeed, it can be supposed that substituting Fe to Cr in CrN would occur at a constant M/N ratio. This would result in a Cr/N ratio lower than 1 and iron would thus contribute to nitrogen excess. In other words, if the Cr/N ratio does not change, it means that iron in nitrides cannot explain excess nitrogen, either in whole or in part.

As our experimental results show that the Cr/N ratio is never smaller than 1, (equal to 1 in the case of EELS analysis), it can be concluded that, contrary to what had been suggested [15], the presence of iron in chromium nitrides does not contribute to excess nitrogen in nitrided Fe–Cr alloys.

4.3.2. Effect of Elastic Accommodation of the Misfit between Nitride and Matrix

TEM and APT analyses show that nitrides grow along three orthogonal directions of space with a disk morphology. A 3D representation of this configuration corresponding to a 5% volume fraction is shown in Figure 11.

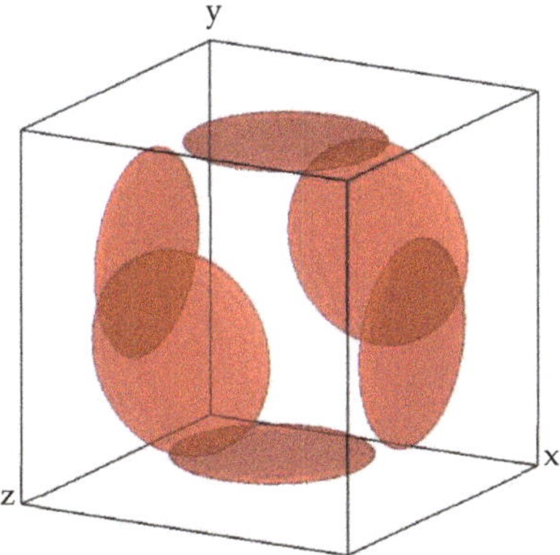

Figure 11. Representation of the 3D configuration of chromium nitride platelets at a volume fraction of 5%.

The initial strain of the platelets was obtained from the lattice mismatch between CrN and ferrite based on electron diffraction patterns. The anisotropic elastic strain related to the morphology of the particles can be written as the strain tensor below:

$$\varepsilon = \begin{pmatrix} \varepsilon_E & 0 & 0 \\ 0 & \varepsilon_{L1} & 0 \\ 0 & 0 & \varepsilon_{L2} \end{pmatrix} \quad (3)$$

with $\varepsilon_E = -O.45$ in the thickness direction and $\varepsilon_{L1} = \varepsilon_{L2} = -0.023$ in the main axes of particles. These data are fully coherent with the misfits measured along the [100] and [110] directions.

The properties of the nitrides and the matrix used for the mechanical calculations are summarized in Table 2.

Table 2. Physical properties of ferrite (bcc-iron) and chromium nitride (fcc-CrN) used as input for the mechanical calculations.

Physical Properties	α-Fe	CrN
Young modulus (GPa)	178	230
Poisson coefficient	0.3	0.29
Molar volume (cm^3/mol)	7.16	11.30
Volume weight (g/cm^3)	7.8	5.9

The calculation of the initial elastic strain relaxation, leads to the equilibrium energy configuration that can be visualized as stress and strain energy distribution maps. For example, pressure and total elastic strain energy in characteristic planes of the system for a precipitate volume fraction of 5% are shown in Figure 12.

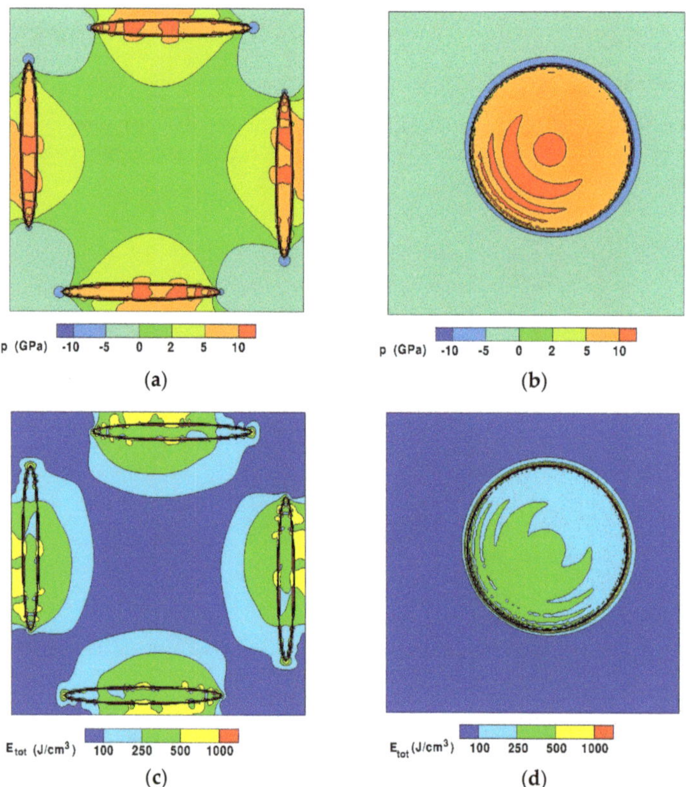

Figure 12. Results of mechanical calculations for a 5% volume fraction of precipitates. (**a**) Pressure map in plane (X = 0.5; Y; Z); (**b**) Pressure map in plane (X = 0.065; Y; Z); (**c**) Total elastic strain energy map in plane (X = 0.5; Y; Z); (**d**) Total elastic strain energy map in plane (X = 0.065; Y; Z).

Overall, it can be noted that the matrix is under strong pressure gradients, from compression in the centre of the system to tension at the edges. As expected, the particles are strongly compressed.

In order to determine the influence of the stress field on the thermodynamic equilibrium, the evolution of the elastic energy stored in the system can be calculated as a function of the volume (or atomic) fraction of precipitates. In that case, from the work of [53] and if the surface tension contribution is neglected, it can be shown that the solubility product is affected in the following manner [54]:

$$K_{el}^{\Phi} = K^{\Phi} exp\left(\frac{V^{\Phi}}{RT}\frac{\partial G^{el,\Phi}}{\partial f^{\Phi}}\right) \tag{4}$$

where $G^{el,\Phi}$ corresponds to the volumetric elastic energy density in the system due to the formation of phase Φ in the ferritic matrix and V^{Φ} is the molar volume of nitride.

In the same way as for the Gibbs-Thomson effect [55], an additional term to the chemical potential of precipitate, equal to $V^{\Phi}\frac{\partial G^{el,\Phi}}{\partial f^{\Phi}}$, must be added to account for the elastic strain energy stored in the system.

These mechanical calculations lead to the results shown in Figure 13, where the elastic strain energy is plotted as a function of the atomic fraction of precipitates. To that end, the volume fraction of particles was converted into an atomic fraction.

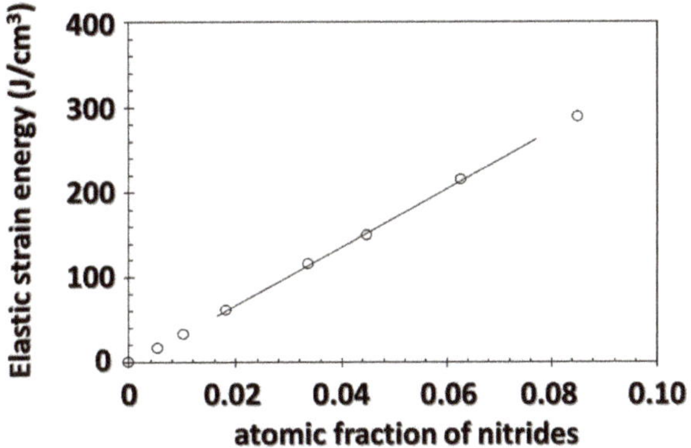

Figure 13. Evolution of the stored elastic strain energy as a function of the atomic fraction of nitrides, based on mechanical calculations.

It should be noted that the first derivative of this curve $\frac{\partial G^{el,\Phi}}{\partial f^{\Phi}}$ is always positive, which indicates that the stress field resulting from the misfit between the particle and its matrix causes an increase in its solubility, in accordance with Equation (4). For atomic fractions below 8% (volume fractions below 10%), the elastic strain energy changes linearly with the atomic fraction of precipitates, with a slope of 3300 J·cm^{-3}. This type of evolution has already been observed in the case of coherent equilibria if one assumes a low precipitate fraction and that the elastic properties of the matrix and the precipitate are similar, homogeneous and isotropic. Three important observations should be made about this analysis. First, in a ternary system, it is not only the equilibrium concentration of nitrogen that is affected but also the product of the activities of both chromium and nitrogen, contrary to what is commonly stated [19,56]. Second, as the elastic strain energy changes linearly with atomic fraction of precipitates in the considered range, the increase in nitrogen content induced by the stress field is not expected to be related to the volume fraction of precipitates (see Equation (4)). Finally, this increase cannot be ascribed to the Gibbs-Thomson effect, which can only cause a maximum increase—for precipitate radius the order of interplanar distances—of the solubility product by a factor of 20.

Further analyses can be formulated by comparing the above results to the isothermal section at 590 °C of the Fe–Cr–N ternary diagram, shown in Figure 14. It can first be seen that the composition of the alloy, represented in Figure 14a by a black dot, lies in the two-phase α + CrN region. The nitrogen and chromium contents in the matrix in equilibrium with CrN are set by the tie-line running through the alloy composition as illustrated in Figure 14a. They are found to be 0.18 at.% and 0.61 at.% for N and Cr, respectively. The nitrogen content measured in the matrix is found to be 0.55 ± 0.02 at.%, or three times the predicted equilibrium content. The solvi of CrN with and without the stress field are shown in the iron rich corner of the section in Figure 14b. From this diagram, it is clear that the stress field increases the solubility of CrN in α and thus enlarges the single phase α region. In the presence of the stress field, the solvus is almost a vertical line. However, the predicted N content under stress is 0.43 at.% at most, which is still lower than the measured content. It is reminded here that the calculated solvus under stress is an upper limit of the solubility increase due to elastic strain, as it assumes that the elastic stress is entirely relaxed through equilibrium shift. In reality, only a certain part of stress can be relaxed by modification of the equilibrium state. According to [57], many mechanisms of stress relaxation can co-exist simultaneously. As an example, one can mention crystal shape optimization and plastic deformation. Therefore, the measured excess nitrogen can only be partially explained by stress relaxation. After annealing for 24 h at 590 °C in vacuum, the nitrogen content in the matrix decreases from 0.55 at.% to 0.32 at.%. Part of the mobile nitrogen leaves the sample during the heat treatment. It is interesting to observe that the matrix concentrations now lie closer to the stress-free solvus. However, they do not match the point given by the tie-line corresponding to the alloy, suggesting the presence of immobile excess nitrogen trapped at interfaces.

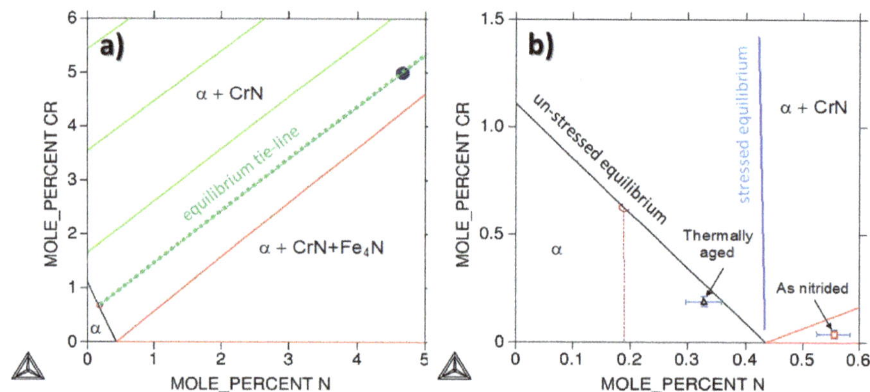

Figure 14. Isothermal sections of the ternary Fe–Cr–N phase diagram at 590 °C. (**a**) Nitrogen and chromium contents in the ferritic matrix in equilibrium with CrN are set by the tie-line running through the alloy composition. (**b**) The solvi of CrN in ferrite, under stress and stress-free, are represented along with nitrogen and chromium concentrations in ferrite after nitriding and after annealing in vacuum for 24 h at 590 °C.

5. Conclusions

In summary, advanced local analysis techniques such as APT and TEM-EELS have been used in the present work to study original aspects of the precipitation of alloying element nitrides in iron based systems during nitriding. Their high analytical resolution at the nanometer scale or below provides invaluable information regarding the precipitation of such nitrides and brings new information regarding issues that had been identified years ago, such as excess nitrogen in Fe–Cr.

In the case of Fe–Si alloys, APT analyses showed, early during the treatment, nitrogen clusters with trace elements with which it has a greater affinity, Al here. As the treatment progresses, stoichiometric Si_3N_4 precipitates systematically nucleate on those small but numerous clusters as they provide lower energy barrier sites. As a result, controlling the distribution of those clusters by changing the clustering element or its content, or by using a clustering treatment at a different temperature, should permit, in accordance with the classical precipitation theory, tailoring of the population of subsequent silicon nitride precipitates.

In the case of Fe–Cr alloys, independent measurements using APT and TEM-EELS confirmed the presence of iron in CrN particles precipitated during nitriding at 590 °C of Fe–Cr alloys. Both techniques yielded the same content, approximately 6 at.%. This content proved to be an equilibrium value as further annealing for 24 h lead to coarsening of the precipitate without change in their iron concentration. This presence of iron in CrN is qualitatively consistent with the thermodynamic properties of FeN and CrN. It has implications on the phase diagram of the Fe–Cr–N ternary system, which should be reassessed in order to better understand the nitriding process and better model the precipitation kinetics of CrN. Iron in CrN, however, did not affect the Cr/N ratio in the precipitates and, thus, cannot explain excess nitrogen, either in whole or in part. It was further shown using 3D calculations that only part of this excess nitrogen can be due to misfit-induced elastic stresses and that further investigation remains necessary to provide a complete explanation.

Author Contributions: Software, S.B.; Validation, P.J.; Formal Analysis, M.G.; Investigation, R.D.; Data Curation, S.B.; A.M.; F.D.; T.E.; Writing-Original Draft Preparation, H.V.; Writing-Review & Editing, A.R.; Project Administration, B.H.

Funding: This research received no external funding

Acknowledgments: This work was supported by the French State through the program "Investment in the future" operated by the National Research Agency (ANR) and referenced by ANR-11-LABX-0008-01 (LabEx DAMAS) (www.labex-damas.com). Part of the TEM work was performed at the CLYM platform in Lyon (www.clym.fr) with the support of the French CNRS and CEA network METSA (www.metsa.fr).

Conflicts of Interest: The authors declare no conflict of interest.

References

1. Dalton, J.C. Thermodynamics of Paraequilibrium Carburization and Nitridation. Master Thesis, Case Western Reserve University, Cleveland, OH, USA, 2014.
2. Jack, D.H. The structure of nitride Iron-Titanium alloys. *Acta Metall.* **1976**, *24*, 137–146. [CrossRef]
3. Goune, M.; Belmonte, T.; Redjaïmia, A.; Weisbecker, P.; Fiorani, J.M.; Michel, H. Thermodynamic and structural studies on nitrided Fe-1.62% Mn and Fe-0.56% V alloys. *Mater. Sci. Eng. A* **2003**, *351*, 23–30. [CrossRef]
4. Van Landeghem, H.P.; Goune, M.; Bordère, S.; Danoix, F.; Redjaïmia, A. Competitive precipitation of amorphous and crystalline silicon nitride in ferrite: Interaction between structure, morphology, and stress relaxation. *Acta Mater.* **2015**, *93*, 218–234. [CrossRef]
5. Xiong, X. Elaboration et genèse des microstructures dans les aciers fer-azote. Ph.D. Thesis, University of Lorraine, Lorraine, France, 2008.
6. Van Landeghem, H.P.; Goune, M.; Redjaïmia, A. Investigation of a Ferrite/Silicon Nitride Composite Concept Aimed at Automotive Applications. *Steel Res. Int.* **2012**, *83*, 590–593. [CrossRef]
7. Mittemeijer, E.J.; Biglari, M.H.; Böttger, A.J.; Van der Pers, N.M.; Sloof, W.G.; Tichelaar, F.D. Amorphous precipitates in a crystalline matrix: Precipitation of amorphous Si_3N_4 in α-Fe. *Scr. Mater.* **1999**, *41*, 625–630. [CrossRef]
8. Meka, S.R.; Jung, K.S.; Bischoff, E.; Mittemeijer, E.J. Unusual precipitation of amorphous silicon nitride upon nitriding Fe-2at.% Si alloy. *Philos. Mag.* **2012**, *92*, 1435–1455. [CrossRef]
9. Van Landeghem, H.P.; Goune, M.; Epicier, T.; Redjaïmia, A. Unexpected low temperature crystallization of amorphous silicon nitride into α-Si_3N_4 in a ferritic matrix. *Scr. Mater.* **2013**, *68*, 187–190. [CrossRef]
10. Cundy, S.L.; Grundy, P.J. Combined electron microscopy and energy analysis of an internally oxidized Ni + Si alloy. *Philos. Mag. A* **1966**, *14*, 1233–1242. [CrossRef]
11. Nolan, P.J.; Grundy, P.J. Morphology and stability of precipitates in internally oxidised silicon-bearing nickel and cobalt alloys. *J. Mater. Sci.* **1971**, *6*, 1143–1150. [CrossRef]
12. Sato, H.; Ota, I.; Fujii, T.; Onaka, S.; Kato, M. Morphological evolution of grain-boundary SiO_2 in internally oxidized Cu–Si bicrystals. *Mater. Trans.* **2004**, *45*, 818–823. [CrossRef]

13. Yanagihara, K.; Yamazaki, S. Characterization of oxidation behavior at Fe–Si alloy surface. *Nippon Steel Tech. Rep.* **2011**, *2*, 27–32.
14. Danoix, R.; Legras, L.; Hanoyer, B.; Dulcy, J.; Danoix, F. Atom probe tomography and transmission electron microscopy investigations of chromium nitrides in FeCr alloys. In Proceedings of the International Conference on Solid to Solid Phase Transformations in Inorganic Materials (PTM) 2005, Phoenix, AZ, USA, 29 May–3 June 2005; Howe, J.M., Laughlin, D.E., Lee, J.K., Dahmen, U., Soffa, W.A., Eds.; TMS (The Minerals, Metals and Materials Society): Warrendale, PA, USA, 2005; Volume 1, pp. 351–356.
15. Ginter, C.; Torchane, L.; Dulcy, J.; Gantois, M.; Malchère, A.; Esnouf, C.; Turpin, T. A new approach to hardening mechanisms in the diffusion layer of gas nitrided α-alloyed steels. Effects of chromium and aluminium: Experimental and simulation studies. *Metall. Ital.* **2006**, *98*, 29–35.
16. Hosmani, S.S.; Schacherl, R.E.; Mittemeijer, E.J. Nitrogen uptake by an Fe-V alloy: Quantitative analysis of excess nitrogen. *Acta Mater.* **2006**, *54*, 2783–2792. [CrossRef]
17. Jung, K.S.; Schacherl, R.E.; Bischoff, E.; Mittemeijer, E.J. Nitriding of ferritic Fe-Cr-Al alloys. *Surf. Coat. Technol.* **2010**, *204*, 1942–1946. [CrossRef]
18. Schacherl, R.E.; Graat, P.C.J.; Mittemeijer, E.J. The nitriding kinetics of Iron-Chromium Alloys; The Role of Excess Nitrogen: Experiments and Modelling. *Metall. Mater. Trans. A* **2004**, *35A*, 3388–3398. [CrossRef]
19. Hekker, P.M.; Rozendall, H.C.F.; Mittemeijer, E.J. Excess nitrogen and discontinuous precipitation in nitride iron-chromium alloys. *J. Mater. Sci.* **1985**, *20*, 718–729. [CrossRef]
20. Jessner, P.; Danoix, R.; Hannoyer, B.; Danoix, F. Investigations of the nitrided subsurface layers of an Fe-Cr-model alloy. *Ultramicroscopy* **2009**, *109*, 530–534. [CrossRef] [PubMed]
21. Jessner, P. Phénomène de précipitation dans les alliages nitrurés: contribution tomographique atomique et de la microscopie électronique à transmission. Ph.D. Thesis, Univ ersity of Rouen, Rouen, France, 2010.
22. Martinavičius, A.; Van Landeghem, H.P.; Danoix, R.; Redjaïmia, A.; Gouné, M.; Danoix, F. Mechanism of Si_3N_4 precipitation in nitrided Fe-Si alloys: A novel example of particle-stimulated-nucleation. *Mater. Lett.* **2017**, *189*, 25–27. [CrossRef]
23. DaCosta, G.; Vurpillot, F.; Bostel, A.; Bouet, M.; Deconihout, B. Design of a delay-line position-sensitive detector with improved performance. *Rev. Sci. Instrum.* **2005**, *76*, 013304. [CrossRef]
24. Bas, P.; Bostel, A.; Deconihout, B.; Blavette, D. A general protocole for the reconstruction of 3D atom probe data. *Appl. Surf. Sci.* **1995**, *87*, 298–304. [CrossRef]
25. Vaumousse, D.; Cerezo, A.; Warren, P.J. A procedure for quantification of precipitate microstructures from three-dimensional atom probe data. *Ultramicroscopy* **2003**, *95*, 215–221. [CrossRef]
26. Lefebvre, W.; Danoix, F.; Da Costa, G.; De Geuser, F.; Hallem, H.; Deschamps, A. 3DAP measurements of Al content in different types of precipitates in aluminium alloys. *Surf. Interface Anal.* **2007**, *39*, 206–212. [CrossRef]
27. Ayache, J.; Beaunier, L.; Boumendil, J.; Ehret, G.; Laub, D. *Sample Preparation Handbook for Transmission Electron Microscopy—Techniques*; Springer: New York, NY, USA, 2010.
28. Egerton, R.F. *Electron Energy-Loss Spectroscopy in the Electron Microscopy*, 2nd ed.; Plenum Press: New York, NY, USA; London, UK, 1996.
29. Mitterbauer, C.; Hébert, C.; Kothleitner, G.; Hofer, F.; Schattschneider, P.; Zandbergen, H.W. Electron energy loss-near edge structure as a fingerprint for identifying chromium nitrides. *Solid State Commun.* **2004**, *130*, 209–213. [CrossRef]
30. Courtois, E.; Epicier, T.; Scott, C. EELS study of niobium carbo-nitride nano-precipitates in ferrite. *Micron* **2006**, *37*, 492–502. [CrossRef] [PubMed]
31. Bordère, S.; Caltagirone, J.-P. A unifying model for fluid flow and elastic solid deformation: A novel approach for fluid–structure interaction. *J. Fluids Struct.* **2017**, *51*, 344–353. [CrossRef]
32. Miller, M.K.; Hetherington, M.G. Local magnification effects in the atom probe. *Surf. Sci.* **1991**, *246*, 442–449. [CrossRef]
33. Jessner, P.; Gouné, M.; Danoix, R.; Hannoyer, B.; Danoix, F. Atom probe tomography evidence of nitrogen excess in the matrix of nitrided Fe-Cr. *Philos. Mag. Lett.* **2010**, *90*, 793–800. [CrossRef]
34. Larson, D.J.; Gault, B.; Geiser, B.P.; De Geuser, F.; Vurpillot, F. Atom probe tomography spatial reconstruction: Status and directions. *Curr. Opin. Solid State Mater. Sci.* **2013**, *17*, 236–247. [CrossRef]
35. Danoix, F.; Grancher, G.; Bostel, A.; Blavette, D. Standard deviations of composition measurements in atom probe analyses–Part II: 3D atom probe. *Ultramicroscopy* **2007**, *107*, 739–743. [CrossRef] [PubMed]

36. Hellman, O.C.; Vandenbroucke, J.A.; Rusing, J.; Isheim, D.; Seidman, D.N. Analysis of three-dimensional atom-probe data by the proximity histogram. *Microsc. Microanal.* **2000**, *6*, 437–444. [PubMed]
37. Schwarz, B.; Rossi, P.J.; Strassberger, L.; Jörg, F.; Meka, S.R.; Bischoff, E.; Schacherl, R.E.; Mittemeijer, E.J. Coherency strain and precipitation kinetics: Crystalline and amorphous nitride formation in ternary Fe–Ti/Cr/V–Si alloys. *Philos. Mag.* **2014**, *94*, 3098–3119. [CrossRef]
38. Van Landeghem, H.P.; Gouné, M.; Redjaïmia, A. Nitride precipitation in compositionally heterogeneous alloys: Nucleation, growth and coarsening during nitriding. *J. Cryst. Growth* **2012**, *341*, 53–60. [CrossRef]
39. Phillips, V.A.; Seybolt, A.U. A Transmission Electron Microscopic Study of some Iron-Nitrided Binary Iron Alloys and Steels. *Trans. Metall. Soc. AIME* **1968**, *242*, 2415–2422.
40. Locquet, J.-N.; Soto, R.; Barrallier, L.; Charaï, A. Complete investigation of a nitrided layer for Cr alloy steel. *Microsc. Microanal. Microstruct.* **1998**, *8*, 335–352. [CrossRef]
41. Sennour, M.; Jacq, C.; Esnouf, C. TEM and EBSD investigation of continuous and discontinuous precipitation of CrN in nitrided pure Fe-Cr alloys. *J. Mater. Sci.* **2004**, *39*, 4521–4531. [CrossRef]
42. Vollstädt, H.; Ito, E.; Akaishi, M.; Akimoto, S.; Fukunaga, O. High Pressure synthesis of Rocksalt type of AlN. *Proc. Jpn. Acad.* **1990**, *66*, 7–9. [CrossRef]
43. Porter, D.A.; Easterling, K.E. *Phase Transformations in Metals and Alloys*, 2nd ed.; Chapman & Hall: London, UK, 1992.
44. Vurpillot, F.; Danoix, F.; Gilbert, M.; Koelling, S.; Dagan, M.; Seidman, D.N. True atomic-scale imaging in three-dimensions: A review of the rebirth of field-ion microscopy. *Microsc. Microanal.* **2017**, *23*, 210–220. [CrossRef] [PubMed]
45. Akré, J.; Danoix, F.; Leitner, H.; Auger, P. The morpholology of secondary-hardening carbides in a martensitic steel at the peak hardness by 3D-FIM. *Ultramicroscopy* **2009**, *109*, 518–523. [CrossRef] [PubMed]
46. Blum, I.; Portavoce, A.; Chow, L.; Hoummada, K.; Mangelinck, D. Diffusion and redistribution of boron in nickel silicides. *Defect Diffus. Forum* **2012**, *322*, 129–150. [CrossRef]
47. Massardier, V.; Voron, L.; Esnouf, C.; Merlin, J. Identification of the nitrides formed during the annealing of a low-carbon low-aluminium steel. *J. Mater. Sci.* **2001**, *36*, 1363–1371. [CrossRef]
48. Sennour, M.; Jacq, C.; Esnouf, C. Mechanical and microstructural investigations of nitrided Fe-Cr layers. *J. Mater. Sci.* **2004**, *39*, 4533–4541. [CrossRef]
49. Catteau, S.D.; Van Landeghem, H.P.; Teixeira, J.; Dulcy, J.; Dehmas, M.; Denis, S.; Redjaïmia, A.; Courteaux, M. Carbon and nitrogen effects on microstructure and kinetics associated with bainitic transformation in a low-alloyed steel. *Journal of Alloys and Compounds* **2016**, *658*, 832–838. [CrossRef]
50. Frisk, K. A Thermodynamic evaluation of the Cr-N, Fe-N, Mo-N and Cr-Mo-N systems. *Calphad* **1991**, *15*, 79–106. [CrossRef]
51. Hillert, M.; Staffansson, L.I. The regular solution model for stoeichiometric phases and ionic melts. *Acta Chem. Scand.* **1970**, *24*, 3618–3626. [CrossRef]
52. Tessier, F.; Navrotsky, A.; Niewa, R.; Leineweber, A.; Jacobs, H.; Kikkawa, S.; Takahashi, M.; Kanamaru, F.; DiSalvo, F.S. Energetic of binary iron nitrides. *Solid State Sci.* **2000**, *2*, 457–462. [CrossRef]
53. Cahn, J.W.; Larché, F. A simple model for coherent equilibrium. *Acta Metall.* **1984**, *32*, 1915–1923. [CrossRef]
54. Williams, R. The calculation of coherent phase equilibria. *Calphad* **1984**, *8*, 1–14. [CrossRef]
55. Maugis, P.; Gouné, M. Kinetics of vanadium carbonitride precipitation in steel: A computer model. *Acta Mater.* **2005**, *53*, 3359–3367. [CrossRef]
56. Somers, M.A.J.; Lankreijer, R.M.; Mittemeijer, E.J. Excess nitrogen in ferrite matrix of nitride binary iron-based alloys. *Philos. Mag. A* **1989**, *59*, 353–378. [CrossRef]
57. Hazotte, A. Transformations et contraintes de cohérence dans les superalliages et les intermétalliques de base TiAl. *Matér. Tech.* **2010**, *97*, 23–31. [CrossRef]

Article

Key Parameters to Promote Granularization of Lath-Like Bainite/Martensite in FeNiC Alloys during Isothermal Holding

Meriem Ben Haj Slama [1,2,3,*], Nathalie Gey [2,3], Lionel Germain [2,3], Kangying Zhu [4] and Sébastien Allain [1,3]

1 Institut Jean Lamour UMR Université de Lorraine, CNRS 7198 Nancy, France; sebastien.allain@univ-lorraine.fr
2 Laboratoire d'Etude des Microstructures et de Mécanique des Matériaux (LEM3), UMR, Université de Lorraine, F-57000 Metz, France; nathalie.gey@univ-lorraine.fr (N.G.); lionel.germain@univ-lorraine.fr (L.G.)
3 Laboratory of Excellence on Design of Alloy Metals for low-mAss Structures (DAMAS), Université de Lorraine, F-57000 Metz, France
4 ArcelorMittal Maizières Research SA, Automotive Product Center, Voie Romaine, BP30320, 57283 Maizières-les-Metz CEDEX, France; kangying.zhu@arcelormittal.com
* Correspondence: meriem.ben-haj-slama@univ-lorraine.fr; Tel.: +33-640-547-335

Received: 31 August 2018; Accepted: 21 September 2018; Published: 24 September 2018

Abstract: The stability of lath-like microstructures during low-temperature isothermal ageing was analyzed in a Fe5Ni0.33C (in wt %) steel. The microstructures were characterized using Scanning Electron Microscopy (SEM) coupled with Electron Backscatter Diffraction (EBSD). Advanced orientation data processing was applied to quantify the hierarchical and multiscale organization of crystallographic variants subdividing Prior Austenite Grains (PAG) into packets/blocks/sub-blocks. The result shows that ferrite laths of martensite or lower bainite are stable, whatever the ageing temperature (up to 380 °C). On the contrary, a granularization process is triggered when microstructures contain a fraction of upper bainite. This metallurgical evolution corresponds to a rapid and significant change of the ferrite matrix involving a disappearance of 60° disoriented blocks. The phenomenon affects in turn the mechanical properties. The final microstructures obtained after isothermal holding look like granular bainite, which raises some questions about the classification of bainite.

Keywords: steel; bainite; martensite; isothermal treatment; mechanical properties; austenite reconstruction; variant

1. Introduction

Similar to martensite, bainite is a phase transformation product of austenite appearing at low temperature in conventional carbon steels. In most cases, its microstructure can be described as a non-polygonal ferrite matrix containing or not second phases. Depending on the chemical composition of the studied steel and the thermomechanical treatment, a wide range of bainite microstructures can be observed at the micrometer scale. That explains why numerous classifications were proposed in the past [1–5]. The first classifications were derived from Optical Microscopy (OM) or Scanning Electron Microscopy (SEM) observations and thus, mainly based on the apparent morphology and spatial distribution of the constituting phases after etching.

From the late 1990s, the progress in orientation-based microscopy [6–8] have permitted investigating in a statistical way and with a high accuracy the crystallographic disorientations between features constituting bainite. As bainite exhibits an orientation relationship (OR) with its austenite parent phase [9–12] due to its associated shear transformation strain, the analysis of disorientations between ferrite variants has been the ground for new crystallography-based classifications [13–15]. They have permitted to rationalize the mechanical properties of lath-like bainite, depending on their transformation temperatures. Thus, crystal orientation-based techniques have become essential to study bainite steels.

These later classifications focus on the crystallographic variant arrangement within PAGs (Prior Austenite Grains). This arrangement is inherited from the mechanism of the transformation with respect of an OR between the austenite and ferrite phases, similar to martensite steels [5,9–12]. The crystallographic variants are grouped in packets (neighboring variants sharing the same habit plane), further divided into highly disoriented blocks (>45°, mainly 60°) themselves divided in low disoriented sub-blocks (<20°) (or variants) [16]. The different ORs have been reported depending on the chemical composition of the steel and the manufacturing conditions, but the multiscale and hierarchical organization of the structure discussed above has unanimous support.

Thanks to their good balance between ductility and strength, bainite/martensite steels are products widely used in industry [17]. Their strength in particular results from their refined microstructure (block/lath sizes), the dense carbide distribution (if any), the carbon in solid solution and the high density of dislocations inherited from the shear transformations at low temperature [18–20]. However, all these microstructure features are prone to evolve during further thermal treatments. In the case of martensite steels, these phenomena are known as tempering processes and have been widely studied in the past [21–23]. This versatility offers a fantastic lever to metallurgists to tune the final properties of steels. On the contrary, very few published studies focus on these processes in bainite steels; the ferrite matrix is thought to be stable during further thermal treatments, and only the lath structure is thought to thicken to some extent [24]. Significant evolution of lath-bainite microstructure was reported only after very long isothermal holding at 500 °C for ~1 year, where mechanisms such as carbide coarsening and ferrite recrystallization were observed [20]. Only severe thermal treatment seems able to induce significant evolution of bainite microstructures, contrary to martensite steels and despite their numerous similarities.

To rationalize the comparison of different isothermal treatments at different temperatures, Hollomon and Jaffe [25] have defined a tempering parameter H_{HJ} to describe time/temperature equivalence such as:

$$H_{HJ} = T\,(C + \log(t)) \tag{1}$$

where C is a dimensionless constant proposed by the authors to be ~20 for steels with carbon contents of 0.25–0.4 wt %. T is the temperature in kilokelvin and t is the time in hour. The severity of the thermal treatment increases with H_{HJ}. The authors reported that the mechanical strength of bainite steels dropped with further tempering over a threshold H_{HJ} value of 18, whatever the bainite microstructure [25]. This value corresponds for instance to a heat treatment of 80 days at 500 °C. The same behavior was confirmed by different authors for alloyed steels (Fe-0.5C-0.8Mn-0.2Si-0.9Ni-1.1Cr-0.5Mo and Fe-0.42C-0.7Mn-0.5Si-0.8Ni-2.5Cr-1Mo in wt % [26], and Fe-0.20C-0.3Si-1.4Mn-0.7Ni and Fe-0.13C-0.2Si-0.6Mn-2.4Cr-0.9Mo in wt % [27]). The main mechanism of softening was carbide coarsening and in some cases a stress-release process.

However, our recent work on a model alloy Fe-5Ni-0.13C (wt %) reported a softening of bainite for an unexpected low value of H_{HJ} (12.8) [28]. The recorded drop in hardness and yield strength was 45 HV and 160 MPa, respectively. According to this result, microstructural evolution of bainite can be largely faster than what was reported in the previous literature, and could occur at lower temperatures (Ms + 20 K; Ms is the martensite start temperature of the alloy). In fact, a fast granularization of lath-like bainite was observed during the studied isothermal holding.

The present work aimed at a better understanding of this lath granularization process in a similar Ni-based steel (Fe-5Ni-0.33C in wt %). Different microstructures (lower or upper bainite, martensite and mixed microstructures) and different ageing time/temperatures were considered. They were characterized by high resolution EBSD analysis coupled with crystallographic reconstruction of PAGs using software developed by our group.

The result of this research is of major importance for various industrial applications (pipelines, nuclear reactor vessels, car-to-ground connecting components, engine components, etc.); it helps to better understand and predict the evolution of the properties of bainite/martensite steels during manufacturing and in service conditions.

2. Materials and Methods

2.1. Material

The composition of the investigated steel was Fe-0.33C-5Ni (wt %). It is a model alloy compared to industrial steels. In particular, Mn has been substituted with Ni to avoid composition heterogeneities and segregations. A 15 kg ingot was cast in a vacuum induction furnace. After a 2 h homogenization treatment at 1200 °C, it was hot and then cold rolled, respectively, to the thicknesses of 3 mm and 1 mm. Dilatometric samples of 10 mm × 3 mm were cut from this as-rolled plate for further controlled heat treatments. Vickers Hardness on Zwick-Roell/Indentec machine (Zwick France, Metz, France), as well as SEM/EBSD analysis were performed on the treated samples. All microstructure analyses were performed at quarter thickness of the sample to avoid possible decarburized layers at the surface.

2.2. Controlled Heat Treatments

Three families of microstructures were obtained by controlled heat treatment performed in a laboratory-made rapid-cycle dilatometer (RCD) (Institut Jean Lamour, Nancy, France): (1) high temperature bainite; (2) low temperature bainite; and (3) martensite-based microstructures (either fully martensite or mixed bainite/martensite). For each family of microstructures, two metallurgical states were generated: (i) the so-called "transformed" state (T) was obtained by quenching the samples directly after the bainite transformation was completed at the considered temperature (according to the dilatometric signal), or directly from the austenite domain to obtain fully martensite microstructure; and (ii) the so-called "Aged" state (A) was obtained by further holding (either at the transformation temperature or at higher/lower temperatures). This "aged" state corresponds in fact to a tempering in the case of a martensite-based microstructure.

The different heat treatments associated to each type of microstructures are further described in the coming subsections and summarized in Figure 1 and Table 1. This table also indicates Vickers hardness results (ten measurements were averaged for each reported values) and H_{HJ} parameters calculated according to Equation (1).

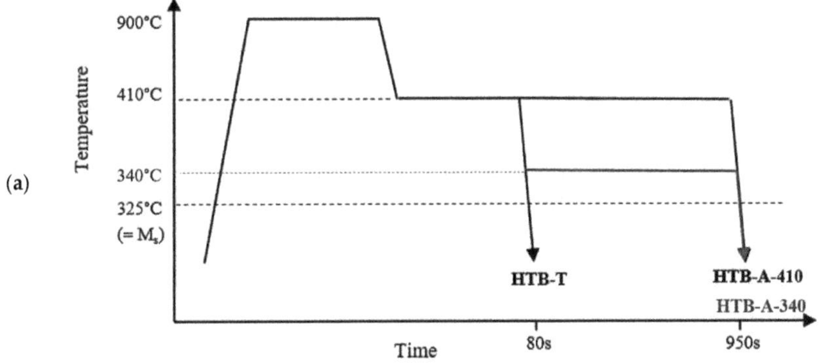

Figure 1. *Cont.*

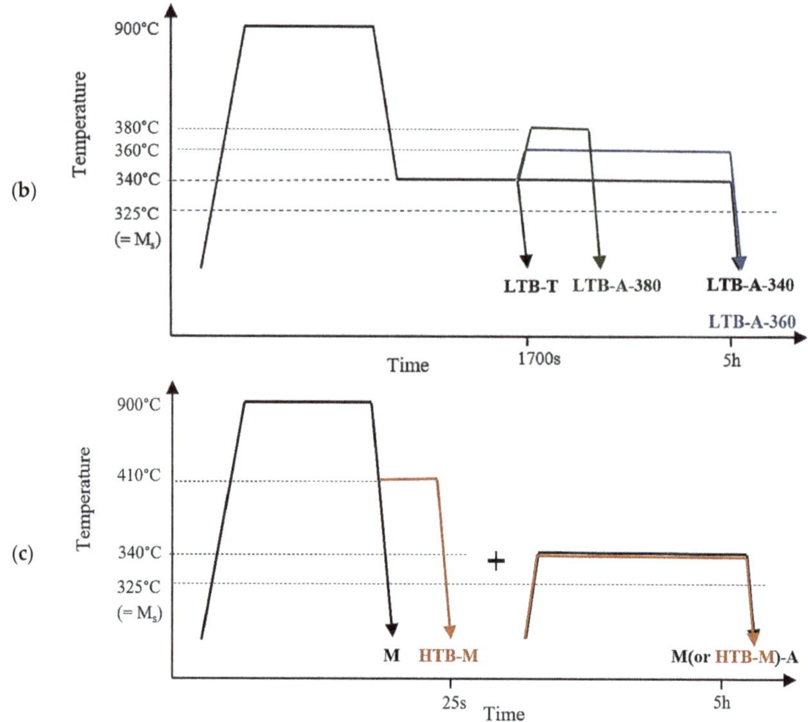

Figure 1. Heat treatments to obtain: (**a**) High Temperature Bainite HTB-type microstructures; (**b**) Low Temperature Bainite LTB-type microstructures; and (**c**) martensite-based microstructures (T stands for as-transformed and A for as-aged conditions).

Table 1. Main parameters of the heat treatments to obtain: (**a**) HTB-type microstructures; (**b**) LTB-type microstructures; and (**c**) martensite (M)-based microstructures (T stands for as-transformed and A for as-aged conditions).

	Microstructure		Heat Treatment Temperature/Time	State	Hardness (HV)	H_{HJ}
(a)	High Temperature Bainite	HTB-T	410 °C/80 s	Transformed	302	12.5
		HTB-A-410	410 °C/80 s + 410 °C/870 s	Aged	242	13.3
		HTB-A-340	410 °C/80 s + 340 °C/870 s	Aged	264	11.9
(b)	Low Temperature Bainite	LTB-T	340 °C/1700 s	Transformed	347	12.0
		LTB-A-340	340 °C/80 s + 340 °C/5 h	Aged	326	12.7
		LTB-A-360	340 °C/80 s + 360 °C/5 h	Aged	315	13.1
		LTB-A-380	340 °C/80 s + 380 °C/1800 s	Aged	320	12.9
(c)	Martensite based microstructure	M	Quenching	Transformed	520	0
		M-A	Quenching + 340 °C/5 h	Aged	460	12.7
		HTB-M	410 °C/25 s + Quenching	Transformed	400	-
		HTB-M-A	410 °C /25 s + Quenching + 340 °C/5 h	Aged	352	12.7

Prior to bainite and martensite transformations, all samples were first austenitized for 5 min at 900 °C (1173 K) and rapidly cooled at 70 °C/s to:

- the transformation temperature in the case of bainite containing microstructures; and
- room temperature for fully martensite microstructures.

The austenite holding time was optimized to ensure a homogenous austenite composition with a conventional average grain size of 17 µm. Moreover, the cooling was fast enough to avoid the ferrite transformation, as confirmed by the dilatometry measurements and the microstructure investigations.

2.2.1. High Temperature Bainite (HTB)

HTB was isothermally formed at 410 °C (683 K), i.e., Ms + 85 °C. At this temperature, the transformation was completed after 80 s. The sample in the "transformed" state (HTB-T) was directly quenched after the transformation was finished. Two additional samples were further aged without any intermediate quench (In contrast to standard definition, the "aged" condition is here obtained without intermediate cooling): sample HTB-A-410 was maintained at the transformation temperature for 870 s and sample HTB-A-340 at 340 °C (613 K) for 870 s (see Figure 1a and Table 1a).

2.2.2. Low Temperature Bainite

LTB was isothermally formed at 340 °C, i.e., Ms + 15 °C. At this temperature, the transformation was completed after 1700 s. The sample in the "transformed" state (LTB-T) was directly quenched after the transformation was finished. Three additional samples were further aged at different temperatures, without any intermediate quench: sample LTB-A-340 was maintained at the transformation temperature for 5 h, sample LTB-A-360 at 360 °C for 5 h and sample LTB-A-380 at 380 °C for 1800 s (see Figure 1b and Table 1b).

2.2.3. Martensite-Based Microstructures

A fully martensite structure (sample M) was also obtained by fast quenching from the austenite domain. Finally, a mixed bainite–martensite microstructure (sample HTB-M) was formed by an isothermal holding at 410 °C for 25 s followed by a quench. In the latter case, the resulting microstructure was supposed to contain 60% of HTB and 40% of fresh martensite.

For each type of microstructure, an additional sample was further tempered at 340 °C, i.e., Ms + 15 °C. They are referred to as M-A and HTB-M-A samples (Figure 1c and Table 1c).

2.3. Microstructure and Microtexture Characterization

The microstructures were characterized by electron microscopy with the JEOL 6500F Field Emission Gun-SEM (JEOL, Peabody, MA, USA) equipped with the Oxford-instrument EBSD system (Nordlys II CCD camera (Oxford-instrument, Gometz-la-Ville, France) and AZTec acquisition software) (AZTecHKL, Oxford-instrument, Gometz-la-Ville, France). For the EBSD measurements, the samples were mechanically polished down to 1-µm diamond paste and finished with colloidal silica. Orientation maps were acquired at an acceleration voltage of 15 kV, with a current of 3.3 nA, allowing a signal of acceptable quality for a camera resolution of 336 × 256 pixels (i.e., with a 4 × 4 binning) and with an integration time of 48 ms per shot without averaging. Under these conditions, it is estimated that the spatial resolution of the diffraction signal is about 50/70 nm for steels (http://lionelgermain.free.fr/merengue2.htm) [29,30]. For a 1500 × 1150 pixel field with a 0.1 µm step size, the acquisition time is approximately of 24 h.

The disorientation maps were plotted from EBSD data as well as the pixel-to-pixel disorientation histogram. The maps highlight low disorientations (<20°) corresponding to sub-block boundaries, high disorientations (>45°, mainly 60°) corresponding to block and packet boundaries and random disorientations (between 20° and 45°) corresponding to PAG boundaries. Notice that only the disorientation angles were considered to characterize the boundaries of the bainite/martensite microstructures, leaving out disorientation axes.

After EBSD studies, some samples were etched with Picral solution to reveal carbides and characterized by SEM using an in-lens detector.

MERENGUE2 in-house software was applied to the ESBD maps to identify the prior austenite grains [30]. The PAG size was about 17 µm after five minutes at 900 °C. DECRYPT (Direct Evaluation of CRYstallographic Phase Transformation) in-house software was then applied to further analyze the variant arrangement in packets and blocks inside the PAGs and quantify the number of blocks and packets per PAG. Using this approach, two crystallographic types of variant arrangements inside PAGs were identified on the fully transformed microstructures (bainite or martensite) and called Type A and Type B microstructures. Figure 2 shows the typical disorientation map obtained on two PAGs mainly transformed into Type A and Type B microstructures. For the sake of readability, a simplified schematic illustration is also provided. Type A is characterized by a small number of blocks per packet. This number is statistically found to be less than three in the majority of the studied experimental configurations. On the contrary, Type B microstructure is characterized by a higher number of blocks per packet, largely more than four statistically (21 on average). Type A is thus representative of upper bainite microtexture while Type B is representative of lower bainite or martensite microtexture. This crystallographic classification based on the number of blocks per packet is in agreement with recent bainite classifications; in particular the ones suggested by Furuhara et al. [14] and Takayama et al. [15].

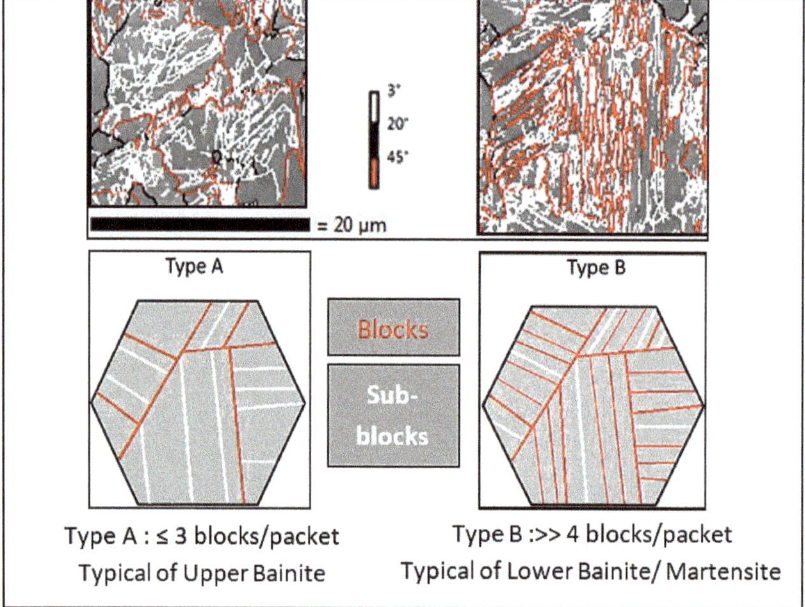

Figure 2. Crystallographic variant organizations of Type A and Type B microtextures (from the top to the bottom: experimental disorientation maps of a single PAG and corresponding schematic organization of blocks and packets inside a PAG)—PAG, packet/block, and sub-block boundaries are plotted in black, red and white respectively.

3. Results

3.1. Evolution of Bainite Microstructures during Extended Isothermal Holding

Figure 3 compares the EBSD results for a transformed (T) and an aged (A) microstructure respectively for LTB (a,b,c) and HTB (d,e,f) microstructures. For this first comparison, the ageing temperatures correspond to the transformation temperatures (340 °C and 410 °C respectively for LTB and HTB).

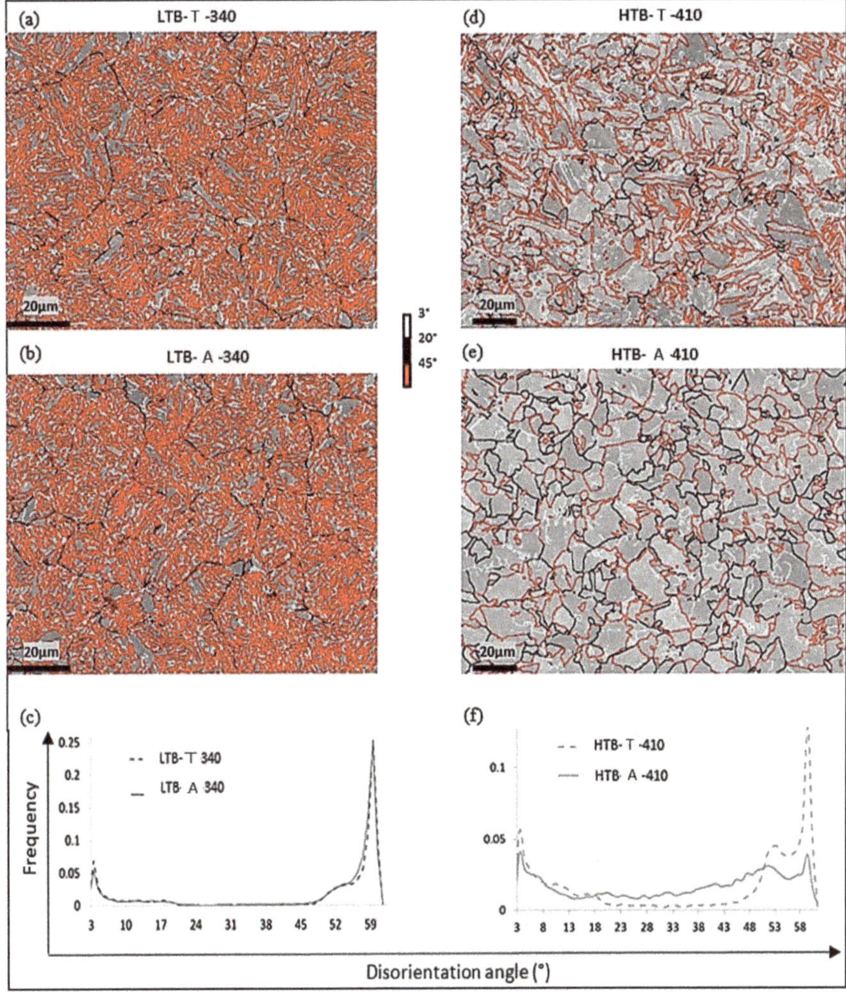

Figure 3. Angular disorientation maps (with the associated color code shown on the figure), comparing LTB-T (**a**) with LTB-A-340 (**b**) and HTB-T (**d**) with HTB-A-410 (**e**), (**c**) and(**f**) angular disorientation histograms associated with the EBSD data.

The LTB-T sample exhibited a lath microstructure with predominance of block boundaries (high disoriented boundaries, mainly of 60°, in red on the EBSD map). The coupled MERENGUE2/DECRYPT analysis indicated a 100% Type B variant organization which is typical of a lower bainite. Inter- and intra-lath carbides, characteristic of lower bainite according to Mehl [5] were also observed. The extended isothermal ageing (5 h at 340°C—H_{HJ} = 12.7) induced no substantial change in the ferrite matrix. Figure 3b shows that the LTB-A-340 microstructure is still homogeneous and presents a very pronounced lath structure. The disorientation distributions before and after ageing presented in Figure 3c are remarkably similar and confirm that the spatial organization of variants was not affected by the prolonged thermal treatment at 340 °C. Nevertheless, a meaningful decrease in hardness between LTB-T and LTB-A-340 was observed (from 347 HV to 326 HV). It can be attributed to the dislocation recovery in the ferrite matrix.

In contrast, the HTB microstructures exhibited a decomposition of the ferrite matrix into a granular structure only after 950 s at 410 °C (H_{HJ} = 13.3). The coupled MERENGUE2/DECRYPT analysis indicated that the HTB-T microstructure was a mixture of Type A and B. The respective proportion of Type A and B was 70% and 30% in the transformed state. Moreover, domains containing only inter-lath carbides (characteristic of upper bainite according to Mehl classification) and domains containing both inter and intra-lath carbides (characteristic of lower bainite according to Mehl classification) were observed in the microstructure. It can be thus concluded that the as-transformed HTB-T microstructure is a mixture of upper and lower bainite.

After 950 s of ageing (H_{HJ} = 13.3), the HTB-A-410 microstructure was significantly different from the HTB-T one: the acicular blocks (essentially those disoriented by 60°, in red in Figure 3d) disappeared in favor of a granular structure (see Figure 3e). This evolution is also detected in the disorientation histogram of Figure 3f, by the significant decrease in the frequencies of 60° disorientations (against an increase in the relative frequencies of random disorientations between 20° and 45°). This ferrite decomposition resulted in a granular structure containing a network of low disoriented sub-boundaries. Since very low disorientations down to 0.3° were detectable, such sub-boundaries were visible on the EBSD map as shown in Figure 4. The final mean size of the granular structure was close to the packet size of the as-transformed microstructure. It must be emphasized here that the entire microstructure, was affected by this ageing evolution, even Type B domains. This microstructure evolution produced a drop in hardness of 60HV, which is far higher than in the case of the sole recovery at 340 °C. The same phenomenon was already observed in Fe-Ni-C steel with lower carbon content [28].

Figure 4. Enlarged area taken from the EBSD map of the HTB-A-410 highlighting in green very low disoriented sub-boundaries (between 0.3° and 3°) present in the granular structure. The lath structure has completely disappeared after this granularization process.

3.2. Evolution of Martensite-Type Microstructures during Ageing

The martensite microstructure M obtained by direct quenching (shown in Figure 5a) had a crystallographic variant arrangement of Type B, in accordance with the prior works of Morito et al. [16]. After 5 h ageing at 340 °C, it remained unaffected, as seen with the microtextures of the M-A sample in Figure 5b and the corresponding disorientation histograms in Figure 5c.

As expected, the as-transformed HTB-M microstructure was made of both coarse domains containing low disorientated boundaries (formed presumably at 410 °C) and very fine lath structures (formed during quenching) (as seen in Figure 5d). The coupled MERENGUE2/DECRYPT analysis confirmed that the structure was composed of 42% Type A and 58% Type B. This ratio is consistent with the considered heat treatment. As a rule of thumb, HTB-M microstructure was in fact expected to contain 60% of high temperature bainite (HTB-T)—made, respectively, of 70% Type A and 30% Type B—and 40% of martensite (100% Type B).

After ageing at 340 °C for 5 h, the HTB-M-A microstructure exhibited a remarkable decrease in the number of block boundaries compared to HTB-M, as seen in Figure 5d,e. Again, it is mostly due to the decrease in frequency of 60° boundaries, as evidenced by the superposed disorientation histograms in Figure 5f.

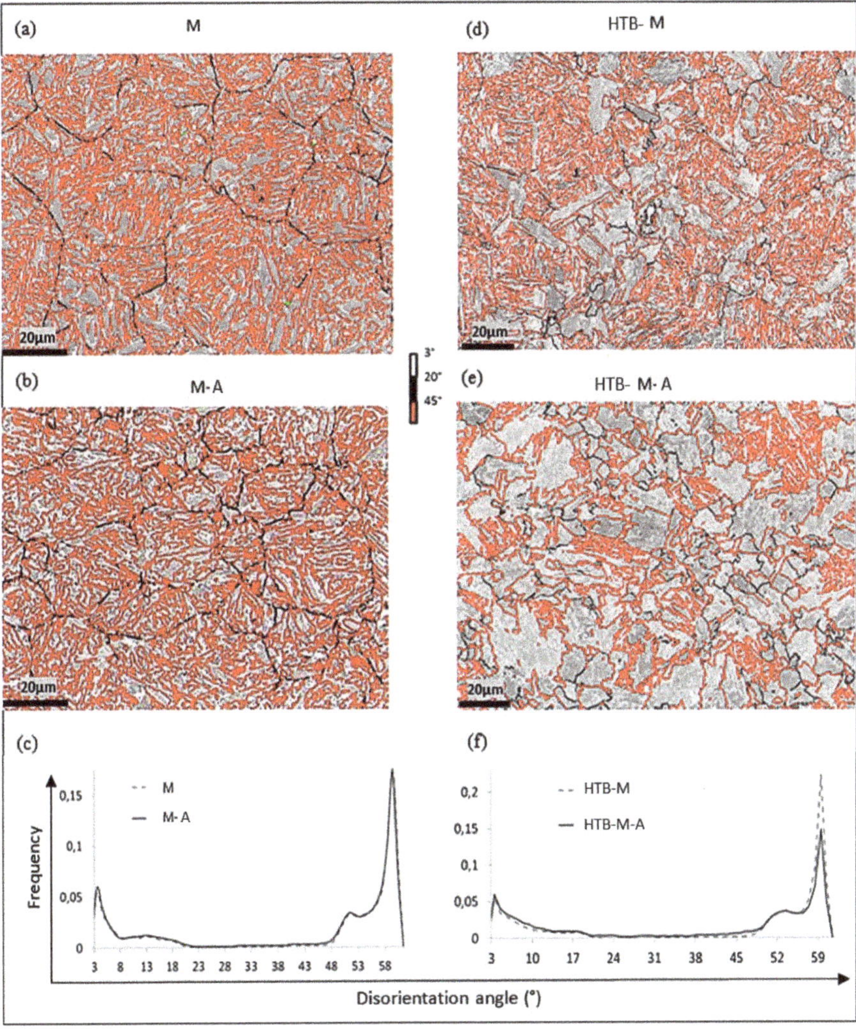

Figure 5. Angular disorientation maps (with the associated color code shown on the figure), comparing: M (**a**) with M-A (**b**); and HTB-M (**d**) with HTB-M-A (**e**). (**c**,**f**) Angular disorientation histograms associated with EBSD data.

3.3. *Influence of the Ageing Temperature on the Stability of As-Transformed Bainite Microstructures*

At this step, it cannot be excluded that the microstructure evolution observed for HTB-A-410 and HTB-M-A could be triggered by the ageing temperature thanks to a thermally activated mechanism regardless of the microstructure. Consequently, additional treatments were performed by varying the holding temperatures once the transformation was finished (see Table 1a,b).

First, the LTB-T microstructure was aged at higher temperatures than 340 °C. Figure 6a shows the disorientation histograms obtained on the LTB-A-360 and LTB-A-380 samples which were isothermally held after transformation, at 360 °C for 5 h and at 380 °C for 30 min, respectively. In both cases, the H_{HJ} parameters of 12.9 and 13.1 were close to the ones applied to HTB-A-410 (H_{HJ} = 13.5) and HTB-M-A (H_{HJ} = 12.7) for which granularization was observed. Whatever the applied heat treatments, the disorientation histograms are strictly similar. It means that, in both cases, the ferrite matrix does not evolve despite a severe treatment. The corresponding hardness decrease after ageing at 360 °C was similar to the one observed after ageing at 340 °C. The mechanisms behind this decrease were probably similar, i.e., the recovery of the ferrite matrix.

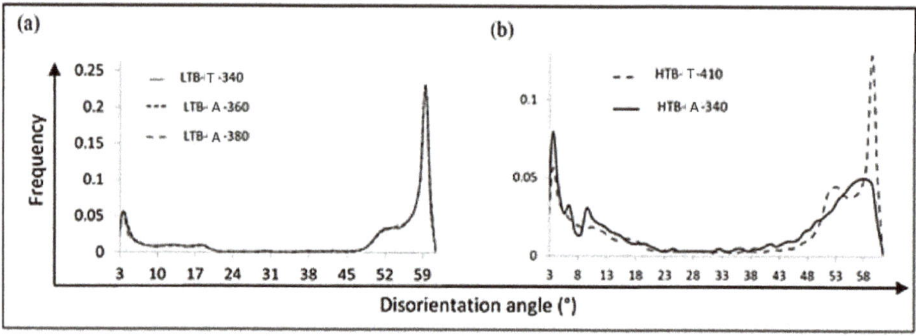

Figure 6. Angular disorientation histograms of: (**a**) LTB-T, LTB-A-360 and LTB-A-380; and (**b**) HTB-T and HTB-A-340 samples.

Secondly, the isothermal ageing of the HTB-T sample was performed at a lower temperature. The bainite microstructure fully transformed at 410 °C after 80 s was held at 340 °C for 870 s (sample HTB-A-340). Figure 6b compares the disorientation histogram obtained for both samples. One can notice a remarkable decrease in the 60° peak. This result highlights the fact that, in presence of Type A microstructure, the lath decomposition starts at this low temperature (340 °C, i.e., Ms + 15 °C) and after a short equivalent time (H_{HJ} = 11.9 only).

4. Discussion

4.1. *Key Role of the Initial Microstructure Regardless of the Ageing Temperature*

This paper demonstrates that, in Fe-5Ni-0.33C (wt %) steel, lower bainite (Type B microstructure) is stable during isothermal holdings at both low and high temperatures. Yet, some microstructure evolutions are still possible at atomistic or nm-scale since the microstructural transformations presented were performed by SEM. However, lower bainite (Type B microstructure) evolves as soon as it contains a fraction of upper bainite (Type A microstructure). This evolution consists mainly in the decomposition of the block structures (disappearance of 60° disoriented boundaries) in favor of a polygonal structure containing a network of sub-boundaries. This is the reason we have qualified this process as "granularization". Martensite behaves similarly as lower bainite under low temperature tempering—its ferrite matrix is stable—while, in the presence of upper bainite, it decomposes under the same tempering conditions as lower bainite. This decomposition is not limited to the fraction of Type A bainite; the entire microstructure is affected by the process.

This result suggests that the as-transformed microstructure plays a key role to trigger the granularization process. Temperature is thus not the sole decisive parameter promoting the evolution. For the studied steel, as well as the previous one with slightly lower carbon content [28], the decrease in mechanical properties can be observed for H_{HJ} tempering parameters in the range of 11.8–13.2, far smaller than the value of 18 expected by the seminal work of Hollomon and Jaffe [25]. Consequently, their approach is questionable for two aspects:

(1) Critical tempering conditions that degrade the steel mechanical properties are not valid for all steel compositions, particularly for FeNiC alloys.
(2) Equivalent tempering parameters without considering the microstructure type seems not relevant.

4.2. *Effect of the Steel Composition*

In the studied FeNiC systems (Fe-0.13C-5Ni previously [28] and Fe-0.33C-5Ni in this paper), the carbon content seems to have an impact only on the phase transformation kinetic. It is a known effect of carbon: the higher the carbon content, the slower the bainite transformation [20]. For our studies, fully lower bainite was obtained after only 50 s in the 0.13 C steel, whereas 1700 s were necessary for the 0.33 C steel. However, the granularization process seems not highly sensitive to the carbon content. Indeed, for both steels, the microstructure became granular for ageing conditions with similar low Hollomon—Jaffe parameters (between 12.5 and 13).

On the contrary, substitutional elements seem to have a remarkable effect on the granularization kinetics. In fact, Mn based alloys studied in the literature are reported to undergo a granularization only after far longer ageing and at higher temperatures, i.e., in more severe conditions (typically for H_{HJ} higher than 16–18). It is thus probable that Mn is an alloying element reducing the kinetics of granularization contrarily to Ni. This difference in the behavior between both substitutional elements is not surprising as they do not affect the mobility of interfaces in the same way during ferrite transformation and conventional recrystallization [31,32]. Consequently, it is likely that the mobility of high angle boundaries in the granularization process is controlled by a solute drag mechanism, as in a more conventional recrystallization process and despite the low temperature [33–35]. Uhm et al. [35] showed that Mn and Ni affect differently the activation energy of α/α interface mobility. Compared to Ni, Mn slows further this mobility down. We predict that granularization can occur in Mn-containing alloys, but with probably slower kinetics than in Mn-free alloys.

4.3. *Granularization and Bainite Classifications*

According to the literature, granular bainite formation has only been observed during continuous cooling treatments. In fact, Bhadeshia reported that granular bainite occurs only in continuously cooled low-carbon steels and explained that it cannot be produced by isothermal transformations [19]. A similar conclusion was drawn by Krauss et al. [36].

Interestingly, we obtained granular bainite from the granularization of lath-like structures by isothermal treatment. The final microstructure is described as equiaxed ferrite grains having the average size of prior packets, presenting low disoriented sub-grain boundaries and containing inter- and intra-granular carbides (cementite). The isothermally aged and granularized microstructures thus show many similarities with granular bainite.

In our opinion, it suggests that the formation of granular bainite can also be explained by the granularization of an intermediate lath structure (mainly of Type A, of course). This assumption opens new roads for discussing granular bainite formation mechanisms.

5. Conclusions

It has been proven that, for Fe-0.33C-5Ni steel, mixed upper and lower bainite microstructures show a remarkable evolution when aged. They evolve quickly toward a granular structure. This evolution is also observed in model microstructures made of upper bainite and fresh martensite.

On the contrary, lower bainite and martensite do not undergo any morphological or microtextural evolution when they are subjected to isothermal holding (even at temperatures above their transformation temperatures), at the studied micrometer scale.

This granularization process on the ferrite matrix consists in the disappearance of all acicular block boundaries (high disoriented) and lath boundaries (low disoriented). This process is triggered by the presence of upper bainite and affects the entire microstructure, even that made of lower bainite or martensite.

This phenomenon is accompanied by a significant drop in hardness, higher than the sole recovery effect.

The process seems to be controlled by both boundary energies and sizes of the involved features (packets, blocks, etc.) and requires a deeper investigation. The results give new paths for discussing the formation of granular bainite.

Author Contributions: Investigation, M.B.H.S.; Methodology, M.B.H.S., S.A. and N.G.; Project administration, S.A.; Resources, K.Z.; Software, L.G.; Supervision, S.A. and N.G.; Writing—original draft, M.B.H.S.; and Writing—review and editing, M.B.H.S., S.A., N.G., L.G. and K.Z.

Funding: This work was supported by the French State through the program "Investment in the future" operated by the National Research Agency (ANR) and referenced by ANR-11-LABX-0008-01 (LabEx DAMAS) and was also supported by the Région Grand Est. The material was supplied by ArcelorMittal Research Center.

Conflicts of Interest: The authors declare no conflict of interest.

References

1. Mehl, R.F. *Hardenability of Alloy Steels*; ASM: Cleveland, OH, USA, 1939; pp. 1–54.
2. Spanos, G.; Fang, H.S.; Aaronson, H.I. A mechanism for the formation of lower bainite. *Metall. Trans. A* **1990**, *21*, 1381–1390. [CrossRef]
3. Ohmori, Y.; Ohtani, H.; Kunitake, T. The bainite in low carbon low alloy high strength steels. *Trans. ISIJ* **1971**, *11*, 250–259.
4. Bramfitt, B.L. Speer, J.G. Perspective morphology of bainite. *Metall. Trans. A* **1990**, *21*, 817–829. [CrossRef]
5. Zajac, S.S.; Schwinn, V.; Tacke, K.H. Characterisation and quantification of complex bainitic microstructures in high and ultra-high strength linepipe steels. *Mater. Sci. Forum* **2005**, *500*, 387–394. [CrossRef]
6. Kelly, P.M.; Jostsons, A.; Blake, R.G. The orientation relationship between lath martensite and austenite in low carbon, low alloy steels. *Acta Metall. Mater.* **1990**, *38*, 1075–1081. [CrossRef]
7. Bunge, H.J.; Weiss, W.; Klein, H.; Wcislak, L.; Garbed, U.; Richard Schneider, J. Orientation relationship of Widmannst ätten plates in an iron meteoritemeasured with high-energy synchrotron radiation. *J. Appl. Crystall.* **2003**, *36*, 137–140. [CrossRef]
8. Zaefferer, S.; Ohlert, J.; Bleck, W. A study of microstructure, transformation mechanisms and correlation between microstructure and mechanical properties of a low alloyed TRIP steel. *Acta Mater.* **2004**, *52*, 2765–2778. [CrossRef]
9. Bain, E.C.; Dunkirk, N.Y. The nature of martensite. *Trans. AIME* **1924**, *70*, 25–46.
10. Kurdjumov, G.; Sachs, G. About the mechanism of steel hardening. *ZeitschriftfürPhysik* **1930**, *64*, 325–343.
11. Nishiyama, Z. Mechanism of transformation from face-centred to body-centred cubic lattice. *Sci. Rep. Tohoku Imp. Univ. Tokyo* **1934**, *23*, 637–664.
12. Greninger, A.B.; Troiano, A.R. The mechanism of martensite formation. *Metals Trans.* **1949**, *185*, 5–15. [CrossRef]
13. Gourgues, A.F.; Flower, H.M.; Lindley, T.C. Electron backscattering diffraction study of acicular ferrite, bainite, and martensite steel microstructures. *Mater. Sci. Technol.* **2000**, *16*, 26–40.
14. Furuhara, T.; Kawatab, H.; Morito, S.; Maki, T. Crystallography of upper bainite in Fe–Ni–C alloys. *Mater. Sci. Eng. A* **2006**, *431*, 228–236. [CrossRef]

15. Takayama, N.; Miyamoto, G.; Furuhara, T. Effects of transformation temperature on variant pairing of bainitic ferrite in low carbon steel. *Acta Mater.* **2012**, *60*, 2387–2396. [CrossRef]
16. Morito, S.; Tanaka, H.; Konishi, R.; Furuhara, T.; Maki, A.T. The morphology and crystallography of lath martensite in Fe-C alloys. *Acta Mater.* **2003**, *51*, 1789–1799. [CrossRef]
17. Caballero, F.G.; Poplawsky, J.D.; Yen, H.W.; Rementeria, R.; Morales-Rivas, L.; Yang, J.R.; García-Mateo, C. Complex Nano-Scale Structures for Unprecedented Properties in Steels. *Mater. Sci. Forum* **2017**, *879*, 2401–2406. [CrossRef]
18. Zhu, K.; Bouaziz, O.; Oberbillig, C.; Huang, M. An approach to define the effective lath size controlling yield strength of bainite. *Mater. Sci. Eng. A* **2010**, *527*, 6614–6619. [CrossRef]
19. Bush, M.E.; Kelly, P.M. Strengthening mechanisms in bainitic steels. *Acta Metall.* **1971**, *19*, 1363–1371. [CrossRef]
20. Bhadeshia, H. *Bainite in Steels*, 2nd ed.; IOM Communications: London, UK, 2001.
21. Badinier, G.; Sinclair, C.W.; Allain, S.; Danoix, F.; Gouné, M. The Mechanisms of Transformation and Mechanical Behavior of Ferrous Martensite, Encyclopedia. In *Reference Module in Materials Science and Materials Engineering*; Elsevier: Dublin, Ireland, 2017.
22. Ghosh, G.; Olson, G.B. Precipitation of paraequilibrium cementite: Experiments, and thermodynamic and kinetic modeling. *Acta Mater.* **2002**, *50*, 2099–2119. [CrossRef]
23. Winchell, P.G.; Cohen, M. The strength of martensite. *Trans. ASM* **1962**, *55*, 347–361.
24. Singh, S.B.; Bhadeshia, H.K.D.H. Estimation of bainite plate-thickness in low-alloy steels. *Mater. Sci. Eng. A* **1998**, *245*, 72–79. [CrossRef]
25. Hollomon, J.H.; Jaffe, L.D. Time-Temperature Relations in Tempering Steel. *Trans. TMS-AIME* **1945**, *162*, 223–249.
26. Virtanen, E.; Van Tyne, C.J.; Levy, B.S.; Brada, G. The tempering parameter for evaluating softening of hot and warm forging die steels. *J. Mater. Processing Technol.* **2013**, *213*, 1364–1369. [CrossRef]
27. Enami, T.; Sato, S.; Tanaka, T.; Funakoshi, T. Effects of cooling rates and tempering conditions on the strength and toughness of Mn-Ni-Mo, Cr-Mo steel plates. *Kawasaki Steel Giho* **1974**, *6*, 145–161.
28. Ben Haj Slama, M.; Gey, N.; Germain, L.; Hell, J.C.; Zhu, K.; Allain, S. Fast granularization of lath-like bainite in FeNiC alloys during isothermal holding at Ms + 20 K (+20° C). *Metall. Mater. Trans. A* **2016**, *47*, 15–18. [CrossRef]
29. Bordín, S.F.; Limandri, S.; Ranalli, J.M.; Castellano, G. EBSD spatial resolution for detecting sigma phase in steels. *Ultramicroscopy* **2016**, *171*, 177–185. [CrossRef] [PubMed]
30. Germain, L.; Gey, N.; Mercier, R.; Blaineau, P.; Humbert, M. An advanced approach to reconstructing parent orientation maps in the case of approximate orientation relations: Application to steels. *Acta Mater.* **2012**, *60*, 4551–4562. [CrossRef]
31. Bhadeshia, H.K.D.H. Thermodynamic analysis of isothermal transformation diagrams. *Metal Sci.* **1982**, *16*, 159–166. [CrossRef]
32. Hillert, M. Diffusion and Interface Control of Reactions in Alloys. *Metall. Trans. A* **1975**, *6*, 5–19. [CrossRef]
33. Kashif Rehman, M.D.; Zurob, H.S. A novel Approach to model Static Recrystallization of Austenitre During Hot-Rolling of Nb Microalloyed Steel. Part I: Precipitate-Free Case. *Metall. Mater. Trans. A* **2013**, *44*, 1862–1871. [CrossRef]
34. Hamada, J.; Enomoto, M.; Fujishiro, T.; Akatsuka, T. In-Situ Observation of the Growth of Massive Ferrite in Very Low-Carbon Fe-Mn and Ni Alloys. *Metall. Mater. Trans. A* **2014**, *45*, 3781–3789. [CrossRef]
35. Uhm, S.; Moon, J.; Lee, C.; Yoon, J.; Lee, B. Prediction model for the austenite grain size in the coarse grained heat affected zone of Fe-C-Mn steels: Considering the effect of initial grain size on isothermal growth behavior. *ISIJ Int.* **2004**, *44*, 1230–1237. [CrossRef]
36. Krauss, G.; Thompson, S.W. Ferritic microstructures in continuously cooled low-and ultralow-carbon steels. *ISIJ Int.* **1995**, *35*, 937–945. [CrossRef]

Article

Three Dimensional Methodology to Characterize Large Dendritic Equiaxed Grains in Industrial Steel Ingots

Marvin Gennesson [1,2,3,*], Julien Zollinger [1,2], Dominique Daloz [1,2], Bernard Rouat [1,2], Joëlle Demurger [3] and Hervé Combeau [1,2]

1 Laboratory of Excellence on Design of Alloy Metals for low-mAss Structures (DAMAS), Université de Lorraine, 57073 Metz, France; julien.zollinger@univ-lorraine.fr (J.Z.); dominique.daloz@univ-lorraine.fr (D.D.); bernard.rouat@univ-lorraine.fr (B.R.); herve.combeau@univ-lorraine.fr (H.C.)

2 Department of Metallurgy & Materials Science and Engineering, Institut Jean Lamour, 2 allée André Guinier Campus Artem, 54000 Nancy, France

3 Ascométal CREAS, Avenue de France, BP 70045, 57301 Hagondange, France; joelle.demurger@ascometal.com

* Correspondence: marvin.gennesson@ascometal.com

Received: 25 May 2018; Accepted: 12 June 2018; Published: 13 June 2018

Abstract: The primary phase grain size is a key parameter to understand the formation of the macrosegregation pattern in large steel ingots. Most of the characterization techniques use two-dimensional measurements. In this paper, a characterization method has been developed for equiaxed dendritic grains in industrial steel castings. A total of 383 contours were drawn two-dimensionally on twelve 6.6 cm^2 slices. A three-dimensional reconstruction method is performed to obtain 171 three-dimensional grains. Data regarding the size, shape and orientation of equiaxed grains is presented and thereby shows that equiaxed grains are centimeter-scale complex objects. They appear to be a poly-dispersed collection of non-isotropic objects possessing preferential orientations. In addition, the volumetric grain number density is 2.2×10^7 grains/m^3, which compares to the 0.5×10^7 grains/m^3 that can be obtained with estimation from 2D measurements. The 2.2×10^7 grains/m^3 value is ten-times smaller than that previously used in the literature to simulate the macrosegregation profile in the same 6.2 ton ingot.

Keywords: industrial ingot; steel; dendritic grain size

1. Introduction

Steel manufacturing requires more homogeneous and cleaner products in terms of inclusions, chemical segregations and porosity. Chemical segregations at the product scale cannot be reduced once the product is fully solidified and are responsible for the differences of mechanical properties in rolled products. Those macrosegregations arise because of the relative motion between the sedimenting solid grains and the thermosolutal convected liquid phase. Consequently, the number, shape and size of the grains are expected to have a great impact on the relative motion and thus on the final macrosegregation profile. Good characterization methods to measure grain properties are of primary importance. Moreover, improvements in casting processes need multi-scale and multi-physics numerical modeling. To be realistic, these models need an extensive description of the grain properties at the beginning of solidification [1–10]. Those numbers are not accessible for industrial castings, and postmortem techniques need to be developed.

Conventional metallography uses 2D measurements for equiaxed dendritic grain size measurements [11–15]. More recently, computer-aided methods that take binary images as the input can be found in the literature. On these binary images, combined segmentation/erosion methods to isolate individual dendrites have also been applied to equiaxed dendritic structures [11,12], as well as intercept techniques, [13–15]. They only measure primary and secondary dendrite arm spacings. Moreover, these image processing techniques give only qualitative and non-comparative results as they are very sensitive to the segmentation and filtering steps, which remove very small objects. Finally, the major drawback of these techniques is that for non-convex 3D objects, no stereological relationship exists to estimate the 3D properties of the grain distribution. It is then necessary to move to 3D in order to characterize equiaxed dendrites.

Advanced techniques like computed X-ray tomography [16–19] or serial cutting/polishing have been used to assess the real 3D grain size of dendritic structures [20,21]. The large size of equiaxed grain structures, which is in the order of magnitude of 1 cm, in addition to the low chemical contrast as encountered in heavy steel ingots makes using X-ray computed tomography quite impossible. The same problem exists with commercially available serial polishing microscopes for steel. For such large structures, conventional cutting remains the best solution. An example was given by Laren and Fredriksson in 1972 to identify the shape of one columnar grain (five slices) and one equiaxed grain (three slices) [22]. However, the small number of cutting planes and number of grains prevented drawing a general conclusion about dendritic equiaxed grains.

The current study in this paper proposes a macro serial cutting method to characterize 3D centimetric equiaxed grains and to estimate the 3D volumetric equiaxed grain density for industrial steel ingots. The method used along with its advantages and its drawbacks are presented with one example of an industrial ingot. It is depicted that the grain density, the characteristic size, the shape and the orientation of the 3D equiaxed grain distribution can be determined with this method.

2. Materials and Methods

2.1. Material and Microscopy

The steel grade used in this study is 100Cr6 (ASTM 52100). 100Cr6 is a through hardening steel used for bearing steel. It is industrially cast as ingots, and its nominal composition is given in Table 1. A 6.2 ton 100Cr6 bottom poured ingot was cast in an Asco Industries plant at Fos-sur-Mer, France, for use in this study. as seen in Figure 1a.

Table 1. Composition of 100Cr6 steel (wt %).

Grade	C	Mn	Si	Cr	Cu	Ni
100Cr6	0.98	0.32	0.22	1.4	0.12	0.14

As a part of conventional metallographic characterization, one central slice was cut along the main axis of the ingot and divided into thirty sections. Polishing followed by macroetching in a warm 15 wt % HCl aqueous solution was performed on each section to reveal the solidification structure. It must be noted that electron back-scattering diffraction cannot be used for characterization of low-alloyed steel primary structures. 100Cr6 fully solidifies in γ-austenite. At lower temperatures, primary austenite undergoes solid state phase transformations, and all the primary crystallographic structure is transformed. In the as-cast state, 100Cr6's microstructure is mostly fully pearlitic. Dendrites exist at room temperature only as chemical segregations that can be revealed with an appropriate reagent.

Macrostructures were imaged on a high resolution Expression 12000XL EPSON scanner (EPSON, Suwa, Japan). Of these etched structures, one sample has been chosen in the equiaxed zone and outlined in Figure 1. The sample is located 60 cm from the top of the ingot and 1.5 cm from the central axis of the ingot (drawn as the green dotted line). The distance from the edge of the sample to the columnar-to-equiaxed transition (drawn as a red dashed line) is 7 cm. The top and the bottom of the ingot are located at the left and right side of the figure, respectively. The sample dimensions are 1.75 cm × 3.8 cm × 5.1 cm.

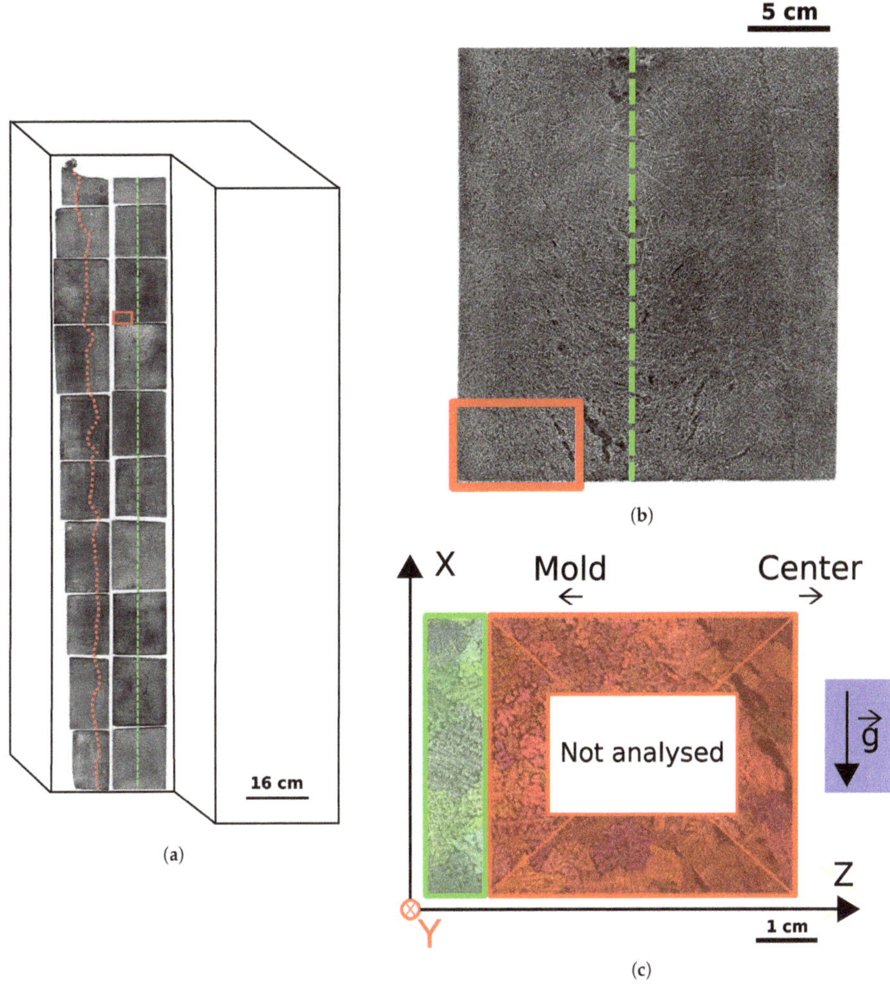

Figure 1. Description of the sample position in a 6.2 ton 100Cr6 ingot. The total height of the ingot is 267 cm. The red dotted line is the columnar-to-equiaxed transition. The green dashed line is the central axis of the ingot. (**a**) Macroetched central slice inside a 3D schematic view of the ingot. (**b**) Position of the sample for serial cutting. (**c**) Zoom on the red rectangle with the position of the serial cut metal in the green rectangle. The normal vector to the slices is directed along the Z axis.

2.2. Serial Cutting

On the sample taken from the ingot in Figure 1a, 12 slices of 0.7 mm in thickness were cut by wire erosion, the diameter of which is 300 µm. Those slices are taken in the green part of Figure 1c with their normal vectors parallel to $\overrightarrow{z}$. The total analyzed thickness was 1 cm, which is the width of the green area along the ingot radial direction, $\overrightarrow{z}$. Marks were made before the cutting to ensure that each slice was properly tracked and the final slice had dimensions of 1.75 cm × 3.8 cm × 0.07 cm. Optical microscopy was used to control the slice widths, where no defect in parallelism was observed.

2.3. Manual Outlining of Individual Grains

All the slices were ground up to 4000 SiC paper, and the final polishing was performed with a 3 µm diamond suspension. The solidification structure was revealed by warm etching in a supersaturated picric acid aqueous solution (Bechet–Beaujard reagent). Micrographs were taken with a Axioplan 2/Axiocam MRc5 optical microscope (Carl Zeiss, Oberkochen, Germany). The motorized platform combined with Axiovision software (version 4.6) allowed macroscopic tile images to be created.

The freeware Fiji was used for image analysis [23]. Manual identification of each 2D grain was performed after histogram equalization. Orientations of the dendrite primary axis and secondary arms were used to outline one grain on the first slice. Once a grain was outlined on its first slice of appearance, primary axis and secondary axis of the same grain were once again used as indicators for outlining on the other slices. The process was repeated until the grain could not be seen on the other slices or until the last slice was reached. Other grains were then outlined with the same procedure. A total of 390 2D contours were drawn manually.

A custom Fiji macro was used to store all the data about each 3D grain as a Fiji region of interest. The macro also produced a colored image of each individual slice, where each grain remained the same color on each slice. The final output is presented in Figure 2.

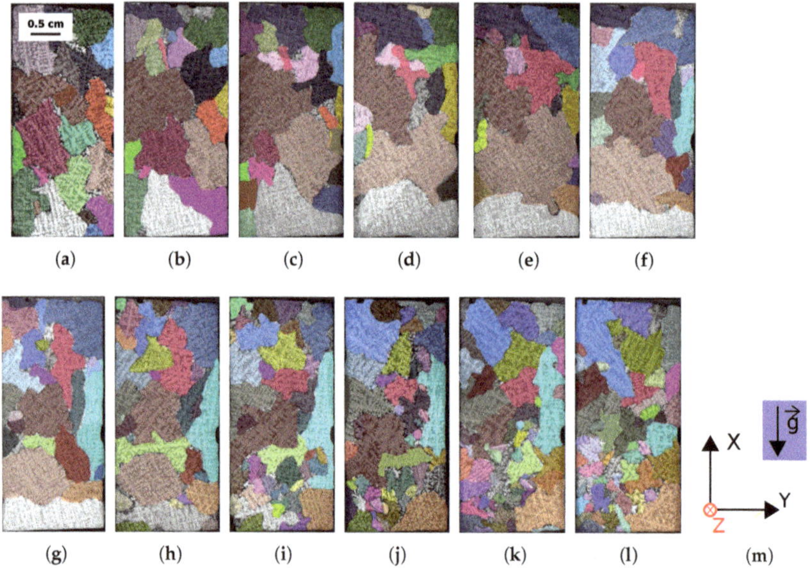

Figure 2. Solidification microstructure after FiJi processing. In the web version, each grain maintains the same color through all slices. (**a**) Z = 0 cm; (**b**) Z = 0.11 cm; (**c**) Z = 0.22 cm; (**d**) Z = 0.33 cm; (**e**) Z = 0.44 cm; (**f**) Z = 0.55 cm; (**g**) Z = 0.66 cm; (**h**) Z = 0.77 cm; (**i**) Z = 0.88 cm; (**j**) Z = 0.99 cm; (**k**) Z = 1.1 cm; (**l**) Z = 1.2 cm; (**m**) Associated coordinated system.

2.4. 3D Reconstruction

A program was written in Python 3 to reconstruct each grain in 3D as shown in Figure 3. After retrieving the data per grain from Fiji (as shown in Figure 3a) using the read-roi functions developed by Mary et al. [24], each 2D contour belonging to the same 3D grain needed to be discretized.

As the main goal is surface reconstruction, polar discretization was done on each contour with the same predefined array of angles. Henceforth, each angle is represented as a line, which is depicted in Figure 3b. If random points were taken on two successive contours, vertices could cross each other, and the surface would not be properly reconstructed. The chosen grain in Figure 3 is the white grain at the bottom of each slice in Figure 2a–i.

To ensure that the solution is unique for each angle, the discretization must be performed on a convex surface. The convex hull of the contour, i.e., the external blue contour in Figure 3b, was used. Each angle was then computed with respect to the center of mass of the new convex contour. This point is the origin of all the blue lines, and each line represents one angle.

The intersection between the lines and the convex Hull contour are the points used for the reconstruction of the surface in 3D between two sets of contours. The final result for one grain is shown in Figure 3c. The number of 3D grains retrieved in the sample volume from the twelve slices is 171. This 3D reconstruction is only used for graphical purposes.

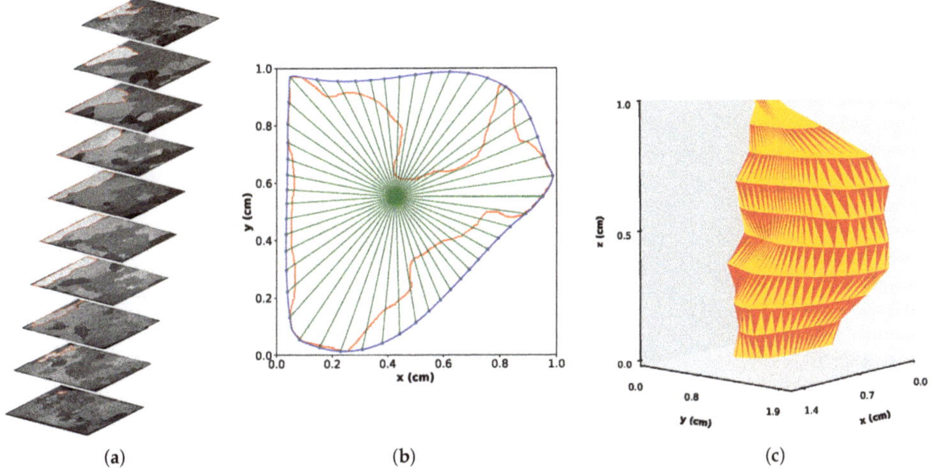

Figure 3. Principle of 3D reconstruction with 2D outlining followed by polar discretization. (**a**) Same grain outlined on each slice. (**b**) Concave contours (red) resulting from the 2D outlining in one slice of (**a**) with its convex Hull contour (blue). Green lines shows the polar discretization of the blue contour. (**c**) 3D reconstructed equiaxed grain. This the same grain as in (**a**). This grain is also colored in light grey at the bottom of each slice in Figure 2.

For the analysis, the use of grain area and grain volume is inconvenient in the international unit system. Image analysis studies often compare lengths by using equivalent circle and sphere diameters. Those equivalent diameters do not make any sense in the present case because the 2D dendrite morphology is very far from being circular.

The Feret diameter, which is defined as the longest distance between two points in a cloud of points, can be considered as a more physical length to describe the grains. For 2D objects, Feret diameters are found from the 2D point clouds created by the manually-drawn concave contours. An example of Feret estimation is shown in Figure 4a for the grain contour from Figure 3a. It can be seen that for concave contours, the Feret diameter may not lie in the drawn contour. For 3D objects, the point cloud for Feret calculation is the sum of all the 2D contours that belong to an individual grain.

In our dataset, it has been verified that grain area and grain volume are respectively proportional to the square and cube of the 2D and 3D Feret diameters. As a consequence, these two parameters are relevant to describe the geometry of equiaxed dendrites. They are also quite easy to handle, to compute elongation factors and also to define orientation angles. Feret diameters (2D or 3D) alone are sufficient to precisely describe a 3D object because they cannot describe if an object is isotropic, and other size parameters need to be introduced in other directions of the space.

Secondary and tertiary Feret diameters are defined similarly to 3D Feret diameter (also called 3D principal Ferret diameter): the secondary Feret diameter is defined as the longest distance found in a direction perpendicular to the principal Feret diameter, whereas the tertiary Feret diameter is the longest distance perpendicular to the principal Feret diameter and the secondary Feret diameters. To ensure that those other Feret diameters always numerically exist, the perpendicularity condition needs to be relaxed with a tolerance of $\pm 5^{\circ}$. This is because of the lack of data in between the different slices that creates non-existing z positions and angles in the 3D cloud point. Examples of the 3D principal, secondary and tertiary Feret diameters are given on Figure 4b for the 3D grain of Figure 3c. It has to be noted that with those definitions, nothing ensures that the three 3D Feret cross all at the same unique point.

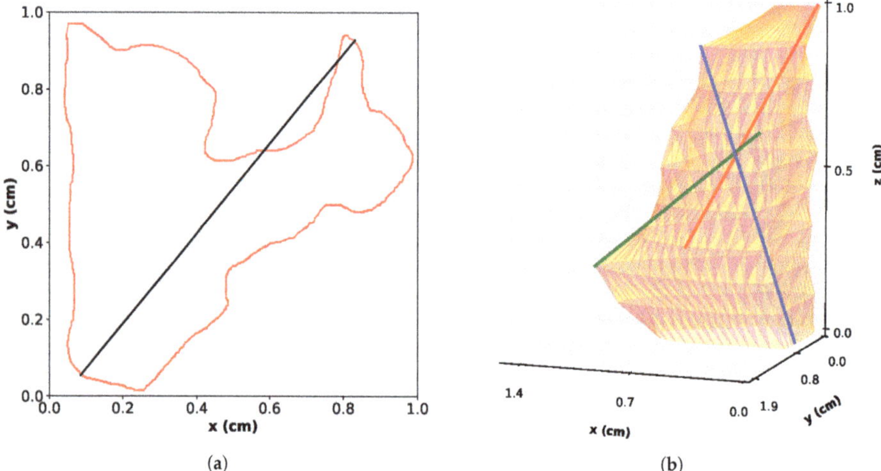

Figure 4. Principle of Feret calculation. (**a**) 2D Feret for the contour of Figure 3b. The 2D Feret diameter is drawn as the black line. (**b**) 3D principal, secondary and tertiary Feret for the grain of Figure 3c. The 3D principal, secondary and tertiary Feret are drawn as the red, green and blue lines, respectively.

3. Results

3.1. 2D Identification of 3D Grains

As said in Section 1, common metallographic measurements are two dimensional. With our dataset, it is possible to retrieve this kind of measurement using only the 2D contours. If each slice were considered individually, the average 2D Feret diameters, $< F_{2D} >$, would vary through the slices, as shown in Figure 5. The represented error bars are 50 % of the standard deviation ($0.5 \cdot \sigma$). The mean value, of all the grains on all slices, is 0.67 cm with a standard deviation of 0.15 cm. This mean value is not consistent with the variation observed through all the slices. On the first slice, the mean 2D Feret is 0.45 cm, whereas on the last seven slices, the mean 2D Feret length is close to 0.85 cm $\pm$ 0.1 cm. In other words, it means that grain size measurements can give results varying up to a factor of two just by changing the sampling position for the 2D measurement by 1 cm, i.e., measuring on the 0 cm slice depth or measuring on the 1 cm slice depth. It is the proof that gradients of structure sizes exist on very short scales in industrial ingots.

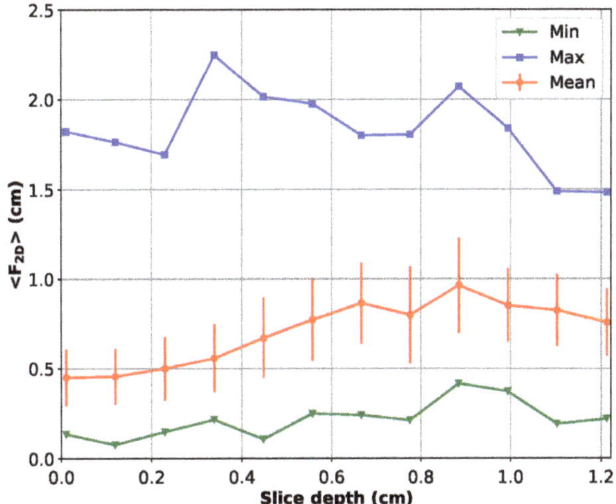

Figure 5. Evolution of the mean 2D Feret diameter on all the slices.

Figure 6a shows the distribution in terms of 2D Feret diameter (F_{2D}). The distribution follows a classical log-normal law that has been fitted with the blue line. The low probability of cutting 3D grains in a plan containing the principal Feret diameter is well described by a log-normal distribution law. The principal mode and the maximum Feret diameter are 0.72 cm and 2.2 cm, respectively. However, there is no information regarding the real shape of the grain. It is therefore impossible to understand what really is the grain size and also, for example, to infer if the grain distribution is uniform, bimodal or even more complex.

The volumetric grain number density can be at first approximated by dividing the surface grain number density by the mean 2D Feret diameter for each slice [25]. Although it is only valid for monodisperse and convex objects, it is a fast way of providing a lower boundary for the real 3D value. With this workflow, values from 0.5×10^7–2×10^7 m^{-3} are retrieved. Those values are strongly under-evaluated because the calculation performed averages the values of the Feret diameter on each slice.

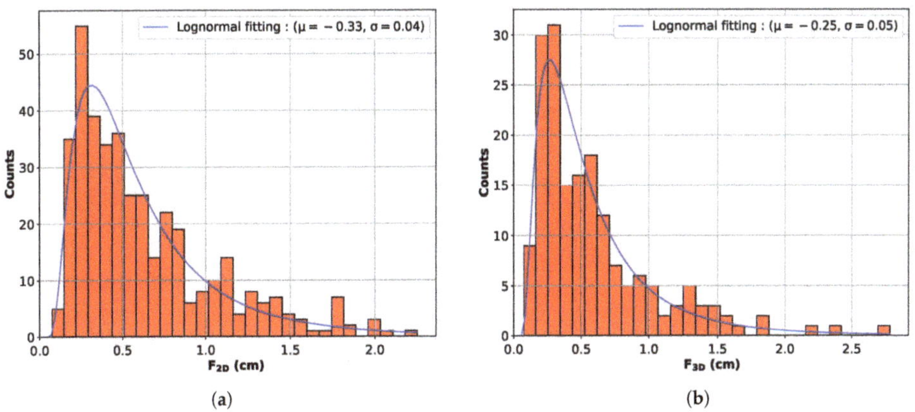

Figure 6. Histogram of measured Feret. (**a**) 2D Feret diameters. (**b**) 3D Feret diameters.

3.2. 3D Grain Structures

The final 3D result is thereby presented in Figure 7. Each reconstructed grain envelope was given a random color. As for 2D measurements, it is also possible to observe the distribution of 3D Feret diameters as shown in Figure 6b. Consequently, the number of 3D Feret diameters is much lower than the number of 2D contours and equals the number of grains, i.e., 171 grains. The shape of the distribution is the same, but the characteristic values are slightly different. The principal mode and maximum values are respectively 0.8 cm and 2.75 cm. The ratio between 2D and 3D mean Feret diameters is 0.9. A short Monte Carlo study on 1000 random slices on monodisperse cubes and spheres gives values of 0.7 and 0.94, respectively, for the same ratio. The volumetric grain number density is 2.2×10^7 grains/m^3.

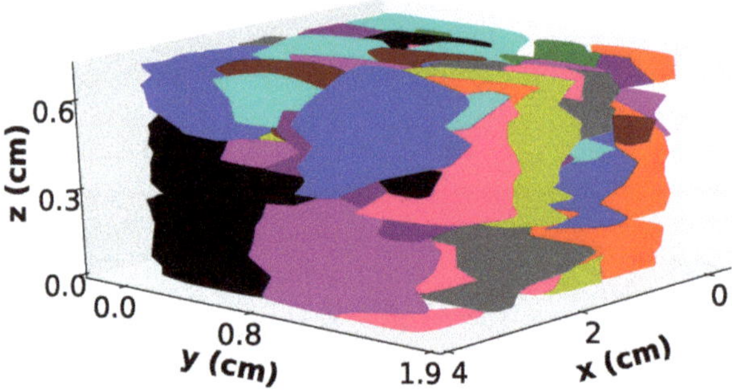

Figure 7. 3D reconstructed envelope of the dendritic microstructure. Each color in the web version accounts for one grain.

3.3. 3D Grain Properties

In 3D, the Feret diameter is only a partial indicator of grain size. Further insight can be gained by finding the longest size in directions perpendicular to the principal Feret diameter. After calculations, three Feret diameters (principal, secondary and tertiary) are obtained in the direction perpendicular to each other with a decrease in lengths. For isotropic objects like spheres, the secondary Feret diameter equals the principal Feret diameter. For non-isotropic objects, they provide a quick way to determine if the objects are elongated. Elongation factors, represented by secondary or tertiary Feret diameters divided by the principal Feret diameter values, are plotted in Figure 8. From the two elongation factors, it can be seen that the dendritic grains are non-isotropic objects with complex shapes. The mean elongation factors are equivalent to a 5:3:2 ellipsoid. The most symmetrical case that exists is a 5:5:4 ellipsoid, and the most elongated one is equivalent to a 10:5:2 ellipsoid. There is no isotropic case in the analyzed volume. In comparison, the 2D elongation mean case is a 5:3 ellipse. The 2D value is close in comparison to the 3D one, but does not give any information regarding the 3D shape.

The orientation of dendritic grains is also of interest. The orientation of the principal Feret diameter is represented in Figure 9. To obtain a density plot, kernel density estimation was applied to the data with the use of the seaborn python library, [26]. This library is very useful because it allows one to directly see the density function for each variable along with the two-dimensional estimated density function. The colorscale values are directly proportional to the number of points by unit of area in the 2D space (θ_x, θ_z). It can be seen that most of the principal Feret diameters are perpendicular to the slicing direction, $\overrightarrow{z}$, which is also the radial direction of the ingot. Moreover, the distribution of the principal Feret diameter orientation has two main characteristic orientations with respect to the x direction ($\overrightarrow{x}$ is oriented along $-\overrightarrow{g}$, which is the opposite direction of gravity). Those characteristic values are located at the center of the darker spots in Figure 9 at 10° and 65° with respect to $\overrightarrow{x}$. Those characteristic orientations do not appear to be related in any way to the principal Feret diameters.

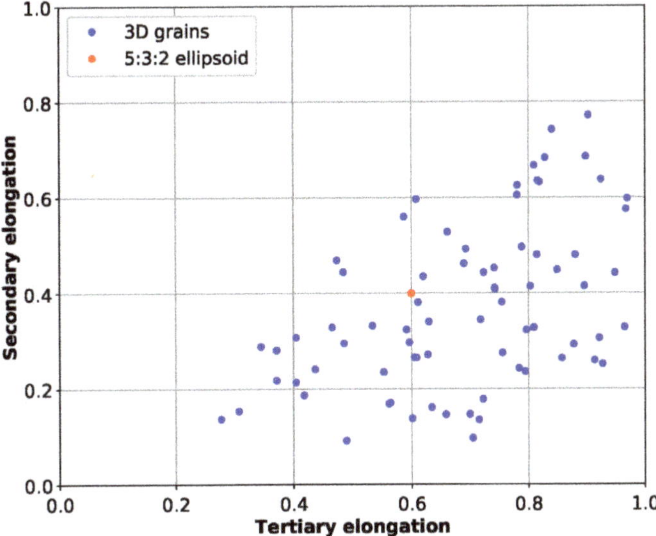

Figure 8. Comparisons of secondary and tertiary equiaxed dendritic grain elongation factors.

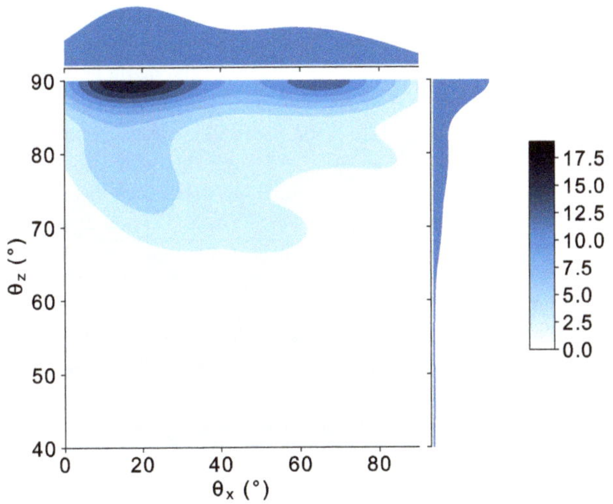

Figure 9. Density map for 3D principal Feret orientations with respect to the x and z directions. Density function shape for θ_x and θ_z are respectively plotted parallel to the corresponding axis. The z direction is a radial direction of the ingot, and the x direction is opposite gravity.

4. Discussion

The main drawback of the method is that all the grains are kept during the analysis. Usually in image analysis, objects that intercept picture sides are removed [27]. This is not the case in the present analysis because of the small size of each slice and also the large grain size. It is a source of errors in the estimation of grain size, orientation and shape factors. However, it can be corrected in the evaluation of the volumetric grain number density, which is the most important parameter in this study. This can be done by adding factors regarding the number of picture sides that the grain is crossing [27]. In the same way, some grains are only observed on one slice. The size of these grains is only estimated in 2D and thus under-evaluated. Last, but not the least, grains finer than the total distance between two observation planes are lost.

As previously stated, common practice for equiaxed grain size determination is 2D measurement, which is a fast way to obtain rudimentary results. In this case, the 2D grain size measurements allow us to establish a mean grain size of 0.67 cm. Previous measurements reported by Kumar et al. on the same kind of ingots gave experimental grain size values on the order of 0.1 cm via the ASTM E-112 intercepts technique [28]. This kind of measurement gives values related to the secondary dendrite arm spacing and is by nature not adapted to dendritic equiaxed grain size measurements. Secondary arms that intercept the measurement line will be counted as the intercept and will decrease the mean intercepted length. The norm would only be applicable to the manually outlined grain contours that will be 2D convex contours. This case would be equivalent to the usual measurements of austenitic grain size.

In this study, 3D measurements show that grain size heterogeneities can arise in an industrial ingot on a scale less than 1 cm. On the first slice, the mean 2D Feret is 0.45 cm, whereas on the last seven slices, the mean 2D Feret length is close to 0.85 cm $\pm$ 0.1 cm. The 25 % of the largest grains fill 80 % of the analyzed volume. Equiaxed grains appear to be a polydisperse collection of objects, which was not possible to prove with two-dimension analysis alone. The ratio of 0.94 between 2D and 3D mean Feret diameters is a direct consequence of the polydispersity. Random cuttings of each 3D dendrite alone would not give results as close to unity because of the low shape isotropy (i.e., secondary and tertiary elongation factors far from one).

Three-dimensional analysis indicated that grains are elongated and not fully isotropic. The real shape of dendritic grains can be estimated by ellipsoids at the first order. This means that no good estimation, in terms of volumetric grain number density, can be made with only 2D measurements. For example, by neglecting the concave shape of equiaxed dendrites, it is possible to estimate the volumetric grain number density by dividing the mean 2D surface grain number density by the 2D mean Feret diameter, which is $0.7 \times 10^7\,\mathrm{m}^{-3}$. Though the order of magnitude is correct, the numerical value is four-times lower than the 3D value. It has to be noted that 2D and 3D Feret diameter calculations are performed on concave contours. The 2D and 3D Feret diameters as defined in this study may not be fully contained respectively in the surface or volume of interest. This underestimation accounts for one part of the difference between the volumetric density estimated from 2D measurement and the value estimated from the number of 3D grains.

This parameter is crucial to validate macrosegregation numerical models currently under development like finite volume or finite element models [4,29]. Numerical predictions by Kumar et al. for the ingot investigated in this study, consider three different cases with varying initial volumetric densities: $1 \times 10^8\,\mathrm{m}^{-3}$, $5 \times 10^9\,\mathrm{m}^{-3}$ and $5 \times 10^{10}\,\mathrm{m}^{-3}$, respectively [1]. According to the authors, those initial values produce final castings with structures that are fully-dendritic, mixed globular/dendritic and fully-globular, respectively. The experimental value, $2.2 \times 10^7\,\mathrm{m}^{-3}$, is one order of magnitude less than the smallest value chosen for the simulation. The dendritic morphology as shown in Figure 2 clearly demonstrates that there is a good agreement between the experimental and the predicted morphology.

Finally, equiaxed grains are not randomly oriented. Their Feret diameters are always very close to the slicing planes. The most probable angle is perpendicular to the radial direction (relatively to the ingot) and less than 15° disoriented from the direction of gravity. Another characteristic angle exists, perpendicular to the slicing direction, at an angle of 65 degrees from the direction of gravity. These two characteristic angles probably come from the settling period in the liquid prior to packing and from the shape of the liquid pool. This kind of textural effect cannot be retrieved from 2D characterization.

5. Conclusions

A method for 3D characterization in industrial castings with large equiaxed dendritic structures has been proposed. Slicing of one sample was performed by wire erosion, and outlines of each grain were made precisely. This manual method allows the accurate determination of 2D properties along with 3D estimation of grain size. Equiaxed grains appear to be complex, non-isotropic and non-randomly-oriented objects. The volumetric grain number density as calculated is close to 2.2×10^7 grains/m^3. This value is ten-times smaller than the currently used value in the literature so as to simulate this kind of ingot. Numerical calculations need to be performed to identify the underlying cause of the determined textural effect. As a part of future work, new measurements should be carried out at different heights on the central line of an industrial ingot to see the evolution in terms of equiaxed structures. The number of slices has also to be increased to improve the precision of the technique. Such data could really help to understand what is the origin of equiaxed grains and to provide better inputs for the numerical models.

Author Contributions: Conceptualization, J.D. and H.C. Data curation, M.G. Formal analysis, M.G., J.Z., D.D., B.R., J.D. and H.C. Funding acquisition, J.D. and H.C. Investigation, M.G. and B.R. Methodology, M.G. and B.R. Project administration, D.D., J.D. and H.C. Software, M.G. Supervision, D.D., J.D. and H.C. Validation, M.G. Visualization, M.G. and H.C. Writing, original draft, M.G. Writing, review and editing, M.G., J.Z., D.D., B.R., J.D. and H.C.

Funding: This work was supported by ASCO Industries and ANRTFrance under a CIFREPh.D. fellowship (Grant Number 2015-1286). This work was supported by the French State through the program "Investment in the future" operated by the National Research Agency (ANR) and referenced by ANR-11-LABX-0008-01 (LabEx DAMAS).

Acknowledgments: The authors would also like to thank Jacob Roman Kennedy and Manoj Joishi improving the use of English in the manuscript.

Conflicts of Interest: The authors declare no conflict of interest.

References

1. Kumar, A.; Založnik, M.; Combeau, H. Prediction of equiaxed grain structure and macrosegregation in an industrial steel ingot: Comparison with experiment. *Int. J. Adv. Eng. Sci. Appl. Math.* **2010**, *2*, 140–148. [CrossRef]
2. Combeau, H.; Založnik, M.; Hans, S.; Richy, P.E. Prediction of macrosegregation in steel ingots: Influence of the motion and the morphology of equiaxed grains. *Metall. Mater. Trans. B* **2009**, *40*, 289–304. [CrossRef]
3. Combeau, H.; Založnik, M.; Bedel, M. Predictive Capabilities of Multiphysics and Multiscale Models in Modeling Solidification of Steel Ingots and DC Casting of Aluminum. *JOM* **2016**, *68*, 2198–2206. [CrossRef]
4. Heyvaert, L.; Bedel, M.; Založnik, M.; Combeau, H. Modeling of the Coupling of Microstructure and Macrosegregation in a Direct Chill Cast Al-Cu Billet. *Metall. Mater. Trans. A* **2017**, *48*, 4713–4734. [CrossRef]
5. Bousquet-Melou, P.; Goyeau, B.; Quintard, M.; Fichot, F.; Gobin, D. Average momentum equation for interdendritic flow in a solidifying columnar mushy zone. *Int. J. Heat Mass Transf.* **2002**, *45*, 3651–3665. [CrossRef]
6. Ni, J.; Beckermann, C. A volume-averaged two-phase model for transport phenomena during solidification. *Metall. Trans. B* **1991**, *22*, 349. [CrossRef]
7. Goyeau, B.; Gobin, D.; Benihaddadene, T.; Quintard, M. Numerical calculation of the permeability in a dendritic mushy zone. *Metall. Mater. Trans. B* **1999**, *30*, 613–622. [CrossRef]
8. Beckermann, C. Modelling of macrosegregation: Applications and future needs. *Int. Mater. Rev.* **2002**, *47*, 243–261. [CrossRef]
9. Gu, J.; Beckermann, C. Simulation of convection and macrosegregation in a large steel ingot. *Metall. Mater. Trans. A* **1999**, *30*, 1357–1366. [CrossRef]
10. Ahmadein, M.; Wu, M.; Ludwig, A. Analysis of macrosegregation formation and columnar-to-equiaxed transition during solidification of Al-4 wt.% Cu ingot using a 5-phase model. *J. Cryst. Growth* **2015**, *417*, 65–74. [CrossRef] [PubMed]

11. Watt, T.J.; Taleff, E.M.; Lopez, L.; Beaman, J.; Williamson, R. Solidification mapping of a nickel alloy 718 laboratory VAR ingot. In *Proceedings of the 2013 International Symposium on Liquid Metal Processing & Casting*; The Minerals, Metals & Materials Society (TMS); Springer: Cham, Switzerland, 2013; pp. 261–270.
12. Ivanoff, T.A.; Watt, T.J.; Taleff, E.M. Microstructure Characterization of Nickel Alloy 718 with Automated Optical Image Processing. In *Characterization of Minerals, Metals, and Materials 2016*; Springer: Cham, Switzerland, 2016; pp. 19–25.
13. Zlatičanin, B.V.; Durić, S.; Jordović, B.M.; Radonjić, B. Characterization of microstructure and properties of AlCuMg alloys. *J. Min. Metall. B Metall.* **2003**, *39*, 509–526. [CrossRef]
14. Wang, Q.; Knight, J.W.; Hess, D.R. Method for Automatic Quantification of Dendrite Arm Spacing in Dendritic Microstructures, 2013. U.S. Patent 9500594 B2, 22 November 2016.
15. Wang, Q.; Knight, J.W. Method for Automatic Quantification of Dendrite Arm Spacing in Dendritic Microstructures, 2015. U.S. Patent 8,942,462, 27 January 2015.
16. Fuloria, D.; Lee, P.D.; Bernard, D. Microtomographic characterization of columnar Al–Cu dendrites for fluid flow and flow stress determination. *Mater. Sci. Eng. A* **2008**, *494*, 3–9. [CrossRef]
17. Tolnai, D.; Townsend, P.; Requena, G.; Salvo, L.; Lendvai, J.; Degischer, H. In situ synchrotron tomographic investigation of the solidification of an AlMg4,7Si8 alloy. *Acta Mater.* **2012**, *60*, 2568–2577. [CrossRef] [PubMed]
18. Yang, M.; Xiong, S.M.; Guo, Z. Characterisation of the 3-D dendrite morphology of magnesium alloys using synchrotron X-ray tomography and 3-D phase-field modelling. *Acta Mater.* **2015**, *92*, 8–17. [CrossRef]
19. Wang, M.; Xu, Y.; Jing, T.; Peng, G.; Fu, Y.; Chawla, N. Growth orientations and morphologies of α-Mg dendrites in Mg–Zn alloys. *Scr. Mater.* **2012**, *67*, 629–632. [CrossRef]
20. Alkemper, J.; Voorhees, P. Three-dimensional characterization of dendritic microstructures. *Acta Mater.* **2001**, *49*, 897–902. [CrossRef]
21. Madison, J.; Spowart, J.; Rowenhorst, D.; Fiedler, J.; Pollock, T. Characterization of three-dimensional dendritic structures in nickel-base single crystals for investigation of defect formation. In Proceedings of the Superalloys Conference, Champion, PA, USA, 14–18 September 2008.
22. Laren, I.; Fredriksson, J. Relations between ingot size and microsegregations. *Scand. J. Metall.* **1972**, *1*, 59–68.
23. Schindelin, J.; Arganda-Carreras, I.; Frise, E.; Kaynig, V.; Longair, M.; Pietzsch, T.; Preibisch, S.; Rueden, C.; Saalfeld, S.; Schmid, B.; et al. Fiji: An open-source platform for biological-image analysis. *Nat. Methods* **2012**, *9*, 676–682. [CrossRef] [PubMed]
24. Mary, H. Read-roi. Available online: https://github.com/hadim/read-roi (accessed on 31 August 2017).
25. Underwood, E. Stereology, or the quantitative evaluation of microstructures. *J. Microsc.* **1969**, *89*, 161–180. [CrossRef] [PubMed]
26. Waskom, M.; Botvinnik, O.; Hobson, P.; Cole, J.B.; Halchenko, Y.; Hoyer, S.; Miles, A.; Augspurger, T.; Yarkoni, T.; Megies, T.; et al. Seaborn: v0.5.0 (November 2014). *Zenodo* **2014**. [CrossRef]
27. Coster, M.; Chermant, J.L. *Précis D'analyse D'images*; Presses du CNRS: Paris, France, 1989.
28. Kumar, A.; Založnik, M.; Combeau, H.; Demurger, J.; Wendenbaum, J. Experimental and Numerical Studies of the influence of hot top conditions on Macrosegregation in an industrial steel ingot. In Proceedings of the First International Conference on Ingot Casting, Rolling and Forging (IRCF), Aachen, Germany, 3–7 June 2012.
29. Gandin, C.A.; Combeau, H.; Založnik, M.; Bellet, M. Finite Element Multi-scale Modeling of Chemical Segregation in Steel Solidification Taking into Account the Transport of Equiaxed Grains. *Metall. Mater. Trans. A* **2018**, *49*, 1725–1748.

Article

In Situ Investigation of the Iron Carbide Precipitation Process in a Fe-C-Mn-Si Q&P Steel

Sébastien Y. P. Allain [1,*], Samy Aoued [2,3], Angéline Quintin-Poulon [2], Mohamed Gouné [2], Frédéric Danoix [3], Jean-Christophe Hell [4], Magali Bouzat [4], Michel Soler [4] and Guillaume Geandier [1]

1 Institut Jean Lamour, UMR CNRS-Université de Lorraine 7198, 54000 Nancy, France; guillaume.geandier@univ-lorraine.fr
2 Institut de Chimie de la Matière Condensée de Bordeaux, UPR CNRS 9048, 33608 Pessac, France; samy.aoued@gmail.com (S.A.); Angeline.Poulon@icmcb.cnrs.fr (A.Q.-P.); mm.goune@gmail.com (M.G.)
3 Normandie Université, UNIROUEN, INSA Rouen, CNRS, Groupe de Physique des Matériaux, 76000 Rouen, France; frederic.danoix@univ-rouen.fr
4 Maizières Automotive Products, Arcelormittal Maizières Research SA, 57283 Maizières les Metz, France; jean-christophe.hell@arcelormittal.com (J.-C.H.); jean-christophe.hell@arcelormittal.com (M.B.); michel.soler@arcelormittal.com (M.S.)
* Correspondence: sebastien.allain@univ-lorraine.fr; Tel.: +33-670-608-831

Received: 28 May 2018; Accepted: 21 June 2018; Published: 26 June 2018

Abstract: Quenching and Partitioning (Q&P) steels are promising candidates for automotive applications because of their lightweight potential. Their properties depend on carbon enrichment in austenite which, in turn, is strongly influenced by carbide precipitation in martensite during quenching and partitioning treatment. In this paper, by coupling in situ High Energy X-Ray Diffraction (HEXRD) experiments and Transmission Electron Microscopy (TEM), we give some clarification regarding the precipitation process of iron carbides in martensite throughout the Q&P process. For the first time, precipitation kinetics was followed in real time. It was shown that precipitation starts during the reheating sequence for the steel studied. Surprisingly, the precipitated fraction remains stable all along the partitioning step at 400 °C. Furthermore, the analyses enable the conclusion that the iron carbides are most probably eta carbides. The presence of cementite was ruled out, while the presence of several epsilon carbides cannot be strictly excluded.

Keywords: steel; Q&P; transition carbide; precipitation; HEXRD; TEM

---

1. Introduction

Quenching and Partitioning (Q&P) is an annealing process proposed in 2003 to elaborate a new generation of advanced high strength steel (AHSS) for automotive construction [1,2]. The steels manufactured according to this new route show high yield strengths thanks to a refined microstructure as well as good ductility provided by the large fraction of austenite retained after processing, which enables a transformation induced plasticity (TRIP) effect. They are thus seen as possible solutions for automotive makers to lighten their body-in-white structures and to improve their crash resistance.

This processing route comprises an incomplete quenching step after an austenitic soaking in order to partially transform austenite into martensite followed by an isothermal partitioning step during which carbon is supposed to diffuse out from martensite (α') to austenite (γ). This mechanism of retained austenite enrichment and stabilization is made possible if carbon does not remain trapped in the martensite. The final step consists generally in a rapid cooling during which a final martensitic transformation in stabilized austenite could occur. Usually, the Q&P heat treatment is thus defined by three key parameters, its quenching temperature (QT), its partitioning temperature (PT), and the partitioning time (Pt) [3].

The soaking conditions in the austenitic range have been chosen to suppress any trace of ferrite and to dissolve pre-existing cementite carbides. Ferrite is to be avoided as it causes a decrease in the yield strength of the steels. Residual carbides reduce the available carbon content to stabilize the retained austenite. Nevertheless, too severe austenitic soaking (too long or at too high temperature) leads to abnormal growth of austenite grains. Too large grains of austenite are unfavorable for the toughness of the steel and preclude in situ investigations using the considered diffraction setup (individual diffraction spots instead of Debye–Scherrer rings).

QT permits the initial fraction of martensite to set and thus the maximum fraction of austenite that can be stabilized during partitioning. If the initial fraction of martensite is high, the carbon content in martensite is sufficient to stabilize all the available austenite. On the contrary, if the fraction of martensite is low, the available carbon content in martensite is not sufficient to stabilize the austenite during the final quench. As suggested by the pioneering work of Speer et al. [1–3], the fraction of retained austenite can thus be maximized as a function of QT. PT and Pt serve to control the partitioning conditions, i.e., the diffusion of carbon from martensite to austenite. As the process is thermally activated, the higher the PT, the faster complete partitioning is achieved. As the partitioning mechanism is fast, the Pt parameter has in practice little effect except if other physical mechanisms take place. The so-called partitioning process can in fact be hindered by a possible carbide free bainitic transformation in retained austenite [3–6], by carbon segregations on martensite defects [7], and also by carbide precipitation in martensite [6,8–11]. These two last processes also occur commonly when tempering as-quenched martensitic steels [11,12]. In the Q&P treatment, the specificity is that they compete with carbon diffusion from martensite into retained austenite.

The main indirect effect of carbide precipitations is to trap carbon in martensite and thus block the enrichment of carbon in austenite. According to our knowledge, the direct effect of carbides on the mechanical properties of tempered martensite has not been demonstrated so far. Nevertheless, the direct and reliable measurement of the mass fraction of carbides is challenging in steels because of their low fractions and their nanometer sizes. Mossbauer spectroscopy or technique based on selective dissolutions allow accurate post mortem measures but carbide precipitation kinetics to our knowledge has never been determined in situ.

HajyAkbary et al. [6] reported the presence of epsilon (ε) carbides at the very early stage of the partitioning step in a Fe-0.3C-3.6Mn-1.6Si steel (in wt %). They claim that these transition carbides appear during the quenching step and then tend to dissolve slowly during partitioning step (PT = 400 °C). Their thermodynamic calculations exclude the occurrence of eta carbides. Pierce et al. [8,9] characterized by transmission electron microscopy (TEM) and Mössbauer spectroscopy (MS) the transition carbides in a Fe-0.4C-1.5Mn-1.5Si (in wt %) steel along different Q&P and Q&T (Quenching and Tempering) processes. In both cases, they report the precipitation of eta (η) carbides after tempering or partitioning at 400 °C. They also showed that eta carbides are essentially non-stoichiometric with compositions close to Fe_3C, instead of Fe_2C. This observation can explain why Toji et al. [11] identified their transition carbides as cementite (θ) particles, as they were only based on local composition measurement by atom probe tomography (APT). In addition, their observations of cementite are sustained by an original reassessed thermo-kinetic model. More recently, [9] Pierce et al. also observed cementite by TEM after Q&P at PT = 450 °C but attribute its occurrence to the decomposition of austenite at long partitioning time (Pt). Carbon clustering in martensite was also reported by Thomas et al. [7] in a highly alloyed system. In the literature, the nature, the composition of carbides, as well as their precipitation sequences thus all remain a bone of contention.

All these recent studies however lead to the conclusion that a certain fraction of carbon must also be trapped in martensitic laths, preventing a complete carbon partitioning between martensite and austenite, and thus limiting austenite enrichment. In a previous paper [13], we measured for instance that the carbon content in austenite of a Fe-0.3C-2.5Mn-1.5Si (in wt %) steel after quenching at 200 °C and 200 s partitioning at 400 °C was 1.05% (QT = 200 °C; PT = 400 °C; Pt = 200 s). As the final fraction of retained austenite is 12%, a simple mass balance shows that the overall residual carbon content in tempered martensite is 0.19%. The carbon depletion of martensite expected from the original theory of Speer et al. is thus far from being completed [3]. In addition, the lattice tetragonality of the martensite gives a mean carbon content of 0.09%C, meaning that 0.10% of carbon is segregated or precipitated in martensite. Even if the use of the body centered tetragonal (BCT) lattice of martensite is not considered as an absolute method, it shows that a large fraction of carbon is trapped in martensite. This value of 0.10% is consistent with the findings of Pierce et al. [8], which show that between 24% and 41% of bulk carbon content can be trapped in η carbides in their alloys.

The carbon balance was also sustained by the identification of a third phase on synchrotron X-ray diffractograms which can be indexed most probably as eta carbides according to the literature [13]. Nevertheless, their precise nature as well as their precipitation kinetics have not been analyzed and discussed so far. In the present paper, in the light of novel coupled SEM/TEM observations and diffraction experiments, we aim to confirm the nature of the third phase reported in our previous work. We show that particles observed by microscopy in tempered martensite are definitely transition carbides, most probably eta carbides but without excluding the presence of epsilon carbides. Based on these results, HEXRD experiments conducted on synchrotron beamlines was used to determine for the first time in situ the precipitation kinetics of the transition carbides throughout the studied Q&P schedule.

2. Materials and Methods

The studied steel is a model alloy produced at laboratory scale, with composition Fe-0.3C-2.5Mn-1.5Si (in wt %). For more details about the manufacturing condition of the sample, please refer to [13]. This alloy is very close to the alloy studied by HajyAkbari et al. [6].

2.1. High Energy X-Ray Diffraction

High Energy X-Ray Diffraction (HEXRD) experiments conducted on synchrotron beamlines offer opportunities to follow in real-time the complex phase transformation processes and their possible interactions taking place in a thermomechanical treatment [4,13–17]. Our recent in situ experiments revealed for instance that intense second-order internal stresses at phase scale are produced all along the processing route of Q&P steels [13,14]. The stresses affect the way carbon enrichment in austenite during partitioning should be measured, and the apparent stability of retained austenite at room temperature. Such in situ experiments are thus the sole indirect method based on XRD to measure reliably carbon enrichment in austenite as they allow deconvoluting unambiguously the chemical and mechanical contribution in the austenite lattice parameter evolution.

The experiment analyzed in this paper was conducted at Deutsches Elektronen-Synchrotron (DESY, Hambourg, Germany). The experimental set-up is extensively described in our previous paper [13]. Petra P-07 beamline was operated at 100 keV using a powder diffraction configuration in transmission and a high throughput 2D detector which enables an acquisition rate higher than 10 Hz. The studied cylindrical sample (Φ = 4 mm; 10 mm height) was heated and cooled using a commercial Bähr dilatometer (TAinstruments, New Castle, DE, USA) available on the line. The in situ experiment was permitted to follow the phase transformations during a model Q&P schedule characterized by a quench temperature (QT) of 200 °C and a partitioning temperature (PT) of 400 °C. The detailed thermal schedule is represented in Figure 1. Different points of interest are reported in the cycle to simplify future discussions.

As shown in our previous papers [13,14], the low signal/noise ratio permitted by high energy enables the presence of transition carbides to be detected on diffraction spectra even at the beginning of partitioning (point 5 in Figure 1). These transition carbides were interpreted most probably as eta carbides on the sole basis of HEXRD results. In this paper, we aim to analyze and discuss further the nature of these carbides thanks to electron microcopy observations, and to appraise the precipitation kinetics.

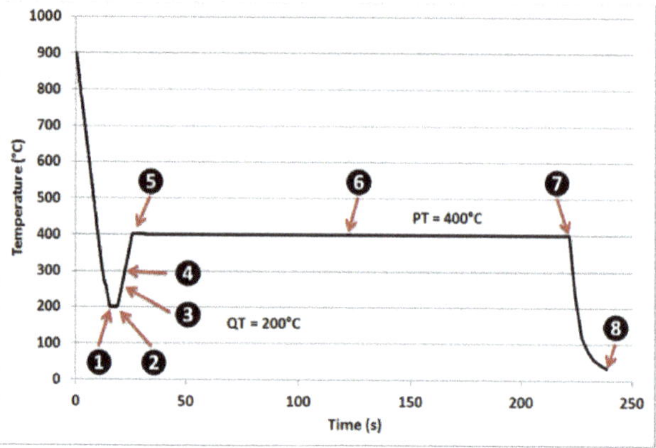

Figure 1. Quenching and Partitioning (Q&P) thermal treatment applied on the studied steel after a 5 min austenitization at 900 °C. The experiment was carried out in situ on a synchrotron beamline. 2D diffraction patterns were acquired every 0.1 s all along this cycle. Eight points of interest are highlighted.

2.2. Electron Microscopy

The HEXRD sample was also analyzed post mortem (state 8 in Figure 1) using conventional scanning electron microscopy (SEM) and TEM. By coupling these different techniques, it was possible to measure the size and the morphology of precipitates but above all to assess some additional information related to the nature of the carbides.

After a standard metallographic (polishing) preparation, the sample was etched with a combination of 1% Nital and Picral etchants for SEM observations using a JEOL 7001F Field Emission Gun (JEOL, Tokyo, Japan) operating at 10 kV. Thin foils were also prepared at low temperature using an EM-09100 JEOL cryo-ion slicer system (with a liquid nitrogen tank, JEOL, Tokyo, Japan). TEM observations were carried out on a JEOL 2100 (JEOL, Tokyo, Japan) operated at 200 kV.

3. Results and Discussions

3.1. Morphology and Size of Carbides by SEM Observations

Figure 2 shows a SEM micrograph of the studied steel after Q&P heat treatment (state 8). The microstructure is obviously duplex with a fibrous martensitic matrix (dark contrast) containing interwoven fine austenite films and coarse austenite islands (clear contrast). According to our HEXRD measurements [13], the fraction of retained austenite is about 12% and the fraction of fresh martensite is lower than 1%. This latter fraction is determined during final cooling but the phase cannot be distinguished from austenite on the SEM micrograph.

Tempered martensitic matrix contains also numerous intralath carbides which appear in bright contrast. Carbides show a platelet morphology. Their mean length is about 400 nm but the resolution of SEM does not permit a reliable estimate of their thickness (approx. a few nanometers).

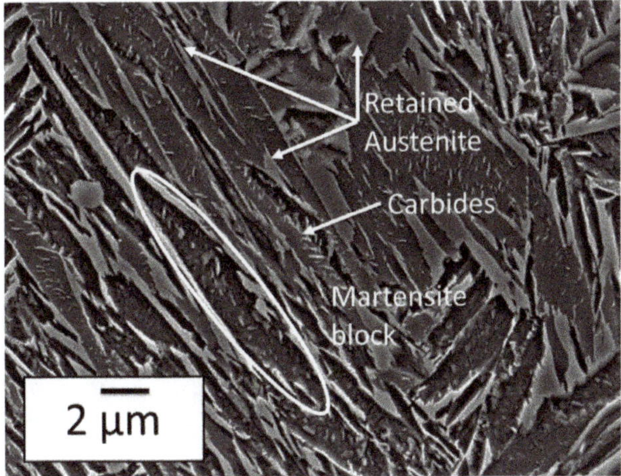

Figure 2. Scanning electron microscopy (SEM) micrograph of the studied steel after Q&P heat treatment (state 8). Etching has dissolved the tempered martensitic matrix preferentially. Intralath carbides and interlath retained austenite thus appear in clear contrast. A martensite block is also highlighted.

3.2. *Crystallographic Nature of Carbides by TEM Observations*

TEM observations were conducted first to confirm SEM observations regarding the morphology and the size of precipitates, but also to investigate their crystallographic structure. Figure 3a shows a bright field (BF) micrograph obtained on a wide block of tempered martensite. In that case, two variants of carbides were observed. In the literature, the majority of the linear precipitates are reported to align along martensite [010] and [001] crystallographic directions [6,18].

According to the literature, three types of carbides have been reported in Q&P steels: θ, ε, and η. The crystallographic structures of ε and η carbides are very close (η carbide structure is in fact pseudo-hexagonal and similar to the hexagonal ε carbide one) and are often difficult to distinguish from TEM experiments, especially when carbides are non-stoichiometric, and thus with vanishing super-lattice diffraction spots [8,17]. In this study, to determine the type of carbides (θ, ε, or η), we chose to work with those aligned along martensite $[-2\text{-}11]_{\alpha'}$ (Figure 3b). Selected area electron diffraction (SAED) observations were conducted in the same area with different zone axes for carbides. It was possible to interpret unambiguously all diffraction patterns using an orthorhombic structure and with lattice parameters consistent with the η structure. Figure 3c shows a dark field (DF) image using the $(111)_\eta$ reflexion of the SAED pattern. Figure 3d shows for instance, a diffraction pattern of the carbides observable in Figure 3b and indexed considering the η carbide structure (Pnnm). This last zone axis is specific of the η structure.

The presence of the highly distorted and magnetic ferritic matrix make it difficult to obtain clear diffraction conditions, which jeopardize the possibility to confirm for certain the nature of all the precipitates using TEM, and in particular to exclude the presence of epsilon carbides. Nevertheless, the presence of cementite even at lath boundaries was ruled out.

The direct observation of samples by SEM and TEM after heat treatment shows a uniform distribution of carbides with platelet morphology (about 400 nm wide and 5 nm thick). The resulting microstructure is thus very similar to those reported by Pierce et al. [8,10] and by HajyAkbary et al. [6]. The techniques used do not permit however a reliable density of precipitate to be measured and thus a correct estimate of the carbide fraction.

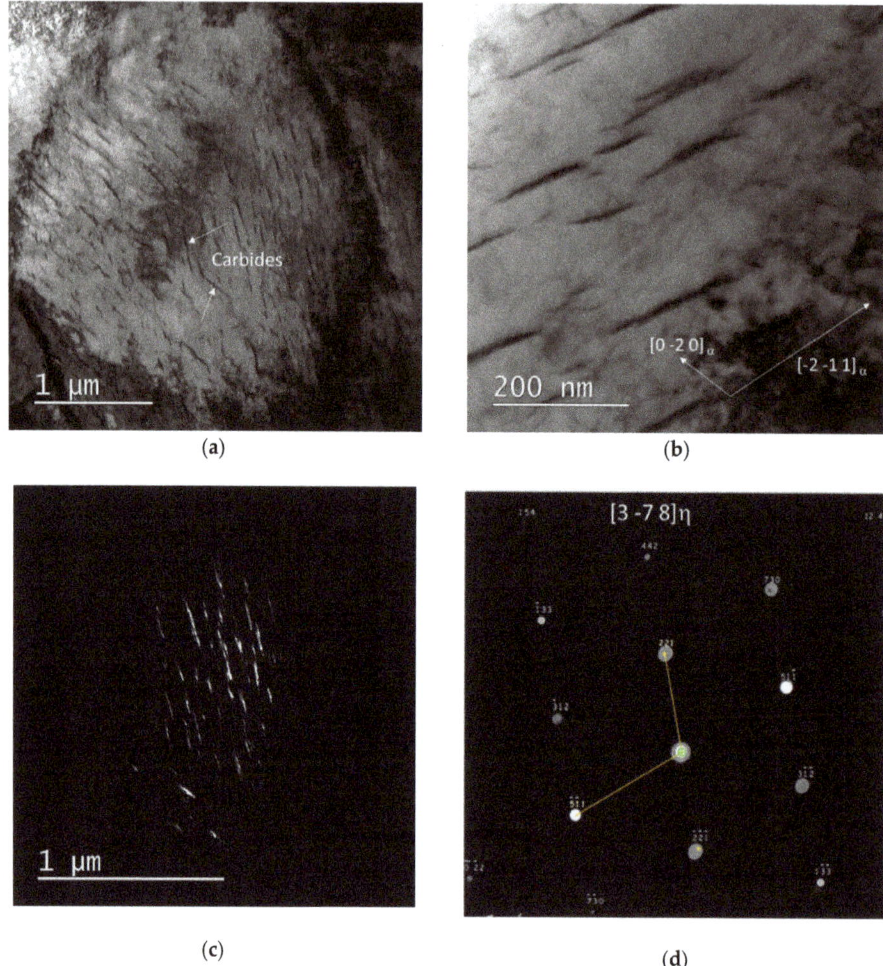

Figure 3. (**a**) Bright field transmission electron microscopy (BF TEM) micrograph of the studied steel after Q&P heat treatment (state 8). Arrows highlight the presence of 2 variants of carbides (dark contrast) in the tempered martensitic matrix (wide block). (**b**) BF TEM micrograph of the studied carbides aligned along martensite $[\text{-}2\text{-}11]_{\alpha'}$ (zone axis $[210]_{\alpha'}$). (**c**) Dark field (DF) image using $(111)_{\eta}$ spot. (**d**) Indexed diffraction pattern (zone axis $[3\text{-}78]_{\eta}$) of the η carbides aligned along martensite $[\text{-}2\text{-}11]_{\alpha'}$.

3.3. *Crystallographic Nature of Carbides by HEXRD*

All 2D X-ray diffraction patterns obtained all along the heat-treatment (one experimental pattern every 0.1 s as shown in Figure 1) were integrated circularly to obtain 1D diffractogram [19]. Such a procedure permits in particular statistical fluctuations and texture effects to be minimized, as discussed in [14]. Figure 4a shows an example of such a 1D diffractogram obtained in final state 8 (full intensity scale). The two families of diffraction peaks from the major constituents (martensite and austenite) are highlighted. Austenite (γ) diffraction peaks can be modelled considering a fcc lattice (space group Fm-3m) and martensite (α′) peaks considering a bct lattice (space group I4/mmm) as explained in our previous work [13,17]. The fraction of bainite formed during the considered heat treatment is low (less than 3%). This latter fraction was measured by HEXRD considering that the increase in bct phase fraction during partitioning is solely due to the bainitic transformation (no interface mobility between martensite and austenite).

The presence of carbides can only be revealed by enlarging the scale and by examining carefully the shoulders of the main peaks. Figure 4b evidences minor diffraction peaks emerging significantly from the statistical fluctuations (highlighted by a red circle) using a log scale. These diffraction peaks are not possible to observe with a conventional laboratory set-up from the author's knowledge.

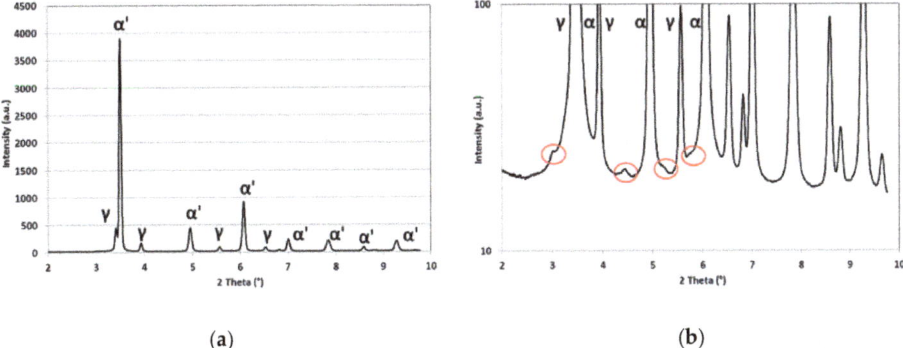

(a) (b)

Figure 4. (**a**) 1D full scale diffractogram after circular integration of 2D Debye–Scherrer pattern (state 8 at room temperature after Q&P treatment). Major diffraction peaks are indexed and attributed to austenite (γ) and martensite (α′) respectively. (**b**) Enlargement of (**a**) (log scale) reveals the presence of minor diffraction peaks in between major peaks' shoulders (red circles). These minor diffraction peaks are attributed to carbides.

The similarities of η and ε crystallographic structures raise the same issues as in TEM when identifying these additional diffraction peaks. X-ray diffraction (XRD) patterns of the three possible structures were simulated using Fullprof [20] assuming their compositions and crystallographic structures. Table 1 shows the space groups, lattice parameters, and compositions used for the calculations. Other values can be found in the literature for the lattice parameters [6,21] but the differences are minor and do not affect the conclusions.

Table 1. Space groups and lattice parameters of carbides used for X-ray diffraction (XRD) pattern simulations with Fullprof.

Carbide	Space Group	a (nm)	b (nm)	c (nm)	Composition	Ref.
θ	Pnma	5.0696	6.7671	4.5159	Fe_3C	[22]
η	Pnnm	4.7040	4.3100	2.8300	Fe_2C	[23]
ε	$P6_322$	4.7670	4.7670	4.3540	Fe_3C	[24]

Figure 5a,b show the expected positions of diffraction peaks of the three studied structures in the range of the two first minor diffraction peaks highlighted in Figure 4b. For each given structure, the relative heights of the peaks were calculated using structure and Lorentz polarization factors, and thus can be compared. Nevertheless, as the maximum intensities for each structure were chosen arbitrarily, it is useless to compare their absolute heights. θ appears in blue, η carbide in red, and ε carbide in green respectively.

All minor diffraction peaks observed in Figure 4b can thus be accounted for either by η or by ε structures. On the contrary, cementite cannot explain in particular the wide peak observed at 4.45° and would have produced extra peaks at 4.22° and at 2.35° which are obviously not visible. Hence, as by

TEM, the presence of cementite can be ruled out within the resolution limit of the technique (typically 0.1% mass fraction if the size of precipitates is sufficient).

Nevertheless, it is not possible to conclude with confidence on the nature of the observed transition carbides using our HEXRD set-up. When comparing expected diffractograms of η and ε, the sole discriminative and isolated fundamental peak is the $(002)_\eta$ expected at 5.02°. Unfortunately, if this particular peak exists, it is convoluted with a major peak of the diffraction pattern associated to martensite and thus undetectable. Nevertheless, the width and the asymmetry of the minor diffraction peaks observed at 3.00° and 4.45° plead in favor of an η structure as suggested by Pierce et al. [8–10]. This result is also sustained by our own TEM observations.

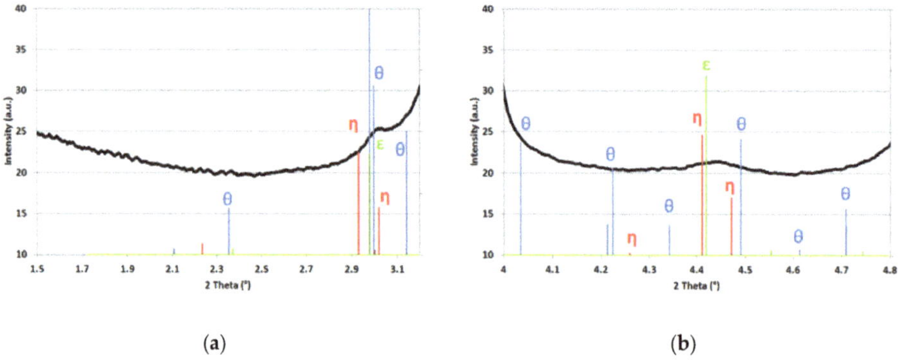

(**a**) (**b**)

Figure 5. 1D experimental diffractogram (black continuous curve) and position of simulated diffraction peaks of possible carbides (cementite in blue, eta carbide in red, and epsilon carbide in green); (**a**,**b**) diffraction angle windows corresponding to the first and second minor diffraction peaks in Figure 4b respectively.

3.4. Precipitation Kinetics

Whatever the considered structure (η or ε), the mass fraction of transition carbide was estimated around 0.45% using a Rietveld simulation of peak height (at the end of the thermal treatment, i.e., at point of interest 8). Even if our in situ diffraction set-up does not permit with certainty the nature of the transition carbide to be confirmed, it enables following of the carbide precipitation sequence.

Figure 6a,b shows the evolution of the experimental diffraction patterns in the angular window corresponding to the distinctive diffraction peak of transition carbides at 4.45°. They are plotted for the points of interest indicated in Figure 1 and correspond respectively to the quenching step (1 and 2), the reheating step (3, 4, and 5) and the partitioning step (5, 6, and 7). The color code refers to the temperature. For the sake of readability, a 6th order polynomial fit has been adjusted on each experimental curve to highlight the main variations out from the statistical fluctuations.

At high and low diffraction angles in the studied range; the increases in intensity correspond to the shoulders of main diffraction peaks, to the $(200)_\gamma$ peak on the left, and to the $(200)_{\alpha'}$ peak on the right. In between, the minor diffraction peaks appears already at 250 °C (point 3) as a small deviation from the flat signal observed after quenching (no evolution between points 1 and 2). The hump becomes more and more pronounced during reheating (between points 3 and 5). The mean intensity of the diffraction pattern increases during the reheating sequence because of the Debye–Waller overall isotropic displacement. During partitioning, the mean intensity decreases probably owing to a recovery process in major phases which leads to a narrowing of diffraction peaks and shoulders. Nevertheless, during this stage (points 5 to 7) at constant temperature, the relative height of the minor peak does not evolve significantly.

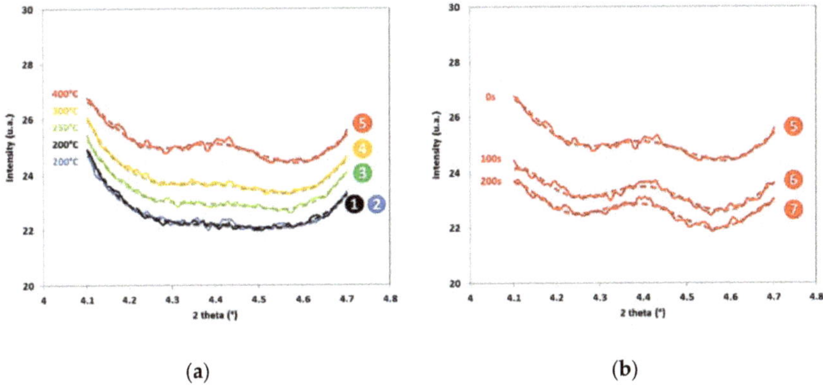

(a) (b)

Figure 6. Experimental 1D diffractograms in a reduced angular windows corresponding to the second minor diffraction peak in Figure 4b obtained at different steps of Q&P treatment (**a**) points 1 to 5 according to Figure 1, i.e., after quenching and during reheating (**b**) points 5 to 7 according to Figure 1, i.e., during partitioning.

It is difficult to quantify with a reasonable precision the fraction of transition carbides all along the partitioning step, but our experiment shows undoubtedly that precipitation of carbide starts between 200 °C and 250 °C in the studied case, i.e., at the very beginning of the reheating step even if the reheating rate is high (50 °C/s). The comparison between patterns at points 1 and 2 shows no evolution during the 5 s pause before reheating at QT. This absence of evolution is the proof that transition carbide precipitation does not start during the quenching step after austenitization for the studied cooling rate, contrary to the post mortem observations of HajyAkbary et al. [6] or Pierce et al. [9] at room temperature.

This precipitation occurs even before the carbon enrichment process in austenite estimated at 270 °C for the studied case [13]. During partitioning, the fraction of transition carbides seems to remain constant.

4. Conclusions

As a conclusion, the experiments presented in this paper uphold the carbon mass balance established by HEXRD in our previous papers to explain the carbon enrichment in austenite after Q&P in a Fe-0.3C-2.5Mn-1.5Si steel [13]. Post mortem SEM and TEM observations after the heat treatment revealed the presence of numerous carbides in martensite, which prevents a complete partitioning of carbon into austenite in the studied conditions (QT = 200 °C, PT = 400 °C, Pt = 200 s). Carbon content in austenite (1.05%) is thus far lower than the composition expected from the sole constrained para-equilibrium (CPE) condition (about 1.9%) [3]. These carbides have a platelet morphology with a mean diameter of 400 nm and a thickness of approx. 5 nm. TEM observations ruled out the presence of cementite in the studied condition (even at martensite lath boundaries) but were not able to determine unambiguously the nature of the observed transition carbide (η or ε).

Our synchrotron experiments also permitted investigation for the first time the nature of carbides in Q&P steels by XRD. Diffraction patterns revealed the presence of a third phase in addition to austenite and martensite, which can be indexed as η or ε carbides. Despite a very low noise/signal ratio, it was not possible to conclude on the nature of the transition carbides. The asymmetry of minor diffraction peaks however pleads in favor of the η structure, as claimed by Pierce et al. [8–10]. No cementite was detected on the contrary with a high level of certainty, supporting the TEM observations.

As the experiment was carried out in situ with a high acquisition rate, real time precipitation kinetics was followed. Contrary to HajyAkbary et al. [6] and Pierce et al. [8–10], it was observed that transition carbide precipitation does not start during quenching, but during reheating below 250 °C, before carbon enrichment in austenite at 270 °C. In the studied conditions, the maximum fraction of precipitate is reached at the beginning of partitioning (about 0.5%). During the partitioning step (200 s at 400 °C), the fraction of precipitate does not evolve significantly (no dissolution), as already observed by Pierce et al. [8].

Author Contributions: Data curation, S.Y.P.A. and G.G.; Funding acquisition, S.Y.P.A., M.G., F.D., and M.S.; Investigation, S.Y.P.A., S.A., A.Q.-P., M.G., J.-C.H., M.B., M.S., and G.G.; Project administration, S.Y.P.A.; Writing-original draft, S.Y.P.A.; Writing-review and editing, S.Y.P.A., F.D., A.Q.-P., M.G., and M.S.

Funding: This work was supported by the French State through the project "CAPNANO" referenced by ANR-14-CE07-0029 and also through the program "Investment in the future" operated by the National Research Agency (ANR) and referenced by ANR-11-LABX-0008-01 (LabEx DAMAS).

Acknowledgments: The synchrotron experiments were realized in December 2016, under the P160 grant at DESY PETRA P-07 in Hamburg, which is fully acknowledged. The authors would like also to thank MATERALIA cluster (Region Grand Est) for their support.

Conflicts of Interest: The authors declare no conflict of interest. The funders had no role in the design of the study; in the collection, analyses, or interpretation of data; in the writing of the manuscript, and in the decision to publish the results.

References

1. Speer, J.G.; Edmonds, D.V.; Rizzo, F.C.; Matlock, D.K. Partitioning of carbon from supersaturated plates of ferrite, with application to steel processing and fundamentals of the bainite transformation. *Curr. Opin. Solid State Mater. Sci.* **2004**, *8*, 219–237. [CrossRef]
2. Speer, J.; Matlock, D.K.; De Cooman, B.C.; Schroth, J.G. Carbon partitioning into austenite after martensite transformation. *Acta Mater.* **2003**, *51*, 2611–2622. [CrossRef]
3. Clarke, A.J.; Speer, J.G.; Miller, M.K.; Hackenberg, R.E.; Edmonds, D.V.; Matlock, D.K.; Rizzo, F.C.; Clarke, K.D.; De Moor, E. Carbon partitioning to austenite from martensite or bainite during the quench and partition (Q&P) process: A critical assessment. *Acta Mater.* **2008**, *56*, 16–22.
4. Allain, S.Y.P.; Geandier, G.; Hell, J.C.; Soler, M.; Danoix, F.; Gouné, M. In-Situ investigation of quenching and partitioning by High Energy X-Ray Diffraction experiments. *Scr. Mater.* **2017**, *131*, 15–18. [CrossRef]
5. Nishikawa, A.S.; Santofimia, M.J.; Sietsma, J.; Goldenstein, H. Influence of bainite reaction on the kinetics of carbon redistribution during the quenching and partitioning process. *Acta Mater.* **2018**, *142*, 142–151. [CrossRef]
6. HajyAkbary, F.; Sietsma, J.; Miyamoto, G.; Furuhara, T.; Santofimia, M.J. Interaction of carbon partitioning, carbide precipitation and bainite formation during the Q&P process in a low C steel. *Acta Mater.* **2016**, *104*, 72–83.
7. Thomas, G.A.; Danoix, F.; Speer, J.G.; Thompson, S.W.; Cuvilly, F. Carbon Atom Re-Distribution during Quenching and Partitioning. *ISIJ Int.* **2014**, *54*, 2900–2906. [CrossRef]
8. Pierce, D.T.; Coughlin, D.R.; Williamson, D.L.; Clarke, K.D.; Clarke, A.J.; Speer, J.G.; De Moor, E. Characterization of transition carbides in quench and partitioned steel microstructures by Mössbauer spectroscopy and complementary techniques. *Acta Mater.* **2015**, *90*, 417–430. [CrossRef]
9. Pierce, D.T.; Coughlin, D.R.; Williamson, D.L.; Kähkönen, J.; Clarke, A.J.; Clarke, K.D.; Speer, J.G.; De Moor, E. Quantitative investigation into the influence of temperature on carbide and austenite evolution during partitioning of a quenched and partitioned steel. *Scr. Mater.* **2016**, *121*, 5–9. [CrossRef]
10. Pierce, D.T.; Coughlin, D.R.; Clarke, K.D.; De Moor, E.; Poplawsky, J.; Williamson, D.L.; Mazumder, B.; Speer, J.G.; Hood, A.; Clarke, A.J. Microstructural evolution during quenching and partitioning of 0.2C-1.5Mn-1.3Si steels with Cr or Ni additions. *Acta Mater.* **2018**, *151*, 454–469. [CrossRef]
11. Toji, Y.; Miyamoto, G.; Raabe, D. Carbon partitioning during quenching and partitioning heat treatment accompanied by carbide precipitation. *Acta Mater.* **2015**, *86*, 137–147. [CrossRef]

12. Badinier, G.; Sinclair, C.W.; Allain, S.; Danoix, F.; Gouné, M. The mechanisms of transformation and mechanical behavior of ferrous martensite. In *Reference Module in Materials Science and Materials Engineering*; Elsevier: New York, NY, USA, 2017; ISBN 9780128035818.
13. Allain, S.Y.P.; Gaudez, S.; Geandier, G.; Hell, J.C.; Gouné, M.; Danoix, F.; Soler, M.; Aoued, S.; Poulon-Quintin, A. Internal stresses and carbon enrichment in austenite of Quenching and Partitioning steels from high energy X-ray diffraction experiments. *Mater. Sci. Eng. A* **2018**, *710*, 245–250. [CrossRef]
14. Allain, S.Y.P.; Geandier, G.; Hell, J.C.; Soler, M.; Danoix, F.; Gouné, M. Effects of Q&P Processing Conditions on Austenite Carbon Enrichment Studied by In Situ High-Energy X-ray Diffraction Experiments. *Metals* **2017**, *7*, 232. [CrossRef]
15. Ariza, E.A.; Nishikawa, A.S.; Goldenstein, H.; Tschiptschin, A.P. Characterization and methodology for calculating the mechanical properties of a TRIP-steel submitted to hot stamping and quenching and partitioning (Q&P). *Mater. Sci. Eng.* **2016**, *671*, 54–69.
16. Epp, J.; Hirsch, T.; Curfs, C. In Situ X-Ray diffraction analysis of carbon partitioning during quenching of low carbon steel. *Metall. Mater. Trans. A* **2012**, *43*, 2210–2217. [CrossRef]
17. Rementeria, R.; Jimenez, J.A.; Allain, S.Y.; Geandier, G.; Poplawsky, J.D.; Guo, W.; Urones-Garrote, E.; Garcia-Mateo, C.; Caballero, F.G. Quantitative assessment of carbon allocation anomalies in low temperature bainite. *Acta Mater.* **2017**, *133*, 333–345. [CrossRef]
18. Thompson, S.W. Structural characteristics of transition-iron-carbide precipitates formed during the first stage of tempering in 4340 steel. *Mater. Charact.* **2015**, *106*, 452–462. [CrossRef]
19. The FIT2D Home Page. Available online: http://www.esrf.eu/computing/scientific/FIT2D/ (accessed on 11 April 2017).
20. Rodríguez-Carvajal, J. Recent advances in magnetic structure determination by neutron powder diffraction. *Phys. B Condens. Matter* **1993**, *192*, 55–69. [CrossRef]
21. Fang, C.M.; Van Huis, M.A.; Zandbergen, H.W. Stability and structures of the ε-phases of iron nitrides and iron carbides from first principles. *Scr. Mater.* **2011**, *64*, 296–299. [CrossRef]
22. Wood, I.G.; Vočadlo, L.; Knight, K.S.; Dobson, D.P.; Marshall, W.G.; Price, G.D.; Brodholt, J. Thermal expansion and crystal structure of cementite, Fe3C, between 4 and 600 K determined by time-of-flight neutron powder diffraction. *J. Appl. Crystallogr.* **2004**, *37*, 82–90. [CrossRef]
23. Hirotsu, Y.; Nagakura, S. Crystal structure and morphology of the carbide precipitated from martensitic high carbon steel during the first stage of tempering. *Acta Metall.* **1972**, *20*, 645–655. [CrossRef]
24. Jang, J.H.; Kim, I.G.; Bhadeshia, H.K.D.H. ε-Carbide in alloy steels: First-principles assessment. *Scr. Mater.* **2010**, *63*, 121–123. [CrossRef]

Article

Variations of the Elastic Properties of the CoCrFeMnNi High Entropy Alloy Deformed by Groove Cold Rolling

Paul Lohmuller [1,2,3], Laurent Peltier [1,3], Alain Hazotte [1,3], Julien Zollinger [2,3], Pascal Laheurte [1,3] and Eric Fleury [1,3,*]

1 Laboratoire d'Etude des Microstructures et de Mécanique des Matériaux, UMR 7239 CNRS, Arts et Métiers ParisTech Campus de Metz, Université de Lorraine, 7 rue Félix Savart, F-57073 Metz, CEDEX 03, France; paul.lohmuller@univ-lorraine.fr (P.L.); laurent.peltier@ensam.eu (L.P.); alain.hazotte@univ-lorraine.fr (A.H.); pascal.laheurte@univ-lorraine.fr (P.L.)

2 Institut Jean Lamour, UMR 7198 CNRS, Université de Lorraine, Campus ARTEM, F-54011 Nancy, CEDEX 01, France; julien.zollinger@univ-lorraine.fr

3 LABoratory of EXcellence Design of Alloy Metals for low-mAss Structures, Université de Lorraine, F-54011 Nancy, CEDEX 01, France

* Correspondence: eric.fleury@univ-lorraine.fr; Tel.: +33-372-747-772

Received: 29 June 2018; Accepted: 27 July 2018; Published: 2 August 2018

Abstract: The variations of the mechanical properties of the CoCrFeMnNi high entropy alloy (HEA) during groove cold rolling process were investigated with the aim of understanding their correlation relationships with the crystallographic texture. Our study revealed divergences in the variations of the microhardness and yield strength measured from samples deformed by groove cold rolling and conventional cold rolling processes. The crystallographic texture analyzed by electron back scattered diffraction (EBSD) revealed a hybrid texture between those obtained by conventional rolling and drawing processes. Though the groove cold rolling process induced a marked strengthening effect in the CoCrFeMnNi HEA, the mechanical properties were also characterized by an unusual decrease of the Young's modulus as the applied groove cold rolled deformation increased up to about 0.5 before reaching a stabilized value. This decrease of the Young's modulus was attributed to the increased density of mobile dislocations induced by work hardening during groove cold rolling processing.

Keywords: high entropy alloy; crystallographic texture; groove rolling; elastic properties

1. Introduction

High entropy alloys (HEAs), a new class of metallic materials in which the stability of the solid solution is explained by contribution of the configurational entropy, have recently received particular attention, owing to the prospect of developing new systems with tailored properties particularly suitable for a large range of different applications from dies and molding materials to corrosion-resistant coatings in manufacturing, energy, transport, and aeronautical industries [1–4]. The stabilization of the solid solution is also associated to special features such as low diffusion kinematics, lattice distortion and "cocktail effect" whose effects have been discussed in the literature [1,3,5].

The equi-atomic CoCrFeMnNi composition developed by Cantor and colleagues [6] is currently one of the most studied HEA. Though its mechanical properties are not outstanding [7–9], this alloy characterized by a single solid solution of face centered cubic structure has appeared attractive as a model material to gain a better understanding on characteristics particular to solid solution HEA alloys against those of conventional solid solution alloys. Some key studies, such as those published by Wu et al. [10] and Laurent-Brocq and co-workers [11,12], already helped in the understanding the formation of these solid solutions, particularly the influence of each constituent element on the stability of the microstructure. The mechanical properties of HEAs, as those of conventional metallic materials, are also controlled by the microstructure. If several studies have revealed the dependency of the mechanical properties with the grain size [7,13,14], a few have recently reported the variation of the crystallographic texture during mechanical processing of HEAs. For example, conventional cold rolling [15–18], rotary swaging [19], or severe plastic deformation [20] have been investigated and the results of these studies show characteristics very similar to those obtained with low stacking fault energy (SFE) materials such as brass alloys [21,22]. Examinations of the deformation mechanisms have demonstrated that deformation occurred mainly by dislocation glide and sub-cell creation [8,18]. However, under particular conditions, such as tests performed under high strain level or at cryogenic temperature [16,18,23,24], the CoCrFeMnNi also presented deformation twinning, which can be seen as a secondary deformation mechanism.

Though most of studies are concerned about the mechanical properties, the modifications of the Young's modulus induced by work hardening have not been intensively studied. However, as seen earlier in the 1950′s, in their theory of work hardening, both Mott [25] and Friedel [26] took into account a network of dislocations in face-centered cubic crystals to explain the changes in the elastic modulus upon work hardening. More recently, others authors have proposed to account for the increase of internal stresses during deformation, the development of crystallographic texture and change in the dislocation arrangement [27] to describe the variation of solid solution alloys.

This study was thus undertaken to investigate the evolution of the hardness and tensile properties of CoCrFeMnNi HEA alloy with the increasing area reduction achieved by grooved cold rolling. Though this process resulted in a marked strengthening effect, it was also accompanied by a reduction of the Young's modulus. The variation of the crystallographic texture with the increasing groove rolling deformation was found to be slightly different from that obtained during conventional cold rolling. This study attempted to provide elements enabling the understanding of the observed changes in the mechanical properties.

2. Materials and Methods

2.1. Samples Production

A 5 kg ingot of the equi-atomic CoCrFeMnNi alloy of composition $Co_{20.1}Cr_{19.9}Fe_{20.6}Mn_{19.5}Ni_{19.9}$ (in at.%) as analyzed by electron dispersive spectroscopy (EDS) was produced by vacuum arc melting at KIST, Korea, using initial elements with purity above 99.9%. Hot forging treatment and hot rolling at 1000 °C were performed to homogenize the microstructure and to reduce the initial ingot thickness down to 12 mm. After another thickness reduction by conventional cold rolling, 4 samples of dimension $9.3 \times 9.3 \times 80$ mm^3 were machined perpendicularly to the rolling direction. Finally an annealing heat treatment was then applied during 3600 s at 900 °C under a protective Argon atmosphere. Following this heat treatment, samples were considered, for this study, in a reference state in term of microstructure and mechanicals properties.

Room temperature groove cold rolling process was performed on the annealed square bars. In this process, rollers present V-grooves of different dimensions as illustrated in Figure 1A. These grooves result in a bidirectional loading on the square sections of the samples, which prevents transversal displacement. This transversal displacement, though moderate, is inherent to the conventional rolling process (Figure 1B). From this viewpoint, groove cold rolling may be seen as a hybrid process between conventional rolling and swaging or extrusion.

The samples were respectively rolled down to 14% and 33% area reduction ratios, i.e., rolling deformation $\varepsilon = 0.2$ and $\varepsilon = 0.4$, respectively. The obtained square wire-type samples were cut off in order to investigate the tensile properties (minimal length about 90 mm). The rests of the samples were rolled to successively up to 89% area reduction ratio (i.e., rolling deformation $\varepsilon = 2.2$), corresponding to a final section of about 3.1 × 3.1 mm^2. For each deformation level, a wire was extracted with a minimal length of about 120 mm for tensile test and microhardness measurements. In order to determine the tensile properties of the initial state, the last sample was rolled until 38% area reduction ratio (i.e., rolling deformation $\varepsilon = 0.7$) and annealed under the same conditions as described above. Table 1 summarizes the thickness of each sample, the associated reduction state calculated by $\varphi = 100 \times \left(1 - {e_f}^2/{e_i}^2\right)$ and the true deformation obtained by $\varepsilon = 2 \times \ln(e_i/e_f)$ [28], where e_i and e_f correspond to the initial and final thicknesses, respectively.

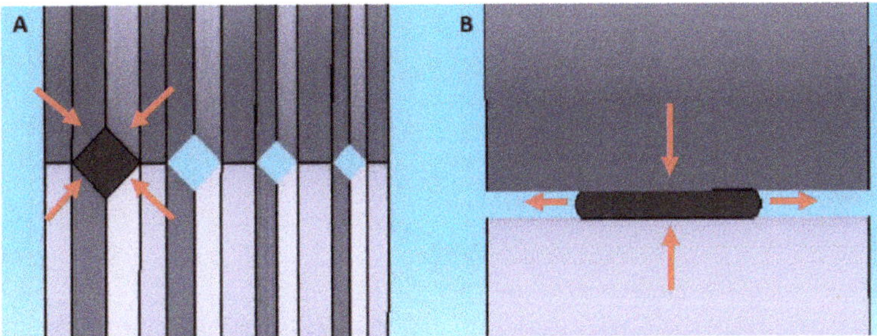

Figure 1. Grooved cold rolling (**A**) versus conventional rolling (**B**), the red arrows correspond to the in plane displacement direction induced by the process.

Table 1. Different grooved cold rolling deformation level produced for microhardness and tensile test, grayed background correspond to samples used for electron back scattered diffraction (EBSD) analyses.

e_i (mm)	e_f (mm)	Section Reduction (%)	True Deformation
9.3	9.3	0.0	0.0
-	8.6	14.5	0.2
-	7.6	33.2	0.4
-	7.3	38.4	0.5
-	6.6	49.6	0.7
-	5.4	66.3	1.1
-	4.2	79.6	1.6
-	3.1	88.9	2.2

2.2. Mechanical Testing

Uniaxial tensile tests were carried out on 100 kN universal Zwick tensile test machine, using a 50 kN load cell. With the aim of determining the Young's modulus, samples were first loaded to 1% strain, then 5 charge/discharge cycles were performed. Strains were measured using an Epsylon extensometer with a gage length of 25 mm, and the samples were deformed at a constant strain rate of 1×10^{-3} s^{-1}. Values of the Young's modulus were determined as the mean of the initial slope of each discharging portion of the curves. Values of the yield strength were obtained from the conventionally 0.2% of plastic deformation. A typical obtained curve is shown in the right-hand side inset in Figure 2.

For microstructural observations, samples were polished (along the rolling direction) using SiC papers with grit size about 80 in order to reduce the thickness and to reach approximately the center part of the samples. Then surfacing polishing was performed with grit size from 1000 to 4000. Hardness values were measured by micro-indentation Vickers tests carried out using a Zwick Rowel machine with a load of 0.3 gf, leading to indentation size of about 35 to 70 μm. Distance between adjacent measurements was at least 200 μm. For each sample, the hardness value was averaged from 15 measurements, and the errors bars correspond to the measured standard deviation.

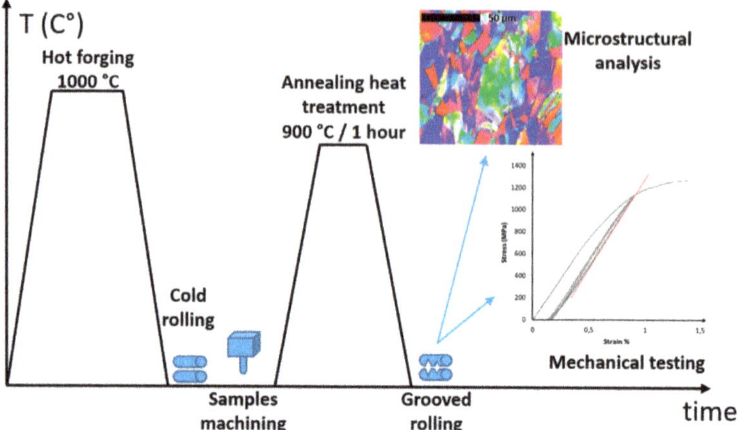

Figure 2. Graphical representation of samples production procedure and characterization.

2.3. Microstructural Analysis

Before tensile testing, pieces of rolled materials were cut off for microstructural analyses. Samples were polished using SiC paper with grit size down to 4000 followed by a Al_2O_3 polishing with final particle size of 0.1 µm. Microstructural analyses were carried out on SEM Jeol 6490, at 20 kV with a working distance of 21 mm. Electron back scattered diffraction (EBSD) analyses were performed by Oxford instrument detector with a minimal number of detected Kikuchi bands equal to 8 and a step size of 0.4 µm. EBSD were studied in four cases: after annealing and after section reduction of $\varepsilon = 0.5$, 1.1 and 2.2. EBSD maps, Figure pole (FP), inverse figure pole (IFP) and orientation density function (ODF) were obtained using ATEX software developed at LEM3 [29]. Texture components were also performed using a spread of 15° from their ideal orientation to compute their volume fractions. All the obtained texture components were summed in order to determine the randomly oriented components by evaluating the difference between this sum and the totality [16,18]. The different analyzed samples are indicated in Table 1. The whole procedure from hot forging to EBSD analysis is summarized in Figure 2.

3. Results

3.1. Microhardness and Tensile Results

Figure 3a shows the variations of the microhardness values and associated standard deviations for the different deformation states applied by groove cold rolling and data obtained by Otto et al. [15] on CoCrFeMnNi HEA processed by conventional cold rolling have been added for comparison. In the as-annealed state, the value of the microhardness was about 165 $HV_{0.3}$, which is in good agreement with values obtained in published articles under the same annealing treatment [8,23]. The variation of the microhardness can be described by two different linear curves with a breakdown at a level of deformation of about 0.5. From the initial annealing state to a condition corresponding to the groove rolling deformation ε ~0.5, the microhardness displayed a marked increased from 165 $HV_{0.3}$ to about 310 $HV_{0.3}$, characterized by a larger value of the slope (284 kgf/mm^2) than those reported by Otto et al. [8] (92 kgf/mm^2) represented by the green dashed line in the Figure 3a. For values of the groove rolling deformation beyond 0.5, the microhardness continued to increase however with a reduced slope until reaching a value of 380 $HV_{0.3}$ for a groove rolling deformation $\varepsilon = 2.2$. It is remarkable that the microhardness values followed the same trend under both groove rolled and conventional rolling processes, though the initial slope was larger and that the breakdown in the slope occurred earlier for samples deformed by groove rolled process.

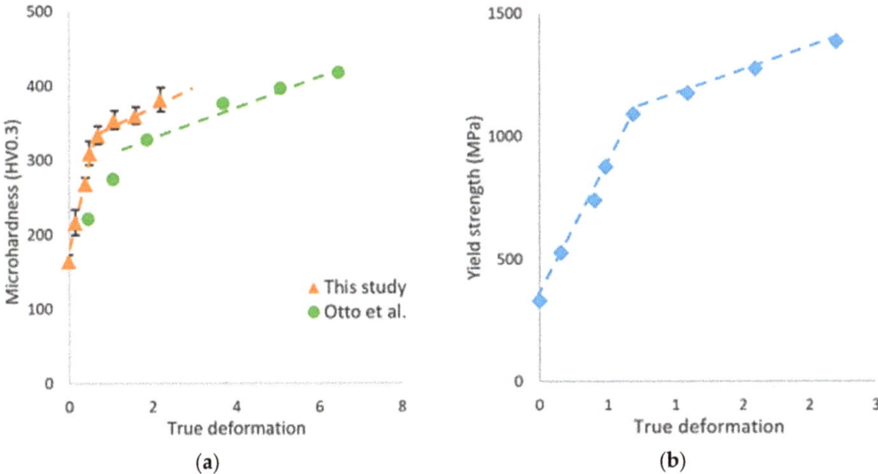

Figure 3. (**a**) Microhardness Vickers ($HV_{0.3}$) evolution with the surface reduction, for this study (orange triangle) and comparison with results published by Otto et al. [15] (green dot), (**b**) yield strength evolution during grooved cold rolling process.

The evolution of the yield strength, as shown in Figure 3b, followed a similar trend to that of the microhardness. Values of the yield strength obtained in the annealed condition are in good agreement with those reported in references [7,8]. The high values of the yield strength obtained after large area reduction were also found to be in accordance with the values reported by Sun et al. for CoCrFeMnNi HEA samples with fine-grained microstructure [7].

As mentioned in the introduction, a part of this study focused on understanding the evolution of the Young's modulus as function of the groove rolling deformation. Figure 4 indicates that the value obtained in the as-annealed sample, i.e., 210 GPa, is also in agreement with those reported in the literature [9,30]. It also shows that the value decreased down to about 175 GPa, as the groove rolling deformation reached 0.5, which is corresponding to approximately a 15% decrease in the Young's modulus. As far as we know, no similar results have to be reported on deformed HEA. To the authors' knowledge, only a few studies have attempted to correlate the variations of the Young's modulus of engineering materials with the increase of cold plastic deformation [27,31,32]. As mentioned by Benito et al. [27], four possible contributions have been proposed to explain the decrease in Young's modulus: (i) debonding between a matrix and a second-phase or inclusions [33,34]; (ii) increase in the internal stresses during plastic deformation [35]; (iii) introduction and development of texture during deformation [27,31]; and (iv) change in dislocation arrangement and kinematics [27,31].

By considering the single-phase character of CoCrFeMnNi HEA, the first mentioned reason appeared to be irrelevant. In their study, Benito et al. have suggested that the internal stresses generated during the early stage of cold rolling were not sufficient to describe the variation of the Young's modulus [27]. Consequently, the decrease in the Young's modulus during the early stage of groove cold rolling could result from either a crystallographic texture effect or a change in the dislocation arrangement and deformation mechanism.

The different mechanisms in the CoCrFeMnNi HEA are now well documented in the literature for different deformation modes and different degrees of deformation [7,8,18,23,36,37]. From these studies, it appears that the deformation mechanisms are governed by dislocation glide upon low strain value, and by dislocation glide combined with twinning upon high level of plastic deformation. In an attempt to elucidate the decrease of the Young's Modulus, a thorough study of the evolution of the crystalline texture as a function of the section reduction has thus been undertaken.

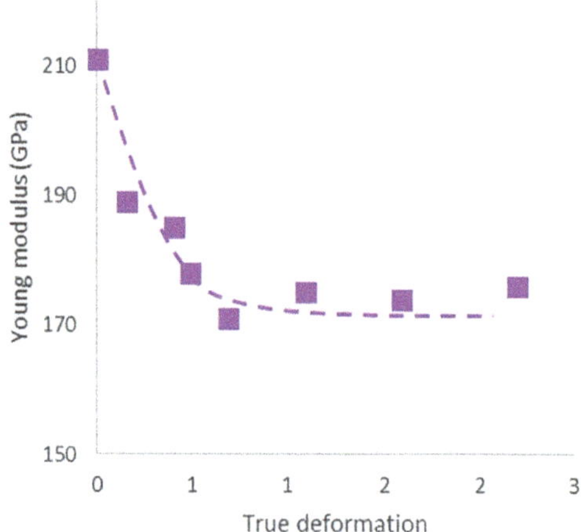

Figure 4. Young's modulus variation with the thickness reduction during grooved cold rolling (maximal incertitude ± 6 GPa).

3.2. Texture Analysis

Figure 5 shows the EBSD maps obtained for the annealed state and for samples after three different values of the area reduction, i.e., ε = 0.5, 1.1, and 2.2. The initial state microstructure consisted of equiaxed grains with a mean grain diameter about 27 μm (as computed by ATEX). Figure 5b,d show the evolution of the microstructure with increasing strain induced by the groove cold rolling process. It can be observed that grains were progressively elongated along the rolling direction (RD) and compressed in the normal direction (ND). For samples processed under low deformation level (Figure 5b), some deformation features oriented at 45° to the rolling direction appeared and their fraction progressively increased with the increasing level of groove rolling deformation (Figure 5c,d). Figure 5e shows a focused area of the Figure 5d obtained with a step size of 0.05 μm and a high indexation rate. These features resulted from different deformation mechanisms operating in CoCrFeMnNi HEA that are dislocation glide and twinning as already described in different studies [7,8,19,36].

Figure 6 shows the evolution of the ODF section intensity and IFP intensity along the rolling direction. The map of the sample in the as-annealed state shows a slight recrystallization texture of the cube on face Cb {001} <100> type as well as a partial α-fiber (<110> along ND). Such texture has been reported for CoCrFeMnNi HEA deformed by conventional cold rolling or hot forging prior to an annealing treatment at 900 °C during 1 h [11,18]. One may also note that recrystallized twinning occurred during the annealing treatment as reported in [16,18].

The Figure 6, showing the IFP of the deformed state, indicates a strong increase of the <111> fiber texture along RD as the level of deformation induced by groove cold rolling increased. This fiber texture is associated to the <001> fiber, which is a minor component of the texture, as indicated by local maximal intensity. Both fiber textures are frequently observed for FCC materials [38,39], and have also been reported for CoCrFeMnNi HEA [19] deformed to swaging or drawing deformation. These observations based on IFP may be balanced with the analyses of the ODF sections. In the ODF map, the α-fiber component (<110> along ND, not shown in the IFP (Figure 6) is represented by the red dashed line. The <111> and <001> directions along RD are highlighted by, respectively, the (Copper (Cu), A) and (Cube (Cb), Goss (G)) texture components in the ODF with $\varphi_2 = 45°$ (referred in Table 2) [40].

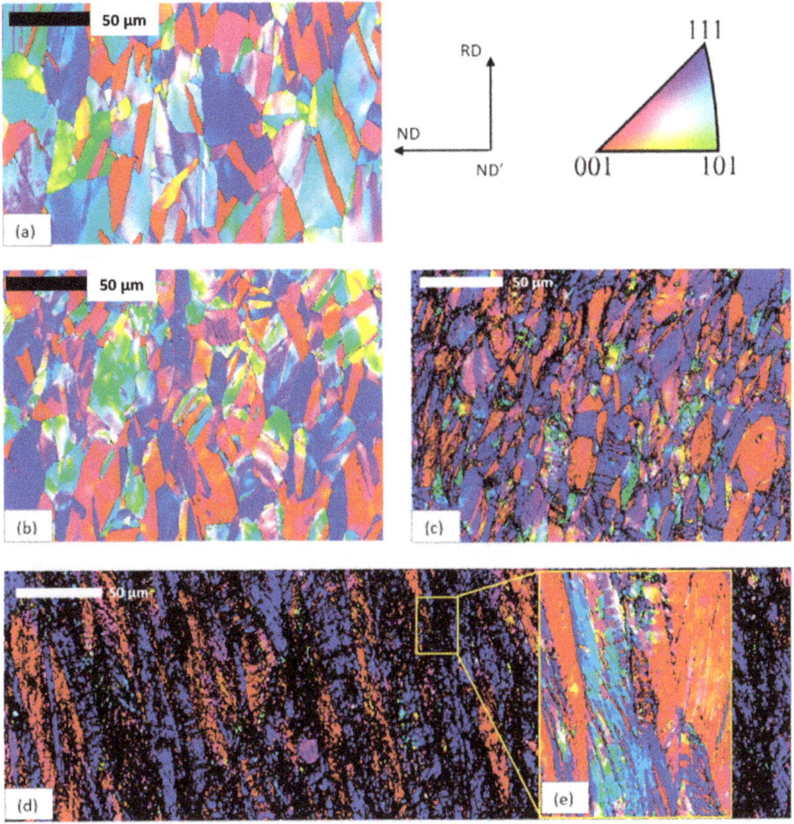

Figure 5. EBSD-band contrast maps described by inverse figure pole (IFP) maps (upper right corner): (**a**) as-annealed; (**b**) after $\varepsilon = 0.5$ by groove cold rolling; (**c**) after $\varepsilon = 1.1$ groove cold rolling; (**d**) $\varepsilon = 2.2$ groove cold rolling; and (**e**) focused area of the image (**d**). Note that magnification is the same for all maps (maps (**d**) corresponding to a stack of 2 EBSD maps with a step size of about 0.2 µm).

Table 2. Definition of the texture components in the alloy.

Texture Component	Symbol	Euler Angle φ_1, Φ, φ_2	Miller Indices
Cb (Cube)	■	0, 0, 0	{001}<100>
G (Goss)	⊙	0, 45, 0	{110}<100>
B (Brass)	▰	35, 45, 0	{110}<112>
G/B (Goss/Brass)	⊗	74, 90, 45	{110}<115>
Cu (Copper)	<>	90, 35, 45	{112}<111>
A	▲	35, 90, 45	{110}<111>
F	⬢	30/90, 55, 45	{111}<112>
P	⬟	30, 90, 45	{011}<211>
CuT	♦	90, 74, 45	{552}<115>
S	●	59, 37, 63	{123}<634>

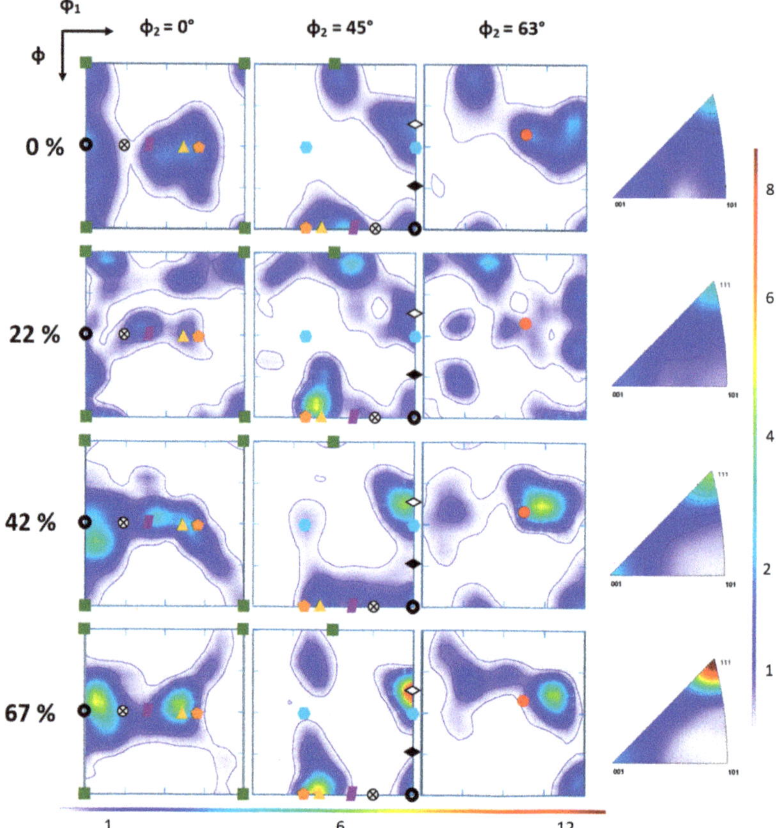

Figure 6. Orientation density function (ODF) section at $\varphi_2 = 0°$, 45° and 63° (right) at each analyzed deformation state and their respective IFP along rolling direction, red arrows indicate maximal intensity for orientation {111} and {001}. Horizontal and vertical bars, respectively, correspond to ODF and IFP maximal intensity. The red dashed line represent the α-fiber. The symbols for each texture component are, respectively, Cb: ■, G: ●, B: ▮, G/B: ⊗, Cu: <>, A: ▲, F: ⬢, P: ⬟, CuT: ♦, S: ●, as defined in Table 2.

For the sample processed to a groove rolling deformation ε = 0.5, a slight partial α-fiber appeared between the G and P components, and it increased till ε = 1.1. As the deformation level increased, this fiber seemed to be "broken" for higher deformation states as indicated by the ODF with $\varphi_2 = 45°$, in which the Brass (B) and Rotated Goss (RtG) texture components were not detected. The S component also seemed to follow the same trend as the α-fiber texture, with a weaker decrease for deformation up to 2.2 in comparison to the B and RtG components. This first features observed during the conventional rolling process, with the α-fiber was found to continuously increased as the deformation increased, as it was reported by Sathiaraj et al. [16] and Haase et al. [18]. The <111> fiber (Cu and A) texture component displayed the strongest increase with the grooved rolling deformation, which firstly confirms the existence of this fiber but may also indicate the presence of twinning deformation mechanism. This deformation by twinning may also occur at lower level of deformation in comparison to conventional rolling. Haase et al. have shown that for conventional rolling, the Cu texture component appeared after a rolling deformation ε = 1.6 [18]. In our study, for a deformation of about 1.1, this component appeared to be the major one in the ODF section and it was found to continuously increase. However, the two components of the <001>RD appeared to have different behavior: as the deformation by groove cold rolling increased, the Cb component tended to progressively disappear, while the G component slightly increased without been affected by the rupture of the α fiber. It is important to note that the G components are both appertaining to <110> along ND and <001> along ND fiber texture, so that their variations may be correlated with the evolution of these two fiber textures.

Figure 7a shows the variation of the volume fraction of each texture component with the increasing deformation by groove cold rolled process, the texture components being obtained from the ODF with a maximum deviation around the ideal orientation of 10° [18]. In addition to the grain elongation as a function to the groove rolling strain (Table 3), a global tendency is a decrease of the random orientation, which could be correlated to the formation of a strong crystallographic texture. It can also be observed that at least 4 texture components reached a volume fraction of 8% after a groove rolling deformation $\varepsilon = 2.2$, which can be directly correlated to the complex texture described above. Figure 7b shows the different volume fraction associated to fiber textures reported previously. The aim here is to compare conventional cold rolling to grooved cold rolling by summing different texture component associated to each fiber texture at different area reduction. The <110>ND component was obtained by summing G, G/B, and B components, <111>RD with Cu and A texture component and <001>RD with Cb and G. One may firstly note that the divergence from the as-annealed state for the two deformation processes may be explained by the different annealing temperatures and by slightly different mechanical post-processing conditions applied to the HEA (see ref. [16]). As expected by the previous analysis, the evolutions of the volume fraction of each fiber texture are clearly different for the two conventional and groove rolling processes. As previously reported by Sathiaraj et al. [16], <110>ND is the main fiber-texture by conventional cold rolling, while the two other studied fibers show a continuous decrease. In the case of groove cold rolling, the <001>RD fiber remained relatively constant from the initial state up to the high deformation state. The two other fibers conserved the same evolution until intermediate deformation state, with a simultaneous increase. For higher deformation level, the rupture of α-fiber shown in Figure 6 (between $\varepsilon = 1.1$ and 2.2) is responsible for the observed decrease of this fiber. In contrast, the <111>RD fiber texture continued to evolve and it became predominant over the other texture components. These final states are closely corresponding to the textures found in FCC materials deformed by swaging or drawing processes, as already published by Otto et al. [15].

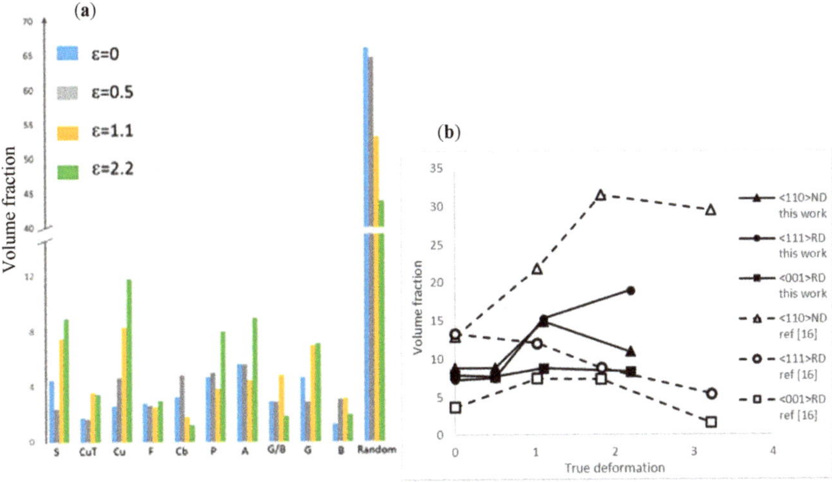

Figure 7. (**a**) Volume fraction of the main principal different detected texture component and (**b**) fiber-texture comparison between groove cold rolling (this work) and conventional cold rolling (ref. [16]).

Table 3. Values of the length/width aspect ratio of the grains, *r*, and Young's modulus, *E*, measured along the rolling direction as a function of the groove rolling strain, ε.

ε	0	0.5	1.1	2.2
r	1.9 ± 0.7	2.3 ± 0.7	3.4 ± 1.0	5.8 ± 2.0
E (GPa)	262.2 ± 5.8	263.0 ± 13.1	262.8 ± 13.1	279.0 ± 12.3

To summarize, the texture of CoCrFeMnNi HEA alloy by groove cold rolling may be described as a hybrid texture between the texture obtained by conventional cold rolling and that obtained by swaging or drawing processes. Under high level of deformation, the <111>RD fiber tended to become predominant due to the decrease of the <110>ND fiber. This final state appeared to be very similar to the texture of FCC materials deformed by swaging process.

4. Discussion

This study firstly focused on the evolution of mechanical properties of the CrCoFeMnNi HEA during deformation by groove cold rolling. Our results revealed two stages in the evolution of the microhardness, yield strength, and Young's modulus, with a change of slopes for a groove rolling deformation of about ε = 0.5. Both microhardness and yield strength displayed a strong increase with the strain up to ε = 0.5 by groove cold rolling followed by a moderate increase, to reach values of, respectively, around 400 HV and 1400 MPa for an applied groove rolling deformation ε equal to 2.2. In contrast, the Young's modulus displayed firstly a decrease of about 15% from its initial value as the groove rolling deformation increased to about ε = 0.5, followed by a slight increase from 175 to 179 GPa as the groove rolling deformation was further increased till ε = 2.2.

The texture analysis revealed that the as-annealed and deformed microstructures presented similar features to those reported in previous studies on CoCrFeMnNi HEA processed by conventional cold rolling. From these considerations, it has been assumed that the deformation mechanisms involved during groove cold rolling were the same as those already reported for CoCrFeMnNi HEA alloy, i.e., dislocation glide and twinning. In order to interpret the variation of the mechanical properties, identification of the main texture components and their evolutions tended to confirm the emergence of a specific crystallographic texture responsible for the change in the microhardness and yield strength as described above. It has been shown that these textures could be interpreted as a hybrid texture between conventional rolling and swaging texture. Also, under high deformation level, our analyses suggested that the swaging type texture tended to become predominant.

The evolution of hardness and yield strength may be regarded through the evolution of the active deformation mechanism. Otto et al. [15], as well as Hasse et al. [18], had reported that plastic deformation was mainly governed by dislocation glide for reduction less than 50% (i.e., ε ~2), while both twinning and dislocation glide are active deformation mechanisms beyond this limit for CrCoFeMnNi HEA processed by conventional cold rolling. For alloy processed by groove rolling process, the similar variation in the properties suggested that the deformation mechanisms might also be operating, and that the breakdown occurred for lower value of rolling deformation owing to the biaxial state of stress applied during groove cold rolling. It might also indicate that the work hardening, and the change in the density of dislocation, in CoCrFeMnNi HEA may be more important when deformed under groove cold rolling, i.e., bi-axial compression, than during conventional cold rolling.

The variation of the Young's modulus followed a different pattern that might be more complicated and less intuitive at this current stage of our investigation. Owing to the strong anisotropy of FCC alloys, such as austenitic Fe and Ni alloys, values of the Young's modulus are expected to be strongly dependent upon the crystallographic texture [41]. Since the elastic constants of the CoCrFeMnNi HEA have not been reported yet, the Young's modulus was calculated from the fourth-order elastic compliance tensor S_{ijkl} of the quaternary CoCrFeNi alloy [42]. Without any texture, these elastic compliances give values of 96.1, 232.9, and 442.8 GPa for the Young's modulus calculated, respectively, along the <001>, <101>, and <111> directions. Taking into account the crystallographic texture expressed by the ODF measured at the different levels of groove rolling deformation, the calculated values of the Young's modulus have been reported in Table 3. Though the calculated values are larger than those reported in the literature and measured in this study, these results tend to suggest that the Young's modulus do not change significantly for CoCrFeMnNi HEA with texture corresponding to groove rolling deformation varying between $\varepsilon = 0$ and 1.1. Beyond $\varepsilon = 1.1$, the crystallographic texture oriented toward the orientation <111>, as seen in the IPF (Figure 6) and identified as the <111>RD fiber-texture (Figure 7), induced a marked increase of the Young modulus. Consequently, from this point of view, the crystallographic texture cannot provide a satisfactory explanation for the change in the Young's modulus detected in our study, which is in agreement with earlier studies [27] undertaken on conventional steels.

As it has been advanced in the introduction, changes in the dislocation arrangement could be responsible for the variation of the elastic properties. Otto et al. [8] explained that the deformation was essentially the results of dislocations gliding at low deformation level and they reported the formation of dislocation cell-structures under higher degrees of deformation. Laplanche et al. [23] have also shown that the dislocation density increased continuously as the deformation increased. Joo et al. [36] have shown that twinning had an effect in term of dynamic Hall-Petch relationship but also led to an accumulation of stacking fault. In addition Benito et al. [27] have demonstrated a strong connection between dislocation structure developed during deformation and decrease in the Young's modulus. By considering all of these features, and assuming that they are also operating in solid solution deformed by groove cold rolling, it seems reasonable to consider that the change in the Young's modulus could be mainly governed by the structure of dislocation developed under level of deformation and the gliding of mobile (non-pinned) dislocations within the cells. The stabilized value of the Young's modulus observed for larger value of the rolled deformation could be explained by modification in the dislocation arrangement resulting from the contribution of the twinning deformation.

5. Conclusions

In this study, samples of CoCrFeMnNi HEA were successively deformed by groove cold rolling at different strain levels. The mechanical properties and crystallographic texture were evaluated as a function of the increasing cold rolling deformation, leading to the following conclusions:

- Both the microhardness and the yield strength exhibited similar evolution as the area reduction ratio applied by groove cold rolling increased. A first marked increase until a groove rolling deformation of about 0.5 was followed by a moderate increase under higher strain level. This two-step behavior has been attributed to change in the activated deformation mechanisms, and characterized by a transition from dislocation mechanism and dislocation + twinning deformation mechanism as reported by Otto et al. [8] and Haase et al. [18]. For the groove rolling process, these breakdowns occurred at lower levels of deformation in comparison to conventional cold rolling.
- The Young's modulus also displayed a two-step variation as the groove cold rolled deformation increased. An approximately 15% decrease from 210 GPa to 175 GPa was observed between annealed HEA samples and those deformed up to about 0.5 by groove rolled process. For higher level of the groove cold rolling deformation, the Young's modulus value stabilized around 175 GPa. This evolution of the elastic property, which could not be attributed by a modification of the crystallographic texture, has been explained by the transition of operating deformation mechanisms between gliding of non-pinned dislocations within dislocation cell in the first stage, and contribution of twinning for larger values of the groove rolling deformation.
- The texture analysis highlighted a complex crystallographic texture induced by groove cold rolling deformation. This texture has been described as a hybrid texture between the texture developed during conventional rolling and that from swaging or drawing process. This texture

was found to consist in a combination of <111> fiber texture along the RD, <110> fiber texture along ND and ND′ and <001> texture along RD. At higher strain level, the <110> fiber texture tended to disappear for the benefit of <111> fiber texture along the RD. This texture was found to be consistent with the bi-axial deformation induced by groove cold rolling but could not account for the decrease in the Young's modulus.

Author Contributions: Conceptualization, E.F., J.Z., and A.H.; Methodology, L.P. and P.L. (Pascal Laheurte); Investigation, P.L. (Paul Lohmuller); Data Curation, P.L. (Paul Lohmuller); Writing-Original Draft Preparation, P.L. (Paul Lohmuller); Writing-Review & Editing, E.F., J.Z., and A.H.; Funding Acquisition, E.F. and J.Z.

Funding: The authors acknowledge the financial support from French National Research Agency ANR (LabEx DAMAS, Grant no.ANR-11-LABX-0008-01).

Acknowledgments: The authors are thankful to Jin-Yoo Suh researcher at KIST, Seoul, Korea, who provided the ingot. A special thank is also addressed at Charbonnier and Lecoz, for their assistance and advices. Eric Fleury acknowledges the France-Korea Partenariat Hubert Curien (PHC) STAR program.

Conflicts of Interest: The authors declare no conflict of interest.

References

1. Gao, M.C.; Yeh, J.; Liaw, P.K.; Zhang, Y. *High-Entropy Alloys: Fundamentals and Applications*; Springer Internationnal Publishing: New York, NY, USA, 2016.
2. Yeh, J.W.; Chen, S.K.; Lin, S.J.; Gan, J.Y.; Chin, T.S.; Shun, T.; Tsau, C.H.; Chang, S.Y. Nanostructured high-Entropy alloys with multiple principal elements: Novel alloy design concepts and outcomes. *Adv. Eng. Mater.* **2004**, *6*, 299–303. [CrossRef]
3. Yeh, J.W. Alloy design strategies and future trends in high-Entropy alloys. *JOM* **2013**, *65*, 1759–1771. [CrossRef]
4. Zhang, Y.; Yang, X.; Liaw, P.K. Alloy design and properties optimization of high-Entropy. *JOM* **2013**, *63*, 830–838. [CrossRef]
5. Miracle, D.; Senkov, O. A critical review of high entropy alloys and related. *Acta Mater.* **2017**, *122*, 448–511. [CrossRef]
6. Cantor, B.; Chang, I.; Knight, P.; Vincent, A. Microstructural development in equiatomic multicomponent alloys. *Mater. Sci. Eng. A* **2004**, *375*, 213–218. [CrossRef]
7. Sun, S.; Tian, Y.; Lin, H.; Dong, X.; Wang, Y.; Zhang, Z.; Zhang, Z. Enhanced strength and ductility of bulk CoCrFeMnNi high entropy alloy having fully recrystallized ultrafine-grained structure. *Mater. Des.* **2017**, *133*, 122–127. [CrossRef]
8. Otto, F.; Dlouhy, A.; Somsen, C.; Bei, H.; Eggeler, G.; George, E. The influences of temperature and microstructure on the tensile properties of a CoCrFeMnNi high-entropy alloy. *Acta Mater.* **2013**, *61*, 5743–5755. [CrossRef]
9. Laplanche, G.; Gadaut, P.; Horst, O.; Otto, F.; Eggeler, G.; George, E.P. Temperature dependencies of the elastic young moduli and thermal expansion coefficient of an equiatomic, single-phase CoCrFeMnNi high-entropy alloy. *J. Alloys Compd.* **2015**, *623*, 348–353. [CrossRef]
10. Wu, Z.; Bei, H.; Otto, F.; Pharr, G.; George, E. Recovery, recrystallization, grain growth and phase stability of a family of FCC-Structured multi-Component equiatomic solid solution alloys. *Intermetallics* **2014**, *46*, 131–140. [CrossRef]
11. Bracq, G.; Laurent-Brocq, M.; Perrière, L.; Pirès, R.; Joubert, J. The fcc solid solution stability in the Co-Cr-Fe-Mn-Ni multi-Component high-Entropy alloy. *Acta Mater.* **2017**, *128*, 327–336. [CrossRef]
12. Laurent-Brocq, M.; Akhatova, A.; Perrière, L.; Chebini, S.; Sauvage, X.; Leroy, E.; Champion, Y. Insights into the phase diagram of CrMnFeCoNi high entropy alloy. *Acta Mater.* **2015**, *88*, 355–365. [CrossRef]
13. Bae, J.; Moon, J.; Jang, M.; Yim, D.; Kim, D.; Lee, S.; Kim, H. Trade-off between tensile property and formability by partial recrystallization of CrMnFeCoNi high-Entropy alloy. *Mater. Sci. Eng. A* **2017**, *703*, 324–330. [CrossRef]
14. Pradeep, K.; Tasan, C.; Yao, M.; Deng, Y.; Springer, H.; Raabe, D. Non-Equiatomic high entropy alloys: Approach towards rapid alloy screening and property-Oriented design. *Mater. Sci. Eng. A* **2015**, *648*, 183–192. [CrossRef]

15. Otto, F.; Hanold, N.; George, E. Microstructural evolution after thermomechanical processing in an equiatomic, single-Phase CoCrFeMnNi high-Entropy alloy with special focus on twin boundaries. *Intermetallics* **2014**, *54*, 39–48. [CrossRef]
16. Dan Sathiaraj, G.; Bhattacharjee, P. Effect of cold-rolling strain on the evolution of annealing texture of equiatomic CoCrFeMnNi high entropy alloy. *Mater. Charact.* **2015**, *109*, 189–197. [CrossRef]
17. Dan Sathiaraj, G.; Bhattacharjee, P. Effect of starting grain size on the evolution of microstructure and texture during thermo-mechanical processing of CoCrFeMnNi high entropy alloy. *J. Alloys Compd.* **2015**, *647*, 92–96. [CrossRef]
18. Haase, C.; Barrales-Mora, L. Influence of deformation and annealing twinning on the microstructure and texture evolution of face-centered cubic high-entropy alloy. *Acta Mater.* **2018**, *150*, 88–103. [CrossRef]
19. Laplanche, G.; Horst, O.; Otto, F.; Eggeler, G.; George, E. Microstructural evolution of a CoCrFeMnNi high-Entropy alloy after swaging and annealing. *J. Alloys Compd.* **2015**, *647*, 548–557. [CrossRef]
20. Skrotzki, W.; Pukenas, A.; Joni, B.; Odor, E.; Ungar, T.; Hohenwarter, A.; Pippan, R. Microstructure and texture evolution during severe plastic deformation of CrMnFeCoNi high-entropy alloy. In Proceedings of the IOP Conference Series: Materials Science and Engineering, Sydney, Australia, 2–7 July 2017; Volume 194.
21. Dan Sathiaraj, G.; Bhattacharjee, P. Analysis of microstructure and microtexture during grain growth in low stacking fault energy equiatomic CoCrFeMnNi high entropy and Ni–60 wt.%Co alloys. *J. Alloys Compd.* **2015**, *637*, 267–276. [CrossRef]
22. Huang, S.; Li, W.; Lu, S.; Tian, F.; Shen, J.; Holmström, E.; Vitos, L. Temperature dependent stacking fault energy of FeCoCrNiMn high entropy alloy. *Scr. Mater.* **2015**, *108*, 44. [CrossRef]
23. Laplanche, G.; Kostka, A.; Horst, O.; Eggeler, G.; George, E. Microstructure evolution and critical stress for twinning in the CrMnFeCoNi high-entropy alloy. *Acta Mater.* **2016**, *118*, 152–163. [CrossRef]
24. Kang, M.; Won, J.; Kwon, J.; Na, Y. Intermediate strain rate deformation behavior of a CoCrFeMnNi high entropy alloy. *Mater. Sci. Eng. A* **2017**, *707*, 16–21. [CrossRef]
25. Mott, N. A theory of work-hardening of metal crystals. *Philos. Mag.* **1952**, *43*, 1151–1178. [CrossRef]
26. Friedel, F. Anomaly in the rigidity modulus of copper alloys for small concentrations. *Philos. Mag.* **1953**, *44*, 444–448. [CrossRef]
27. Benito, J.; Manero, J.; Jorba, J.; Roca, A. Change of Young's Modulus of Cold-Deformed Pure Iron in a Tensile Test. *Metall. Mater. Trans. A* **2005**, *36*, 3317–3324. [CrossRef]
28. Callister, W.D. *Materials Science and Engineering: An Introduction*, 8th ed.; John Wiley & Sons: Hoboken, NJ, USA, 2010.
29. Beausir, B.; Fundenberger, J.J. ATEX-Analysis Tools for Electron and X-Ray Diffraction. Available online: http://atex-software.eu/ (accessed on 28 June 2018).
30. Ge, H.; Song, H.; Shen, J.; Tian, F. Effect of alloying on the thermal-elastic properties of 3d high-Entropy alloys. *Mater. Chem. Phys.* **2018**, *210*, 302–326. [CrossRef]
31. Ledbetter, H.; Kim, S. Low Temperature Elastic constants of deformed polycrystalline copper. *Mater. Sci. Eng. A* **1988**, *101*, 87–92.
32. Morestin, F.; Boivin, M. On the necessity of taking into account the variation in the young modulus with plastic strain in elastic-Plastic software. *Nucl. Eng. Des.* **1996**, *162*, 107–116. [CrossRef]
33. Elmay, W.; Prima, F.; Gloriant, T.; Bolle, B.; Zhong, Y.; Patoor, E.; Laheurte, P. Effects of thermomechanical process on the microstructure and mechanical properties of a fully martensitic titanium-Based biomedical alloy. *J. Mech. Behav. Biomed. Mater.* **2013**, *18*, 47–56. [CrossRef] [PubMed]
34. Yamaguchi, K.; Adachi, H.; Takakura, N. Effects of plastic strain and strain path on young's modulus of sheet metals. *Met. Mater.* **1998**, *4*, 420–425.
35. Böcker, W.; Bunge, H.J.; Reinert, T. Anomalies of young modulus in Fe-Cu composites after high degree of deformation. *Mater. Sci. Forum* **1994**, *157*, 1551–1558. [CrossRef]
36. Joo, S.; Kato, H.; Jang, M.; Moon, J.; Tsai, C.; Yeh, J.; Kim, H. Tensile deformation behavior and deformation twinning of an equimolar CoCrFeMnNi high-Entropy alloy. *Mater. Sci. Eng. A* **2017**, *689*, 122–133. [CrossRef]
37. Kireeva, I.; Chumlyakov, Y.; Pobedennaya, Z.; Vyrodova, K.I.; Karaman, I. Orientation dependence of twinning in single crystalline CoCrFeMnNi high-Entropy alloy. *Mater. Sci. Eng. A* **2017**, *705*, 176–181. [CrossRef]
38. Kalua, P.; Brandao, L.; Ortiz, F.; Egungwu, O.; Igeb, F. On the texture evolution in swaged Cu-Based wires. *Acta Metall.* **1998**, *38*, 1755–1761.

39. Kauffman, A.; Freudenberger, J.; Klauss, H.; Klemm, V.; Schillinger, W.; Subramanya-Sarma, V.; Schultz, L. Properties of cryo-Drawn copper with severely twinned microstructure. *Mater. Sci. Eng. A* **2013**, *588*, 132–141. [CrossRef]
40. Bunge, H. *Texture Analysis in Materials Science*; Mathematical Methods; Butterworth-Heinemann: Oxrord, UK, 1968.
41. Nye, J. *Physical Properties of Crystal: Their Representation by Tensors and Matrices*; Oxford Science Publications; Oxford University Press: Oxford, UK, 1957.
42. Tian, F.Y.; Varga, L.K.; Chen, N.X.; Delczeg, L.; Vitos, L. Ab initio investigation of high-entropy alloys of 3d elements. *Phys. Rev. Lett. B.* **2013**, *87*, 075144. [CrossRef]

Article

The Effect of Crystal Defects on 3D High-Resolution Diffraction Peaks: A FFT-Based Method

Komlavi Senyo Eloh [1,2,3], Alain Jacques [1,2,*], Gabor Ribarik [1,4] and Stéphane Berbenni [1,3]

1 Laboratory of Excellence on Design of Alloy Metals for low-mAss Structures (DAMAS), Université de Lorraine, f-57073 Metz, France; komlavi-senyo.eloh@univ-lorraine.fr (K.S.E.); ribarik@elte.hu (G.R.); stephane.berbenni@univ-lorraine.fr (S.B.)
2 Institut Jean Lamour (IJL), Université de Lorraine, CNRS, f-54000 Nancy, France
3 Laboratoire d'Etude des Microstructures et de Mécanique des Matériaux (LEM3), Université de Lorraine, CNRS, f-57073 Metz, France
4 Department of Materials Physics, Eötvös University, h-1117 Budapest, Hungary
* Correspondence: alain.jacques@univ-lorraine.fr; Tel.: +33-3-7274-2679

Received: 28 June 2018; Accepted: 31 August 2018; Published: 9 September 2018

Abstract: Forward modeling of diffraction peaks is a potential way to compare the results of theoretical mechanical simulations and experimental X-ray diffraction (XRD) data recorded during in situ experiments. As the input data are the strain or displacement field within a representative volume of the material containing dislocations, a computer-aided efficient and accurate method to generate these fields is necessary. With this aim, a current and promising numerical method is based on the use of the fast Fourier transform (FFT)-based method. However, classic FFT-based methods present some numerical artifacts due to the Gibbs phenomenon or "aliasing" and to "voxelization" effects. Here, we propose several improvements: first, a consistent discrete Green operator to remove "aliasing" effects; and second, a method to minimize the voxelization artifacts generated by dislocation loops inclined with respect to the computational grid. Then, we show the effect of these improvements on theoretical diffraction peaks.

Keywords: dislocations; diffraction; fast Fourier transform (FFT)-based method; discrete green operator; voxelization artifacts; sub-voxel method; simulated diffraction peaks; scattered intensity

1. Introduction

X-ray diffraction (XRD) is one of the most powerful non-destructive tools to investigate materials, as their wavelength is commensurate with the distance between atoms within a crystal [1–9]. Successive improvements of both the X-ray sources (from X-ray tubes to third generation synchrotrons) and detectors (from photographic plates and gas counters to fast two-dimensional arrays) have led to a tremendous increase in the quantity of data recorded per unit time, allowing real time in situ or in operando measurements [10,11]. It is now possible to determine the 3D grain microstructure of a bulk material with a submicron resolution (using topo-tomography), to follow the evolution of the elastic strain state of the grains of a polycrystal during mechanical tests (3D-XRD, far field diffractometry), or to measure the distribution of strains within a few grains in real time (2D diffractometry) [12,13]. Such experiments result in terabytes of data recorded within a few days, which need to be analyzed efficiently. In fact, only a low fraction of those data is actually treated because scientists lack both time and numerical tools (software) for further analysis [14].

The classical techniques used to analyze the 1D or 2D diffraction patterns recorded during tests performed on polycrystalline specimens such as the Rietveld method, the square sines method to measure internal stresses, or CMWP (convolutional multiple whole profile) fitting for dislocations content often rely on simplified and mathematically tractable models of a microstructure.

Calculations which may involve a simplifying hypothesis lead to a general formula which can be used to fit one or several parameters of the microstructure (dislocation densities and type, internal stress tensor etc.) to the diffraction pattern (peak profiles, variation of the $2\theta_B$ angle with orientation etc.).

During the last 10 years, several authors proposed the opposite approach: forward modeling [15–22]. This requires the design of a microstructure and the simulation of its behavior (often under process or thermo-mechanical solicitation), and the computation of the elastic strain field or the displacement field. The last step is the generation through a 'virtual diffractometer' of a theoretical diffraction pattern (different $\boldsymbol{G}$ vectors and different orientations of the lattice planes), which can be compared with the experimental one. Depending on the size of the simulated representative volume of matter and the experimental conditions such as the X-ray beam coherence, different assumptions can be made such as a coherent beam (where the amplitudes scattered by different points add) or an incoherent beam (scattered intensities add), or for a partially coherent beam where a full calculation may be necessary. Such modeling can be quite successful and can be used to validate the different steps involved, mainly the microstructure and the constitutive law used to simulate the material's behavior.

However, as diffraction peaks contain information on different scales of a specimen: from average quantities such as Type I (average) stresses related to the peaks' positions, Type II (at grain level) stresses related to its width, and Type III stresses (near the core of defects such as dislocations) related to the peaks' tails, a realistic simulation of a diffraction peak requires a description of a material's representative volume element with a very fine mesh, i.e., a huge amount of CPU time with classical methods used for simulations such as the finite element method.

Numerical approaches based on the fast Fourier transform (FFT) for calculating the stress and strain fields within a composite material received a surge of interest since the pioneering work of Moulinec and Suquet [23,24]. They were first developed to compute effective properties and mechanical field of linear elastic composites [23–26] and were extended to heterogeneous materials with eigenstrains (dislocations, thermal strains etc.) [14,27–30]. They are also used for conductivity problems [31], non-linear materials [25,27], and viscoplastic or elasto-viscoplastic polycrystals [32–36]. Today, FFT-based approaches represent an attractive alternative to the finite element method because of lower computation time [32].

However initial tests indicate that the displacement field computed (essential for diffraction pattern generation) with FFT algorithms presents some numerical artifacts. These numerical artifacts are due to Gibbs phenomenon or "aliasing" and to voxelization. The accuracy of the calculated strain or displacement field is strongly influenced by these shortcomings and the simulated peaks may provide wrong information on mechanical behavior or material characteristics. Therefore, it is important to control these artifacts in order to simulate correct diffraction pattern in the case of a microstructure containing different phases, grains, and crystal defects.

The aim of this paper is to improve the accuracy of the displacement field for diffraction peak generation. This improvement is based on the introduction of a consistent discrete periodized Green operator associated with the displacement field in order to take explicitly into account the discreteness of the discrete Fourier transform method [37]. The improvement of the voxelization in FFT-method is performed through a sub-voxelization method described for inclined dislocation loops. These improvements are reported and discussed in the present contribution. In the Section 2, the FFT-based method to compute the displacement field in a periodic medium is described. In Section 3, the treatment of voxelization problems in FFT-based approaches by a sub-voxelization method is detailed in the case of slip plane not conforming to FFT grid. In the Section 4, simulation of diffraction peaks is reported and discussed.

2. Fast Fourier Transform (FFT)-Based Numerical Calculation of the Displacement Field and Periodized Green Operators

2.1. FFT-Based Algorithm and Mechanical Fields

Let us consider a homogeneous elastic medium with eigenstrain assuming a periodic unit cell discretized in $N \times N \times N$ voxels and subjected to a uniform overall strain tensor denoted $\boldsymbol{E}$. Here, this overall strain is the spatial average of the strain field in the unit cell (with external loading and a given eigenstrain field). The unit cell may contain voxel size defects (0D) such as chemical inhomogeneities, line defects (1D) with arbitrary shape and distribution, planar (2D) defects such as stacking faults, or 3D precipitates which are all modeled with an eigenstrain tensor. (See [14].) These defects create a displacement field and thus generate strain and stress fields [38].

The displacement vector is denoted $\boldsymbol{u}$ and in the forthcoming equations, $\boldsymbol{x}$ denotes any position vector within the unit cell. All vector and tensor fields will be written using bold characters.

Starting from the equation for mechanical equilibrium, $div\ \boldsymbol{\sigma}(\boldsymbol{x}) = 0$, and using field equations (strain compatibility, generalized Hooke's law, decomposition of total strain compatible strain into elastic strain and eigenstrain), the displacement field is given at every position by the Green's function technique [39]:

$$\boldsymbol{u}(x) = \left(\boldsymbol{B} * \boldsymbol{c}^0 : \boldsymbol{\varepsilon}^*\right)(x) \tag{1}$$

where the symbol $*$ denotes the spatial convolution product, $\boldsymbol{c}^0$ is the homogeneous linear elastic stiffness, $\boldsymbol{\varepsilon}^*$ is the eigenstrain field and $\boldsymbol{B}$ is a third order Green operator defined in Fourier space as:

$$\widehat{B_{ijk}}(\boldsymbol{\xi}) = \frac{i}{2}\left(\widehat{G_{ij}}\xi_k + \widehat{G_{ik}}\xi_j\right) \tag{2}$$

in which $\widehat{\boldsymbol{B}}$ is the Fourier transform of $\boldsymbol{B}$ and $\widehat{\boldsymbol{G}}$ is the Fourier transform of the elastic Green tensor [39]. Therefore, using the Fourier transform of spatial convolution product, Equation (1) can be written in Fourier space as:

$$\widehat{\boldsymbol{u}}(\boldsymbol{\xi}) = \widehat{\boldsymbol{B}}(\boldsymbol{\xi}) : \boldsymbol{c}^0 : \widehat{\boldsymbol{\varepsilon}^*}(\boldsymbol{\xi}) \tag{3}$$

Several numerical results showed that the use of the third order operator $\widehat{\boldsymbol{B}}$ derived from the classic Green $\widehat{\boldsymbol{G}}$ leads to spurious oscillations on the computed displacement field near materials discontinuities and dislocations [37]. The discrete Fourier transform (DFT) used in this algorithm indeed transforms a periodic function in real space into a periodic function in reciprocal space. However, the operator $\widehat{\boldsymbol{B}}$ commonly used is the continuous analytic operator truncated to the size of the unit cell of the reciprocal space: it is not periodic function. To fix this problem, we very recently developed a periodized consistent discrete Green operator using the DFT. The mathematical derivations of this discrete Green operator are given elsewhere [37]. The Fourier transform of this discrete Green operator denoted $\widehat{\boldsymbol{B}}'$ is written as function of $\widehat{\boldsymbol{B}}$ and reads:

$$\widehat{\boldsymbol{B}}'\left(\xi_{ijk}\right) = A_{ijk} \sum_{m,n,p=-\infty}^{+\infty} \frac{(-1)^{m+n+p}}{(mN+i)} \frac{1}{(nN+j)} \frac{1}{(pN+k)} \widehat{\boldsymbol{B}}\left(\xi_{mN+i,nN+j,pN+k}\right)$$

$$\text{With } A_{ijk} = \left(\frac{N}{\pi}\right)^3 \sin\left(\frac{i\pi}{N}\right) \sin\left(\frac{j\pi}{N}\right) \sin\left(\frac{k\pi}{N}\right) \tag{4}$$

Discrete frequencies appearing in this equation are given when N is even by (T is the period of the unit cell):

$$\xi = \left(-\frac{N}{2}+1\right)\frac{1}{T}, \left(-\frac{N}{2}+2\right)\frac{1}{T}, \ldots, -\frac{1}{T}, 0, \frac{1}{T}, \ldots \left(\frac{N}{2}-1\right)\frac{1}{T}, \left(\frac{N}{2}\right)\frac{1}{T}$$

Here, the sum on the $\hat{B}$ operator is extended to the whole reciprocal space (in practice for m, n, p up to a few tens) and folded up onto the unit cell of the DFT with suitable coefficients. The inverse transform of $\hat{u}(\xi)$ gives the displacement field at the center of each voxel.

We can also compute the displacement field at each voxel's corner with a shifted operator using the shift theorem:

$$\hat{B}'\left(\xi_{ijk}\right) = A_{ijk}e^{p\pi\frac{i+j+k}{N}} \sum_{m,n,p=-\infty}^{+\infty} \frac{1}{(mN+i)} \frac{1}{(nN+j)} \frac{1}{(pN+k)} \hat{B}\left(\xi_{mN+i,nN+j,pN+k}\right) \quad (5)$$

2.2. Numerical Examples

Let us consider a homogeneous material with isotropic elastic constants: Young's modulus $E = 333.4$ GPa and Poisson ration $\vartheta = 0.26$. This approximately corresponds to the room temperature elastic constants of single crystalline Ni-based superalloys. The unit cell (Figure 1a) is discretized in $128 \times 128 \times 128$ voxels and contains a square-shaped inclusion discretized in $32 \times 32 \times 1$ voxels corresponding to an Eshelby-like square prismatic loop perpendicular to the z-axis. In order to generate a shift of the upper surface of the inclusion relative to its lower surface by a Burgers vector $\boldsymbol{b}(0,\ 0,\ b_3)$, only the voxels within the inclusion are submitted to a non-zero eigenstrain tensor defined as: $\varepsilon^*_{ij} = 0$ except $\varepsilon^*_{33} = 1$. Then, we have $b_3 = t \times \varepsilon^*_{33}$ where t the thickness of the inclusion in the z-direction (i.e., the voxel size). This displacement field computed with the FFT algorithm using the different Green operators defined in Section 2.1 is represented along z-axis in Figure 1.

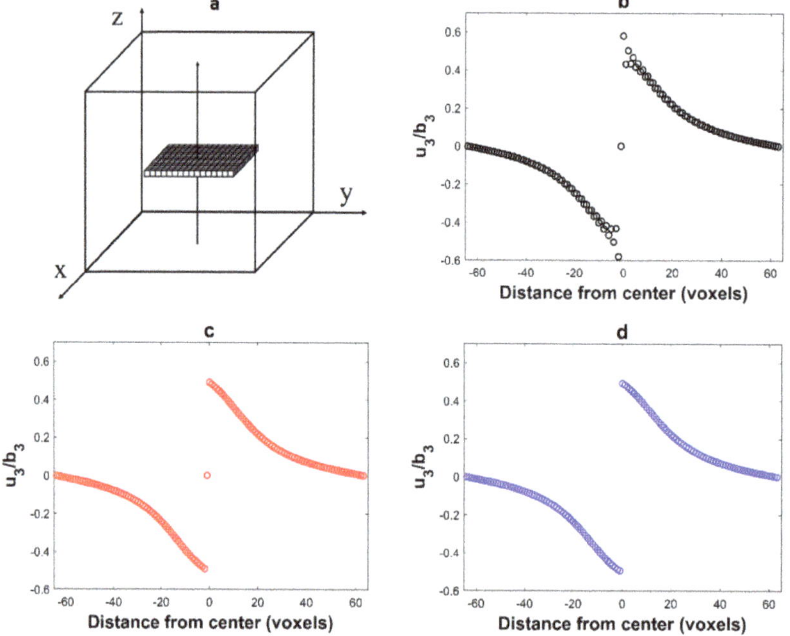

Figure 1. (**a**) Simulation of a square dislocation loop in plane (001) by a platelet with eigenstrain; (**b**) component u_3 of the displacement field (normalized by b_3) along the z axis (arrow) computed with the Green operator $\boldsymbol{B}$ and showing spurious oscillations; (**c**) same component u_3 computed with $\boldsymbol{B}'$. The displacement at voxel (64,64,64) is zero in the center of the inclusion (**c**) and $b_3/2$ on its surface; (**d**) same component u_3 computed with $\boldsymbol{B}''$.

When computed along a line crossing a dislocation loop, the displacement field exhibits a discontinuity with a jump equal to Burgers vector $\boldsymbol{b}$. This is indeed observed in Figure 1. However, the displacement field computed with the usual Green operator $\boldsymbol{B}$ (Figure 1b) also shows spurious oscillations as soon as the discontinuity is approached. These oscillations (numerical artifacts) are not observed with the periodized operators $\boldsymbol{B}'$ and $\boldsymbol{B}''$. An artificial damping of the oscillations in Figure 1b (such as a low pass filtering) might smooth these oscillations, but it would also smooth the discontinuity, which is not searched.

2.3. Voxelization Effect on the Displacement Field

While the displacement field computed for dislocation loops having their planes parallel to the faces of the simulated volume is correctly given with discrete Green operators $\boldsymbol{B}'$ or $\boldsymbol{B}''$, voxelization artifacts appear for inclined loops, as shown with $\boldsymbol{B}'$ in Figure 2 for a dislocation loop with a $\left[0\bar{1}1\right]$ Burgers vector lying in a (111) slip plane of a fcc crystal. The eigenstrain tensor is constrained in the region occupied by the dislocation loop (transformed voxels) and is given by:

$$\varepsilon^*_{ij} = \frac{A_s}{2V}\left(n_i b_j + n_j b_i\right) \tag{6}$$

where A_s is the area on which planes with normal $\boldsymbol{n}(n_1, n_2, n_3)$ has slipped by a relative amount $\boldsymbol{b}(b_1, b_2, b_3)$ and V is the volume occupied by the dislocation loop [40,41]. As before, the dislocation loop is 32 voxels wide in the x and y directions, and 1 voxel thick but now with a z position such that $x + y + z = constant$. The displacement has been computed at the center of voxels with the periodized operator $\boldsymbol{B}'$ along z (Figure 2b). As in Figure 1c, the displacement in the center of a voxel belonging to the loop plane (black dot in the reddish transformed voxel in Figure 2b) is zero. The displacement in the first neighboring voxels (red dot in Figure 2b) are shifted relative to the expected position, so that the displacement difference between these voxels is significantly lower than $\boldsymbol{b}$, see Figure 2c. It can be checked in Figure 2b that each of these voxels shares three faces with a transformed voxel. A more detailed analysis shows that the second neighbors (which share three edges with transformed voxels) are also slightly shifted in the opposite direction. The result is shown in Figure 2d with: a strong localized oscillation of the phase (taken here as the displacement modulo $\boldsymbol{b}$).

Although the amplitude of this shift is small (less than 10% of the Burgers vector) it has unwanted consequences on the diffraction peak simulation:

The dislocation loops are surrounded by four impaired layers of voxels: As the scattered X-ray amplitude is proportional to the Fourier transform of $\boldsymbol{G}\cdot\boldsymbol{u}$ (see Equation (8) in Section 4), we can expect a phantom streak in the intensity in a direction perpendicular to the loop plane.

The displacement field near the edges of the loop (near the dislocation line) will be quite different from its expected value, and the strain field will not vary with the distance r to the dislocation line as $1/r$. This will strongly affect the tails of the diffraction peaks.

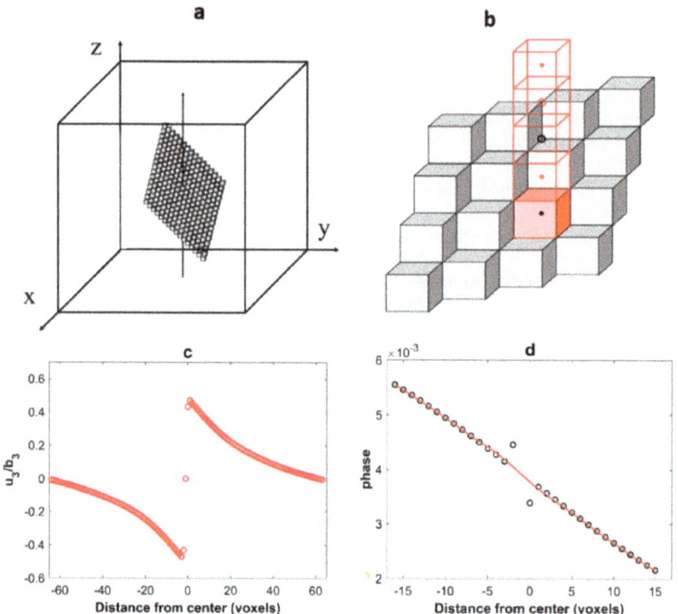

Figure 2. (**a**) Modeling of a dislocation loop in a (111) plane as a layer of voxels with eigenstrain; (**b**) position of the computed points relative to the transformed voxels with eigenstrains; (**c**) plot of the displacement field u_3 (normalized by b_3) along the z direction for the dislocation loop of Figure 2; (**d**) local oscillation of the phase due to the voxelization of the dislocation loop (the representation is made for 32 voxels centered in the unit cell along z direction). The red line is approximately equal to the phase expected for this displacement field.

3. Sub-Voxelization Method to Correct Voxelization Artifacts

3.1. Sub-Voxelization Method

The (conceptually) simplest way to remove this voxelization artifact would be to work on a multiple grid (to multiply the number of voxels along each direction by 2, 4, or more), then to downsample the displacement field data. In that case, FFT algorithms would lose much of their interest due to these more demanding computational efforts. We show below that this can be done in a more "economical"—and simple—way by applying a patch to the FFT-computed displacement field. The basic method is to compute, on the same grid, the difference vector:

$$\Delta_i(\boldsymbol{x}) = u_i^{sub}(\boldsymbol{x}) - u_i^{hom}(\boldsymbol{x}) \tag{7}$$

where $u_i^{sub}(\boldsymbol{x})$ is the displacement vector calculated for voxels where this eigenstrain is concentrated on a single plane of sub voxels (Figure 3b) and u_i^{hom} the displacement field in direction i of voxels with a uniform eigenstrain (Figure 3a). For the sake of clarity, we use 2D diagrams in Figure 3, but here the technique is applied to real 3D problems.

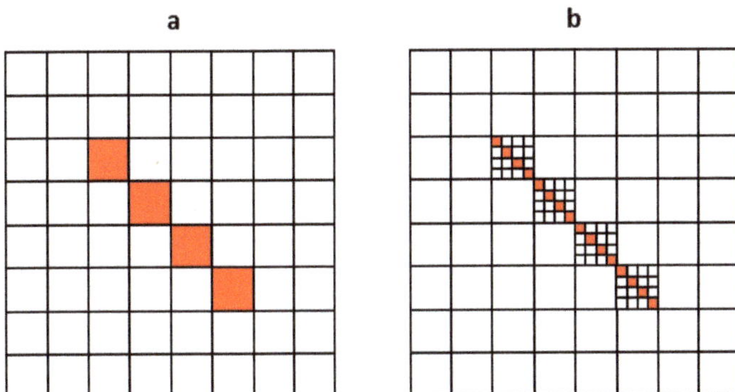

Figure 3. 2D representation of a dislocation loop in a tilted plane on a (8 × 8) fast Fourier transform (FFT) grid: (**a**) with a homogeneous eigenstrain ε^* in the voxels occupied by the dislocation loop; (**b**) with each voxel subdivided into 4 × 4 sub-voxels, only 4 of which have a 4 ε^* eigenstrain field.

In order to compute the displacement due to sub-voxels, we use a $N \times N$ ($N \times N \times N$) grid for 2D (resp. 3D) problems where each voxel can be subdivided into $n \times n$ ($n \times n \times n$) sub voxels. Only n ($n \times n$) sub voxels are submitted to an eigenstrain field. At a point $\boldsymbol{A}$ of the grid (black dots, Figure 4a), we need to compute the sum of the displacements $\boldsymbol{u}_i^j$ due to the n ($n \times n$) sub voxels j (center $\boldsymbol{B}_j$) within a voxel centered at point $\boldsymbol{O}$. This sum is equivalent to the sum of the displacements due to a strained sub voxel at point $\boldsymbol{O}$ on the grid points $\boldsymbol{A}_j$ such as $\boldsymbol{OA}_j = \boldsymbol{B}_j\boldsymbol{A}$ (Figure 4b). It is also equivalent to the sum of the displacements $\boldsymbol{u'}_i^j$ due to a *full* voxel at point O on the initial grid on points $\boldsymbol{A'}_j$ such as $\boldsymbol{OA'}_j = \boldsymbol{nB}_j\boldsymbol{A}$ (Figure 4c). The only difference between these last two sums is due to the long-range strain field, and approximately results in a linear drift of the displacement. As the end of the vectors $\boldsymbol{OA'}_j$ does not lie on the grid points (voxel centers) but on the corners of the voxels, the $\boldsymbol{u'}_i^j$ displacements must be calculated with the shifted operator $\boldsymbol{B''}$ (Equation (5)). A last point is the scaling of the $\boldsymbol{u}_i^j$ and $\boldsymbol{u'}_i^j$ sums during the operations of Figure 4. To keep the one Burgers vector jump between both sides of the sub voxels plane in Figure 4a, the eigenstrain in the sub voxels must be multiplied by n. The backwards change of scale requires a division by n: there is no scaling factor between u_i^{hom} and $u_i^{sub} = \sum u'^j_i$.

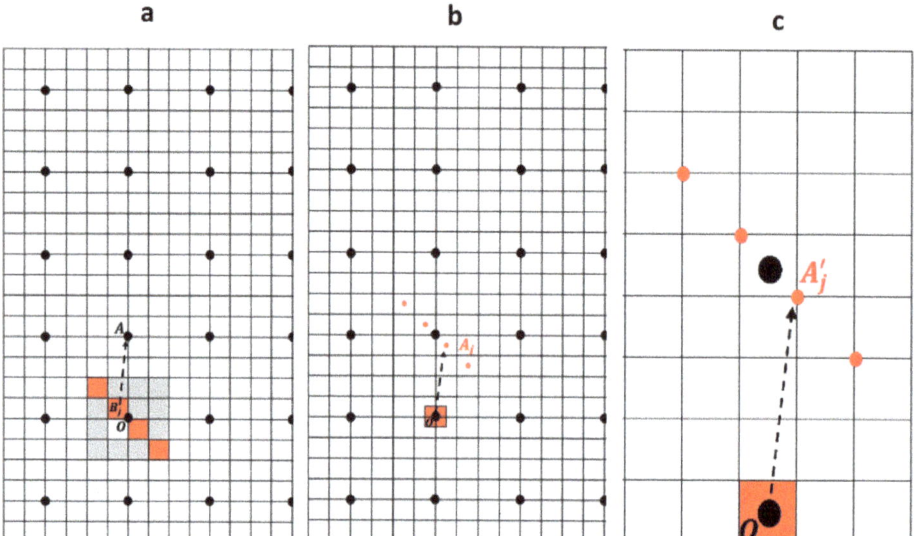

Figure 4. (**a**) 2D representation of the computational grid. The black dots correspond to the voxels centers. A voxel with center O is discretized in 4 × 4 in 2D (4 × 4 × 4 in 3D) sub-voxels. The red sub-voxels have a non-zero eigenstrain. We want to compute the displacement field at point A, due to these deformed sub-voxels centered at B_j. (**b**) Displacement field generated by a deformed sub-voxel centered at O on a row of sub-voxels centered at B_j such as $\boldsymbol{OA_j} = \boldsymbol{B_jA}$. The sum of these displacements is equal to the previous displacement field. (**c**) Displacement field generated by a deformed voxel centered at O on a row of voxels (computed at the corners A'_j using Green operator $\boldsymbol{B''}$) such as $\boldsymbol{OA'_j} = n\boldsymbol{B_jA}$. This sum is equal to the previous sum.

We need to compute $\Delta_{ij\_pl}(\boldsymbol{x})$ the difference in displacement in direction i due to a voxel which belongs to the plane "pl" (for fcc "pl" is equal to (111), $(\bar{1}11)$, $(1\bar{1}1)$, $(11\bar{1})$) of a dislocation loop with a Burgers vector j at a position x relative to the transformed voxel. In practice, in a material with cubic symmetry, it is sufficient to compute $\Delta_{13\_(111)}(\boldsymbol{x})$ and $\Delta_{33\_(111)}(\boldsymbol{x})$, and to use the symmetries of the cube (fourfold [001] axis, threefold [111] axis, and $(1\bar{1}0)$ symmetry plane) (and suitable exchanges of the components of x) to obtain the required components. As can be seen in Figure 2b, $\Delta_{ij\_pl}(\boldsymbol{x})$ is non-zero only for the neighbors of the transformed voxel, except the drift due to the long range strain alluded above. The final computational procedure to determine $\Delta_{ij\_pl}(\boldsymbol{x})$ and use the patch is now detailed:

Compute the field $\boldsymbol{c^0} : \boldsymbol{\varepsilon^*}$ defined in Equation (1) for an isolated voxel with the eigenstrain associated to a dislocation loop (Equation (6)) with a Burgers vector [001] in a (111) plane (see Figure 2).

Compute the displacement field in directions x ($\boldsymbol{u_1^{hom}}$) and z ($\boldsymbol{u_3^{hom}}$) at the voxels' center around the transformed voxel by convolution with the discrete periodized operator $\boldsymbol{B'}$ (Equation (4)).

Compute the displacement field in directions x and z at the voxels' corners around the transformed voxel by convolution with the shifted operator $\boldsymbol{B''}$ (Equation (5)).

Calculate the $\boldsymbol{u_1^{sub}} = \sum \boldsymbol{u'^j_1}$ and $\boldsymbol{u_3^{sub}} = \sum \boldsymbol{u'^j_3}$ sums ($n \times n$ terms for each sum) as in Figure 4c, then the raw $\Delta_{13\_(111)}(\boldsymbol{x})$ and $\Delta_{33\_(111)}(\boldsymbol{x})$ for (x_1, x_2, x_3) going from -3 to 3 times the voxel size t.

Use the farthest voxels to correct the drift of the components so that all terms for large $\boldsymbol{x}$ are zero, and keep non zero only the terms for the first three neighbors.

The patch can then be applied on the raw (FFT-based) displacement field by adding the convolution of all transformed voxels of the different slip systems by the relevant $\Delta_{ij\_pl}(\boldsymbol{x})$.

3.2. Results

For numerical tests, we used the same 128 × 128 × 128 grid as before, and the transformed voxel was divided into 8 × 8 × 8 sub voxels (using a reference medium with the same elastic constants as before). Only the final values in units of b (after drift correction) of $\Delta_{13\_(111)}(x)$ and $\Delta_{33\_(111)}(x)$ are used and other components are obtained by symmetries (Appendix A).

The patch was used on the same configuration as in Figure 2. Figure 5a shows the resulting displacement field and Figure 5b the phase (i.e., the displacement modulo a Burgers vector) in Burgers vector units. As it can be observed from Figure 5a, the voxelization artifacts of the displacement field are removed thanks to the sub-voxelization method. In addition, the resulting phase varies smoothly even during the crossing of the dislocation loop which is more realistic, see Figure 5b.

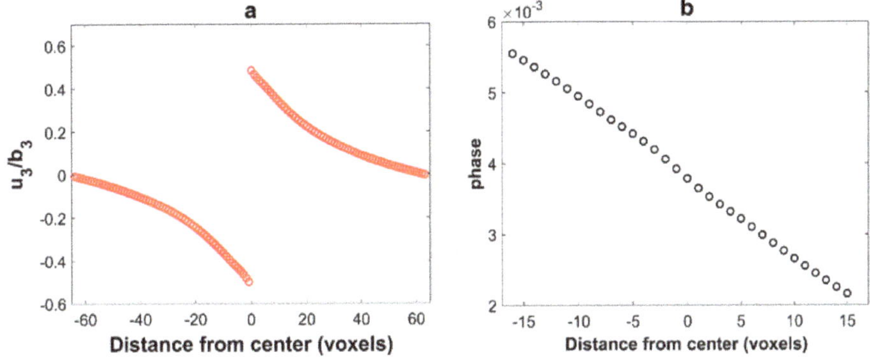

Figure 5. (**a**) Plot of the displacement field u_3 (normalized by b_3) along the z direction for dislocation loop illustrated on Figure 2. The voxelization artifacts are removed by the sub-voxel method described above. (**b**) The phase (i.e., the displacement modulo a Burgers vector). With this correction, the phase is almost continuous.

4. Application on Diffraction Peak Simulation

In this section, we show simulated diffraction peaks in order to point the effects of voxelization artifacts and of the patch on numerical results. Under kinematical conditions and assuming a coherent beam, the amplitude of a diffracted wave at a position $\boldsymbol{q}$ in the vicinity of a reciprocal $\boldsymbol{G}$ lattice vector is [14,18,42–44]:

$$A(\boldsymbol{q}) = FT[A_0(\boldsymbol{x}) \times F(\boldsymbol{G},\boldsymbol{x}) \times exp(-2i\pi\, \boldsymbol{G}{\cdot}\boldsymbol{u}(\boldsymbol{x}))] \tag{8}$$

where $\boldsymbol{x}$ is the position of the scattering atom, $A_0(\boldsymbol{x})$ is the amplitude of the incidence wave, $F(\boldsymbol{G},\boldsymbol{x})$ is the local structure factor, and $\boldsymbol{u}(\boldsymbol{x})$ the displacement field. The scattered intensity is $I(\boldsymbol{q}) = |A(\boldsymbol{q})|^2$. For a face-centered cubic crystal, this intensity is non zero when $\boldsymbol{G}$ (h, k, l) is such as h, k, and l have the same parity. Here two diffraction vectors $\boldsymbol{G}$ (200) and $\boldsymbol{G}$ (002) are used. They respectively correspond to $\boldsymbol{G}{\cdot}\boldsymbol{b} = 0$ and $\boldsymbol{G}{\cdot}\boldsymbol{b} = 1$. The 3D diffracted intensity has been calculated using the FFT instead of the continuous Fourier transform, then summed in the planes perpendicular to the $\boldsymbol{G}$ vector to obtain a linear plot along $\boldsymbol{G}$ equivalent to a $I(2\theta)$ plot. In Figure 6a ($\boldsymbol{G}$ (200)) and Figure 6c ($\boldsymbol{G}$(002)), we show the diffracted intensity (logarithmic scale) as a function of the pixel position i, and in Figure 6b,d a logarithmic/logarithmic plot of the intensity vs. $|i - i_0|$ where i_0 is the center of the peak. In order to only study the effect of the displacement fields, we set $A_0(\boldsymbol{x}) = 1$ and $F(\boldsymbol{G},\boldsymbol{x}) = 1$ for these simulations.

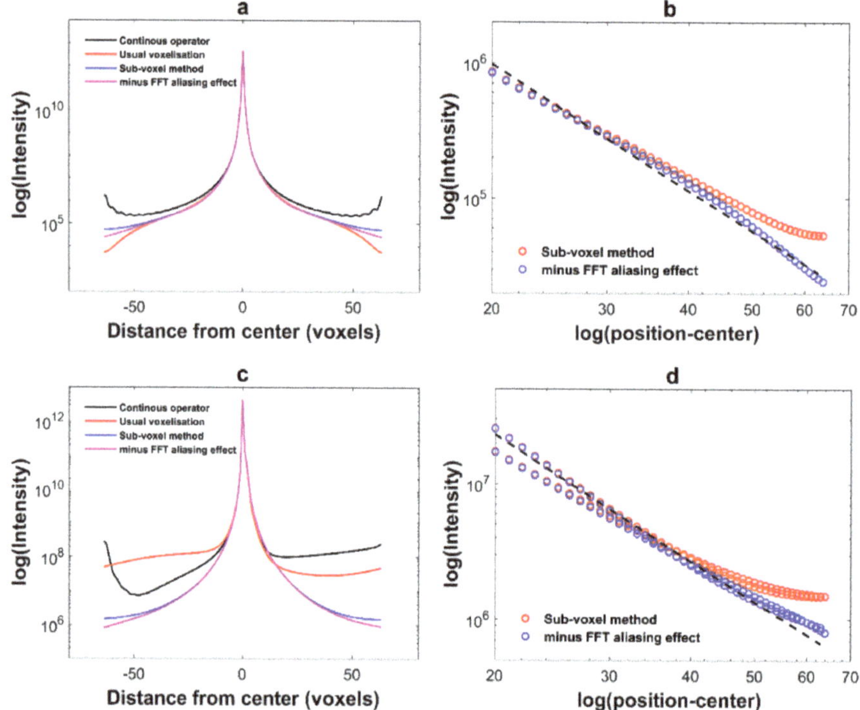

Figure 6. Simulated diffracted intensity as a function of the pixel position (logarithmic scale). 3D configuration is represented in a 1D plot by making the sum in each plane along an x-axis. Different way for computing the displacement fields are studied for a dislocation loop with a $\frac{a}{2}[0\bar{1}1]$ Burgers vector lying in a (111) slip plane. (**a**) Diffracted vector studied is $\boldsymbol{G}$ (200) corresponding to $\boldsymbol{G}{\cdot}\boldsymbol{b} = 0$. (**b**) Log/log representation of the intensity vs. $|i - i_0|$. (**c**) and (**d**) same as (**a**) et (**b**) but the studied diffracted vector is $\boldsymbol{G}$ (002) ($\boldsymbol{G}{\cdot}\boldsymbol{b} = 1$).

The peak shape near the top of the peaks is the same for both computing methods. It is perfectly symmetric in the $\boldsymbol{G}{\cdot}\boldsymbol{b} = 0$ case and exhibits a bump on the right side for $\boldsymbol{G}{\cdot}\boldsymbol{b} = 1$. The long-range behavior is, however, quite different. When the displacement field has been calculated with the usual truncated operator (black line), a phantom peak is observed at large $|i - i_0|$ (at large q), which is due to the short period oscillations near the displacement field discontinuity (Figure 1a). The behavior of the peak calculated with the modified Green operator (red curve) is only slightly better: the intensity at large q is underestimated in one case and overestimated in the other. When the intensity has been calculated with the sub voxel patch (dark blue curve) the long range intensity follows the expected $I_0|i - i_0|^{-3}$ law [45,46]: the peak tails are indeed related to the highly distorted zones near the dislocations' cores. However, the dark blue curve saturates at very large $\boldsymbol{q}$. We suppose this is due to the use of the FFT instead of the continuous Fourier transform in the calculation of the scattered amplitude (Equation (8)). The plot of Figure 6 represents only one period in Fourier space, and is repeated over and over on all Fourier space. We can now calculate the intensity of the tails of these repetitions:

$$I_{neib.} = \sum I_0|i - i_0 - 128m|^{-3} \tag{9}$$

where m varies from −5 to 5 (zero excluded). If we now plot the difference between the dark blue intensity curve and this background line, we obtain the pink curve. On the log./log. plots, Figure 6b,d, it can be checked that this curve follows the $I_0|i - i_0|^{-3}$ law to the end. Thus, the residual error in the intensity computed by FFT results of the FFT itself, and not from an error on the sub voxel-corrected displacement field. If the number of voxels is increased to 512^3 or 1024^3 while keeping the physical size of the representative volume constant, this residual error should fall down to undetectable levels.

5. Conclusions

In this paper, we have shown that although the use of a periodized Green operator in the FFT-based method improves the final displacement field solution in a representative volume containing discontinuities (dislocation loops), artifacts due to the voxelization of the dislocation loop planes are still present with respect to analytical solutions. These artifacts have unwanted consequences on the tails of diffraction peaks simulated by using this displacement field as input data.

We have introduced a patch through a sub-voxelization method, which corrects these artifacts by simulating the displacement field without employing a finer grid resolution. A simple construction method for this patch has been given and the patch can be used in a single post-processing step to modify the initial FFT-based displacement field.

The modified displacement field has been used to simulate one-dimensional diffraction peaks. The procedure strongly improves the shape of the peaks' tails, i.e., it gives a good description of the displacement field and the phase near the dislocation lines.

Author Contributions: For the present paper, K.E. performed the simulation work and wrote the first draft of the paper; A.J. analyzed numerical results and helped in the simulation; G.R. helped in constructing the computer code and in analyzing results; S.B. helped in developing the first theoretical part of the paper. The paper was written by K.E., A.J. and S.B.

Funding: This research was funded by the french National Research Agency (ANR), grant number ANR-11-LABX-0008-01 (LabEx DAMAS).

Acknowledgments: Gabor Ribarik greatly acknowledges the support of the Janos Bolyai Research Fellowship of the Hungarian Academy of Sciences.

Conflicts of Interest: The authors declare no conflict of interest.

Appendix A

$\Delta_{33\_(111)}(x_1, x_2, x_3)$ and $\Delta_{13\_(111)}(x_1, x_2, x_3)$ have these symmetries due to the permutation properties of the plane (111):

$$\Delta_{33\_(111)}(x_1, x_2, x_3) = \Delta_{33\_(111)}(x_2, x_1, x_3)$$

$$\Delta_{33\_(111)}(x_1, x_2, x_3) = -\Delta_{33\_(111)}(-x_1, -x_2, -x_3)$$

$$\Delta_{13\_(111)}(x_1, x_2, x_3) = -\Delta_{13\_(111)}(-x_1, -x_2, -x_3)$$

The others values of $\Delta_{ij\_(111)}(\boldsymbol{x})$ are given as function of $\Delta_{33\_(111)}(x_1, x_2, x_3)$ and $\Delta_{13\_(111)}(x_1, x_2, x_3)$:

$$\Delta_{11\_(111)}(x_1, x_2, x_3) = \Delta_{33\_(111)}(x_3, x_2, x_1) \qquad \Delta_{12\_(111)}(x_1, x_2, x_3) = \Delta_{13\_(111)}(x_1, x_3, x_3)$$

$$\Delta_{22\_(111)}(x_1, x_2, x_3) = \Delta_{33\_(111)}(x_1, x_3, x_2) \qquad \Delta_{31\_(111)}(x_1, x_2, x_3) = \Delta_{13\_(111)}(x_3, x_2, x_1).$$

$$\Delta_{32\_(111)}(x_1, x_2, x_3) = \Delta_{13\_(111)}(x_3, x_1, x_2) \qquad \Delta_{21\_(111)}(x_1, x_2, x_3) = \Delta_{13\_(111)}(x_1, x_3, x_2)$$

$$\Delta_{23\_(111)}(x_1, x_2, x_3) = \Delta_{13\_(111)}(x_2, x_1, x_3)$$

The value of $\Delta_{33}(x_1, x_2, x_3)$ and $\Delta_{13}(x_1, x_2, x_3)$ for the remaining plane $(\bar{1}11), (1\bar{1}1), (11\bar{1})$ are obtained using these symmetries:

$$\begin{aligned}
\Delta_{33_(\bar{1}11)}(x_1, x_2, x_3) &= \Delta_{33_(111)}(x_2, -x_1, x_3) \\
\Delta_{13_(\bar{1}11)}(x_1, x_2, x_3) &= \Delta_{13_(111)}(-x_1, x_2, x_3) \\
\Delta_{33_(1\bar{1}1)}(x_1, x_2, x_3) &= \Delta_{33_(111)}(-x_2, x_1, x_3) \\
\Delta_{13_(1\bar{1}1)}(x_1, x_2, x_3) &= \Delta_{13_(111)}(x_1, -x_2, x_3) \\
\Delta_{33_(11\bar{1})}(x_1, x_2, x_3) &= -\Delta_{33_(111)}(x_2, x_1, -x_3) \\
\Delta_{13_(11\bar{1})}(x_1, x_2, x_3) &= -\Delta_{13_(111)}(x_1, x_2, -x_3)
\end{aligned}$$

References

1. Graverend, J.-B.L.; Dirand, L.; Jacques, A.; Cormier, J.; Ferry, O.; Schenk, T.; Gallerneau, F.; Kruch, S.; Mendez, J. In Situ Measurement of the γ/γ' Lattice Mismatch Evolution of a Nickel-Based Single-Crystal Superalloy during Non-isothermal Very High-Temperature Creep Experiments. *Metall. Mater. Trans. A* **2012**, *43*, 3946–3951. [CrossRef]
2. Robinson, I.; Harder, R. Coherent X-ray diffraction imaging of strain at the nanoscale. *Nat. Mater.* **2009**, *8*, 291–298. [CrossRef] [PubMed]
3. Pfeifer, M.A.; Williams, G.J.; Vartanyants, I.A.; Harder, R.; Robinson, I.K. Three-dimensional mapping of a deformation field inside a nanocrystal. *Nature* **2006**, *442*, 63. [CrossRef] [PubMed]
4. Ungár, T. Strain Broadening Caused by Dislocations. Available online: https://www.scientific.net/MSF.278-281.151 (accessed on 16 June 2018).
5. Ungár, T.; Gubicza, J.; Ribárik, G.; Borbély, A. Crystallite size distribution and dislocation structure determined by diffraction profile analysis: Principles and practical application to cubic and hexagonal crystals. *J. Appl. Crystallogr.* **2001**, *34*, 298–310. [CrossRef]
6. Ribárik, G.; Ungár, T.; Gubicza, J. MWP-fit: A program for multiple whole-profile fitting of diffraction peak profiles by ab initio theoretical functions. *J. Appl. Crystallogr.* **2001**, *34*, 669–676. [CrossRef]
7. Ribárik, G.; Gubicza, J.; Ungár, T. Correlation between strength and microstructure of ball-milled Al–Mg alloys determined by X-ray diffraction. *Mater. Sci. Eng. A* **2004**, *387*, 343–347. [CrossRef]
8. Balogh, L.; Ribárik, G.; Ungár, T. Stacking faults and twin boundaries in fcc crystals determined by X-ray diffraction profile analysis. *J. Appl. Phys.* **2006**, *100*, 023512. [CrossRef]
9. Groma, I. X-ray line broadening due to an inhomogeneous dislocation distribution. *Phys. Rev. B* **1998**, *57*, 7535. [CrossRef]
10. Tréhorel, R.; Ribarik, G.; Schenk, T.; Jacques, A. Real time study of transients during high temperature creep of a Ni-based superlloy by far field high energy synchrotron X-rays diffraction. *J. Appl. Crystallogr.* **2018**, under review.
11. Tréhorel, R. *Comportement Mécanique Haute Température Du Superalliage Monocristallin Am1: Etude In-Situ Par Une Nouvelle Technique De Diffraction En Rayonnement Synchrotron*; Université de Lorraine: Nancy, France, 2018.
12. Bernier, J.V.; Park, J.-S.; Pilchak, A.L.; Glavicic, M.G.; Miller, M.P. Measuring Stress Distributions in Ti-6Al-4V Using Synchrotron X-Ray Diffraction. *Metall. Mater. Trans. A* **2008**, *39*, 3120–3133. [CrossRef]
13. Miller, M.P.; Bernier, J.V.; Park, J.-S.; Kazimirov, A. Experimental measurement of lattice strain pole figures using synchrotron X rays. *Rev. Sci. Instrum.* **2005**, *76*, 113903. [CrossRef]
14. Jacques, A. From Modeling of Plasticity in Single-Crystal Superalloys to High-Resolution X-rays Three-Crystal Diffractometer Peaks Simulation. *Metall. Mater. Trans. A* **2016**, *47*, 5783–5797. [CrossRef]
15. Weisbrook, C.M.; Gopalaratnam, V.S.; Krawitz, A.D. Use of finite element modeling to interpret diffraction peak broadening from elastic strain distributions. *Mater. Sci. Eng. A* **1995**, *201*, 134–142. [CrossRef]
16. Miller, M.P.; Dawson, P.R. Understanding local deformation in metallic polycrystals using high energy X-rays and finite elements. *Curr. Opin. Solid State Mater. Sci.* **2014**, *5*, 286–299. [CrossRef]
17. Demir, E.; Park, J.-S.; Miller, M.P.; Dawson, P.R. A computational framework for evaluating residual stress distributions from diffraction-based lattice strain data. *Comput. Methods Appl. Mech. Eng.* **2013**, *265*, 120–135. [CrossRef]

18. Vaxelaire, N.; Proudhon, H.; Labat, S.; Kirchlechner, C.; Keckes, J.; Jacques, V.; Ravy, S.; Forest, S.; Thomas, O. Methodology for studying strain inhomogeneities in polycrystalline thin films during in situ thermal loading using coherent X-ray diffraction. *New J. Phys.* **2010**, *12*, 035018. [CrossRef]
19. Song, X.; Xie, M.; Hofmann, F.; Illston, T.; Connolley, T.; Reinhard, C.; Atwood, R.C.; Connor, L.; Drakopoulos, M.; Frampton, L. Residual stresses and microstructure in powder bed direct laser deposition (PB DLD) samples. *Int. J. Mater. Form.* **2015**, *8*, 245–254. [CrossRef]
20. Hofmann, F.; Song, X.; Jun, T.-S.; Abbey, B.; Peel, M.; Daniels, J.; Honkimäki, V.; Korsunsky, A.M. High energy transmission micro-beam Laue synchrotron X-ray diffraction. *Mater. Lett.* **2010**, *64*, 1302–1305. [CrossRef]
21. Hofmann, F.; Abbey, B.; Liu, W.; Xu, R.; Usher, B.F.; Balaur, E.; Liu, Y. X-ray micro-beam characterization of lattice rotations and distortions due to an individual dislocation. *Nat. Commun.* **2013**, *4*, 2774. [CrossRef] [PubMed]
22. Suter, R.M.; Hennessy, D.; Xiao, C.; Lienert, U. Forward modeling method for microstructure reconstruction using X-ray diffraction microscopy: Single-crystal verification. *Rev. Sci. Instrum.* **2006**, *77*, 123905. [CrossRef]
23. Moulinec, H.; Suquet, P. A numerical method for computing the overall response of nonlinear composites with complex microstructure. *Comput. Methods Appl. Mech. Eng.* **1998**, *157*, 69–94. [CrossRef]
24. Moulinec, H.; Suquet, P. Fast numerical method for computing the linear and nonlinear properties of composites. *C. R. Acad. Sci. Ser. II* **1994**, *318*, 1417–1423.
25. Michel, J.C.; Moulinec, H.; Suquet, P. A computational scheme for linear and non-linear composites with arbitrary phase contrast. *Int. J. Numer. Methods Eng.* **2001**, *52*, 139–160. [CrossRef]
26. Müller, W. Mathematical vs. Experimental Stress Analysis of Inhomogeneities in Solids. *J. Phys. IV Colloq.* **1996**, *6*, C1-139–C1-148. [CrossRef]
27. Vinogradov, V.; Milton, G.W. An accelerated FFT algorithm for thermoelastic and non-linear composites. *Int. J. Numer. Methods Eng.* **2008**, *76*, 1678–1695. [CrossRef]
28. Anglin, B.S.; Lebensohn, R.A.; Rollett, A.D. Validation of a numerical method based on Fast Fourier Transforms for heterogeneous thermoelastic materials by comparison with analytical solutions. *Comput. Mater. Sci.* **2014**, *87*, 209–217. [CrossRef]
29. Graham, J.T.; Rollett, A.D.; LeSar, R. Fast Fourier transform discrete dislocation dynamics. *Model. Simul. Mater. Sci. Eng.* **2016**, *24*, 085005. [CrossRef]
30. Berbenni, S.; Taupin, V.; Djaka, K.S.; Fressengeas, C. A numerical spectral approach for solving elasto-static field dislocation and g-disclination mechanics. *Int. J. Solids Struct.* **2014**, *51*, 4157–4175. [CrossRef]
31. Eyre, D.J.; Milton, G.W. A fast numerical scheme for computing the response of composites using grid refinement. *Eur. Phys. J. AP* **1999**, *6*, 41–47. [CrossRef]
32. Prakash, A.; Lebensohn, R.A. Simulation of micromechanical behavior of polycrystals: Finite elements versus fast Fourier transforms. *Model. Simul. Mater. Sci. Eng.* **2009**, *17*, 064010. [CrossRef]
33. Lebensohn, R.A. N-site modeling of a 3D viscoplastic polycrystal using Fast Fourier Transform. *Acta Mater.* **2001**, *49*, 2723–2737. [CrossRef]
34. Lebensohn, R.A.; Rollett, A.D.; Suquet, P. Fast fourier transform-based modeling for the determination of micromechanical fields in polycrystals. *JOM* **2011**, *63*, 13–18. [CrossRef]
35. Lebensohn, R.A.; Kanjarla, A.K.; Eisenlohr, P. An elasto-viscoplastic formulation based on fast Fourier transforms for the prediction of micromechanical fields in polycrystalline materials. *Int. J. Plast.* **2012**, *32–33*, 59–69. [CrossRef]
36. Suquet, P.; Moulinec, H.; Castelnau, O.; Montagnat, M.; Lahellec, N.; Grennerat, F.; Duval, P.; Brenner, R. Multi-scale modeling of the mechanical behavior of polycrystalline ice under transient creep. *Procedia IUTAM* **2012**, *3*, 76–90. [CrossRef]
37. Eloh, K.S.; Jacques, A.; Berbenni, S. Development of a new consistent discrete Green operator for FFT-based methods to solve heterogeneous problems with eigenstrain. *Int. J. Plast.* **2018**. submitted.
38. Hirth, J.P.; Lothe, J. *Theory of Dislocations*; Krieger Publishing Company: Malabar, FL, USA, 1982; ISBN 978-0-89464-617-1.
39. Mura, T. *Micromechanics of Defects in Solids*, 2nd ed.; Mechanics of Elastic and Inelastic Solids; Springer: Dordrecht, The Netherlands, 1987; ISBN 978-90-247-3256-2.
40. Li, Q.; Anderson, P.M. A Compact Solution for the Stress Field from a Cuboidal Region with a Uniform Transformation Strain. *J. Elast.* **2001**, *64*, 237–245. [CrossRef]

41. Anderson, P.M. Crystal-based plasticity. In *Fundamentals of Metal Forming*; Wagoner, R.H., Chenot, J.-L., Eds.; Wiley: New York, NY, USA, 1997; pp. 263–293.
42. Takagi, S. A Dynamical Theory of Diffraction for a Distorted Crystal. *J. Phys. Soc. Japan* **1969**, *26*, 1239–1253. [CrossRef]
43. Vartanyants, I.A.; Yefanov, O.M. Coherent X-ray Diffraction Imaging of Nanostructures. *arXiv* **2013**, arXiv:1304.5335.
44. Takagi, S. Dynamical theory of diffraction applicable to crystals with any kind of small distortion. *Acta Crystallogr.* **1962**, *15*, 1311–1312. [CrossRef]
45. Ungár, T. Microstructural parameters from X-ray diffraction peak broadening. *Scr. Mater.* **2004**, *51*, 777–781. [CrossRef]
46. Krivoglaz, M.A. *Theory of X-Ray and Thermal Neutron Scattering by Real Crystals*; Springer: New York, NY, USA, 1969; ISBN 978-1-4899-5584-5.

Article

Dislocation Densities and Velocities within the γ Channels of an SX Superalloy during In Situ High-Temperature Creep Tests

Thomas Schenk [1,2,*], Roxane Trehorel [1], Laura Dirand [3] and Alain Jacques [1,2]

1 Institute Jean Lamour, CNRS UMR 7198, 54011 Nancy, France; roxane.trehorel@univ-lorraine.fr (R.T.); alain.jacques@univ-lorraine.fr (A.J.)
2 Laboratory of Excellence on Design of Alloy Metals for low-mAss Structures (DAMAS), Université de Lorraine, 57073 Metz, France
3 Fives Stein Bilbao, S.A., 48011 Bilbao, Spain; laura.dirand@gmail.com
* Correspondence: thomas.schenk@univ-lorraine.fr; Tel.: +33-372-742-668

Received: 29 June 2018; Accepted: 10 August 2018; Published: 24 August 2018

Abstract: The high-temperature creep behavior of a rafted [001] oriented AM1 Ni-based single crystal superalloy was investigated during in situ creep tests on synchrotrons. Experiments were performed at constant temperatures under variable applied stress in order to study the response (plastic strain, load transfer) to stress jumps. Using two different diffraction techniques in transmission (Laue) geometry, it was possible to measure the average lattice parameters of both the γ matrix and the γ' rafts in the [100] direction at intervals shorter than 300 s. The absolute precision with both diffraction techniques of the constrained transverse mismatch (in the rafts' plane) is about 10^{-5}. After stress jumps, special attention is given to the evolution of plastic strain within the γ channels. The relaxation of the Von Mises stress at leveled applied stress shows evidence of dislocation multiplication within the γ channels. From the analysis, we showed an interaction between plastic stress and dislocation density of the γ phase.

Keywords: nickel-based single crystal superalloy; lattice mismatch; in situ experiments; X-ray diffractometry; creep

---

1. Introduction

Ni-based superalloys are used primarily for manufacturing turbines in power plants and jet and helicopter engines [1]. Their high creep resistance is mainly due to precipitation hardening of the solid solution fcc γ phase by $L1_2$ γ' precipitates [2]. The best creep resistance therefore shows single crystals being used for parts exposed to very high temperatures and stresses.

Under these conditions, cuboidal γ' precipitates evolve into semi-coherent platelets embedded in the softer γ matrix. This results in a quasi-lamellar structure perpendicular to the applied tensile stress σ_a of alternating γ and γ' successive layers. This directional coarsening process is widely known as rafting [3–6].

At high temperatures the hard γ' phase deforms by dislocation climb [7–9], while the softer γ channels are plastically deformed by a glide of the usual fcc $a/2$ <110> {111} dislocations [10]. The high symmetry of the [001] tensile axis results in eight of the 12 available slip systems being equally active. However, the long-term aim of understanding the high-temperature creep behavior of superalloys remains elusive as the behavior of each phase within the rafted microstructure is different to the behavior of single crystals of the same materials and the actual stress and strain state of these phases is generally unknown.

While in the rafted microstructure we can assume that the σ_{zz} component of the stress tensor in each phase is equal to the σ_a applied stress, the $\sigma_{xx} = \sigma_{yy}$ components result from coherence stresses and from the difference in plastic strain between both phases. The coherence stresses are due to a known [11] temperature-dependent natural $\gamma/\gamma\prime$ lattice mismatch defined as:

$$\delta = 2\frac{a_{\gamma'} - a_{\gamma\prime}}{a_{\gamma'} + a_{\gamma}} \quad (1)$$

where a_{γ} and $a_{\gamma'}$ are the free lattice parameters of the unconstrained γ and $\gamma\prime$ phases, respectively. As mobile dislocation segments glide within the channels of the rafted microstructure, they leave straight segments at the intersections of their slip planes and the $\gamma/\gamma\prime$ interface. These dislocations react with each other to form an interface network. This results in a constrained lattice parameter misfit perpendicular to the tensile axis (i.e., in the plane of the interfaces) $\delta_{\perp}$:

$$\delta_{\perp} = 2\frac{a_{\gamma'200} - a_{\gamma 200}}{a_{\gamma'200} + a_{\gamma 200}} = -\frac{b}{d}. \quad (2)$$

This so-called perpendicular misfit $\delta_{\perp}$ (typical values of -3×10^{-3}) equals to $-b/d$ where b is the magnitude of the Burgers vector and d is the average distance between interface dislocations [12]. Depending on the experimental conditions the magnitude of $\delta_{\perp}$ can be lower than, equal to, or larger than the magnitude of δ. The stresses generated by the interface dislocation network can thus decrease, cancel out, or overcompensate for the stresses due to the natural lattice mismatch. As $\delta_{\perp}$ results from dislocations that moved within the γ channels and did not cross the interface, it is also related to the difference between the plastic strain of the two phases.

The perpendicular misfit $\delta_{\perp}$ can be measured by analyzing the diffraction profile of a (200) or (020) reflection using synchrotron X-ray beams in transmission [13]. Such measurements have been carried out by Three-Crystal Diffractometry, and also by Far-Field Diffractometry (i.e., a set-up in Laue geometry, but with the detector put at far field [14]). The high precision of the relative positions of diffraction angles (10^{-5}) was possible due to these techniques. Both the (200) γ and (200) $\gamma\prime$ peaks can be recorded with a single scan/image in a short time span, and it is possible to follow the variations of $\delta_{\perp}$. Using a multilayer model, thoroughly detailed in [15], we can calculate the thermomechanical response of both phases to changes in loading or temperature conditions [15–21].

The aim of the experiments described below was to study the effect of the size of the microstructure (thickness of a γ channel + thickness of a γ' raft) on the mechanical behavior and to use a new diffraction technique with an improved time resolution in order to follow fast transients (a few hundred seconds) in the behavior of the rafts. However, in this paper we shall focus on the behavior of the γ channels. We use the variations with time of $\delta_{\perp}$ during in situ, high-temperature mechanical tests with stepwise loading to show evidence of dislocation multiplication within the γ channels and build a constitutive law for the γ channels.

2. Materials and Methods

2.1. Materials

The experiments were performed using specimens of the AM1 Ni-based single crystal superalloy provided by SNECMA-SAFRAN group (Courcouronnes, France) and by ONERA (Châtillon, France) [22]. Specimens having a [001] orientation within a few degrees were machined out from single crystalline AM1 rods. The samples diameters for samples were chosen as a function of the used X-ray energies to obtain the optimum ratio of diffraction intensity versus X-ray absorption. TCD samples have been submitted to a "*standard*" heat treatment [23], a solution treatment at 1300 °C for 3 h and air quenched with a speed of 10 °C per second, followed by two aging steps. This leads to a structure of well-defined cuboidal regularly arranged $\gamma\prime$ precipitates with an average edge length in the 0.4–0.5 µm range and a maximum creep life. TCD samples are of 3.4 mm in diameter and 29 mm in gauge length.

Samples for Far Field Diffractometry are 2 mm in diameter and 12 mm in gauge length. Both sample geometries are depicted in Figure 1 and verify the DIN 50125 standard [24], since the parallel length of the sample is ≥6 times the diameter, and the neck radius is larger than 0.75 times the diameter of the respective specimen. To obtain a different periodic $\gamma/\gamma\prime$ height of the *"lamellar"* microstructure, casted AM1-samples just annealed at 1300 °C for 3 h and air quenched with a speed of 10 °C per second, but not aged. The resulting cubic $\gamma\prime$ precipitates are more disordered and have an average edge length of 0.25 µm. After rafting, the mean value of the period (sum of the height of one raft and one channel) is lower (623 nm) than that of the *"standard"* heat treatment (893 nm).

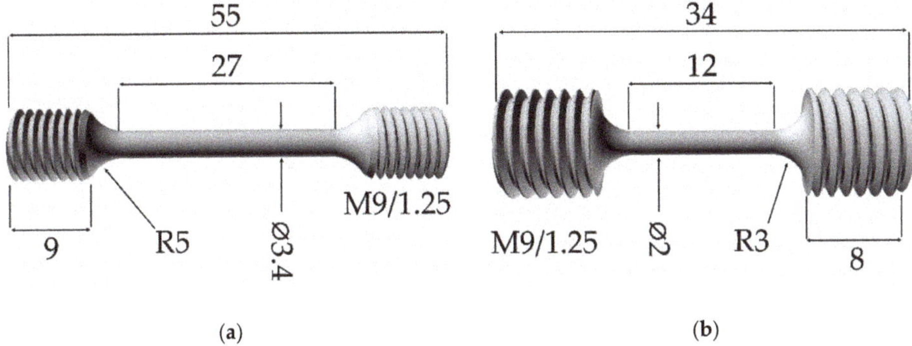

Figure 1. TCD specimen (**a**) and Far-Field Diffractometry specimen (**b**).

These values were measured for a *"standard"* specimen crept at 1000 °C (TCD measurements, on BW5, HASYLAB, Hamburg, Germany) and a *"small"* microstructure specimen crept at 970 °C (Far-Field Diffractometry, on ID11, ESRF, France), which we shall mainly use for illustration.

2.2. High-Temperature Straining Device

The specimens were tested in a "home-made" high-temperature straining device [25]. The sample is radiation-heated by two separate heating zones in a vacuum. The combination of radiation heating and heat losses (via the water-cooled rods, for instance) leads to a near-constant (slightly parabolic) temperature distribution within 5 mm of the sample center. The elongation of the sample is measured by a LVDT extensometer with a resolution of 0.5 µm and the applied force is recorded by a 5 kN force cell.

Temperatures are measured with 3 K-type thermocouples: As shown in Figure 2, two are positioned on the upper and lower grip, and are for temperature regulation, and the third is on the center of the sample. A stepper motor with a high gear ratio is coupled to the mobile rod, driving the upper grip of the straining device. All parameters (heating, force, strain rate of the stage) are continuously recorded via a LabVIEW^{TM} program.

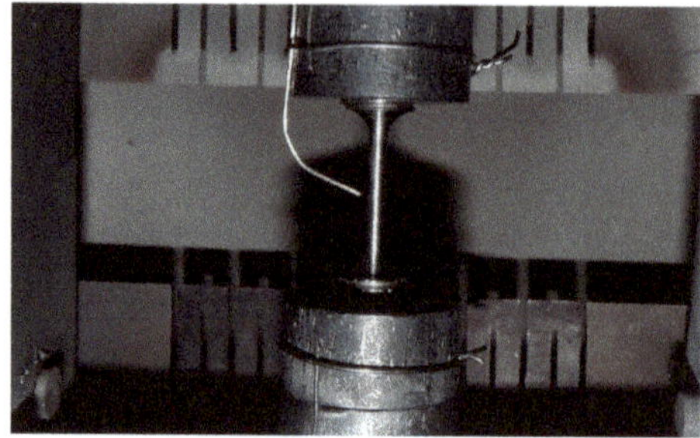

Figure 2. Position of the three thermocouples.

The parabolic temperature profile is systematically checked by diffraction at the beginning of each test. The beam position along z corresponds to the hottest part (lowest Θ_B) of the specimen.

2.3. In Situ XRD

TCD and Far-Field Diffractometry have been thoroughly compared in [14,26].

For in situ creep tests (in the "home-made" high-temperature straining device), the bulk sample is placed with the high-temperature straining device on the diffractometer. The monochromatic beam travels through the specimen and the single crystal samples are rotated along (the vertical [001] tensile axis) to bring the (200) lattice plane into Bragg condition.

TCD experiments using three crystals in a non-dispersive (+, −, +) geometry were done at 120 KeV ($\lambda = 1.033 \times 10^{-11}$ m) at DeSy BW5 [13] using an energy selective Ge detector. A 1D intensity profile of a (200) peak scan of the *"standard"* sample (Figure 3) is typically obtained in 300 s.

Furthermore, we investigated creep of samples using Far-Field Diffractometry at ESRF ID11, at 67 KeV ($\lambda = 1.851 \times 10^{-11}$ m) with a 2D 2048 × 2048 pixel FReLoN Camera with a pixel size of 50 μm × 50 μm as detector. The camera was placed (as far as possible) at 8.5 m from the specimen. From the recorded images, we obtained 1D integrated intensity profiles along 2θ by adding up the pixels' intensities of the detector along the direction perpendicular to the diffraction plane. Acquisition times for a (200) peak using Far-Field Diffractometry is reduced to seven seconds permitting to follow fast transients.

For both experiments the gauge volume is large enough to probe both dendritic and interdendritic zones.

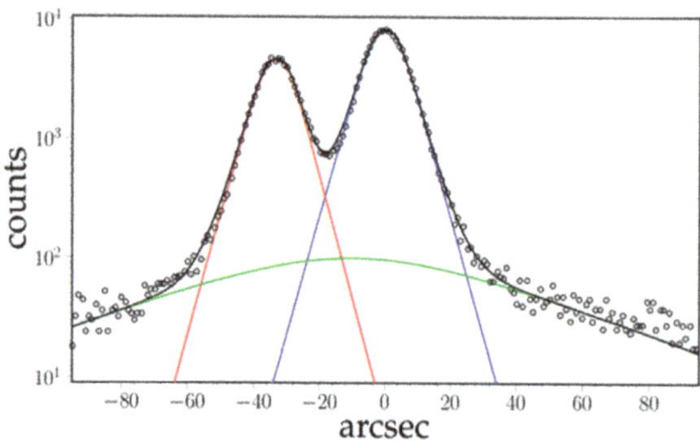

Figure 3. Typical X-ray data and fit of TCD intensity profile for the two peaks of phases γ (red) and $\gamma\prime$ (blue).

In Figure 3 a typical TCD intensity plot is shown together with the profile fitting functions. All diffraction profiles (independently of the applied diffraction set-up) were fitted as the sum of three symmetric peaks, corresponding to the $\gamma\prime$ rafts (blue in Figure 3), the γ channels (red in Figure 3), and a wide tail for the background (supposed to be the contribution of the distorted zones near dislocation cores).

An automated procedure using a dedicated ad hoc fitting function (Equation (3), [18]) for both γ and $\gamma\prime$ peak-shapes is used:

$$P_i(\alpha) = A_i \exp\left[1 + \left[1 - \left(\frac{|\alpha - \alpha_{0i}|}{w_i}\right)^{\beta_i}\right]^{\frac{1}{\beta_i}}\right]. \tag{3}$$

The fit parameters are the peak positions α_{0i}, the peak heights A_i, powers $\beta_i > 2$ that define the sharpness of the peaks, and the profile parameters w_i defining the slope of Napier logarithm of the peaks. The centers of the fitted γ and $\gamma\prime$ peaks are used to calculate the perpendicular misfit $\delta_\perp$, and the area under the peaks can be used to evaluate the phase fraction f'_{XRD} of the $\gamma\prime$ rafts (assuming the same structure factors for both phases).

2.4. Post Mortem TEM and SEM Studies

At the end of the in situ X-ray experiments, crept samples were cooled down fast, cutting the heating under the load to freeze their microstructure (thickness of rafts and channels, dislocation densities). TEM (FEI/Philips CM200, Eindhoven, The Netherlands) please add the information and SEM (Philips/FEI XL30 S-FEG, Eindhoven, The Netherlands) micrographs were taken of both samples. All samples exhibit a classical rafted microstructure (Figure 4). The thicknesses of channels and rafts were measured with an intercept method.

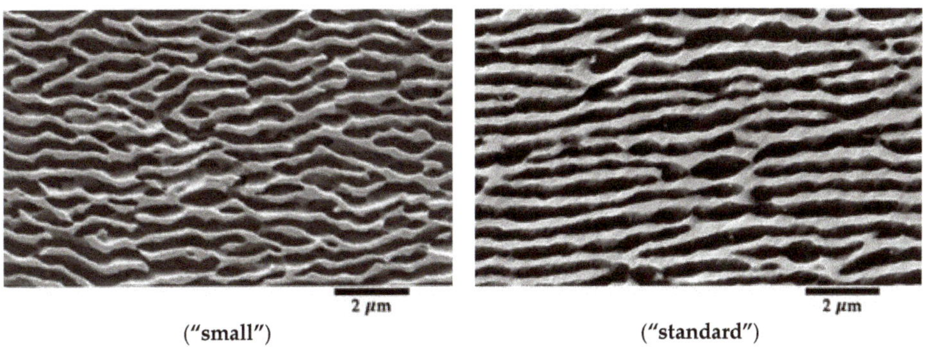

Figure 4. SEM micrographs of two specimens with different initial microstructure at different temperatures: 970 °C ("**small**" microstructure), 1000 °C ("**standard**").

Since the equilibrium phase fraction $f\prime$ of γ' at room temperature is larger than at high temperatures, small γ' cubes precipitate within the γ-channels during cooling. The $\gamma/\gamma\prime$ interfaces are also expected to migrate during cooling. The thicknesses of the channels measured *post mortem* samples thus slightly underestimate its value at high temperature. However, the average period of the microstructure (one raft and one channel) does not change during cooling.

2.5. Data Analysis

The plasticity of the γ-phase (matrix) of an AM1 superalloy is studied by evaluating stress/plastic strain curves obtained by similarly following temperature, total elongation, and (2D) diffraction patterns of the (200) γ/γ' planes. The raw relative precision in peaks shifts due to changes in temperature or stresses, as well as the difference of the parameters of γ and γ' peaks after fitting the intensity profiles is 10^{-5}, much better than the precision of the applied force, the temperature, and the elongation of the specimen.

Assuming that the specimen's volume remains constant during plastic strain we correct the applied stress σ_a using the total plastic deformation $\varepsilon^{pl}_{zz,tot}$ of the sample:

$$\sigma_a = \left(1 + \varepsilon^{pl}_{zz,tot}\right)\frac{F_a}{S_0}, \tag{4}$$

where F_a is the applied force and S_0 the initial section of the specimen.

The evolution of the lattice parameters in the [100] direction allows us to deduce the elastic strains of the γ channels and the γ' rafts and also the average elastic strain in the [100] direction. With the multilayer model given in [15], it is possible from this recorded creep data to calculate with a good precision both the main components of the stress tensor for each phase and the strain of each phase, provided that the actual value of the perpendicular misfit $\delta_\perp$ is known.

The following approximations are used:

- homogenous and isotropic phases
- same Poisson coefficient $\nu = 0.42$ [4]
- distinct Young's moduli: E (γ), E$'$ (γ')
- symmetric stress tensors (i.e., $\sigma_{xx} = \sigma_{yy}$).

The transverse stress σ_{xx} within the channels is given by [15]:

$$\sigma_{xx} = -\frac{\chi E}{1-\nu}\left[(\delta_\perp - \delta) + \left(\frac{1}{E'} - \frac{1}{E}\right)\nu\sigma_a\right] = A\cdot(\delta_\perp - \delta) - B\cdot\sigma_a, \tag{5}$$

$$\text{where } \chi = \frac{E'f'}{E(1-f') + E'f'}. \tag{6}$$

Equation (5) allows us to calculate the von Mises stress:

$$\sigma_{VM} = \sigma_a - \sigma_{xx}, \tag{7}$$

which is a function of parameters $(\nu, \sigma_a, f', \delta_\perp - \delta, E, E')$. ν, δ are known, $\delta_\perp$, $f' = f'_{XRD}$ can be measured and the evolution of E, E' is given in [15].

The plastic strain of the whole specimens $\varepsilon^{pl}_{zz,tot} = (1-f').\varepsilon^{pl}_{zz} + f'.\varepsilon'^{pl}_{zz}$ was deduced from the variations of $\Delta\updownarrow/\updownarrow_0$ of the instantaneous elastic variation during loading and unloading at constant temperature. Still assuming a constant volume and as discussed above, the perpendicular misfit $\delta_\perp = \left(\varepsilon'^{pl}_{zz} - \varepsilon^{pl}_{zz}\right)/2$ resulting from the difference of plastic strain between both phases, we obtain the strain of the γ channels as follows:

$$\varepsilon^{pl}_{zz} \approx \varepsilon^{pl}_{zz,tot} - 2f'\delta_\perp. \tag{8}$$

3. Results

3.1. Real-Time Experiment with Far-Field Diffractometry

The aim of the experimental procedure described below was first to obtain well-rafted microstructures with a different wavelength, then to study the response of both phases to changes in their stress state. Oriented coalescence of the γ' precipitates (i.e., rafting) results from the difference in stored elastic energy between the different γ channels in the initial cuboid microstructure [5,6]. In superalloys with a negative lattice mismatch, during tensile creep tests, the 'horizontal' channels perpendicular to the tensile axis are plastically strained while the 'vertical' channels remain nearly dislocation-free. As the coherence stresses are partly relaxed by plastic strain in the 'horizontal' channels their stored energy becomes lower than that of the 'vertical' channels: the horizontal channels widen, while the vertical channels become thinner and disappear. The pre-strain load must be chosen so that dislocations can glide within all the horizontal channels, i.e., the total stress on dislocations (applied stress + coherence stresses) must be larger than the Orowan stress. As the Orowan stress increases when the thickness of the channels decreases (about 160 MPa for the *"small"* microstructure vs. 95 MPa for the standard one), the pre-strain load was increased from 150 MPa to 200 MPa. As the horizontal channels widen during rafting, the final Orowan stress is lower. Last, former experiments [15] showed a strong dependence of the strain rate of the γ' rafts vs. the internal stress $-\sigma\prime_{xx}$. In order to investigate this behavior and its dependence on the microstructure, the experimental conditions were chosen so that $-\sigma\prime_{xx}$ would vary between 0 MPa and about 100 MPa. The specimen with an initial *"small"* microstructure was first heated to 970 °C, then put under a 200 MPa load. The (200) diffraction profile was continuously followed during the test. It evolved from the one peak distribution (mainly $\gamma\prime$) with a bump on the lower 2θ side (mainly γ) observed for the cuboids microstructure [17] into a well-defined two peaks profile as seen in Figure 3 for a rafted microstructure. After 13 h the rafting was deemed to be complete, and the tests began using the same procedure as in [15] (Figure 5). The specimen was submitted to a stepwise loading first to 260 MPa, then by 10 MPa positive steps up to 300 MPa (each step lasting 10 min), then unloaded to 200 MPa (corresponding to a corrected applied stress of 212 MPa). Those first loadings are reported in Figure 4a with greater detail). After a ~1 h relaxation further upwards and downwards steps under different loads followed by a relaxation were done. The specimen was then cooled down to room temperature under a 230 MPa stress.

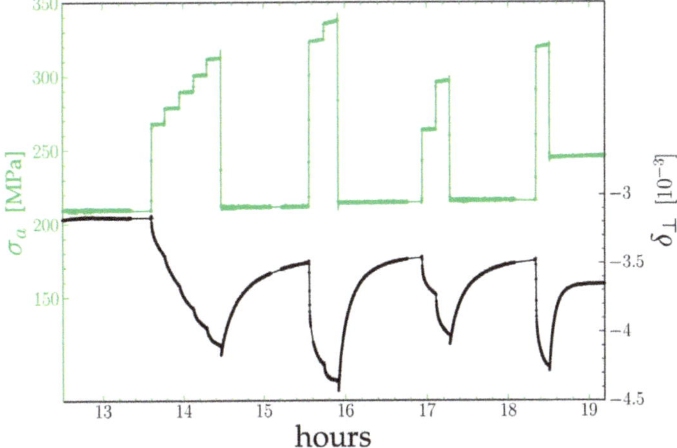

Figure 5. Evolution of the perpendicular misfit $\delta_\perp$ (black) with the corrected applied stress σ_a (green) of the "small" AM1 sample at 970 °C. Each dot corresponds to a seven seconds diffraction peak recording.

At each positive stress jump both γ and $\gamma\prime$ peaks shift to a larger 2θ angle by the same amount ($\delta_\perp$ does not change.), as the transverse lattice parameters of both phases decrease instantaneously according to the material's Poisson's ratio: This is the elastic response. Following this shift, the distance between both peaks increases rapidly, then at a slower pace: $\delta_\perp$ becomes increasingly negative. Each downwards stress jump first results in a dip of $\delta_\perp$: as the length of the tensile rod decreases the specimen moves downwards along the z axis by a few tens of micrometers and the specimen's volume illuminated by the 50×50 μm^2 X-ray beam changes (This effect is also present during positive steps but is masked by the fast decrease of $\delta_\perp$). The misfit then increases in algebraic value until it settles after one hour.

At each measurement point the σ_{xx} stress component has been calculated using Equation (5), then the Von Mises stress $\sigma_a - \sigma_{xx}$. The plastic strain was obtained using Equation (8). Both are plotted on Figure 6b. As it can be seen, each upwards step of the applied stress results in an equivalent step in the Von Mises stress. The Von Mises stress then decreases first fast, then more slowly as the plastic strain of the rafts increases, $\delta_\perp$ decreases algebraically, and the σ_{xx} stress component (i.e., the back stress due to interface dislocations) increases: The Von Mises stress is obviously larger than the Orowan stress. The opposite is observed after a downwards stress step, but as the plastic strain of the channels decreases only slightly the increase of $\delta_\perp$ results from the plastic strain of the rafts. Following a downwards stress step, the strain rate of the channels is first negative (arrow), then zero, and later becomes positive even before the next positive stress jump.

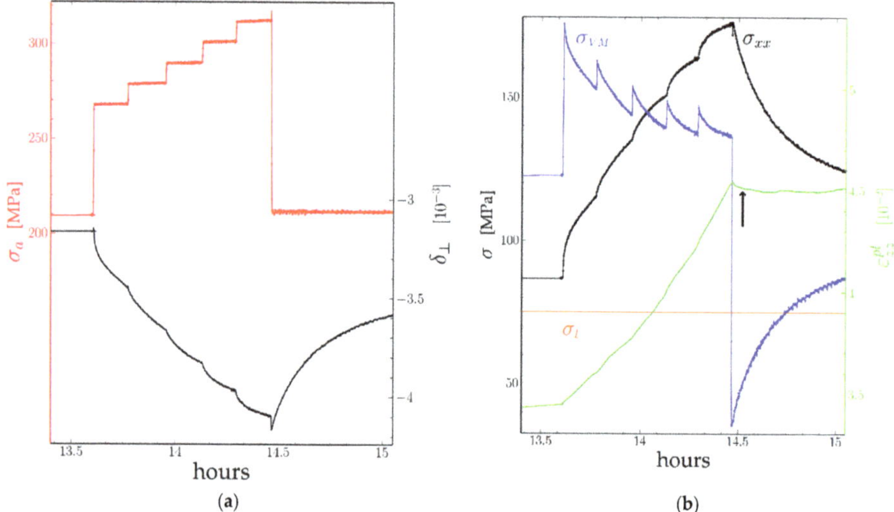

Figure 6. Detail between 13 h 30 and 15 h of the creep test of a sample with an initial "**small**" microstructure at 970 °C: (**a**) applied stress σ_a (red), $\delta_\perp$, (black), **b**) σ_{xx} (black) σ_{VM} (blue) and plastic strain ε_{zz}^{pl} in green.

An interesting point is the evolution of the stress and strain rates during the initial stepwise loading between 13 h 30 and 14 h 30. While each step results in an increase then a decrease of the Von Mises stress, the global tendency is a decrease of the Von Mises stress. On the contrary, the plastic strain rate increases. This point will be discussed below.

3.2. *Successive Relaxation Tests*

Figure 7 shows the evolution of the Von Mises stress of a "**standard**" AM1 sample recorded by Three-Crystal Diffractometry (with peak recording lasting 300 s instead of 7 s).

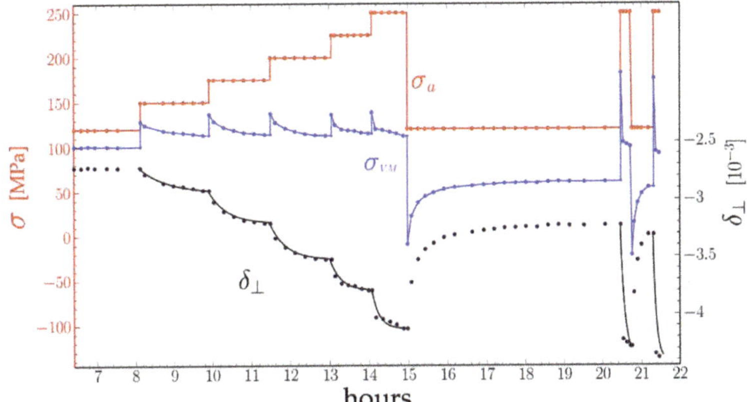

Figure 7. Creep test of a "standard" sample at 1000 °C: applied stress σ_a (red), calculated σ_{VM} (blue) and in black, $\delta_\perp$ measured (points) and calculated (lines) for an exponential decay of $\delta_\perp$ for the different decay times given in Table 1.

The same procedure as before was qualitatively followed: heating to 1000 °C then loading to 120 MPa until complete rafting, and stepwise loading and unloading. The evolution of the applied stress σ_a, the perpendicular misfit $\delta_\perp$ and the calculated Von Mises stress σ_{VM} are shown in Figure 7. The general shape of these functions follows the same behavior as that described in Section 3.1 with some differences detailed below.

The first five 25 MPa stress jumps at levels from 120 MPa up to 250 MPa were performed within 7 h. During the one-hour span following a loading step, $\delta_\perp$ had time to settle at increasingly negative levels and the Von Mises to return each time to a 113 MPa level. The strain rate of the channels, as deduced from the leveling of $\delta_\perp$, then fell to 0. The evolution of $\delta_\perp$ was approximately exponential, but as can be seen from the figure the evolution was initially quite slow then increasingly fast: the characteristic relaxation times τ_i decrease with each step i.

4. Analysis of the Experimental Results

4.1. Base of Modelling

The strain rate $\dot{\varepsilon}_p$ of a material is commonly given by the Orowan equation:

$$\dot{\varepsilon}_p = s.\rho_m.b.v,$$

where s is the Schmid factor (here: $1/\sqrt{6}$), ρ_m the density of mobile dislocation segments, b the magnitude of the Burgers vector, and v the velocity of the dislocation segments. In the following, we shall try to separate the respective contributions of the density and the velocity of dislocations.

Dislocation glide in γ channels will take place, if the Von Mises stress σ_{VM} is large enough to overcome a threshold stress stress σ_t, generally assumed to be the Orowan stress. The glide velocity v of the dislocations should be a unique increasing function of the effective resolved shear stress τ^* they can feel:

$$v = f(\tau^*) = f\left(\frac{1}{\sqrt{6}}(\sigma_{VM} - \sigma_t)\right) for\ \sigma_{VM} > \sigma_t. \quad (9)$$

The Orowan stress [11] results from the necessity for a dislocation segment to leave one segment at each γ/γ' interface on both sides of a channel. If the Von Mises stress becomes lower than the threshold stress the moving segment may bow in the opposite direction, and perhaps move backwards on a limited distance by reabsorbing these segments. However, as these are supposed to climb within the interface and react with the preexisting network of dislocations [27], only a short length can be reabsorbed. Further backwards glide would require the creation of new interface dislocations with the opposite sign, i.e., a Von Mises stress with the opposite sign and a magnitude larger than the threshold stress. As can be seen from our measurements (Figure 6b), the backwards strain is quite limited (in the 5×10^{-4} range).

We may remark that as different channels have different thicknesses they also have different Orowan stresses. However, the plastic strain should stop when the local Von Mises stress (the applied load minus the local σ_{xx} stress due to the local density of interface dislocations) is equal to the local Orowan stress. One hour after a first step, as in Figure 7, the local equilibrium has been reached for all channels, and Equation (9) will remain correct at the local level.

4.2. Determination of the Threshold Stress

In Equation (9) we define the Orowan stress as the stress for which the dislocation velocity is zero and above which it is positive. In the case of the TCD tests (Section 3.2) this is obviously the 113 MPa level at which the Von Mises settles after a load step. An estimation of this stress in [11], taking into account the average thickness of the channels and the elastic constants determined at high temperature indeed gives about 100 MPa. The case of the Far-Field Diffractometry test (Section 3.1) is more complicated, as the relaxation of the Von Mises stress following a load step is never complete. However, during a downwards load step, the Von Mises stress decreases by the same amount and the strain rate becomes negative (Figures 6b and 8). As the Von Mises stress re-increases the strain rate becomes positive again. However, following a downwards step the Von Mises stress increases, while the strain rate becomes positive. If we determine the moment at which this change of sign takes place then look at the value of the Von Mises stress at this moment, we obtain 75 MPa. (It may be noticed that, without the applied load correction (Equation (4)) the result would have been decreasing stress.)

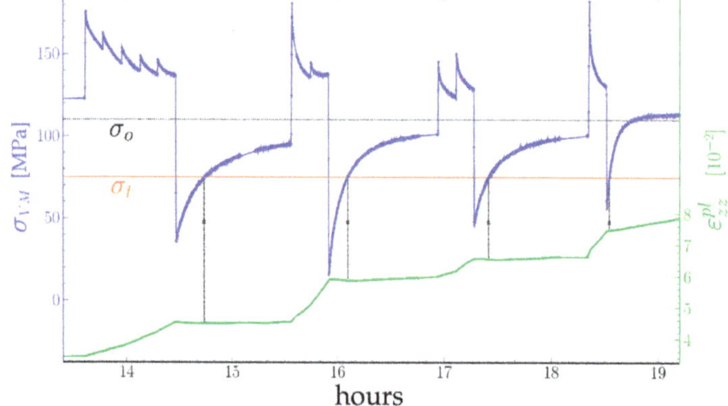

Figure 8. Creep test of the AM1 sample with short period at 970 °C σ_{VM} stress (blue), expected Orowan stress σ_O (black), apparent threshold stress σ_t (orange) and plastic strain ε_{zz}^{pl} in green.

This 75 MPa threshold stress σ_t is much lower than the expected Orowan stress, being evaluated to 101 MPa for this "**small period**" sample if compared to σ_O of the "standard" sample. We can exclude such a large error in the measurement of σ_{VM}. Hence σ_t cannot be σ_O and we expect a different deformation mechanism than pure Orowan-limited dislocation glide for those low stresses.

4.3. Dislocation Velocities and Densities

During the successive relaxations of Figure 8 the Von Mises stress had several times the same value. At these points, the velocity of dislocations should be the same. However, as can be seen in Figure 9, the strain rate changes.

In Figure 9, we plot the value of the strain rate for different levels of the Von Mises stress as a function of the plastic strain. As it can be seen, the points fall along straight lines. The slope S of these lines (obtained by linear regression) increases with the level of the Von Mises stress. The first two lines (140 MPa and 130 MPa, measurements between 13 h 30 and 14 h 30) intersect near the origin of the plot, while the latter lines (130 MPa, 124 MPa and 95 MPa for measurements after 16 h) intersect the plastic strain axis at $\varepsilon_{zz}^{pl} \approx 0.02$. If we admit that the dislocation velocity was the same for all points along a line, the only explanation is that the dislocation density increases linearly with the strain, and the initial dislocation density (at the beginning of the test) was quite low. The shifted lines could then be explained by a drop of the dislocation density, probably at (15 h 55) the instant of the deepest drop of the Von Mises stress during the test.

We can thus rewrite the Orowan law as:

$$\dot{\varepsilon}_{zz}^{pl} = S(\sigma_{VM}) \cdot \left(\varepsilon_{zz}^{pl} - \varepsilon_0^{pl}\right). \tag{10}$$

The constant ε_0^{pl} would be very small before (15 h 55), and 0.02 afterwards. The slope $S(\sigma_{VM})$ should then be proportional to the dislocation velocity and should be a unique function of the Von Mises stress.

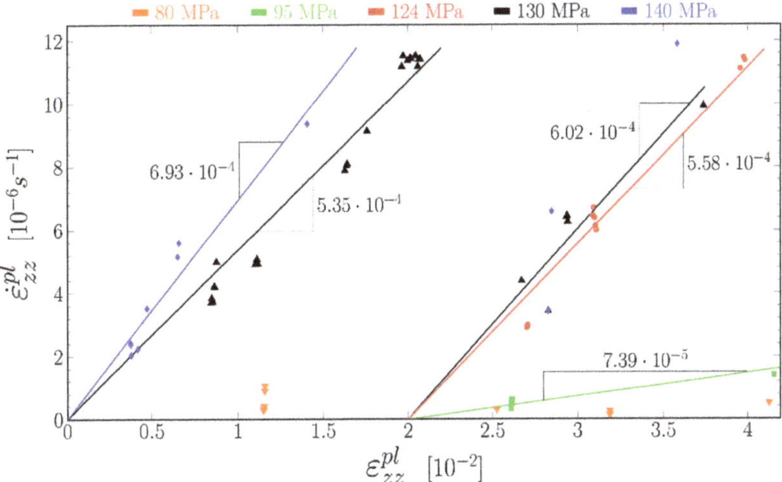

Figure 9. Strain rate of the γ channels as function of their strain for equal Von Mises stresses.

In Figure 10 we plot the ratio between the plastic strain rate and the $\varepsilon_{zz}^{pl} - \varepsilon_0^{pl}$ factor (i.e., $S(\sigma_{VM})$) as a function of the Von Mises stress. Despite some time-averaging (60 s) there is some scatter in the data, especially for very low stresses (<50 MPa) and very high stresses. As stated above this noise results mainly from the noise in the measure of the specimen elongation. The points at the lowest and highest stresses were recorded during specimen loading or unloading when we expected the largest errors on our experimental data. However, the whole set of points clearly settles along two lines (black). At low stresses (from 50 to 110 MPa) the behavior is linear, with a 2.3×10^{-6} $s^{-1} \cdot MPa^{-1}$ slope. The best fit goes through the stress axis for σ_t = 78 MPa. All these low stress points were recorded within the hour following a load drop, as the Von Mises stress was increasing again. The high Von Mises stress points (>120 MPa) were recorded during the stepwise upwards loading. They follow a line with a 2.02×10^{-5} s^{-1}·MPa^{-1} slope. The transition between these two behaviors cannot be seen as we lack data points in the 110 MPa < σ_{VM} < 120 MPa domain.

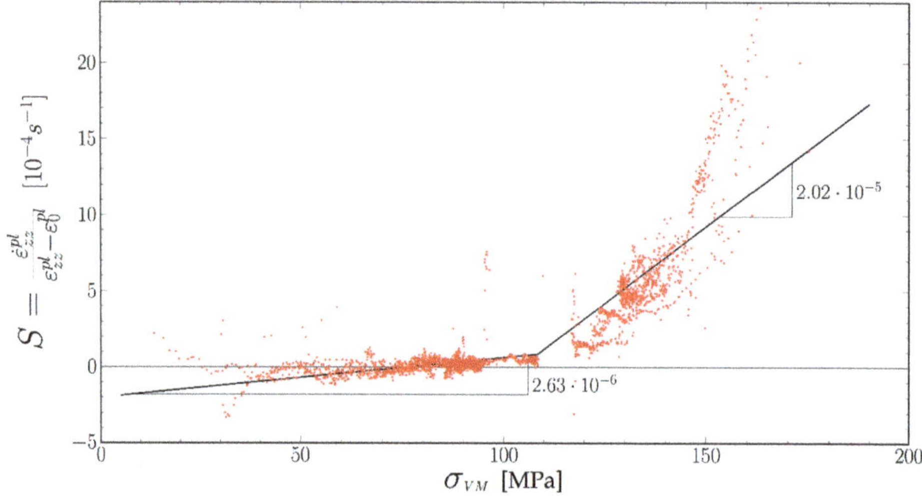

Figure 10. Strain rate divided by strain as function of Von Mises stresses.

In order to check the accuracy of the above modelling, we performed a simulation of the evolution of the plastic strain during the whole test. The cumulative plastic strain was calculated by integrating stepwise for an integration time of 10 s the strain rate and the plastic strain, using the two slopes S-factor adjusted from Figure 10, and an initial strain-rate of 0 before 13 h. Each time segment between a load drop and the next was taken independently, with the same $S(\sigma_{VM})$ law but different initial ε_0^{pl} factors in order to allow for losses of dislocations during the stress drops. Such a drop of ρ_m calculated as $\rho_m = \varepsilon_{zz}^{pl} - \varepsilon_0^{pl}$ is indicated with an arrow on Figure 11, corresponding to an increase of the ε_0^{pl} constant of 0.02 at 15 h 55 during the most important σ_{VM} release. The resulting plastic strain curve is compared with the experimental one in Figure 11.

As can be seen in Figure 11, the evolution of the strain is well reproduced both in the low and high Von Mises stress regimes.

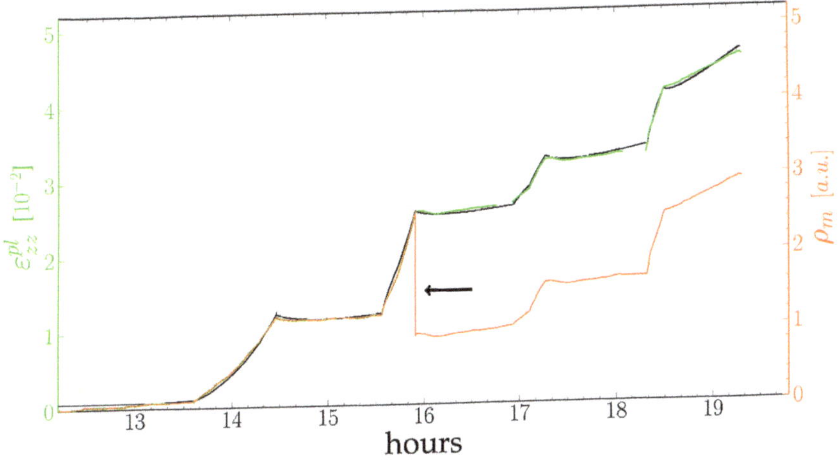

Figure 11. Plastic strain measured (green) and calculated (black) and dislocation density (orange). Arrow: drop of the dislocation density at 15 h 55.

4.4. Successive Stress Relaxations and Activation Energy

The long recording time of TCD peaks does not allow us to follow the variations of the strain rate during a single loading step as in the above paragraph. However, as a full set of data is available between 1000 °C and 1125 °C it can be used to observe variations due to temperature. The evolution of the perpendicular mismatch recorded in Figure 7 was put into equations in [28].

As during the tests, the difference between the Von Mises stress and the Orowan stress varies first from zero to 25 MPa during a load step then from 25 MPa to zero during the following relaxation. The dislocations' velocities are assumed to vary with the dislocation mobility μ as: $v = \mu \cdot (\sigma_{VM} - \sigma_O)$ (we neglect the low-stress regime). The relaxation time should be inversely proportional to μ and to ρ_m. The results of the successive exponential fits are given in Table 1 and Figure 12 shows the variation of τ^{-1} the inverse of the relaxation time vs. the plastic strain of the rafts ε_{zz}^{pl} for the TCD specimen. As can be seen, the variation is linear with a 0.065 slope: the increase in dislocation density is proportional to the strain increment.

Table 1. Adjusted parameters for calculating $\delta_\perp$ (Figure 12) of "standard" specimen.

Jump n°	σ_a (MPa)	$-\delta_f$ (10^{-3})	ε_{zz}^{pl} (10^{-3})	$\dot{\varepsilon}_{zz}^{pl}$ (s^{-1})	τ (s)
1	150	2.9406	0.4	3.0×10^{-8}	2700
2	175	3.2091	1	7.0×10^{-8}	1700
3	200	3.52515	2.2	3.0×10^{-7}	1500
4	225	3.79425	5	1.24×10^{-6}	1000
5	250	4.12654	14	5.0×10^{-6}	650
6	250	4.40814	28	1.2×10^{-5}	420
7	250	4.40813	44	1.50×10^{-5}	300

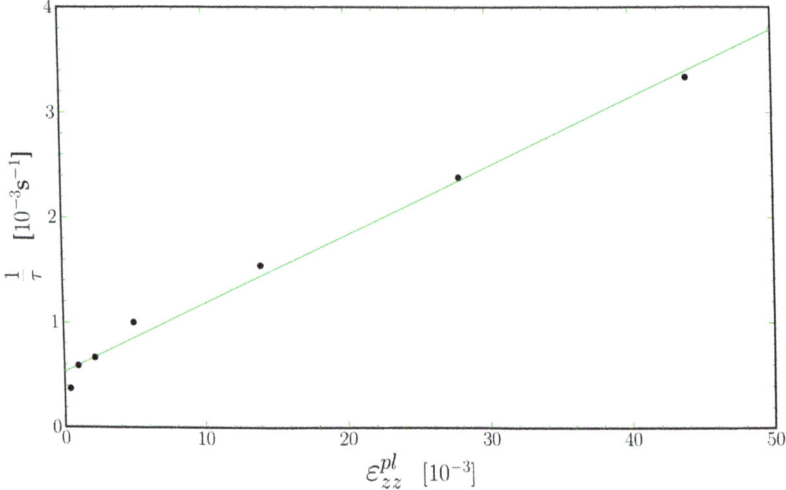

Figure 12. Variation of the relaxation time τ as function of the plastic deformation ε_{zz}^{pl} of the γ phase during a creep test of "standard" sample.

The slope should be proportional to the dislocations' mobility μ. If we now plot the slopes obtained for different specimens at different testing temperatures, reported in Table 2 on an Arrhenius plot (Figure 13), they seem to fit on a straight line.

Table 2. Slope $\Delta(1/\tau)/\Delta\varepsilon_{zz}^{pl}$ for different "standard" specimens [28].

Sample	5A1	4L2	5C1	1C2	4R2
T	1000	1049	1066	1119	1125
slope	0.0651	0.121	0.43	1	0.55

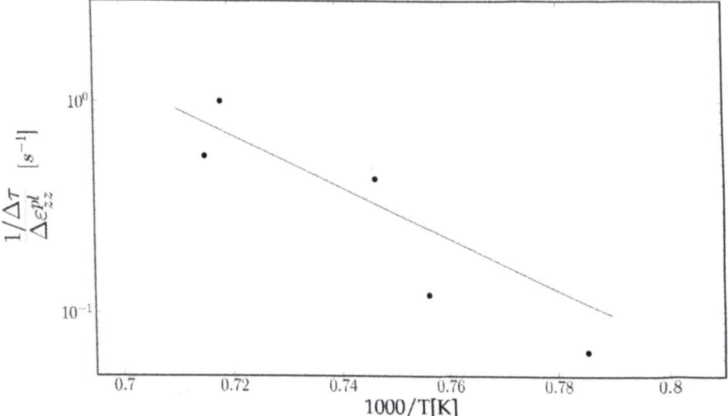

Figure 13. Arrhenius plot of $\Delta(1/\tau)\Delta\varepsilon_{zz}^{pl}$ as function of the inverse temperature for deformations after stress releases following small stress jumps.

The corresponding activation energy is (with some incertitude) in the 3 eV range. This value is of the same magnitude as the 3.26 eV apparent activation energy given in [29] and 3.7 eV [30] for dislocation glide within the channels during Stage I of creep.

5. Discussion

Two points will be discussed in the following: the shape of the dislocation velocity vs. Von Mises stress (Figure 7), and the mechanisms of dislocation multiplication and annihilation.

5.1. Dislocation Velocity Law

Even if it gave the best fit to our data (much better than a power law or a hyperbolic sine) the apparent velocity law with two slopes and two thresholds found in Figure 10 raises a question: the 78 MPa Orowan stress does not fit with its expected value (101 MPa) and is much lower than the values found in [15], while the 108 MPa threshold is more in line with those. The strain rate of the channels measured after each unloading to 200 MPa was in the 10^{-7} range. This is low, but as the same stress level was kept for nearly one hour, the specimen elongation was about 20 μm, much larger than the error bars. Besides, the dislocation density increases slightly during that time and is consistent with that of the next high stress step: we can exclude a large error in the measurement of the strain rate.

The theoretical Orowan stress has been calculated assuming that a moving dislocation segment needs to leave one segment at each interface and that the work of the force on the gliding dislocation segment is equal to the energy needed to create these two dislocation segments (Figure 14a).

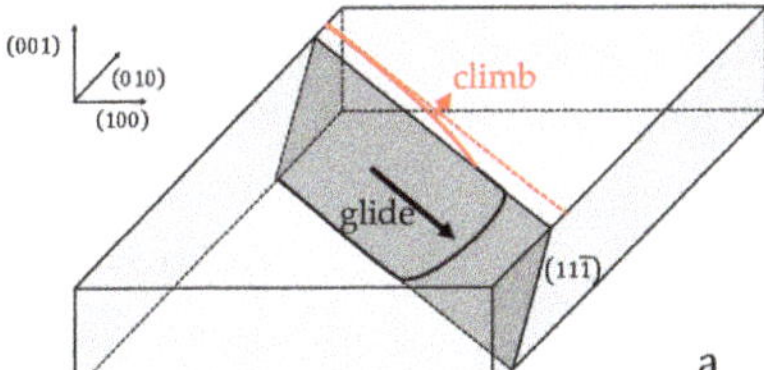

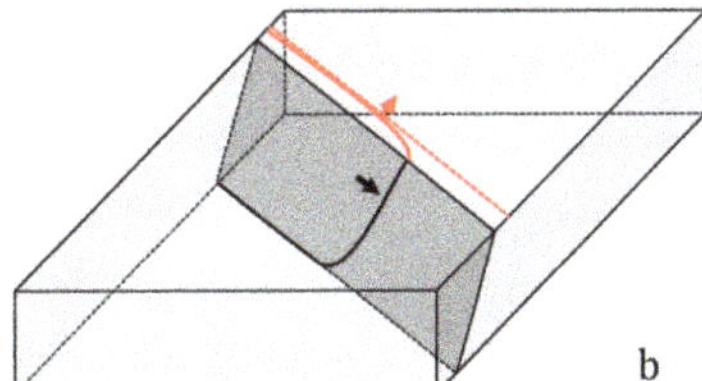

Figure 14. A dislocation glides within (**a**) $(11\bar{1})$ plane of a γ channel leaving a dislocation segment at each intersection of its glide plane and the γ/γ' interface. The segment at the top interface (red line) later climbs within the interface to react with another interface dislocation (red dashed line). (**b**) If the glide velocity is low the interface segment immediately bows into the interface and no extra energy is needed to create it.

These segments later climb within the γ/γ' interface until they react with other dislocations of the interface network and decrease their line energy. However, if the glide velocity of the dislocation is of the same order of magnitude as its climb velocity (Figure 14b) the interface segment can bow immediately therein: it is no longer necessary to provide its line energy.

The dislocation can thus glide slowly within the rafts under a Von Mises stress much lower than the theoretical Orowan stress. If then the Von Mises stress goes past the Orowan stress, the mobile dislocation segment can glide freely within a channel and leave the reacting segments far behind.

5.2. Dislocation Multiplication and Annihilation

The layered microstructure of a superalloy with a rafted microstructure presents some analogies with pearlite [31] and the Persistent Slip Bands (PSBs) observed in fcc single crystals during oligocyclic fatigue tests [32,33], and we may expect some common features in their behavior. During fatigue tests under single slip conditions, dislocations are supposed to bulge out of the PSBs upon stress reversal and screw segments to glide within the channels until they get trapped by another screw segment gliding in the opposite direction in a parallel slip plane. Both segments then annihilate by cross slip. In pearlite, a specific mechanism of dislocation multiplication ("scolopendra" sources) was observed in [31]: dislocations reaching the end of a channel between two cementite islands can propagate into neighboring channels. As in the present case we have no stress reversal and dislocations segments from the interface network are probably unable to bulge back into the channels. The multiplication mechanism should thus be of the scolopendra type. As the dislocations gliding within the channels belong to all of the eight active slip systems we also may expect different kinds of annihilation reactions to occur: reactions between dislocations of the same slip systems as in fatigue, reactions between dislocations from coplanar slip systems as well as between dislocations gliding in secant planes. However, we shall assume here that two dislocations annihilate if they pass within a distance shorter than a capture radius y of each other. (This capture distance is about 50 nanometers for screw dislocations in parallel slip planes. It should be of the order of magnitude of the channel width for dislocations moving in secant planes.) The balance equation for dislocation density as a dislocation moves by a distance dx [31] should thus be:

$$d\rho = \rho\frac{dx}{L} - 2y\rho^2 dx = \frac{d\varepsilon}{sbL} - 2y\rho\frac{d\varepsilon}{sb}, \quad (11)$$

where L should be of the same order of magnitude as the rafts' length (about 5 μm), s is the Schmidt factor for all active slip systems and b the magnitude of the Burgers vector.

As in the beginning of our experiment the initial dislocation density is low, the annihilation term should be negligible and the dislocation density proportional to the strain:

$$\rho = \frac{\varepsilon}{sbL} \sim 3.10^{14}\varepsilon. \quad (12)$$

Using Equation (10), we thus find a relation between the dislocation velocity and factor $S(\sigma_{VM})$:

$$v(\sigma_{VM}) = L \cdot S(\sigma_{VM}).$$

The range of dislocation velocities in the γ channels is $\sim 5 \times 10^{-9}$ m·s^{-1}. After a 1% strain, the dislocation density within the channels should be 3×10^{12} m^{-2}: this is one order of magnitude higher than actually observed by TEM [28]. However, at the end of the rafts a larger density of dislocations is observed at points where several channels interconnect: these dislocations did apparently glide along a channel and are waiting to enter the next ones. The dislocation velocity law given above is thus an average velocity law, much lower than the actual glide velocity within the channels.

These dislocations queuing up to enter the channels might also explain an apparent paradox concerning dislocation annihilation. Following Equation (11), the dislocation density should saturate for

$$\rho_{sat} = \frac{1}{2yL}.$$

Assuming the capture radius y in the 10^{-7} m range saturation should occur at $\rho_{sat} \sim 10^{12}$ m^{-1}, for a 0.3% strain. This was not observed here (Figure 11). However, if the dislocation distribution is uneven, and the actual dislocation density within the channels is lower than the average density, the probability of meeting of two dislocations gliding within a channel is small and the rate of mutual annihilation lower than expected from Equation (12). Here, we have to assume that the dislocations waiting at the entry of a raft do not annihilate each other, either because their distance is larger than y or because they are bending at the entry of a channel. Under a load drop, as the Von Mises stress is near 0, these dislocations can move backwards under their mutual stress field and annihilate each other (Figure 11 at 15 h 55).

6. Conclusions

Using a combination of high-resolution X-ray diffraction measurements and in situ, high-temperature creep tests, it is possible to determine both the stress state and the strain rate of the γ channels under complicated stress paths. The main results of the analysis of experimental data are:

- The mobile dislocation density increases proportionately with the strain. This is consistent with a multiplication mechanism of the scolopendra type. However, as the dislocation distribution is uneven (there is probably a waiting time before dislocations enter the channel) usual annihilation laws do not apply and the main annihilation events take place during drops in the Von Mises stress.
- Two different velocity laws for high and low Von Mises stresses seem to operate. In the first case (Von Mises stress larger than the Orowan stress), the velocity is proportional to $\sigma_{VM} - \sigma_O$ and the dislocation mobility is thermally activated a ~3 eV activation energy. At Von Mises stresses lower than the Orowan stress, dislocations can still glide within the rafts without needing to trail one dislocation at each γ/γ' interface. This mechanism is probably controlled by dislocation climb within the interface.

Author Contributions: Conceptualization, A.J. and T.S.; Methodology, A.J.; Software, A.J. and R.T.; Validation, T.S., A.J. and R.T.; Formal Analysis, R.T., L.D. and A.J.; Investigation, R.T., L.D., T.S. and A.J.; Resources, ONERA and SNECMA-SAFRAN.; Data Curation, T.S.; Writing—Original Draft Preparation, T.S.; Writing—Review & Editing, A.J. and T.S.; Supervision, A.J.; Project Administration, A.J.; Funding Acquisition, A.J.

Funding: This research received no external funding.

Acknowledgments: This work was supported by the French State through the program "Investment in the future", operated by the National Research Agency (ANR) and referenced by ANR-11-LABX-0008-01 (LabEx DAMAS). The authors would like to acknowledge Jonathan Wright, Vadim Diadkin, and the ESRF staff for their help with the ID 11 beamline. They would like to thank Martin von Zimmermann for help with the DeSy BW5 beamline. They are grateful to ONERA and Jonathan Cormier for providing and treating specimens, and to Olivier Ferry for help with the high-temperature mechanical testing device. Gabor Ribarik, Lucile Dezerald, and Pierre Bastie are gratefully acknowledged for their help during the execution of in situ tests.

Conflicts of Interest: The authors declare no conflict of interest.

References

1. Reed, R.C.; Rae, C.M.F. 22—Physical Metallurgy of the Nickel-Based Superalloys. In *Physical Metallurgy*, 5th ed.; Laughlin, D.E., Hono, K., Eds.; Elsevier: Oxford, UK, 2014; pp. 2215–2290, ISBN 978-0-444-53770-6.
2. Pollock, T.M.; Argon, A.S. Creep resistance of CMSX-3 nickel base superalloy single crystals. *Acta Metall. Mater.* **1992**, *40*, 1–30. [CrossRef]
3. Miyazaki, T.; Nakamura, K.; Mori, H. Experimental and theoretical investigations on morphological changes of γ' precipitates in Ni-Al single crystals during uniaxial stress-annealing. *J. Mater. Sci.* **1979**, *14*, 1827–1837. [CrossRef]
4. Pollock, T.M.; Argon, A.S. Directional coarsening in nickel-base single crystals with high volume fractions of coherent precipitates. *Acta Metall. Mater.* **1994**, *42*, 1859–1874. [CrossRef]
5. Buffiere, J.Y.; Ignat, M. A dislocation based criterion for the raft formation in nickel-based superalloys single crystals. *Acta Metall. Mater.* **1995**, *43*, 1791–1797. [CrossRef]
6. Véron, M.; Bréchet, Y.; Louchet, F. Directional coarsening of Ni-based superalloys: Computer simulation at the mesoscopic level. *Acta Mater.* **1996**, *44*, 3633–3641. [CrossRef]
7. Srinivasan, R.; Eggeler, G.; Mills, M. γ'-cutting as rate-controlling recovery process during high-temperature and low-stress creep of superalloy single crystals. *Acta Mater.* **2000**, *48*, 4867–4878. [CrossRef]
8. Huang, M.; Zhuo, L.; Xiong, J.; Li, J.; Zhu, J. Core structure of a<100> superdislocations in a single-crystal superalloy during high-temperature and low-stress creep. *Philos. Mag. Lett.* **2015**, *95*, 496–503. [CrossRef]
9. Jacques, A.; Tréhorel, R.; Schenk, T. High temperature dislocation climb in the γ' rafts of Single Crystal Superalloys: The hypothesis of a control by dislocation entry into the rafts. *Metall. Mater. Trans. A* **2018**, *49*, 4110–4125. [CrossRef]

10. Nabarro, F.R.N.; Frank, R.N.; De Villiers, H.L.; Heidi, L. *The Physics of Creep: Creep and Creep-Resistant Alloys*; Taylor & Francis: London, UK, 1995; ISBN 0850668522.
11. Jacques, A.; Dirand, L.; Chateau, J.P.; Schenk, T.; Ferry, O.; Bastie, P. In Situ Measurement of Internal Stresses and Strain Rates by High Energy X-ray Diffraction during High Temperature Mechanical Testing. *Adv. Mater. Res.* **2011**, *278*, 48–53. [CrossRef]
12. Hantcherli, M.; Pettinari-Sturmel, F.; Viguier, B.; Douin, J.; Coujou, A. Evolution of interfacial dislocation network during anisothermal high-temperature creep of a nickel-based superalloy. *Scr. Mater.* **2012**, *66*, 143–146. [CrossRef]
13. Bouchard, R.; Hupfeld, D.; Lippmann, T.; Neuefeind, J.; Neumann, H.B.; Poulsen, H.F.; Rütt, U.; Schmidt, T.; Schneider, J.R.; Süssenbach, J.; et al. A triple-crystal diffractometer for high-energy synchrotron radiation at the HASYLAB high-field wiggler beamline BW5. *J. Synchrotron Radiat.* **1998**, *5*, 90–101. [CrossRef] [PubMed]
14. Tréhorel, R.; Ribarik, G.; Jacques, A.; Schenk, T. Real time study of transients during high temperature creep of a Ni-based superalloy by far field high energy synchrotron X-rays diffraction. *J. Appl. Crystallogr.* **2018**. submitted. [CrossRef]
15. Dirand, L.; Jacques, A.; Chateau-Cornu, J.P.; Schenk, T.; Ferry, O.; Bastie, P. Phase-specific high temperature creep behaviour of a pre-rafted Ni-based superalloy studied by X-ray synchrotron diffraction. *Philos. Mag.* **2013**, *93*, 1384–1412. [CrossRef]
16. Royer, A.; Bastie, P.; Veron, M. In situ determination of γ' phase volume fraction and of relations between lattice parameters and precipitate morphology in Ni-based single crystal superalloy. *Acta Mater.* **1998**, *46*, 5357–5368. [CrossRef]
17. Diologent, F.; Caron, P.; D'Almeida, T.; Jacques, A.; Bastie, P. The γ/γ' mismatch in Ni based superalloys: In situ measurements during a creep test. *Nucl. Instrum. Methods Phys. Res. Sect. B Beam Interact. Mater. Atoms.* **2003**, *200*, 346–351. [CrossRef]
18. Jacques, A.; Bastie, P. The evolution of the lattice parameter mismatch of a nickel-based superalloy during a high-temperature creep test. *Philos. Mag.* **2003**, *83*, 3005–3027. [CrossRef]
19. Jacques, A.; Diologent, F.; Bastie, P. In situ measurement of the lattice parameter mismatch of a nickel-base single-crystalline superalloy under variable stress. *Mater. Sci. Eng. A* **2004**, *387–389*, 944–949. [CrossRef]
20. Jacques, A.; Diologent, F.; Caron, P.; Bastie, P. Mechanical behavior of a superalloy with a rafted microstructure: In situ evaluation of the effective stresses and plastic strain rates of each phase. *Mater. Sci. Eng. A* **2008**, *483*, 568–571. [CrossRef]
21. Dirand, L.; Cormier, J.; Jacques, A.; Chateau-Cornu, J.P.; Schenk, T.; Ferry, O.; Bastie, P. Measurement of the effective gamma/gamma' lattice mismatch during high temperature creep of Ni-based single crystal superalloy. *Mater. Charact.* **2013**, *77*, 32–46. [CrossRef]
22. Caron, P.; Khan, T. Improvement of Creep strength in a nickel-base single-crystal superalloy by heat treatment. *Mater. Sci. Eng.* **1983**, *61*, 173–184. [CrossRef]
23. Fredholm, A.; Khan, T.; Theret, J.M.C.F.; Davidson, J.H. Alliage Monocristallin à Matrice à Base de Nickel. EP0149942, 1985.
24. Testing of Metallic Materials—Tensile Test Pieces. Available online: https://www.din.de/de/mitwirken/normenausschuesse/nmp/normen/wdc-beuth:din21:262241217 (accessed on 7 August 2018).
25. Feiereisen, J.P.; Ferry, O.; Jacques, A.; George, A. Mechanical testing device for in situ experiments on reversibility of dislocation motion in silicon. *Nucl. Instrum. Methods Phys. Res. Sect. B Beam Interact. Mater. Atoms* **2003**, *200*, 339–345. [CrossRef]
26. Tréhorel, R. Comportement Mécanique Haute Température du Superalliage Monocristallin AM1: Étude In Situ Par Une Nouvelle Technique de Diffraction en Rayonnement Synchrotron. Ph.D. Thesis, Université de Lorraine, Vandœuvre-lès-Nancy, France, 2018.
27. Epishin, A.; Link, T. Mechanisms of high-temperature creep of nickel-based superalloys under low applied stresses. *Philos. Mag.* **2004**, *84*, 1979–2000. [CrossRef]
28. Dirand, L. *Fluage à Haute Température d'un Superalliage Monocristallin: Expérimentation In Situ en Rayonnement Synchrotron*; INPL Nancy: Vandœuvre-lès-Nancy, France, 2011.
29. Carry, C.; Strudel, J.L. Apparent and effective creep parameters in single crystals of a nickel base superalloy-I Incubation period. *Acta Metall.* **1977**, *25*, 767–777. [CrossRef]
30. Ma, A.; Dye, D.; Reed, R.C. A model for the creep deformation behaviour of single-crystal superalloy CMSX-4. *Acta Mater.* **2008**, *56*, 1657–1670. [CrossRef]

31. Louchet, F.; Doisneau-Cottignies, B.; Bréchet, Y. Specific dislocation multiplication mechanisms and mechanical properties in nanoscaled multilayers: The example of pearlite. *Philos. Mag. A Phys. Condens. Matter Struct. Defects Mech. Prop.* **2000**, *80*, 1605–1619. [CrossRef]
32. Essmann, U.; Mughrabi, H. Annihilation of dislocations during tensile and cyclic deformation and limits of dislocation densities. *Philos. Mag. A* **1979**, *40*, 731–756. [CrossRef]
33. Brown, L.M. Dislocation bowing and passing in persistent slip bands. *Philos. Mag.* **2006**, *86*, 4055–4068. [CrossRef]

materials

Article

In Situ Stress Tensor Determination during Phase Transformation of a Metal Matrix Composite by High-Energy X-ray Diffraction

Guillaume Geandier [1,2,*], Lilian Vautrot [1,2], Benoît Denand [1,2] and Sabine Denis [1,2]

1. Institut Jean Lamour, UMR 7198, CNRS—Université de Lorraine, Campus ARTEM, 2 allée André Guinier BP 50840, 54011 Nancy CEDEX, France; lilianvautrot@gmail.com (L.V.); benoit.denand@univ-lorraine.fr (B.D.); sabine.denis@univ-lorraine.fr (S.D.)
2. Laboratory of Excellence on Design of Alloy Metals for low-mAss Structures (DAMAS), Université de Lorraine, 54000 Nancy-Metz, France

* Correspondence: guillaume.geandier@univ-lorraine.fr; Tel.: +33-372-742-728

Received: 30 June 2018; Accepted: 6 August 2018; Published: 12 August 2018

Abstract: In situ high-energy X-ray diffraction using a synchrotron source performed on a steel metal matrix composite reinforced by TiC allows the evolutions of internal stresses during cooling to be followed thanks to the development of a new original experimental device (a transportable radiation furnace with controlled rotation of the specimen). Using the device on a high-energy beamline during in situ thermal treatment, we were able to extract the evolution of the stress tensor components in all phases: austenite, TiC, and even during the martensitic phase transformation of the matrix.

Keywords: metal matrix composite; in situ X-ray diffraction; internal stresses; phase transformation; microstructure

1. Introduction

Mass reduction, such as in transportation applications where it is desired to reduce fuel consumption and pollution, can be obtained by using new lighter materials with mechanical properties that are at least equivalent to those of the previous ones. This goal can be reached using Metal Matrix Composites (MMCs) reinforced with ceramic particles. In our study, a steel matrix composite reinforced with TiC particles obtained by powder metallurgy (a mixture of 75% steel powder and 25% TiC powder) allows the density of this composite to be reduced to 7 g/cm^3, compared to 7.8 g/cm^3 for steel alone (i.e., a decrease of 11.4% in mass). The final properties of an MMC depend on the chemical composition, on the nature of the interfaces, on the microstructure of the matrix, and on the stresses in the reinforcements and in the matrix. Many studies deal with load transfer between the phases in composite materials induced by external loading [1,2]. However, large stresses can be generated during the heat treatment, resulting from the differences in the coefficients of thermal expansion between matrix and reinforcements [3–5], and also from the phase transformations [6,7]. The residual stress levels and distributions are a key factor for the final properties of the MMC [8]. In a previous study [9], the evolution of the matrix and the reinforcements were analysed during the heat treatment using in situ high-energy X-ray diffraction focusing on the structural aspects. In this paper, we focus on the internal stress analysis. First, we describe the dedicated original device and the methodology that was developed to allow an in situ determination of the evolutions of internal stresses in a steel matrix composite during cooling. Then, internal stresses in the phases are analysed, emphasising the role of martensitic transformation. Finally, a 3D finite element micromechanical model is used to better understand this role, and a comparison with experimental results is discussed.

2. Material and Thermal Cycle

A metal matrix composite was elaborated using powder metallurgy by Mecachrome (France). Steel powder (32CrMoV13, see Table 1 for the chemical composition of the matrix) at a volume fraction of 75% and TiC powder at a volume fraction of 25% were milled under Argon atmosphere. The powders were consolidated by hot isostatic pressing (HIP) under a stress of 100 MPa at 1120 °C during 4 h. After hot isostatic pressing, the microstructure of the steel matrix composite (see Figure 1a) was non-homogeneous and presented two different areas: a steel area (pearlitic microstructure) without TiC particles, called the unreinforced area (lighter area), and a darker area comprising a mixture of TiC particles and steel.

Table 1. Chemical composition of the steel matrix 32CrMoV13 (data from Mecachrome).

C	Cr	Mo	Mn	Si	V	Ni	N
0.312	3.831	0.721	0.434	0.583	0.136	0.067	152 ppm

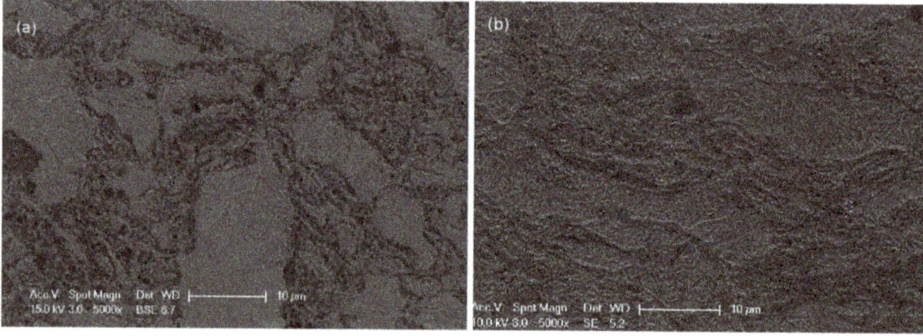

Figure 1. Metal matrix composite (MMC) microstructure (**a**) at initial state and (**b**) after the thermal treatment.

More details on the microstructure and chemical composition can be found in [9]. Figure 2 presents the thermal cycle measured by a thermocouple welded to the sample. The sample was heated to 900 °C at 3 °C·s^{-1}. The temperature was held for 5 min at 900 °C, and then the sample was cooled down. To achieve cooling, the lamps of the furnace were switched off at the end of the dwell, and a controlled gas flow (argon) allowed the cooling rate to be controlled. The cooling rate was fast enough to avoid ferrite or bainite formation before the martensitic transformation at M_S temperature (measured at 180 °C). The microstructure after the thermal treatment is presented in Figure 1b: the unreinforced areas composed of martensite and retained austenite (as we show later in the paper) and the reinforced areas where the matrix was mixed with TiC particles.

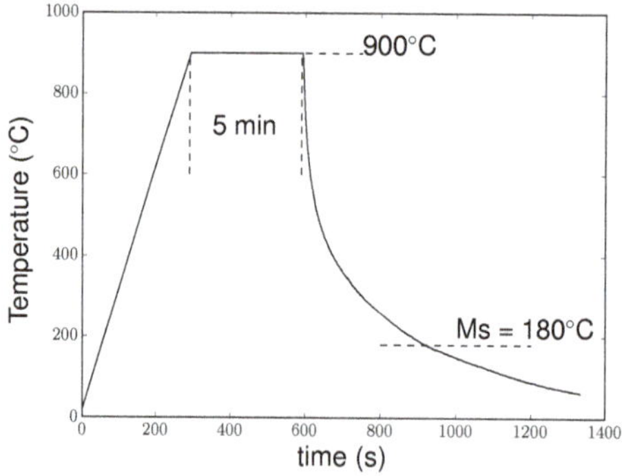

Figure 2. Thermal cycle applied to the MMC.

3. High-Energy X-ray Diffraction

3.1. Experimental Setup

The high-energy X-ray diffraction (HEXRD) experiments were performed at the European Synchrotron Radiation Facility (ESRF, Grenoble, France) on the ID15B beam line. The in situ measurements were conducted with an 87 keV monochromatic beam. The high-energy beam allowed us to analyse a large volume of the sample (due to the low absorption of the sample), thus being representative of the bulk and lessening the surface effect. The transmitted signal was collected by a large-area 2D detector (Perkin Elmer XRD1621) that recorded the whole Debye–Scherrer rings with a maximum 2θ angle of 12°. The high photon flux gives a high-quality diffraction signal, and thus HEXRD frames can be recorded at a high frame rate (up to 10 fps). Using this method, we could follow the evolution of Debye–Scherrer rings during a thermal treatment. For our experiments, the sample was a cylinder of 3 mm diameter and 30 mm length, and the beam size was 0.4×0.4 mm^2.

The sample was heated with a radiation furnace that was specially developed to perform a thermal treatment with a controlled rotation of the sample [10]. Temperature was controlled by a thermocouple welded on the sample surface, placed just below the beam path to avoid diffraction signal from the thermocouple. The sample and furnace configuration ensured that the temperature was homogeneous around the thermocouple and beam path. The rotation of the sample around its vertical axis was controlled by stopping it at fixed positions (0°, 90°, 180°, and 270°) for 0.3 s. The change between each position took 0.2 s. The sample rotation during in situ cooling and the fast acquisition rate allowed the data to be collected for determination of the full-stress tensor within one second, as we obtained all necessary orientations of the sample by combining the rotation of the sample and 2D diffraction in the perpendicular plane within two rotation steps.

3.2. Phase Analysis

As the sample presented small grains and no texture during the whole thermal treatment (see some examples of 2D patterns recorded at high temperature and at room temperature in Figure 3), a Rietveld analysis was conducted to extract phase fractions and mean cell parameters from the HEXRD frames using FullProf software [11,12]. Data were corrected (dark field and flat field) and reduced to (2θ, intensity) patterns using fit2D software [13]. Integration of the intensities was performed all around the rings.

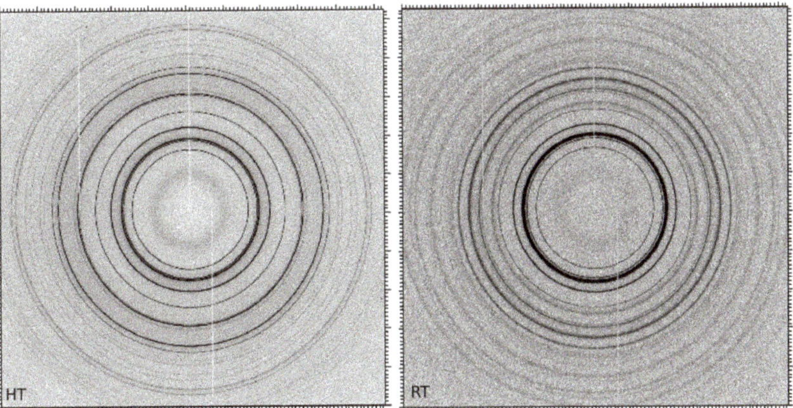

Figure 3. *Cont.*

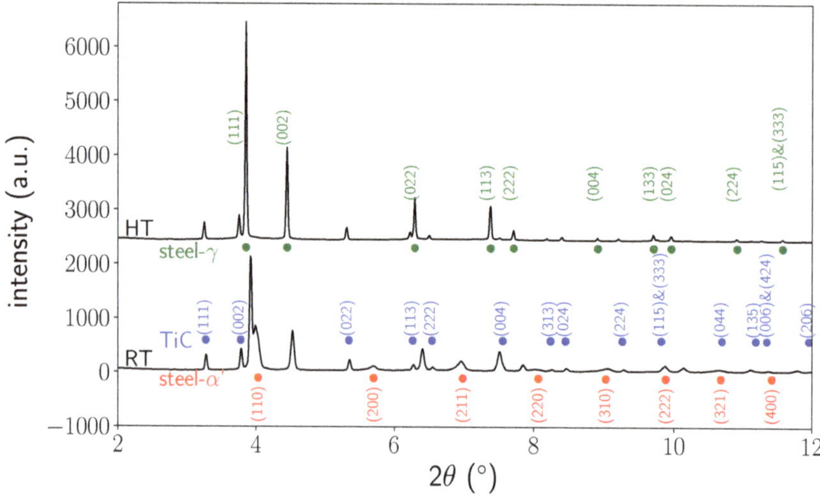

Figure 3. 2D patterns showing austenite and TiC diffraction signals at high temperature (HT) and showing austenite, martensite and TiC diffraction signals at room temperature (RT). Diffractograms from integrated images for the different phases (austenite, martensite (α'), TiC).

4. Internal Stress Determination

From the same corrected frames used for Rietveld analysis, a stress analysis was performed by applying the $sin^2\Psi$ method, as this method is more sensitive to small variations compared to a direct extraction of strain values from peak positions at specific azimuths on the rings. Contrary to Rietveld analysis, an integration was performed only for sectors of 1° of the ring using fit2D software [13]. Thus, we obtained 360 diffractograms (2θ, intensity) for the 360 azimuths (δ) that constitute the ring. In order to extract stresses in each phase from these data, it is necessary to determine the 2θ position of one peak in each phase. We selected $\alpha'(211)$, γ (200), and *TiC* (220), as they did not overlap with other peaks. Diffraction peaks were approximated by a Pearson VII function that allows the shape of our peaks to be reproduced and for their 2θ position, full width at half maximum (FWHM), intensity, shape factor, and background parameters to be obtained. This operation was repeated for the three phases for the 360 azimuths for each image collected during the thermal treatment.An automatic procedure was developed in Python to analyse the data (approximately 3 million peaks were analysed per phase extracted from 150 GB of raw data).

To account for uncertainties in the 2θ position due to the variations of the beam position over time, the average 2θ position for the opposite azimuths (δ and δ + 180°) was calculated. This step allowed us to free ourselves of the exact knowledge of the position of the centre of the ring [14].

In order to apply the $sin^2\Psi$ method, it is necessary to convert our configuration to the classical (Φ, Ψ) configuration (Figure 4 presents a sketch of the setup and describes the angles). Indeed, each azimuth corresponds to a given pair of angles (Φ, Ψ) that can be determined by the following equations [15,16]:

$$cos\Psi = cosc\left((1-\frac{sin^2\theta\, sin^2\delta}{1-cos^2c})(1-\frac{sin^2\omega\, sin^2\delta}{1-cos^2c})\right)^{1/2} - \frac{sin\omega\, sin\theta\, sin^2\delta}{1-cos^2c}, \quad (1)$$

where

$$cosc = cos\theta\, cos\omega + sin\theta\, sin\omega\, cos\delta, \quad (2)$$

$$\Phi = \chi + arccos\left(\frac{sin(\theta - \omega)}{sin\Psi}\right), \tag{3}$$

with θ as the Bragg angle, χ as the sample rotation, and ω as the sample tilt. Thus, angle pairs (Φ, Ψ) were distributed as shown in Figure 4 versus their position on the ring (azimuth δ).

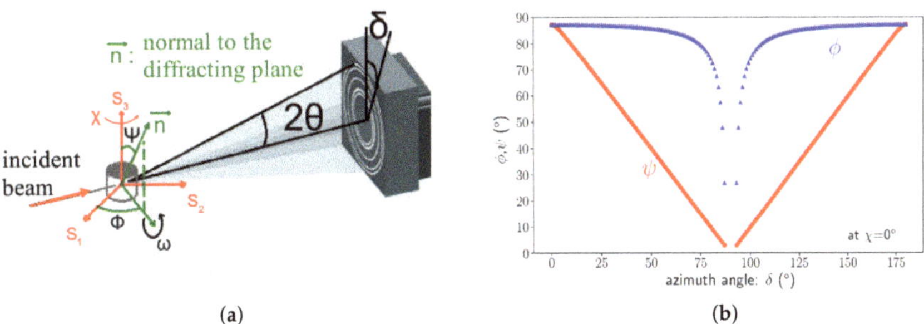

Figure 4. (**a**) Angles definition from setup configuration: (S1, S2, S3) is the sample reference system and (**b**) evolution of (Φ, Ψ) defining the direction normal to the diffracting plane {hkl}.

Finally, the $sin^2\Psi$ method can be applied. The deformation $\epsilon_{\Phi\Psi}$ in the direction normal to the diffracting plane defined by Φ and Ψ angles is expressed as a function of the components of the deformation tensor ϵ_{ij} in the sample reference system:

$$\epsilon_{\Phi\Psi} = ln\left(\frac{sin\theta_0}{sin\theta}\right) = \left[(\epsilon_{11}cos^2\Phi + \epsilon_{12}sin2\Phi + \epsilon_{22}sin^2\Phi) - \epsilon_{33}\right] sin^2\Psi + (\epsilon_{13}cos\Phi + \epsilon_{23}sin\Phi)sin2\Psi + \epsilon_{33}. \tag{4}$$

Introducing Hooke's law, $\epsilon_{\Phi\Psi}$ can be expressed as a function of the components of the stress tensor:

$$\epsilon_{\Phi\Psi} = ln\left(\frac{sin\theta_0}{sin\theta}\right) = \frac{1+\nu_{hkl}}{E_{hkl}}(\sigma_\Phi - \sigma_{33})sin^2\Psi + \frac{1+\nu_{hkl}}{E_{hkl}}, \tag{5}$$

with:

$$\tau_\Phi sin2\Psi + \frac{1+\nu_{hkl}}{E_{hkl}}\sigma_{33} - \frac{\nu_{hkl}}{E_{hkl}}Tr(\overline{\overline{\sigma}}), \tag{6}$$

$$\sigma_\Phi = \sigma_{11}cos^2\Phi + \sigma_{12}sin2\Phi + \sigma_{22}sin^2\Phi, \tag{7}$$

$$\tau_\Phi = \sigma_{13}cos\Phi + \sigma_{23}sin\Phi, \tag{8}$$

$$Tr(\overline{\overline{\sigma}}) = \sigma_{11} + \sigma_{22} + \sigma_{33}. \tag{9}$$

E_{hkl} and ν_{hkl} are the X-ray elastic constants (Young's modulus and Poisson's ratio). θ_0 is the stress-free diffraction angle for the diffracting plane. Then, $ln\left(\frac{1}{sin\theta}\right) = f(sin2\Psi)$ was plotted for each phase. For triaxial stress states, an ellipse is obtained (without texture) for a given Φ angle. The shear stresses are determined from the ellipse opening. The slope of the ellipse axis allows the stress difference $\sigma_\Phi - \sigma_{33}$ to be obtained, where σ_{33} is the stress component along the sample axis. By selecting the rotation position corresponding to a specific orientation of the transverse section of the sample, we can extract the σ_{11}–σ_{33} and σ_{22}–σ_{33} stress differences from positions $(0^\circ, 180^\circ)$ and $(90^\circ, 270^\circ)$ versus temperature. These stress differences are the mean values for the considered phase.

Because the intercept (Equation (10)) of the $sin^2\Psi$ curve (at $\Psi = 0$) is related to ϵ_{33} [17], it is possible to determine all the strain components ϵ_{ii} and then determine the stress components σ_{ii} if θ_0 is known:

$$Intercept = \epsilon_{33} + ln(\frac{1}{sin\theta_0}). \quad (10)$$

Nevertheless, data on the intercept need to be corrected, as the values are sensitive to the sample's tilt (as shown in Section 5.5). Due to the thermal expansion of the sample and its holder and also to the weight of the thermocouple, a tilt was identified during the sample rotation that resulted in a small variation of the peak position over time, correlated with sample rotation. In Figure 5, an example of the evolution of peak position over time is presented, in addition to a zoomed in view of a selected range of image numbers. We could identify the image where the sample position was fixed (i.e., rotation stopped), and thus where the sample orientation was known. As the tilt was reproducible over time, we averaged the intercept values corresponding to equivalent orientations and given temperature.

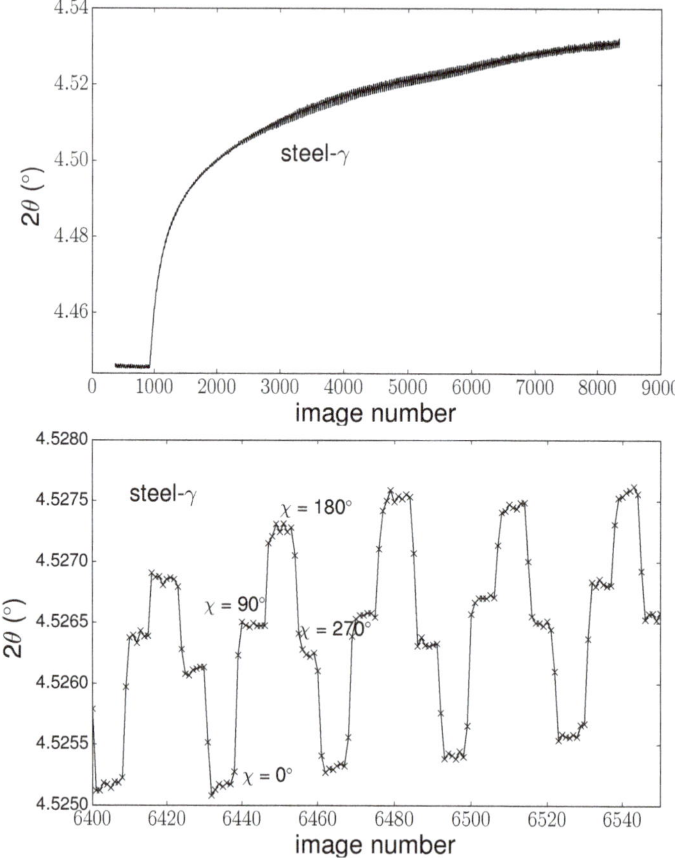

Figure 5. Evolution of peak position (γ (200)) over image number, and a zoomed view of image numbers 6400–6550.

As already seen, to obtain the full strain and stress tensor, we need θ_0 (i.e., the stress-free cell parameters of the phases and their evolution versus temperature). For steel-γ phase and TiC, we used the value of the cell parameters measured at high temperature assuming, that stresses were fully relaxed at 900 °C. For martensite values, we used values from the literature for a steel with composition close to the one in this study at room temperature [18]. The evolution of stress-free cell parameter versus temperature was calculated using the thermal expansion coefficient of phases, extracted from experimental studies for steel [9] or from the literature for TiC [19]. For our analysis, we used the isotropic elastic constants (Young's modulus E and Poisson's ratio ν). For the steel [20] and for the TiC reinforcements [19], the evolution of E and ν with temperature were taken from the literature as well as for TiC reinforcements [19]. Using macroscopic elastic constants we assume that the mechanical anisotropy of all phases can be neglected.

5. Results

5.1. Phase Transformation Kinetics

Figure 6 presents the evolution of the integrated 2D images versus temperature during the dwell and cooling periods. During cooling, we could identify the phase transformation of the matrix by the intensity decrease of the peak corresponding to the γ phase (face-centred cubic structure) and the increase of the α' phase (tetragonal structure). In Figure 7, we present the evolutions of the phase fractions during the cooling from 900 °C to 50 °C, leading to the formation of martensite in the matrix (as mentioned previously in part 2). At the beginning of the cooling, only austenite (γ) and TiC were present, with 81 m% and 19 m% respectively. During cooling, the amount of TiC remained constant. The martensitic transformation began at 180 °C. From 180 °C to 50 °C, we could follow the increase of martensite fraction and decrease of austenite fraction. At 50 °C, the microstructure was composed of 54% martensite and 46% retained austenite.

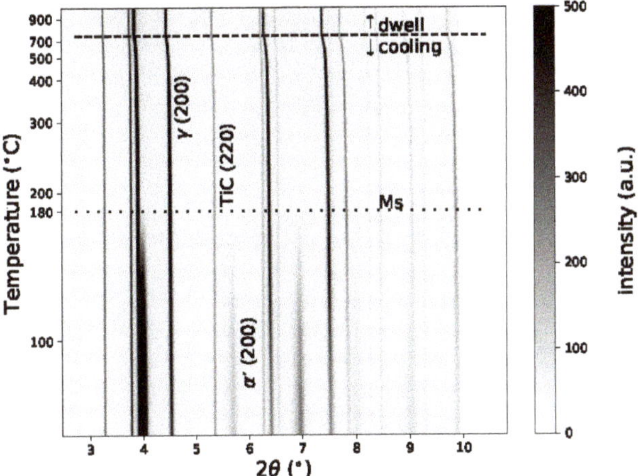

Figure 6. Evolution of diffractograms during cooling.

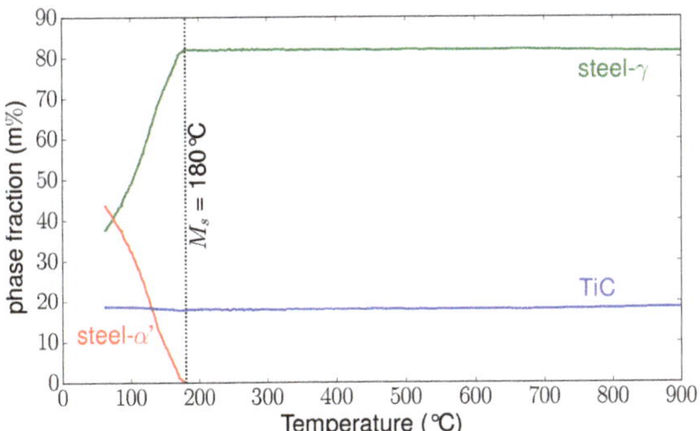

Figure 7. Phase fraction evolutions during cooling.

5.2. Evolution of Mean Cell Parameters

From the beginning of cooling until the beginning of martensitic transformation, the cell parameters of the different phases decreased continuously without change of slope (Figure 8). From a previous analysis [21] of the apparent coefficient of thermal expansion (CTE), it was concluded that thermal stresses are generated in the phases during cooling. TiC was under a mean compression state since the CTE of austenite is about three times larger than that of TiC. Indeed, as shown in Figure 8, the apparent CTE of TiC was much higher than the stress-free CTE. Before the martensitic transformation, the mean stress state in austenite seemed very small.

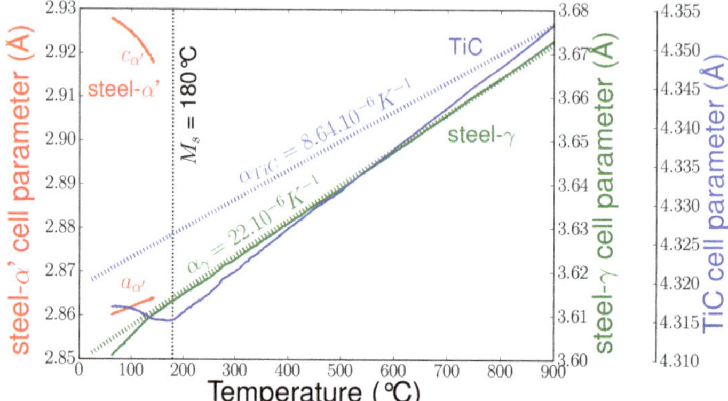

Figure 8. Cell parameter evolutions during cooling (stress-free thermal expansion coefficients of TiC and austenite are also shown).

As the martensitic transformation occurred, the TiC cell parameter increased, indicating a decrease of the mean compression in TiC. As the martensite content reached approximately 15%, the austenite cell parameter decreased, indicating an increase in compression. We could also observe an increase of the tetragonality (c/a ratio) of the martensite (the c parameter increased while the a parameter decreased linearly). Note that as the martensitic transformation began, peak intensities were very low (as can be seen in Figure 6), and thus accurate determination of cell parameters was not possible. Therefore, cell parameters for low fractions of martensite (below 10%) are not presented.

5.3. Full Width at Half Maximum

In addition to the evolution of mean cell parameters, we also characterised the mean full width at half maximum (FWHM) evolutions for each phase during the heat treatment. From the integrated images, a fit was made using a Pearson VII function to extract the FWHM for each phase versus temperature. Figure 9 shows the evolution of the FWHM of selected peaks for the different phases present in the composite. Between 900 °C and the martensitic start temperature (Ms) temperature, the FWHM could be linked to heterogeneities of thermal stresses generated in the phases. At Ms, the FWHM for γ phase increased more sharply. A sharp increase was also observed for the α' phase. As the martensitic transformation occurs without a change in chemical composition, these evolutions of FWHM can be mainly attributed to heterogeneous transformation stresses due to the deformation associated with martensitic transformation. Below Ms, the FWHM for TiC remained nearly constant, indicating low stress gradients in the particles.

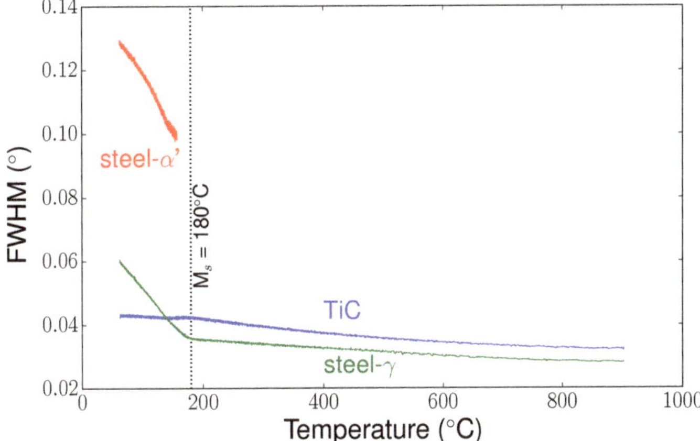

Figure 9. Evolution during cooling of the full width at half maximum (FWHM) for the different phases (steel-γ phase (200), TiC (220), and steel-α' phase (211)) versus temperature.

5.4. Stress Evolutions

Using $sin^2\Psi$ method, we were able to determine all the components of the stress tensors in the different phases during cooling. Figure 10 shows some examples of $sin^2\Psi$ curves for the three phases at different temperatures. We can see that the curves were close to straight lines, meaning that shear stresses were negligible, and the slopes were very small. This is confirmed in Figure 11, which presents the evolutions of the slopes of $sin^2\Psi$ curves (i.e., the stress differences) versus temperature for the different phases. Thus, the mean stress states remained hydrostatic in all phases during cooling, even during the martensitic transformation. In the following, we only present the results for σ_{ii} components, as the other components $\sigma_{ij,i\neq j}$ were negligible. Figure 12 shows the evolution of the intercept of the $sin^2\Psi$ curve versus temperature for the different phases. As mentioned previously, the intercept evolution was sensitive to the tilt of the sample during rotation. This effect was clearly seen for temperatures lower than 300 °C for TiC particles and austenite, where the different curves corresponded to different χ angles. In our approach, we corrected this tilt artefact by using average values of the intercepts.

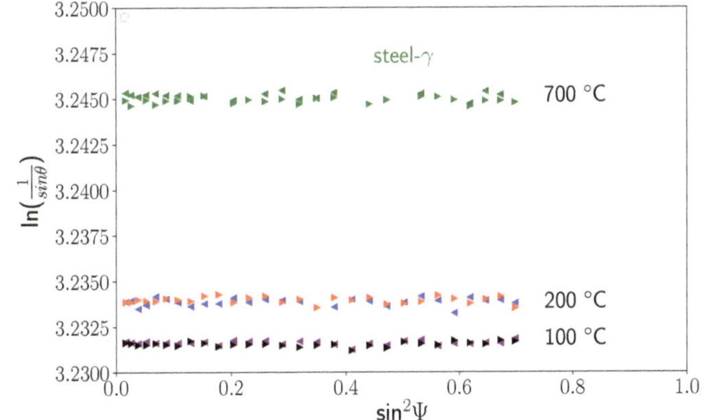

Figure 10. *Cont.*

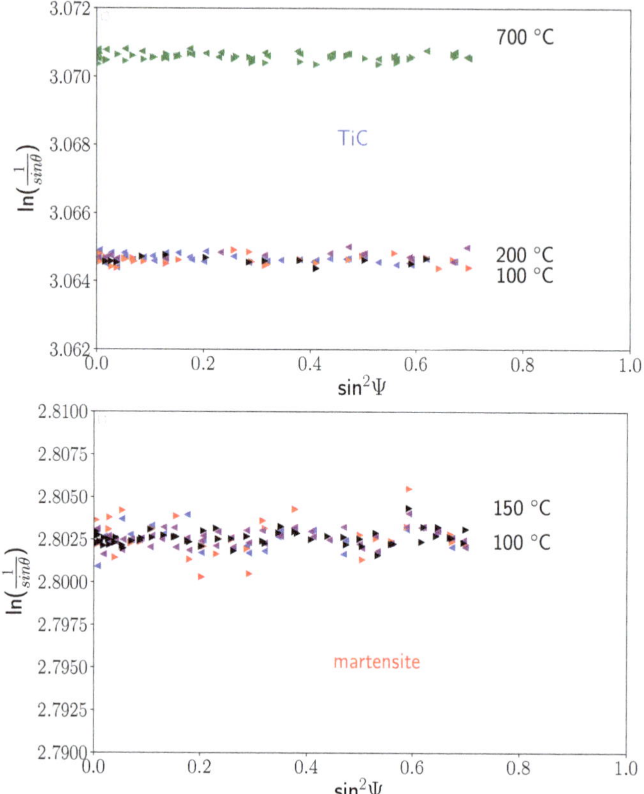

Figure 10. Evolution of the $sin^2\Psi$ curves at different temperatures for steel-γ phase, TiC, and steel-α' phase.

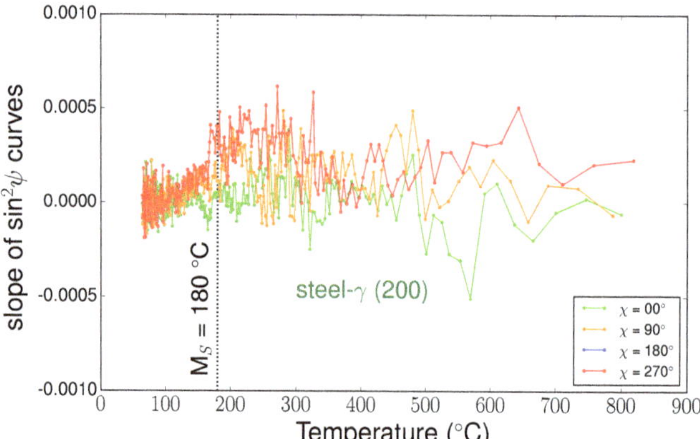

Figure 11. *Cont.*

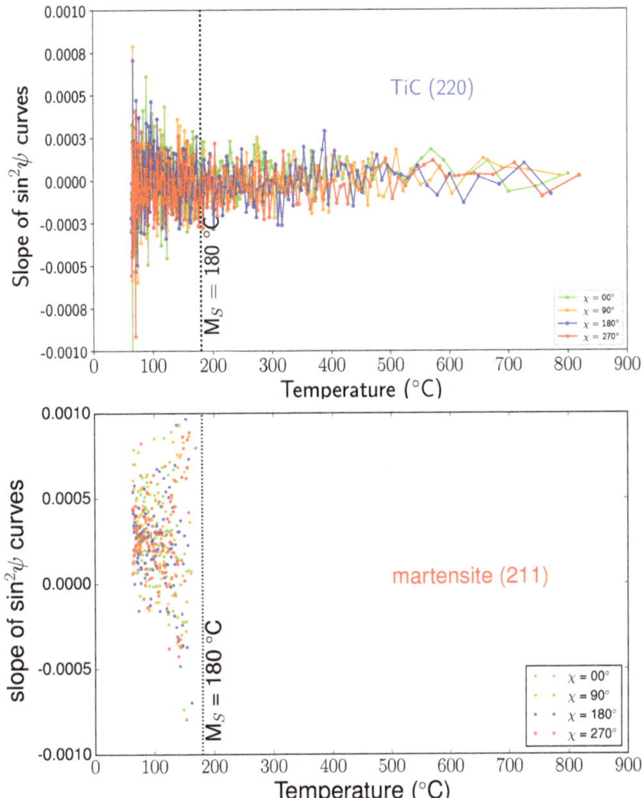

Figure 11. Evolution of the slopes of the $sin^2\Psi$ curve versus temperature for steel-γ phase, TiC, and steel-α' phase.

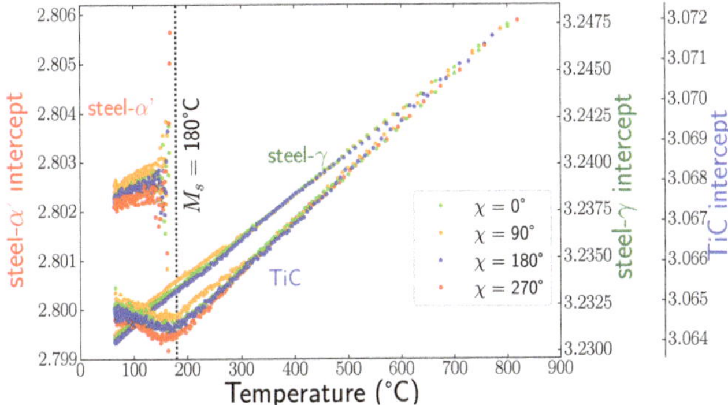

Figure 12. Evolution of the intercepts of the $sin^2\Psi$ curves versus temperature for different χ angle for steel-γ phase, TiC, and steel-α' phase.

Finally, Figure 13 presents the determined stress evolutions during cooling from 900 °C to 50 °C.

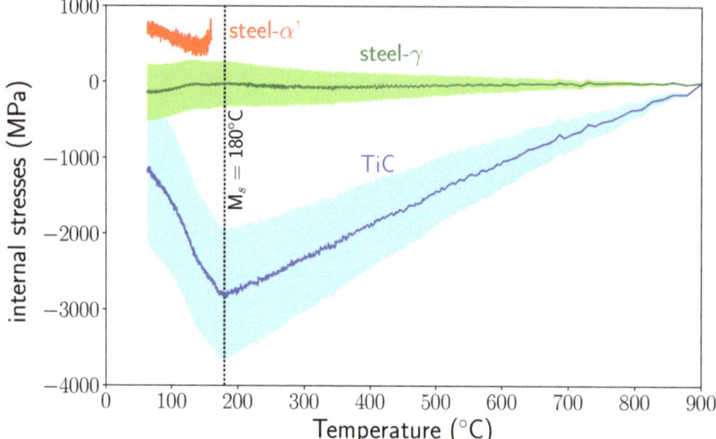

Figure 13. Internal stress evolutions during cooling (because the stress states are hydrostatic, only one component of the stress tensor is presented). Envelopes show the stress variations introduced by uncertainties in thermal expansion coefficients (see Section 5.5).

Note the high level of compressive stress in TiC increasing from the high temperature until the appearance of martensite. The stress level in austenite was relatively low. The martensitic transformation induced a large decrease of the compressive stress in TiC and small compressive stresses in austenite. For martensite, the tensile stress first decreased and then increased as temperature decreased.

5.5. Discussion

From the experimental results, we can say that the calculated macroscopic stress for the composite (using the measured phase fractions) was not zero, even before the phase transformation. Note that stress values in the phases were highly dependent on the stress-free cell parameters and CTE, and to a lesser extent on X-ray elastic constants. In the following, we give an estimation of the possible effect of these parameters on the determined stress levels.

5.5.1. Coefficient of Thermal Expansion

From a literature review on the coefficient of thermal expansion of TiC [19,22,23] and experimental results for steels, we estimated that the error on CTE is around ($\pm 1.10^{-6}$ K^{-1}). In Figure 13, the envelopes present the variations in the stress levels for austenite and TiC, including these experimental uncertainties (for martensite, the stress variations are not shown as they would overlap the stress scattering). We can see that the variation in stress level can be very large. There were even variations large enough to find values of CTE for which the macroscopic stresses were zero.

5.5.2. Stress-Free Parameters

For steel and TiC, the stress-free parameters were estimated by Rietveld refinement on the diffractograms at the end of the dwell at 900 °C, assuming stress relaxation at 900 °C. We compared the value of the cell parameters for a dwell at 950 °C. The Rietveld analysis showed a variation of the cell parameters close to the thermal expansion between 900 and 950 °C, thus justifying our assumption.

Another difficulty came from the fact that values of stress-free parameters of martensite are very difficult to obtain since local stresses are generally generated as martensite forms. As described above, in our approach, we used values from the PDF-4 database, which are close to the ones determined by Roberts [18]. This means that the level of stress in martensite is unknown. To estimate the stress level in martensite, the criterion of macroscopic stress equal to zero in the composite could be used.

5.5.3. Macroscopic Elastic Constants

Different values for the Young's Modulus and Poisson's ratio for TiC can be found in the literature [19,22,24,25], depending on the chemical composition of the TiC (nitrogen can be substituted for carbon and the stoichiometry of the structure is not perfect). Values from Wall et al. [19] used in our study are close to the average values of the different studies. If we take the extreme values found in the literature into account, the maximum level of compressive stress reached before M_S temperature could vary by about 250 MPa.

5.6. Micromechanical Modelling

In order to better understand the experimental results, 3D finite element micromechanical modelling (Code Zebulon [26]) was performed to calculate the internal stress evolutions during cooling.

5.6.1. Description of the Model

Calculations were performed for a 3D simplified microstructure: a periodic distribution of spherical TiC particles (representing 25% volume fraction) embedded in a steel matrix. Due to symmetry, only one eighth of the unit cell needed to be meshed (see Figure 14). Periodicity was imposed through the boundary conditions.

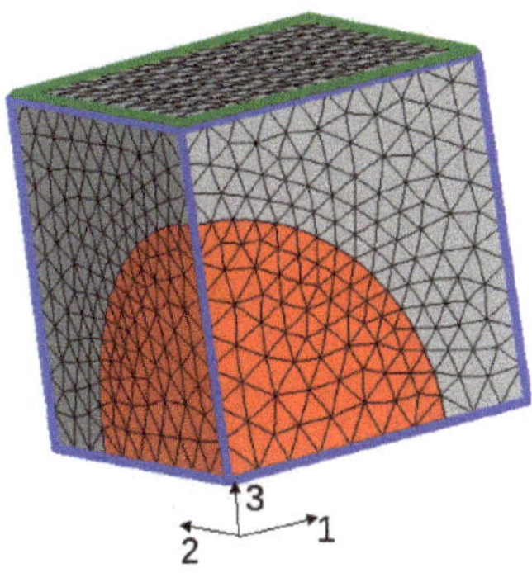

Figure 14. Unit cell used for the 3D finite element micromechanical modelling.

The thermal treatment was the same as that imposed during the experiments: cooling of the composite was simulated from 900 °C to room temperature by imposing cooling in accordance with Figure 2. Phases were assumed to be fully relaxed at 900 °C. Martensitic transformation was described by a Koistinen–Marburger law [27], with parameters determined from experimental results.

$$y_m = (1 - exp^{-\beta(Ms-T)}), \tag{11}$$

where y_m is the martensitic mass fraction, Ms is the martensitic start temperature (Ms = 180°C), and β is a coefficient equal to 0.014 $°C^{-1}$. For the actual model, only two phases were taken into account: the matrix and the particles.

The model considers a a thermo-elasto-visco-plastic behaviour law of the matrix including the phase transformation deformations (volume change and transformation plasticity). This behaviour law derives from previous works on the prediction of internal stresses in metallic alloys [28,29]. This law is written in incremental form by:

$$d\epsilon^t_{ij} = d\epsilon^e_{ij} + d\epsilon^p_{ij} + d\epsilon^{th}_{ij} + d\epsilon^{tr}_{ij} + d\epsilon^{pt}_{ij}, \tag{12}$$

with:

$d\epsilon_{ij}^{e}$: incremental elastic strain related to stress increment by Hooke's law with temperature-dependent Young's modulus and Poisson's ratio.
$d\epsilon_{ij}^{p}$: incremental visco-plastic strain at high temperature and plastic strain at lower temperatures.

These components are calculated using the classical plasticity theory with Von Mises criterion and isotropic hardening.

For each phase *k*, the yield stress σ_k is determined by:

$$\sigma_k = \sigma_{0k} + H_k \epsilon_{vp}^{n_k} + K_k \dot{\epsilon}_{vp}^{m_k}, \tag{13}$$

where σ_{0k} is the threshold stress, $H_k \epsilon_{vp}^{n_k}$ is the hardening by plastic strain, and $K_k \dot{\epsilon}_{vp}^{m_k}$ is the viscous stress. H_k, n_k, K_k, and m_k are temperature-dependent coefficients determined experimentally for each phase.

$d\epsilon_{ij}^{th}$: incremental thermal strain

$$\epsilon_{ij}^{th} = \sum y_k \int \alpha_k(T) dT, \tag{14}$$

where α_k is the temperature-dependent coefficient of thermal expansion of phase *k* and y_k is the volume fraction of phase *k*.

The strain increments due to phase transformation are:

$d\epsilon_{ij}^{tr}$: incremental strain due to volume change

$$\epsilon_{ij}^{tr} = y_k \sum \epsilon_{k,0\,^\circ\mathrm{C}}^{tr}, \tag{15}$$

$d\epsilon_{ij}^{pt}$: incremental strain due to transformation plasticity

$$d\epsilon_{ij}^{pt} = \frac{3}{2} K_k f'(y_k) dy_k s_{ij}, \tag{16}$$

where s_{ij} are the components of the deviatoric stress tensor, and K_k is an experimentally determined coefficient. For martensitic transformation $f(y_m) = (2 - y_m) y_m$.

All data concerning the steel matrix (austenite and martensite) were extracted from previous studies done in our laboratory. For TiC reinforcements, a thermo-elastic behaviour was considered using data from the literature [19].

5.6.2. Calculated Results

Calculations allow the analysis of stress and deformation gradients in the particles and in the matrix during the entire cooling process. Figure 15 presents the stress profiles (components σ_{11}, σ_{22}, σ_{33} and σ_m (hydrostatic stress)) along axis 1 of the cell (see Figure 14) for temperatures ranging from 900 °C to 20 °C.

The profiles of the cumulated equivalent plastic strain in the matrix are represented in Figure 15. Before the martensitic transformation, stress profiles show compressive stresses in the particle, as expected from the differences in thermal expansion coefficients. These stresses increased as temperature decreased, and reached −200 MPa at 180 °C in the centre of the particle. It can be noticed that stress gradients existed in the particle with lower stresses close to the interface, particularly for σ_{11}, and that the stress state was not hydrostatic as would be predicted by Eshelby's model [30]. This is mainly related to the interactions between the particles, as the volume fraction was relatively high (25%).

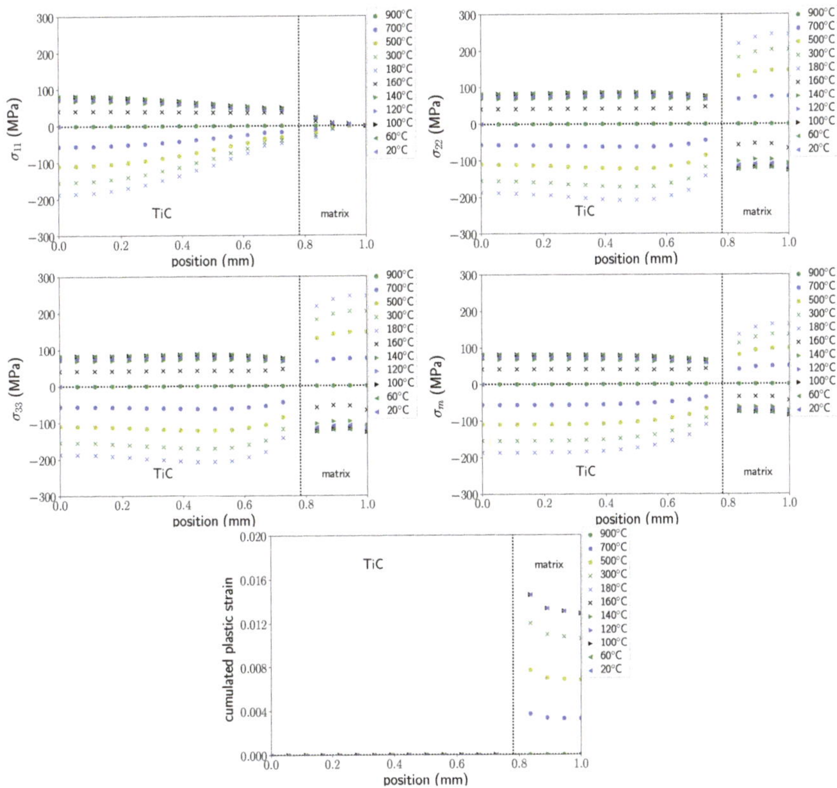

Figure 15. Stresses and cumulated equivalent plastic strain profiles along axis 1 for different temperatures.

In the matrix, σ_{11} was small. σ_{22} and σ_{33} increased in tension as temperature decreased, and showed relatively small gradients. Hydrostatic stress was approximately 160 MPa at the boundary of the cell at 180 °C. Viscoplastic strains developed in the matrix (austenite) from the beginning of cooling and reached a maximum value (about 1.5%) just before the martensitic transformation. As the martensitic transformation occurred, a large stress relaxation occurred in the matrix as well as in the particle that led to a stress reversal—stresses in the matrix and in the particle became respectively compressive and tensile. This stress relaxation was due to the volumetric variation associated with martensitic transformation. Due to this stress relaxation, no more plastic strains developed in the matrix, but as the transformation progressed, transformation plasticity strains (not shown here) were generated. Finally, the residual stress state was tensile in the particle and compressive in the matrix.

5.6.3. Comparison with Experimental Results and Discussion

In order to compare the micromechanical results with the experimental ones, the mean values of the stresses were calculated in the matrix and in the reinforcement during the entire cooling process. The mean values were calculated from the stress tensor components in each finite element weighted by the element volume. The evolution of the mean stresses in the matrix and in TiC particles during cooling are presented in Figure 16.

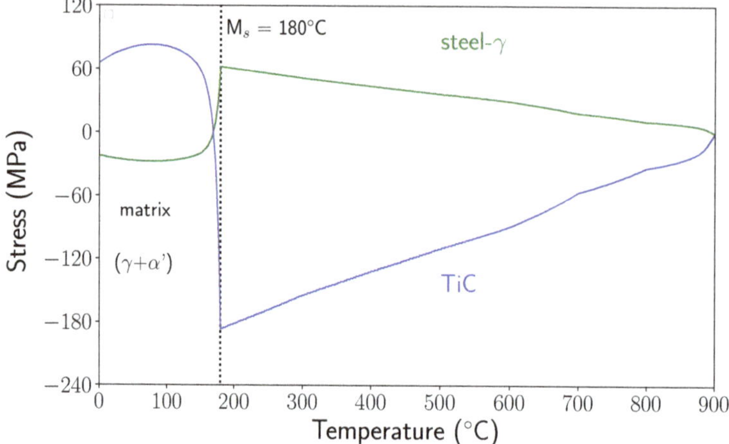

Figure 16. Evolution of the mean calculated stresses in the reinforcements and in the matrix during cooling.

Between 900 °C and Ms temperature, the stresses increased in tension and compression for the matrix (which was austenitic) and in compression in the TiC particles. At Ms, they reached +62 MPa in the matrix and −185 MPa in the particle. As noted above, martensitic transformation induced a large stress relaxation, particularly in the reinforcements. Residual stresses were tensile in TiC and compressive in the matrix, with relatively low levels. These evolutions were in accordance with the evolutions observed experimentally, but the stress levels were very different. In particular, in TiC, just before the matrix phase transformation, the calculated stress level was 10 times smaller than the experimentally determined value.

Note that experimental stress values in TiC particles were much closer to those calculated considering a thermoelastic behaviour of the matrix. Indeed, a calculation performed considering a thermoelastic behaviour of the matrix led to a stress level in the TiC particles just before Ms of about −1600 MPa. Thus, it seems that our calculation overestimated the (visco)plastic deformations in the austenite during cooling before the martensitic transformation. This assumption is reinforced by the fact that only small variations of the FWHM of austenite were observed before the martensitic transformation (see Section 5.3). Thus, questions arise about the accommodation of the thermal stresses generated during cooling (before the transformation). One important point is that the real microstructure was very different from the modelled one. As mentioned previously, the actual microstructure was very heterogeneous and there are reinforced zones where the volume fraction of particles is very high (50% or even more) and zones without particles that could lead to different plastic strain development. In addition, in the reinforced area, interparticle distances were very small (on the order of 1 µm), and would probably lead to less plastic accommodation, as simulated presently.

Except for the questions addressed above, it can be said that the main contribution of our study is to reveal the strong influence of the phase transformation of the matrix on the internal stress evolutions in the composite material. Indeed, in the literature, a number of studies deal with phase transformations and microstructural changes during the processing of composite materials or during heat treatment, either phase transformation of the reinforcements [6,31–33] or within the matrix [34,35], and their consequences on the mechanical behaviour. However, a thorough analysis of the role of the phase transformation in the internal and residual stresses and their in situ quantification has not yet been performed. In previous work on different nonreinforced steels [36], it was shown experimentally through the evolution of cell parameters that martensitic transformation induces compressive stresses in the austenite, once about 15% martensite has been formed. This was an unexpected behaviour, since tensile stresses were expected in the austenite.

Indeed, the free strain associated with the formation of a martensite plate has two contributions: a volumic strain and a shear strain. The order of magnitude of the shear strain is 0.2. The shear strain may be accommodated by the formation of self-accommodating plates, and/or by elastic and plastic deformation. The volumic strain is positive (its order of magnitude for steels varies from 1.5% to 3%), and is accommodated by elastic and plastic deformations. Considering Eshelby's model, the formation of a martensite plate will lead to a mean compressive stress in the martensite, and a mean tensile stress in the infinite medium of austenite. The building up of compression stresses in the parent austenite is still complex to understand, but we explain it by the increasing interactions between the martensite plates as transformation progresses. An increase of FWHM in austenite during the phase transformation was also observed for martensite amounts larger than 15% and was related to elastic strain heterogeneities.

Here, in the case of our reinforced steel, through the cell parameter evolutions (Figure 8), we also revealed compressive stresses in the austenite once the amount of martensite reached 15%. FWHM evolutions also revealed elastic heterogeneities in austenite as well as in martensite resulting from the accommodation of the transformation strain. However, the very new feature is that the martensitic transformation of the matrix induced an unexpected high-stress relaxation in the particles, and this can only be explained by the micromechanical approach that considers the volumic variation associated with the martensitic transformation. Consequently, the phase transformation of the matrix also had a strong influence on the residual stress states after cooling in the composite material. From the micromechanical approach, it was also shown that the accommodation mechanisms of the thermal and transformation stresses (elasticity, plasticity, etc.) were determinant on the level of internal and residual stresses, and in the future deeper studies on that point are necessary: determination of local strains, stresses, and dislocation densities by synchrotron (Laue microdiffraction) and high-resolution TEM.

6. Conclusions

We developed an original experimental device and a methodology for robust and rapid in situ stress analysis starting from a very large number of 2D images (i.e., Debye–Scherrer rings) obtained by synchrotron X-ray diffraction experiments during heat treatment. Indeed, the device allows control of not only heating and cooling, but also the rotation of the sample in order to obtain data in all necessary directions to extract the informations on the evolution of strains and stresses. As far as we know, it is the first time such experiments have been carried out. From the present experiment performed during the heat treatment of a steel matrix composite reinforced by TiC particles, we determined the evolutions of the full-stress tensor in the different phases: austenite, martensite and TiC. We showed that the mean stresses in the phases remained hydrostatic during the entire cooling process. Large stresses were generated in the particles due to the differences in thermal expansion coefficient with austenite. The role of the martensitic transformation of the matrix on the stress states in the particles was clearly demonstrated. It was shown that martensitic transformation led to a high stress relaxation in the TiC particles and to a lesser extent in the matrix.

The 3D finite element micromechanical model allows detailed analysis of the stress gradients in the matrix and in the particles during the entire cooling process. It was shown that internal stresses resulting from the thermal expansion mismatch between the austenite and TiC particles may be high enough to lead to plastic accommodation in the matrix, resulting in relatively low stress levels in the particles. It was demonstrated that the martensitic transformation, due to the volumetric variation and transformation plasticity, led to a high stress relaxation in the particles and in the matrix. Nevertheless, the calculated mean values of the stresses in the different phases, even if the evolutions were similar, showed large discrepancy with the experimental results, especially in the particles. These discrepancies were related on one hand to experimental uncertainties in stress levels due to uncertainties in thermal expansion coefficients and stress-free cell parameters. On the other hand, the present micromechanical calculations on a simplified microstructure probably overestimated the plastic accommodation in the matrix. Work is on course for taking the heterogeneity of the real microstructure into account by performing micromechanical simulations on SEM images of the microstructure.

The results obtained in this paper also address the wider issue of better mastering (i.e., tailoring) the mechanical performance of MMCs. Particularly, the understanding of damage mechanisms (fracture of particles, interface decohesion, etc.) and their modelling needs to consider not only microstructural features, but also the mechanical states of the phases inherited from materials processing (internal stresses, plastic strains, etc.), which is often neglected. For example, the studies on steel–TiC composites obtained by SLM (selective laser melting) [5,37] pointed out the occurrence of cracks resulting from the high internal stresses induced on one hand by the solidification shrinkage and on the other hand by the rapid heating and cooling in the solid state. Experimental knowledge of internal stress evolutions as well as their numerical prediction in order to better control and optimise them is thus a key issue for different processing routes of composite materials and particularly additive manufacturing processes.

Funding: This work was supported by the French State through the program "Investment in the future" operated by the National Research Agency (ANR) and referenced by ANR-11-LABX-0008-01 [38].

Acknowledgments: The authors gratefully acknowledge Mecachrome for supplying the materials, the Direction Générale des Entreprises (DGE) for financial support in the AMETIS program, and the European Synchrotron Radiation Facility (ESRF) for provision of beamtime at beamline ID15B.

Conflicts of Interest: The authors declare no conflicts of interest.

References

1. Meixner, M. Measurement of the evolution of internal strain and load partitioning in magnesium hybrid composites under compression load using in-situ synchrotron X-ray diffraction analysis. *Compos. Sci. Technol.* **2011**, *71*, 167–176. [CrossRef]
2. Yang, F. Microstructure and phase stress partition of Mo fiber reinforced CuZnAl composite. *Mater. Sci. Eng. A* **2015**, *628*, 419–422. [CrossRef]
3. Lee, R.S. Thermal and grinding induced residual stresses in a silicon carbide particle-reinforced aluminium metal matrix composites. *Composites* **1995**, *26*, 425–429. [CrossRef]
4. Pagounis, E. Effect of thermal expansion coefficients on the martensitic transformation in a steel matrix composite. *Scr. Mater.* **1996**, *34*, 407–413. [CrossRef]
5. AlMangour, B. Rapid fabrication of bulk-form TiB2/316L stainless steel nanocomposites with novel reinforcement architecture and improved performance by selective laser melting. *J. Alloy. Compd.* **2016**, *680*, 480–493. [CrossRef]
6. Alexander, K. Internal Stresses and the Martensite Start Temperature in Alumina-Zirconia Composites: Effects of Composition and Microstructure. *J. Am. Ceram. Soc.* **1999**, *78*, 291–296. [CrossRef]
7. Schöbel, M. Experimental Simulation of Thermally Induced Stresses during Cooling of Continuously Cast Steel Slabs. *Steel Res. Int.* **2016**, *87*, 1312–1322. [CrossRef]
8. Wu, Y. The effect of phase transformation on the thermal expansion property in Al/ZrW_2O_8 composites. *J. Mater. Sci.* **2012**, *48*, 2928–2933. [CrossRef]
9. Mourot, M. Transformation Kinetics and Resulting Microstructure in MMC Reinforced with TiC Particles. *Solid State Phenom.* **2011**, *172–174*, 747–752. [CrossRef]
10. Denand, B. Four d'Analyse Portable Pour Ligne De Rayonnement. Patent FR 1759981, June 2018.
11. Rietveld, H.M. A Profile Refinement Method for Nuclear and Magnetic Structures. *J. Appl. Crystallogr.* **1969**, *2*, 65–71. [CrossRef]
12. Rodriguez-Carvajal, J. Recent advances in magnetic structure determination by neutron powder diffraction. *Physica B* **1993**, *192*, 55–69. [CrossRef]
13. Hammersley, A.P. Calibration and correction of distortions in two-dimensional detector systems. *Rev. Sci. Instrum.* **1995**, *66*, 2729–2733. [CrossRef]
14. Le Bourlot, C. Caractérisation de L'hétérogénéité du Champ des Déformations Élastiques dans les Matériaux Polycristallins par Diffraction des Rayons X et des neutrons-application à un acier duplex. Ph.D. Thesis, Université Paris, Paris, France, 2012.
15. Heidelbach, F. Quantitative texture analysis of small domains with synchrotron radiation X-rays. *J. Appl. Crystallogr.* **1999**, *32*, 841–849. [CrossRef]

16. Gelfi, M. X-ray diffraction Debye Ring Analysis for STress measurement (DRAST): A new method to evaluate residual stresses. *Acta Mater.* **2004**, *52*, 583–589. [CrossRef]
17. Dolle, H. The Influence of Multiaxial Stress States, Stress Gradients and Elastic Anisotropy on the Evaluation of (Residual) Stresses by X-rays. *J. Appl. Crystallogr.* **1979**, *12*, 489–501. [CrossRef]
18. Roberts, C.S. Effect of carbon on the volume fractions and lattice parameters of retained austenite and martensite. *J. Appl. Crystallogr.* **1953**, *5*, 203–204. [CrossRef]
19. Wall, J. Thermal residual stress evolution in a TiC-50 vol.% Ni_3Al cermet. *Mater. Sci. Eng. A* **2006**, *421*, 40–45. [CrossRef]
20. Vidal, G. Influence de la température sur les constantes élastiques des métaux et alliages. *Rev. Métall.* **1951**, *11*, 864–874. [CrossRef]
21. Geandier, G. In situ structural evolution of steel-based MMC by high energy X-ray diffraction and comparison with micromechanical approach. *J. Heat Treat. Mater.* **2014**, *69*, 54–59. [CrossRef]
22. Winkler, B. In situ observation of the formation of TiC from the elements by neutron diffraction. *J. Alloys Compd.* **2007**, *441*, 374–380. [CrossRef]
23. Dartigues, F. La Précipitation à la Solidification du Monoborure de Titane Dans L'alliage de Titane Ti-6Al-4V) Peut-Elle Modifier sa Microstructure et Son Comportement Mécanique? Ph.D. Thesis, University of Bordeaux, Bordeaux, France, 2004.
24. Yang, Y. First-principles calculations of mechanical properties of TiC and TiN. *J. Alloys Compd.* **2009**, *485*, 542–547. [CrossRef]
25. Cheng, L. Densification and mechanical properties of TiC by SPS-effects of holding time, sintering temperature and pressure condition. *J. Eur. Ceram. Soc.* **2012**, *32*, 3399–3406. [CrossRef]
26. Besson, J. Object-oriented programming applied to the finite element method part II. application to material behaviors. *Rev. Eur. Élém.* **1998**, *7*, 567–588. [CrossRef]
27. Koistinen, D.P. A general equation prescribing the extent of the austenite-martensite transformation in pure iron-carbon alloys and plain carbon steels. *Acta Crystallogr.* **1959**, *7*, 60–61. [CrossRef]
28. Denis, S. *Models for Stress-Phase Transformation Couplings in Metallic Alloys in Lemaître Handbook of Materials Behaviour Models*; Academic Press: Cambridge, MA, USA, 2001; pp. 896–904, ISBN 012-443341-3.
29. Denis, S. Prediction of residual stress and distortion of ferrous and non-ferrous metals: Current status and future developments. *J. Mater. Eng. Perform.* **2002**, *11*, 92–102. [CrossRef]
30. Mura, T. *Micromechanics of Defects in Solids*; Springer: Berlin, Germany, 1987; ISBN 978-94-009-3489-4.
31. Yilmaz, S. Phase transformations in thermally cycled Cu/ZrW_2O_8 composites investigated by synchrotron X-ray diffraction. *J. Phys. Condens. Matter* **2002**, *14*, 365–275. [CrossRef]
32. Zwigl, P. Transformation-mismatch plasticity of $NiAl/ZrO_2$ composites-finite-element modeling. *Mater. Sci. Eng. A* **2002**, *335*, 128–136. [CrossRef]
33. Armstrong, W.D. Fiber phase transformation and matrix plastic flow in a room temperature tensile strained NiTi phase shape memory alloy fiber reinforced 6082 Aluminium matrix composite. *Scr. Mater.* **1997**, *36*, 1037–1043. [CrossRef]
34. Zhang, S. Microstructural evolution and phase transformation during partial remelting of in-situ $Mg_2Si_p/AM60B$ composite. *Trans. Nonferr. Met. Soc. China* **2016**, *26*, 1564–1573. [CrossRef]
35. Luo, X. Effect of quenching on the matrix microstructure of SiCf/Ti-6Al-4V composites. *J. Mater. Sci.* **2018**, *53*, 1922–1932. [CrossRef]
36. Bruneseaux, F. Etude des Transformations de Phases des Alliages de Titane et Aciers Avec et Sans Charge Externe par DRX In Situ. Ph.D. Thesis, Institut National Polytechnique de Lorraine, Nancy, France, 2008.
37. AlMangour, B. Densification behavior, microstructural evolution, and mechanical properties of TiC/316L stainless steel nanocomposites fabricated by selective laser melting. *Mater. Des.* **2018**, *138*, 119–128. [CrossRef]
38. LabEx DAMAS. Available online: labex-damas.univ-lorraine.fr (accessed on 10 August 2018).

Article

A Multiscale Analysis on the Superelasticity Behavior of Architected Shape Memory Alloy Materials

Rui Xu [1,2,3], Céline Bouby [4], Hamid Zahrouni [1,3,*], Tarak Ben Zineb [1,4], Heng Hu [2] and Michel Potier-Ferry [1,3]

1 Laboratory of Excellence on Design of Alloy Metals for Low-Mass Structure (Labex-DAMAS), Université de Lorraine, 57070 Metz, France; ruixu@whu.edu.cn (R.X.); tarak.ben-zineb@univ-lorraine.fr (T.B.Z.); michel.potier-ferry@univ-lorraine.fr (M.P.-F.)
2 School of Civil Engineering, Wuhan University, 8 South Road of East Lake, Wuchang, 430072 Wuhan, China; huheng@whu.edu.cn
3 Université de Lorraine, CNRS, Arts et Métiers ParisTech, LEM3, F-57000 Metz, France
4 Université de Lorraine, CNRS, Arts et Métiers ParisTech, LEM3, F-54000 Nancy, France; Celine.Bouby@univ-lorraine.fr
* Correspondence: hamid.zahrouni@univ-lorraine.fr; Tel.: +33-03-7274-7815

Received: 31 July 2018; Accepted: 12 September 2018; Published: 17 September 2018

Abstract: In this paper, the superelasticity effects of architected shape memory alloys (SMAs) are focused on by using a multiscale approach. Firstly, a parametric analysis at the cellular level with a series of representative volume elements (RVEs) is carried out to predict the relations between the void fraction, the total stiffness, the hysteresis effect and the mass of the SMAs. The superelasticity effects of the architected SMAs are modeled by the thermomechanical constitutive model proposed by Chemisky et al. 2011. Secondly, the structural responses of the architected SMAs are studied by the multilevel finite element method (FE^2), which uses the effective constitutive behavior of the RVE to represent the behavior of the macroscopic structure. This approach can truly couple the responses of both the RVE level and structural level by the real-time information interactions between two levels. Through a three point bending test, it is observed that the structure inherits the strong nonlinear responses—both the hysteresis effect and the superelasticity—of the architected SMAs at the cellular level. Furthermore, the influence of the void fraction at the RVE level to the materials' structural responses can be more specifically and directly described, instead of using an RVE to predict at the microscopic level. Thus, this work could be referred to for optimizing the stiffness, the hysteresis effect and the mass of architected SMA structures and extended for possible advanced applications.

Keywords: shape memory alloys; architected cellular material; numerical homogenization; multiscale finite element method

1. Introduction

Cellular materials are widely used for their high strength-to-weight ratio and high energy absorption performance (Gibson and Ashby [1]; Ashby et al. [2]). For instance, honeycomb, folded cellular materials and foam are usually used as the core of the sandwich structures for dissipating the kinetic energy, damping or reducing the weight of the structure (Ashby et al. [2]; Yazdani et al. [3]; Garcia-Moreno [4]; Hangai et al. [5]; Strano et al. [6]). However, honeycomb and folded cellular materials have high manufacturing costs and moisture problem, as well as buckling problems (Rashed et al. [7]). The mechanical behaviors of foams are too difficult to be accurately measured for their stochastic cells, which always results in the excessive use of materials to satisfy the safety factor. To overcome these shortcomings, partially-ordered foams allowing limited structural control of the pore and spatial distribution of pore levels, such as metal syntactic foams (Taherishargh et al. [8–10];

Broxtermann et al. [11]; Linul et al. [12]; Luong et al. [13]) are studied. Furthermore, architected cellular materials with an ordered structure were designed and studied during these years (Rashed et al. [7]; Pingle et al. [14]; Schaedler et al. [15]; Schaedler and Carter [16]; Lehmhus et al. [17]). Thanks to the highly developed additive manufacturing techniques, such as 3D printing (Ngo et al. [18]; Mostafaei et al. [19]) and selective laser welding (Mehrpouya et al. [20]; Rashed et al. [7]), the manufacturing of architected cellular materials is no longer impossible. Users can design a cellular material with a certain behavior by tuning its cellular parameters, such as the geometry, components, local mechanical properties, etc.

Architected cellular materials' functionality could be extend by combining the features of various materials, such as shape memory alloys. It is well known that shape memory alloys, such as NiTi, can endure large deformation and recover their initial shape after unloading (see for example the reviews of Lagoudas [21], Patoor et al. [22], Lagoudas et al. [23], Tobushi et al. [24] and Cisse et al. [25]). This superelasticity effect brings high performance to SMA in energy absorption. When the given load reaches a critical level in a superelastic test, SMA will apparently soften due to its inner phase transformation. This behavior enables SMA to absorb the external energy as much as possible and prevents material from crushing or buckling. Such a kind of response is very similar to the ideal response of the cellular material designed by Schaedler et al. [15]. Meanwhile, the hysteresis effects of SMA can dissipate a large amount of energy. All mentioned features of SMAs meet the requirements of an architected structure for energy absorption applications very well. In addition, taking advantage of the lightweight and shape memory effect, architected SMAs may be designed for advanced applications in aerospace, civil engineering, etc.

To investigate the behavior of architected cellular SMAs, rare, but valuable works have been proposed (Machado et al. [26]; Ravari et al. [27]; Ashrafi et al. [28]). Machado et al. [26] proposed an experimental and modeling study on the cellular NiTi tube-based materials. In order to design and optimize the architected SMA tube materials, the authors studied the effective behavior of the thin-walled NiTi cellular materials by carrying out a study based on experiment and numerical simulation. The influences of SMAs' material properties and cellular architecture on the effective behavior were investigated. To reduce the high cost of fabrication, Ravari et al. [27] focused mainly on numerical modeling for designing and optimizing SMA cellular lattice structures. The effects of the geometry and cellular imperfections on the effective behavior of the material were investigated by unit cell and multi-cell methods. This work was later developed by Ashrafi et al. [28], who proposed an efficient unit cell model with modified boundary conditions for SMA cellular lattice structures. The shape memory effect was also simulated by this model, which had good agreement with the experiment.

The above works mainly focused on the effective cellular response in order to represent or predict the behavior of architected SMA structures. Considering the scale separation between the microscopic cellular scale and macroscopic structural scale, however, it is difficult to predict the structural response of a unit cell without certain assumed boundary conditions, because the stress-strain states of the macroscopic structure are usually not uniform and the deformations at the microscopic level could be totally different. Thus, in order to directly simulate the structural responses of architected SMA, more appropriate numerical methods should be used. During the past few decades, multiscale modeling approaches have been developed and widely used (Kanoute et al. [29]; Geers et al. [30]; El Hachemi et al. [31]; Kinvi-Dossou et al. [32]). As one of the most popular and effective multiscale methods, the multilevel finite element method (FE^2, see Feyel [33]) to describe the response of high nonlinear structures using generalized continua shows good performance in various applications, such as fiber buckling (Nezamabadi et al. [34]), composite shells (Cong et al. [35]), rate-dependent response (Tikarrouchine et al. [36]) and SMA-based fiber/matrix composites (Kohlhaas and Klinkel [37]; Chatzigeorgiou et al. [38]; Xu et al. [39]). In this approach, both the structural level and the RVE level are simulated by the finite element method (FEM). Two levels are fully coupled and computed simultaneously, where the unknown constitutive behaviors on the structure level are represented by the effective behaviors of homogenized RVEs, and the strain states of the RVEs are given by the associated integration points. The FE^2 method shows good performance on multiscale modeling of SMA-based materials. Xu et al. [39] proposed a 3D FE^2 model for simulating composites with stiff SMA fibers embedded in a soft matrix. This model was validated by the literature and showed a good ability in modeling the superelasticity and the shape memory effects of SMA composites.

Following previous works, it is therefore necessary and feasible to investigate the behavior of architected SMAs with a multilevel finite element model. Towards a better understanding of the studied architected SMA, unit cells (RVEs) with different void fractions are introduced to study the superelasticity effect of the materials and structures. In Section 2, a constitutive model for SMA is introduced briefly. Then, a parametric analysis with different void fractions at the cellular level is performed to predict the relations between the void fraction, the total stiffness, the hysteresis effect and the mass of the architected cellular SMAs. In Section 3, after a short presentation for the multilevel finite element model, multiscale modelings on architected SMA structures are carried out to simulate the structural and the cellular responses simultaneously.

2. Cellular Response

To investigate the architected SMA structures, RVEs with different geometrical parameters are studied firstly, then the structural response is investigated further by the computational homogenization method. First of all, a short presentation of the SMA constitutive model is given in the next subsection.

2.1. SMA Constitutive Model

The SMA model, proposed by Chemisky et al. [40], is adopted for the thermomechanical behavior modeling of the architected SMA. This model follows the work of Peultier et al. [41], who proposed a macroscopic phenomenological SMA approach based on the Gibbs free energy. This model was implemented on ABAQUS via user-defined materials (UMAT) and validated by experiments. It was later improved by Chemisky and Duval; see Chemisky et al. [40]; Duval et al. [42]. Here, only a brief introduction is presented for the readers due to the limited length of the paper.

The SMA constitutive model used here is able to describe four different strain mechanisms: the elastic strain $\boldsymbol{\varepsilon}^{e}$, the thermal expansion strain $\boldsymbol{\varepsilon}^{th}$, the martensitic transformation strain $\boldsymbol{\varepsilon}^{tr}$ and the twin accommodation strain $\boldsymbol{\varepsilon}^{tw}$. To this end, the total strain is decomposed in the following form:

$$\boldsymbol{\varepsilon} = \boldsymbol{\varepsilon}^{e} + \boldsymbol{\varepsilon}^{th} + \boldsymbol{\varepsilon}^{tr} + \boldsymbol{\varepsilon}^{tw}, \tag{1}$$

where the above strains are formulated as:

$$\begin{cases} \boldsymbol{\varepsilon}^{e} = \mathbb{S} : \boldsymbol{\sigma}, \\ \boldsymbol{\varepsilon}^{th} = \boldsymbol{\alpha}(T - T_{ref}), \\ \boldsymbol{\varepsilon}^{tr} = f \bar{\boldsymbol{\varepsilon}}^{tr}, \\ \boldsymbol{\varepsilon}^{tw} = f^{FA} \bar{\boldsymbol{\varepsilon}}^{tw}. \end{cases} \tag{2}$$

where $\mathbb{S}$ denotes the isotropic fourth order compliance tensor, $\boldsymbol{\alpha}$ represents the isotropic thermal expansion tensor, T_{ref} gives the temperature without expansion strain, f gives the martensitic volume fraction in SMA, the transformation strain $\bar{\boldsymbol{\varepsilon}}^{tr}$ describes the mean strain over the martensite related to martensite reorientation and the mean strain $\bar{\boldsymbol{\varepsilon}}^{tw}$ and self-accommodated martensitic volume fraction f^{FA} are introduced to describe the twin accommodation over the martensite.

The potential energy of the SMA model is based on the following Gibbs free energy expression:

$$
\begin{aligned}
G = &(U^A - TS^A)(1-f) + (U^M - TS^M)f - \frac{1}{2}\sigma : \mathbf{S} : \sigma - \sigma : \alpha\Delta T - \sigma : \bar{\varepsilon}^{tr} f - \sigma : \bar{\varepsilon}^{tw} f^{FA} \\
&+ \frac{1}{2} f H_{tr} \bar{\varepsilon}^{tr} : \bar{\varepsilon}^{tr} + \frac{1}{2} H_f f^2 + \frac{1}{2} f^{FA} H_{tw} \bar{\varepsilon}^{tw} : \bar{\varepsilon}^{tw} + C_v \left[(T - T_0) - T \log \frac{T}{T_0} \right] \\
= &U^A - TS^A + B(T - T_0)f - \frac{1}{2}\sigma : \mathbf{S} : \sigma - \sigma : \alpha\Delta T - \sigma : \bar{\varepsilon}^{tr} f - \sigma : \bar{\varepsilon}^{tw} f^{FA} + \frac{1}{2} f H_{tr} \bar{\varepsilon}^{tr} : \bar{\varepsilon}^{tr} \\
&+ \frac{1}{2} H_f f^2 + \frac{1}{2} f^{FA} H_{tw} \bar{\varepsilon}^{tw} : \bar{\varepsilon}^{tw} + C_v \left[(T - T_0) - T \log \frac{T}{T_0} \right].
\end{aligned} \tag{3}
$$

where U^A and U^M are the austenitic and the martensitic internal energy and S^A and S^M are the austenitic and the martensitic entropy. Several terms, such as $\Delta U = U^M - U^A$, $\Delta S = S^M - S^A$ and $\Delta T = T - T_{ref}$, are also introduced. Considering $T_0 = \frac{\Delta U}{\Delta S}$ the equilibrium temperature of transformation, a linear variation of entropy around T_0 is defined as $B = -\Delta S$. The terms H_{tr}, H_f and H_{tw} comprise, respectively, a set of material parameters characterizing interactions between grains of the microscopic RVE, between variants inside grains and between twins. C_v is the transformation latent heat coefficient. The control equation of the thermodynamic system is established by introducing the Clausius–Duhem inequality:

$$
-\dot{G} - \varepsilon : \dot{\sigma} - S\dot{T} - \vec{q} \cdot \frac{\overrightarrow{grad}\, T}{T} \geq 0. \tag{4}
$$

Substituting Equation (3) to Equation (4) and considering the thermo-elastic balance conditions, the Clausius–Duhem inequality is reduced to:

$$
\phi = -\frac{\partial G}{\partial f}\dot{f} - \frac{\partial G}{\partial \bar{\varepsilon}^{tr}} : \dot{\bar{\varepsilon}}^{tr} - \frac{\partial G}{\partial \bar{\varepsilon}^{tw}} : \dot{\bar{\varepsilon}}^{tw} - \vec{q} \cdot \frac{\overrightarrow{grad}\, T}{T} \geq 0. \tag{5}
$$

Here, driving forces related to this dissipation expression are introduced:

1. Transformation driving force related to f:

$$
\begin{aligned}
A_f = -\frac{\partial G}{\partial f} = &\sigma : \bar{\varepsilon}^{tr} + \zeta^{FA} \sigma : \bar{\varepsilon}^{tw} - B(T - T_0) - \frac{1}{2} f H_{tr} \bar{\varepsilon}^{tr} : \bar{\varepsilon}^{tr} \\
&- H_f f - \frac{1}{2} \zeta^{FA} H_{tw} \bar{\varepsilon}^{tw} : \bar{\varepsilon}^{tw}.
\end{aligned} \tag{6}
$$

   where the definition of $\dot{f}^{FA} = \zeta^{FA} \dot{f}$ is introduced here.
2. Orientation force related to $\bar{\varepsilon}^{tr}$:

$$
A_{\bar{\varepsilon}^{tr}} = -\frac{1}{f} \frac{\partial G}{\partial \bar{\varepsilon}^{tr}} = \sigma' - H_{tr} \bar{\varepsilon}^{tr}. \tag{7}
$$

   where σ' denotes the deviatoric part of the stress tensor σ.
3. Twin accommodation related to $\bar{\varepsilon}^{tw}$:

$$
A_{\bar{\varepsilon}^{tw}} = -\frac{1}{f^{FA}} \frac{\partial G}{\partial \bar{\varepsilon}^{tw}} = \sigma' - H_{tw} \bar{\varepsilon}^{tw}. \tag{8}
$$

By introducing a series of predefined critical thresholds related to these driving forces, the evolution of phase transformation is controlled. For more clarity of the relations between the main control equations, the constitutive equations for this model are summarized in Table 1.

The implementation of this model in a finite element code is realized; for more details about the implementation process, see Chemisky et al. [40] and Duval et al. [42]. Moreover, the evolution of tension-compression asymmetry and internal loops during the partial loadings can also be well simulated with this adopted model. In the following subsection, architected SMA RVEs with different geometric parameters are studied to see the cellular response of the material.

Table 1. Summary of the control equations for the SMA model.

Strain mechanisms:
$\varepsilon = \varepsilon^e + \varepsilon^{th} + \varepsilon^{tr} + \varepsilon^{tw}$.
Thermodynamical potential:
$G = U^A - TS^A + B(T - T_0)f - \frac{1}{2}\sigma : \mathbb{S} : \sigma - \sigma : \alpha\Delta T - \sigma : \bar{\varepsilon}^{tr} f - \sigma : \bar{\varepsilon}^{tw} f^{FA} + \frac{1}{2} f H_{tr} \bar{\varepsilon}^{tr} : \bar{\varepsilon}^{tr} +$
$\frac{1}{2} H_f f^2 + \frac{1}{2} f^{FA} H_{tw} \bar{\varepsilon}^{tw} : \bar{\varepsilon}^{tw} + C_v \left[(T - T_0) - T \log \frac{T}{T_0}\right]$.
Clausius–Duhem inequality:
$-\dot{G} - \varepsilon : \dot{\sigma} - S\dot{T} - \vec{q} \cdot \frac{\overrightarrow{grad}\, T}{T} \geq 0$.
Thermo-elastic balance conditions:
$\mathbb{S} : \sigma + \alpha\Delta T + f\bar{\varepsilon}^{tr} + f^{FA}\bar{\varepsilon}^{tw} - \varepsilon = 0$,
$S^A - Bf + \sigma : \alpha - C_v \left(\log \frac{T}{T_0}\right) - S = 0$.
Thermodynamic forces:
$A_f = -\frac{\partial G}{\partial f} = \sigma : \bar{\varepsilon}^{tr} + \zeta^{FA}\sigma : \bar{\varepsilon}^{tw} - B(T - T_0) - \frac{1}{2} f H_{tr} \bar{\varepsilon}^{tr} : \bar{\varepsilon}^{tr} - H_f f - \frac{1}{2}\zeta^{FA} H_{tw} \bar{\varepsilon}^{tw} : \bar{\varepsilon}^{tw}$,
$A_{\bar{\varepsilon}^{tr}} = -\frac{1}{f}\frac{\partial G}{\partial \bar{\varepsilon}^{tr}} = \sigma' - H_{tr}\bar{\varepsilon}^{tr}$,
$A_{\bar{\varepsilon}^{tw}} = -\frac{1}{f^{FA}}\frac{\partial G}{\partial \bar{\varepsilon}^{tw}} = \sigma' - H_{tw}\bar{\varepsilon}^{tw}$.
Criterion functions:
$F_f^{crit} = F_f^{max} + (B_f - B) \cdot (T - T_0)\ if\ \dot{f} > 0$,
$F_f^{crit} = -F_f^{max} + (B_r - B) \cdot (T - T_0)\ if\ \dot{f} < 0$.
Physical limitations:
$0 \leq f \leq 1$.

2.2. Convergence Analysis for the RVE Mesh

An architected SMA RVE is introduced to describe the microscopic structure; see Figure 1. It is formed by excavating cylindrical holes through the center of each face of a cube SMA. The size of the cube is given by 1 mm × 1 mm × 1 mm, while the radius of the cylindrical holes is given as 0.38 mm. According to computational homogenization theory, periodic boundary conditions are introduced into the RVE by the multi-point constraints (MPCs) on ABAQUS:

$$\Delta\mathbf{u}^+ - \Delta\mathbf{u}^- = \Delta\bar{\varepsilon}\cdot(\mathbf{x}^+ - \mathbf{x}^-)\ on\ \partial\omega, \tag{9}$$

where $\mathbf{u}$ is the displacement vector, $\mathbf{x}$ is the coordinates of nodes and $\bar{\varepsilon}$ is the strain load applied on the RVE. The notations $+$ and $-$ denote the nodes on opposite boundaries; the notation Δ represents the incremental case. Thus, a set of reference points (RPs) for three different directions is introduced in MPCs and the meshes on the opposite boundaries must be the same in order to create periodic boundaries; see Figure 1. Both the continuum 3D solid element with full integration (labeled C3D8 on ABAQUS) and continuum 3D solid element with reduced integration (labeled C3D8R on ABAQUS) for the mesh in Figure 1a are studied in order to optimize the computation efficiency on the RVE. Following the constitutive model introduced in the last subsection, the material parameters of a conventional NiTi alloy are given in Table 2, which are identified with the experimental data of Sittner et al. [43].

Table 2. Material parameters for SMA.

E (MPa)	39,500	B_r (MPa °C^{-1})	7	H_f (MPa)	2
ν	0.3	B_f (MPa °C^{-1})	7	H_{tr} (MPa)	1635
ε^T_{trac}	0.056	M_s (°C)	−80	H_{tw} (MPa)	25,000
ε^{TFA}_{trac}	0.053	A_f (°C)	−2	H_s (MPa)	68.5
ε^T_{comp}	0.044	F_ε (MPa)	220		

To check the effective response of the RVE, a tensile strain up to 10% in the Y direction is applied on the RVE at a constant temperature of 30 °C.

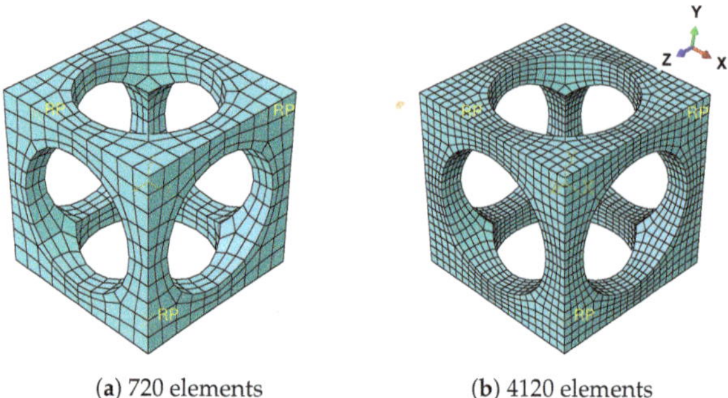

(**a**) 720 elements (**b**) 4120 elements

Figure 1. Architected RVE with fine meshes and reference points (RPs).

To describe the effective behavior of RVE, the averaged values, such as stress $\bar{\sigma}$, strain $\bar{\varepsilon}$ and martensitic volume fraction $\bar{f}_t$, are introduced. The averaged values are computed by volume averaging over the whole 1 mm^3 cube from Element 1 to N, where N denotes the total element number. Note the averaged strain is equal to the prescribed strain due to the periodic boundary conditions, Equation (9). Figure 2a shows the stress-strain relations simulated with different meshes and element types. The hysteresis effect is observed during the loading/unloading cycle. This nonlinear response is caused by the inner forward phase transformation of SMA. The averaged martensitic volume fraction $\bar{f}_t$ increases along with the loads until reaching a maximum value; see Figure 2b. When the unloading begins, $\bar{f}_t$ decreases immediately until recovering to zero. From the comparison of the curves, it is observed that the mesh with 720 elements is fine enough to model this architected SMA RVE. Furthermore, the RVEs simulated by the C3D8R and C3D8 elements have exactly the same responses; see Figure 2.

Generally speaking, the mesh with 720 C3D8R elements has relatively enough accuracy and lower computational cost than other meshes. Thus, meshes like density and C3D8R elements are used for the RVE in the following work.

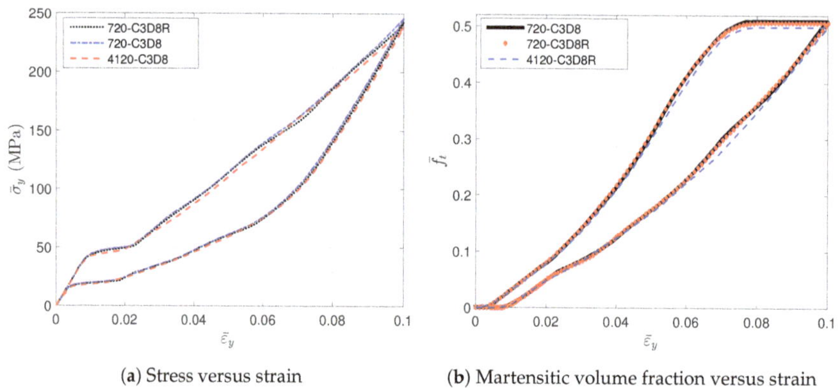

(**a**) Stress versus strain (**b**) Martensitic volume fraction versus strain

Figure 2. The evolution of the averaged stress and the averaged martensitic volume fraction simulated by RVEs with different meshes.

2.3. Cells with Different Geometries

In this subsection, five types of RVEs with different void volume factions ξ are studied, as illustrated in Figure 3. The size of the cube is given by 1 mm × 1 mm × 1 mm, while the radius of the cylindrical hole varying along with ξ is given in Table 3. The material parameters of SMA remain the same as those in the last subsection.

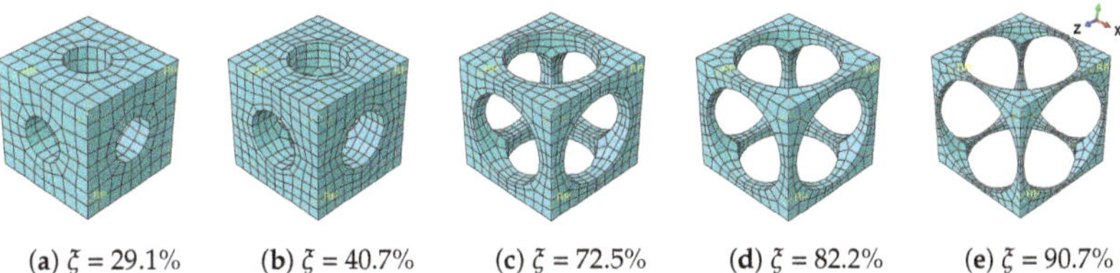

(a) ξ = 29.1% (b) ξ = 40.7% (c) ξ = 72.5% (d) ξ = 82.2% (e) ξ = 90.7%

Figure 3. Meshes for architected SMA RVEs with different void volume fractions ξ.

Table 3. Geometrical parameters for RVEs.

Radius (mm)	**0.2**	**0.24**	**0.38**	**0.42**	**0.47**
Void volume fraction ξ	29.1%	40.7%	72.5%	82.2%	90.7%

A tensile strain up to 10% and a compressive strain up to −10% in the Y direction are respectively applied on the RVEs to simulate effective responses for different RVEs. Figure 4 gives the curves of averaged stress versus the averaged stain along the loading direction, simulated by the above RVEs respectively. The absolute value of stress in different RVEs at a certain strain level increases along with the decreasing of the void volume fraction, as depicted in Figure 4. In more detail, Figure 5 gives the relation of maximum stress and material volume fraction (1-ξ) when the absolute value of strain is up to 10%. The trend of the curve shows that the higher the material volume fraction is, the faster the stress increases. This trend could be explained by exploring the stress distribution on the RVE at the maximum strain level 10%, as shown in Figures 6 and 7. The high stresses are mainly distributed over the middle of the pillars along the loading direction. Lower material volume fraction results in a stronger stress concentration effect. Thus, the stiffness of the RVE decreases more rapidly than the material volume fraction. This test could be a reference for designers to balance the stiffness and the mass of the architected structure.

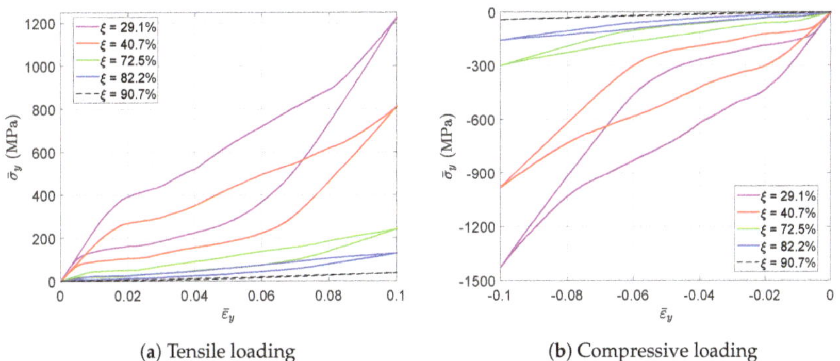

(a) Tensile loading (b) Compressive loading

Figure 4. The evolution of the averaged stress-strain curves simulated by RVEs with different void fractions.

Figure 8 gives the martensitic volume fraction averaged over SMA versus the strain averaged over the cube. The martensitic volume fraction $\bar{f}_t$ increases along with the increasing of the strain level, but stops before it reaches one. More specifically, each RVE has a certain maximum $\bar{f}_t$ value in this loading case. The maximum martensitic volume fraction versus the material volume fraction at the maximum absolute strain level of 10% is depicted in Figure 9. It shows that the lower the material volume fraction is, the more rapidly the martensitic volume fraction decreases. This also means the usage rate of SMA is at a very low level when the material volume fraction is low, and it could result in a waste of SMA. For example, the hysteresis effect of the SMA architected structure is not proportional to the SMA mass, but to the product of the SMA mass and the SMA usage rate. Thus, the hysteresis effect could be very weak when the SMA usage rate is very low. Like the stiffness discussed before, this phenomenon could also be explained by the stress concentration where only the high stress area transforms into martensite; see Figures 10 and 11. This also means SMA cannot completely transform from austenite into martensite. Thus, as long as the unloading begins, the reverse transformation starts immediately, resulting in the reduction of $\bar{f}_t$.

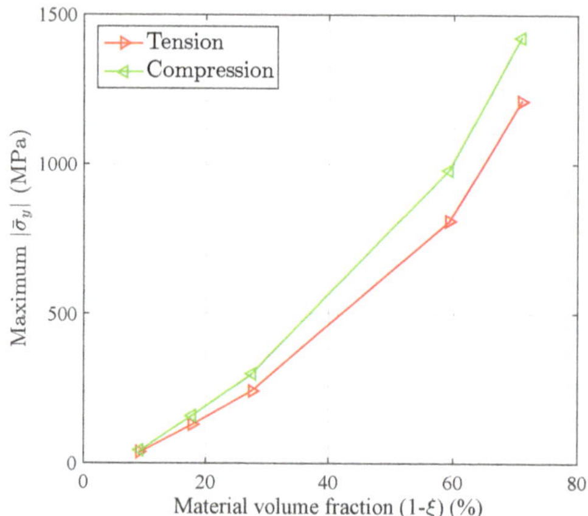

Figure 5. The relation between the stiffness and the material volume fraction (1-ξ) of the RVE when the absolute value of strain is up to 10%.

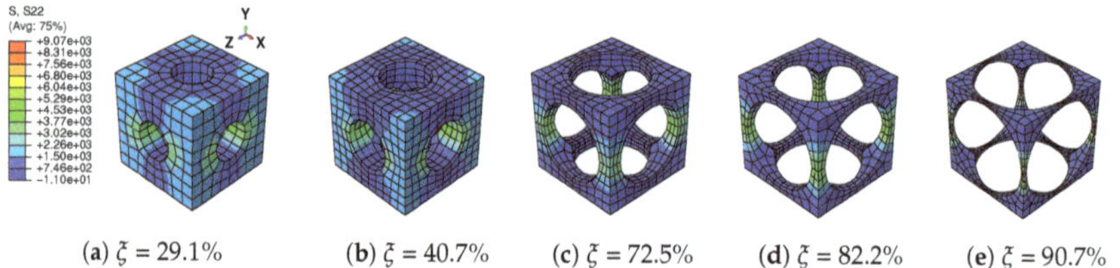

(a) ξ = 29.1% (b) ξ = 40.7% (c) ξ = 72.5% (d) ξ = 82.2% (e) ξ = 90.7%

Figure 6. Distribution of stress in Y direction (S22) on the RVEs with different void volume fractions ξ in tension loading at strain level 10%.

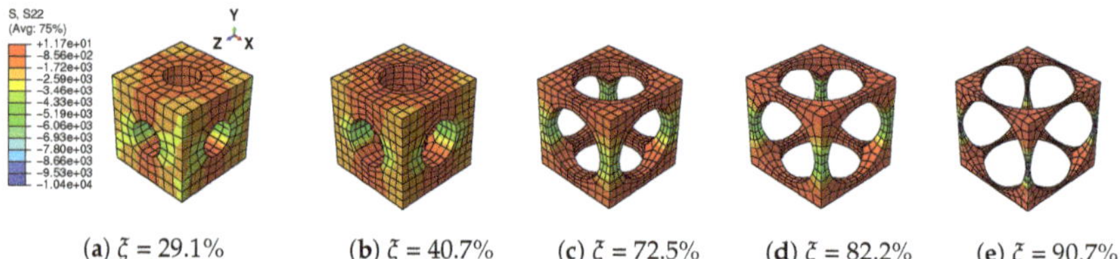

(a) ξ = 29.1% (b) ξ = 40.7% (c) ξ = 72.5% (d) ξ = 82.2% (e) ξ = 90.7%

Figure 7. Distribution of stress in Y direction (S22) on the RVEs with different void volume fractions ξ in compression loading at strain level −10%.

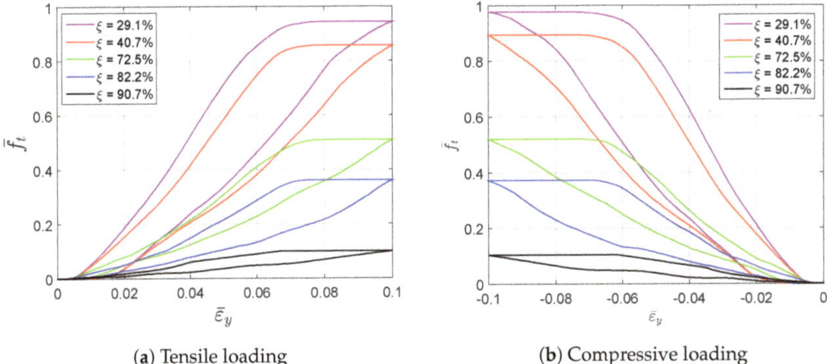

(**a**) Tensile loading (**b**) Compressive loading

Figure 8. The evolution of the martensitic volume fraction versus the averaged strain curves simulated by RVEs with different void volume fractions.

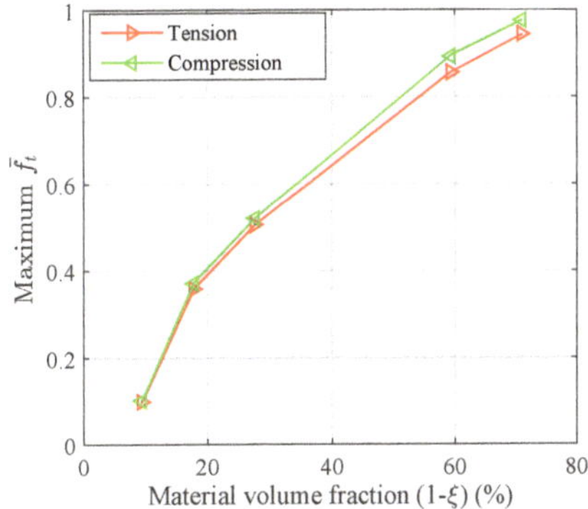

Figure 9. The relation between maximum martensitic volume fraction versus the material volume fraction when the absolute value of strain is up to 10%.

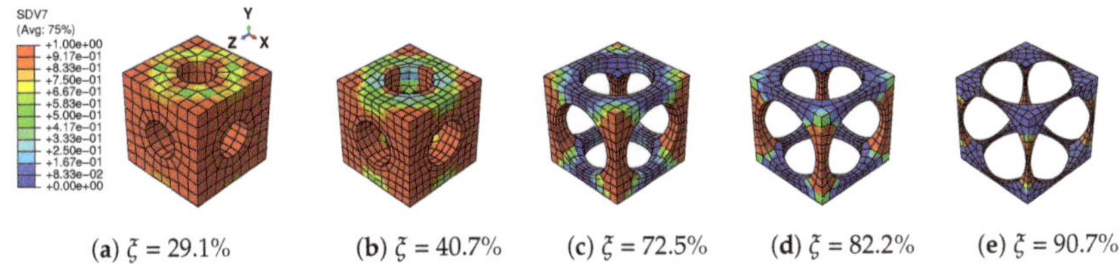

(**a**) ξ = 29.1% (**b**) ξ = 40.7% (**c**) ξ = 72.5% (**d**) ξ = 82.2% (**e**) ξ = 90.7%

Figure 10. Distribution of martensitic volume fraction $\bar{f}_t$ (named SDV7 in colorbar) on the RVEs with different void volume fractions ξ in tension loading at strain level 10%.

To design a light weight architected SMA structure, it is required to minimize the material volume fraction in order to decrease the weight and cost of the structure. However, the necessary stiffness of the structure should be firstly satisfied. It should be noticed that the waste of SMA should also be avoided as much as possible due to SMA's high price. Therefore, the results given in Figures 5 and 9 can be a reference for the designers to balance the stiffness, mass and SMA usage according to their specific requirements.

Following the tests at the RVE level, the structural response of the architected SMAs is studied in the next section.

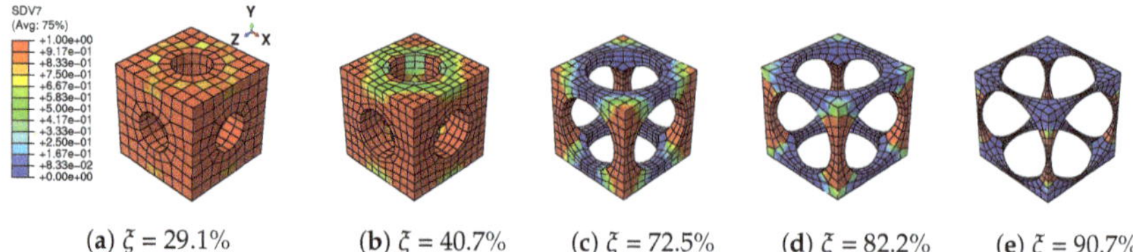

Figure 11. Distribution of martensitic volume fraction $\bar{f}_t$ (named SDV7 in colorbar) on the RVEs with different void volume fractions ξ in compression loading at strain level -10%.

3. Structural Response

In this section, the structural responses of the architected SMAs are focused on. The constitutive behavior of the macroscopic structure is represented by the computational homogenized RVEs in the last subsection. To simulate the structural response, a short introduction of the multilevel finite element method (FE^2) is presented firstly.

3.1. FE^2 Formulation

Generally speaking, an architected SMA with a periodic microscopic structure can be divided into two scales and be simulated by the multiscale homogenization method. As depicted in Figure 12, each point at the macroscopic level is represented by a periodic RVE after homogenization. The constitutive behavior and the stress of each macroscopic point are transferred from the RVE by the computational homogenization technique, while the macroscopic strain is applied on the RVE with periodic boundary conditions. Both levels are simulated by FEM, which can capture their mechanical fields accurately. This approach is implemented on ABAQUS via the user subroutine UMAT. The real-time interaction of the two levels is realized by the iteration of the Newton–Raphson method. This method is able to compute both the macroscopic and microscopic responses of the structures simultaneously. The authors have developed this FE^2 approach on the ABAQUS platform for SMA-based composites; see Xu et al. [39] for more detailed formulations. Related valuable works on multiscale modeling of SMA composite could be also referred to, such as Kohlhaas and Klinkel [37], Chatzigeorgiou et al. [38], Chatzigeorgiou et al. [44] and Fatemi Dehaghani et al. [45].

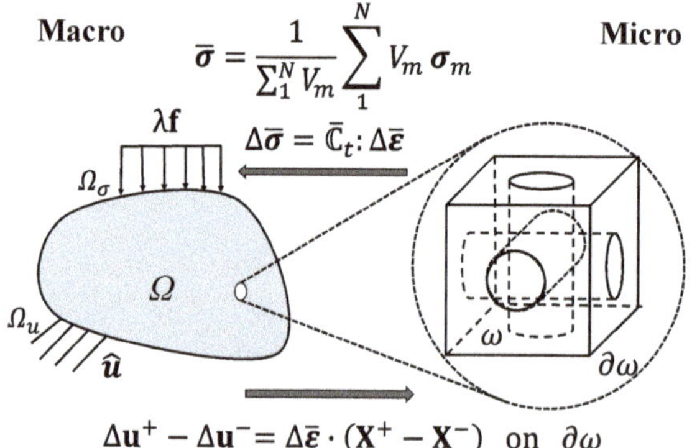

Figure 12. Basic concept of the finite element squared method.

3.2. Beam with Three Kinds of Cells Subjected to Three-Point Bending

The multiscale finite element method mentioned above has been validated by the reference and shows good ability in modeling the superelasticity and the shape memory effect of the SMA composites [39]. Therefore, this multiscale model is used to simulate the multiscale response of the architected SMA structure herein. A 3D beam subjected to three-point bending load is shown in Figure 13.

This beam is composed of architected SMA. The width, length and height of the beam are 5 mm, 20 mm and 5 mm, respectively. Edge Y = 5 mm, Z = 10 mm is fixed in the Y and Z directions. A displacement load up to 0.5 mm in the Y direction is applied on the edges Y = Z = 0 mm and Z = 20 mm, Y = 0 mm. Node X = 0 mm, Y = 5 mm, Z = 10 mm is fixed in the X direction in order to eliminate the rigid body displacement in the X direction. Considering the symmetry of the structure and the boundaries, only the left half of the structure is simulated in order to reduce the computation cost. To do this, an additional displacement constraint in the Z direction is given on face Z = 10 mm. Since the deformation in the X direction is not obvious in the three-point bending test, one element is used in this direction. Each edge in the Y direction is meshed by two elements and in the Z direction by four elements; see Figure 14. The continuum 3D solid element with incompatible modes (labeled C3D8I in ABAQUS) is adopted for the modeling of the macroscopic beam since it is enhanced by incompatible modes to improve its bending behavior. The RVEs studied in the last section are used herein with void volume fractions ξ of 40.7%, 72.5% and 82.2%, respectively.

As the macroscopic constitutive behavior on each integration point is not clear, it has to be represented by the effective behavior of the associated RVE at each macroscopic increment. A brief flow diagram, showing how this multiscale problem is solved, is illustrated in Figure 15. Specifically, the effective behavior of the RVE is computed by seeking the relations between the averaged stresses and averaged strains over the RVE via a series of loading tests. Once the effective behavior is obtained, the macroscopic problem is to be solved. Considering the nonlinear response of the RVE during loading, the macroscopic convergence has to be checked in each macroscopic iteration of the Newton–Raphson framework. In an iteration, the strain states of RVEs are updated with the macroscopic strains, and in return, the macroscopic stresses are renewed by updating the averaged stress of the RVE at the new strain states.

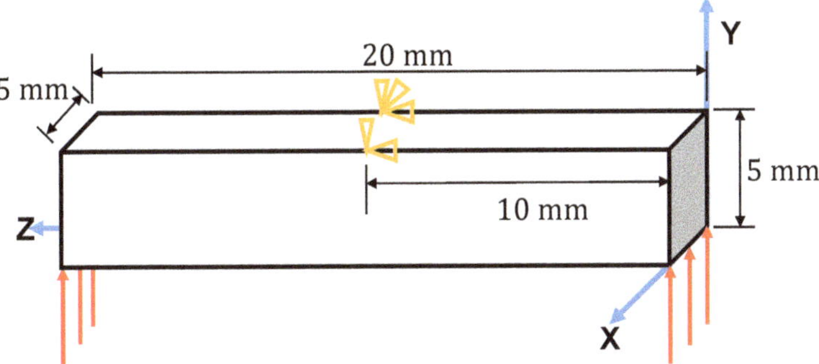

Figure 13. Geometry and boundary conditions for the three-point bending beam.

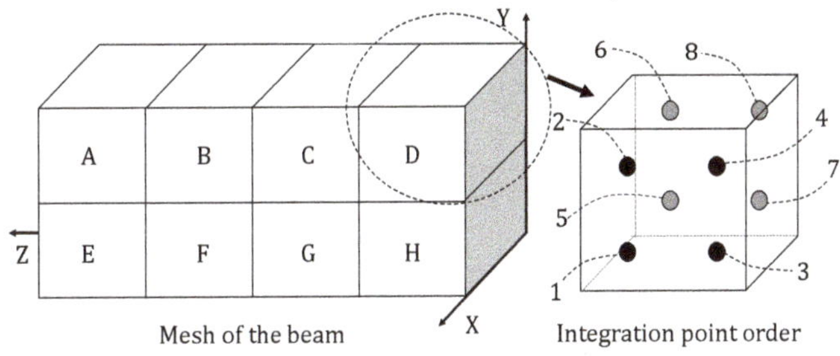

Figure 14. The meshes for the left half beam and the integration points in each C3D8I element.

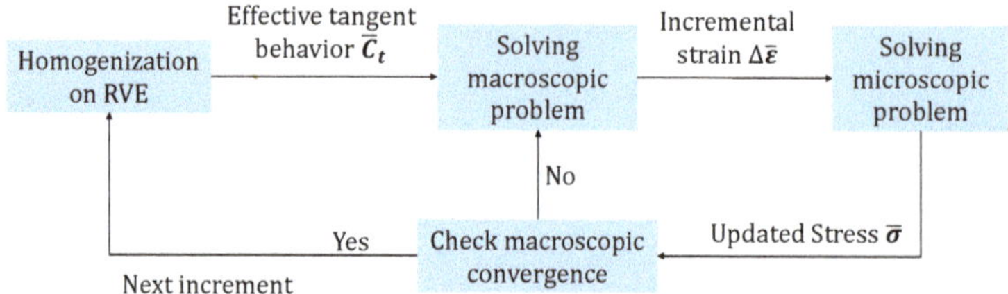

Figure 15. The nonlinear interaction between two scales in the Newton–Raphson framework.

3.3. Stress Distributions at the Macroscopic and Microscopic Levels

Let us consider the case with volume fraction $\xi = 40.7\%$. The stress distribution of the bending beam with the prescribed displacement in the Y direction reaching 0.5 mm is illustrated in Figure 16. The deformations of both the macroscopic and microscopic structures are magnified by three times. The high compressive stresses in the Z direction are observed in the elements above the middle plane Y = 2.5 mm, while in contrast, the compressive stresses are observed in the elements below the middle plane. In the meantime, for different macroscopic points, the RVEs have different stress states in correspondence with their associated macroscopic strain states. To specify the microscopic structure more clearly, we introduce RVE_{kl} to denote an RVE corresponding to the integration point l of the macroscopic element k. The RVEs above the middle plane, such as RVE_{A4} and RVE_{D4}, are mainly subjected to compression, while the RVEs below the middle plane, such as RVE_{E3} and RVE_{H3}, are mainly subjected to tension. The stresses in the RVEs far from face Z = 10 mm, such as RVE_{A4} and RVE_{E3}, are at a relatively low level compared to those RVEs near face Z = 10 mm, such as RVE_{D4} and RVE_{H3}.

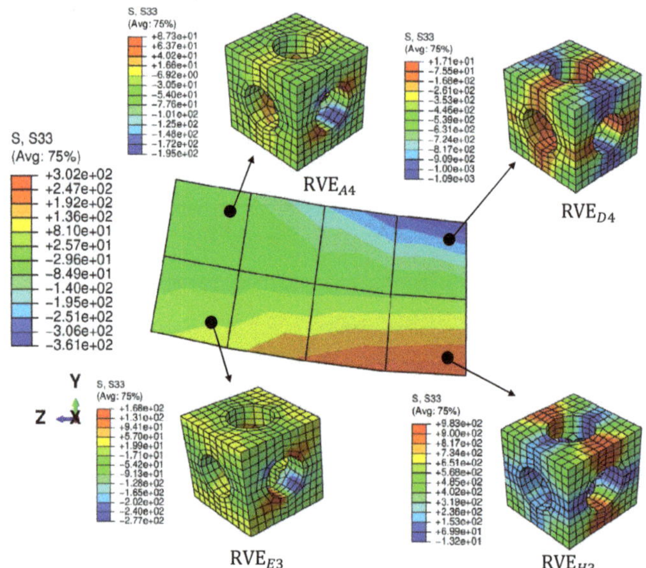

Figure 16. The distribution of stress in Z direction (S33) for the macroscopic structure and microscopic structure with $\xi = 40.7\%$, where the deformations of both levels are magnified by three times.

3.4. Evolution of the Loading

Figure 17 shows the nonlinear response of the macroscopic structure with three different void fractions. The linear response is observed at the very beginning, and the microscopic structures deform without any phase transformation. As the loading increases, the forward phase transformation begins over the high stress area of the beam. For example, Figure 18 gives the stress strain relations of integration points in element D and element H with ξ = 40.7%, which also represent the effective behavior of the associated RVEs. Note that Integration Points 5 to 8 in each element are not illustrated considering the symmetry of the structure in the middle plane X = 2.5 mm. It is observed that the hysteresis effects in RVE_{D3} and RVE_{H3} are relatively much more obvious than those in RVE_{D4} and RVE_{H4}; because RVE_{D3} and RVE_{H3} are located in the high stress area. The martensite transformation states of the RVEs with the prescribed displacement in the Y direction reaching 0.5 mm, shown in Figure 19, also give a reasonable confirmation from the RVE level. This figure depicts the RVEs corresponding to these integration points closest to the face Z = 10 mm. Only the RVEs far from the middle plane Y = 2.5 mm have an obvious phase transformation. Simultaneously, the phase transformation at the RVE level is accompanied by the softening of macroscopic stiffness. This softening goes gradually, because the thermomechanical phase transformations at the microscopic level are not synchronous, considering that the stress states of the macroscopic structure are not uniform over the whole beam and that the microscopic structure is architected. Once the unloading begins, the behaviors of the structure and RVEs immediately turn into the linear case. Later, the reverse transformation follows, and finally, all the phases transform back into austenite.

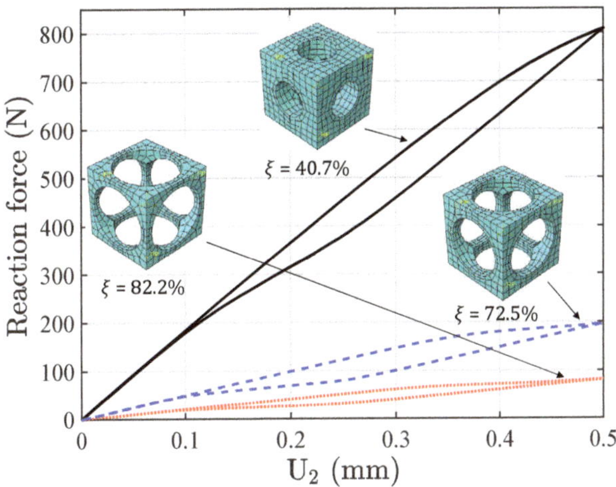

Figure 17. Force-displacement curve of the boundary Z = 20 mm, Y = 0 mm on the bending beam.

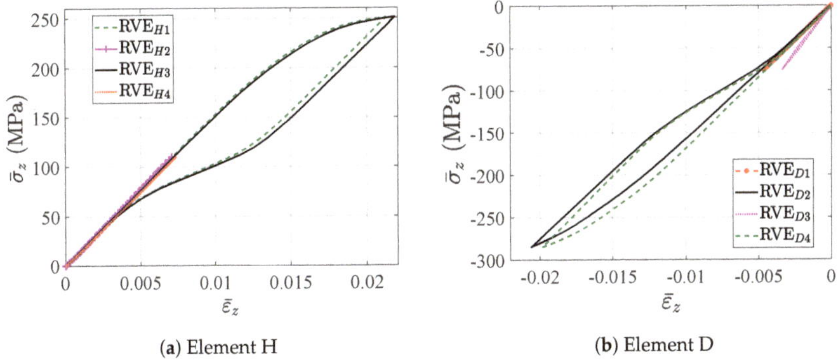

(**a**) Element H (**b**) Element D

Figure 18. The stress-strain relations on the macroscopic integration points with ξ = 40.7%.

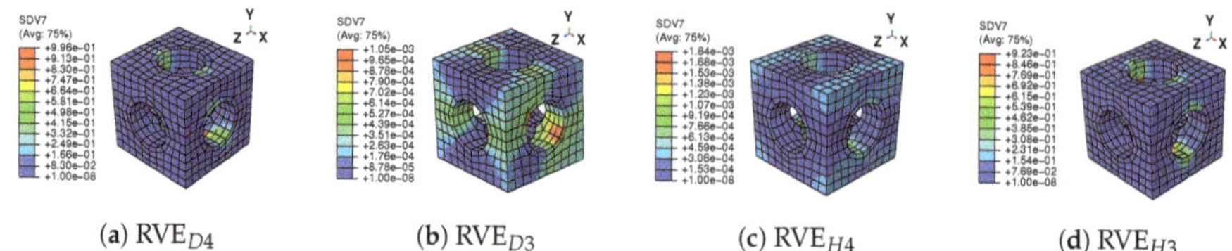

(a) RVE_{D4} (b) RVE_{D3} (c) RVE_{H4} (d) RVE_{H3}

Figure 19. Distribution of martensitic volume fraction $\bar{f}_t$ (named SDV7 in colorbar) on the RVEs with the prescribed displacement in the Y direction reaching 0.5 mm and ξ = 40.7%.

3.5. Structural Response with Different Microscopic Structures

The differences between structural responses with three void fractions are observed in Figure 17. With a higher void fraction, the macroscopic structure shows lower stiffness and a weaker hysteresis effect, because the macroscopic behavior is represented by the mean behavior of the microscopic structures. The stress-strain relations of Integration Points *H3* and *D4*, which are also the averaged stress-strain relations of RVE_{H3} and RVE_{D4}, are depicted in Figure 20 to investigate the microscopic response of the structure. Figure 21 gives the stress distributions in the Z direction for associated RVEs at maximum loads corresponding to the displacement of the beam ends U_2 = 0.5 mm. It is observed that the averaged stiffness and the hysteresis effect of the RVE with a higher void fraction are lower than those of the RVE with a lower void fraction, which is consistent with the response at the macroscopic level. As we have discussed in Section 2, the high void fraction results in less SMA, which can provide low stiffness for the structure. For this architected structure, the stresses are concentrated over the pillars along the main loading direction when the void fraction is high. This also results in the martensite phase being mainly concentrated over the pillars.

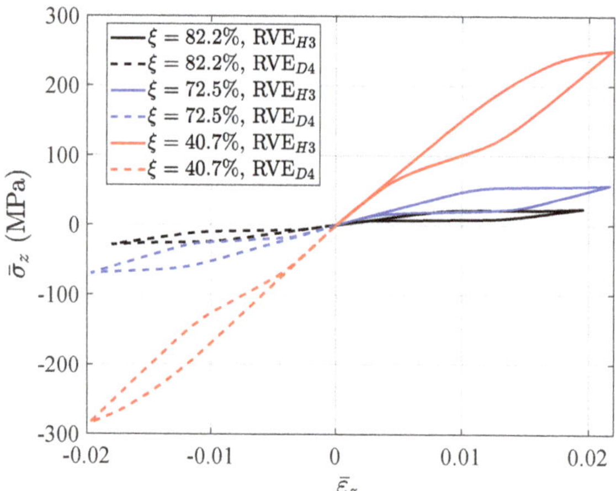

Figure 20. The comparison of the stress-strain relations on the macroscopic Integration Points *H3* and *D4* simulated by three kinds of RVEs respectively.

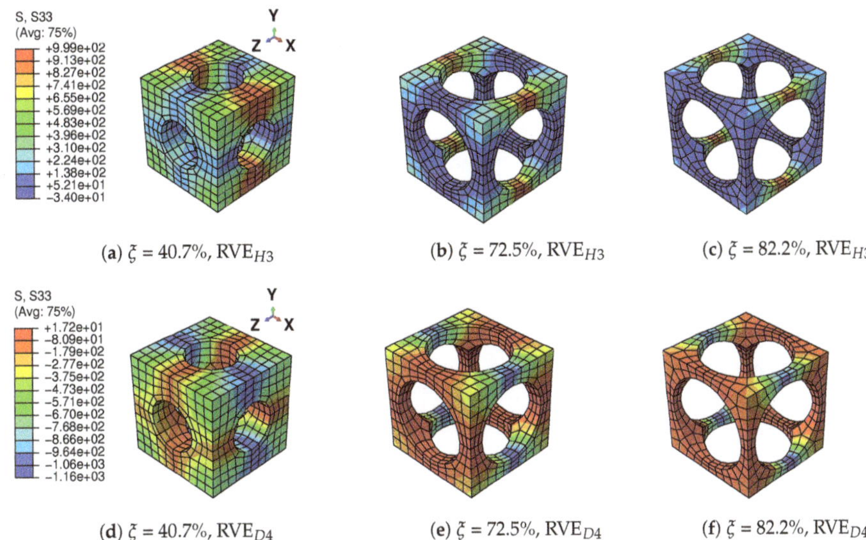

(**a**) ξ = 40.7%, RVE$_{H3}$ (**b**) ξ = 72.5%, RVE$_{H3}$ (**c**) ξ = 82.2%, RVE$_{H3}$

(**d**) ξ = 40.7%, RVE$_{D4}$ (**e**) ξ = 72.5%, RVE$_{D4}$ (**f**) ξ = 82.2%, RVE$_{D4}$

Figure 21. Distribution of stress in Z direction (S33) on RVE$_{H3}$ and RVE$_{D4}$ of different beams corresponding to the displacement of the beam ends U_2 = 0.5 mm.

3.6. Comments on the Computational Efficiency

The proposed numerical tool contributes to the design of innovative SMA applications, which can optimize the quantity of material with the required functionalities (actuation, recovery force, damping, energy absorption, etc.), lighten the structure and consider alternative manufacturing processes such as additive manufacturing. The latter can consider more complex geometries of the unit cell, which has a direct influence on the overall behavior of the structure. Note that this numerical multiscale procedure is very time consuming. To reduce the CPU time, the authors propose to associate this method with model reduction techniques. Among these techniques, we can introduce the proper orthogonal decomposition (POD) to build up a small number of basis functions where the solution can be computed (Yvonnet et al. [46]). When this technique is used in an optimization procedure requiring several repetitive calculations, it can significantly reduce the computation time. We can also consider the proper generalized decomposition (PGD) technique (Ammar et al. [47]; Kpogan et al. [48]), which can reduce the dimensionality of the problem to be solved and which is well adapted for unsteady problems, as well as parametric studies. Another technique that has proven its efficiency in solving non-linear problems and that was developed by our group is the asymptotic numerical method (ANM). This technique is based on the development of variables in the form of Taylor series truncated at large order, which allows one to reduce the computation time significantly by reducing the number of computation steps. This technique can be considered as a high order predictor algorithm without the need for any iteration phase. It is particularly suited for strong nonlinearities and for thin-structure problems involving instability phenomena (Nezamabadi et al. [49]; Assidi et al. [50]; Aggoune et al. [51]). In addition, bridging techniques (Hu et al. [52]; Yu et al. [53]) can also be considered to combine the advantages of different reduced finite element models. These tools will be associated with the present algorithm to reduce the computation time and to deal with complex structural geometry and complex response curves.

4. Conclusions

In this paper, the superelasticity behavior of the architected SMA structure is studied with a 3D multiscale finite element model. Firstly, RVEs with architected SMA are built to simulate the cellular responses of the structure. The behavior of SMA is described by the constitutive model proposed by Chemisky et al. [40]. Both tension and compression loading cycles are applied on the RVEs with different void fractions. The superelasticity responses and the hysteresis effects are observed in the RVEs. The effect of changing the void fraction on the stiffness, the maximum martensitic volume fraction and hysteresis effect are discussed in detail. Moreover, a multiscale approach is carried out to model the structural response, as well as the cellular response. The relations between the structural and cellular responses are studied in a three-point bending test. It is observed that the macroscopic response is related to the phase transformation in the RVE, which changes the effective constitutive behavior of the structure. The stress-strain state of the RVE directly depends on the stress strain state of the associated macroscopic point. Thus, the multiscale model is necessary and successful for simulating the nonlinear behavior of the architected SMA structure. Furthermore, structural responses with different void fractions are studied, which gives a good reference of the void fractions' influences on structural stiffness and hysteresis.

Author Contributions: Conceptualization, C.B., H.Z., T.B.Z., H.H. and M.P.-F. Formal analysis, R.X. and C.B. Funding acquisition, H.Z., T.B.Z. and H.H. Investigation, R.X. and T.B.Z. Methodology, R.X., C.B., H.Z., T.B.Z. and M.P.-F. Project administration, H.Z. and T.B.Z. Resources, H.Z. and H.H. Supervision, H.Z., T.B.Z., H.H. and M.P.-F. Visualization, R.X. Writing, original draft, R.X. Writing, review and editing, C.B., H.Z., T.B.Z., H.H. and M.P.-F.

Funding: This research received no external funding.

Acknowledgments: This work has been supported by the National Natural Science Foundation of China (Grant No. 11772238). The authors also gratefully acknowledge the financial support of the French National Research Agency ANR (LabEx DAMAS, Grant No. ANR-11-LABX-0008-01).

Conflicts of Interest: The authors declare no conflict of interest.

Abbreviations

The following abbreviations are used in this manuscript:

SMA	Shape memory alloy
RVE	Representative volume element
FE^2	Multilevel finite element method
FEM	Finite element method
UMAT	User-defined materials
MPCs	Multi-point constraints
RP	Reference point
C3D8	Continuum 3D solid element with full integration
C3D8R	Continuum 3D solid element with reduced integration
C3D8I	Continuum 3D solid element with incompatible modes

References

1. Gibson, L.J.; Ashby, M.F. *Cellular Solids: Structure and Properties;* Cambridge University Press: Cambridge, UK, 1997.
2. Ashby, M.F.; Evans, T.; Fleck, N.A.; Hutchinson, J.; Wadley, H.; Gibson, L. *Metal Foams: A Design Guide;* Elsevier: Amsterdam, The Netherland, 2000.
3. Yazdani Sarvestani, H.; Akbarzadeh, A.H.; Niknam, H.; Hermenean, K. 3D printed architected polymeric sandwich panels: Energy absorption and structural performance. *Compos. Struct.* **2018**, *200*, 886–909. [CrossRef]
4. Garcia-Moreno, F. Commercial applications of metal foams: Their properties and production. *Materials* **2016**, *9*, 85.

[CrossRef] [PubMed]
5. Hangai, Y.; Nakano, Y.; Koyama, S.; Kuwazuru, O.; Kitahara, S.; Yoshikawa, N. Fabrication of aluminum tubes filled with aluminum alloy foam by frictionwelding. *Materials* **2015**, *8*, 7180–7190. [CrossRef] [PubMed]
6. Strano, M.; Marra, A.; Mussi, V.; Goletti, M.; Bocher, P. Endurance of damping properties of foam-filled tubes. *Materials* **2015**, *8*, 4061–4079. [CrossRef] [PubMed]
7. Rashed, M.G.; Ashraf, M.; Mines, R.A.W.; Hazell, P.J. Metallic microlattice materials: A current state of the art on manufacturing, mechanical properties and applications. *Mater. Des.* **2016**, *95*, 518–533. [CrossRef]
8. Taherishargh, M.; Belova, I.V.; Murch, G.E.; Fiedler, T. Pumice/aluminium syntactic foam. *Mater. Sci. Eng. A* **2015**, *635*, 102–108. [CrossRef]
9. Taherishargh, M.; Sulong, M.A.; Belova, I.V.; Murch, G.E.; Fiedler, T. On the particle size effect in expanded perlite aluminium syntactic foam. *Mater. Des.* **2015**, *66*, 294–303. [CrossRef]
10. Taherishargh, M.; Belova, I.V.; Murch, G.E.; Fiedler, T. The effect of particle shape on mechanical properties of perlite/metal syntactic foam. *J. Alloys Compd.* **2017**, *693*, 55–60. [CrossRef]
11. Broxtermann, S.; Taherishargh, M.; Belova, I.V.; Murch, G.E.; Fiedler, T. On the compressive behaviour of high porosity expanded Perlite-Metal Syntactic Foam (P-MSF). *J. Alloys Compd.* **2017**, *691*, 690–697. [CrossRef]
12. Linul, E.; Movahedi, N.; Marsavina, L. On the Lateral Compressive Behavior of Empty and Ex-Situ Aluminum Foam-Filled Tubes at High Temperature. *Materials* **2018**, *11*, 554. [CrossRef] [PubMed]
13. Luong, D.; Lehmhus, D.; Gupta, N.; Weise, J.; Bayoumi, M. Structure and compressive properties of invar-cenosphere syntactic foams. *Materials* **2016**, *9*, 115. [CrossRef] [PubMed]
14. Pingle, S.M.; Fleck, N.A.; Deshpande, V.S.; Wadley, H.N.G. Collapse mechanism maps for the hollow pyramidal core of a sandwich panel under transverse shear. *Int. J. Solids Struct.* **2011**, *95*, 3417–3430. [CrossRef]
15. Schaedler, T.A.; Ro, C.J.; Sorensen, A.E.; Eckel, Z.; Yang, S.S.; Carter, W.B.; Jacobsen, A.J. Designing metallic microlattices for energy absorber applications. *Adv. Eng. Mater.* **2014**, *16*, 3. [CrossRef]
16. Schaedler, T.A.; Carter, W.B. Architected cellular materials. *Annu. Rev. Mater. Res.* **2016**, *46*, 187–210. [CrossRef]
17. Lehmhus, D.; Vesenjak, M.; Schampheleire, S.; Fiedler, T. From stochastic foam to designed structure: Balancing cost and performance of cellular metals. *Materials* **2017**, *10*, 922. [CrossRef] [PubMed]
18. Ngo, T.D.; Kashani, A.; Imbalzano, G.; Nguyen, K.T.Q.; Hui, D. Additive manufacturing (3D printing): A review of materials, methods, applications and challenges. *Compos. Part B Eng.* **2018**, *143*, 172–196. [CrossRef]
19. Mostafaei, A.; Vecchis, P.R.D.; Stevens, E.L.; Chmielus, M. Sintering regimes and resulting microstructure and properties of binder jet 3D printed Ni-Mn-Ga magnetic shape memory alloys. *Acta Mater.* **2018**, *154*, 355–364. [CrossRef]
20. Mehrpouya, M.; Gisario, A.; Elahinia, M. Laser welding of NiTi shape memory alloy: A review. *J. Manuf. Process.* **2018**, *31*, 162–186. [CrossRef]
21. Lagoudas, D.C. *Shape Memory Alloys: Modeling and Engineering Applications*; Springer: New York, NY, USA, 2008.
22. Patoor, E.; Lagoudas, D.C.; Entchev, P.B.; Catherine Brinson, L.; Gao, X. Shape memory alloys, Part I: General properties and modeling of single crystals. *Mech. Mater.* **2006**, *38*, 391–429. [CrossRef]
23. Lagoudas, D.C.; Entchev, P.B.; Popov, P.; Patoor, E.; Catherine Brinson, L.; Gao, X. Shape memory alloys, Part II: Modeling of polycrystals. *Mech. Mater.* **2006**, *38*, 430–462. [CrossRef]
24. Tobushi, H.; Hayashi, S.; Sugimoto, Y.; Date, K. Two-way bending properties of shape memory composite with SMA and SMP. *Materials* **2009**, *2*, 1180–1192. [CrossRef]
25. Cisse, C.; Zaki, W.; Ben Zineb, T. A review of constitutive models and modeling techniques for shape memory alloys. *Int. J. Plast.* **2016**, *76*, 244–284. [CrossRef]
26. Machado, G.; Louche, H.; Alonso, T.; Favier, D. Superelastic cellular NiTi tube-based materials: Fabrication, experiments and modeling. *Mater. Des.* **2015**, *65*, 212–220. [CrossRef]
27. Ravari Karamooz, M.R.; Nasr Esfahani, S.; Taheri Andani, M.; Kadkhodaei, M.; Ghaei, A.; Karaca, H.; Elahinia, M. On the effects of geometry, defects, and material asymmetry on the mechanical response of shape memory alloy cellular lattice structures. *Smart Mater. Struct.* **2016**, *25*, 025008. [CrossRef]

28. Ashrafi, M.J.; Amerinatanzi, A.; Saebi, Z.; Moghaddam, N.S.; Mehrabi, R.; Karaca, H.; Elahinia, M. Shape memory response of cellular lattice structures: Unit cell finite element prediction. *Mech. Mater.* **2018**, *125*, 26–34. [CrossRef]
29. Kanouté, P.; Boso, D.P.; Chaboche, J.L.; Schrefler, B.A. Multiscale methods for composites: A review. *Arch. Comput. Method E* **2009**, *16*, 31–75. [CrossRef]
30. Geers, M.G.D.; Kouznetsova, V.G.; Brekelmans, W.A.M. Multi-scale computational homogenization: Trends and challenges. *J. Comput. Appl. Math.* **2010**, *234*, 2175–2182. [CrossRef]
31. El Hachemi, M.; Koutsawa, Y.; Nasser, H.; Giunta, G.; Daouadji, A.; Daya, E.M.; Belouettar, S. An intuitive computational multi-scale methodology and tool for the dynamic modeling of viscoelastic composites and structures. *Compos. Struct.* **2016**, *144*, 131–137. [CrossRef]
32. Kinvi-Dossou, G.; Matadi Boumbimba, R.; Bonfoh, N.; Koutsawa, Y.; Eccli, D.; Gerard, P. A numerical homogenization of E-glass/acrylic woven composite laminates: Application to low velocity impact. *Compos. Struct.* **2018**, *200*, 540–554. [CrossRef]
33. Feyel, F. A multilevel finite element method (FE^2) to describe the response of highly non-linear structures using generalized continua. *Comput. Methods Appl. Mech. Eng.* **2003**, *192*, 3233–3244. [CrossRef]
34. Nezamabadi, S.; Potier-Ferry, M.; Zahrouni, H.; Yvonnet, J. Compressive failure of composites: A computational homogenization approach. *Compos. Struct.* **2015**, *127*, 60–68. [CrossRef]
35. Cong, Y.; Nezamabadi, S.; Zahrouni, H.; Yvonnet, J. Multiscale computational homogenization of heterogeneous shells at small strains with extensions to finite displacements and buckling. *Int. J. Numer. Meth. Eng.* **2015**, *104*, 235–259. [CrossRef]
36. Tikarrouchine, E.; Chatzigeorgiou, G.; Praud, F.; Piotrowski, B.; Chemisky, Y.; Meraghni, F. Three-dimensional FE^2 method for the simulation of non-linear, ratedependent response of composite structures. *Compos. Struct.* **2018**, *193*, 165–179. [CrossRef]
37. Kohlhaas, B.; Klinkel, S. An FE^2 model for the analysis of shape memory alloy fiber-composites. *Comput. Mech.* **2014**, *55*, 421–437. [CrossRef]
38. Chatzigeorgiou, G.; Chemisky, Y.; Meraghni, F. Computational micro to macro transitions for shape memory alloy composites using periodic homogenization. *Smart Mater. Struct.* **2015**, *24*, 035009. [CrossRef]
39. Xu, R.; Bouby, C.; Zahrouni, H.; Ben Zineb, T.; Hu, H.; Potier-Ferry, M. 3D modeling of shape memory alloy fiber reinforced composites by multiscale finite element method. *Compos. Struct.* **2018**, *200*, 408–419. [CrossRef]
40. Chemisky, Y.; Duval, A.; Patoor, E.; Ben Zineb, T. Constitutive model for shape memory alloys including phase transformation, martensitic reorientation and twins accommodation. *Mech. Mater.* **2011**, *43*, 361–376. [CrossRef]
41. Peultier, B.; Ben Zineb, T.; Patoor, E. Macroscopic constitutive law of shape memory alloy thermomechanical behaviour. Application to structure computation by FEM. *Mech. Mater.* **2006**, *38*, 510–524. [CrossRef]
42. Duval, A.; Haboussi, M.; Ben Zineb, T. Modelling of localization and propagation of phase transformation in superelastic SMA by a gradient nonlocal approach. *Int. J. Solids Struct.* **2011**, *13*, 1879–1893. [CrossRef]
43. Sittner, P.; Heller, L.; Pilch, J.; Sedlak, P.; Frost, M.; Chemisky, Y.; Duval, A.; Piotrowski, B.; Zineb, T.B.; Patoor, E.; et al. Roundrobin SMA modeling. In Proceedings of the ESOMAT 2009, Prague, Czech Republic, 7–11 September 2009.
44. Chatzigeorgiou, G.; Charalambakis, N.; Chemisky, Y.; Meraghni, F. Periodic homogenization for fully coupled thermomechanical modeling of dissipative generalized standard materials. *Int. J. Plast.* **2016**, *81*, 18–39. [CrossRef]
45. Fatemi Dehaghani, P.; Hatefi Ardakani, S.; Bayesteh, H.; Mohammadi, S. 3D hierarchical multiscale analysis of heterogeneous SMA based materials. *Int. J. Solids Struct.* **2017**, *118–119*, 24–40. [CrossRef]
46. Yvonnet, J.; Zahrouni, H.; Potier-Ferry, M. A model reduction method for the post-buckling analysis of cellular microstructures. *Comput. Methods Appl. Mech. Eng.* **2007**, *197*, 265–280. [CrossRef]
47. Ammar, A.; Chinesta, F.; Cueto, E.; Doblaré, M. Proper generalized decomposition of time-multiscale models. *Int. J. Numer. Meth. Eng.* **2012**, *90*, 569–596. [CrossRef]
48. Kpogan, K.; Tri, A.; Sogah, A.; Mathieu, N.; Zahrouni, H.; Potier-Ferry, M. Combining MFS and PGD methods to solve transient heat equation. *Numer. Methods Part. Differ. Equ.* **2018**, *34*, 257–273. [CrossRef]

49. Nezamabadi, S.; Yvonnet, J.; Zahrouni, H.; Potier-Ferry, M. A multilevel computational strategy for handling microscopic and macroscopic instabilities. *Comput. Methods Appl. Mech. Eng.* **2009**, *198*, 2099–2110. [CrossRef]
50. Assidi, M.; Zahrouni, H.; Damil, N.; Potier-Ferry, M. Regularization and perturbation technique to solve plasticity problems. *Int. J. Mater. Form* **2009**, *2*, 1–14. [CrossRef]
51. Aggoune, W.; Zahrouni, H.; Potier-Ferry, M. Asymptotic numerical methods for unilateral contact. *Int. J. Numer. Meth. Eng.* **2006**, *68*, 605–631. [CrossRef]
52. Hu, H.; Damil, N.; Potier-Ferry, M. A bridging technique to analyze the influence of boundary conditions on instability patterns. *J. Comput. Phys.* **2011**, *230*, 3753–3764. [CrossRef]
53. Yu, K.; Hu, H.; Chen, S.; Belouettar, S.; Potier-Ferry, M. Multi-scale techniques to analyze instabilities in sandwich structures. *Compos. Struct.* **2013**, *96*, 751–762. [CrossRef]

Article

Stress Concentration and Mechanical Strength of Cubic Lattice Architectures

Paul Lohmuller [1,2,3,*], Julien Favre [4], Boris Piotrowski [1], Samuel Kenzari [2] and Pascal Laheurte [1,3]

1. Laboratoire d'Etude des Microstructures et de Mécanique des Matériaux LEM3 UMR CNRS 7239, Arts et Métiers ParisTech Campus de Metz, Université de Lorraine, F-57078 Metz, France; boris.piotrowski@ensam.eu (B.P.); pascal.laheurte@univ-lorraine.fr (P.L.)
2. Institut Jean Lamour, UMR 7198 CNRS-Université de Lorraine, Campus Artem, F-54011 Nancy, France; samuel.kenzari@univ-lorraine.fr
3. Laboratory of Excellence on Design of Alloy Metals for Low-mAss Structures (DAMAS), Université de Lorraine, F-54011 Nancy, France
4. Mines Saint-Etienne, Univ Lyon, CNRS, UMR 5307 LGF, Centre SMS, Departement PMM, F-42023 Saint-Etienne, France; julien.favre@emse.fr

* Correspondence: paul.lohmuller@univ-lorraine.fr

Received: 30 May 2018; Accepted: 3 July 2018; Published: 5 July 2018

Abstract: The continuous design of cubic lattice architecture materials provides a wide range of mechanical properties. It makes possible to control the stress magnitude and the local maxima in the structure. This study reveals some architectures specifically designed to reach a good compromise between mass reduction and mechanical strength. Decreased local stress concentration prevents the early occurrence of localized plasticity or damage, and promotes the fatigue resistance. The high performance of cubic architectures is reported extensively, and structures with the best damage resistance are identified. The fatigue resistance and S–N curves (stress magnitude versus lifetime curves) can be estimated successfully, based on the investigation of the stress concentration. The output data are represented in two-dimensional (2D) color maps to help mechanical engineers in selecting the suitable architecture with the desired stress concentration factor, and eventually with the correct fatigue lifetime.

Keywords: lattice structures; porous materials; 3D surface maps; finite element; fatigue; plasticity

---

1. Introduction

Recent progress in manufacturing technologies based on additive manufacturing [1,2] and welded assemblies [3] open new possibilities for the production of architectured materials with open porosities. Lattice periodic structures illustrate the appearance of new hybrid materials [4] combining high specific strength with ultra-low specific density, usually out of reach for any bulk materials. These porous structures combine properties that are usually considered as incompatible, and can be positioned in the blank areas on the strength/density material selection chart [3,5].

As a result of the promising specific properties of these materials, significant efforts have been made to highlight the importance of choosing the optimal architecture offering the best specific mechanical strength. The research procedure is usually iterated by trial and error; small alterations are operated on a well-known structure, and the benefits of each design are presented one by one. For instance, the well-known octet-truss [6,7] has been edited to make a new structure called an IsoTruss [8], with a superior specific stiffness and improved isotropy. To accelerate the optimization procedure, new design rules have emerged. The use of Bravais lattices provides simple rules to produce periodic structures [9]. Some recent works have shown that it is possible to use crystallography formalism to design very versatile complex architectures by just using a point group and two architectural parameters [10]. It was possible to define a large range of structures with variable stiffness and Poisson ratio. New solutions for the mass reduction have emerged from this formalism by developing structures with a high specific stiffness. The current work proposes an investigation of the stress concentration in a wide range of topologies. It provides an indication on the plasticity onset and thus on the associated fatigue resistance.

The specific stiffness of the lattice structures varies with the square of the specific density [10,11]. However, it was shown that the power exponent could deviate significantly from its nominal value of two, depending on the architecture [10]. Likewise, the yield strength and fatigue strength are expected to follow a power law of density. According to the early works of Ashby et al. [11], the power law coefficient is expected to be three/two for the dependence of yield stress on the specific density for a stochastic foam. Here again, some discrepancies on the power exponent are expected because of the effect of architecture. Some architectures could be more prone to stress concentration, leading to an early onset of plasticity and damage. Therefore, any architecture minimizing the stress concentration with a low density would be of high interest for mechanical engineers. This work aims to identify these optimal structures by evaluating the stress concentration coefficient as a function of density. Once the stress concentration is known, it is possible to discuss the onset of plasticity and damage. By transposing the data from the bulk materials into the case of the porous architectures, it becomes possible to give a first estimation of the possible fatigue lifetime. The ultimate goal of this work is to provide guidelines for mechanical engineers on ways to save mass and material consumption by producing architectured materials from known bulk materials, while preserving high mechanical strength and a controlled lifetime.

The second section will explain the architecture generation procedure and the associated numerical simulations. The third section details the determination of the stress concentration factor to discuss the heterogeneities of the stress field. The fourth part focuses on the application for mechanical design, and on the capability to predict the fatigue lifetime from the stress concentration factors.

2. Materials and Methods

The cubic lattice structures are generated using crystallographic rules and symmetry operations. A cluster of points defines the architecture, and the struts correspond to the shortest possible segments between the nodes. This definition of struts is consistent with the concept of chemical bonding in crystals. The lattice design is defined by a set of two initial points belonging to the lattice network. Point A is set at the origin, and the position of point B is set by two parameters, x and y. Figure 1 illustrates some structures defined by points A and B. The structures are represented with a strut radius equal to 0.05 times the lattice parameter. The position along the z-axis must be set to 0.5 times the lattice parameter to produce a continuous network. Then the use of symmetry operations of the $m\bar{3}m$ space group on points A and B results in an infinite network of points corresponding to the possible positions of atoms in a crystal. The whole procedure is not described in this study, but interested readers should turn to the work of Favre et al. [10]. The case x = 0.0 and y = 0.0 results in a primitive cubic lattice (Figure 1), and the progressive increase of both x and y forms an auxetic re-entrant structure, called a hexatruss [12,13]. The further increase results in complex structures with a high struts density (for example, x = 0.25 y = 0.25 in Figure 1), and finally reaches the upper bound x = 0.5 and y = 0.5, corresponding to the centered cubic lattice. This technique provides a continuum between a known cubic Bravais lattice, such as primitive, face-centered, or cube-centered, and new complex architectures.

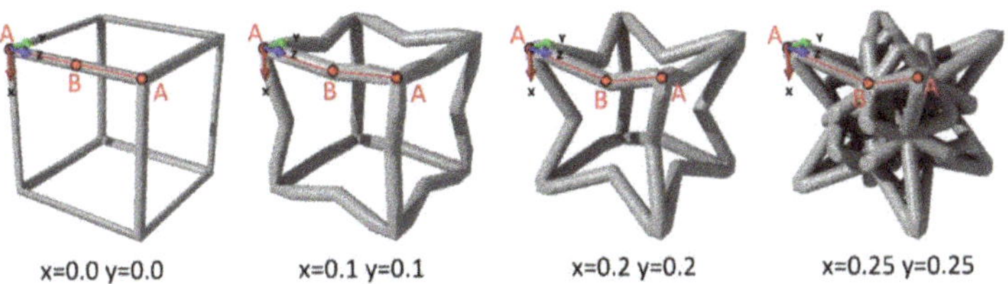

Figure 1. Cubic lattices designed by setting the (x, y) coordinates of the B point.

The one-dimensional (1D) struts lattice network is converted into a three-dimensional (3D) object with struts, with a circular section. The radius of the struts is given as a fraction of the lattice parameter. For instance, if the lattice parameter is 1 mm, then the radius would be a fraction of 1 mm, and the computed stress unit will be in megapascals. The generation of 3D structures is operated using Openscad (version 2015.03-2) and Autodesk Meshmixer software (version 3.2.37). The 3D object is exported in '.stl' format, and then it is post-processed by Tetgen, a meshing tool that is used to produce a model suitable for finite elements analysis. The 3D model is imported to Abaqus v6.13. The deformation applied on the lattice is a small compressive strain of $\varepsilon = 1\%$. The modeling technique is further detailed by Favre et al. [10]. The opposite facets of the representative volume are symmetry planes, corresponding to a frictionless sliding. The material has homogeneous isotropic elastic properties. Simulations were conducted on an elastic material with a Young's modulus E = 1 GPa, representative of the ABS polymer (Acrylonitrile Butadiene Styrene) that is used in commercial 3D printers. The output data were post-treated by Python scripts to determine the volume and the stress of each element. The statistical analysis of the stress distribution over the population of the elements provides some reliable values of the maximal stress, denoted σ_m, and of the average stress, denoted $\overline{\sigma}$. Furthermore, the macroscopic compression stress, denoted $\widetilde{\sigma}$, is computed from the sum of all of the resulting forces applied on the outer surface of the cube bearing the external compressive pressure.

During the elastic compression of the lattice structures, the macroscopic stress applied on the outer surface of the cube $\widetilde{\sigma}$ is proportional to the macroscopic strain, and the proportionality coefficient is the apparent Young's modulus E, reported by Favre et al. [10]. The lattice is considered as a homogeneous equivalent material, and the mechanical design relies on the value of $\widetilde{\sigma}$, applied on the representative elementary volume. The stress concentration is defined by the ratio of the maximal stress to the macroscopic stress $\widetilde{\sigma}$. It becomes a useful information for mechanical engineers, as it connects the effective macroscopic stress to the microscopic stress localizations.

Thereby, the stress concentration factor K is defined as the ratio between the maximal stress σ_m and the external stress $\widetilde{\sigma}$ applied on the lattice, as follows: $K = \sigma_m/\widetilde{\sigma}$. This factor provides an estimation of the localized stress maximum. A second stress concentration factor $\overline{K}$ is computed by the ratio of the average stress in the lattice to the external stress: $\overline{K} = \overline{\sigma}/\widetilde{\sigma}$. This factor illustrates the average stress concentration over the whole volume of the lattice, regardless of the localized stress maximum.

3. Result

3.1. Stress Field and Maximum

The finite element model provides a large set of information on the heterogeneous stress field in the lattices. Figure 2 illustrates the stress field in the case of an octet-truss lattice (FCC, x = 0.0 and y = 0.5). The stress concentration is the highest at the outer surface of the intersections of the beams. The magnitude of the Von Mises stress is about ten times larger than in the median part of the beams. This stress concentration at the surface is likely to result in localized plasticity and crack initiation.

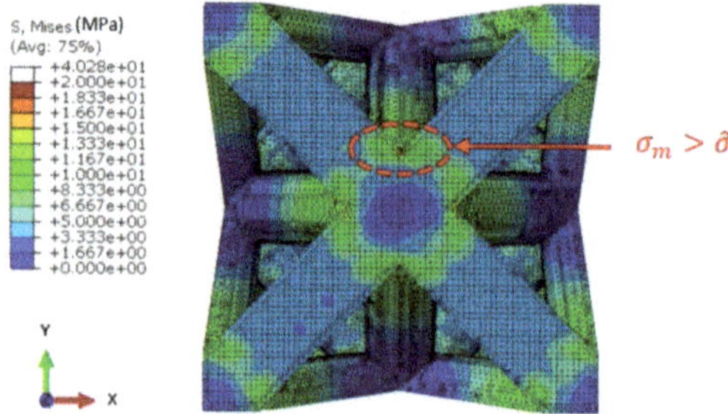

Figure 2. Von Mises stress field in an FCC lattice (x = 0.0 and y = 0.5), and stress concentration at nodes.

Unfortunately, the analysis of the maximum stress is not straightforward, and it is preferable to avoid a simplistic search of the maximal stress value. Indeed, the finite element method often results in singularities in the stress field because of the possible distortion of some elements. Consequently, the evaluation of the maximal stress cannot rely on a single element, and a statistical approach must be set. The stress was analyzed over the complete population of the elements, and the distribution as a function of the volume fraction was plotted in Figure 3a. Depending on the architecture, the distribution exhibits a series of peaks for the low stress (<20 MPa for ε = 1% and E = 1 GPa). For the larger stress values, the fraction tends to zero, and the stress value is not representative of the effective stress field. For such distributions, taking the maximal value would result in poor accuracy. Therefore, the maximal stress σ_m was defined as the mean value of 1% of the elements having the largest stress value. In other words, the maximal stress is the mean value within the tail of the distribution.

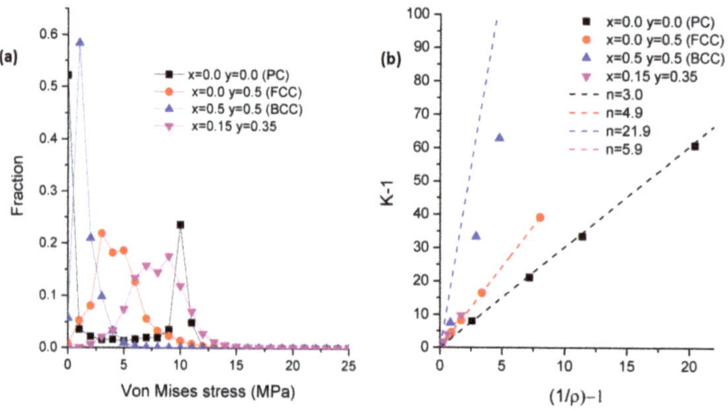

Figure 3. (**a**) Von Mises stress distribution in the set of elements for different architecture parameters; (**b**) stress concentration factor K as a function of density ρ.

The factor K varies as a power law of density [11], as follows:

$$K = 1/\rho^m = \rho^{-m} \tag{1}$$

However, Equation (1) does not represent properly the variation of K with density. Instead, K follows a linear relation with $1/\rho$. Then, the variation of K is represented by the following:

$$K = n_1 \times \frac{1}{\rho} + n_2 \tag{2}$$

with n_1 and n_2 being two parameters depending on the architecture. For $\rho = 1$, there is naturally no stress concentration, and $K = 1$. Therefore, Equation (2) becomes the following:

$$K = 1 + n \times \left(\frac{1}{\rho} - 1\right) \quad (3)$$

with n a parameter to be determined. Figure 3b illustrates $K - 1$ as a function of $1/\rho - 1$, and the relation is linear with a slope equal to n. The fit of K values using the Equation (3) yields to a correlation coefficient r^2 above 0.98.

The coefficient denoted n has been determined for every structure in the x–y parametric plan. The same relationship holds for the global stress concentration factor $\overline{K}$. It results in a set of two parameters, n and $\overline{n}$, defining the variation of the stress concentration with the density. These two parameters directly give an estimation of the evolution of K and $\overline{K}$, because of the different architectures described in the previous section.

One must remember that the density is not set only by the value of the struts radius, but also by the architecture defined by x and y. Indeed, depending on the architecture, the connectivity and the number of struts in the representative volume are affected, and the density varies. The density follows a polynomial relation with the radius value, as follows:

$$\rho = k_1 \times r^2 + k_2 \times r \quad (4)$$

where k_1 and k_2 are two parameters specific to a given architecture, defined by the x and y parameters. It is reminded that the struts radius must be expressed as a fraction of the lattice parameter, and therefore, it has the same unit as the lattice parameter. For instance, if the lattice parameter is 1 mm, then the unit of k_1 is mm^{-2} and the unit of k_2 is mm^{-1}. Figure 4 illustrates the variation of the k_1 and k_2 parameters with x and y. Using these two maps, it is possible to estimate the density of any structure from the value of the strut radius. Then, by introducing the value of ρ into Equation (3), it is possible to estimate the stress concentration factors K and $\overline{K}$.

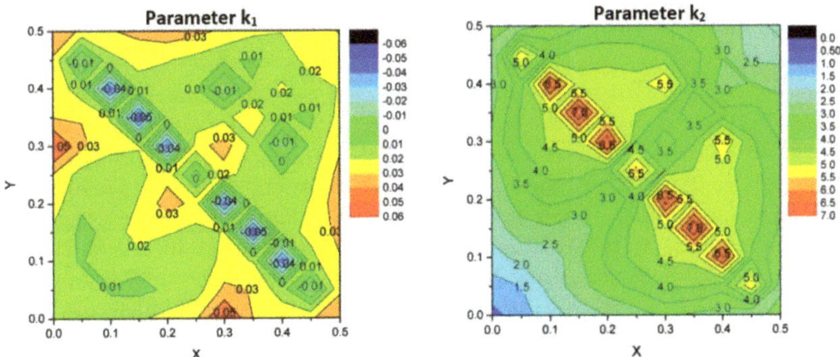

Figure 4. Polynomial parameters k_1 and k_2 for the dependence of density on the struts radii.

3.2. Stress Concentration Factors

The stress concentration factor K has to be minimized to ensure a lower risk of plastic deformation and a maximal fatigue lifetime. Figure 5 illustrates the variation of K with x and y, for two strut radii. A global trend is visible. For low or high values of x + y, the stress concentration is quite high. This means that the structures similar to that of the primitive cube (x + y = 0) or to the BCC structure (x + y = 1) are prone to stress concentration. On the contrary, structures on the diagonal of the map (x + y = 0.5) exhibit a very low concentration factor K, and the mechanical strength is very promising.

Figure 1 illustrates the progressive change in the architecture responsible for this trend; the increase of x + y leads to an increase in the density, and ultimately to a decrease of *K*.

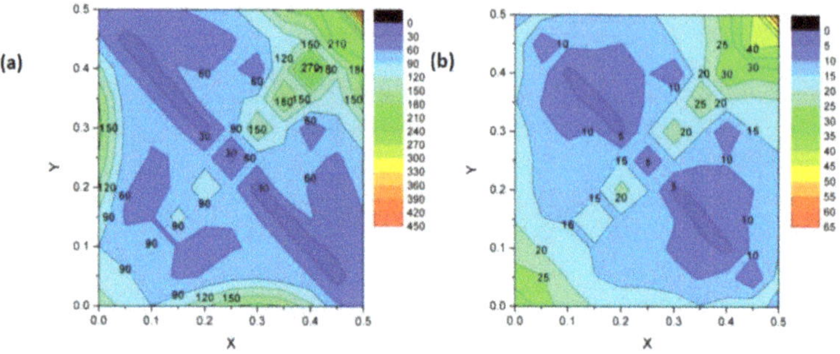

Figure 5. Stress concentration factor *K* as a function of (x, y) for two radii: (**a**) $r = 0.05$ a and (**b**) $r = 0.1$ a.

The change in the architecture affects the density, and it leads to the variations of *K* on Figure 5. The decrease of *K* at intermediate values of x + y is mostly due to an increase of ρ, resulting in a decrease of the stress variations. Therefore, Figure 5 is useful to set a specific value for the *K* factor, for situations where the density variation is of low importance.

On the other hand, mechanical engineers are often willing to decrease the mass of parts to reduce the material consumption and the product weight. Then, ρ has to be minimized together with *K*. It becomes possible by examining the n parameter of Equation (3). It directly gives an indication of the stress concentration variation by the variation of the density for each topology.

Figure 6a illustrates the variations of n parameter with the x and y parameters. The lower the n, the faster *K* decreases with increasing density (Equation (3)). Therefore, having low values of n is attractive for structural applications. When the density is preserved at a small value, the *K* factor is minimized and the risks for struts failure due to stress localization is reduced. The structures with a minimal n value would have a better resistance to localized plasticity and fatigue damage, even with decreased density. Having this in mind, it appears that the optimal structure for mechanical engineering is x = 0.0 and y = 0.0, because it minimizes the n value. The compression along the <100> direction results in pure compression stresses on the struts, and the stress concentration at the nodes remains at a very low level. The octet-truss structure (x = 0.0 and y = 0.5, FCC) and other structures located on the diagonal (x + y = 0.5) are also very interesting candidates for mass reduction because of their low n value. For a given pair of density and material, this architecture is minimizing the local stress concentration, and it should result in a higher fatigue lifetime and a higher resistance to localized damage, compared with other architectures. On the contrary, the structure x = 0.4 and y = 0.4 has the maximal n value, and belongs to the family of complex lattices with a high strut connectivity. The *K* factor is maximized because of the large volume fraction of the material localized at nodes where the stress is maximal.

A similar analysis was conducted on $\overline{K}$, illustrating the evolution of the average stress concentration with the global lattice volume. Consequently, $\overline{n}$ is the sensitivity parameter of $\overline{K}$, with respect to density, and it provides information only on the global magnitude of the stress field. Using $\overline{K}$, one can discuss the occurrence of generalized plasticity or damage within an extended fraction of the part. $\overline{K}$ is especially relevant to determine the optimal structure with the maximized macroscopic yield strength, because only the average stress magnitude is considered. A low value of $\overline{n}$ means that there is great potential to preserve a low mean stress when the density is decreased. Again, the optimal architecture for a maximal yield stress at low density is x = 0.0 and y = 0.0. It allows a good compromise between the decreased density and the increased mean stress field magnitude. For a given density and a given material, this architecture results in a maximal macroscopic yield stress. The structure x = 0.2 and y = 0.3 shows a local minimum of $\overline{n}$, and it is also a suitable candidate to minimize the global stress magnitude. However, the structure with x = 0.35 and y = 0.45 maximizes the magnitude of the stress field at a given density, and the generalized plasticity is expected to occur quickly. This structure may be beneficial for some applications, for example to form the lightweight parts by plastic deformation, and it may be useful for packaging or shaping models.

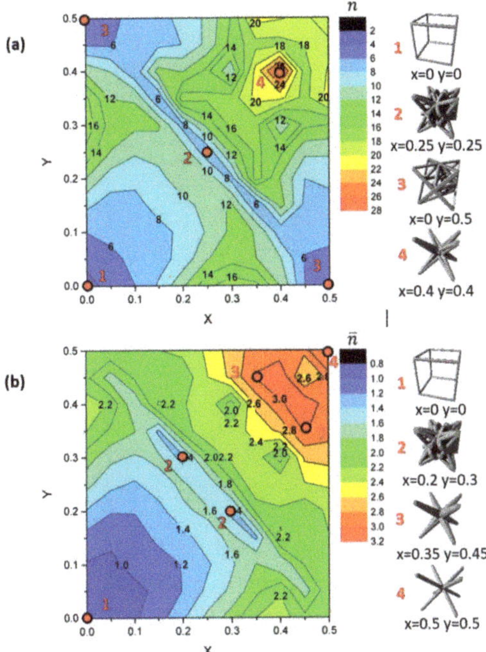

Figure 6. Parameter n indicating the dependence of the stress concentration factor on density for: (**a**) $K = \sigma_m/\tilde{\sigma}$ and (**b**) $\overline{K} = \overline{\sigma}/\tilde{\sigma}$. The optimal solutions to maximize the yield stress are highlighted in red and the solutions to minimize it are indicated in blue.

4. Discussion

4.1. Case Study and Applications

Equation (3) relates three variables n, ρ, and K. To determine the optimal parameters for a workpiece, mechanical engineers must set two of these variables, and determine the third one. When (n, ρ) or (n, K) are set, then the architecture has already been selected, and thus it is not a variable. This corresponds to the conventional mechanical design technique, the structure is drawn into the Computer Aided Design software, and then the strut section is set to get the suitable stress. Alternatively, the stress magnitude is set in the specifications, and the strut section is determined. For a given structure, it is not possible to choose both K and ρ, because they are both dependent on n, which is a constant.

However, there is a third way, by considering n as a variable, and setting (K, ρ). It is far more convenient because it becomes possible to choose any value of K and ρ separately, which is impossible in conventional mechanical design. Let us consider the following specification: the target of a mechanical engineer is a value of $\rho = 0.1$ and $K = 100$. Then, using Equation (3), the n parameter must be set to 11 to fulfil this requirement. Finally, reading Figure 6a, one can deduce relevant architectures, such as x = 0.20 and y = 0.25, or x = 0.3 and y = 0.4, for instance. All of the families of the equivalent structures are on the iso-values line n = 11 on the map, and the engineer can choose deliberately the most convenient one to manufacture. Of course, if the specification implies a joint minimization of (K, ρ), then n has to be minimized as mentioned previously, and the optimal choice is the trivial solution x = 0.0 and y = 0.0.

An example of application is the control of the deformation mode by selective plasticity. It becomes interesting to set different levels of K to control the exact area undergoing plasticity, and finally, to generate strain and stress gradients. An interesting illustration of this phenomenon is presented in Figure 7. If a bulk part is constituted of different domains called A and B, with two different lattice structures, and different n parameters, then the stress field will be different. Assuming domain B has a larger K value, it is prone to be deformed plastically at an early stage. After the plastic deformation and unloading, domain B has a larger persistent plastic strain, and it leads to the shape change of the object due to the residual stress. This unusual phenomenon is typical of functionally graded materials, and it can be tailored at will using Figure 6 to design domains A and B.

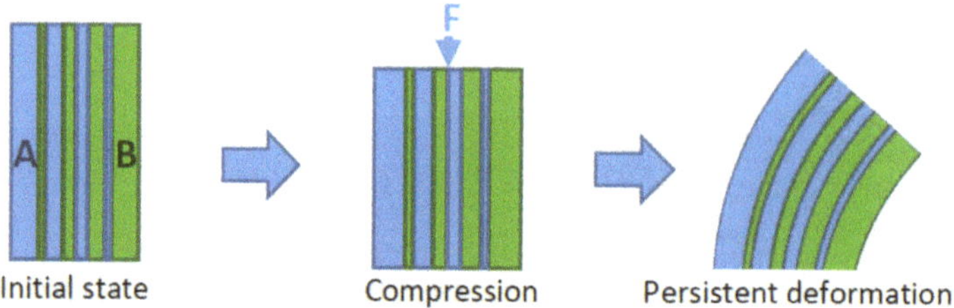

Figure 7. Illustration of a part composed of two different domains, A (blue, low K) and B (green, high K), with different architectures, resulting in residual stress, with F the applied force.

Considering now the damage, some attractive applications can be considered to control the cracking path within the domains of the specific high K value. It would be an extension of the pre-cutting usually seen on paper, but transposed to 3D volumes. When the ultimate tensile strength or the fatigue strength of the bulk is known, the optimal choice of K by adjusting n is possible, while keeping a low density in any situation.

4.2. Fatigue Strength

The fatigue strength has been studied for specific cases in the literature by experimental and modeling works. The most commonly investigated structures are primitive cubic and its variants [14], the diamond structure [15–18] and rhombic dodecahedron [16,19,20]. Specific attention has been paid to the rhombic dodecahedron structure, and its mechanical properties are now well known [21]. A very large set of experimental data conducted by Yavari et al. [22] and Zhang et al. [23,24] provides a large panel of S–N curves for different levels of porosities, in the case of structures made with Ti-6Al-4V by laser or electron beam melting. The experimental dataset will be compared with the first-order estimations resulting from the estimation of the stress concentration in lattices.

The rhombic dodecahedron was generated using the symmetries of the $m\bar{3}m$ space group, as described previously. However, it cannot be produced with only two initial points, as illustrated in Figure 1. This structure requires an initial set of three points, namely: A (0, 0); B (0, 0.5), and C (0.25, 0.25). It is therefore similar to an FCC structure, having all of its tetrahedral sites filled with a node. The resulting structure is illustrated in Figure 8. The structure was generated for different struts radii corresponding to the relative densities between 0.16 and 0.38, in agreement with the authors of [22,23]. The n parameter was found to be $n = 9.2$ for this structure. Therefore, the stress concentration factor K is within the range of 16–49, when the relative density is in the range of 0.16–0.38.

The fatigue strength of rhombic dodecahedron is illustrated in Figure 8 [22,23]. In addition, the fatigue experimental data for the bulk Ti-6Al-4V produced by selective laser melting is superimposed in the same plot [25–28]. The S–N curves follow the Basquin equation during the high cycle fatigue, as follows:

$$\Delta\sigma \cdot N^{A} = C \tag{5}$$

where A and C are the material parameters, $\Delta\sigma$ is the amplitude of stress, and N the number of cycles to failure. The parameters A and C were identified by fitting the experimental dataset for the rhombic dodecahedron lattices and the Ti-6Al-4V alloy.

The decrease of density from ρ = 0.38 to 0.16 results in a decrease of the failure stress. This is mainly due to the stress concentration at the nodes in the lattice, resulting in an early fracture of the material compared to the bulk material. There are two levels of stress concentration. The first one is the K factor, examined previously. The second one, denoted K_{strut}, is as a result of the surface roughness on the struts. The size of the defects is denoted a in Figure 9, and results in a local strut radius R−a. Therefore, the local stress near a cavity increases because of the local section reduction. The concentration factor is then given by Equation (6), as follows:

$$K_{strut} = \left(\frac{R}{R-a}\right)^{2} \tag{6}$$

As a result, the failure stress σ is given by the following:

$$\sigma = \frac{\sigma_B}{K \cdot K_{strut}} \tag{7}$$

with σ_B, the stress in the bulk material for a density ρ = 1. In the specific case of a single cycle N = 1, σ_B corresponds to the parameter C in Equation (5).

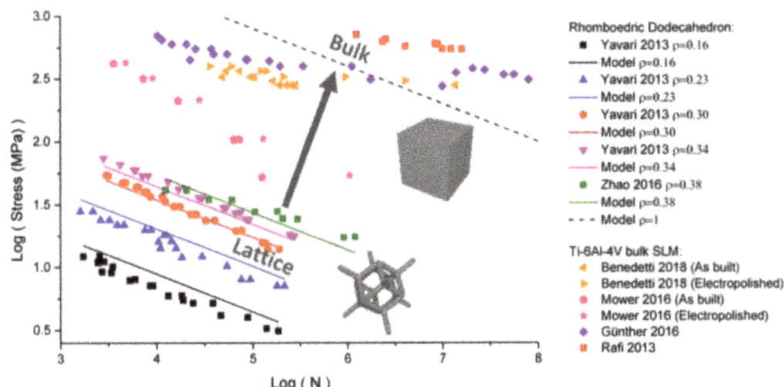

Figure 8. S–N curve of bulk and lattice materials (scatter plot), and the estimated values from the calculation of the stress concentration (solid lines).

According to Yavari et al. [22], the average error between the nominal struts diameter and the effective value is between 20 and 40 µm. For the rhombic dodecahedron with a lattice parameter of 1 mm, the parameters for Equation (4) are k_1 = 12.51 mm^{-2} and k_2 = 1.76 mm^{-1}. Using Equation (4), the radius of the beams varies from 0.06 to 0.11 for a density from 0.16 to 0.38. Using an average defect size, a = 30 µm, the stress concentration K_{strut} due to the defects is within the range of 1.8–3.6. The stress concentration K due to the architecture is obtained from Equation (3), and varies from 16 to 49 for the density range considered. Therefore, the factor K is about one order of magnitude larger than K_{strut}, and the effect of the architecture is predominant for the fatigue strength compared to the effect of the defects. However, the factor K_{strut} cannot be neglected, because it would lead to an incorrect estimation of the S–N curves.

In Figure 8, the experimental points corresponding to lattice structures can be fitted with a slope of −0.3, corresponding to A = 0.3 for all of the values of the relative density. The experimental data for the bulk materials show a similar value of A, but with a large discrepancy due to some additional effects of the defects and processing parameters that become predominant in the bulk parts.

The vertical offset of the lattice S–N curves increases monotonously with density for the lattice structures. Therefore, the vertical offset is mainly driven by the stress concentration due to the occurrence of large open porosities in the lattices. The parameter $n = 9.2$, found by the finite element modeling, helps to set a phenomenological model to predict the S–N curves. Using C = 25 GPa for the bulk material, it becomes possible to predict with a suitable accuracy all of the data set for rhombic dodecahedron for all of the density range (solid lines in Figure 8). The change for the vertical offset is entirely explained by the stress concentration factors K and K_{strut}. The stress σ_B follows the S–N curve for the bulk material (dashed line), and the apparent decrease of the vertical offset with decreasing density is due to the decrease of σ, caused by the stress concentration, following Equation (7).

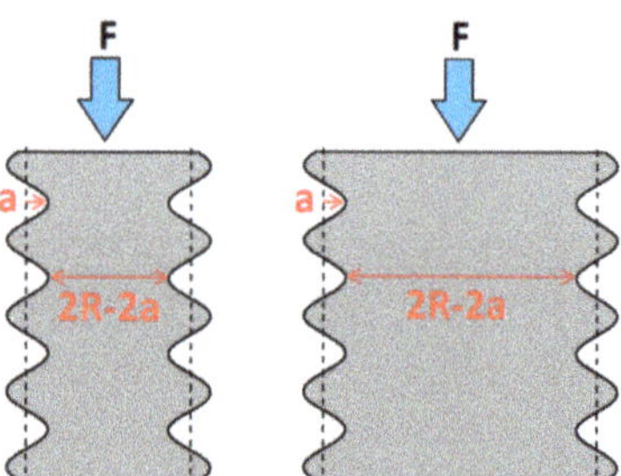

Figure 9. Variation of the effective beam section due to surface roughness, with F the applied force.

Using extrapolation at $\rho = 1$ for the bulk material, the estimation of the bulk fatigue strength σ_B is within an acceptable range between the measurement of Benedetti et al. [25] and that of Gunther et al. [26]. The estimated bulk properties are one order of magnitude higher than the data of Mower et al. [27], and one order lower than the one of Rafi et al. [28]. This illustrates the large variability of the fatigue strength, depending on the processing conditions, and the conclusions concerning the prediction of the S–N curves should be considered carefully. A significant effect of the laser scanning parameters, the powder properties, and of the possible annealing treatments may completely affect the final fatigue lifetime. The stress concentration analysis should be used as a first estimation of the possible fatigue strength so as to determine the optimal architecture. The exact quantification of the S–N curve can be obtained accurately only by an extensive experimental study with conventional fatigue tests.

To conclude, the n parameter, indicated in Figure 6, is a powerful tool to predict the local maximal stress in a lattice structure. This information can be put into the Basquin equation to produce a first-order estimation of the fatigue strength of the lattice structures.

5. Conclusions

Stress concentration was determined during the uniaxial compression of cubic lattice structures by finite element modeling. Maximal stress was identified by statistically analyzing the stress over the population of the elements. This method provides a reliable estimation of the maximal stress independent of the possible occurrence of singularities in the finite element calculation. The stress concentration factor K was deduced, and its variation with the relative density was determined. The parameter n is a relevant indicator of the performance of each structure to resist localized plasticity and damage during fatigue loading. The variation of the fatigue strength with density was successfully predicted in the case of the rhombic dodecahedron using the n parameter. The use of n is therefore applicable to estimate the stress concentration and the fatigue strength when the properties of the bulk material are known from the literature. From now on, the lifetime of a product can be ruled not only by the control of the density, but also by the relevant choice of the x and y architectural parameters. From these considerations, it is possible to independently adjust the weight of the part and the lifetime by adjusting the architecture within the part at will.

Author Contributions: J.F., S.K., and P.L. (Pascal Laheurte) conceived and supervised this research; P.L. (Paul Lohmuller) and B.P. performed numerical modelling; P.L. (Paul Lohmuller) and J.F. analyzed the data; P.L. (Paul Lohmuller), writing the article, J.F., S.K., B.P. and P.L. (Pascal Laheurte) editing the article.

Funding: This research was funded by National Research Agency grant number ANR-11-LABX-0008-0 (LabEx DAMAS).

Acknowledgments: This work was supported by the French State through the program "Investment in the future" operated by the National Research Agency (ANR) and referenced by ANR-11-LABX-0008-01 (LabEx DAMAS).

Conflicts of Interest: The authors declare no conflict of interest.

References

1. Yan, C.; Hao, L.; Hussein, A.; Raymont, D. Evaluations of cellular lattice structures manufactured using selective laser melting. *Int. J. Mach. Tools Manuf.* **2012**, *62*, 32–38. [CrossRef]
2. Lhuissier, P.; de Formanoir, C.; Martin, G.; Dendievel, R.; Godet, S. Geometrical control of lattice structures produced by EBM through chemical etching: Investigations at the scale of individual struts. *Mater. Des.* **2016**, *110*, 485–493. [CrossRef]
3. Dong, L.; Deshpande, V.; Wadley, H. Mechanical response of Ti–6Al–4V octet-truss lattice structures. *Int. J. Solids Struct.* **2015**, *60–61*, 107–124. [CrossRef]
4. Ashby, M.; Bréchet, Y. Designing hybrid materials. *Acta Mater.* **2003**, *51*, 5801–5821. [CrossRef]
5. Suard, M.; Martin, G.; Lhuissier, P.; Dendievel, R.; Vignat, F.; Blandin, J.J.; Villeneuve, F. Mechanical equivalent diameter of single struts for the stiffness prediction of lattice structures produced by Electron Beam Melting. *Addit. Manuf.* **2015**, *8*, 124–131. [CrossRef]
6. Nayfeh, A.H.; Hefzy, M.S. Continuum Modeling of Three-Dimensional Truss-like Space Structures. *AIAA J.* **1978**, *16*, 779–787. [CrossRef]
7. Deshpande, V.S.; Fleck, N.A.; Ashby, M.F. Effective properties of the octet-truss lattice material. *J. Mech. Phys. Solids* **2001**, *49*, 1747–1769. [CrossRef]
8. Messner, M.C. Optimal lattice-structured materials. *J. Mech. Phys. Solids* **2016**, *96*, 162–183. [CrossRef]
9. Méjica, G.F.; Lantada, A.D. Comparative study of potential pentamodal metamaterials inspired by Bravais lattices. *Smart Mater. Struct.* **2013**, *22*, 115013. [CrossRef]
10. Favre, J.; Lohmuller, P.; Piotrowski, B.; Kenzari, S.; Laheurte, P.; Meraghni, F. A continuous crystallographic approach to generate cubic lattices and its effect on relative stiffness of architectured materials. *Addit. Manuf.* **2018**, *21*, 359–368. [CrossRef]
11. Salimon, A.; Bréchet, Y.; Ashby, M.F.; Greer, A.L. Potential applications for steel and titanium metal foams. *Mech. Behav. Cell. Solids* **2005**, *40*, 5793–5799. [CrossRef]
12. Reis, F.D.; Ganghoffer, J.F. Equivalent mechanical properties of auxetic lattices from discrete homogenization. *Comput. Mater. Sci.* **2012**, *51*, 314–321. [CrossRef]
13. Evans, K.E. Molecular Network Design. *Nature* **1991**, *353*, 124. [CrossRef]
14. Dallago, M.; Fontanari, V.; Torresani, E.; Leoni, M.; Pederzolli, C.; Potrich, C.; Benedetti, M. Fatigue and biological properties of Ti-6Al-4V ELI cellular structures with variously arranged cubic cells made by selective laser melting. *J. Mech. Behav. Biomed. Mater.* **2018**, *78*, 381–394. [CrossRef] [PubMed]
15. Van Hooreweder, B.; Appers, Y.; Lietaert, K.; Kruth, J.P. Improving the fatigue performance of porous metallic biomaterials produced by Selective Laser Melting. *Acta Biomater.* **2017**, *47*, 193–202. [CrossRef] [PubMed]
16. Zargarian, A.; Esfahanian, M.; Kadkhodapour, J.; Ziaei-Rad, S. Numerical simulation of the fatigue behavior of additive manufactured titanium porous lattice structures. *Mater. Sci. Eng. C* **2016**, *60*, 339–347. [CrossRef] [PubMed]
17. Van Hooreweder, B.; Kruth, J.P. Advanced fatigue analysis of metal lattice structures produced by Selective Laser Melting. *CIRP Ann. Manuf. Technol.* **2017**, *66*, 221–224. [CrossRef]

18. Hrabe, N.W.; Heinl, P.; Flinn, B.; Körner, C.; Bordia, R.K. Compression-compression fatigue of selective electron beam melted cellular titanium (Ti-6Al-4V). *J. Biomed. Mater. Res.* **2011**, *99B*, 313–320. [CrossRef] [PubMed]
19. Hedayati, R.; Hosseini-Toudeshky, H.; Sadighi, M.; Mohammadi-Aghdam, M.; Zadpoor, A.A. Computational prediction of the fatigue behavior of additively manufactured porous metallic biomaterials. *Int. J. Fatigue* **2016**, *84*, 67–79. [CrossRef]
20. Li, S.J.; Murr, L.E.; Cheng, X.Y.; Zhang, Z.B.; Hao, Y.L.; Yang, R.; Medina, F.; Wicker, R.B. Compression fatigue behavior of Ti–6Al–4V mesh arrays fabricated by electron beam melting. *Acta Mater.* **2012**, *60*, 793–802. [CrossRef]
21. Xiao, L.; Song, W.; Wang, C.; Liu, H.; Tang, H.; Wang, J. Mechanical behavior of open-cell rhombic dodecahedron Ti–6Al–4V lattice structure. *Mater. Sci. Eng. A* **2015**, *640*, 375–384. [CrossRef]
22. Yavari, S.A.; Wauthle, R.; van der Stok, J.; Riemslag, A.C.; Janssen, M.; Mulier, M.; Kruth, J.P.; Schrooten, J.; Weinans, H.; Zadpoor, A.A. Fatigue behavior of porous biomaterials manufactured using selective laser melting. *Mater. Sci. Eng. C* **2013**, *33*, 4849–4858. [CrossRef] [PubMed]
23. Zhao, S.; Li, S.J.; Hou, W.T.; Hao, Y.L.; Yang, R.; Misra, R.D.K. The influence of cell morphology on the compressive fatigue behavior of Ti-6Al-4V meshes fabricated by electron beam melting. *J. Mech. Behav. Biomed. Mater.* **2016**, *59*, 251–264. [CrossRef] [PubMed]
24. Zhao, S.; Li, S.J.; Wang, S.G.; Hou, W.T.; Li, Y.; Zhang, L.C.; Hao, Y.L.; Yang, R.; Misra, R.D.K.; Murr, L.E. Compressive and fatigue behavior of functionally graded Ti-6Al-4V meshes fabricated by electron beam melting. *Acta Mater.* **2018**, *150*, 1–15. [CrossRef]
25. Benedetti, M.; Fontanari, V.; Bandini, M.; Zanini, F.; Carmignato, S. Low- and high-cycle fatigue resistance of Ti-6Al-4V ELI additively manufactured via selective laser melting: Mean stress and defect sensitivity. *Int. J. Fatigue* **2018**, *107*, 96–109. [CrossRef]
26. Günther, J.; Krewerth, D.; Lippmann, T.; Leuders, S.; Tröster, T.; Weidner, A.; Niendorf, T.; Biermann, H. Fatigue life of additively manufactured Ti–6Al–4V in the very high cycle fatigue regime. *Int. J. Fatigue* **2016**, *94*, 236–245. [CrossRef]
27. Mower, T.M.; Long, M.J. Mechanical behavior of additive manufactured, powder-bed laser-fused materials. *Mater. Sci. Eng. A* **2016**, *651*, 198–213. [CrossRef]
28. Rafi, H.K.; Karthik, N.V.; Gong, H.; Starr, T.L.; Stucker, B.E. Microstructures and Mechanical Properties of Ti6Al4V Parts Fabricated by Selective Laser Melting and Electron Beam Melting. *J. Mater. Eng. Perform.* **2013**, *22*, 3872–3883. [CrossRef]

Article

Influence of the Non-Schmid Effects on the Ductility Limit of Polycrystalline Sheet Metals

Mohamed Ben Bettaieb [1,2] and Farid Abed-Meraim [1,2,*]

1 Université de Lorraine, CNRS, Arts et Métiers ParisTech, LEM3, F-57000 Metz, France; Mohamed.BenBettaieb@ensam.eu

2 Laboratory of Excellence on Design of Alloy Metals for Low-mAss Structures (DAMAS)—Université de Lorraine, F-57073 Metz, France

* Correspondence: Farid.Abed-Meraim@ensam.eu; Tel.: +33-3-87-37-54-79

Received: 30 June 2018; Accepted: 2 August 2018; Published: 8 August 2018

Abstract: The yield criterion in rate-independent single crystal plasticity is most often defined by the classical Schmid law. However, various experimental studies have shown that the plastic flow of several single crystals (especially with Body Centered Cubic crystallographic structure) often exhibits some non-Schmid effects. The main objective of the current contribution is to study the impact of these non-Schmid effects on the ductility limit of polycrystalline sheet metals. To this end, the Taylor multiscale scheme is used to determine the mechanical behavior of a volume element that is assumed to be representative of the sheet metal. The mechanical behavior of the single crystals is described by a finite strain rate-independent constitutive theory, where some non-Schmid effects are accounted for in the modeling of the plastic flow. The bifurcation theory is coupled with the Taylor multiscale scheme to predict the onset of localized necking in the polycrystalline aggregate. The impact of the considered non-Schmid effects on both the single crystal behavior and the polycrystal behavior is carefully analyzed. It is shown, in particular, that non-Schmid effects tend to precipitate the occurrence of localized necking in polycrystalline aggregates and they slightly influence the orientation of the localization band.

Keywords: crystal plasticity; non-Schmid effects; Taylor multiscale scheme; localized necking; bifurcation theory

---

1. Introduction

Despite the significant progress accomplished in the modeling of the mechanical behavior of metallic materials, the study of localized necking in thin metal sheets remains an active research topic for both academic and industrial communities. This research area has been initiated in the pioneering contributions of Keeler and Backofen [1], and Goodwin [2], who have introduced the representation by a forming limit diagram (FLD) as a characterization of the initiation of localized necking in thin metal sheets. For each strain path ranging from uniaxial tension to equibiaxial tension, the in-plane principal strains, which are associated with the incipience of localized necking, are reported on the forming limit diagram. Considering the practical complexity related to the experimental determination of FLDs (precise identification of the moment of the onset of strain localization, scatter in the experimental data...), as well as their high cost (a lot of experimental tests required to build a complete FLD...), important efforts have been devoted to the development of several alternative theoretical and/or numerical prediction models. These models are generally based on the coupling between a localization criterion used to predict the onset of localized necking and a constitutive model describing the evolution of the mechanical fields. Among the most known localization criteria used in the literature, one can quote the initial imperfection approach, which was developed initially by Marciniak and Kuczynski [3], the bifurcation theory initiated by Rice in [4,5], and the energy criterion of instability of a deformation process formulated by Petryk et al. [6–8]. Despite its large popularity, the initial imperfection approach has a main drawback: the over-sensitivity of its predictions (in terms of limit strain) to the amount of initial imperfection (which may be viewed as a nonphysical parameter). By contrast, the use of the bifurcation theory does not need any additional parameter (such as the initial imperfection factor). Furthermore, the formulation of the bifurcation theory is based on sound mathematical considerations, and its numerical implementation is relatively easy. For these reasons, the bifurcation theory is used in the current paper to detect the incipience of strain localization. For elastic-plastic constitutive frameworks with an associated plastic flow rule and a smooth yield surface, it has been demonstrated (see, e.g., Reference [5]) that the bifurcation approach is unable to predict material instability at a realistic strain level in the range of positive strain paths. Consequently, the prediction of localized necking at realistic strain levels requires the use of constitutive models exhibiting some destabilizing effects. When phenomenological constitutive models are coupled with the bifurcation theory, destabilizing phenomena may be introduced either by including damage-induced softening effects [9,10] or by deviating the plastic flow rule from normality [11,12]. Despite their popularity, the phenomenological models are not able to accurately capture the effects of some essential physical and microstructural mechanisms (initial and induced crystallographic and morphologic textures, crystallographic structure, dislocation density evolution...) on some important in-use properties (strength, formability...). These limitations have motivated researchers to use multiscale models in the prediction of FLDs. In the current contribution, a multiscale model has been coupled with the bifurcation theory to predict the incipience of localized necking in polycrystalline aggregates. In this multiscale model, the Taylor scheme is used to determine the macroscopic behavior of the polycrystalline aggregates from that of their constituents (single crystals). The mechanical behavior at the single crystal scale follows a finite strain rate-independent formulation. Compared to the rate-dependent formulation that has been widely used to predict strain localization [13,14], the rate-independent one is more appropriate for the simulation of cold forming processes, where viscous effects are limited. In the majority of the previous rate-independent contributions [15–18], the plastic flow is modeled by the classical Schmid law [19]. In this case, the destabilizing mechanism required to predict bifurcation localization at realistic limit strain levels is an obvious consequence of the crystal plasticity multi-slip and the associated yield surface vertex effects, which is taken into account by using this classical Schmid law. The effect of a regularization of this Schmid law (by substituting the vertices at the yield surface by rounded corners) on the prediction of the ductility limit has been recently analyzed in [20]. It has been demonstrated that the limit strains predicted in the range of positive strain paths are unrealistically high when the regularized version of the Schmid law is used to model the plastic flow at the single crystal scale. Although the classical Schmid law is widely accepted, various experimental observations made on BCC (Body Centered Cubic) metallic materials, such as molybdenum [21], tungsten [22], or tantalum [23], have revealed that this classical law is not always able to accurately describe the plastic flow. Indeed, most BCC single crystals have 24 slip systems, but the slip planes are not ideally close-packed. The $1/2\langle 111 \rangle$ screw dislocations in BCC metals have a non-planar core structure that spreads on three $\{1\bar{1}0\}$ planes. This causes non-Schmid effects (effects that are not considered by the classical law), in which stresses developing on planes and directions other than those on the primary slip systems will influence dislocation motion [24]. Note that the non-Schmid effects have been introduced in the modeling of the mechanical behavior of steel materials in at least three recent studies [24–26]. In these contributions, two particular steel grades have been studied: the three-phase QP980 and dual-phase DP980 steels. The reason for the choice of these steels is principally related to the fact that the deformation of the ferrite phase in both grades shows non-negligible non-Schmid behavior. These non-Schmid effects are incorporated in a generalized version of the Schmid law [27]. In such a generalization, the single crystal yield function includes stress components other than the Schmid stress, which results in non-associated plastic flow (i.e., non-normality). The main motivation behind the current investigation is to carefully analyze the implications of deviations from the classical Schmid law (by the incorporation of the non-Schmid effects) on the constitutive modeling of both BCC single crystals and polycrystalline aggregates. Then, the impact of these deviations on the onset of localized necking in thin metal sheets and on the orientation of the associated localization band is investigated. It will be especially demonstrated that the non-Schmid effects tend to precipitate the occurrence of strain localization in polycrystalline aggregates.

The paper is organized as follows. Section 2 gives the theoretical framework on which the prediction of localized necking in BCC polycrystalline aggregates is based. Particular attention is paid in this section to the introduction of non-Schmid effects in the plastic flow. Section 3 outlines the numerical aspects related to the implementation of the equations that govern the prediction of the ductility limit of polycrystalline aggregates. In Section 4, various numerical results are provided, which illustrate the influence of the non-Schmid effects on the mechanical behavior and on the onset of localized necking. The conclusions drawn from this study are provided in Section 5. Finally, Appendix A provides the list of crystallographic slip systems for BCC single crystals.

Standard notations and conventions are used throughout:

- First, second, or fourth-order tensors are represented by bold-face letters and symbols (the order of which is indicated by the context).
- Scalar parameters and variables are designated by thin letters and symbols.
- Macroscopic (resp. microscopic) fields are designated by capital (resp. small) letters and symbols.

$\dot{\bullet}$	time derivative of •.
$\bullet^{\nabla}$	co-rotational derivative of •.
$\bullet^{T}$	transpose of •.
•.•	inner product.
• : •	double contraction product (= $\bullet_{ij}\bullet_{ij}$ for the product between two second-order tensors, and
$\bullet_{ijkl}\bullet_{kl}$	for the product between a fourth-order tensor and a second-order tensor).
• × •	vector product.
det(•)	determinant of tensor •.
sgn(•)	sign of •.

2. Theoretical Framework

In the following theoretical developments, an updated Lagrangian formulation will be used to express the different equations (constitutive equations and localization analysis). Let us consider a polycrystalline aggregate made of N_g single crystals, which are initially randomly oriented. This aggregate is assumed to be representative of the studied thin metal sheet. To predict the ductility limit of the latter, the aggregate is submitted to uniform biaxial straining, where the in-plane components of the macroscopic velocity gradient **G** are known and defined as follows:

$$G_{11} = 1 \quad ; \quad G_{22} = \rho \quad ; \quad G_{12} = G_{21} = 0, \tag{1}$$

where ρ denotes the strain-path ratio, and it ranges between $-1/2$ (uniaxial tensile state) and 1 (equibiaxial tensile state). The out-of-plane components of **G** are unknown and should be deduced from the plane-stress condition in the direction normal to the plane of the sheet:

$$\dot{N}_{13} = \dot{N}_{23} = \dot{N}_{31} = \dot{N}_{32} = \dot{N}_{33} = 0, \tag{2}$$

where $\dot{\mathbf{N}}$ is the macroscopic nominal stress rate, which is related to the macroscopic velocity gradient **G** through the macroscopic tangent modulus **L**:

$$\dot{\mathbf{N}} = \mathbf{L} : \mathbf{G}. \tag{3}$$

The bifurcation criterion is used to numerically determine the onset of localized necking in the polycrystalline aggregate for the whole range of strain-path ratios (comprised between $-1/2$ to 1). This criterion asserts that strain localization occurs when the acoustic tensor becomes singular [5]. Hence, this criterion is defined by the following expression:

$$\det(\vec{\mathcal{N}}.\mathbf{L}^{\mathrm{PS}}.\vec{\mathcal{N}}) = 0, \tag{4}$$

where $\vec{\mathcal{N}}$ is the unit vector (lying in the plane of the sheet) normal to the localization band. Here, $\vec{\mathcal{N}}$ is taken equal to $(\cos\theta, \sin\theta)$, where the orientation θ of the localization band is comprised between $0°$ and $90°$. As to tensor $\mathbf{L}^{\mathrm{PS}}$, it represents the 2D macroscopic tangent modulus relating the in-plane components of the nominal stress rate tensor to the in-plane components of the velocity gradient. The analytical expression for the 2D tangent modulus $\mathbf{L}^{\mathrm{PS}}$ is derived from the general expression of the 3D tangent modulus by the classical relation [20]:

$$\forall \mathrm{i},\mathrm{j},\mathrm{k},\mathrm{l} = 1,2: \quad \mathrm{L}^{\mathrm{PS}}_{\mathrm{ijkl}} = \mathrm{L}_{\mathrm{ijkl}} - \frac{\mathrm{L}_{\mathrm{ij33}}\mathrm{L}_{\mathrm{33kl}}}{\mathrm{L}_{3333}} \tag{5}$$

Then, to check the occurrence of strain localization, the 3D macroscopic tangent modulus $\mathbf{L}$ should be determined by integrating the constitutive equations associated with the polycrystalline aggregate. In the current contribution, the Taylor multiscale scheme was used to determine the macroscopic mechanical behavior from the behavior of the microscopic constituents (the single crystals). This scheme was based on the assumption of the homogeneity of the strain field over the aggregate. Consequently, the microscopic velocity gradient $\mathbf{g}$ is equal to its macroscopic counterpart $\mathbf{G}$. The macroscopic nominal stress rate $\dot{\mathbf{N}}$ is obtained from its microscopic counterpart $\dot{\mathbf{n}}$ by the averaging relation:

$$\dot{\mathbf{N}} = \frac{1}{\mathrm{V}}\int_{\mathrm{V}} \dot{\mathbf{n}}(\mathbf{x})\,\mathrm{d}\mathbf{x}, \tag{6}$$

where V denotes the volume of the polycrystalline aggregate and $\mathbf{x}$ is a material point within this aggregate.

By combining the assumption of homogeneity of the strain field and Equation (6), one can deduce that the macroscopic tangent modulus $\mathbf{L}$ is related to its microscopic counterpart $\mathbf{l}$ (which relates $\dot{\mathbf{n}}$ to $\mathbf{g}$) by a relationship that is similar to Equation (6):

$$\mathbf{L} = \frac{1}{\mathrm{V}}\int_{\mathrm{V}} \mathbf{l}(\mathbf{x})\,\mathrm{d}\mathbf{x} \tag{7}$$

Therefore, to compute the macroscopic tangent modulus $\mathbf{L}$ via Equation (7), the microscopic tangent modulus $\mathbf{l}$ for all individual grains should be first determined. To this end, the subsequent developments are dedicated to the derivation of the analytical expression of the microscopic tangent modulus.

The microscopic velocity gradient $\mathbf{g}$ is additively split into its symmetric and skew-symmetric parts, denoted $\mathbf{d}$ and $\mathbf{w}$, respectively:

$$\mathbf{g} = \mathbf{d} + \mathbf{w} \tag{8}$$

The strain rate tensor $\mathbf{d}$ and the spin tensor $\mathbf{w}$ are decomposed into their elastic and plastic parts:

$$\mathbf{d} = \mathbf{d}^{\mathrm{e}} + \mathbf{d}^{\mathrm{p}} \quad ; \quad \mathbf{w} = \mathbf{w}^{\mathrm{e}} + \mathbf{w}^{\mathrm{p}} \tag{9}$$

The evolution of the rotation $\mathbf{r}$ of the lattice frame of the single crystal is expressed as a function of the elastic spin tensor $\mathbf{w}^{\mathrm{e}}$:

$$\dot{\mathbf{r}}.\mathbf{r}^{\mathrm{T}} = \mathbf{w}^{\mathrm{e}} \tag{10}$$

The elastic part of the behavior law is defined by the relation between the co-rotational derivative $\mathbf{k}^{\nabla}$ of the Kirchhoff stress tensor and the elastic strain rate $\mathbf{d}^{e}$:

$$\mathbf{k}^{\nabla} = \mathbf{C}^{e} : \mathbf{d}^{e} \tag{11}$$

where $\mathbf{C}^{e}$ is the fourth-order elasticity tensor. Here, isotropic and linear elasticity is assumed. We recall that the Kirchhoff stress tensor $\mathbf{k}$ is defined by the following relation:

$$\mathbf{k} = \mathrm{j}\,\boldsymbol{\sigma} \tag{12}$$

where j is the Jacobian of the deformation gradient and $\boldsymbol{\sigma}$ is the Cauchy stress tensor.

The combination of Equations (11) and (12) gives:

$$\boldsymbol{\sigma}^{\nabla} + \boldsymbol{\sigma}\,\mathrm{Tr}(\mathbf{d}) = \mathbf{C}^{e} : \mathbf{d}^{e} \tag{13}$$

The use of the co-rotational derivative $\mathbf{k}^{\nabla}$ instead of the simple time derivative $\dot{\mathbf{k}}$ of the Kirchhoff stress aims to satisfy the objectivity principle. These two derivatives are related by the following relation:

$$\dot{\mathbf{k}} = \mathbf{k}^{\nabla} - \mathbf{k}.\mathbf{w}^{e} + \mathbf{w}^{e}.\mathbf{k} \tag{14}$$

The plastic deformation of the single crystal is assumed to be only due to the slip on the crystallographic planes. Thus, the plastic strain rate $\mathbf{d}^{p}$ and the plastic spin $\mathbf{w}^{p}$ can be expressed as follows:

$$\mathbf{d}^{p} = \sum_{\alpha=1}^{N_s} \dot{\gamma}^{\alpha}\,\mathrm{sgn}(\tau^{\alpha})\,\mathbf{R}^{\alpha} \quad ; \quad \mathbf{w}^{p} = \sum_{\alpha=1}^{N_s} \dot{\gamma}^{\alpha}\,\mathrm{sgn}(\tau^{\alpha})\,\mathbf{S}^{\alpha} \tag{15}$$

where:

- N_s is the total number of slip systems.
- $\dot{\gamma}^{\alpha}$ is the absolute value of the slip rate of the αth slip system.
- $\mathbf{R}^{\alpha}$ and $\mathbf{S}^{\alpha}$ are respectively the symmetric and skew-symmetric part of the Schmid orientation tensor, which is defined as the tensor product $\vec{\mathbf{m}}^{\alpha} \otimes \vec{\mathbf{n}}^{\alpha}$. Vectors $\vec{\mathbf{m}}^{\alpha}$ and $\vec{\mathbf{n}}^{\alpha}$, corresponding to BCC single crystals that we have used in the current work are listed in Appendix A.
- τ^{α} is the resolved shear stress of the αth slip system, which is equal to $\mathbf{R}^{\alpha} : \boldsymbol{\sigma}$.

The plastic flow of the single crystal is modeled by a generalized version of the Schmid law [21]:

$$\forall \alpha = 1, \ldots, N_s : \quad \left\{ \begin{array}{l} \left| \tau^{\alpha} + \sum\limits_{i=1}^{N_{ns}} a_i^{\alpha} \tau_i^{\alpha} \right| < \tau_c^{\alpha} \Rightarrow \dot{\gamma}^{\alpha} = 0 \\ \left| \tau^{\alpha} + \sum\limits_{i=1}^{N_{ns}} a_i^{\alpha} \tau_i^{\alpha} \right| = \tau_c^{\alpha} \Rightarrow \dot{\gamma}^{\alpha} \geq 0 \end{array} \right. . \tag{16}$$

where:

- τ_c^{α} is the critical shear stress of the αth slip system.
- $\sum\limits_{i=1}^{N_{ns}} a_i^{\alpha} \tau_i^{\alpha}$ is an additional term (compared to the classical Schmid law) used to capture the non-Schmid effects. Here, τ_i^{α} and N_{ns} denote the non-Schmid shear stresses and their number, respectively. As to a_i^{α}, they represent material parameters, which can be determined by experimental tests or atomistic simulations [28]. For simplicity, we assume in the current

contribution that a_i^α are the same for all of the slip systems, and we choose the following expansion for the term $\sum_{i=1}^{N_{ns}} a_i^\alpha \tau_i^\alpha$ [28]:

$$\sum_{i=1}^{N_{ns}} a_i^\alpha \tau_i^\alpha = a_1 \boldsymbol{\sigma} : (\overrightarrow{\mathbf{m}}^\alpha \otimes \overrightarrow{\mathbf{n}}_1^\alpha) + a_2 \boldsymbol{\sigma} : [(\overrightarrow{\mathbf{n}}^\alpha \times \overrightarrow{\mathbf{m}}^\alpha) \otimes \overrightarrow{\mathbf{n}}^\alpha] + a_3 \boldsymbol{\sigma} : [(\overrightarrow{\mathbf{n}}_1^\alpha \times \overrightarrow{\mathbf{m}}^\alpha) \otimes \overrightarrow{\mathbf{n}}_1^\alpha] \tag{17}$$

where vectors $\overrightarrow{\mathbf{n}}_1^\alpha$ are enumerated in Appendix A (for the case of BCC single crystals).

Compared to other approaches given in the literature (see, for instance, References [28–30]) devoted to the study of the impact of the non-Schmid effects on the plastic flow of single crystals, Equation (17) seems to be the simplest expansion that we can consider.

By using the expression of the resolved shear stress τ^α ($= \mathbf{R}^\alpha : \boldsymbol{\sigma}$) and Equation (17), the generalized Schmid law can be rewritten as follows:

$$\forall \alpha = 1, \ldots, N_s : \quad \begin{cases} |\tau^{*\alpha}| < \tau_c^\alpha \Rightarrow \dot{\gamma}^\alpha = 0 \\ |\tau^{*\alpha}| = \tau_c^\alpha \Rightarrow \dot{\gamma}^\alpha \geq 0 \end{cases} \tag{18}$$

where $\tau^{*\alpha}$ is equal to $\mathbf{R}^{*\alpha} : \boldsymbol{\sigma}$, and $\mathbf{R}^{*\alpha}$ is defined as follows:

$$\forall \alpha = 1, \ldots, N_s : \quad \mathbf{R}^{*\alpha} = \mathbf{R}^\alpha + \mathbf{R}^{ns\alpha} \tag{19}$$

with $\mathbf{R}^{ns\alpha}$ being the symmetric part of the non-Schmid orientation tensor $\mathbf{M}^{ns\alpha}$, which can be easily deduced from Equation (17):

$$\forall \alpha = 1, \ldots, N_s : \quad \mathbf{M}^{ns\alpha} = a_1 (\overrightarrow{\mathbf{m}}^\alpha \otimes \overrightarrow{\mathbf{n}}_1^\alpha) + a_2 [(\overrightarrow{\mathbf{n}}^\alpha \times \overrightarrow{\mathbf{m}}^\alpha) \otimes \overrightarrow{\mathbf{n}}^\alpha] + a_3 [(\overrightarrow{\mathbf{n}}_1^\alpha \times \overrightarrow{\mathbf{m}}^\alpha) \otimes \overrightarrow{\mathbf{n}}_1^\alpha] \tag{20}$$

The rate of the critical shear stresses is expressed by the following generic form:

$$\forall \alpha = 1, \ldots, N_s : \quad \dot{\tau}_c^\alpha = \sum_{\beta=1}^{N_s} h^{\alpha\beta} \dot{\gamma}^\beta \tag{21}$$

where $\mathbf{h}$ is a symmetric hardening matrix.

The time derivative of $\tau^{*\alpha}$ can be obtained after some straightforward calculations:

$$\forall \alpha = 1, \ldots, N_s : \quad \dot{\tau}^{*\alpha} = \mathbf{R}^{*\alpha} : \boldsymbol{\sigma}^\nabla \tag{22}$$

The expression of the co-rotational stress rate $\boldsymbol{\sigma}^\nabla$ can be obtained by combining Equations (13) and $(15)_{(1)}$:

$$\boldsymbol{\sigma}^\nabla = [\mathbf{C}^e : \mathbf{d} - \boldsymbol{\sigma}\,\mathrm{Tr}(\mathbf{d})] - \sum_{\alpha=1}^{N_s} \dot{\gamma}^\alpha \mathrm{sgn}(\tau^\alpha)\, \mathbf{C}^e : \mathbf{R}^\alpha \tag{23}$$

Consequently, $\dot{\tau}^{*\alpha}$ can be expressed as follows:

$$\forall \alpha = 1, \ldots, N_s : \quad \dot{\tau}^{*\alpha} = \mathbf{R}^{*\alpha} : [\mathbf{C}^e : \mathbf{d} - \boldsymbol{\sigma}\,\mathrm{Tr}(\mathbf{d})] - \sum_{\beta=1}^{N_s} \dot{\gamma}^\beta\, \mathrm{sgn}(\tau^\beta)\, \mathbf{R}^{*\alpha} : \mathbf{C}^e : \mathbf{R}^\beta \tag{24}$$

Let us now introduce the set $\mathcal{A}$ of active slip systems, which are defined as:

$$\forall \alpha \in \mathcal{A} : \quad \dot{\gamma}^\alpha > 0 \quad ; \quad \dot{\tau}^{*\alpha}\, \mathrm{sgn}(\tau^{*\alpha}) - \dot{\tau}_c^\alpha = 0 \tag{25}$$

By using Equations (21) and (24), Equation (25) can be transformed as follows:

$$\forall \alpha \in \mathcal{A}: \quad \sum_{\beta \in \mathcal{A}} \left(\mathrm{sgn}(\tau^{*\alpha})\,\mathrm{sgn}(\tau^{\beta})\,\mathbf{R}^{*\alpha} : \mathbf{C}^{e} : \mathbf{R}^{\beta} + \mathrm{h}^{\alpha\beta}\right) \dot{\gamma}^{\beta} = \mathrm{sgn}(\tau^{*\alpha})\,\mathbf{R}^{*\alpha} : [\mathbf{C}^{e} : \mathbf{d} - \sigma\,\mathrm{Tr}(\mathbf{d})] \tag{26}$$

The absolute values for the slip rates of the active slip systems can then be obtained from Equation (26):

$$\forall \alpha \in \mathcal{A}: \quad \dot{\gamma}^{\alpha} = \sum_{\beta \in \mathcal{A}} \mathrm{M}^{\alpha\beta}\,\mathrm{sgn}(\tau^{*\beta})\mathbf{R}^{*\beta} : [\mathbf{C}^{e} : \mathbf{d} - \sigma\,\mathrm{Tr}(\mathbf{d})] \tag{27}$$

where **M** is the inverse of matrix **P** defined by the following index form:

$$\forall \alpha, \beta \in \mathcal{A}: \quad \mathrm{P}^{\alpha\beta} = \mathrm{sgn}(\tau^{*\alpha})\,\mathrm{sgn}(\tau^{\beta})\,\mathbf{R}^{*\alpha} : \mathbf{C}^{e} : \mathbf{R}^{\beta} + \mathrm{h}^{\alpha\beta} \tag{28}$$

The nominal stress tensor **n** and the Cauchy stress tensor $\boldsymbol{\sigma}$ are related by the following relation:

$$\mathbf{n} = \mathrm{j}\,\mathbf{f}^{-1}\boldsymbol{\sigma} \tag{29}$$

where j is the Jacobian of the microscopic deformation gradient **f**. The rate $\dot{\mathbf{n}}$ of the microscopic nominal stress is determined by computing the time derivative of Equation (29):

$$\dot{\mathbf{n}} = \mathrm{j}\,\mathbf{f}^{-1}(\dot{\boldsymbol{\sigma}} + \boldsymbol{\sigma}\,\mathrm{Tr}(\mathbf{d}) - \mathbf{g}.\boldsymbol{\sigma}) \tag{30}$$

As an updated Lagrangian approach is used in the current investigation, Equation (30) can be reduced to the following form:

$$\dot{\mathbf{n}} = \dot{\boldsymbol{\sigma}} + \boldsymbol{\sigma}\,\mathrm{Tr}(\mathbf{d}) - \mathbf{g}.\boldsymbol{\sigma} \tag{31}$$

The latter can be related to $\dot{\gamma}^{\alpha}$ by combining Equations (9), (14) and (15):

$$\dot{\mathbf{n}} = \mathbf{C}^{e} : \mathbf{d} - \boldsymbol{\sigma}.\mathbf{w} - \mathbf{d}.\boldsymbol{\sigma} - \sum_{\alpha \in \mathcal{A}} (\mathbf{C}^{e} : \mathbf{R}^{\alpha} + \mathbf{S}^{\alpha}.\boldsymbol{\sigma} - \boldsymbol{\sigma}.\mathbf{S}^{\alpha})\,\dot{\gamma}^{\alpha}\mathrm{sgn}(\tau^{\alpha}). \tag{32}$$

The microscopic tangent modulus **l** can be obtained from Equations (27) and (32) after some lengthy but straightforward algebraic manipulations [16]:

$$\mathbf{l} = \mathbf{C}^{e} - {}^{1}_{\sigma}\mathbf{l} - {}^{2}_{\sigma}\mathbf{l} - \sum_{\alpha \in \mathcal{A}} \left(\mathrm{sgn}(\tau^{\alpha})\,(\mathbf{C}^{e} : \mathbf{R}^{\alpha} + \mathbf{S}^{\alpha}.\boldsymbol{\sigma} - \boldsymbol{\sigma}.\mathbf{S}^{\alpha}) \otimes \sum_{\beta \in \mathcal{A}} \mathrm{M}^{\alpha\beta}\,\mathrm{sgn}(\tau^{*\beta})\mathbf{R}^{*\beta} : (\mathbf{C}^{e} - \boldsymbol{\sigma} \otimes \mathbf{Id})\right) \tag{33}$$

where **Id** stands for the second-order identity tensor, while ${}^{1}_{\sigma}\mathbf{l}$ and ${}^{2}_{\sigma}\mathbf{l}$ are fourth-order tensors, which are defined by convective terms of Cauchy stress components:

$${}^{1}_{\sigma}\mathbf{l}_{ijkl} = \frac{1}{2}\left(\delta_{lj}\sigma_{ik} - \delta_{kj}\sigma_{il}\right) \quad ; \quad {}^{2}_{\sigma}\mathbf{l}_{ijkl} = \frac{1}{2}\left(\delta_{ik}\sigma_{lj} + \delta_{il}\sigma_{kj}\right) \tag{34}$$

By analyzing Equations (18) and (25), one can easily deduce that $\dot{\gamma}^{\alpha}$ is not determined by τ^{α} and $\dot{\tau}^{\alpha}$ alone. In this case, the plastic flow is non-associated with the yield criterion. Indeed, stresses other than the Schmid stress (which is parallel to the slip) on that slip system enter the flow rule. Thereby, the normality rule is not respected. This phenomenon can be easily checked by analyzing Equation $(15)_{(1)}$ and the definition of $\tau^{*\alpha}$. Indeed, we have:

$$\mathbf{d}^{\mathrm{p}} \neq \sum_{\alpha=1}^{\mathrm{N_s}} \dot{\gamma}^{\alpha}\,\mathrm{sgn}(\tau^{*\alpha})\,\frac{\partial\,\tau^{*\alpha}}{\partial\,\boldsymbol{\sigma}} = \sum_{\alpha=1}^{\mathrm{N_s}} \dot{\gamma}^{\alpha}\,\mathrm{sgn}(\tau^{*\alpha})\,\mathbf{R}^{*\alpha} \tag{35}$$

Equation (35) implies that the plastic strain rate deviates from the direction of the outward normal to the yield surface (represented in stress space).

The effect of the non-associativity of the plastic flow on the decrease of the ductility limit has been largely studied in the literature with phenomenological models [11,31]. The aim of the current contribution is to numerically explore the effect of this non-associativity, considered at the single crystal scale, on the ductility limit of polycrystalline aggregates.

3. Algorithmic Aspects

The algorithm for the prediction of forming limit diagrams by the bifurcation approach is based on the two following nested loops:

- For each strain-path ratio ρ comprised between $-1/2$ and 1 (with typical intervals of 0.1).
    - For each time increment $I^{\Delta} = [t_0, t_0 + \Delta t]$:
        - ✓ Compute the plane-stress tangent modulus $\mathbf{L}^{PS}$ from the 3D tangent modulus $\mathbf{L}$ by using an iterative procedure similar to the one developed in [16]. On the other hand, $\mathbf{L}$ is determined from the microscopic tangent moduli $\mathbf{l}$ of the different single crystals by Equation (7). Some indications on the method used to compute $\mathbf{l}$ are given after this algorithm.
        - ✓ For $\theta = 0°$ to $90°$, at user-defined intervals (with typical increments of $1°$):
            - compute the determinant of the acoustic tensor $\det(\vec{\mathcal{N}}.\mathbf{L}^{PS}.\vec{\mathcal{N}}) = 0$. We recall that the components of $\vec{\mathcal{N}}$ are equal to $(\cos\theta, \sin\theta)$.
        - ✓ Search for the orientation that minimizes $\det(\vec{\mathcal{N}}.\mathbf{L}^{PS}.\vec{\mathcal{N}}) = 0$, over the different values of θ. If $\det(\vec{\mathcal{N}}_{\min}.\mathbf{L}^{PS}\vec{\mathcal{N}}_{\min}) \leq 0$,, then localized necking is reached. The corresponding angle θ is the orientation of the localization band, while the corresponding limit strain E_{11} is equal to $\int_0^{t_0+\Delta t} G_{11} dt = t_0 + \Delta t$ (as G_{11} is equal to 1). The computation is then stopped. Otherwise, the integration is continued for the next time increment.

The microscopic tangent modulus $\mathbf{l}$ should be computed at $t_0 + \Delta t$ by using Equation (33). As clearly demonstrated in Equations (33) and (34), the different microscopic variables should be updated to compute $\mathbf{l}$. The update of these variables is based on an implicit integration scheme over I^{Δ}. This integration scheme belongs to the family of ultimate algorithms and is similar to the one developed in [32] for the classical Schmid law. The novelty in the algorithm used here, compared to the algorithm of [32], consists in the addition of the non-Schmid effects. By carefully analyzing the constitutive equations at the single crystal scale, one can deduce easily that the determination of the set of active slip systems, from the set of potentially active slip systems $\mathcal{P}$ $(= \{\alpha = 1, \ldots, N_s : |\tau^{*\alpha}(t_0)| = \tau_c^{\alpha}(t_0)\})$, as well as the corresponding slip rates $\dot{\gamma}^{\alpha}$ allows the determination of all other mechanical variables. In this objective, a combinatorial search strategy, analogous to the one proposed in [32], is adopted to determine the set of active slip systems from the set of potentially active slip systems. This search strategy is performed iteratively and, at each iteration, a subset of the set of potentially active slip systems is chosen to be the set of active slip systems. The slip rates corresponding to the assumed set of active slip systems are calculated by using Equation (27). If matrix $\mathbf{P}$ (Equation (28)) is singular (which corresponds to the widely-known indetermination problem), the pseudo-inversion technique is adopted to invert it and then to compute the slip rates of the active slip systems. For the other slip systems, their slip rates are assumed to be equal to zero. After this step, the generalized Schmid law defined by Equation (18) is assessed for all of the potentially active slip systems. If at least one constraint of this generalized Schmid law is not satisfied, then the assumed set is not an effective set of active slip systems and another set is chosen. It must be noted that, due to the introduction of the non-Schmid effects, matrix $\mathbf{P}$ is not symmetric. The asymmetry of matrix $\mathbf{P}$ leads to a slight increase in the computational effort spent by the solver. Once the slip rates of the active slip systems are computed, the other mechanical variables (the Cauchy stress tensor $\boldsymbol{\sigma}$, the rotation of the lattice frame $\mathbf{r}$) should be updated, and the microscopic tangent modulus $\mathbf{l}$ is computed.

4. Numerical Results

4.1. Material Data

In this section, we examine the impact of the non-Schmid effects on the constitutive response at the single crystal scale (Section 4.2) and on the prediction of the onset of strain localization at the polycrystal scale (Section 4.3). A polycrystalline aggregate made of 1000 single crystals is considered. The initial crystallographic texture corresponding to this polycrystalline aggregate is randomly generated, as shown in Figure 1.

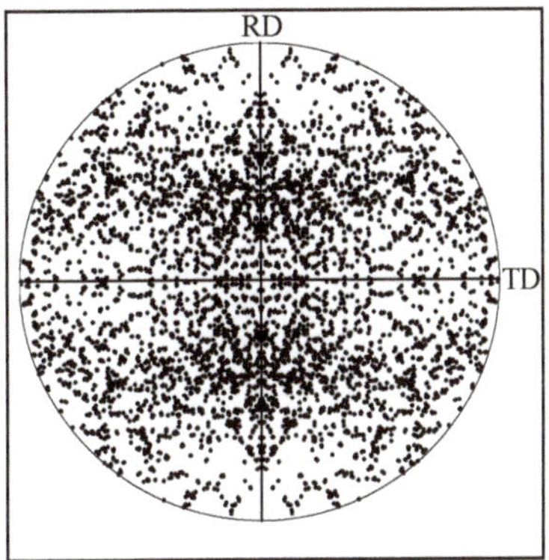

Figure 1. Initial crystallographic texture: {111} pole figure.

The initial state for each single crystal, in terms of stress and crystallographic slip, is defined as:

$$\boldsymbol{\sigma}(t=0) = \mathbf{0} \quad ; \quad \forall \alpha = 1, \ldots, 12 : \quad \gamma^{\alpha}(t=0) = 0 \tag{36}$$

The hardening law used in this work has been initially introduced in [33]. In this law, the hardening matrix $\mathbf{h}$ involved in Equation (21) is given by the following index expression:

$$\forall \alpha, \beta = 1, \ldots, 24 : \quad h^{\alpha\beta} = \hat{h}(A)\left[q + (1-q)\,\delta^{\alpha\beta}\right] \quad ; \quad \hat{h}(A) = h_0 \left(1 + \frac{h_0\, A}{n\, \tau_0}\right)^{n-1} \tag{37}$$

where $\delta^{\alpha\beta}$ is the Kronecker symbol and A is the sum of the accumulated plastic slip on all slip systems. However, n, τ_0, q and h_0 are material parameters. The values of the parameters relating to elasticity and hardening are reported in Table 1.

Table 1. Material parameters.

Elasticity		Hardening			
E	ν	n	τ_0	q	h_0
65 GPa	0.3	0.15	40 MPa	1.4	390 MPa

To extensively analyze the influence of the non-Schmid effects on the material response and formability limit prediction, it is desirable to set up a sensitivity study where the non-Schmid parameters a_1, a_2, and a_3 (see Equation (17)) would be varied simultaneously or separately in the range of their admissible values. Here, and for the sake of brevity, attention is restricted to the effect of parameter a_1 on the numerical predictions.

4.2. Significance of the Non-Schmid Effects on the Microscale Constitutive Response

To analyze the significance of the non-Schmid effects on the microscopic mechanical behavior, let us apply a plane-stress loading to a single crystal. This loading is defined by the following microscopic velocity gradient:

$$\mathbf{g} = \begin{pmatrix} 1 & 0 & ? \\ 0 & -0.5 & ? \\ ? & ? & ? \end{pmatrix} \tag{38}$$

where the unknown components of $\mathbf{g}$ ($g_{13}, g_{23}, g_{31}, g_{32}$, and g_{33}) are deduced by the following plane-stress condition:

$$\dot{n}_{13} = \dot{n}_{23} = \dot{n}_{31} = \dot{n}_{32} = \dot{n}_{33} = 0 \tag{39}$$

The initial orientation matrix of the studied single crystal is assumed to be equal to the identity tensor.

The impact of the non-Schmid effects on the evolution of the accumulated plastic slip of the activate slip systems $\gamma^\alpha (= \int_0^t \left|\dot{\gamma}^\alpha\right| dt)$, as a function of $\varepsilon_{11} (= \int_0^t g_{11}\, dt)$, is depicted in Figure 2. For the case of classical Schmid law, the parameters a_1, a_2, and a_3 are obviously set to 0. This figure shows that the activity of the slip systems and the evolution of the corresponding accumulated plastic slips are strongly influenced by the value of parameters a_1, a_2, and a_3. Indeed, when the non-Schmid effects are not considered, systems 2, 5, 6, 8, 9, 10 are activated during loading. By contrast, systems 3, 5, 6, 7, 11, 12 are activated when the non-Schmid effects are considered in the constitutive modeling. This result is expected, considering the influence of parameters a_1, a_2 and a_3 on the values of the components of matrix $\mathbf{P}$ (see Equation (28)). This matrix clearly affects the activity of the slip systems and the evolution of the corresponding accumulated plastic slips (see Equation (27)).

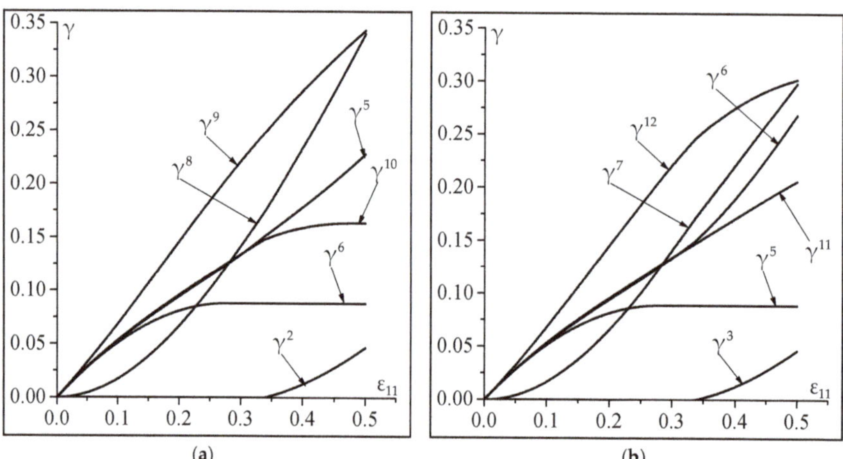

Figure 2. Significance of the non-Schmid effects on the slip system activity: (**a**) Results without non-Schmid effects ($a_1 = a_2 = a_3 = 0$); (**b**) Results with non-Schmid effects ($a_1 = 0.2; a_2 = a_3 = 0$).

The above-observed difference in the activity of the slip systems induces a difference in the evolution of the in-plane components of the Cauchy stress tensor, as shown in Figure 3. The abrupt changes observed on the evolution of the components of the stress tensor are explained by the change in the activity of the slip systems.

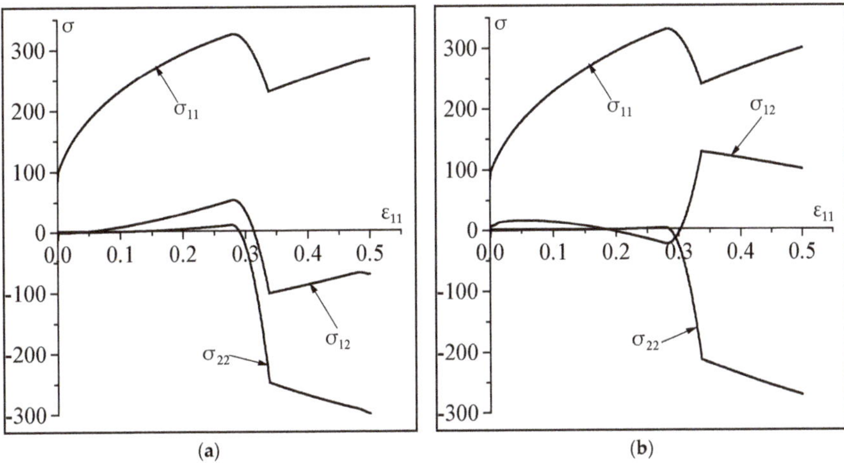

Figure 3. Impact of the non-Schmid effects on the evolution of the in-plane components of the Cauchy stress tensor: (**a**) Results without non-Schmid effects ($a_1 = a_2 = a_3 = 0$); (**b**) Results with non-Schmid effects ($a_1 = 0.2; a_2 = a_3 = 0$).

4.3. *Influence of the Non-Schmid Effects on Localized Necking*

As previously mentioned, the occurrence of plastic strain localization is predicted by the bifurcation theory. This theory states that bifurcation takes place when the determinant of the acoustic tensor becomes equal to zero (Equation (4)). Hence, to analyze the impact of the non-Schmid effects on plastic strain localization, let us first analyze their impact on the evolution of some components of the 2D macroscopic tangent modulus $\mathbf{L}^{PS}$. To this aim, the evolution of the components L^{PS}_{1111}, L^{PS}_{1122}, L^{PS}_{2211} and L^{PS}_{1212}, as function of $E_{11}(=\int_0^t G_{11}\,dt)$ for the plane-strain state, is plotted in Figure 4. This figure clearly shows that the value of the non-Schmid coefficient a_1 has a significant effect on the evolution of the components of $\mathbf{L}^{PS}$. Furthermore, the consideration of the non-Schmid effects leads to an important fluctuation of the evolution of the components of $\mathbf{L}^{PS}$. This observation is attributed to the fact that the normality of the plastic flow to the yield function is not respected when the non-Schmid effects are accounted for. As clearly shown in Figure 4, all of the components of $\mathbf{L}^{PS}$ decrease during loading including also, in particular, the shearing component L^{PS}_{1212}. This feature is an obvious consequence of the multi-slip nature of crystal plasticity, which results in the formation of vertices at the current points of the single crystal yield surfaces. Note that the decrease of the shearing components of the tangent modulus is the most important destabilizing factor responsible for bifurcation, thus leading to early plastic strain localization. It is also important to notice that the difference between L^{PS}_{1122} and L^{PS}_{2211} does not exceed 1%, which explains why the curves representing these tangent modulus components are indistinguishable, as clearly observed in Figure 4.

The influence of the non-Schmid effects on the initiation of plastic strain localization for two strain paths (uniaxial tension and plane-strain tension) is investigated in Figure 5, where the minimum of the determinant of the acoustic tensor is plotted as a function of E_{11}. The results of this figure clearly show that the non-Schmid effects tend to precipitate the occurrence of localized necking. Indeed, the limit strains predicted when the non-Schmid effects are considered are lower than those determined by the classical Schmid law, by about 3% of deformation.

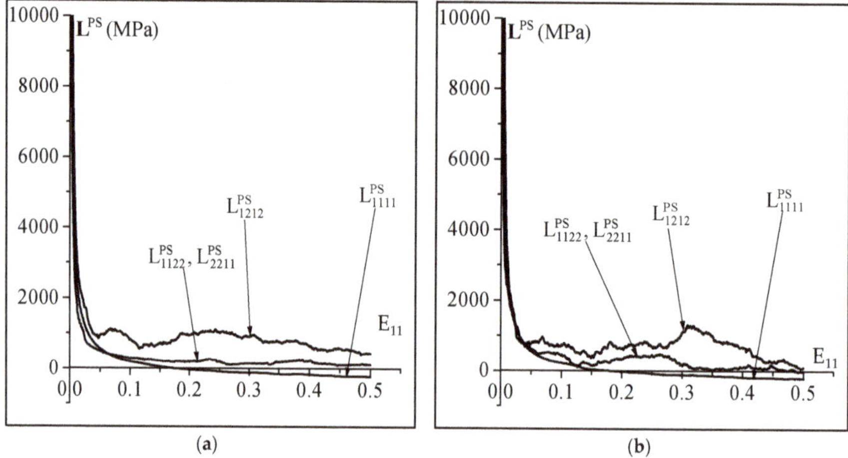

Figure 4. Significance of the non-Schmid effects on the evolution of some representative components of the macroscopic tangent modulus for the plane-strain state: (**a**) Results without non-Schmid effects ($a_1 = a_2 = a_3 = 0$); (**b**) Results with non-Schmid effects ($a_1 = 0.2; a_2 = a_3 = 0$).

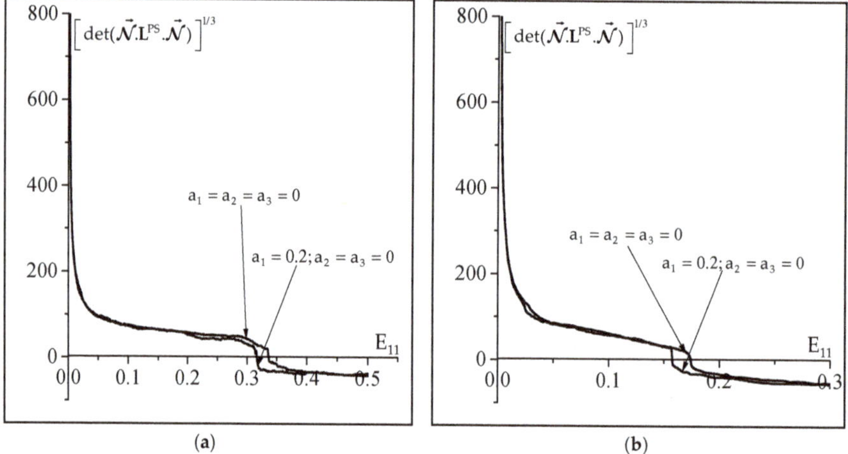

Figure 5. Impact of the non-Schmid effects on the evolution of the minimum of the determinant of the acoustic tensor as a function of E_{11}: (**a**) Uniaxial tensile state ($\rho = -0.5$); (**b**) Plane-strain state ($\rho = 0$).

The results obtained in Figure 5 are generalized to the whole range of strain paths in Figure 6a. This latter figure shows that the forming limit diagram predicted by the classical Schmid law is higher than its counterpart, which is predicted when the non-Schmid effects are accounted for. The impact of the non-Schmid effects on the necking band orientation is analyzed in Figure 6b. This latter figure shows that the non-Schmid effects have a slight influence on the necking band orientation. Also, the band inclination angle is found to be equal to 0 for strain-path ratios larger than 0.1, for both plastic flow rules (i.e., the classical Schmid law and the one which accounts for non-Schmid effects).

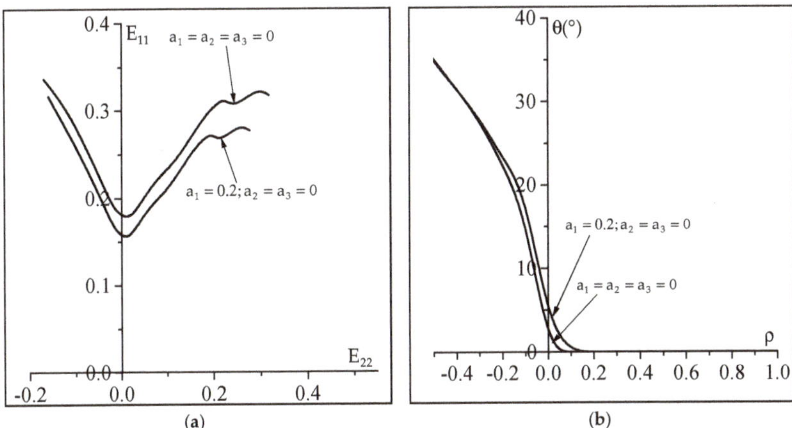

Figure 6. Impact of the non-Schmid effects on: (**a**) the shape and the level of the forming limit diagrams; (**b**) the evolution of the necking band orientation as a function of the strain-path ratio.

The effect of the crystallographic texture on the ductility of polycrystalline sheet metals has been analyzed in several contributions (see for instance, references [16,34]). It has been demonstrated in earlier contributions that the predicted limit strains remain almost insensitive to the initial texture in the range of negative strain paths. By contrast, in the range of positive strain paths, the opposite trend is observed. In fact, both the shape and the overall level of the predicted FLDs are highly sensitive to the initial crystallographic texture. To further investigate the influence of the non-Schmid effects on the ductility limit, we enlarge the analysis by adopting a new crystallographic texture (see Figure 7a), in order to compare the corresponding FLDs, predicted with and without considering the non-Schmid effects. The FLDs predicted using this new texture are reported in Figure 7b. The material parameters used to obtain the predictions of Figure 7b are the same as those used for Figure 6a (except for the initial crystallographic texture). The conclusion of these new comparisons, associated with the new crystallographic texture, is the same as that revealed by Figure 6: the non-Schmid effects tend to reduce the ductility limit for the whole range of strain paths. A more detailed study will be conducted in future works to further investigate and to analyze the combined influence of the crystallographic texture and the non-Schmid effects.

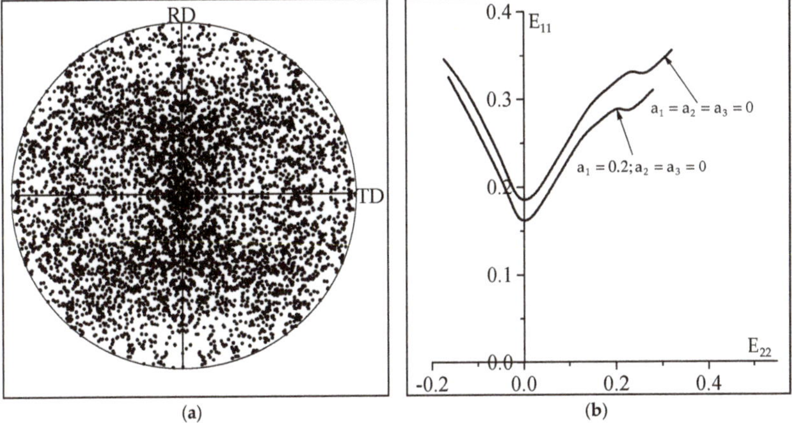

Figure 7. (**a**) Initial crystallographic texture ({111} pole figure); (**b**) Predicted forming limit diagrams.

5. Concluding Remarks

In the current contribution, we have analyzed the impact of the consideration of the non-Schmid effects on the mechanical behavior and on the onset of plastic strain localization in BCC polycrystalline materials. In the proposed model, the non-Schmid effects are accounted for at the single crystal scale by introducing additional terms in the expression of the resolved shear stress. Three scalar parameters (a_1, a_2 and a_3) have been used to account for these non-Schmid effects. The ductility limit is predicted by coupling the full-constraint Taylor scale-transition scheme to the bifurcation theory. The influence of the non-Schmid effects, through the variation of parameter a_1 on the prediction of plastic strain localization is particularly analyzed. The numerical simulations reveal that the consideration of non-Schmid effects ($a_1 \neq 0$) tends to precipitate the onset of localized necking. The present investigation will be extended in future works to other multiscale schemes, which are more relevant than the Taylor multiscale scheme, such as the self-consistent approach and the periodic homogenization technique. An extensive study will be carried out in future works, which will adopt more elaborate non-Schmid models, to better analyze the impact of the non-Schmid effects on the prediction of plastic strain localization.

Author Contributions: Conceptualization, M.B.B. and F.A.-M.; Methodology: M.B.B. and F.A.-M.; Software: M.B.B.; Writing: M.B.B. and F.A.-M. All of the authors participated in the discussion and reviewed the paper.

Funding: This research received no external funding.

Conflicts of Interest: The authors declare no conflict of interest.

Appendix A

Slip Systems for the BCC Single Crystals Used in the Non-Schmid Crystal Plasticity Model

For BCC single crystals, the number of crystallographic slip systems is equal to 24:12 slip systems of the family $\{110\}\langle 111\rangle$ and 12 slip systems of the family $\{112\}\langle 111\rangle$. However, attention is focused here on the $\{110\}\langle 111\rangle$ slip systems with non-Schmid effects, since the geometry of $1/2\langle 111\rangle$ screw dislocations on the $\{110\}$ planes has been better understood in experiments and atomistic simulations at low temperatures, which is of particular interest in this work. Each slip system is characterized by three orthonormal vectors $(\vec{\mathbf{m}}^{\alpha}, \vec{\mathbf{n}}^{\alpha}, \vec{\mathbf{n}}_1^{\alpha})$ in the deformed configuration. $\vec{\mathbf{m}}^{\alpha}$ is the vector parallel to the slip line, $\vec{\mathbf{n}}^{\alpha}$ is the vector normal to the slip plane and $\vec{\mathbf{n}}_1^{\alpha}$ is the normal to the plane $\{1\bar{1}0\}$ in the slip direction zone, which also forms an angle of $-60°$ with $\vec{\mathbf{n}}^{\alpha}$. In the intermediate configuration, these vectors are defined by the following relations:

$$\forall \alpha = 1, \ldots, 24: \quad \|\vec{\mathbf{m}}_0^{\alpha}\| = \|\vec{\mathbf{n}}_0^{\alpha}\| = \|\vec{\mathbf{n}}_{10}^{\alpha}\| = 1 \quad ; \quad \vec{\mathbf{m}}_0^{\alpha} . \vec{\mathbf{n}}_0^{\alpha} = 0 \tag{A1}$$

Vectors $\vec{\mathbf{m}}_0^{\alpha}$, $\vec{\mathbf{n}}_0^{\alpha}$, and $\vec{\mathbf{n}}_{10}^{\alpha}$ are given in Table A1.

Table A1. List of slip systems for a BCC crystallographic structure [28].

α	**1**	**2**	**3**	**4**	**5**	**6**
$\sqrt{3}\vec{\mathbf{m}}_0^{\alpha}$	$[1\bar{1}1]$	$[\bar{1}\bar{1}1]$	$[111]$	$[\bar{1}11]$	$[\bar{1}11]$	$[\bar{1}\bar{1}1]$
$\sqrt{2}\vec{\mathbf{n}}_0^{\alpha}$	$[011]$	$[011]$	$[0\bar{1}1]$	$[0\bar{1}1]$	$[101]$	$[101]$
$\sqrt{2}\vec{\mathbf{n}}_{10}^{\alpha}$	$[\bar{1}10]$	$[0\bar{1}1]$	$[10\bar{1}]$	$[\bar{1}\bar{1}0]$	$[101]$	$[01\bar{1}]$
α	**7**	**8**	**9**	**10**	**11**	**12**
$\sqrt{3}\vec{\mathbf{m}}_0^{\alpha}$	$[111]$	$[1\bar{1}1]$	$[\bar{1}11]$	$[\bar{1}1\bar{1}]$	$[111]$	$[11\bar{1}]$
$\sqrt{2}\vec{\mathbf{n}}_0^{\alpha}$	$[\bar{1}01]$	$[\bar{1}01]$	$[110]$	$[110]$	$[\bar{1}10]$	$[\bar{1}10]$
$\sqrt{2}\vec{\mathbf{n}}_{10}^{\alpha}$	$[1\bar{1}0]$	$[011]$	$[\bar{1}0\bar{1}]$	$[110]$	$[\bar{1}01]$	$[0\bar{1}\bar{1}]$

References

1. Keeler, S.P.; Backofen, W.A. Plastic instability and fracture in sheets stretched over rigid punches. *Trans. ASM* **1963**, *56*, 25–48.
2. Goodwin, G.M. Application of strain analysis to sheet metal forming problems in press shop. *Metall. Ital.* **1968**, *60*, 767–774.
3. Marciniak, Z.; Kuczynski, K. Limit strains in processes of stretch-forming sheet metal. *Int. J. Mech. Sci.* **1967**, *9*, 609–620. [CrossRef]
4. Stören, S.; Rice, J.R. Localized necking in thin sheets. *J. Mech. Phys. Solids* **1975**, *23*, 421–441. [CrossRef]
5. Rice, J.R. The localization of plastic deformation. In Proceedings of the 14th International Congress of Theoretical and Applied Mechanics, Delft, The Netherlands, 30 August–4 September 1976; pp. 207–220.
6. Petryk, H. On energy criteria of plastic instability. In *Plastic Instability, Proc. Considere Memorial*; Ecole Nationale des Ponts et Chaussees: Paris, France, 1985; pp. 215–226.
7. Petryk, H.; Thermann, K. Post-critical plastic deformation of biaxially stretched sheets. *Int. J. Solids Struct.* **1996**, *33*, 689–705. [CrossRef]
8. Petryk, H.; Thermann, K. Post-critical plastic deformation in incrementally nonlinear materials. *J. Mech. Phys. Solids* **2002**, *50*, 925–954. [CrossRef]
9. Haddag, B.; Abed-Meraim, F.; Balan, T. Strain localization analysis using a large deformation anisotropic elastic–plastic model coupled with damage. *Int. J. Plasticity* **2009**, *25*, 1970–1996. [CrossRef]
10. Mansouri, L.Z.; Chalal, H.; Abed-Meraim, F. Ductility limit prediction using a GTN damage model coupled with localization bifurcation analysis. *Mech. Mater.* **2014**, *76*, 64–92. [CrossRef]
11. Kuroda, M.; Tvergaard, V. A phenomenological plasticity model with non-normality effects representing observations in crystal plasticity. *J. Mech. Phys. Solids* **2004**, *49*, 1239–1263. [CrossRef]
12. Ben Bettaieb, M.; Abed-Meraim, F. Investigation of localized necking in substrate-supported metal layers: Comparison of bifurcation and imperfection analyses. *Int. J. Plasticity* **2016**, *65*, 168–190. [CrossRef]
13. Signorelli, J.W.; Bertinetti, M.A.; Turner, P.A. Predictions of forming limit diagrams using a rate-dependent polycrystal self-consistent plasticity model. *Int. J. Plasticity* **2009**, *25*, 1–25. [CrossRef]
14. Schwindt, C.; Schlosser, F.; Bertinetti, M.A.; Signorelli, J.W. Experimental and Visco-Plastic Self-consistent evaluation of forming limit diagrams for anisotropic sheet metals: An efficient and robust implementation of the M-K model. *Int. J. Plasticity* **2015**, *73*, 62–99. [CrossRef]
15. Franz, G.; Abed-Meraim, F.; Berveiller, M. Strain localization analysis for single crystals and polycrystals: Towards microstructure-ductility linkage. *Int. J. Plasticity* **2013**, *48*, 1–33. [CrossRef]
16. Akpama, H.K.; Ben Bettaieb, M.; Abed-Meraim, F. Localized necking predictions based on rate-independent self-consistent polycrystal plasticity: Bifurcation analysis versus imperfection approach. *Int. J. Plasticity* **2017**, *91*, 205–237. [CrossRef]
17. Miehe, C.; Schröder, J.; Schotte, J. Computational homogenization analysis in finite plasticity Simulation of texture development in polycrystalline materials. *Comput. Method Appl. Mech. Eng.* **1999**, *171*, 387–418. [CrossRef]
18. Watanabe, I.; Setoyama, D.; Nagasako, N.; Iwata, N.; Nakanishi, K. Multiscale prediction of mechanical behavior of Ferrite-Pearlite steel with Numerical Material Testing. *Int. J. Numer. Meth. Eng.* **2012**, *89*, 829–845. [CrossRef]
19. Schmid, E.; Boas, W. *Plasticity of Crystals*; Chapman and Hall: London, UK, 1935.
20. Akpama, H.K.; Ben Bettaieb, M.; Abed-Meraim, F. Influence of the yield surface curvature on the forming limit diagrams predicted by crystal plasticity theory. *Lat. Am. J. Solids Struct.* **2016**, *13*, 1250–1269. [CrossRef]
21. Bassani, J.; Ito, K.; Vitek, V. Complex macroscopic plastic flow arising from non-planar dislocation core structures. *Mater. Sci. Eng. A* **2001**, *319–321*, 97–101. [CrossRef]
22. Gröger, R.; Vitek, V. Multiscale modeling of plastic deformation of molybdenum and tungsten. III. Effects of temperature and plastic strain rate. *Acta Mater.* **2008**, *56*, 5426–5439. [CrossRef]
23. Wang, Z.; Beyerlein, I. An atomistically-informed dislocation dynamics model for the plastic anisotropy and tension-compression asymmetry of BCC metals. *Int. J. Plasticity* **2011**, *27*, 1471–1484. [CrossRef]
24. Mapar, A.; Ghassemi-Armaki, H.; Pourboghrat, F.; Kumar, K.S. A differential-exponential hardening law for non-Schmid crystal plasticity finite element modeling of ferrite single crystals. *Int. J. Plasticity* **2017**, *91*, 268–299. [CrossRef]

25. Chen, P.; Ghassemi-Armaki, H.; Kumar, S.; Bower, A.; Bhat, S.; Sadagopan, S. Microscale-calibrated modeling of the deformation response of dual-phase steels. *Acta Mater.* **2014**, *65*, 133–149. [CrossRef]
26. Srivastava, A.; Ghassemi-Armaki, H.; Sung, H.; Chen, P.; Kumar, S.; Bower, A.F. Micromechanics of plastic deformation and phase transformation in a three-phase TRIP-assisted advanced high strength steel: Experiments and modeling. *J. Mech. Phys. Solids* **2015**, *78*, 46–69. [CrossRef]
27. Qin, Q.; Bassani, J.-L. Non-associated plastic flow in single crystals. *J. Mech. Phys. Solids* **1992**, *40*, 835–862. [CrossRef]
28. Gröger, R.; Bailey, A.; Vitek, V. Multiscale modeling of plastic deformation of molybdenum and tungsten: I. Atomistic studies of the core structure and glide of 1/2111 screw dislocations at 0K. *Acta Mater.* **2008**, *56*, 5401–5411. [CrossRef]
29. Qin, Q.; Bassani, J.-L. Non-Schmid yield behavior in single crystals. *J. Mech. Phys. Solids* **1992**, *40*, 813–833. [CrossRef]
30. Koester, A.; Ma, A.; Hartmaier, A. Atomistically informed crystal plasticity model for body-centered cubic iron. *Acta Mater.* **2012**, *60*, 3894–3901. [CrossRef]
31. Kuroda, M. A phenomenological plasticity model accounting for hydrostatic stress-sensitivity and vertex-type of effect. *Mech. Mater.* **2004**, *36*, 285–297. [CrossRef]
32. Akpama, H.K.; Ben Bettaieb, M.; Abed-Meraim, F. Numerical integration of rate-independent BCC single crystal plasticity models: Comparative study of two classes of numerical algorithms. *Int. J. Numer. Meth. Eng.* **2016**, *108*, 363–422. [CrossRef]
33. Chang, Y.W.; Asaro, R.J. An experimental study of shear localization in aluminum-copper single crystals. *Acta Metall.* **1981**, *29*, 241–257. [CrossRef]
34. Barlat, F. Crystallographic texture, anisotropic yield surfaces and forming limits of sheet metals. *Mater. Sci. Eng.* **1987**, *91*, 55–72. [CrossRef]

 materials

Article

Microsegregation Model Including Convection and Tip Undercooling: Application to Directional Solidification and Welding

Thomas Billotte [1,2], Dominique Daloz [1,2,*], Bernard Rouat [1,2], Guillaume Tirand [3], Jacob R. Kennedy [1,2], Vincent Robin [4] and Julien Zollinger [1,2]

1 Department of Metallurgy and Materials Science and Engineering, Institut Jean Lamour, UMR CNRS 7198, Université de Lorraine, 54000 Nancy, France; t.billotte@isgroupe.com (T.B.); bernard.rouat@univ-lorraine.fr (B.R.); jacob-roman.kennedy@univ-lorraine.fr (J.R.K.); julien.zollinger@univ-lorraine.fr (J.Z.)

2 Laboratory of Excellence on Design of Alloy Metals for Low-mAss Structures ('LabEx DAMAS'), Université de Lorraine, 57073 Metz, France

3 AREVA, Technical Center, 71100 Saint-Marcel, France; guillaume.tirand@framatome.com

4 AREVA, Engineering & Projects, 69006 Lyon, France; vincent.robin@edf.fr

* Correspondence: dominique.daloz@univ-lorraine.fr; Tel.: +33-372-742-669

Received: 29 June 2018; Accepted: 15 July 2018; Published: 20 July 2018

Abstract: The microsegregation behavior of alloy filler metal 52 (FM 52) was studied using microprobe analysis on two different solidification processes. First, microsegregation was characterized in samples manufactured by directional solidification, and then by gas tungsten arc welding (GTAW). The experimental results were compared with Thermo-Calc calculations to verify their accuracy. It was confirmed that the thermodynamic database predicts most alloying elements well. Once this data had been determined, several tip undercooling calculations were carried out for different solidification conditions in terms of fluid flow and thermal gradient values. These calculations allowed the authors to develop a parametrization card for the constants of the microsegregation model, according to the process parameters (e.g., convection in melt pool, thermal gradient, and growth velocity). A new model of microsegregation, including convection and tip undercooling, is also proposed. The Tong–Beckermann microsegregation model was used individually and coupled with a modified Kurz-Giovanola-Trivedi (KGT) tip undercooling model, in order to take into account the convection in the fluid flow at the dendrite tip. Model predictions were compared to experimental results and showed the microsegregation evolution accurately.

Keywords: microsegregation; gas tungsten arc welding; directional solidification; FM52 filler metal; ERNiCrFe-7; tip undercooling

1. Introduction

The first pressurized water reactor (PWR) components were manufactured from stainless steel and nickel A600 alloys. Welding of the reactors was performed using FM82 filler metal in order to maintain both mechanical and anti-corrosion properties. Since 1990, A690 nickel alloy has replaced the A600 alloy in order to improve stress corrosion cracking resistance [1]. In addition, a new filler metal, FM52, was developed. A690 alloy mainly consists of an austenitic matrix containing $M_{23}C_6$ (MC) carbides [2–4]. MC carbides are also observed in the filler metal FM52 [5–7], as well as some titanium nitrides, as reported in the literature [8,9].

A690 and FM52 satisfy the specification for corrosion resistance [10], but suffer from a tendency for ductility dip cracking (DDC) during welding [11–17]. Different factors may affect the DDC sensitivity, such as alloy composition, element segregation (phosphorous and sulfur), secondary phase precipitation at grain boundaries, grain boundary sliding, or grain boundary orientation [12]. Since 2000, this has led to many developments to improve the DCC resistance, focusing mainly on the chemical composition [18]. FM52 can now be reinforced using boron and zirconium for FM52M [19], and using molybdenum and niobium for FM52MSS, to improve its mechanical properties [20,21]. However, segregation of boron at the grain boundaries promotes the formation of low melting point compounds, generating liquation cracks [22]. This cracking behavior can be further exacerbated by the presence of oxide bifilms [23]; the same effect is noted for zirconium [24]. The importance of back-diffusion to reduce solidification cracking has also recently been shown [25], along with the fact that microsegregation in gas tungsten arc welding (GTAW) can be somewhat controlled by the use of pulsed current [26]. It is critical then, for development and application, to thoroughly understand how segregation occurs and how to model it. It is useful to simulate solidification structures using a model in which a cellular automaton is coupled with a finite element mesh (CAFE) [27,28].

When such predictive modeling tools are applied to welding or additive manufacturing, the Scheil rule (or truncated Scheil [29])—based on the hypothesis that no solid diffusion occurs—is employed, due to the relatively rapid solidification compared to other casting processes [30]. Time allowed for back-diffusion in the solid is low in rapid solidification, which correlates well with the Scheil model hypothesis. However, the structures formed are much finer than those obtained with slower solidification processes. This means that less time is needed for elements to back-diffuse into the solid and, thus, the applicability of the Scheil rule becomes more problematic [31]. Additionally, the intense convection that occurs in the weld pool may affect the solutal build-up associated with solidification, and thereby modify the dendrite tip undercooling. This is not taken into account in the Scheil description, and may affect the initial conditions for microsegregation.

Currently, there is no published work which specifically regards dendrite tip undercooling and convection in the melt pool [32]. Convection is in the order of tens of centimeters per second during welding. Thus, it is relevant to quantify the convection effect on undercooling and include it in a microsegregation model. The ability of the Scheil model to correctly represent the microsegregation in such rapid growth conditions may also be investigated. This paper will address both of these points.

In this work, quenched directional solidification (QDS) and gas tungsten arc welding (GTAW) of FM52 were carried out. From the experiments, the microsegregation was characterized. A KGT [33] tip undercooling model—which took fluid flow into account [34]—was applied, in order to quantify the importance of convection on the undercooling. The Tong–Beckermann [35] microsegregation model was also used, which could account for the importance of back-diffusion. This model, together with the classical lever rule and Gulliver–Scheil model, are compared with the experimental results. These results are discussed, and rules for better use of microsegregation models are proposed.

2. Experimental Section

The FM52 alloy was provided by AREVA. The chemical composition of the alloy was measured using the electron probe micro-analysis (EPMA) on a JEOL JXA 8530F apparatus (JEOL, Tokyo, Japan). Table 1 shows the results of this analysis, compared with the manufacturer's specifications.

Table 1. Composition of alloy filler metal 52 (FM52) in weight percent.

	Ni	Cr	Fe	Al	Ti	Mn	Si	C
EPMA	57.98	29.98	10.06	0.65	0.52	0.29	0.13	0.03
Specification [36]	≈60	28–31.5	7–11	<1.1	<1	<0.5	<0.5	<0.04

The solidus and liquidus temperatures were determined using a SETARAM Setsys 16/18 differential thermal analysis (DTA) (SETARAM, Lyon, France). Four cooling and heating rates were applied, ranging from 2 to 20 K/min. The results were examined using the Boettinger [37] recommendations and the Bobadilla observations [38]. The DTA results for solidus and liquidus temperatures are summarized in Table 2, and compared with thermodynamic calculations performed using the Thermo-Calc software (Thermo-Calc 2018b, Thermo-Calc software, Solna, Sweden) with the TTNi8 database [39]. The solidification range (ΔT_0) found by DTA analysis was 10.6 °C, which is very close to the value (13 °C) found by Wu and Tsai [40]. As seen in Table 2, the values given by the TTNi8 database for the FM52 alloy (considering all alloying elements) does not reflect the experimental DTA measurements, especially for the solidification range (36 °C vs. 10.6 °C). If only the major constituent elements of the alloy are considered (Cr, Fe), the calculated solidification range is 8.4 °C, showing much better agreement with the DTA measurements.

Table 2. Results of the differential thermal analyses (DTA) and thermodynamic calculations with the TTNi8 database.

	DTA (°C)	Thermo-Calc (°C, FM52)	Thermo-Calc (°C, Ni-29.98Cr-10.06Fe)
Solidus	1379.2	1364	1409.4
Liquidus	1389.8	1400	1417.8

Experiments involving quenching during directional solidification (QDS) were performed with a specially designed Bridgman type furnace. This method allows the thermal gradient (G) and the solidification velocity (V) to be controlled independently. The quench imposes a cooling rate of 100 °C/s, and is able to freeze the microstructure and microsegregations during solidification [41,42]. Alloy FM52 was melted and cast into cylindrical rods, 5 mm in diameter, which were used in the Bridgman furnace. After the QDS experiments, transverse cross-sections were cut along the length of the mushy zone. Each section corresponds to a given temperature before the quench and, for a given solid fraction, the morphology and the amount of segregation in the solidifying alloy.

The welding experiments were performed using gas tungsten arc welding (GTAW) under argon. The filler metal was FM52 alloy and the welding plates were A690 nickel alloy. Contrary to the QDS experiments, the solidification operating parameters are not controlled in the GTAW process. For this study, the growth velocity was considered equal to the welding speed, and the thermal gradient was determined using the SYSWELD software (2017, ESI, Paris, France). The solidification parameters, and G and V for the QDS and welding experiments are given in Table 3.

Table 3. Solidification parameters used in the quenched directional solidification (QDS) and welding experiments.

Test Type	Thermal Gradient (K/m)	V (mm/s)	G/V (K·s/m^2)	$G{\cdot}V$ (K/s)
QDS	5300	0.09	5.89×10^7	0.477
		0.03	1.77×10^8	0.159
		0.013	4.08×10^8	0.0689
	3000	0.09	3.33×10^7	0.27
		0.03	1×10^8	0.09
		0.013	2.31×10^8	0.039
GTAW	≈300,000	1.66	1.81×10^8	498

To characterize microsegregation, EPMA measurements were performed using regular analysis grids [41–44]. The parameters for the systematic sampling were determined using a previously developed method [44]. The dimensions of the analysis grids were the same for all the analyzed samples, e.g., 196 measurements with a 100 µm step size, so that the sampling parameter, r, defined by $r = \frac{step}{\lambda_1}$, was always less than or equal to 10 [44].

The sorting strategy used was based on the element which was known to have the greatest tendency to segregate in the liquid [41]. In this work, titanium was selected, as Ti is known to segregate in the liquid during solidification [9]. All other elements were arranged following the Ti values in ascending order. In order to compare them, the microsegregation profiles were normalized to the mean composition calculated from all 196 points.

3. Microstructure and Microsegregation

3.1. Secondary Dendrite Arm Spacing (SDAS) Law

The scale of microsegregation was found using secondary dendrite arm spacing (SDAS) measurements. It was necessary to know the SDAS evolution, with respect to the solidification parameters, in order to utilize the data in the Tong–Beckermann (TB) model. For this reason, several QDS experiments were performed with different solidification conditions. For each experiment, several micrographs from the longitudinal section of the QDS samples were taken to measure the SDAS. This was also carried out for the GTAW experiments.

Figure 1 illustrates the different sizes of the solidification structures from the experiments. As shown in Figure 1a, the solidification front is clearly visible on the QDS micrographs (Figure 1a–c). The dendrites grow following the thermal gradient, however, in the case of GTAW, it was less clear, since the thermal gradient vector was not constant. Nevertheless, a dendritic structure can still be observed in Figure 1d. The evolution of the size of the solidification structures is also visible in Figure 1. As the product of *G* and *V* increased, the size of the structures decreased, as can be seen by comparing the QDS experiments. This is even more evident with the GTAW experiments, where the cooling rate is 1000 times higher than the maximum cooling rate obtained with QDS.

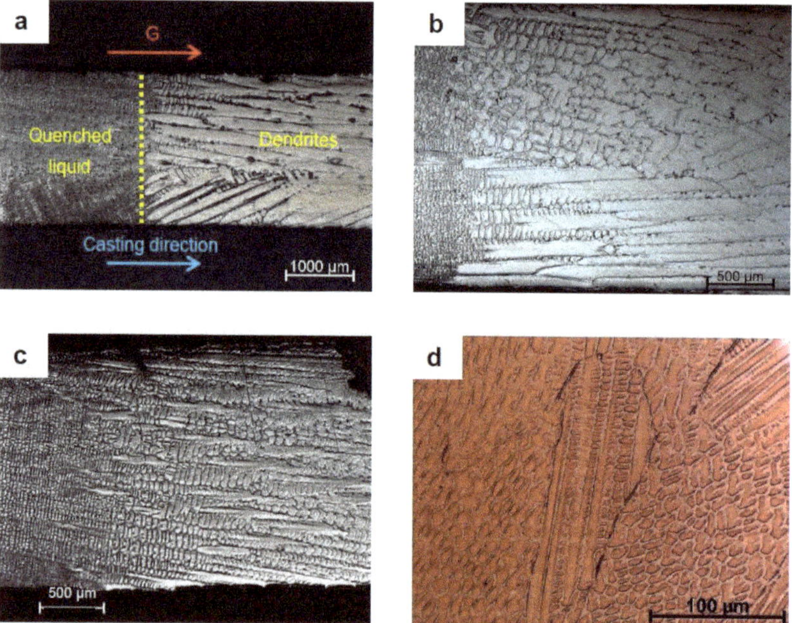

Figure 1. Micrographs for the secondary dendrite arm spacing (SDAS) measurement after electrolytic etching, using a 10 g $H_2C_2O_4$ with 100 mL H_2O reagent, under the applied potential 4 V DC for 30–40 s: (**a**) QDS for a solidification velocity $V = 0.013$ mm/s and thermal gradient (G) = 3000 K/m; (**b**) QDS for $V = 0.03$ mm/s and $G = 5300$ K/m; (**c**) QDS for $V = 0.09$ mm/s and G = 5300 K/m; and (**d**) gas tungsten arc welding (GTAW).

This data was supplemented by the measurements of Blecher et al. [45] in an A690 alloy, which had a composition very similar to the FM52 alloy. The SDAS evolution, with respect to the cooling rate, is plotted in Figure 2, using the Frenk and Kurz method [46]. The equation of the fitted curve, presented as a dashed line in Figure 2, is given by Equation (1). This law is very similar to the Blecher et al. law [45], and is very close to the well-known SDAS law for metallic alloys, $\lambda_2 = A \times (G \times V)^{\frac{1}{3}}$, where A is a material constant [47].

$$\lambda_2 = 23.92 \times (G \times V)^{-0.323} \text{ in microns.} \tag{1}$$

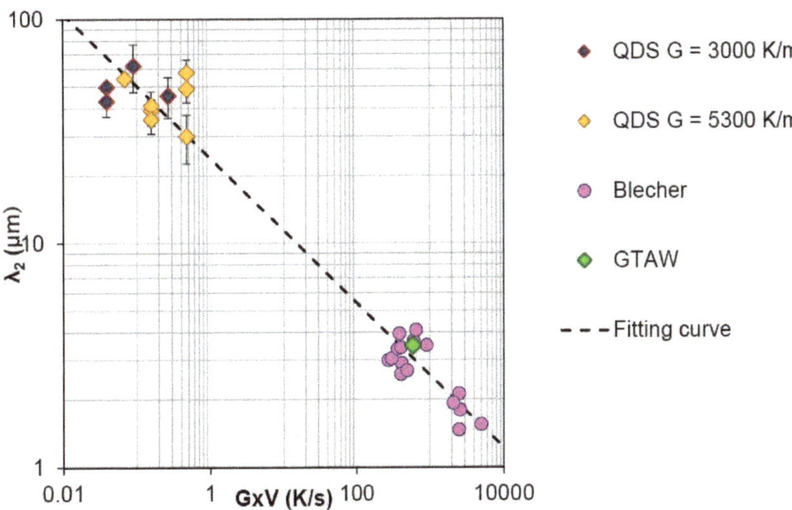

Figure 2. SDAS evolution of FM52 alloy as function of the cooling rate in K/s for QDS, GTAW experiments and Blecher et al.'s [45] measurements.

3.2. Microsegregation Results

The microsegregation profiles for the QDS and GTAW experiments are given for Fe, Cr, Ti and Al in Figure 3. Even though the cooling rate was a thousand times larger in QDS than GTAW, the microsegregation profiles do not differ significantly. In fact, they are identical for all elements (Figure 3), despite the cooling rates differing by three orders of magnitude.

The trends in the segregations shown in Figure 3 are comparable to others reported for different nickel alloys containing chromium, titanium and aluminum [48,49]. These elements segregate in the liquid during solidification, as opposed to iron which does not.

Two methods of estimating the partition coefficient were considered. From the phase diagram, the partition coefficient, denoted k_1 in the following, was estimated using Equation (2). QDS offers a second possibility of estimating the partition coefficient if the microsegregation is analyzed on a transverse section, where the solid fraction is less than 1. This method was used to calculate the partition coefficient, denoted k_2, defined in Equation (3). The second method was interesting as it was able to evaluate the influence of the composition at the interface and the temperature on the partition coefficient.

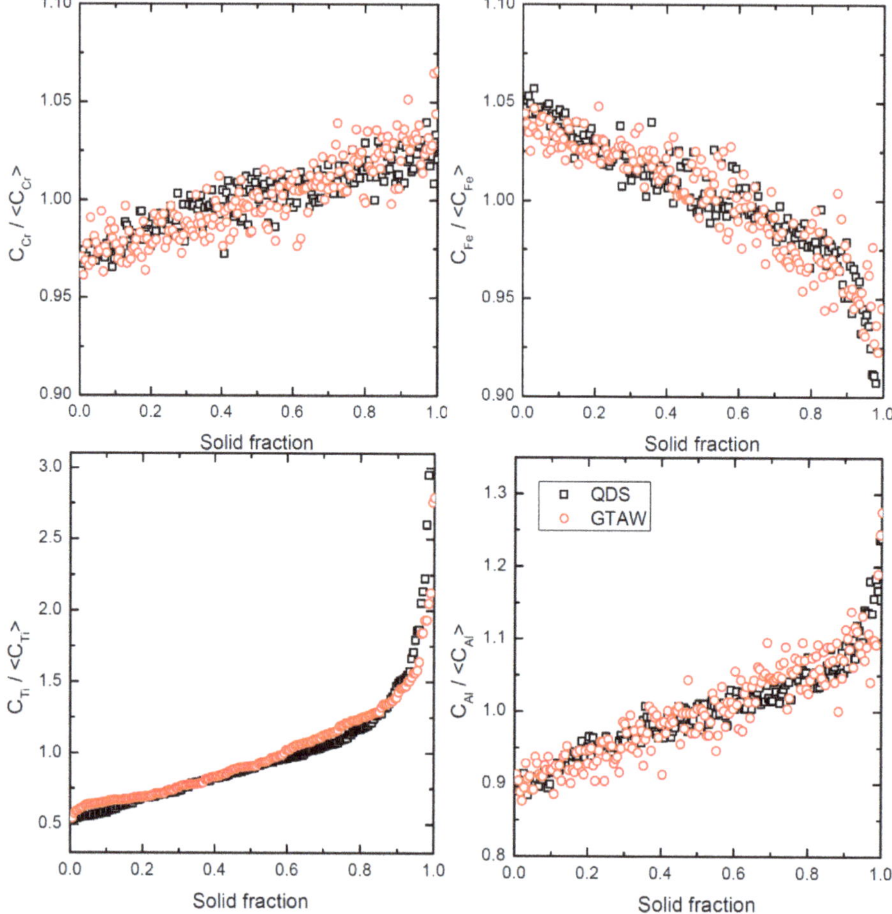

Figure 3. Comparison of microsegregation profiles between QDS, GV = 0.477 K/s and GTAW experiments.

Due to the small solidification range of the FM52 alloy, the length of the mushy zone was only ~2 mm. It was therefore difficult to accurately cut. Nevertheless, it was successfully sectioned on the QDS test for GV = 0.039 K/s. In the analyzed region, the solid fraction was estimated to be approximately 80%.

$$k_1 = \frac{\text{Solid composition for } f_s = 0}{\text{Mean composition}} \tag{2}$$

$$k_2 = \frac{\text{Solid composition at quench moment}}{\text{Mean composition of quenched liquid}} \tag{3}$$

The values of the partition coefficients are presented in Table 4. Over three cooling rates and two solid fractions, the values did not have significant variation. This indicates that the partition coefficients do not depend on the composition and temperature changes at the interface.

Table 4. Partition coefficients QDS, GTAW, and Thermo-Calc calculations.

	k	Cr	Fe	Al	Ti	Mn	Si
QDS, GV = 0.477 K/s	k_1	0.96	1.05	0.87	0.5	0.75	0.59
QDS, GV = 0.16 K/s	k_1	0.96	1.07	0.86	0.46	0.73	0.66
QDS, GV = 0.039 K/s	k_1	0.98	1.06	0.87	0.5	0.76	0.66
	k_2	0.98	1.05	0.94	0.62	0.71	0.59
GTAW	k_1	0.97	1.04	0.91	0.54	x	x
Thermo-Calc for FM52 alloy	k_1	0.93	1.13	0.99	0.41	0.54	0.62
Thermo-Calc for ternary Ni-29.98Cr-10.06Fe	k_1	0.92	1.11	x	x	x	x

The experimental analysis was concluded with partition coefficients determined for alloy FM52 and the ternary alloy approximation from the thermodynamic TTNi8 database. The small effect of composition on the majority element partition coefficients should be noted. The data for the ternary and FM52 alloys were similar. The values presented in Table 4 differ somewhat between the measured and calculated. In the Thermo-Calc computation, the undercooling was not considered, so the first solid that formed had the same composition as was predicted by the phase diagram, i.e., the product between the nominal composition and the partition coefficient. These results supported the assumption that the FM52 alloy behaved like the considered ternary alloy, as was assumed for the following model.

In addition to the thermodynamic data, the most surprising result was that the microsegregation profiles were identical in QDS and GTAW, which could not be explained by a size effect (Fourier number was of the order of 0.3 in QDS, compared to 0.025 in GTAW). Several factors had to be considered in order to explain this. First, it was necessary to clarify the role of convection (important in welding) on undercooling, and by extension on microsegregation. Next, attention had to be paid to the size of the structures and to the experimental determination of the microsegregation. These two questions were not considered by Liang and Chen [50].

4. Modelling Microsegregation: Including Fluid Flow

4.1. Microsegregation Model

The most common microsegregation models are the well-known lever rule and Scheil model [51]. The lever rule assumes an infinite diffusion in both the liquid and the solid phases, while the Scheil model assumes no diffusion in the solid phase and infinite diffusion in the liquid phase. They are usually used to evaluate the possible microsegregation range, but cannot be considered fully realistic. Many more elaborate microsegregation models have been proposed over the years, most based on the Clyne–Kurz (CK) model [52], the Kobayashi model [53], the Wang–Beckermann model [54] or the Tong–Beckermann (TB) model [35]. In the present work, only the TB model was considered.

Compared to the Scheil model, the TB model accounts for solute diffusion in the solid phase, through a solutal Fourier number, α, defined in the CK model as [52]:

$$\alpha_i = \frac{D_i^s t_f}{\lambda_2^2} \tag{4}$$

where D_i^s is the diffusion coefficeient for the solute i in the solid phase, t_f is the local solidification time, defined as $\Delta T_0/G_T V$, and λ_2 is the secondary dendrite arm spacing. α is also known as the back-diffusion parameter.

In addition to the CK model, the TB model considers limited diffusion both into the solid and liquid phase, which can be relevant when the solidification velocity is important, such as is the case in welding operations. Similar to the CK model, the TB model introduces a solutal Fourier number β for diffusion in the liquid phase:

$$\beta = \frac{D_i^l t_f}{\lambda_2^2} \tag{5}$$

where D_i^l is the solute diffusion coefficient in the liquid phase. This Fourier number may also be used to define a second adimensional parameter called β', introducing a "tuning constant", σ, defined as:

$$\beta' = \sigma\beta. \tag{6}$$

This constant is included in the model to compensate for the error introduced by the solute diffusion of the moving solid/liquid interface, which does not satisfy the zero flux condition at the symmetry line separating two dendrites [35]. In practical terms, this tuning constant accounts for the dendrite tip undercooling and changes the composition of the first solid, accordingly. The microsegregation evolution during solidification can be obtained by solving the following differential equation for solid fraction f_s ranging from 0 to 1:

$$\frac{2\beta' f_s}{k_i}(1-\delta)\frac{dw_i^{s*}}{df_s} = (1+6\alpha)\left(w_i^0 - w_i^{s*}\right) + \left(\frac{w_i^{s*}}{k_i} - w_i^0\right)\left(\frac{\delta}{f_s} - 2\beta'(1+6\alpha)(1-\delta)\right) \tag{7}$$

where w_i^0 is the nominal mass fraction, w_i^{s*} is the interfacial solid mass fraction at a given f_s, k_i is the partition coefficient for the solute i, and:

$$\delta = \exp\left(-\frac{1-f_s}{2\beta' f_s}\right). \tag{8}$$

Note that Equation (4) has corrected a typo found in Equation (14), published in Reference [35]. In order to solve Equation (7), the initial solid composition $w_i^s(f_s \to 0)$ must be determined. This initial solid composition actually corresponds to the dendrite tip composition.

4.2. Tip Undercooling

To establish the dendrite tip composition, the tip operating point must be determined by ascertaining the dendrite tip undercooling:

$$\Delta T = \Delta T_T + \Delta T_R + \Delta T_C + \Delta T_k \tag{9}$$

where ΔT_T is the thermal undercooling, ΔT_R is the curvature undercooling, ΔT_C is the chemical or solutal undercooling, and ΔT_k the kinetic undercooling. In the processing conditions considered in this paper, ΔT_T and ΔT_k were neglected. The curvature undercooling is equal to Γ/r where Γ is the Gibbs–Thomson coefficient and r is the dendrite tip radius. The determination of the chemical undercooling can be achieved using the well-known KGT model [33]. In this work, to account for fluid flow that can be significant in the GTAW process, a modified formulation of the KGT model derived by Appolaire et al. was chosen [34]. The KGT model considers an isolated dendrite with a tip described by a circular paraboloid. At the interface, the liquid composition w_l^* is assumed to be homogeneous along the surface of the paraboloid, and at infinity from the nominal solute composition w_0. Accounting for fluid flow, the supersaturation for each solute i $\Omega_i = \left(w_i^{l*} - w_i^0\right)/\left(w_i^{l*}(1-k_i)\right)$ is related to two Péclet numbers: (i) the solutal Péclet number $P_i = rV/2D_i^l$ and the flow Péclet number $P_{u_i} = rU/2D_i^l$, where U the relative velocity between solid and liquid:

$$\Omega_i = \mathcal{F}(P_i, P_{u_i}). \tag{10}$$

For the range of tip radii and fluid flow velocities encountered in this work, the Stokes regime was considered appropriate [55], and thus function $\mathcal{F}$ could be given by:

$$\mathcal{F}(P_i, P_{u_i}) = 2P_i \int_1^\infty \exp\{-\ln\eta + (1-\eta^2)P_i - 2P_{u_i}/E_1(Re_t)[1-\eta^2 + (1-\eta^2)\ln\eta] - 2\varepsilon P_i \ln\eta\}d\eta \tag{11}$$

where η is the coordinate in the paraboloid frame. The surface of the tip is located at η = 1. $Re_t = (P_i + P_{u_i})/Sc_i$; Sc_i is the Schmidt number of the solute i, found by the equation $Sc_i = v/D_i^l$, where ν is the kinematic viscosity of the liquid; and ε is the relative density difference between the solid and liquid phases, $\varepsilon = (\rho_s/\rho_l) - 1$. This term plays an insignificant role on the supersaturation, and so it was neglected here. The chemical undercooling is then related to the interfacial mass fractions:

$$\Delta T_C = m\left(w_i^0 - w_i^{l*}\right) \tag{12}$$

where m is the liquidus slope of the element. The final expression for the total undercooling becomes:

$$\Delta T = \frac{\Gamma}{r} + \sum_i w_i^0 m_i \left[1 - \frac{1}{1-(1-k_i)\mathcal{F}(P_i, P_{u_i})}\right] \tag{13}$$

and the associated value of $w_i^s(f_s \to 0)$ is:

$$w_i^s(f_s \to 0) = \frac{k_i w_i^0}{1-(1-k_i)\mathcal{F}(P_i, P_{u_i})} \tag{14}$$

$$w_i^s(f_s \to 0) = \frac{1+2\beta'}{1+\frac{2\beta'}{k}} \times w_i^0. \tag{15}$$

By coupling Equation (15) from Reference [35] with Equation (11) from this work, the following expression of σ was obtained for each solute i:

$$\sigma_i = \frac{1}{\beta}\left(\frac{1-\mathcal{F}(P_i, P_{u_i})}{2\mathcal{F}(P_i, P_{u_i})}\right) \tag{16}$$

The obtained value for σ_i was then used to determine β' and thus Equation (7) was solved, giving the evolution of the solute mass fraction for solid fractions ranging from 0 to 1.

4.3. Consequence of Convection on Microsegregation

Gabathuler and Weinberg [56] showed that fluid flow does not penetrate the dendritic skeleton above $f_s > 0.2$, while Appolaire et al. [55] showed that fluid flow led to a faster increase of the solid fraction at low solid fractions. At higher solid fractions, the flow is assumed to be described by Darcy's equation, which relates the flow velocity to the pressure gradient in the mushy zone. Defining the local convection time in the mushy zone as $t_{conv} = 2\lambda_2/U_{MZ}$, where U_{MZ} is the flow velocity in the mushy zone, it can be compared to the local solidification time, t_f, to determine whether or not the flow has an impact on microsegregation. In the present work $t_f \ll t_{conv}$ for all the investigated processing conditions. It should be noted that this would be different in the case where cellular solidification occurs, as this would involve a change in the characteristic length from the secondary dendrite arm spacing to the primary dendrite arm spacing, and a sharp increase in the permeability. In this case, it is assumed that convection will not affect the microsegregation, except for the initial solid composition.

4.4. Evaluation of Tip Undercooling in the Presence of Convection

Table 5 shows the data used for the calculations, wherein a ternary Ni-Cr-Fe alloy was considered and the liquidus and solidus planes were linearized [33]. The Gibbs–Thompson coefficient was approximated with the equation of Magnin and Trivedi [57]. The diffusion coefficients in the liquid were evaluated using the Stokes–Einstein relation. The dynamic viscosity was given by Reference [58]. The Thermo-Calc software, in conjunction with the TTNi8 database, were used to determine the phase equilibrium data. The latent heat of fusion was obtained experimentally by DTA. The value we obtained was close to that used in Reference [40].

Table 5. Physical and thermodynamic data used for the calculations.

Physical Quantities	
Gibbs-Thompson coefficient (Km)	$\Gamma = 1.82 \times 10^{-7}$
Diffusion Coefficient for Fe and Cr in molten nickel (m^2/s)	$D_l = 1.57 \times 10^{-9}$
Heat capacity of liquid ($J \cdot kg^{-1} \cdot K^{-1}$) [59,60]	$Cp_l = 700$
Thermal conductivity of liquid ($W \cdot m^{-1} \cdot K^{-1}$) [59,61]	$\lambda_l = 30$
Dynamic viscosity of liquid at 1400 °C ($N \cdot s \cdot m^{-2}$) [58]	$\mu = 0.00483$
Liquid density at T_l ($kg \cdot m^{-3}$) [58]	$\rho_l = 7160$
Latent heat of fusion (J/kg)	$L = 1.3 \times 10^5$
Thermodynamic Quantities	
Liquidus slope for Cr ($K \cdot (\%w)^{-1}$)	$m_{Cr} = -2.27$
Liquidus slope for Fe ($K \cdot (\%w)^{-1}$)	$m_{Fe} = 4.55$
Partition coefficient of Cr	$k_{Cr} = 0.93$
Partition coefficient of Fe	$k_{Fe} = 1.11$
Composition of Cr in weight percent	29.98
Composition of Fe in weight percent	10.06
Fictive reference temperature in °C [33]	1439.5
Solidus temperature	1409.4
Liquidus temperature	1417.8

Figure 4 shows the undercooling evolution as a function of the tip velocity for different liquid convections. The black line is the classical KGT computation. The blue and red lines are computations with convection in the melt for flow velocities of 0.025 and 0.05 m/s, respectively. These speeds are of the typical order of magnitude for convection at the bottom of the molten bath in GTAW, and can reach several tens of centimeters per second at the surface of the bath [32]. The analyzed region for microsegregation in GTAW is located at the bath bottom, so 0.025 m/s was the value used. For QDS, the convection was considered equal to zero. As can be seen in Figure 4, convection reduces the tip undercooling in GTAW nearer to values encountered in QDS.

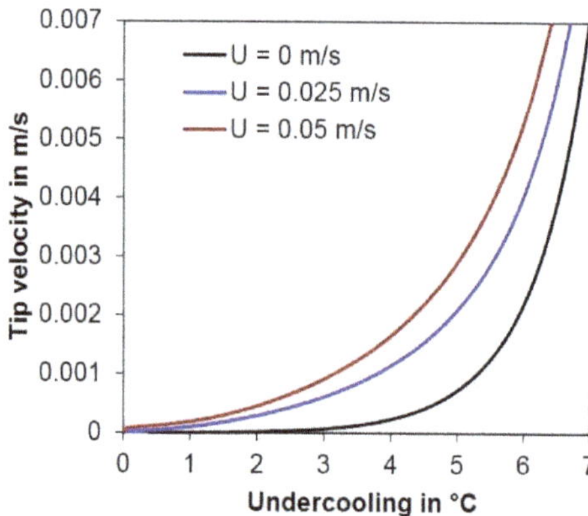

Figure 4. Undercooling as a function of the tip velocity for several fluid flow intensity convections.

5. Comparison between Microsegregation Model and Experimental Results

5.1. QDS Experiment

Figure 5 shows the different microsegregation profiles for one QDS experiment, along with the microsegregation predicted by the TB model, lever rule and Scheil rule.

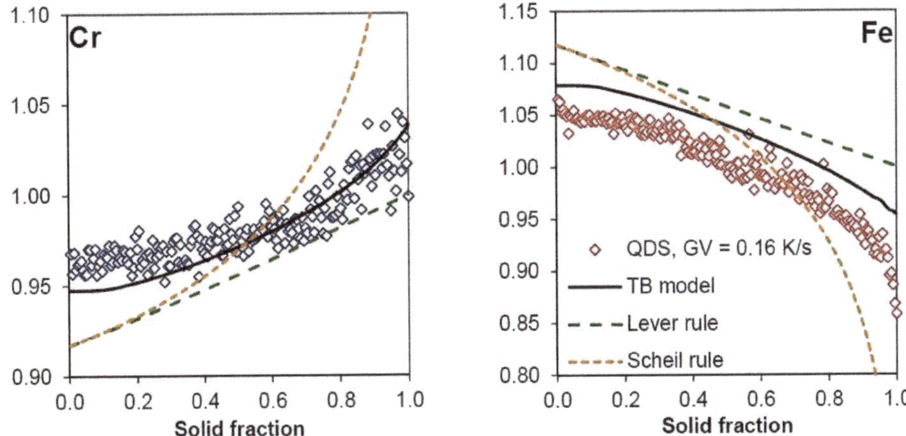

Figure 5. Comparison between microsegregation profile and Tong–Beckermann (TB) model for QDS experience for chromium and iron for GV = 0.16 K/s.

The model was able to predict the observed trend at low solid fractions (between 0 and 0.2) in the case of iron, as shown in Figure 5. The trend was less pronounced for chromium, but still present. The importance of taking into account the undercooling, in particular for the small solid fractions (up to 0.3 fs) on the micro-segregation, may be recognized from this data.

The experimental data points and the TB model are slightly shifted apart from one another, in particular for the first solid formed. This could be due to the fact that the TB model computes the composition of solid at the interface with homogenization during the local solidification time. Post homogenization, occurring during subsequent cooling at the dendrite scale, was not taken into account by the model, although it does occur in the experiment. Simplification when computing the data, by assuming the alloy FM52 to be a ternary alloy, was also carried out.

Nevertheless, the curves given by the TB model compare well with the experimental data despite the slight shift previously noted. It should be remembered that the difference between the compositions of the first solid—detected by EPMA and calculated by the TB model—was small. This deviation corresponds to 0.45% and 0.25% by mass, in the case of chromium and iron, respectively. The TB model, coupled with the undercooling calculation, was then validated by the experiments in order to predict the microsegregation profiles.

5.2. TB Model and GTAW

Figure 6 shows the Cr and Fe microsegregation profiles for the GTAW experiments, and the microsegregation predicted by the TB model. The Lever and Scheil rule are not shown for the sake of clarity. A good fit between the model and the experiments can be observed, except at high solid fractions. However, the effect of microstructure size should also be considered, not only in the modeling parameters—which take into account back-diffusion and finite liquid diffusion—but also in the experiments, because of the fineness of the microstructures. The EPMA probe smoothed the composition gradient due to the interaction volume of the electrons with the analyzed material. In the case of welding, the interdendritic spacing investigated by the microprobe was SDAS. Following the Castaing relation [62] to determine the depth of penetration of the electron beam, using a classical voltage of 20 kV, a 0.850 μm depth interaction zone was predicted. Generally, 1 micron is often used. Thus, the solid fraction analyzed at each microprobe pointed could be approximated by a $\frac{\text{depth penetration}}{\text{SDAS}}$ ratio.

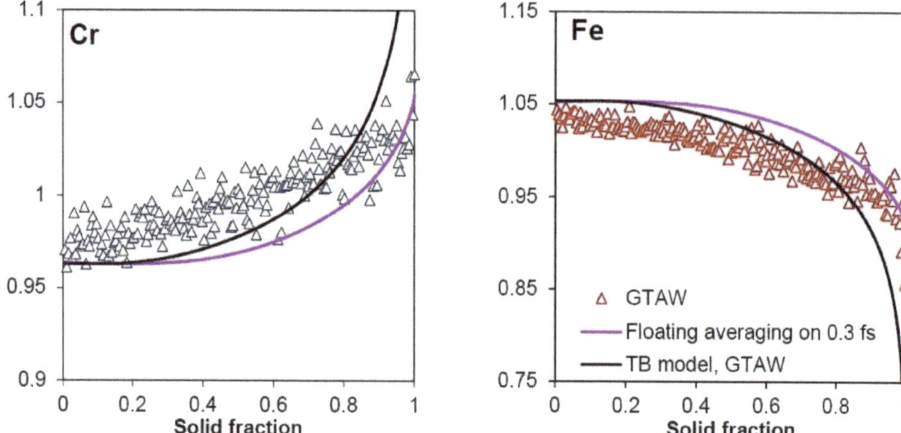

Figure 6. Comparison between the microsegregation profiles given by the TB model, and the TB model with floating averaging in the case of the GTAW experiments.

In the case of welding, the SDAS was approximately 3 or 4 microns. Thus, the primary phase solid fraction over this distance must vary from 0 to 1. If the diameter of the microprobe interaction volume is one micron, each analysis point accounts for one third or one quarter of the solid fraction. This must be taken into consideration when comparing the predicted microsegregation profiles to the experimental. In order to do so in this study, a floating average method was used on the model output concentration, ranging from +/−15% solid fraction, for each output point. The results of this method are also plotted in Figure 6.

With the floating average, the TB curves correspond relatively well to the experimental measurements, especially in the case of iron. The slight difference between the microprobe points and the floating averaging TB model is likely caused by the use of the magnitude approximation for the convection in the molten pool.

6. Conclusions

The microsegregation behaviors and undercooling of the alloy FM52 were investigated using quenched directional solidification and gas tungsten arc welding. The partition coefficients were confirmed to be close to unity for most alloy elements. Microsegregation was experimentally characterized. Despite the large range of solidification conditions used, it was found that microsegregation does not change drastically. The Tong–Beckerman microsegregation model was then modified to take into account the effect of fluid flow on dendrite tip undercooling. With such modification, the TB model fits well with the experimental data, especially when the interaction volume of the experimental analysis was considered. This modified TB model can be used for microsegregation predictions in all solidification conditions, which, when compared with other analytical models, improves the precision without drastically increasing the complexity.

Author Contributions: T.B., D.D., J.Z. and B.R. conceived, designed and carried out the experiments; T.B., D.D. and J.-R.K. wrote and edited the main manuscript. T.B., D.D., B.R., G.T., J.-R.K., V.R. and J.Z. participated in the analysis, discussion and reviewed the manuscript.

Funding: This work is included in the MUSICAS (Méthodologie Unifiée de la Simulation pour l'Intégrité et la Contrôlabilité des Assemblages Soudés) multi-partner project. Special thanks to AREVA NP who realized the welding experiences. This work was also supported by the French State through the program "Investment in the future" operated by the National Research Agency (ANR) and referenced by ANR-11-LABX-0008-01 (LabEx DAMAS).

Acknowledgments: The assistance of Christine Gendarme (Centre de Compétences Microscopie Electronique Microsonde, CC MEM, Institut Jean Lamour, Parc de Saurupt, 54011, NANCY, FRANCE) for microprobe analysis was greatly appreciated.

Conflicts of Interest: The authors declare no conflicts of interest.

References

1. Suh, J.-H.; Shin, J.-K.; Kang, S.-J.L.; Lim, Y.-S.; Kuk, I.-H.; Kim, J.-S. Investigation of IGSCC behaviour of sensitized and laser surface melted Alloy 600. *Mater. Sci. Eng.* **1998**, *A254*, 67–75. [CrossRef]
2. Kai, J.J.; Liu, M.N. The effects of heat treatment on the carbide evolution and the chromium depletion along grain boundary of Inconel 690 alloy. *Scr. Metall.* **1989**, *23*, 17–22. [CrossRef]
3. Noecker, F.F.; DuPont, J.N. Metallurgical investigation into ductility dip cracking in Ni-based alloys: Part II. *Weld. J.* **2009**, *88*, 62s–77s.
4. Wang, J.D.; Gan, D. Effects of grain boundary carbides on the mechanical properties of Inconel 600. *Mater. Chem. Phys.* **2001**, *70*, 124–128. [CrossRef]
5. Mo, W.; Lu, S.; Li, D.; Li, Y. Effects of filler metal composition on the microstructure and mechanical properties for ERNiCrFe-7multi-pass weldments. *Mater. Sci. Eng. A* **2013**, *582*, 326–337. [CrossRef]
6. Jeng, S.L.; Lee, H.T.; Rehbach, W.P.; Kuo, T.Y.; Weirich, T.E.; Mayer, J.P. Effects of Nb on the microstructure and corrosive property in the Alloy 690–SUS 304L weldment. *Mater. Sci. Eng. A* **2005**, *397*, 229–238. [CrossRef]
7. Ramirez, A.J.; Sowards, J.W.; Lippold, J.C. Improving the ductility-dip cracking resistance of Ni-base alloys. *J. Mater. Process. Technol.* **2006**, *179*, 212–218. [CrossRef]
8. Meng, F.; Wang, J.; Han, E.-H.; Ke, W. The role of TiN inclusions in stress corrosion crack initiation for alloy 690TT in high-temperature and high-pressure water. *Corros. Sci.* **2010**, *52*, 927–932. [CrossRef]
9. Kuo, T.Y.; Lee, H.T. Effects of filler metal composition on joining properties of alloy 690 weldments. *Mater. Sci. Eng. A* **2002**, *338*, 202–212.
10. Crum, J.R.; Gosnay, S.M.; Kiser, S.D.; Zhang, R. Corrosion acceptance testing of weld metals for nuclear service, Materials Selection and Design. *NACE Int.* **2011**, *50*, 70–75.
11. Young, G.A.; Capobianco, T.E.; Penik, M.A.; Morris, B.W.; McGee, J.J. The mechanism of Ductility Dip cracking in Nickel-Chromium Alloys. *Weld. J.* **2008**, *87*, 31s–43s.
12. Collins, M.G.; Lippold, J.C. An investigation of ductility Dip Cracking in Nickel-Based Filler Materials—Part 1. *Weld. J.* **2003**, *82*, 288–295.
13. Collins, M.G.; Ramirez, A.J.; Lippold, J.C. An investigation of ductility Dip Cracking in Nickel-Based Filler Materials—Part 2. *Weld. J.* **2003**, *82*, 348–354.
14. Collins, M.G.; Ramirez, A.J.; Lippold, J.C. An investigation of ductility Dip Cracking in Nickel-Based Filler Materials—Part 3. *Weld. J.* **2004**, *83*, 39–49.
15. Noecker, F.F.; DuPont, J.N. Metallurgical investigation into ductility dip cracking in Ni-based alloys: Part 1. *Weld. J.* **2009**, *8*, 7–20.
16. Nissley, N.E.; Lippold, J.C. Ductility-Dip Cracking Susceptibility of Nickel-Based Weld Metals: Part 1—Strain to Fracture Testing. *Weld. J.* **2008**, *87*, 257–264.
17. Nissley, N.E.; Lippold, J.C. Ductility-Dip Cracking Susceptibility of Nickel-Based Weld Metals: Part 2—Microstructural Charaterization. *Weld. J.* **2009**, *88*, 131–140.
18. Kiser, S.D.; Zhang, R.; Baker, B.A. A New Welding Material for Improved resistance to Ductility Dip Cracking. 2008. Available online: http://www.specialmetalswelding.com/papers/trend%20paper%20-2008.pdf (accessed on 6 December 2013).
19. INCONEL Filler Metal 52M, Special Metals. Available online: http://www.specialmetals.com/assets/smc/documents/smw/fm52m.pdf (accessed on 6 December 2013).
20. INCONEL Filler Metal 52MSS, Special Metals. Available online: http://www.specialmetals.com/assets/smc/documents/smw/fm52mss.pdf (accessed on 6 December 2013).
21. Lee, H.T.; Jeng, S.L.; Yen, C.H.; Kuo, T.Y. Dissimilar welding of nickel-based Alloy 690 to SUS 304L with Ti addition. *J. Nucl. Mater.* **2004**, *335*, 59–69. [CrossRef]
22. Ramirez, A.J.; Lippold, J.C. High temperature behaviour of Ni-base weld metal Part II—Insight into the mechanism for ductility deep cracking. *Mater. Sci. Eng. A* **2004**, *380*, 245–258. [CrossRef]

23. Campbell, J.; Tiryakioglu, M. Bifilm defects in Ni-based alloy castings. *Metall. Mater. Trans. B* **2012**, *43*, 902–914. [CrossRef]
24. Lachowicz, M.; Dudzinski, W.; Haimann, K.; Podrez-Radziszewska, M. Microstructure transformations and cracking in the matrix of γ-γ' superalloy Inconel 713C melted with electron beam. *Mater. Sci. Eng. A* **2008**, *479*, 269–276. [CrossRef]
25. Liu, J.; Duarte, H.P.; Kou, S. Evidence of back diffusion reducing cracking during solidification. *Acta Mater.* **2017**, *122*, 47–59. [CrossRef]
26. Srikanth, A.; Manikandan, M. Development of welding technique to avoid the sensitization in the alloy 600 by conventional Gas Tungsten Arc Welding method. *J. Manuf. Process.* **2017**, *30*, 452–466. [CrossRef]
27. Rappaz, M.; Gandin, C.-A.; Desbiolles, J.-L.; Thévoz, P. Prediction of Grain Structures in Various Solidification Processes. *Metall. Mater. Trans. A* **1996**, *27*, 695–705. [CrossRef]
28. Gandin, C.-A.; Desbiolles, J.L.; Rappaz, M.; Thévoz, P. A Three-Dimentional Cellular Automaton-Finite Element Model for the Prediction of Solidification Grain Structures. *Metall. Mater. Trans. A* **1999**, *30*, 3153–3165. [CrossRef]
29. Flood, S.C.; Hunt, J.D. Columnar and equiaxed growth I. A model of a columnar front with a temperature dependent velocity. *J. Cryst. Growth* **1987**, *82*, 543–551. [CrossRef]
30. Chen, S.; Guillemot, G.; Gandin, C.-A. 3D Coupled Cellular Automaton (CA)–Finite Element (FE) Modeling for Solidification Grain Structures in Gas Tungsten Arc Welding (GTAW). *ISIJ Int.* **2014**, *54*, 401–407. [CrossRef]
31. Krauss, H.G. Experimental Measurement of Thin Plate 304 Stainless Steel GTA Weld Pool Surface Temperatures. *Weld. J.* **1987**, *66*, 353s–359s.
32. Tanaka, M.; Yamamoto, K.; Tashiro, S.; Nakata, K.; Yamamoto, E.; Yamazaki, K.; Suzuki, K.; Murphy, A.B.; Lowke, J.J. Time-dependent calculations of molten pool formation and thermal plasma with metal vapour in gas tungsten arc welding. *J. Phys. D Appl. Phys.* **2010**, *43*, 1–11. [CrossRef]
33. Kurz, W.; Giovanola, B.; Trivedi, R. Theory of microstructural development during rapid solidification. *Acta Metall.* **1985**, *34*, 823–830. [CrossRef]
34. Appolaire, B.; Combeau, H.; Lesoult, G. Modeling of equiaxed growth in multicomponent alloys accounting for convection and for the globular/dendritic morphological transition. *Mater. Sci. Eng. A* **2008**, *487*, 33–45. [CrossRef]
35. Tong, X.; Beckermann, C. A diffusion boundary layer model of microsegregation. *J. Cryst. Growth* **1998**, *187*, 289–302. [CrossRef]
36. INCONEL Filler Metal 52, Special Metals. Available online: http://www.specialmetals.com/assets/smc/documents/smw/fm52.pdf (accessed on 6 December 2013).
37. Boettinger, W.J.; Kattner, U.R. On Differential Thermal Analyzer Curves for the Melting and Freezing of Alloys. *Metall. Mater. Trans. A* **2002**, *33*, 1779–1794. [CrossRef]
38. Bobadilla, M.; Lacaze, J.; Lesoult, G. Influence des conditions de solidification sur le déroulement de la solidification des aciers inoxydables austénitiques. *J. Crist. Growth* **1988**, *89*, 531–544. [CrossRef]
39. Anderson, J.O.; Helander, T.; Höglund, L.; Shi, P.; Sundman, B. Thermo-calc and Dictra, Computation Tools for Materials Science. *Calphad* **2002**, *26*, 273–312. [CrossRef]
40. Wu, W.; Tsai, C.H. Hot Cracking Susceptibility of Fillers 52 and 82 in Alloy 690 Welding. *Metall. Mater. Trans. A* **1999**, *30*, 417–426. [CrossRef]
41. Charpentier, M.; Daloz, D.; Hazotte, A.; Gautier, E.; Lesoult, G.; Grange, M. Study of Microstructure and Solute Partitioning in a CastTi-48Al-2Cr-2Nb Alloy by Quenching during Directional Solidification Technique. *Metall. Mater. Trans. A* **2003**, *34*, 2139–2149. [CrossRef]
42. Zollinger, J.; Lapin, J.; Daloz, D.; Combeau, H. Influence of oxygen on solidification behaviour of cast TiAl-based alloys. *Intermetallics* **2007**, *15*, 1343–1350. [CrossRef]
43. Lacaze, J.; Lesoult, G. Experimental investigation of the development of microsegregation during solidification of an Al-Cu-Mg-Si aluminium alloy. *Mater. Sci. Eng. A* **1993**, *173*, 119–122. [CrossRef]
44. Zollinger, J.; Daloz, D. On the sampling methodology to characterize microsegregation. *Mater. Charact.* **2011**, *62*, 1058–1065. [CrossRef]
45. Blecher, J.J.; Palmer, T.A.; Debroy, T. Solidification Map of a Nickel-Base Alloy. *Metall. Mater. Trans. A* **2014**, *45*, 2142–2151. [CrossRef]

46. Frenk, A.; Kurz, W. High speed laser cladding: Solidification conditions and microstructure of a cobalt-based alloy. *Mater. Sci. Eng. A* **1993**, *173*, 339–342. [CrossRef]
47. Dantzig, J.A.; Rappaz, M. *Solidification, Engineering Sciences, Materials*; EPFL Press: Lausanne, Switzerland, 2009.
48. Ganessan, M.; Dye, D.; Lee, P.D. A Technique for characterizing Microsegregation in Multicomponent Alloys and Its Application to Single-Crystal Superalloy Castings. *Metall. Mater. Trans. A* **2005**, *36*, 2191–2204. [CrossRef]
49. Seo, S.M.; Jeong, H.W.; Ahn, Y.K.; Yun, D.W.; Lee, J.H.; Yoo, Y.S. A comparative study of quantitative microsegregation analyses performed during the solidification of the Ni-base superalloy CMSX-10. *Mater. Charact.* **2014**, *89*, 43–55. [CrossRef]
50. Liang, Y.-J.; Cheng, X.; Wang, H.-U. A new microsegregation model for rapid solidification condition multicomponent alloys and its application to single-crystal nickel-base superalloys of laser rapid directional solidification. *Acta Materilia* **2016**, *118*, 17–27. [CrossRef]
51. Scheil, E. Bemerkungen zur Schichtkristallbildung (Retrograde saturation curves). *Z. Metallkunde* **1942**, *34*, 70–72.
52. Clyne, T.W.; Kurz, W. Solute redistribution during solidification with rapid solid state diffusion. *Metall. Trans. A Phys. Metall. Mater. Sci.* **1981**, *12*, 965–971. [CrossRef]
53. Kobayashi, S. Mathematical analysis of solute redistribution during solidification based in a columnar dendrite model. *Trans. ISIJ Int.* **1988**, *28*, 728–735. [CrossRef]
54. Wang, C.Y.; Beckermann, C. A unified solute diffusion model for columnar and equiaxed dendritic solidification. *Mater. Sci. Eng. A* **1993**, *171*, 199–211. [CrossRef]
55. Appolaire, B.; Albert, V.; Combeau, H.; Lesoult, G. Free growth of equiaxed crystals setting in undercooled NH4Cl-H2O melts. *Acta Mater.* **1998**, *46*, 5851–5862. [CrossRef]
56. Gabathuler, J.P.; Weinberg, F. Fluid Flow into a dendritic Array under Forced Convection. *Metall. Trans. B* **1983**, *14*, 733–741. [CrossRef]
57. Magnin, P.; Trivedi, R. Eutectic growth, a modification of the Jackson and Hunt theory. *Acta Metall. Mater.* **1991**, *39*, 453–467. [CrossRef]
58. Kobatake, H.; Brillo, J. Density and viscosity of ternary Cr-Fe-Ni liquid alloys. *J. Mater. Sci.* **2013**, *48*, 6818–6824. [CrossRef]
59. Gao, Z.; Ojo, O.A. Modeling analysis of hybrid laser arc welding of single criystal nickel-based superalloys. *Acta Mater.* **2012**, *60*, 3153–3167. [CrossRef]
60. Bezençon, C. Recouvrement Monocristallin par procédé laser d'un Superalliage à base Nickel. Ph.D. Thesis, Materials science & Eng. Institut, Ecole Polytechnique Fédérale de Lausanne, Lausanne, Switzerland, 2002.
61. Hamouda, H.B. *Modélisation et Simulation de la Structure de Solidification dans les Superalliages Base-Nickel: Application AM1*; Autre; Ecole Nationale Supérieure des Mines de Paris: Paris, France, 2012; NNT: 2012ENMP0040.
62. Castaing, R. *Advances in Electronics and Electron Physics*, 13th ed.; Masson, C., Ed.; Academic Press: New York, NY, USA, 1960; p. 317.

Article

Mechanical Modelling of the Plastic Flow Machining Process

Viet Q. Vu [1,2], Yan Beygelzimer [1,3], Roman Kulagin [4] and Laszlo S. Toth [1,2,*]

1 Université de Lorraine, CNRS, Arts et Métiers ParisTech, LEM3, F-57073 Metz, France
2 Laboratory of Excellence on Design of Alloy Metals for low-mAss Structures (DAMAS), Université de Lorraine, F-57073 Metz, France; vqviet@tnut.edu.vn
3 Donetsk Institute for Physics and Engineering Named after O.O. Galkin, National Academy of Sciences of Ukraine, 03680 Kyiv, Ukraine; yanbeygel@gmail.com
4 Institute of Nanotechnology (INT), Karlsruhe Institute of Technology (KIT), 76021 Karlsruhe, Germany; kulagin_roma@mail.ru
* Correspondence: laszlo.toth@univ-lorraine.fr

Received: 24 May 2018; Accepted: 12 July 2018; Published: 16 July 2018

Abstract: A new severe plastic deformation process, plastic flow machining (PFM), was introduced recently to produce sheet materials with ultrafine and gradient structures from bulk samples in one single deformation step. During the PFM process, a part of a rectangular sample is transformed into a thin sheet or fin under high hydrostatic pressure. The obtained fin is heavily deformed and presents a strain gradient across its thickness. The present paper aims to provide better understanding about this new process via analytical modelling accompanied by finite element simulations. PFM experiments were carried out on square commercially pure aluminum (CP Al) billets. Under pressing, the material flowed from the horizontal channel into a narrow 90° oriented lateral channel to form a fin sheet product, and the remaining part of the sample continued to move along the horizontal channel. At the opposite end of the bulk sample, a back-pressure was applied to increase the hydrostatic pressure in the material. The experiments were set at different width sizes of the lateral channel under two conditions; with or without applying back-pressure. A factor called the lateral extrusion ratio was defined as the ratio between the volume of the produced fin and the incoming volume. This ratio characterizes the efficiency of the PFM process. The experimental results showed that this ratio was greater when back-pressure was applied and further, it increased with the rise of the lateral channel width size. Finite element simulations were conducted in the same boundary conditions as the experiments using DEFORM-2D/3D software, V11.0. Two analytical models were also established. The first one used the variational principle to predict the lateral extrusion ratio belonging to the minimum total plastic power. The second one employed an upper-bound approach on a kinematically admissible velocity field to describe the deformation gradient in the fin. The numerical simulations and the analytical modelling successfully predicted the experimental tendencies, including the deformation gradient across the fin thickness.

Keywords: lateral extrusion ratio; Finite Element (FE) simulation; analytical modelling; plastic flow machining; back pressure

1. Introduction

Ultrafine-grained (UFG) material structure exhibits significantly high strength together with satisfactory ductility, and other attractive service properties such as superplasticity, good fatigue strength, and high wear resistance, among others [1]. The possibility of producing UFG materials has been successfully established for various severe plastic deformation (SPD) techniques. SPD processes are defined as metal forming processes that impose a very large plastic strain on a processed sample to produce UFG materials under very high hydrostatic pressure. Some examples of SPD processes are high pressure torsion (HPT), equal-channel angular pressing (ECAP), multi-axial forging, and twist extrusion, among others [2]. The use of gigantic hydrostatic pressure permits to reach extremely large strains because it reduces the tendencies to form cracks [3].

Concerning sheet materials with UFG structure, several SPD techniques have been proposed. They are the following: accumulated roll bonding [4], asymmetric rolling [5], continuous confined strip shearing [6], equal channel angular sheet extrusion [7], con-shearing process (CSP) [8], repetitive corrugation and strengthening [9], and equal channel angular rolling [10]. However, all these techniques require a sheet form as the initial sample and only transform their microstructure. The production of UFG sheet materials directly from bulk coarse-grained blanks will expand the range of technologies and open new possibilities. Thus, methods that allow for this are of considerable interest. Traditional metal forming approaches, such as rolling or direct extrusion, require considerable energy and very powerful machines for forming UFG sheet materials from bulk billets by cold plastic deformation. Processes where the applied power is reduced are of high interest.

A recently developed process, named large strain extrusion machining (LSEM, [11,12]), is a process developed in this technological direction. It is a modified machining process in which continuous metal chips are produced with UFG structure because of the large imposed strain. In this process, the chip formation is controlled by a constraining channel that is placed on the chip. Yet another new process, called plastic flow machining (PFM), for producing UFG sheet materials, was introduced by the present authors in the literature [13,14]. This process is able to produce sheet materials with ultrafine and gradient structures from bulk samples in a single extrusion step. During the PFM process, a surface layer of a bulk sample is severely deformed under a large hydraulic pressure to transform it into a thin sheet or fin with large imposed shear strain with strain gradient across its thickness. The advantages of PFM are manifolds. First, thanks to the large hydrostatic stress in the deformation zone; shear bands, segmentation, micro-cracks, and voids are suppressed during the fin formation. Second, a UFG structure and simple shear textures are obtained in the PFM-produced fin, providing a significant increase in strength and formability. For example, commercial pure aluminum (CP Al) fins produced by PFM have tensile strength increased by a factor of three and produce a Lankford coefficient of 0.92, which is greater than that in conventional rolling (where it varies between 0.5 and 0.85 [14]). Obtaining a high value of this coefficient is very important for improving the formability of aluminum sheets.

In this paper, we present the results of a further study on the PFM process. Numerical simulations and new analytical modelling are presented and their outcomes are compared with experimental results. The objective of this study is to provide a deeper understanding about this new process, such as the dependence of the fin formation efficiency on the die geometry and the reasons for the formation of a gradient structure within the fin.

2. Fundamental Principles of PFM and Experiments

The fundamental principles of the PFM operation are given in the previous work [14]; they are illustrated here in Figure 1. The initial square billet sample with height H_0 is fit into the horizontal square channel die. In the middle part of the horizontal channel, the height of the channel is reduced slightly from H_0 to H_1. Right at the reduction point, a lateral channel with width size h is opened. The die edge connecting the horizontal and lateral channels is inclined at an angle α with respect to the horizontal channel. A small step is made along the lateral channel to reduce friction of the outgoing fin with the channel's surface. The sample is pressed on its left side by a pressing punch, which moves at a constant velocity U_0. On the right side of the sample, a constant back-pressure P_{BP} is applied by another punch. Because of the geometry of the die, there are two extrusion processes: lateral and forward extrusions. The lateral extrusion produces the thin sheet or fin, and the forward extrusion moves the material horizontally.

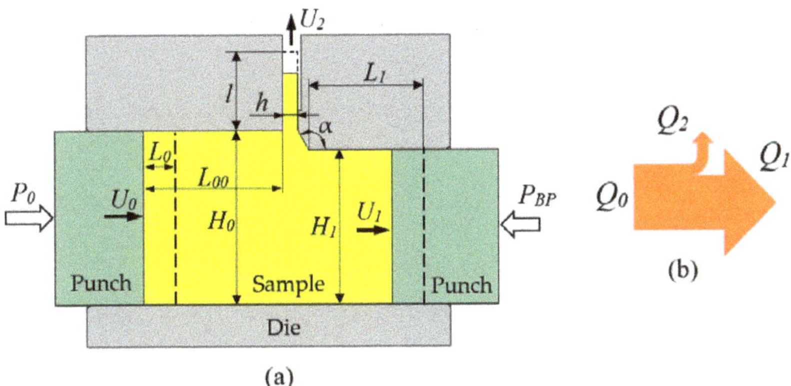

Figure 1. Fundamental principles of the plastic flow machining (PFM) process. (**a**) Die geometry; (**b**) Metal flow.

One of the key features of a lateral extrusion process—pointed out already by Hill (see Figure 33 in the literature [15])—is that a high-pressure zone is generated at the intersection zone between the lateral and horizontal channels, provided that the thickness reduction $r = (H_0 - H_1)/H_0$ is sufficiently small (between about 0.01 and 0.15). This high pressure drives the material into the lateral channel and produces heavy deformation. During processing, the incoming flow Q_0 is separated into two flows: $Q_0 = Q_1 + Q_2$. These flows can be defined as the volumes passing the cross sections per unit time:

Pressing flow:

$$Q_0 = U_0.D.H_0 \tag{1}$$

Forward extrusion flow:

$$Q_1(t) = U_1(t).D.H_1(t) \tag{2}$$

Lateral extrusion flow:

$$Q_2(t) = U_2(t).D.h(t) \tag{3}$$

Here, U_0, $U_1(t)$, and $U_2(t)$ are the respective material flow velocities and D is the transverse dimension of the workpiece. Note that the Q_1 and Q_2 material flows are not necessarily constant during the whole extrusion process, so we consider them as a function of time. Indeed, while Q_0 is imposed to be constant by the pressing punch, which moves with a constant speed, there is no such condition for the other two flows, Q_1 and Q_2. In order to define the relative amount of material moving into the lateral channel, a parameter called lateral extrusion ratio x is introduced:

$$x(t) = Q_2(t)/Q_0 \tag{4}$$

Therefore,

$$Q_1(t) = (1 - x(t))Q_0,\ Q_2(t) = x(t)Q_0 \tag{5}$$

Combining Equations (1), (3), and (4), we obtain the following:

$$x(t) = \frac{U_2(t).h}{U_0 H_0} \tag{6}$$

The U_0 and $U_2(t)$ velocities are defined as follows:

$$U_0 = dL_0(t)/dt,\ U_2(t) = dl(t)/dt \tag{7}$$

Which results in the following:

$$x(t) = \frac{l(t).h}{L_0(t).H_0} \tag{8}$$

The maximum possible value of x is 1, which occurs when the back-pressure punch is fixed. In this case, PFM transforms into a kind of non-equal channel angular pressing process (NECAP [16,17]).

It is important to investigate the dependence of the lateral extrusion ratio on the die geometry and back-pressure, because this ratio determines the efficiency of the fin formation. A larger value of this ratio means a larger amount of material flowing into the lateral channel to form a longer fin. In industrial applications, one should design a proper die geometry and back-pressure so that the lateral extrusion can be operated with high efficiency.

Several experiments were carried out on commercial pure (CP) Al-1050 at room temperature to examine the variations in the x parameter. As the evolution of $l(t)$ was not possible to follow during the test, the following x value was defined from the experiments:

$$X = \frac{l(T).h}{L_0(T).H_0} \tag{9}$$

where T is the total time of the extrusion operation.

The experiments were conducted in a modified equal channel angular pressing machine equipped with two punches that were controlled by a hydraulic system. The force capacity of each hydraulic punch was 72 tons. Dies and punches of the machine were made of high-alloy tool Z160CDV12 steel subjected to heat treatment to have a yield limit about 2 GPa and a hardness of 58 HRC. The pressing punch on the left side of the sample was controlled to move at a constant speed of 1 mm/s. When back-pressure was applied, the punch on the right side was loaded at a constant pressure of 110 MPa. In order to examine the effect of back-pressure, testing was also carried out without applying back-pressure. Only the gap-width h was varied, between 0.2 and 1.5 mm, all other geometrical parameters of the die were kept constant; $H_0 = 20$ mm, $H_1 = 18$ mm, and die angle $\alpha = 120°$. The results obtained for the lateral extrusion ratio X for eight testing conditions are shown in Figure 2.

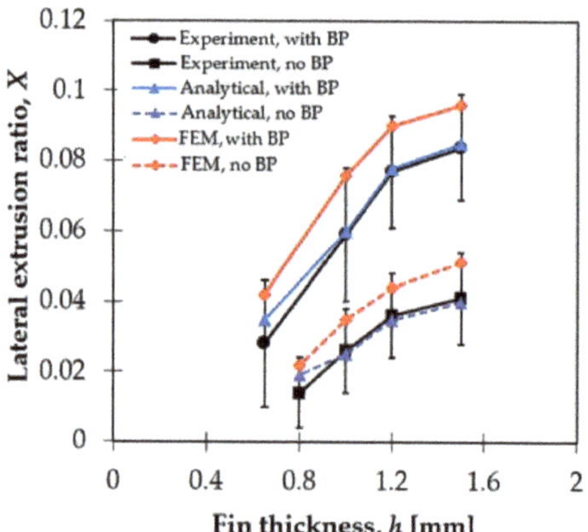

Figure 2. The dependence of the lateral extrusion ratio on the gap-width, for a back-pressure (BP) of 110 MPa, as well as without back-pressure. FEM—finite element modeling.

It can be seen in Figure 2 that the lateral extrusion ratio increased with the increase of the gap-width in both cases, with and without back-pressure, and x was larger with back-pressure. The fin was not formed for h less than 0.8 mm without BP and 0.6 mm with BP, so only direct extrusion took place for small h values. In these cases, a dead-metal-zone adjacent to the lateral channel entry was formed, which hindered the material flow into the lateral channel. The results in Figure 2 clearly show that the fin formation was improved by conducting PFM under back-pressure and that the gap-width size had to be above a minimum value.

As in the modeling below, we will obtain x as a function of time (or L_1), X can be derived as follows from the modeling. The total length of the fin is as follows:

$$l(T) = \int_0^T U_2(t)dt \tag{10}$$

Now, using the following relations:

$$U_2(t) = x(t)\frac{H_0 U_0}{h},\ dt\frac{dL_1(t)}{U_1(t)} = \frac{H_1 dL_1(t)}{(1 - x(t))H_0 U_0} \tag{11}$$

We obtain the following:

$$l(T) = \frac{H_1}{h}\int_0^{L_1} \frac{x(L_1)}{1 - x(L_1)} dL_1 \tag{12}$$

Replacing this relation into Equation (9), we obtain the following:

$$X = \frac{H_1}{L_0(T).H_0}\int_0^{L_1} \frac{x(L_1)}{1 - x(L_1)} dL_1 \tag{13}$$

Therefore, in the modeling, we will determine $x(L_1)$, then apply the integral (13) to derive the X value, which can be directly compared to the experimental X.

3. Numerical Simulation of the PFM Process

The commercial finite element code, DEFORM-2D/3D V11.0 FE, was employed to perform finite element (FE) simulations to obtain velocity fields and the distributions of stress and strain in the sample during processing. The FE simulations were performed in an incremental manner using the updated Lagrangian approach together with the Newton–Raphson iteration method. The die components were set as rigid bodies. The sample was constructed by 12,000 four-node quadrilateral deformable elements. Adaptive meshing was employed for accommodating large strains in the deformation zone near the entry of the lateral channel. The boundary conditions were set similar to the experiments. The pressing punch velocity was set at a constant value of 1 mm/s. The back pressure was set at two different values: 30 MPa and 110 MPa. The material behavior was taken von Mises isotropic type using Hollomon's power law [18]: $\sigma = 180\varepsilon_{vM}^{0.23}$MPa, where σ and ε_{vM} are the von Mises equivalent stress and equivalent strain, respectively. The hardening parameters were identified using the experimental stress-strain curve obtained for Al1050 in the literature [19]. The frictional shear stress τ was modeled by the Siebel friction law [18]: $\tau = \mu\sigma$, where σ is the flow stress, employing a friction coefficient of $\mu = 0.2$.

Regarding possible thermal effects, in general, during SPD processes at room temperature, there is no significant increase in temperature. The maximum observed temperature increase was about 5 °C for aluminum. For example, Alexander and Langdon [20] pointed out that the temperature increase during HPT, which is the most severe SPD technique, was only about 5 °C in aluminum. The main reason for this is that the processing velocities are not high in SPD; in the present study, the pressing punch moved only 1 mm/s. Therefore, thermal effects are small and can be neglected in the modeling.

Figure 3 presents the results of the FE simulations obtained for the velocity field, the strain rate, and the accumulative strain distribution for two back pressure values (30 MPa and 110 MPa) and for two gap-width sizes (h = 0.65 mm and 1.5 mm). The velocity fields for both gap-widths show that an increase in back pressure leads to an increase in the material velocity within the lateral channel. Therefore, the fin is longer with back-pressure. This is in agreement with the experimental results in Figure 2.

For each case in Figure 3, there is a low velocity zone near the die edge, where the velocity is very small, but non-zero (the blue color code indicates the low velocity). One can consider such zones as "dead metal zones" (DMZ), which are commonly seen in extrusion processes because of the friction between the metal and the die walls. If the gap-width for the lateral channel is too small, then the DMZ can be large enough to prevent the occurrence of the lateral material flow.

The effective strain distributions in the fin in Figure 3 show deformation gradients across the fin thickness. The reasons for this strain gradient will be revealed by an analytical model in the next section. The strain rate distributions in Figure 3 show the zones where high deformation takes place; the obtained features are useful to construct the analytical model.

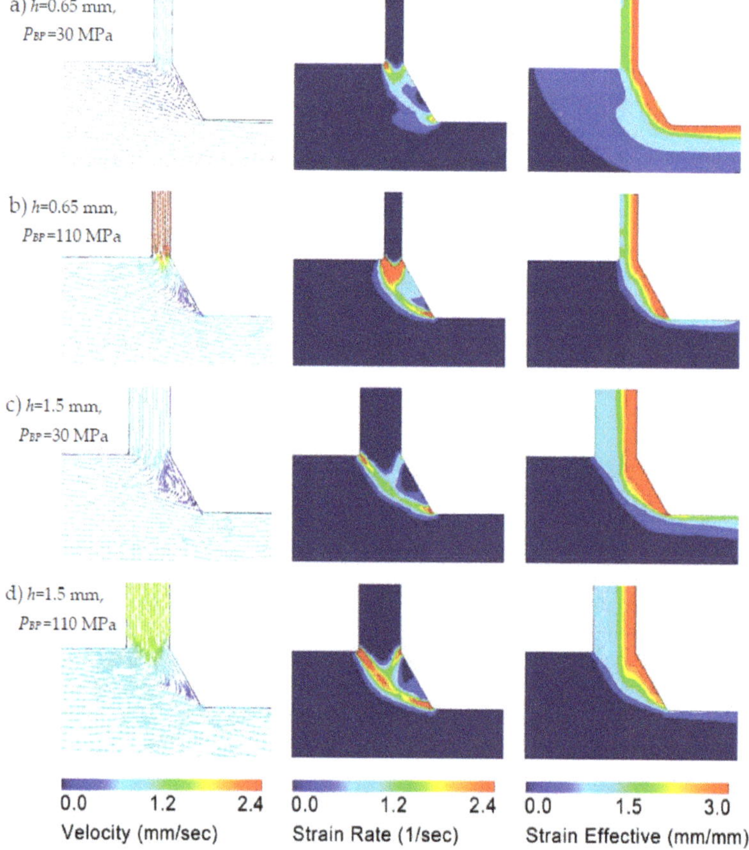

Figure 3. Results of finite element simulations of the PFM process for the parameters: H_0 = 20 mm; H_1 = 18 mm; α = 120°. The magnitudes of the velocity, strain rate, and effective strain (equivalent strain) are indicated by the respective color codes (the maximum value of the equivalent strain is about 4 in all cases).

4. Analytical Modeling of the PFM Process

For a better understanding of the PFM process, two analytical models were established. In these models, the material is considered rigid-plastic, for simplicity. Indeed, our CP Al material was already in a hardened state, so only small further strain hardening took place during PFM. As the sample is constrained by the die walls in the transverse direction (perpendicular to the plane of Figure 1), there is no strain along this direction and this permits us to adopt a plane strain state. In the first model, we aim to model the dependence of the lateral extrusion ratio x on the back-pressure and the width of the slit; we used the "minimum properties of an actual velocity field" principle (p. 332 in the literature [21]). The actual value of x must minimize the total rate of work as we employed the variational principle of mechanics. For the second model, where we study the deformation gradient within the fin, we used the upper-bound approach applied on a kinematically admissible velocity field, formed by rigid blocks [22].

4.1. Model for the Lateral Extrusion Ratio

In this section, a model is presented for predicting the lateral extrusion ratio as a function of the back-pressure and the gap-width.

With the help of the upper bound theorem [21], the x value, which expresses the fraction of the material of the workpiece that flows to form the fin in a time increment, can be determined by minimizing the total power generated during the PFM process, it is given by the following:

$$W = W_1 + W_2 + W_{BP} + W_{fr} \tag{14}$$

where W_1 and W_2 are the plastic powers for the forward and lateral extrusions, respectively; W_{fr} is the friction power caused by the contact between the sample and the die; and W_{BP} is the power produced by the back pressure.

$$W_{BP} = P_{BP}H_1DU_1 \tag{15}$$

Assuming that the whole sample volume is in plastic state during the extrusion process, W_1 and W_2 are defined as follows:

$$W_1 = a_1Q_1,\ W_2 = a_2Q_2 \tag{16}$$

Here, a_1 and a_2 are the plastic work per unit volume, which, for isotropic material, can be obtained from the flow stress σ_0 and the equivalent plastic strain e_{eq} as follows [21]:

$$a = \sigma_0 e_{eq} \tag{17}$$

According to Equations (4) and (5), the PFM process can be represented as a lateral extrusion of the workpiece, where the surface layer with thickness xH_0 flows into the lateral channel, and a simultaneous forward extrusion of the workpiece with a thickness $(1-x)H_0$. The width of the lateral channel is h, and the direct channel's thickness is H_1. If $(1-x)H_0 > H_1$, an elongation of the specimen takes place in the direct extrusion channel, and the equivalent plastic strain e_{eq1} can be determined by the plain strain extrusion formula, adapted from the literature [22].

$$e_{eq1} = \frac{2}{\sqrt{3}}\ln\frac{(1-x)H_0}{H_1} \tag{18}$$

Employing the above-defined r parameter to simplify the expression, we obtain the following:

$$e_{eq1} = \frac{2}{\sqrt{3}}\ln\frac{1-x}{1-r} \tag{19}$$

Note that there is no direct extrusion strain if $x = r$. This happens when all material contained in the upper part of the workpiece (the $H_0 - H_1$ thickness layer) goes into the fin. When $x > r$, then the material in the Q_1 flow is under compression strain. That requires a very high back-pressure. The most probable case is, however, when $x < r$, for which case the direct extruded material is under tensile strain. In order to keep the strain constantly positive for the plastic work calculation, we use the absolute value in Equation (19):

$$e_{eq1} = \frac{2}{\sqrt{3}}\left|\ln\frac{1-x}{1-r}\right| \tag{20}$$

Plugging this relation into Equation (12), one obtains the following:

$$a_1 = \frac{2\sigma_0}{\sqrt{3}}\left|\ln\frac{1-x}{1-r}\right| \tag{21}$$

Taking into account this relation, we obtain from Equation (16) the following formula for the plastic power of the forward extrusion:

$$W_1 = \frac{2\sigma_0}{\sqrt{3}}Q_1\left|\ln\frac{1-x}{1-r}\right| \tag{22}$$

To estimate the average equivalent strain for the lateral extrusion, we simplify the complex material flow obtained from the FE simulations by a simple non-equal channel angular pressing flow process [16,17], by defining a dead metal zone as indicated in Figure 4. The entry dimension is the OD distance, and the exit dimension is OA = h. As mentioned before, during lateral extrusion, the material flows from the surface layer of the bulk sample into the lateral channel have the thickness xH_0, this leads to $OD = xH_0$. In this simplified kinematically admissible velocity field, there is only one velocity discontinuity line, the OC segment.

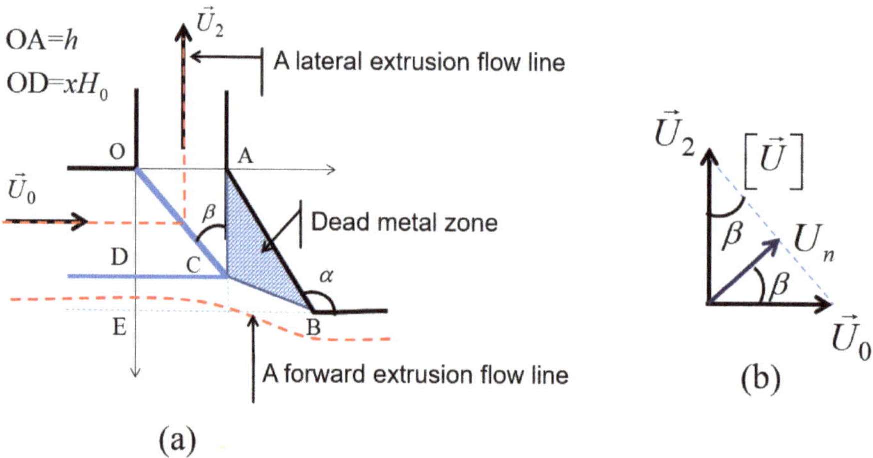

Figure 4. Schematic of a kinematically admissible velocity field for the flow into the lateral channel showing the dead metal zone (**a**); the velocity hodograph is shown in (**b**).

When crossing the velocity discontinuity segment, the material acquires an equivalent plastic strain equal to the following (see [22]):

$$e_{eq} = \frac{1}{\sqrt{3}} \left| \frac{\left[\vec{U}\right]}{U_n} \right| \tag{23}$$

where U_n is the velocity component normal to the OC segment, and $\left[\vec{U}\right]$ is the velocity discontinuity vector. With the help of Figure 4b, one can write the following:

$$\left[\vec{U}\right] = U_n \tan\beta + \frac{U_n}{\tan\beta} \tag{24}$$

Because, in the considered kinematic field, all material flows into the fin within the upper zone defined by the OD segment, the ratio OD/H_0 is equal to the x parameter. Therefore, it follows from the geometry in Figure 4a that,

$$\tan\beta = \frac{OA}{AC} = \frac{OA}{OD} = \frac{h}{xH_0} = \frac{\overline{h}}{x} \tag{25}$$

where $\overline{h} = h/H_0$. Finally, from Equations (23)–(25), we get the following:

$$e_{eq2} = \frac{1}{\sqrt{3}} \left(\frac{x}{\overline{h}} + \frac{\overline{h}}{x} \right) \tag{26}$$

Note that this formula is the same as the one developed for the non-equal channel angular extrusion process for a 90° die [16], where the incoming channel thickness is xH_0 (the OD segment in Figure 4) and the outgoing one is h. Plugging this relation into Equation (17) one can get the following:

$$a_2 = \frac{\sigma_0}{\sqrt{3}} \left(\frac{x}{\overline{h}} + \frac{\overline{h}}{x} \right) \tag{27}$$

Taking into account this relation, from relation (16), we obtain the following formula for the plastic power for the lateral extrusion:

$$W_2 = \frac{\sigma_0}{\sqrt{3}} Q_2 \left(\frac{x}{\overline{h}} + \frac{\overline{h}}{x} \right) \tag{28}$$

With the help of Equations (15), (22), and (28), Equation (14) can be developed as follows:

$$W = \frac{2\sigma_0}{\sqrt{3}} Q_1 \left| \ln \frac{1-x}{1-r} \right| + \frac{\sigma_0}{\sqrt{3}} Q_2 \left(\frac{x}{\overline{h}} + \frac{\overline{h}}{x} \right) + P_{BP} H_1 D U_1 + W_{fr} \tag{29}$$

It is useful to obtain dimensionless quantities, so for this purpose, we divide both sides of this equation by $\sigma_0 D H_0 U_0$:

$$\frac{W}{\sigma_0 D H_0 U_0} = \frac{2\sigma_0}{\sqrt{3}} \frac{Q_1}{\sigma_0 D H_0 U_0} \left| \ln \frac{1-x}{1-r} \right| + \frac{\sigma_0}{\sqrt{3}} \frac{Q_2}{\sigma_0 D H_0 U_0} \left(\frac{x}{\overline{h}} + \frac{\overline{h}}{x} \right) + \frac{P_{BP} H_1 U_1}{\sigma_0 H_0 U_0} + \frac{W_{fr}}{\sigma_0 D H_0 U_0} \tag{30}$$

Taking into account Equations (1), (2), (4), and (5), we obtain the following:

$$\frac{W}{\sigma_0 D H_0 U_0} = \frac{2}{\sqrt{3}} (1-x) \left| \ln \frac{1-x}{1-r} \right| + \frac{x}{\sqrt{3}} \left(\frac{x}{\overline{h}} + \frac{\overline{h}}{x} \right) + \overline{P}_{BP} (1-x) + \frac{W_{fr}}{\sigma_0 D H_0 U_0} \tag{31}$$

where $\overline{P}_{BP} = \frac{P_{BP}}{\sigma_0}$.

According to the principle of the minimum plastic power of an actual velocity field [21], the value of x is provided by a minimum of the right-hand side of Equation (31). We consider the case where $x << 1$ and $r << 1$. For this case, one can apply a first order approximation in Equation (31):

$$\frac{W}{\sigma_0 D H_0 U_0} = \frac{2}{\sqrt{3}}(1-x)|r-x| + \frac{x}{\sqrt{3}}\left(\frac{x}{\overline{h}} + \frac{\overline{h}}{x}\right) + \overline{P}_{BP}(1-x) + \frac{W_{fr}}{\sigma_0 D H_0 U_0} \tag{32}$$

Now, we determine the friction-power. W_{fr} can be divided into three parts:

$$W_{fr} = W_{fDMZ} + W_{f0} + W_{fBP} \tag{33}$$

where W_{frDMZ} is generated by the friction produced by the contact between the flowing material and the edge AC on the dead metal zone (Figure 4):

$$W_{fDMZ} = \mu_{DMZ}.P_{fin}.AC.D.U_2 \tag{34}$$

Combining Equations (3), (4), and (34) using $AC = OD = xH_0$, we obtain the following:

$$W_{fDMZ} = \mu_{DMZ} P_{fin} x H_0 D \frac{Q_2}{hD} = \mu_{DMZ} P_{fin} x H_0 \frac{x Q_0}{h} = \mu_{DMZ} P_{fin} \frac{x^2{}_0}{\overline{h}} D H_0 U_0 \tag{35}$$

W_{f0} and W_{fBP} in Equation (33) are the friction powers for the left and right parts of the samples with respect to the fin's position. These powers were calculated by employing the method of slices (see Appendix A for details):

$$W_{f0} = U_0 H_0 D P_0 \left[1 - \exp\left(\frac{-2\mu L_0 (H_0 + D)}{H_0 D}\right)\right] \tag{36}$$

$$W_{fBP} = U_1 H_1 D P_{BP} \left[\exp\left(\frac{2\mu L_1 (H_1 + D)}{H_1 D}\right) - 1\right] \tag{37}$$

where μ is the friction coefficient for sample-die contact surfaces. This is not unknown, however, because it can be obtained from the equality of the pressures at the position of the fin, calculated from the pressure equations from the left and the right sides within the horizontal extrusion channel. The analytic expression of this friction parameter is as follows (Equation (A5) in Appendix A):

$$\mu = \frac{D}{2} \ln\left(\frac{P_0}{P_{BP}}\right) \left[\frac{L_1(H_1 + D)}{H_1} + \frac{L_0(H_0 + D)}{H_0}\right]^{-1} \tag{38}$$

Combining Equations (1), (2), (5) and (37), one can obtain the following:

$$W_{fBP} = U_0 H_0 D (1-x) P_{BP} \left[\exp\left(\frac{2\mu L_1 (H_1 + D)}{H_1 D}\right) - 1\right] \tag{39}$$

Now, combining Equations (32), (33), (35), (36) and (39), and by considering that the friction part W_{f0} in Equation (36) does not depend on x, the actual x value can be obtained by minimizing the function:

$$f(x) = \frac{2}{\sqrt{3}}(1-x)|r-x| + \frac{x}{\sqrt{3}}\left(\frac{x}{\overline{h}} + \frac{\overline{h}}{x}\right) + \overline{P}_{BP}(1-x) + \mu_{DMZ}\overline{P}_{fin}\frac{x^2}{\overline{h}} + \overline{P}_{BP}(1-x)\left[\exp\left(\frac{2\mu L_1(H_1+D)}{H_1 D}\right) - 1\right] \tag{40}$$

where $\overline{P}_{fin} = P_{fin}/\sigma_0$.

The minimum of $f(x)$ depends on the applied back-pressure. Three intervals of the back-pressure can be distinguished, where the solutions are as follows (see Appendix B for the mathematical details):

Mode 1:

$$x = \overline{h}\left[(1+r) + A\overline{P}_{BP}\right]B^{-1}, \text{ when } \overline{P}_{BP} < \overline{P}_{BP1} \quad (41)$$

Mode 2:

$$x = r, \text{ when } \overline{P}_{BP1} \leq \overline{P}_{BP} \leq \overline{P}_{BP2} \quad (42)$$

Mode 3:

$$x = \overline{h}\left[-(1+r) + A\overline{P}_{BP}\right]C^{-1}, \text{ when } \overline{P}_{BP} > \overline{P}_{BP2} \quad (43)$$

Here,

$$\overline{P}_{BP1} = \left[\frac{r}{\overline{h}}B - (1+r)\right]A^{-1},\ \overline{P}_{BP2} = \left[\frac{r}{\overline{h}}C + (1+r)\right]A^{-1} \quad (44)$$

With the following:

$$A = \frac{\sqrt{3}}{2}\exp\left(\frac{2\mu L_1(H_1+D)}{H_1 D}\right);\ B = 1 + 2\overline{h} + \sqrt{3}\mu_{DMZ}\overline{P}_{fin};\ C = 1 - 2\overline{h} + \sqrt{3}\mu_{DMZ}\overline{P}_{fin} \quad (45)$$

x depends on the back-pressure in Modes 1 and 3, while in Mode 2, x is independent of the back-pressure.

The analytical calculation was performed for the fin thicknesses of h = 0.65, 1.0, 1.2, and 1.5 mm by Equation (38) and Equations (41)–(45) using the measured experimental parameters together with the die geometry parameters: H_0 = 20 mm and H_1 = 20; flow stress: $\sigma_0 = 120$ MPa; and friction coefficient $\mu_{DMZ} = 0.3$. Figure 5 shows the variation of the x parameter as a function of the displacement of the back-pressure punch, L_1. As can be seen in the figure, x is generally increasing during the extrusion process. The deformation mode is generally Mode 1, except for the largest h value, where it reaches Mode 2.

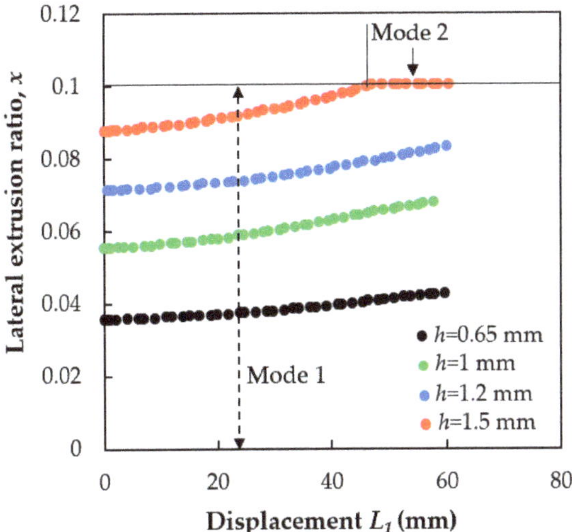

Figure 5. The lateral extrusion ratio x as the function of the displacement L_1 of the back-pressure punch for four values of the gap-width h.

Now, using the varying x parameter displayed in Figure 5 in the integral in Equation (13), we obtain the X value, which can be directly compared with the experiment. The results are shown in Figure 2 and Figure 8. The model reproduces the results that when back pressure is not applied,

X decreases by about half of its value with back-pressure. The modeling results are in excellent agreement with the experiments.

4.2. Model for the Strain Gradient

The strain rate distribution results obtained by the FE simulations (Figure 3) suggest still another, improved kinematically admissible velocity field for an analytical estimation of the equivalent strain in PFM. We use the upper bound theory [22] to construct a kinematically admissible velocity field with rigid blocks for Mode 2 ($\overline{P}_{BP1} \leq \overline{P}_{BP} \leq \overline{P}_{BP2}$) (Figure 6).

The dissipation power W_d for the proposed field is given by the following formula:

$$W_d = k(OC|[U]|_1 + CB|[U]|_2 + AC|[U]|_3) + mk\,ABU' \quad (46)$$

where $|[U]|_1$, $|[U]|_2$, and $|[U]|_3$ are the magnitudes of the velocity discontinuity vectors at boundaries 1, 2, and 3, respectively; U' is the velocity along the AB boundary; and OC, CB, AC, and AB are the lengths of the corresponding segments. The last term in Equation (46) is the friction with the die-wall along the AB segment, where $k = \sigma_0/2$, and m is the plastic friction coefficient.

The proposed kinematically admissible velocity field has one free parameter ξ, which divides the fin into two deformation zones (see Figure 6). According to the upper bound theorem [22], the ξ value is determined by minimizing the dissipation power W_d.

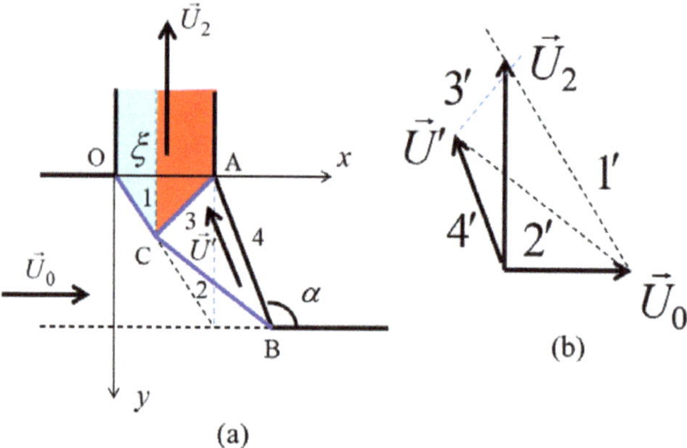

Figure 6. A kinematically admissible velocity field composed of three rigid blocks for the analysis of the strain distribution in the fin. (**a**) The rigid blocks with the velocity discontinuity line segments are identified by 1, 2, and 3. Left to the OCB line, the material moves with the velocity $\vec{U}_0$. The ABC triangle is moving with the velocity $\vec{U}'$. Above the OCA segment, the material moves with velocity $\vec{U}_2$. α is the die angle and ξ is the abscissa of the point C. (**b**) The velocity hodograph along the CB segment.

The material points at the left part of the fin (Zone I) undergo shear deformation only, which is taking place at the boundary 1, while the material arriving at the right part of the fin (Zone II) undergoes shear deformations at both boundaries 2 and 3. When crossing a velocity discontinuity segment, the material acquires an equivalent plastic strain equal to the following (see [22]):

$$e_{eqi} = \frac{1}{\sqrt{3}}\left|\frac{\left[\vec{U}\right]_i}{U_{ni}}\right| \quad (47)$$

where U_{ni} is the velocity component normal to the segment i, and $\left[\vec{U}\right]_i$ is the velocity discontinuity vector on the segment i $(i = 1,2,3)$. Applying this formula to the first zone with the thickness ξ, the equivalent plastic strain is as follows:

$$e_{eqI} = \frac{1}{\sqrt{3}}\left|\frac{\left[\vec{U}\right]_1}{U_{n1}}\right|. \tag{48}$$

In the second zone, with the thickness of $(h - \xi)$, the von Mises strain is as follows:

$$e_{eqII} = \frac{1}{\sqrt{3}}\left|\frac{\left[\vec{U}\right]_2}{U_{n2}}\right| + \frac{1}{\sqrt{3}}\left|\frac{\left[\vec{U}\right]_3}{U_{n3}}\right|. \tag{49}$$

The final analytical forms of Equations (47)–(49) are long to develop, yet nevertheless, quite straightforward. Here, we only present the final results that can be obtained when the formulas are fully developed and the minimization process is done.

Figure 7a shows the solution for ξ/h that characterizes the partition of the two deformation zones as a function of the geometry parameter of the process $\overline{h}/r$. ξ/h varies between 0.5 and 0.62 in the range of 0.5–0.75 for the $\overline{h}/r$ parameter. Higher $\overline{h}/r$ values were not considered because they are not realistic for an experimental PFM process. The strain values in the two zones are plotted in Figure 7b,c, and as a function of the geometry parameter $\overline{h}/r$. As can be seen, the strain in Zone II is more than twice as high as that in Zone I. While the strain in Zone I increases from 1.1 to 1.3, that in Zone 2 rises from 2.8 to 3.0 when the geometry parameter $\overline{h}/r$ changes in the range of 0.5–0.75.

The strain values in Zones 1 and 2 were also determined experimentally using a master curve established for the high angle grain boundary fraction as a function of strain (see Figure 4 in the literature [23]). They are also plotted in Figure 7b,c.

Finite element simulations were also carried out to examine the strain gradient in the fin. Three cases were considered; the obtained results are displayed in Figure 7b,c. As can be seen, the results obtained by the analytical model agree with those obtained from the experiments, as well as with the FE simulations.

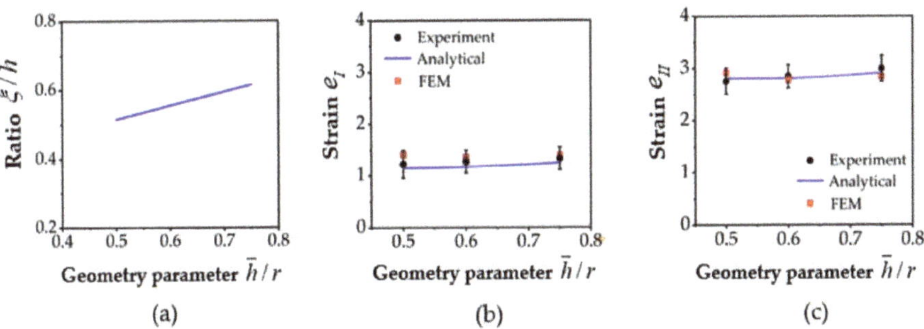

Figure 7. The characteristics of the strain distribution for Mode 2, obtained by the analytical model (continuous lines), and by finite element (FE) simulations (red dots) for the die angle $\alpha = 120$ and for the friction value of $m = 0.2$. (**a**) The predicted width of Zone I; (**b**) the von Mises strain in Zone I; and (**c**) the von Mises strain in Zone II.

5. Discussion

In this section, the different factors that control the PFM process are analyzed using the results obtained in both the analytical and FE calculations. The three modes of the PFM process are also examined.

5.1. The Lateral Extrusion Ratio

Figure 2 shows clearly that the lateral extrusion ratio X achieved in the case with the back-pressure assistance is about double when compared with the case without back-pressure assistance. This means that in practice, PFM should be conducted always with back-pressure so that the process is more efficient. Physically, the back-pressure applied on the right hand side of the sample (Figure 1) restricts the occurrence of the forward extrusion flow Q_1. This results in the increase of the lateral extrusion flow Q_2, because $Q_0 = Q_1 + Q_2$ is constant. Thus, it leads to a greater X value compared with the case without back-pressure.

We now discuss the dependence of the lateral extrusion ratio on the PFM die geometry, represented by the ratio $\overline{h}/r$ in the case where back-pressure is applied (Figure 8). Note that in Figure 2, the lateral extrusion ratio is shown as the function of the gap with size h, which does not describe the overall geometry of the PFM die, and thus the dependence of this ratio on $\overline{h}/r$ (Figure 8) will provide a clearer view of its relationship with the die geometry. The results show that at a constant back-pressure of 110 MPa, X increases with the rise of the ratio $\overline{h}/r$. The results achieved from the analytical model are in excellent agreement with those received from the experiments.

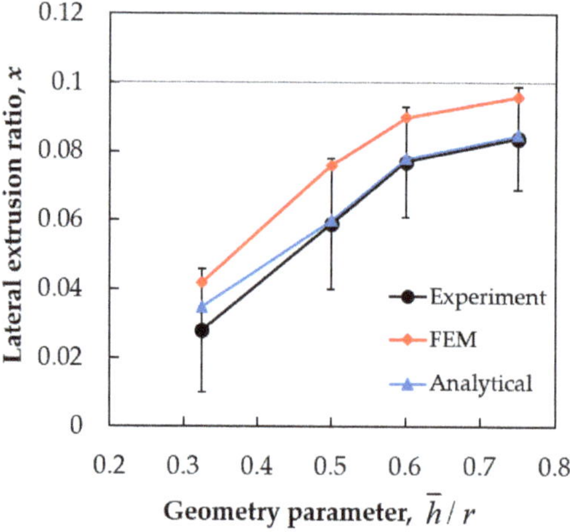

Figure 8. The dependence of the lateral extrusion ratio on the $\overline{h}/r$ parameter in the experiments with back-pressure of 110 MPa, by finite element simulations, as well as by the analytical model.

There is a small difference in the values between the analytical model and the experiments in the case of small values of $\overline{h}/r$. This is because the DMZ is taken into account in a simplified way. Namely, its shape and size depend on the level of the applied back-pressure. At very small $\overline{h}/r$ values, the material flowing into the lateral channel is more restricted by the DMW. This results in a slightly smaller lateral extrusion ratio in the experiment compared with that predicted by the analytical model.

Compared with the FE simulations, the analytical model is in better agreement with the experiments. This can be attributed partly to the high strain hardening parameters used in FE simulations. In fact, the commercially pure aluminum that was used in the experiment was practically rigid-plastic, because it was already in a hardened state, so its hardening after the yield limit was very small. Therefore, the approach of the analytical model with the condition that the material is considered as rigid-plastic is appropriate and the model provides good results. In the FE simulations, treating the as rigid-plastic is difficult as it leads to non-convergences of the FE code. The other reason for the differences could be that a different friction law was used in the FE simulations (Siebel), while the analytical model used the Coulomb-law. Nevertheless, the FE results are also useful as they show that for strain hardening material, the PFM process is more efficient.

5.2. The Three Extrusion Modes

It is possible to give a geometrical interpretation of the three modes that were identified in Section 4.1 above. Simply, they differ from each other by the position of the boundary that separates the forward and lateral metal flows (Figure 9).

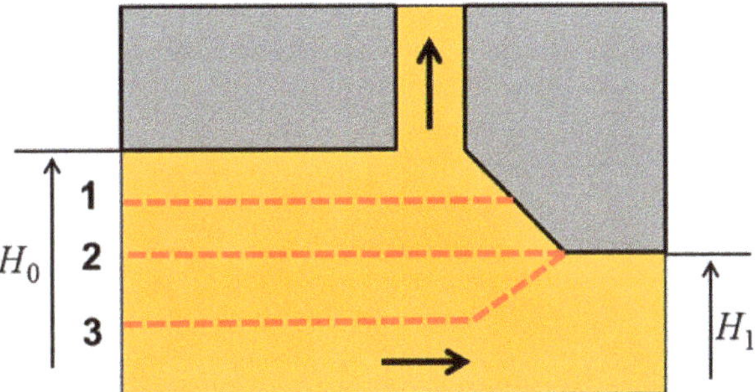

Figure 9. The conditional boundaries of the metal flows for the three possible PFM modes: Mode 1: $\overline{p}_{BP} < \overline{p}_{BP1}$; Mode 2: $\overline{p}_{BP1} \leq \overline{p}_{BP} \leq \overline{p}_{BP2}$; Mode 3: $\overline{p}_{BP} > \overline{p}_{BP2}$.

The separation between Modes 1 and 3 is Mode 2, for which $X = r$ (Equation (42)). In this case, all material within the surface layer of the workpiece of thickness rH_0 flows into the lateral channel, so the position of the boundary line is the same as the top exit line of the workpiece; see Figure 9. This is Mode 2. In Mode 1, only a part of the top layer with thickness $(H_0 - H_1)$ (above the dotted line identified by 1) goes into the lateral channel. With increasing back-pressure, the value of this part increases. If the dotted line 1 moves down to level 2, the entire $(H_0 - H_1)$ layer moves into the lateral channel. When $\overline{p}_{BP} = \overline{p}_{BP1}$, Mode 2 starts. In this mode, the metal flow does not depend on the back-pressure. Mode 3 starts when $\overline{p}_{BP} = \overline{p}_{BP2}$. In this mode, the amount of metal flows moving into the lateral channel increases again, like in Mode 1, with an increase of back-pressure. At extremely high-back pressures, the horizontal part of dotted line 3 can go down to the bottom of the workpiece, then all the metal of the workpiece goes into the lateral channel. This extreme case corresponds to the NECAP process [16,17].

From a practical point of view, the most interesting modes of PFM are Modes 1 and 2, where the deformation is mainly concentrated in the surface layer of the workpiece and occurs by simple shear under pressure. In these modes, the power parameters of the process are small, which allows the process to be readily scaled up. The results obtained are promising for an efficient industrial application of the PFM process.

5.3. The Strain Gradient

The model of strain distribution (Figure 6) and the strain values calculated from this model (Figure 7) clearly show that there are two zones in the fin with different strain values in the cross-section. The obtained strain gradients correspond very well to the experimental results in this study, where the strains were estimated from the portion of next-grain large angle boundaries misorientation. There is one major difference though, which is the existence of a narrow zone (about 100 μm thick) in which the strain is very high; it is located at the right side of the fin, and not produced by the model (see in Vu et al. [14]). This zone can be attributed to the shear deformation associated with friction that is produced at the interface between the die and the workpiece. This friction also brings other effects: it enhances the strain gradient across the fin and produces a high strain zone in the top part of the bulk workpiece.

Now we examine in detail the mechanism that produces the high strain zone by friction. Because of the metal flow at the surface of the die, fresh metal comes out from the workpiece and goes into contact with the surface of the die. The sliding between the metal and the die surface can stop if the adhesion stress τ_a between the fresh metal and the die exceeds the shear flow stress of the metal. As a result, an intensive shear occurs at the surface layer of the workpiece. The plastic friction coefficient m is equal to 1 when there is no sliding. According to Prandtl [24], a surface with $m = 1$ contains an envelope of a family of slip-lines and the metal flow in close proximity to the interface is characterized by a very high strain rate (tends to infinity for ideally plastic materials). For this reason, in a narrow layer of Zone II, at the contact with the die, the metal rapidly hardens. After the shear flow stress of the metal reaches the value of τ_a, the sliding of the metal along the wall of the die restarts. Thus, a thin, highly deformed metal layer appears near the surface of the fin. The same mechanism leads to the formation of a highly deformed thin layer on the top part of the bulk workpiece (see in Vu et al. [14]).

6. Conclusions

In the present work, different modeling was applied for a new SPD process, the PFM process. Two analytical models were formulated; one for obtaining the extrusion ratio, and another for the strain gradient within the produced fin. Finite element simulations were also carried out where strain hardening and friction were also taken into account. The results were compared to experimental data obtained on extrusion of commercially pure aluminum, Al 1050. From the results obtained, we can formulate the following main conclusions:

1. An analytical upper bound model was presented for modeling the lateral extrusion ratio of the PFM process. It was able to reproduce the effect of the applied back-pressure and produced results with excellent agreement with the experiment. With the help of this model, three extrusion modes were identified: Modes 1–3. The selection of the extrusion mode is determined by the applied back-pressure and one geometrical parameter, which is defined by the die geometry.
2. Another analytical model was also established to describe the strain gradient found experimentally in the fin. The model was able to produce strain values in two zones, with values near to the experiment. The third, the very high deformation zone, was interpreted with the help of the friction between the metal flow and the die wall.
3. Finite element modeling of the PFM process was carried out, where strain hardening and friction were considered. This modeling gave important information on the dead metal zone, and gave results that were generally in good agreement with the experimental observations.

Author Contributions: Conceptualization: Y.B. and L.S.T.; Methodology: Y.B., V.Q.V., L.S.T.; Validation: L.S.T., Y.B., R.K.; Formal Analysis: V.Q.V., L.S.T., R.K.; Investigation: V.Q.V.; Resources: L.S.T.; Data Curation: V.Q.V.; Writing-Original Draft Preparation: V.Q.V.; Writing-Review & Editing: V.Q.V., Y.B. and L.S.T.; Supervision: L.S.T.; Project Administration: L.S.T.; Funding Acquisition: L.S.T.

Funding: This research was funded by the National Research Agency (ANR) and referenced by ANR-11-LABX-0008-01 (LabEx DAMAS).

Acknowledgments: This work was supported by the French State through the program "Investment in the future" operated by the National Research Agency (ANR) and referenced by ANR-11-LABX-0008-01 (LabEx DAMAS). Viet Q. VU acknowledges the PhD scholarship awarded by the Vietnamese Government (Project 911).

Conflicts of Interest: There is no conflict of interest for this publication.

Appendix A

Power Dissipation by Friction

First, we employ the method of slices for calculating the distribution of the hydrostatic pressure within the workpiece. Then, the powers generated by the frictions will be calculated. A slice of thickness dy is subjected to pressure from both sides and also to the friction from the four walls of the rectangular channel (Figure A1). The equilibrium of forces provides the following:

$$F(y) - F(y + dy) - F_{f0} = 0 \tag{A1}$$

where y is the distance from the left-surface of the sample, $F(y)$ is the force applied by the pressure $P(y)$, and F_{f0} is the friction force applied by the die on the sample by the four walls.

Using the relations $F(y) = P(y)H_0D$ and $F_{f0} = 2P(y)(H_0 + D)\mu dy$, one obtains the following differential equation for the pressure distribution:

$$\frac{dP(y)}{P(y)} = -\frac{2(H_0 + D)}{H_0D}\mu dy \tag{A2}$$

Now, applying the boundary condition $P(y = 0) = P_0$, the pressure distribution within the left part of the sample is as follows:

$$P(y) = P_0 \exp\left(-\frac{2\mu y(H_0 + D)}{H_0D}\right) \tag{A3}$$

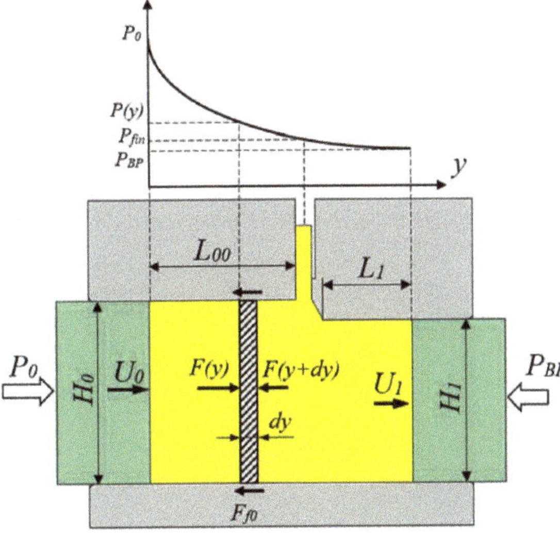

Figure A1. The method of slices schematically showing the stress distribution.

Note that we considered that the pressure applied by the punch is penetrating into the metal as a hydrostatic stress, which can be justified by the fully constrained nature of this extrusion process.

The same above detailed calculation procedure leads to the following formula for the pressure distribution on the right side of the sample:

$$P_{BP}(y) = P_{BP}\exp\left(\frac{2\mu y(H_1 + D)}{H_1 D}\right) \tag{A4}$$

Here, y is measured from the right-hand end of the specimen. Note that the change in sign of the exponent in Equation (A4) compared with Equation (A3) is because of the fact that the friction force is parallel to the force of the applied back-pressure on the right-hand side.

As can be seen from Equations (A3) and (A4), the pressure decreases from the position of the pressing punch and increases from the back-pressure punch within the sample. The two pressures then have to give the same value at the position where the left and right parts of the sample meet; where the fin is formed. This physical condition permits to extract the following formula for the friction:

$$\mu = \frac{D}{2}\ln\left(\frac{P_0}{P_{BP}}\right)\left[\frac{L_1(H_1 + D)}{H_1} + \frac{L_{00}(H_0 + D)}{H_0}\right]^{-1} \tag{A5}$$

Here, L_{00} and L_1 are the lengths of the left and right parts of the sample, respectively. The value of the pressure at the position of the fin can be obtained from Equation (A3), for the y value of $y = L_{00}$:

$$P_{fin} = P_0\exp\left(-\frac{2\mu L_{00}(H_0 + D)}{H_0 D}\right) \tag{A6}$$

By substituting the formula for friction, we get the following:

$$P_{fin} = P_0\exp\left(-\frac{2L_{00}(H_0 + D)}{H_0}\ln\left(\frac{P_0}{P_{BP}}\right)\left[\frac{L_1(H_1 + D)}{H_1} + \frac{L_{00}(H_0 + D)}{H_0}\right]^{-1}\right) \tag{A7}$$

The friction power generated on the left and right sides of the specimen can be readily derived with the help of the equations for the pressure. The increment of the power on a slice is as follows:

$$dW_{f0} = 2\mu(H_0 + D)P(y)U_0 = 2\mu(H_0 + D)P_0\exp\left(-\frac{2\mu y(H_0 + D)}{H_0 D}\right)U_0 \tag{A8}$$

By integrating, we obtain the friction powers for the left and right parts of the sample:

$$W_{f0} = U_0 H_0 D P_0\left[1 - \exp\left(\frac{-2\mu L_{00}(H_0 + D)}{H_0 D}\right)\right] \tag{A9}$$

$$W_{fBP} = U_1 H_1 D P_{BP}\left[\exp\left(\frac{2\mu L_1(H_1 + D)}{H_1 D} - 1\right)\right] \tag{A10}$$

Appendix B

Minimization of the Total Power for Obtaining the x Value

We look for the minimum of the function:

$$f(x) = \frac{2}{\sqrt{3}}(1 - x)|r - x| + \frac{x}{\sqrt{3}}\left(\frac{x}{\overline{h}} + \frac{\overline{h}}{x}\right) + \overline{P}_{BP}(1 - x) + \mu_{DMZ}\overline{P}_{fin}\frac{x^2}{\overline{h}} + \overline{P}_{BP}(1 - x)\left[\exp\left(\frac{2\mu L_1(H_1 + D)}{H_1 D}\right) - 1\right] \tag{A11}$$

Because of the absolute value expression in this function, we examine two cases: $x < r$ and $x \geq r$.

1. $x < r$

$$f(x) = \frac{2}{\sqrt{3}}(1 - x)(r - x) + \frac{x}{\sqrt{3}}\left(\frac{x}{\overline{h}} + \frac{\overline{h}}{x}\right) + \overline{P}_{BP}(1 - x) + \mu_{DMZ}\overline{P}_{fin}\frac{x^2}{\overline{h}} + \overline{P}_{BP}(1 - x)\left[\exp\left(\frac{2\mu L_1(H_1 + D)}{H_1 D}\right) - 1\right] \tag{A12}$$

The minimum is determined by solving the equation $\frac{df(x)}{dx} = 0$:

$$\begin{aligned}\frac{df(x)}{dx} &= -\frac{2}{\sqrt{3}}(r-x) - \frac{2}{\sqrt{3}}(1-x) + \frac{1}{\sqrt{3}}\left(\frac{x}{\overline{h}} + \frac{\overline{h}}{x}\right) + \frac{x}{\sqrt{3}}\left(\frac{1}{\overline{h}} - \frac{\overline{h}}{x^2}\right) - \overline{P}_{BP} + \mu_{DMZ}\overline{P}_{fin}\frac{2x}{\overline{h}} - \overline{P}_{BP}\left[\exp\left(\frac{2\mu L_1(H_1+D)}{H_1 D}\right) - 1\right] \\ &= -\frac{2}{\sqrt{3}}(r-x) - \frac{2}{\sqrt{3}}(1-x) + \frac{2}{\sqrt{3}}\frac{x}{\overline{h}} + \mu_{DMZ}\overline{P}_{fin}\frac{2x}{\overline{h}} - \exp\left(\frac{2\mu L_1(H_1+D)}{H_1 D}\right)\overline{P}_{BP} = 0\end{aligned} \quad (A13)$$

This results in the following:

$$x = \overline{h}\left\{(1+r) + \frac{\sqrt{3}}{2}\exp\left(\frac{2\mu L_1(H_1+D)}{H_1 D}\right)\overline{P}_{BP}\right\}\left(1 + 2\overline{h} + \sqrt{3}\mu_{DMZ}\overline{P}_{fin}\right)^{-1} \quad (A14)$$

Because we examine the case $x < r$, this expression is valid when

$$\overline{h}\left[(1+r) + \frac{\sqrt{3}}{2}\exp\left(\frac{2\mu L_1(H_1+D)}{H_1 D}\right)\overline{P}_{BP}\right]\left(1 + 2\overline{h} + \sqrt{3}\mu_{DMZ}\overline{P}_{fin}\right)^{-1} < r \quad (A15)$$

This implies that $\overline{P}_{BP} < \overline{P}_{BP1}$, where

$$\overline{P}_{BP1} = \left[\frac{r}{\overline{h}}(1 + 2\overline{h} + \sqrt{3}\mu_{DMZ}\overline{P}_{fin}) - (1+r)\right]\left[\frac{\sqrt{3}}{2}\exp\left(\frac{2\mu L_1(H_1+D)}{H_1 D}\right)\right]^{-1} \quad (A16)$$

If $\overline{P}_{BP} = \overline{P}_{BP1}$, then $x = r$.
If $\overline{P}_{BP} > \overline{P}_{BP1}$, Equation (A13) does not have a solution satisfying condition $x < r$.
This means that the minimum of the function (A12) is attained at the point $x = r$.

2. $x \geq r$

$$f(x) = -\frac{2}{\sqrt{3}}(1-x)(r-x) + \frac{x}{\sqrt{3}}\left(\frac{x}{\overline{h}} + \frac{\overline{h}}{x}\right) + \overline{P}_{BP}(1-x) + \mu_{DMZ}\overline{P}_{fin}\frac{x^2}{\overline{h}} + \overline{P}_{BP}(1-x)\left[\exp\left(\frac{2\mu L_1(H_1+D)}{H_1 D}\right) - 1\right] \quad (A17)$$

The minimum is determined by solving equation $\frac{df(x)}{dx} = 0$.

$$\begin{aligned}\frac{df(x)}{dx} &= \frac{2}{\sqrt{3}}(r-x) + \frac{2}{\sqrt{3}}(1-x) + \frac{1}{\sqrt{3}}\left(\frac{x}{\overline{h}} + \frac{\overline{h}}{x}\right) + \frac{x}{\sqrt{3}}\left(\frac{1}{\overline{h}} - \frac{\overline{h}}{x^2}\right) - \overline{P}_{BP} + \mu_{DMZ}\overline{P}_{fin}\frac{2x}{\overline{h}} - \overline{P}_{BP}\left[\exp\left(\frac{2\mu L_1(H_1+D)}{H_1 D}\right) - 1\right] \\ &= \frac{2}{\sqrt{3}}(r-x) + \frac{2}{\sqrt{3}}(1-x) + \frac{2}{\sqrt{3}}\frac{x}{\overline{h}} - \overline{P}_{BP} + \mu_{DMZ}\overline{P}_{fin}\frac{2x}{\overline{h}} - \overline{P}_{BP}\exp\left(\frac{2\mu L_1(H_1+D)}{H_1 D}\right) = 0\end{aligned} \quad (A18)$$

This results in the following:

$$x = \overline{h}\left\{-(1+r) + \frac{\sqrt{3}}{2}\exp\left(\frac{2\mu L_1(H_1+D)}{H_1 D}\right)\overline{P}_{BP}\right\}\left(1 - 2\overline{h} + \sqrt{3}\mu_{DMZ}\overline{P}_{fin}\right)^{-1} \quad (A19)$$

Because we examine the case $x \geq r$, this expression is valid when

$$\overline{h}\left\{-(1+r) + \frac{\sqrt{3}}{2}\exp\left(\frac{2\mu L_1(H_1+D)}{H_1 D}\right)\overline{P}_{BP}\right\}\left(1 - 2\overline{h} + \sqrt{3}\mu_{DMZ}\overline{P}_{fin}\right)^{-1} \geq r \quad (A20)$$

This indicates that $\overline{P}_{BP} > \overline{P}_{BP2}$, where

$$\overline{P}_{BP2} = \left[\frac{r}{\overline{h}}(1 - 2\overline{h} + \sqrt{3}\mu_{DMZ}\overline{P}_{fin}) + (1+r)\right]\left[\frac{\sqrt{3}}{2}\exp\left(\frac{2\mu L_1(H_1+D)}{H_1 D}\right)\right]^{-1} \quad (A21)$$

If $\overline{P}_{BP} = \overline{P}_{BP2}$, then $x = r$.

If $\overline{P}_{BP} < \overline{P}_{BP2}$, Equation (A8) does not have a solution satisfying condition $x > r$. This means that the minimum of the function (2) is attained at the point $x = r$.

Combining cases 1 and 2, we obtain the relations (41)–(45) above.

References

1. Valiev, R.Z.; Estrin, Y.; Horita, Z.; Langdon, T.G.; Zehetbauer, M.J.; Zhu, Y. Producing bulk ultrafine-grained materials by severe plastic deformation: Ten years later. *JOM* **2016**, *68*, 1216–1226. [CrossRef]
2. Estrin, Y.; Vinogradov, A. Extreme grain refinement by severe plastic deformation: A wealth of challenging science. *Acta Mater.* **2013**, *61*, 782–817. [CrossRef]
3. Valiev, R.Z.; Estrin, Y.; Horita, Z.; Langdon, T.G.; Zechetbauer, M.J.; Zhu, Y.T. Producing bulk ultrafine-grained materials by severe plastic deformation. *JOM* **2006**, *58*, 33–39. [CrossRef]
4. Saito, Y.; Utsunomiya, H.; Tsuji, N.; Sakai, T. Novel ultra-high straining process for bulk materials—Development of the accumulative roll-bonding (ARB) process. *Acta Mater.* **1999**, *47*, 579–583. [CrossRef]
5. Cui, Q.; Ohori, K. Grain refinement of high purity aluminium by asymmetric rolling. *Mater. Sci. Technol.* **2000**, *16*, 1095–1101. [CrossRef]
6. Lee, J.C.; Seok, H.K.; Suh, J.Y. Microstructural evolutions of the al strip prepared by cold rolling and continuous equal channel angular pressing. *Acta Mater.* **2002**, *50*, 4005–4019. [CrossRef]
7. Saray, O.; Purcek, G.; Karaman, I. Principles of equal-channel angular sheet extrusion (ecase): Application to if-steel sheets. *Rev. Adv. Mater. Sci.* **2010**, *25*, 42–51.
8. Saito, Y.; Utsunomiya, H.; Suzuki, H.; Sakai, T. Improvement in the r-value of aluminum strip by a continuous shear deformation process. *Scr. Mater.* **2000**, *42*, 1139–1144. [CrossRef]
9. Huang, J.Y.; Zhu, Y.T.; Jiang, H.; Lowe, T.C. Microstructures and dislocation configurations in nanostructured cu processed by repetitive corrugation and straightening. *Acta Mater.* **2001**, *49*, 1497–1505. [CrossRef]
10. Nam, C.Y.; Han, J.H.; Chung, Y.H.; Shin, M.C. Effect of precipitates on microstructural evolution of 7050 al alloy sheet during equal channel angular rolling. *Mater. Sci. Eng. A* **2003**, *347*, 253–257. [CrossRef]
11. Moscoso, W.; Shankar, M.R.; Mann, J.B.; Compton, W.D.; Chandrasekar, S. Bulk nanostructured materials by large strain extrusion machining. *J. Mater. Res.* **2011**, *22*, 201–205. [CrossRef]
12. Efe, M.; Moscoso, W.; Trumble, K.P.; Dale Compton, W.; Chandrasekar, S. Mechanics of large strain extrusion machining and application to deformation processing of magnesium alloys. *Acta Mater.* **2012**, *60*, 2031–2042. [CrossRef]
13. Beygelzimer, Y.; Toth, L.S.; Fundenberger, J.-J. Procédé de Formation D'un Objet Plat Métallique à Grains Ultrafins. U.S. Patent WO2017017341, 2 February 2017.
14. Vu, V.Q.; Beygelzimer, Y.; Toth, L.S.; Fundenberger, J.-J.; Kulagin, R.; Chen, C. The plastic flow machining: A new SPD process for producing metal sheets with gradient structures. *Mater. Charact.* **2018**, *138*, 208–214. [CrossRef]
15. Hill, R. *The Mathematical Theory of Plasticity*; Oxford University Press: Oxford, UK, 1998.
16. Hasani, A.; Tóth, L.S.; Beausir, B. Principles of nonequal channel angular pressing. *J. Eng. Mater. Technol.* **2010**, *132*, 31001–31009. [CrossRef]
17. Tóth, L.S.; Lapovok, R.; Hasani, A.; Gu, C. Non-equal channel angular pressing of aluminum alloy. *Scr. Mater.* **2009**, *61*, 1121–1124. [CrossRef]
18. Hollomon, J.H. Tensile Deformation. *Trans. Metall. Soc. AIME* **1945**, *162*, 268–290.
19. Arzaghi, M.; Fundenberger, J.J.; Toth, L.S.; Arruffat, R.; Faure, L.; Beausir, B.; Sauvage, X. Microstructure, texture and mechanical properties of aluminum processed by high-pressure tube twisting. *Acta Mater.* **2012**, *60*, 4393–4408. [CrossRef]
20. Alexander, P.Z.; Langdon, T.G. Reassessment of temperature increase and equivalent strain calculation during high-pressure torsion. *IOP Conf. Ser. Mater. Sci. Eng.* **2014**, *63*, 012052.
21. Kachanov, L.M. *Fundamentals of the Theory of Plasticity*; North-Holland Publishing Company: Maastricht, The Netherlands, 1971.
22. Johnson, W.; Mellor, P.B. *Engineering Plasticity*; Ellis Horwood Limited, Van Nostrand Reinhold Ltd.: Chichester, UK, 1983.

23. Chen, C.; Beygelzimer, Y.; Toth, L.S.; Fundenberger, J.-J. Microstructure and strain in protrusions formed during severe plastic deformation of aluminum. *Mater. Lett.* **2015**, *159*, 253–256. [CrossRef]
24. Prandtl, L. Anwendungsbeispiele zu einem Henckyschen Satz ueber das plastische Gleichgewicht. *Zeits. Ang. Math. Mech.* **1923**, 3, 401. [CrossRef]

Article

Effects of Processing Conditions on Texture and Microstructure Evolution in Extra-Low Carbon Steel during Multi-Pass Asymmetric Rolling

Satyaveer Singh Dhinwal [1,2,*], Laszlo S. Toth [1,2,*], Peter Damian Hodgson [3] and Arunansu Haldar [4,5]

1. Laboratory of Excellence on Design of Alloy Metals for Low-Mass Structure (Labex-DAMAS), Université de Lorraine, 57070 Metz, France
2. Université de Lorraine, CNRS, Arts et Métiers ParisTech, LEM3, F-57000 Metz, France
3. Institute for Frontier Materials (IFM), Deakin University, Geelong 3216, Australia; peter.hodgson@deakin.edu.au
4. Department of Metallurgy & Materials Engineering, Indian Institute of Engineering Science and Technology, Shibpur-711103, India; arunansuhaldar@gmail.com
5. Tata Steel Ltd., Jamshedpur-831001, India

* Correspondence: satya.rj@gmail.com (S.S.D.); laszlo.toth@univ-lorraine.fr (L.S.T.); Tel.: +33-604-528-208 (L.S.T.)

Received: 22 June 2018; Accepted: 25 July 2018; Published: 31 July 2018

Abstract: Multi-pass rolling was carried out on extra-low carbon steel at room temperature by imposing different ratios of asymmetry in the roll-diameters as well as by conventional mode. The aim of this study is to understand the effect of shear deformation due to the asymmetric conditions on the development of the rolling texture and the possibilities of propagating the shear deformation into the mid-thickness area of the sheet. The trends of the measured texture developments in both symmetric and asymmetric rolling indicate their dependence primarily on the stability and fraction of the Goss {110}<001> and the rotated cube {001}<111> orientations. The effects of asymmetry conditions were further examined on the microstructure evolution and were correlated to the increased orientation inhomogeneity and grain fragmentation. Both texture and microstructure development showed their dependence on the applied thickness reduction per pass, on the total thickness reduction of the sheet as well as on the degree of the imposed asymmetry. It was found that shear textures can be obtained by asymmetric rolling at conditions where all three parameters—asymmetry ratio, strain in one pass, and the total accumulated strain—are as large as possible.

Keywords: rolling; asymmetric ratio; texture; thickness reduction per pass; grain refinement; steel

---

1. Introduction

Asymmetric rolling is one of the most convincing alternatives for producing continuous sheets with through thickness shearing as compared to other existing shearing-based processes in sheet manufacturing [1–10]. The growing interest in this kind of shear based processing is due to the relatively simple modification that existing conventional (symmetric) rolling requires, just by varying a few rolling parameters while keeping the rest of the industrial sheet processing infrastructure almost unchanged. Various studies report that when an asymmetry is introduced in the roll-diameters of the rolling process leads not only to an increase in the productivity of sheet manufacturing but also enhances the formability of the product [11–14]. Such an increase in formability was related to the tilting/rotation of the conventional rolling texture by the imposed shearing, resulting from the asymmetric conditions. Investigations about the rotation of conventional rolling texture in asymmetric rolling further show its dependence on factors such as the thickness reduction per pass (TRPP), the type of applied asymmetry (kinematical, mechanical, or tribological), and the ratio of these imposed asymmetries applied to the two rolls [8–10,12,15–22].

Due to the various parameters that can be utilized for imposing asymmetry in rolling, different investigators applied different sets of rolling schedules. Based on the applied amount of thickness reduction per pass (TRPP), they can be classified into two categories. The small to medium type TRPP in the range of 5–50% are applied in multi-pass rolling while, on other hand, TRPP in the range of 60–75% are examined in a single pass. Results from these investigations provide varying ranges of tilting/rotation of the rolling texture and of its heterogeneity through the thickness. Some simulation-based and experimental studies suggest that the rotation of conventional rolling texture can reach close to the ideal shear texture with small TRPP of 5 to 15% [16,22]. Meanwhile, single pass rolling studies with a high TRPP of 60–75% were more concerned with grain refinement and mechanical properties than the possibilities of rolling texture rotation towards the ideal shear texture [13,18,23–25].

The current situation regarding asymmetric rolling schedules does not provide a clear understanding about the connection between TRPP, imposed asymmetry and the rotation of rolling texture or its trends through the thickness of the sheet. In addition, it can be noted from the literature on asymmetric rolling that there is a lack of experimental evidence about the behavior of prominent texture components of the rolling texture regarding their tilting/rotation imposed by the asymmetric conditions. Nevertheless, some simulation based studies predicted the overall nature of rolling textures with assumed shear coefficient values (the ratio of shear strain/rolling strain) [10,22,26,27].

It should also be noted that, apart from a very few studies on BCC materials, most experimental and simulation based studies were carried out on the HCP and FCC materials [16,21,22,28,29]. However, BCC metals form a very important class of structural materials and have a strong tendency to develop favorable textures, well suited for secondary processing when they are produced by conventional rolling. The potential of asymmetric rolling and its effect on this class of materials has not yet been fully addressed. An alteration in conventional texture by asymmetric rolling may lead to the future implications not only for structural applications but also for others. For example, asymmetric rolling can influence the magnetization properties of silicon steels.

Existing knowledge about asymmetric rolling also does not clarify the respective influence of the shearing effect together with the number of passes and the total final thickness reduction of a sheet. A systematic study of multi-pass asymmetric rolling in comparison with symmetric rolling is indeed a requirement for understanding the characteristic texture and microstructural developments which also results in modification of the mechanical properties of the sheet. In the present study, an extra-low carbon steel was selected for the experiments. The following experimental parameters were varied: thickness reduction per pass (TRPP), total reduction, and the degree of applied roll diameter asymmetry. The employed total thickness reductions went up to very high values: to 89%. The resultant texture rotations and microstructural development were examined and related to the degree of shearing through the sheet. Special emphasis was given to the mid-thickness region since it is known that in the sub-surfaces of a sheet, ideal shear texture can readily form even under symmetric rolling conditions [30–33].

2. Materials and Methods

The nominal composition of the as-received extra-low carbon steel can be seen in Table 1. The plates were hot rolled at 1000 °C during which their thickness was reduced from 27 mm to 10 mm. After hot rolling, the sheets were heat treated for one hour also at 1000 °C and subsequently air cooled. This heat treatment was given to facilitate homogenization of the microstructure and to provide a weak initial texture (Figure 1). Apart from the ferrite phase, the co-existence of scattered tiny pearlite colonies can also be seen in Figure 1a. The average ferrite grain size and surface area fraction of pearlite colonies after the heat treatment were about 36 μm and less than 0.1%, respectively, measured by the open source ImageJ™ software on several optical micrographs. Samples with dimensions of

40 mm width, 70 mm length, and 8 mm in thickness were machined from the heat-treated sheets for the rolling schedules. The front ends of these samples were wedged for easier entry into the roll bite.

Table 1. Chemical composition of extra low carbon steel in weight percentage (wt %).

C	Mn	Si	Al	Cr	Ni	Cu	Ti
0.0300	0.15	0.006	0.043	0.020	0.010	0.003	0.001

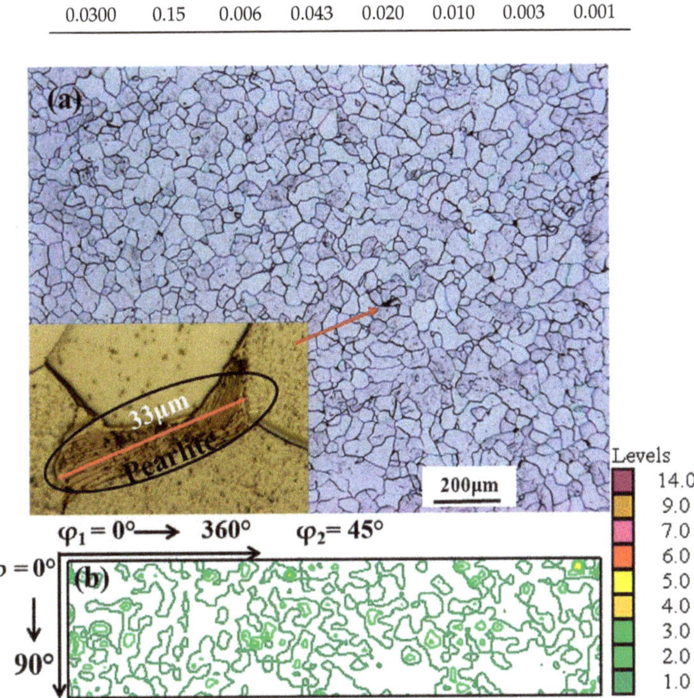

Figure 1. (**a**) Microstructure and (**b**) texture of extra-low carbon steel sheet after heat treatment, before rolling.

Rolling was carried out at room temperature with no lubrication in order to keep similar conditions as it had been in all previous reports on asymmetric rolling. However, it was important that the sheet material should not stick to the rolls. For this purpose, the rolls were lubricated after each pass and later cleaned with oil absorbing paper, so that the maximum amount of lubricant could be removed while retaining the smoothness of the rolls at constant conditions. Three roll diameter ratios 1:1.3, 1:1.6, and 1:2 were used to impose different degrees of asymmetric rolling, as compared with symmetric rolling (1:1). A special roll-set was designed for the experiments in order to facilitate all the ratios adjacent to each other on two shafts, both rotated at 32 revolutions per minute, driven by a single motor. The small diameter roll set was positioned and connected to the upper shaft of the mill while the counterpart large diameter roll set fitted to the lower shaft. TRPP values of 10% were used only for the ratios 1:1 and 1:2 while the 30% and 50% TRPP levels were applied for all roll diameter ratios. Rolling was unidirectional in all cases.

For texture measurements, X-ray diffraction from Panalytical X'pert Pro Instrument (PHILIPS, The Netherlands) was employed using a parallel beam point-focused on the specimen with its surface parallel to the rolling plane at the required thickness depth. The measured bulk texture results were analyzed using the commercially available Labosoft™ 3.0 software. The texture results are presented here in the $\varphi_2 = 45°$ sections of the orientation space where almost all prominent rolling and shear texture components can be seen. Note that positions of shear texture components in the φ_2 = 45° section correspond to simple shear with shear plane in the plane of the sheet and with shear direction parallel to the rolling direction. The identifications of the ideal orientations are given with their Miller indices for rolling in Table 2 and for shear in Table 3. Because of the asymmetric deformation conditions, the texture analysis was carried out without imposing any kind of sample symmetry. This is why the equivalent orientations for symmetric rolling (f1–f2, e1–e2) are specified separately in Table 2 (they are not equivalent when shear is present, due to the loss of orthorhombic symmetry). To quantify the volume fraction of the main texture components, the orientation density function (ODF) was integrated within 10° orientation distance from the ideal position.

Table 2. Ideal deformation rolling texture components of BCC metals and alloys where (hkl)[uvw] represents rolling plane and rolling direction respectively.

Orientation	Miller Indices (hkl)[uvw]	Euler Angles (°) (φ_1, φ, φ_2)
Rotated cube	(001)[110]	0, 0, 45
f1	$(111)[1\bar{2}1]$	30, 54.74, 45
f2	$(111)[\bar{1}\bar{1}2]$	90, 54.74, 45
e1	$(111)[1\bar{1}0]$	0, 54.74, 45
e2	$(111)[0\bar{1}1]$	60, 54.74, 45
i	$(112)[1\bar{1}0]$	0, 35.26, 45

Table 3. Ideal deformation shear texture components of BCC metals and alloys expressed in the rolling reference system with hkl | | ND and uvw | | RD and their FCC rolling system equivalents in parentheses.

Orientation	Miller Indices (hkl)[uvw]	Euler Angles (°) (φ_1, φ, φ_2)
F (Goss)	(110)[001]	(90, 90, 45), (270, 90, 45)
J1 (Brass)	$(0\bar{1}1)[\bar{2}11]$	(125, 90, 45), (305, 90, 45)
J2 (Brass)	$(1\bar{1}0)[\bar{1}\bar{1}2]$	(55, 90, 45), (235, 90, 45)
D1 (Copper)	$(11\bar{2})[111]$	270, 35, 45
D2 (Copper)	$(\bar{1}\bar{1}2)[111]$	90, 35, 45
E1	$(01\bar{1})[111]$	(145, 90, 45), (325, 90, 45)
E2	$(0\bar{1}1)[111]$	(35, 90, 45), (215, 90, 45)

The microstructures were examined by the electron back scattered diffraction (EBSD) technique using a Zeiss LEO 1530 field emission gun scanning electron microscope (FEG-SEM). All samples were scanned in the mid-thickness area of the TD plane. The EBSD data acquisition and post-processing were performed by the AZTEC and HKL channel 5 software provided by Oxford Instruments. For the representation of the microstructures, inverse pole figure (IPF) maps were used where the color key code corresponds to the orientation of the RD axis with respect to the crystal coordinate system. Grain boundaries were identified on the basis of a minimum misorientation angle of 5°. Nevertheless, for a clear visualization, only grain boundaries with ≥15° misorientations were traced on the IPF maps. The misorientations, grain boundaries, and grain size distributions of the EBSD data were analyzed by the ATEX software [34].

There are several parameters that can be varied for asymmetric rolling. As mentioned in the Introduction, three of them were selected for the present experiments: TRPP, total reduction and the degree of applied asymmetry. Figure 2 displays the parameter space in a 3D representation and also in two sections; at constant rolling strains of 1.39 and 2.19. The stars indicate the different combinations of the selected parameters and corresponding rotation of texture achieved by these combinations of parameters. There were 14 cases studied in total, most of them for large parameter values.

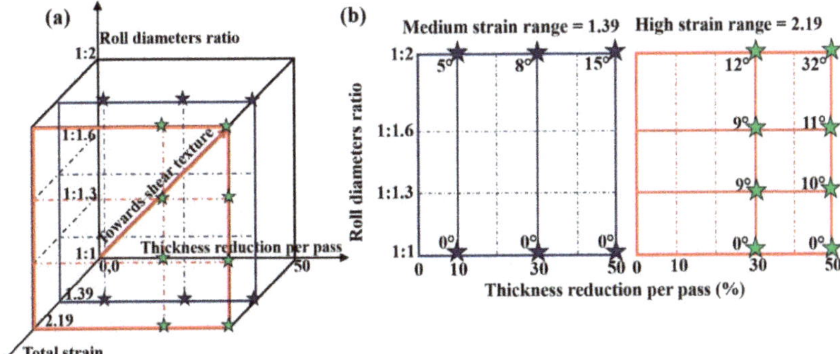

Figure 2. Schematic of the parameter space (**a**) and two sections (**b**) showing the parameter values employed in the present study. The numbers in (**b**) indicate the rotation angle of the rolling texture towards a shear texture, which is identified by the rotation of the f2 rolling component.

3. Results

3.1. Texture

3.1.1. Medium Strain Range (75%)

Figure 3 represents the φ_2 = 45° sections of the ODF for the multi-pass rolling schedules for symmetric (1:1) and asymmetric (1:2) modes with TRPP varying between 10% and 50%. It can be seen in Figure 3 that the resultant texture remains almost identical to the typical symmetric rolling texture after asymmetric rolling with roll diameter ratio of 1:2 and 10% TRPP. Figure 3 also shows that while keeping the same asymmetric rolling ratio, both the intensity of the texture and its fiber nature decreases as the TRPP increases.

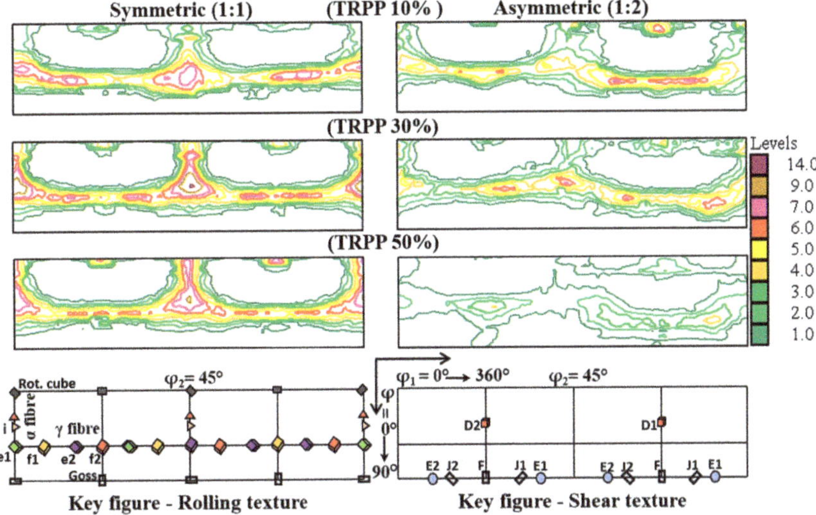

Figure 3. φ_2 = 45° ODF sections at mid-thickness for symmetric (1:1) and asymmetric (1:2) rolling in medium strain range (75%) with TRPP of 10, 30, and 50%.

The volume fractions of the prominent rolling and shear texture components are presented in Figure 4 after 75% total thickness reduction as a function of TRPP. As can be seen, the volume fractions of the rolling texture components sharply decrease with increased TRPP for the asymmetric ratio of 1:2. Nevertheless, they remain higher than the fractions of shear texture components, even at 50% TRPP for the ratio 1:2. Here, among the prominent rolling texture components, the rotated cube component {001}<011> of the α-fiber and the f2 (111)[$\bar{1}\bar{1}$2] component of the γ-fiber emerge as the major texture components. Meanwhile, comparing the shear texture components at 50% TRPP with other lower TRPPs at the same asymmetric ratio, it is clear that their volume fractions increased. It can be also noted in Figure 4a that the volume fractions of the f1 and f2 texture components are no longer equal under asymmetric rolling conditions. Under symmetric rolling conditions, these components are nearly equal. In case of symmetric rolling, most of the rolling texture components show small dependence on the TRPP. Exception is the rotated cube component {001}<011>, which strengthens steadily with the increase in thickness reduction per pass.

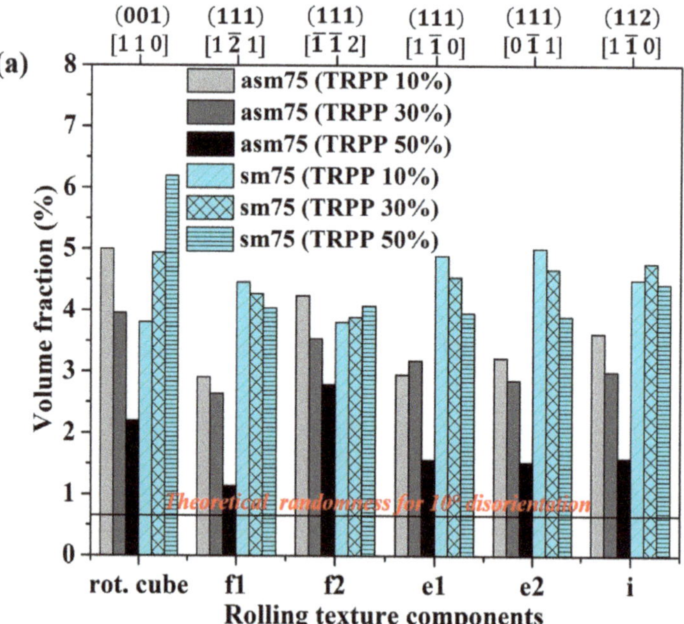

Figure 4. *Cont.*

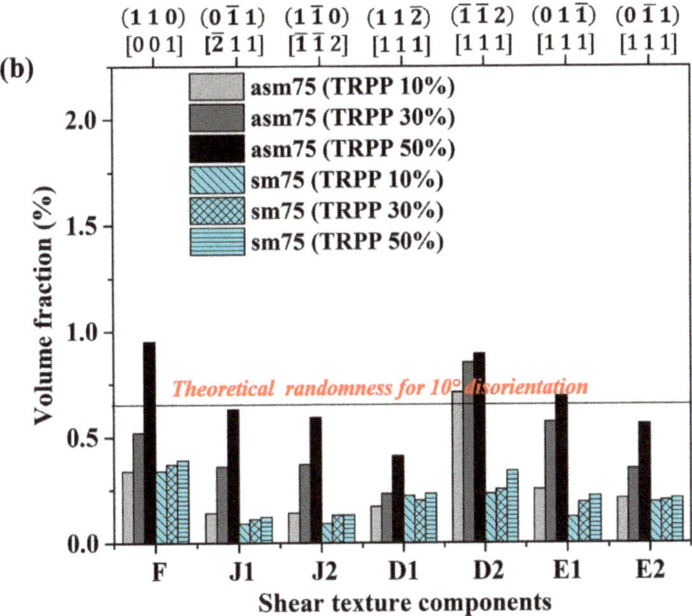

Figure 4. Volume fractions of the prominent (**a**) rolling and (**b**) shear texture components at mid-thickness in multi-pass symmetric and asymmetric (1:2) rolling in medium strain range (75%).

3.1.2. High Strain Range (89%)

The $\varphi_2 = 45°$ sections of ODFs are plotted in Figure 5 for both TRPP; 30% and 50%, while varying the roll diameter ratio from symmetric (1:1) to extreme asymmetric (1:2) rolling. There are more changes in the ODFs for this large strain case for both symmetric and asymmetric rolling compared to the lower rolling reduction. For symmetric rolling, the α and γ fibers are strong for both 30% and 50% TRPP. With increasing asymmetric rolling, these fibers are much weaker and appear in more and more rotated positions (the rotation of the γ fiber can be best seen in the second half of the ODF; from $\varphi_1 = 180°$ to 360°). The rotation is very pronounced for the roll diameters of 1:1.6 and 1:2. The texture is nearly a shear texture for the maximum asymmetry case.

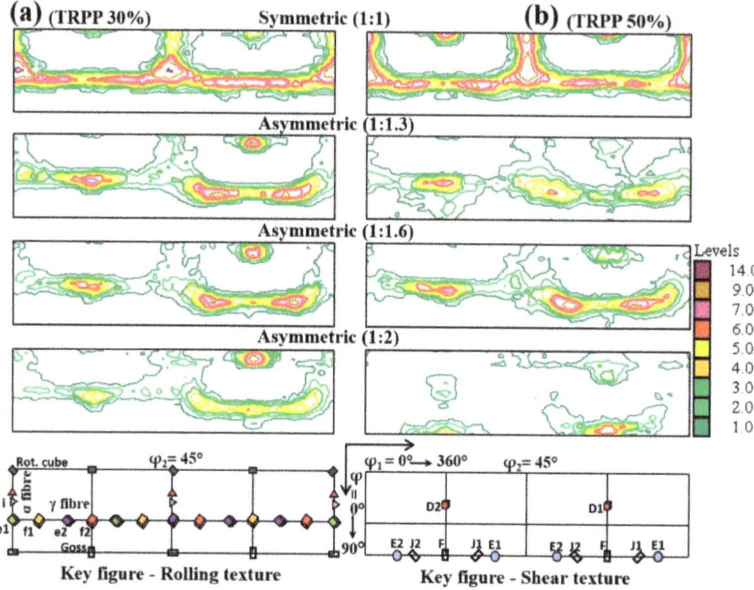

Figure 5. φ_2 = 45° ODF sections in high strain range (89%) for symmetric and asymmetric rolling with TRPP of (**a**) 30% and (**b**) 50% for varying roll diameter ratios from 1:1 to 1:2.

It is also important to emphasize that there is a good through-thickness homogeneity of the textures at this strain level for the asymmetric ratio of 1:2 for the 50% TRPP, see Figure 6. For 30% TRPP, the texture is stronger near the bottom surface of the sheet (at the contact with larger roll).

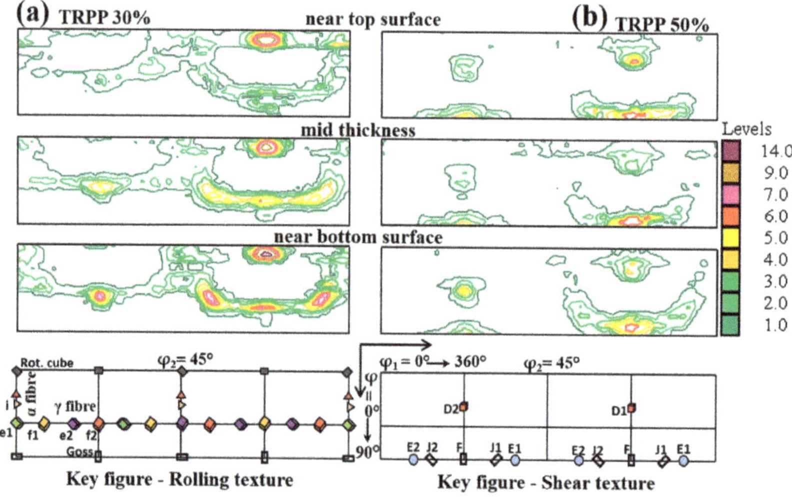

Figure 6. φ_2 =45° ODF sections at three locations across the thickness of the sheet for high strain (89%) asymmetric rolling (1:2) for TRPP of (**a**) 30% and (**b**) 50%.

The computed quantitative volume fraction results presented in Figure 7a,b effectively confirm the general weakening of the rolling texture components with the simultaneous strengthening of the shear texture components for 30% and 50% TRPP. A major rise can be observed for the asymmetry ratio of 1:2, where the volume fractions of the shear texture components jump well above the random intensity level (Figure 7b).

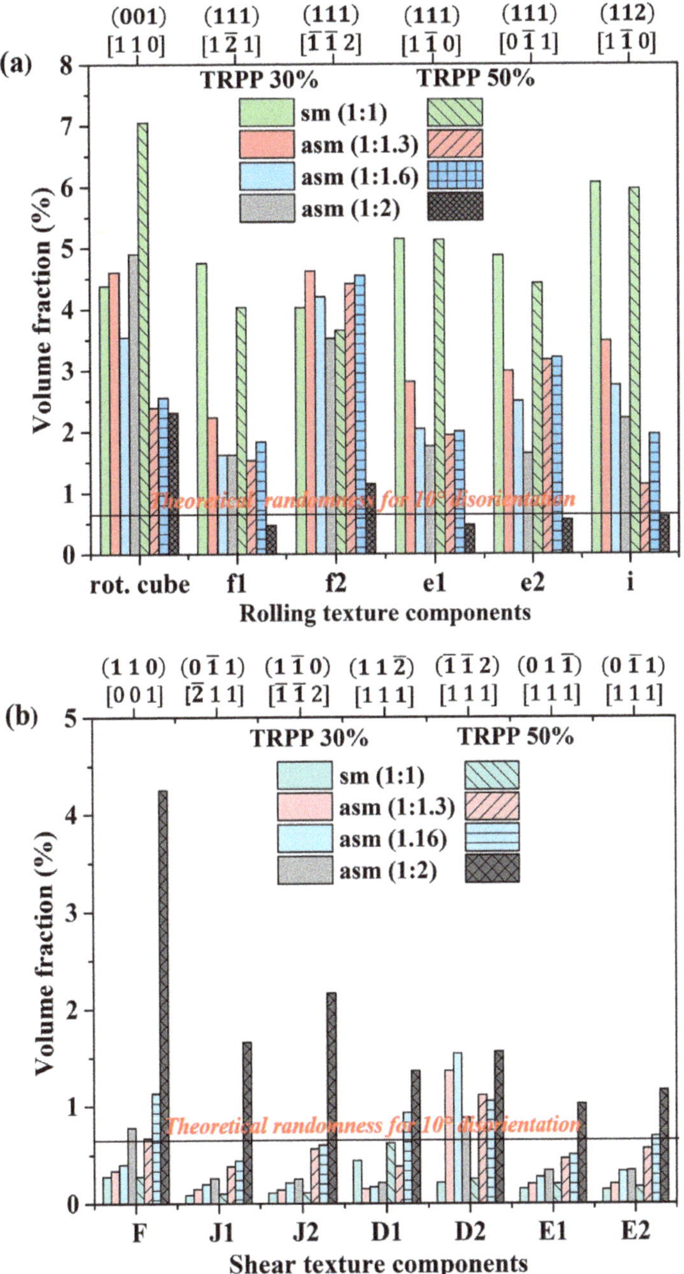

Figure 7. Prominent texture components of (**a**) rolling and (**b**) shear at mid-thickness of the sheet for varying roll diameter ratio from 1:1 to 1:2 at TRPP 30% and 50%.

Concerning the homogeneity of the shear texture components across the thickness, we can see that their strengths are essentially similar throughout the thickness for the asymmetry ratio 1:2 for both 30% and 50% TRPP (Figure 8). It is also evident from Figure 8 that the strengthening in shear texture components is greater for TRPP of 50% than for 30%.

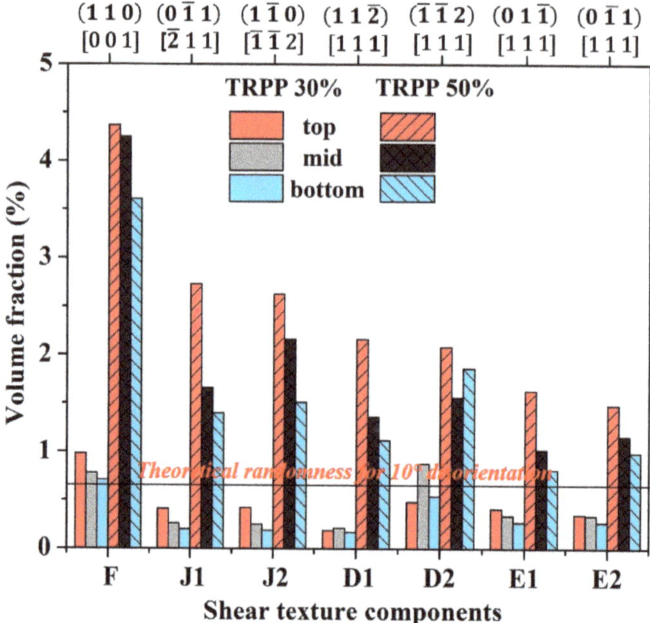

Figure 8. Prominent shear texture components at three locations across the sheet thickness for the asymmetric ratio 1:2 with 30% and 50% TRPP.

3.2. Microstructure

3.2.1. Medium Strain Range (75%)

Representative IPF maps of the microstructure taken from the mid-thickness on TD plane can be seen in Figure 9 for both the symmetric and asymmetric (1:2) cases. As can be seen, the grains show their usual tendency of becoming flattened for all TRPP values. The IPF maps at the asymmetry ratio 1:2 show that the fragmentation of the flattened grains increases with increasing TRPP. Additionally, it appears from Figure 9 that the original grain structure is less fragmented in symmetric compared to asymmetric rolling when increasing the TRPP from 10% to 50%.

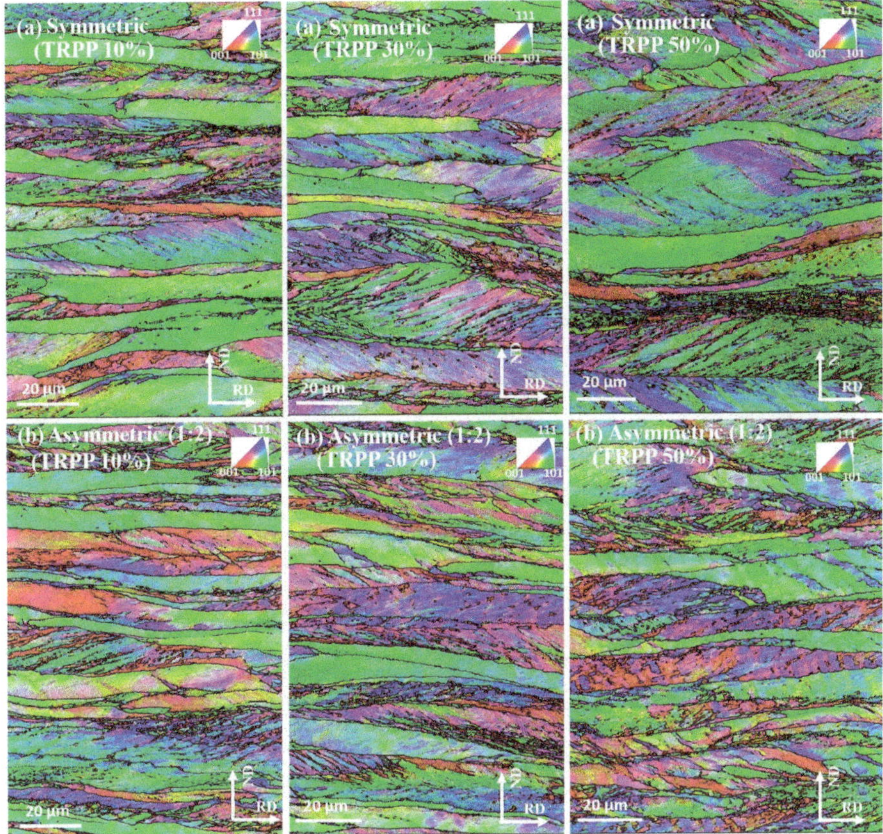

Figure 9. IPF maps (RD axis projection) for medium strain (75%) in (**a**) symmetric and (**b**) asymmetric (1:2) mode with TRPP values of 10, 30, and 50%.

Similar observations can be made in the grain size distributions, see Figure 10a, where the fraction of large grains is greater for the symmetric compared to the asymmetric rolling as TRPP increases to 50%. Clearly, the grain size distributions are bimodal, in all cases. There are many small grains that are below 10 µm size. The large grains size parts of the distribution show parabolic type distributions, which is due to the relatively small numbers of grains found within the selected bin-size of 1 µm [35]. The number of grains corresponding to the different parabolas are indicated in Figure 10a. Another important characteristic of grain fragmentation is the fraction of large angle boundaries. They are illustrated in Figure 10b for three interval values of the misorientations: from 3–5°, 5–15°, and above 15°. As can be seen, the fraction of boundaries having misorientation angles larger than 15° increased more in asymmetric rolling than in the symmetric case, for all TRPP values.

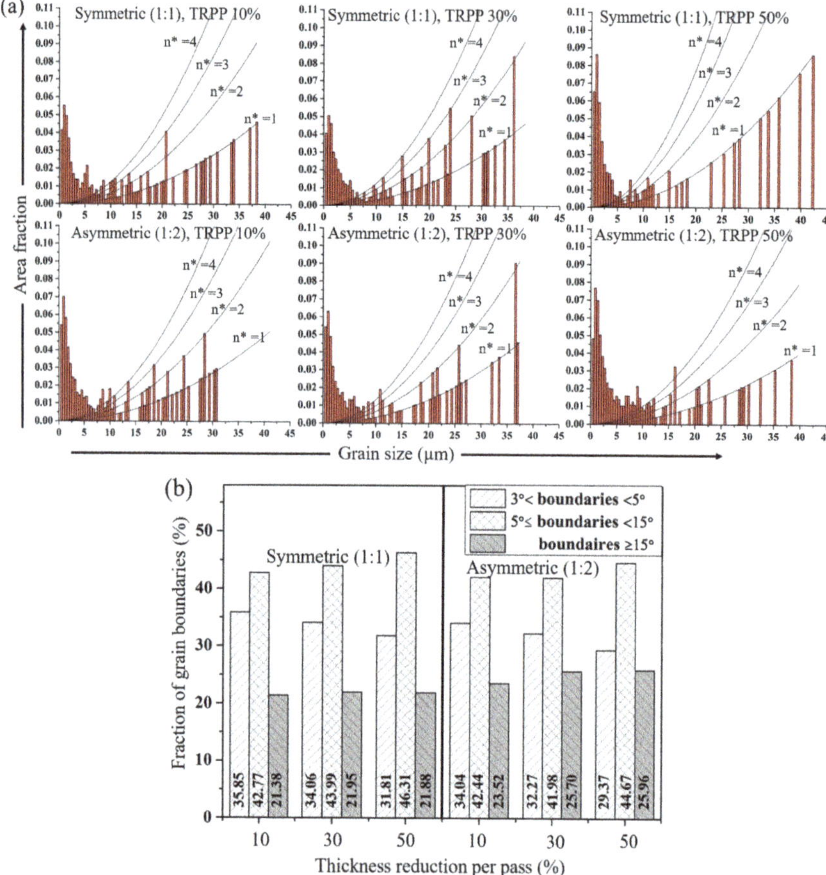

Figure 10. (**a**) Grain size distributions (area fractions) and (**b**) fraction of grain boundaries for symmetric and asymmetric (1:2) rolling in medium strain (75%) for TRPP of 10, 30, and 50%. The added parabolas indicate the number of large grains being in the same bins (n*).

3.2.2. High Strain Range (89%)

At 89% thickness reduction for 30% TRPP, it can be seen from Figure 11 via the green colored grains that the mid-thickness microstructure mainly consisted of elongated grains oriented within the α fiber for both symmetric and asymmetric rolling. However, apart from these α grains, high occurrence of blue colored, severely strained areas are evident, where shear bands and deformation bands intersect; these are γ fiber grains for the asymmetry ratio of 1:2. Additionally, it can also be observed at this stage in the IPF maps that the inclination angle of the shear bands to the rolling direction is smaller as compared to the medium strain range for both symmetric and asymmetric rolling.

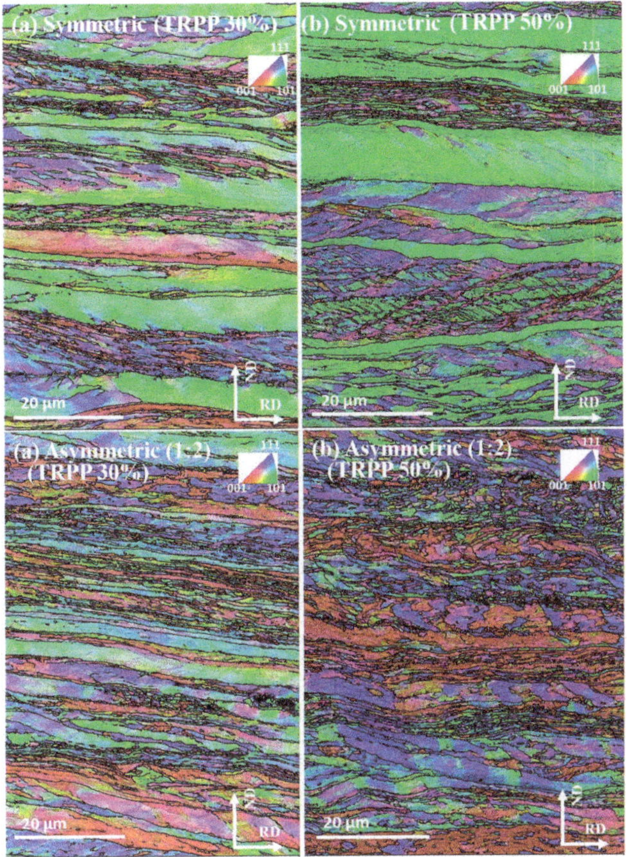

Figure 11. IPF maps (RD axis projection) for high strain (89%) rolled in symmetric and asymmetric (1:2) mode with TRPP of (**a**) 30% and (**b**) 50%.

A significant change in the rolled microstructure can be observed at the asymmetric ratio of 1:2 for 50% TRPP compared to the symmetric case (Figure 11). Indeed, there is more evidence for grain fragmentation. After 89% reduction in asymmetric rolling, the grains are clearly fragmented for both 30% and 50% TRPPs (Figure 11) but it appears that the orientation gradients are greater in the latter case. For symmetric rolling, the microstructures at 89% reduction changed little even when the TRPP was increased from 30% to 50% (Figure 11), although a higher occurrence of fine grain formation can also be seen in severely deformed regions. This is basically similar to the asymmetric case.

Figure 12 presents the distributions for grain boundary fractions (Figure 12a), misorientations (Figure 12b) and for normalized average grain sizes (by area fraction method) (Figure 12c). The grain boundary misorientation fraction values in Figure 12a demonstrate that the high angle boundaries in asymmetric rolling are present in significantly higher fraction with respect to the symmetric rolling case. Consequently, the lower misorientations show lower frequency. Interestingly, the next-neighbor grain-to-grain misorientation distributions are very similar, see Figure 12b. This is not in contradiction with the results of Figure 12a because this distribution is not weighed by the grain size. Indeed, there are further differences in the grain size distribution displayed in Figure 12c, where it is clear that the bimodal nature of the distributions is very much reduced for the asymmetric case.

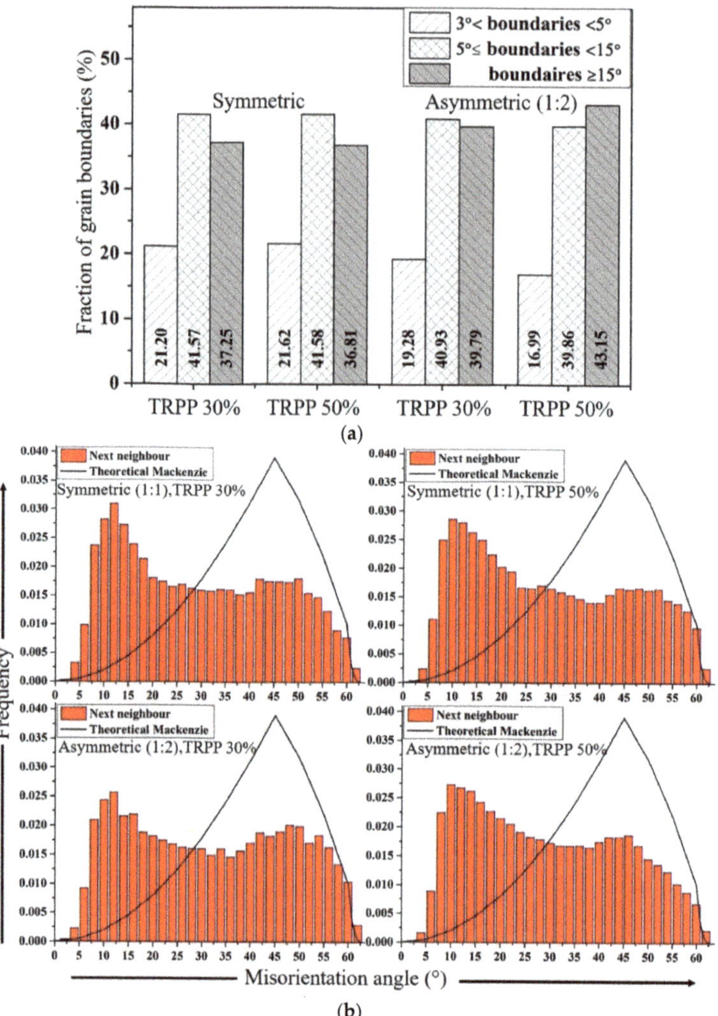

Figure 12. *Cont.*

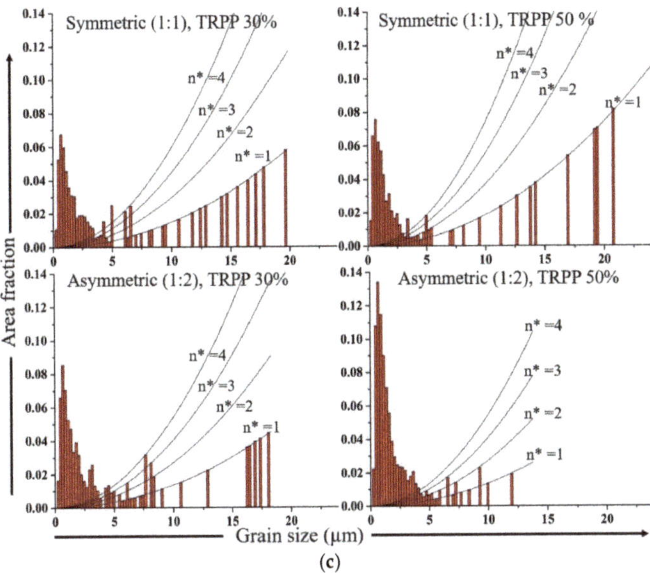

Figure 12. (**a**) Grain boundaries frequencies for symmetric and asymmetric (1:2) rolling for of 30% and 50% in high strain regime (89%); (**b**) next neighbor misorientation distributions of symmetric and asymmetric (1:2) rolling for TRPP of 30% and 50% in the high strain regime (89%); (**c**) grain size distribution (area fraction) for symmetric and asymmetric (1:2) rolling with TRPP of 30% and 50% in high strain regime (89%). The added parabolas indicate the number of large grains being in the same bins (n*).

4. Discussion

The purpose of the present work is to examine the effects of three parameters—one by one, and also combined—on the texture and microstructure developments during asymmetric rolling of steel. These parameters are: the degree of asymmetry defined by the roll diameter ratios, the thickness reduction per pass (TRPP), and the total rolling reduction (75% and 89%). In the following we discuss the obtained results in terms of the evolution of the crystallographic textures and the characteristics of the microstructures.

4.1. Texture

4.1.1. Relation between Rolling and Shear Textures

First we examine the possible relation between the two kinds of textures that were identified in our experiments: the rolling and shear textures. It will be shown that the shear texture can be derived from the rolling one by a simple rotation.

Several previous studies show that the main effect on the texture during asymmetric rolling is a general rotation of the rolling texture in the direction of the shear generated by the asymmetry [9,15–19,22]. The velocity gradient of simple shear $\underline{\underline{L}}$ is composed of the strain rate tensor $\underline{\underline{\dot{\varepsilon}}}$ and the rigid body spin $\underline{\underline{\dot{\beta}}}$ according to the decomposition

$$\underline{\underline{L}} = \begin{pmatrix} 0 & 0 & \dot{\gamma} \\ 0 & 0 & 0 \\ 0 & 0 & 0 \end{pmatrix} = \underline{\underline{\dot{\varepsilon}}} + \underline{\underline{\dot{\beta}}} = \begin{pmatrix} 0 & 0 & \dot{\gamma}/2 \\ 0 & 0 & 0 \\ \dot{\gamma}/2 & 0 & 0 \end{pmatrix} + \begin{pmatrix} 0 & 0 & \dot{\gamma}/2 \\ 0 & 0 & 0 \\ -\dot{\gamma}/2 & 0 & 0 \end{pmatrix} \quad (1)$$

Here the plane of shear is the rolling plane (ND = the plane perpendicular to axis 3), the shear direction is in the rolling direction (RD = axis 1) and $\dot{\gamma}$ is shear strain rate. In the following, we examine only the effect of the rigid body rotation part, which is represented by the second matrix on the right side of Equation (1). This matrix describes a simple rotation around TD (axis 2) with the rotation rate

$$\dot{\omega} = \dot{\gamma}/2 \tag{2}$$

This relation implies a simple relation between the shear strain and the rigid body rotation

$$\gamma = 2\omega \tag{3}$$

Simple rigid body rotations were applied on an experimental rolling texture with increasing rotation angle around the TD axis, up to $\omega = 45°$. The resulting textures are displayed in the $\varphi_2 = 45°$ ODF section in Figure 13 showing the positions of the ideal shear texture components. It can be seen from the figure that an approximate 35° rotation transforms the rolling texture into a shear texture. The corresponding shear strain can be obtained from Equation (3); it is $\gamma = 1.22$. This value is near to the one proposed from polycrystal texture simulations in [22], which is $\gamma = 1.0$. Of course, this analysis is only approximate in the sense that the shear component of the imposed strain, with its rigid body rotation, is decoupled from the rolling strain. During a continuous asymmetric rolling process, the two strain components act at the same time with different amounts. Nevertheless, it is very interesting that a rolling texture appears as a shear texture when rotated by about 35°. This angle is the same as the angle at which shear bands appear during rolling [36–38]. Therefore, our finding also means that the textures that belong to shear bands appear as perfect rolling textures in the rolling reference system. Thus, shear bands do not make a different texture from rolling, they have no distinct signature on the rolling texture.

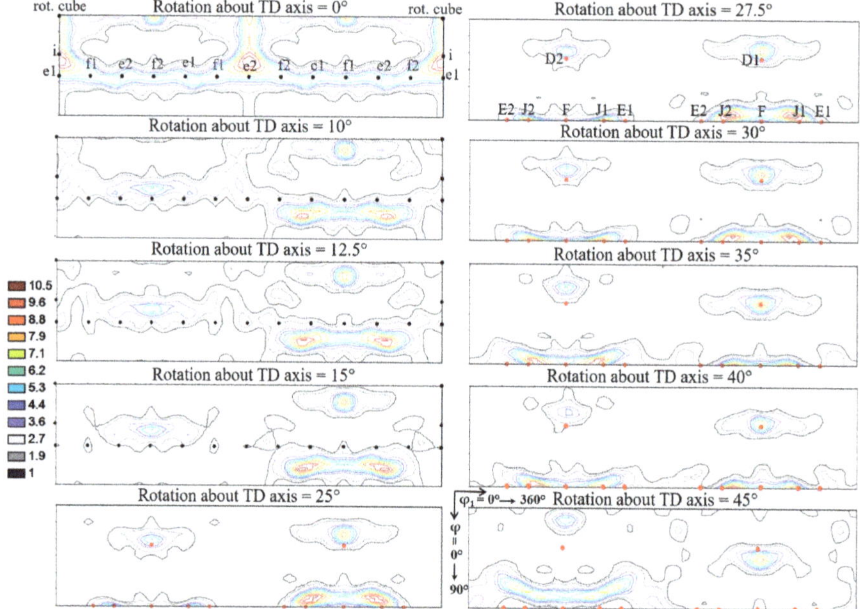

Figure 13. $\varphi_2 = 45°$ sections of ODFs to represent the symmetric rolling texture migration towards shear texture just by a simple rotation about the TD axis.

We obtained in the preceding paragraph that a rolling texture turns into a shear texture when rotating by 35° around the TD axis. Then the question arises: into which components do the rolling ideal components rotate? We have examined this question using the data in Tables 2 and 3, and found the following results:

- The f1 and f2 rolling components are rotating exactly into the F component of the shear texture.
- The i rolling component produces the J1 and J2 shear components.
- The rotated cube produces both the D1 and D2 components of shear.
- The e1 and e2 rolling components are not rotating into any ideal shear component.
- The Goss rolling orientation does not need to be rotated to become stable orientation for shear; it is already the same orientation as the F shear component, so it remains stable for any imposed shear deformation after rolling.

Following the above analysis of the texture rotation due to shear, one can attribute a general rotation angle for a given texture made by asymmetric rolling. For this purpose, we choose the deviation angle of the f2 rolling component from its symmetric rolling position. This component is turning progressively into the F shear texture component during asymmetric rolling. The rotation angle needed for complete transformation of the rolling texture into a shear texture is 35°. The texture rotation angles obtained in this way for the different processing conditions are presented in Figure 2b. As can be seen from the rotation values in the sections of the processing space, they increase as a function of all three variables; the total strain, the TRPP, and the roll diameter ratio. The experimentally obtained maximum rotation angle is 32°, which is nearly as high as the theoretical value (35°) obtained above, so the parameter combination pointed to the vertex of the processing cube presented in Figure 1a is the most effective one in terms of transformation of the rolling texture into a shear texture by asymmetric rolling.

4.1.2. Mixed Nature of the Textures Developing during Asymmetric Rolling

One of the main results of our experimental campaign is that the textures obtained by asymmetric rolling show more similarities to the conventional rolling texture if the TRPP is less or equal to 30%, for all diameter ratios (Figures 3a and 6b). It implies that the plane-strain part of the deformation is more effective than the shear part, even for the high strain regime (89%). Nevertheless, the signature of the shear strain is well visible for 30% TRPP when the asymmetry ratio is 1:2.

After increasing the thickness reduction per pass to 50%, even for medium total thickness reductions (75%), the texture for 1:2 ratio shows 15° rotation at mid-thickness of the sheet, meaning that the superimposed shear is effective and competes well with the plane strain part. However, the amount of deformation is not sufficient to significantly strengthen the shear texture. What is happening is that the rolling texture components weaken considerably, but the shear texture does not become strong (Figure 4). An extension of this deformation condition (50% TRPP and asymmetric ratio 1:2) into the high strain regime (89%) not only rotates the texture further towards the ideal shear texture but also strengthens the shear texture components.

The observed higher strengthening in the F (Goss) $\{110\}\langle 001\rangle$ orientation in high strain regime, as compared to other shear texture components, like J (Brass) $\{1\bar{1}0\}\langle\bar{1}\bar{1}2\rangle$ and D (Copper) $\{\bar{1}\bar{1}2\}\langle 111\rangle$, is in agreement with previous studies concerning the occurrence of shear textures in shear based processes [39–42]. The similarity with these reports indicate that the accumulated shear in asymmetric rolling (1:2) with high total thickness reduction (89%) becomes significantly higher than the plane rolling strain. The formation of Goss (F) preferentially along with the copper (D) $\{\bar{1}\bar{1}2\}\langle 111\rangle$ was also observed in strongly sheared regions of Fe3%Si single crystal sheets [40]. These single crystals were initially rotated cube (001)[110] and cube orientations that were hot rolled to 85% thickness reduction by applying higher TRPPs in an unlubricated condition [40]. The mechanism of this behavior was connected to a rotation about the TD$\parallel$[110] axis for the rotated cube orientation.

Similar to the observations of single crystal behavior in [40], the rotated cube orientation is important in the present investigation for the development of shear texture components. It can be noticed from symmetric rolling that, with increase in thickness reduction per pass, the rotated cube orientation consistently becomes stronger than the other rolling texture components. Nevertheless, conventionally, under small thickness reduction per pass (below <30%), the i {112}<110> and e {111}<110> components prefer to strengthen more than the rotated cube and f {111}<1$\bar{2}$1> orientations. An increased strengthening in the rotated cube component with increase in thickness reduction per pass can be correlated with two factors. Primarily, its high stability is due to the four-fold symmetry of symmetric rolling and its tendency to have greater orientation spread around the rolling direction than the normal and transverse directions [43]. However, earlier studies of symmetric rolling also reported that under the influence of shear near the surface regions of the sheet, components like the rotated cube and f have strong affinity to form Goss and other shear texture components by virtue of rotation about the TD axis [38–40,44–46]. Thereby, simultaneous imposition of higher asymmetric ratio along with high thickness reduction per pass forms favorable condition for the rotation of rolling texture and further strengthening in shear texture components with increase in total thickness reduction.

This kind of behavior in terms of texture rotation with 50% TRPP is also in accord with the elastoplastic FEM modelling studies by Kim et al. [15] and with the convergence–divergence map presented by Toth et al. [22] for asymmetric rolling. It was pointed out in these works that only with a high shear coefficient ($\geq$ 1) the shear texture components can be fully developed. Thus, it seems that asymmetric rolling with a roll diameter ratio of 1:2 and 50% thickness reduction per pass possesses a shear coefficient of $\geq$1.

A comparison of the (1:2) asymmetric rolling texture at high total thickness reduction (89%) with ECAP of IF steel at room temperature shows for ECAP stronger J {110}<112> and D {112}<111> components than the F (Goss) [47–49]. ECAP of IF steel was carried at room temperature for both 90° and 120° die angles in route A, which is a monotonic shearing mode with sudden orientation shifts of the crystal orientations between passes (due to the unavoidable re-insertion of the sample into the die). While in lower strain torsion test, the J {110}<112> component appears sooner than the D {$\bar{1}\bar{1}$2}<111> and F {110}<001> orientations [42]. This difference in texture evolution can be related to the imposed deformation geometry of asymmetric rolling where the strain state depends on the respective contributions of plane strain and shear strain, which controls the crystal rotation. In addition, the thickness of the sample does not remain constant throughout the rolling process, while processes such as ECAP and torsion exert a state of simple shear deformation without any change in the shape and size of the material [42,47–49]. This also means that more passes can be applied in those processes (ECAP and torsion) which results in stronger shear textures as compared to asymmetric rolling.

4.2. Microstructure

It was observed after the total thickness reduction of 75% (medium strain regime) with TRPP up to the 50% that both symmetric and (1:2) asymmetric rolling form elongated and lamellar bands of deformed grains oriented in the rolling direction (Figure 8). However, in the present experiments, the occurrence of orientation inhomogeneity and fragmentation was more pronounced in the asymmetric case when the TRPP was increased to 50%. Increasing the total thickness reduction to the high strain regime (89%) influences not only the asymmetric but also the symmetric rolled microstructures. It is due to the simultaneous activation of both texture and microstructure mechanisms and their respective evolution when deformation increases from medium to high strain regime [50,51]. Thereby, changing the strain regime from medium to high increased the fraction of HAGB from ~22 to ~37% even in the symmetric case with 30% TRPP (Figures 9b and 11a). Although, the fraction of HAGBs achieved in symmetric rolling with further increase in 50% TRPP is still in the same range as observed with 30% TRPP. In fact, it also lies in the same range as reported by Saha et al. for multi-pass rolling of IF steel with small TRPPs [52]. This result indicates that the fraction of HAGBs cannot be increased significantly by increasing TRPP for a given total thickness reduction in symmetric rolling.

In the case of asymmetric rolling (TRPP 30 to 50% and asymmetry ratio 1:2), an increase in deformation and shear bands as a result of fragmentation and inhomogeneity can be anticipated from the imposed deformation condition. There can be two additional effects, which appear in asymmetric rolling compared to the symmetric case. From a geometrical point of view, an increase in macroscopic instabilities, such as macro shear bands, may result from the imposed shear strain as TRPP increases. Whereby, grains can undergo shearing while their flattening/elongation occurs simultaneously. Such external imposed deformation conditions act in addition to the internal constraints of the polycrystalline material which plays an important role in steels and influences the grain orientation stability [53,54]. This effect assists in creating more orientation spread and deviation from the preferred stable end orientations in agreement with the texture results presented in Figures 3 and 6. Thus, it can lead to more orientation splitting and spread in orientations. The other factor could be the strengthening of the shear texture with extension of total thickness reduction from medium to high strain regime. It has been noted that orientations evolving towards their final preferred texture are prone to create deformation induced high angle grain boundaries (HAGBs) [50]. In addition, it is also reported when the process is ideal shear type then the rate of lattice rotation become higher as compared to the symmetric rolling [55]. The consequence of this difference in lattice rotation rate leads to the increased grain refinement and fraction of HAGB in shear assisted processes [55]. The simultaneous occurrence of these two factors along with the microstructural mechanism could be the possible reason for further subdivision of grains and an increased fraction of HAGB in the asymmetric mode, especially for the 50% TRPP compared to symmetric rolling.

The effect of increased thickness reduction from 30 to 50% in high strain regime is also reflected in the misorientation and grain size distribution plots for both types of rolling (Figure 12b,c). Similar to the studies of Liu et al. [56] and Hughes et al. [50,51], a decrease in the fraction of misorientations between 5° and 10° was observed rather than a further increase, in both types of rolling, irrespective of the applied TRPPs. According to these authors, a decrease in the misorientation distribution between 5° and 10° is due to the dominance of the removal process of lamellar boundaries over their generation. However, when higher misorientation angles (≥15°) were taken into consideration, asymmetric rolling showed an increased fraction as compared to the symmetric cases. Hughes et al. suggested that an increase in misorientation angles (≥15°), especially beyond 30°, may be due to the preferred texture formation in subdivided grains which become more prominent after high total thickness reductions [50,51]. According to Toth et al., however, the development of misorientations between 15° and 30° derives from the internal subdivisions of grains rather than from their original grain boundaries which predominantly lie in the range 30–60° [57]. It is more notable in asymmetric rolling (1:2) than in symmetric rolling due to their higher grain fragmentation.

With increased total thickness reduction, it is also noticed that along with increased fraction of fine grains (≤5 μm), a notable fraction of grains between 5 μm and 22 μm remains after symmetric rolling, for both 30% and 50% TRPPs (Figure 12c). Such a trend reaffirms that increased TRPP does not play a significant role in grain fragmentation during symmetric rolling. Asymmetric rolling, on the contrary, reduces the fraction of large residual grains and increases the fraction of fine grains smaller than 5 μm as TRPP is raised from 30% to 50% (Figure 12c). Such a different behavior in asymmetric rolling also highlights how the thickness reduction per pass (TRPP) is an important parameter for achieving high grain fragmentation at high total thickness reductions.

The above-presented argumentation clearly shows that the addition of shear deformation to the rolling strain by asymmetric rolling is beneficial for the grain fragmentation process. The more the texture approaches a shear texture, better the grain fragmentation is. This represents a significant advantage of the asymmetric rolling over the symmetric one. Figure 1a illustrates that a shear texture is approached the best if all three processing parameters are as high as possible, which is represented by the diagonal line in Figure 1a.

5. Conclusions

In the present work, systematic multiple pass rolling has been carried out on extra-low carbon steel in symmetric and asymmetric modes while varying the TRPP between 10–50%. The evolution of the textures and the microstructures were examined at medium (75%) and high (89%) total thickness reductions. The results of our study lead to the following conclusions:

1 The rotation of rolling texture towards the shear texture at the mid-thickness as well as its through-thickness homogeneity are significantly increased by the asymmetry-induced shear strain yielding from the applied TRPP and the final thickness reduction of the sheet with roll diameter ratios $\geq$1:1.6.
2 For TRPP values below about 30%, the resultant texture of asymmetric rolling shows a greater similarity with the plane strain rolling texture compared to a conventional shear texture but prominent asymmetric trends exist in the preferred texture components especially in the f1 and f2 orientations ({111}$<\overline{11}2>$) of the rolling texture.
3 The texture after asymmetric rolling (1:2) resembles to an ideal shear texture when the deformation is extended to the high strain regime and for high TRPP such as 50%.
4 It has been shown that an approximately 35° rigid body rotation of the rolling texture around the TD axis transforms a typical symmetric rolling texture into a shear texture. Form this, it has been estimated that the shear component generated by the asymmetry conditions has to be at least 1.2 in order to get a shear texture.
5 The shear texture that can be obtained by asymmetric rolling inherits several texture components from the symmetric rolling texture. A list of them is provided in this study.
6 Asymmetric rolling is advantageous for grain refinement and for the creation of deformation induced HAGBs as compared to symmetric rolling under similar conditions. However, the significance of this advantage depends primarily on the final total thickness reduction and is associated with the degree of rotation of the rolling texture.

Author Contributions: Conceptualization: L.S.T., S.S.D. and P.D.H. Methodology: L.S.T., S.S.D. and P.D.H. Validation: L.S.T., S.S.D. and A.H. Formal analysis: L.S.T. and S.S.D. Investigation: S.S.D. Data curation: S.S.D. Writing-original draft preparation: S.S.D. Writing-review & editing: S.S.D. and L.S.T. Visualization: S.S.D. Supervision: L.S.T. and P.D.H. Project administration: P.D.H. and A.H. Funding acquisition: L.S.T., P.D.H. and A.H.

Funding: This work was supported by the French State through the program "Investment in the Future" operated by the National Research Agency (ANR) and referenced by ANR-11-LABX-0008-01 (LabEx DAMAS) and by the linkage industrial project LP0989455 of the Australian Research Council.

Acknowledgments: Bevis Hutchison is gratefully acknowledged for valuable advices in bulk texture measurements by X-ray diffraction and in their analysis.

Conflicts of Interest: The authors declare no conflict of interest.

References

1. Saito, Y.; Utsunomiya, H.; Tsuji, N.; Sakai, T. Novel ultra-high straining process for bulk materials—Development of the accumulative roll-bonding (ARB) process. *Acta Mater.* **1999**, *47*, 579–583. [CrossRef]
2. Segal, V.M. Equal channel angular extrusion: From macromechanics to structure formation. *Mater. Sci. Eng. A* **1999**, *271*, 322–333. [CrossRef]
3. Tsuji, N.; Saito, Y.; Lee, S.H.; Minamino, Y. ARB (Accumulative Roll-Bonding) and other new techniques to produce bulk ultrafine grained materials. *Adv. Eng. Mater.* **2003**, *5*, 338–344. [CrossRef]
4. Valiev, R.Z.; Zehetbauer, M.J.; Estrin, Y.; Höppel, H.W.; Ivanisenko, Y.; Hahn, H.; Wilde, G.; Roven, H.J.; Sauvage, X.; Langdon, T.G. The innovation potential of bulk nanostructured materials. *Adv. Eng. Mater.* **2007**, *9*, 527–533. [CrossRef]

5. Tóth, L.S.; Lapovok, R.; Hasani, A.; Gu, C. Non-equal channel angular pressing of aluminum alloy. *Scr. Mater.* **2009**, *61*, 1121–1124. [CrossRef]
6. Efe, M.; Moscoso, W.; Trumble, K.P.; Dale Compton, W.; Chandrasekar, S. Mechanics of large strain extrusion machining and application to deformation processing of magnesium alloys. *Acta Mater.* **2012**, *60*, 2031–2042. [CrossRef]
7. Sagapuram, D.; Efe, M.; Moscoso, W.; Chandrasekar, S.; Trumble, K.P. Controlling texture in magnesium alloy sheet by shear-based deformation processing. *Acta Mater.* **2013**, *61*, 6843–6856. [CrossRef]
8. Lapovok, R.; Tóth, L.S.; Winkler, M.; Semiatin, S.L. A comparison of continuous SPD processes for improving the mechanical properties of aluminum alloy 6111. *J. Mater. Res.* **2009**, *24*, 459–469. [CrossRef]
9. Beausir, B.; Biswas, S.; Kim, D.I.; Tóth, L.S.; Suwas, S. Analysis of microstructure and texture evolution in pure magnesium during symmetric and asymmetric rolling. *Acta Mater.* **2009**, *57*, 5061–5077. [CrossRef]
10. Sidor, J.; Miroux, A.; Petrov, R.; Kestens, L. Microstructural and crystallographic aspects of conventional and asymmetric rolling processes. *Acta Mater.* **2008**, *56*, 2495–2507. [CrossRef]
11. Sakai, T.; Hamada, S.; Saito, Y. Improvement of the r-value in 5052 aluminum alloy sheets having through-thickness shear texture by 2-pass single-roll drive unidirectional shear rolling. *Scr. Mater.* **2001**, *44*, 2569–2573. [CrossRef]
12. Utsunomiya, H.; Ueno, T.; Sakai, T. Improvement in the *r*-value of aluminum sheets by differential-friction rolling. *Scr. Mater.* **2007**, *57*, 1109–1112. [CrossRef]
13. Huang, X.; Suzuki, K.; Chino, Y. Improvement of stretch formability of pure titanium sheet by differential speed rolling. *Scr. Mater.* **2010**, *63*, 473–476. [CrossRef]
14. Kim, H.-K.; Kim, H.-W.; Cho, J.-H.; Lee, J.-C. High-formability Al alloy sheet produced by asymmetric rolling of strip-cast sheet. *Mater. Sci. Eng. A* **2013**, *574*, 31–36. [CrossRef]
15. Kim, K.H.; Lee, D.N. Analysis of deformation textures of asymmetrically rolled aluminum sheets. *Acta Mater.* **2001**, *49*, 2583–2595. [CrossRef]
16. Lee, S.H.; Lee, D.N. Analysis of deformation textures of asymmetrically rolled steel sheets. *Int. J. Mech. Sci.* **2001**, *43*, 1997–2015. [CrossRef]
17. Jin, H.; Lloyd, D.J. Evolution of texture in AA6111 aluminum alloy after asymmetric rolling with various velocity ratios between top and bottom rolls. *Mater. Sci. Eng. A* **2007**, *465*, 267–273. [CrossRef]
18. Huang, X.; Suzuki, K.; Watazu, A.; Shigematsu, I.; Saito, N. Effects of thickness reduction per pass on microstructure and texture of Mg–3Al–1Zn alloy sheet processed by differential speed rolling. *Scr. Mater.* **2009**, *60*, 964–967. [CrossRef]
19. Roumina, R.; Sinclair, C.W. Deformation geometry and through-thickness strain gradients in asymmetric rolling. *Metall. Mater. Trans. A* **2008**, *39*, 2495–2503. [CrossRef]
20. Polkowski, W.; Jóźwik, P.; Polański, M.; Bojar, Z. Microstructure and texture evolution of copper processed by differential speed rolling with various speed asymmetry coefficient. *Mater. Sci. Eng. A* **2013**, *564*, 289–297. [CrossRef]
21. Orlov, D.; Pougis, A.; Lapovok, R.; Toth, L.; Timokhina, I.; Hodgson, P.; Haldar, A.; Bhattacharjee, D. Asymmetric rolling of interstitial-free steel using differential roll diameters. Part I: Mechanical properties and deformation textures. *Metall. Mater. Trans. A* **2013**, *44*, 4346–4359. [CrossRef]
22. Tóth, L.S.; Beausir, B.; Orlov, D.; Lapovok, R.; Haldar, A. Analysis of texture and *R* value variations in asymmetric rolling of IF steel. *J. Mater. Process. Technol.* **2012**, *212*, 509–515. [CrossRef]
23. Kim, W.J.; Lee, B.H.; Lee, J.B.; Lee, M.J.; Park, Y.B. Synthesis of high-strain-rate superplastic magnesium alloy sheets using a high-ratio differential speed rolling technique. *Scr. Mater.* **2010**, *63*, 772–775. [CrossRef]
24. Kim, W.J.; Yoo, S.J.; Lee, J.B. Microstructure and mechanical properties of pure Ti processed by high-ratio differential speed rolling at room temperature. *Scr. Mater.* **2010**, *62*, 451–454. [CrossRef]
25. Kim, W.J.; Lee, Y.G.; Lee, M.J.; Wang, J.Y.; Park, Y.B. Exceptionally high strength in Mg–3Al–1Zn alloy processed by high-ratio differential speed rolling. *Scr. Mater.* **2011**, *65*, 1105–1108. [CrossRef]
26. Choi, C.H.; Kim, K.; Lee, D.N. The effect of shear texture development on the formability in rolled aluminum alloys sheets. *Mater. Sci. Forum* **1998**, *273–275*, 391–396. [CrossRef]
27. Jeong, H.; Park, S.; Ha, T. Evolution of shear texture according to shear strain ratio in rolled FCC metal sheets. *Met. Mater. Int.* **2006**, *12*, 21–26. [CrossRef]

28. Lapovok, R.; Orlov, D.; Timokhina, I.B.; Pougis, A.; Toth, L.S.; Hodgson, P.D.; Haldar, A.; Bhattacharjee, D. Asymmetric rolling of interstitial-free steel using one idle roll. *Metall. Mater. Trans. A* **2012**, *43*, 1328–1340. [CrossRef]
29. Nguyen Minh, T.; Sidor, J.J.; Petrov, R.H.; Kestens, L.A. Texture evolution during asymmetrical warm rolling and subsequent annealing of electrical steel. *Mater. Sci. Forum* **2012**, *702–703*, 758–761. [CrossRef]
30. Truszkowski, W.; Król, J.; Major, B. Inhomogeneity of rolling texture in fcc metals. *MTA* **1980**, *11*, 749–758. [CrossRef]
31. Truszkowski, W.; Król, J.; Major, B. On penetration of shear texture into the rolled aluminum and copper. *MTA* **1982**, *13*, 665–669. [CrossRef]
32. Sakai, T.; Saito, Y.; Kato, K. Texture formation in low carbon Ti bearing steel sheets by high speed hot rolling in ferrite region. *Trans. Iron Steel Inst. Jpn.* **1988**, *28*, 1036–1042. [CrossRef]
33. Mishin, O.V.; Bay, B.; Jensen, D.J. Through-thickness texture gradients in cold-rolled aluminum. *Metall. Mater. Trans. A* **2000**, *31*, 1653–1662. [CrossRef]
34. Beausir, B.; Fundenberger, J.J. ATOM-Analysis Tool for Orientation Maps. Available online: http://www.atex-software.eu/ (accessed on 26 July 2018).
35. Toth, L.S.; Biswas, S.; Gu, C.; Beausir, B. Notes on representing grain size distributions obtained by electron backscatter diffraction. *Mater. Char.* **2013**, *84*, 67–71. [CrossRef]
36. Dillamore, I.L.; Roberts, J.G.; Bush, A.C. Occurrence of shear bands in heavily rolled cubic metals. *Met. Sci.* **1979**, *13*, 73–77. [CrossRef]
37. Chen, Q.Z.; Quadir, M.Z.; Duggan, B.J. Shear band formation in IF steel during cold rolling at medium reduction levels. *Philos. Mag.* **2006**, *86*, 3633–3646. [CrossRef]
38. Yeung, W.Y.; Duggan, B.J. Shear band angles in rolled F.C.C. materials. *Acta Metall.* **1987**, *35*, 541–548. [CrossRef]
39. Mishra, S.; Därmann, C.; Lücke, K. On the development of the goss texture in iron-3% silicon. *Acta Metall.* **1984**, *32*, 2185–2201. [CrossRef]
40. Shimizu, Y.; Ito, Y.; Iida, Y. Formation of the Goss orientation near the surface of 3 pct silicon steel during hot rolling. *MTA* **1986**, *17*, 1323–1334. [CrossRef]
41. Wonsiewicz, B.C.; Chin, G.Y. Inhomogeneity of plastic flow in constrained deformation. *Metall. Trans.* **1970**, *1*, 57–61.
42. Baczynski, J.; Jonas, J.J. Texture development during the torsion testing of α-iron and two IF steels. *Acta Mater.* **1996**, *44*, 4273–4288. [CrossRef]
43. Böcker, A.; Klein, H.; Bunge, H. Development of cross-rolling textures in Armco-iron. *Texture Microstruct.* **1990**, *12*, 103–123. [CrossRef]
44. Nguyen-Minh, T.; Sidor, J.J.; Petrov, R.H.; Kestens, L.A.I. Occurrence of shear bands in rotated Goss ({110}<110>)orientations of metals with bcc crystal structure. *Scr. Mater.* **2012**, *67*, 935–938. [CrossRef]
45. Haratani, T.; Hutchinson, W.B.; Dillamore, I.L.; Bate, P. Contribution of shear banding to origin of Goss texture in silicon iron. *Met. Sci.* **1984**, *18*, 57–66. [CrossRef]
46. Hölscher, M.; Raabe, D.; Lücke, K. Relationship between rolling textures and shear textures in f.c.c. and b.c.c. metals. *Acta Metall. Mater.* **1994**, *42*, 879–886. [CrossRef]
47. Li, S.; Gazder, A.A.; Beyerlein, I.J.; Pereloma, E.V.; Davies, C.H.J. Effect of processing route on microstructure and texture development in equal channel angular extrusion of interstitial-free steel. *Acta Mater.* **2006**, *54*, 1087–1100. [CrossRef]
48. Li, S.; Gazder, A.A.; Beyerlein, I.J.; Davies, C.H.J.; Pereloma, E.V. Microstructure and texture evolution during equal channel angular extrusion of interstitial-free steel: Effects of die angle and processing route. *Acta Mater.* **2007**, *55*, 1017–1032. [CrossRef]
49. De Messemaeker, J.; Verlinden, B.; Van Humbeeck, J. Texture of IF steel after equal channel angular pressing (ECAP). *Acta Mater.* **2005**, *53*, 4245–4257. [CrossRef]
50. Hughes, D.A.; Hansen, N. High angle boundaries formed by grain subdivision mechanisms. *Acta Mater.* **1997**, *45*, 3871–3886. [CrossRef]
51. Hughes, D.A.; Hansen, N. Microstructure and strength of nickel at large strains. *Acta Mater.* **2000**, *48*, 2985–3004. [CrossRef]
52. Saha, R.; Ray, R.K. Microstructural and textural changes in a severely cold rolled boron-added interstitial-free steel. *Scr. Mater.* **2007**, *57*, 841–844. [CrossRef]

53. Inagaki, H.; Suda, T. The development of rolling textures in low-carbon steels. *Texture* **1972**, *1*, 129–140. [CrossRef]
54. Inagaki, H. Fundamental aspect of texture formation in low carbon steel. *ISIJ Int.* **1994**, *34*, 313–321. [CrossRef]
55. Gu, C.F.; Tóth, L.S.; Arzaghi, M.; Davies, C.H.J. Effect of strain path on grain refinement in severely plastically deformed copper. *Scr. Mater.* **2011**, *64*, 284–287. [CrossRef]
56. Liu, Q.; Huang, X.; Lloyd, D.J.; Hansen, N. Microstructure and strength of commercial purity aluminium (AA 1200) cold-rolled to large strains. *Acta Mater.* **2002**, *50*, 3789–3802. [CrossRef]
57. Tóth, L.S.; Beausir, B.; Gu, C.F.; Estrin, Y.; Scheerbaum, N.; Davies, C.H.J. Effect of grain refinement by severe plastic deformation on the next-neighbor misorientation distribution. *Acta Mater.* **2010**, *58*, 6706–6716. [CrossRef]

Article

Toward Better Control of Inclusion Cleanliness in a Gas Stirred Ladle Using Multiscale Numerical Modeling

Jean-Pierre Bellot [1,2,*], Jean-Sebastien Kroll-Rabotin [1,2], Matthieu Gisselbrecht [1,2], Manoj Joishi [1,2], Akash Saxena [3], Sean Sanders [3] and Alain Jardy [1,2]

1 Institut Jean Lamour—UMR 7198 CNRS/Université de Lorraine, CS 50840, 54011 Nancy CEDEX, France; jean-sebastien.kroll-rabotin@univ-lorraine.fr (J.-S.K.R.); matthieu.gisselbrecht@univ-lorraine.fr (M.G.); manoj.joishi@univ-lorraine.fr (M.J.); Alain.Jardy@univ-lorraine.fr (A.J.)
2 Laboratory of Excellence on Design of Alloy Metals for low-mAss Structures (Labex DAMAS), Université de Lorraine, 57073 Metz, France
3 Department of Chemical & Materials Engineering, University of Alberta, Edmonton, AB T6G 1H9, Canada; asaxena@ualberta.ca (A.S.); sean.sanders@ualberta.ca (S.S.)
* Correspondence: jean-pierre.bellot@univ-lorraine.fr; Tel.: +33-372-742-671

Received: 29 May 2018; Accepted: 5 July 2018; Published: 10 July 2018

Abstract: The industrial objective of lowering the mass of mechanical structures requires continuous improvement in controlling the mechanical properties of metallic materials. Steel cleanliness and especially control of inclusion size distribution have, therefore, become major challenges. Inclusions have a detrimental effect on fatigue that strongly depends both on inclusion content and on the size of the largest inclusions. Ladle treatment of liquid steel has long been recognized as the processing stage responsible for the inclusion of cleanliness. A multiscale modeling has been proposed to investigate the inclusion behavior. The evolution of the inclusion size distribution is simulated at the process scale due to coupling a computational fluid dynamics calculation with a population balance method integrating all mechanisms, i.e., flotation, aggregation, settling, and capture at the top layer. Particular attention has been paid to the aggregation mechanism and the simulations at an inclusion scale with fully resolved inclusions that represent hydrodynamic conditions of the ladle, which have been specifically developed. Simulations of an industrial-type ladle highlight that inclusion cleanliness is mainly ruled by aggregation. Quantitative knowledge of aggregation kinetics has been extracted and captured from mesoscale simulations. Aggregation efficiency has been observed to drop drastically when increasing the particle size ratio.

Keywords: steel ladle; non-metallic inclusions; simulation; aggregation

---

1. Introduction

A huge number of inclusions is generated in the secondary steelmaking processes mainly during the deoxidation steps. Non-metallic inclusions such as sulfide and oxide particles may have a detrimental impact on both formability and fatigue life. Furthermore, inclusion size distribution is particularly critical because large aggregates are the most harmful to mechanical properties [1]. Nowadays, the reduction of the weight of high performance steel parts without affecting their mechanical properties is a new challenge, which requires continuous improvement of metal cleanliness.

Gas-stirred ladle treatment of liquid metal has long been recognized as the processing stage responsible for the inclusion population of specialty steels. As seen in Figure 1, argon gas is injected through one or more porous plugs at the bottom of the ladle, which leads to mixing of the liquid metal to achieve chemical homogeneity and entrapment of inclusions by the bubble swarm. The latter is well known as the flotation mechanism. Furthermore, the turbulence generated by the bubble plume enhances the probability of inclusion collision and enhances aggregation, which is an important mechanism for the evolution of the inclusion size distribution.

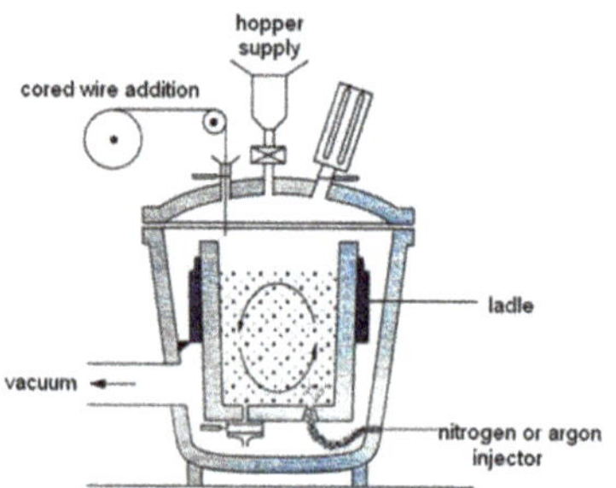

Figure 1. Schematic of a gas-stirred ladle.

The ladle process is a very complex three-phase reactor, which has been the purpose of many modeling studies for the last two decades. Few cold or hot physical models [2,3] and numerous computational fluid dynamics (CFD) simulations have been undertaken to investigate the turbulent flow and mixing phenomena. The modeling of liquid-gas flows is treated in the literature using either a Euler-Euler method [4,5] or a Euler-Lagrange method [6,7]. Each approach has equally important advantages and drawbacks. Among the most sophisticated and recent approaches, the multiphase volume of the fluid (VOF) technique allowed Li et al. [8] to predict the opening of the slag due to bubbles called "open eye" (potentially leading to steel re-oxidation) and slag droplets entrainment (possibly generating large size defects). The behavior of the non-metallic inclusions in a gas-stirred ladle is usually modeled through a population balance method (PBM). The coupling of CFD and PBM techniques is now frequently adopted, which results in promising outcomes in terms of calculating the inclusion removal efficiency and evolution of particle size distribution (PSD) over processing time [9–11]. Recently, Lei et al. [12] proposed the application of a particle size grouping technique for solving the population balance equation across a very broad range of particle sizes (from nuclei of 0.1 nm up to large aggregates of the order of 100 μm in size) but with a noticeable discrepancy regarding the exact solution of Smoluchowski's equation.

Unfortunately, all these sophisticated macroscopic models are based on modeled kinetics of turbulent aggregation and of aggregate restructuring (often described by a single morphological descriptor such as the fractal dimension) that are not capable of covering the exhaustive physics of inclusion interactions and are, in most cases, inadequate for oxide particles in a non-wetting liquid metal.

In the framework of project DAMAS (Design of Alloy Metals for low-mAss Structures), which is dedicated to the light-weighting of metallic materials, particular attention has been paid to the modeling of the behavior of inclusions. Specifically, an original multi-scale approach has been carried out. At the process scale, the evolution of the PSD is simulated by coupling a CFD calculation with a PBM integrating all mechanisms, i.e., flotation, aggregation, sedimentation, and capture at the top layer. This model takes into account the actual geometry of the full-scale industrial ladle and important operating conditions such as the gas flow rate. This 3D numerical tool is used to analyze the effect of successive different gas flow rates on the resulting population of non-metallic inclusions. At the inclusion scale, the simulations of fully resolved inclusions in hydrodynamic conditions represent those found in the ladle, which have been specifically developed. Such simulations are required in order to capture the local aggregation kinetics that impact the macroscopic behavior of inclusion.

2. Methods and Computational Models

2.1. Mesoscale Modeling of Inclusion Behavior

Local dynamics at the inclusion scale are investigated through numerical simulations in which all scales of interest are resolved. The direct numerical simulation (DNS) of the flow uses a lattice Boltzmann method (LBM) [13,14] that is coupled with the dynamics of the solids using an immersed boundary method (IBM) [15], which is detailed by Saxena et al. [16]. Solids are discretized with marker points on their surfaces at which flow properties are interpolated from the surrounding LBM nodes. The momentum exchange between the solid and liquid phases is calculated for each marker point and is distributed back to the LBM nodes as a source term so that the flow undergoes stress from the action of solids. Similarly, the total force and torque acting on each solid object are obtained by adding all the contributions from all the marker points. The motion of the solids is then obtained by solving Newton's equation and integrating it over time.

Such simulations of the flow and solid dynamics are fully deterministic. However, the information required in larger scale simulations is of statistical nature. This is why multiple simulations are required for all studies on inclusions and aggregate behavior.

2.1.1. Aggregation Kinetics

Due to the strong non-wettability of non-metallic inclusions, colliding solids are assumed to form aggregates bonded by gas bridges. Therefore, in the present study, aggregation kinetics are not differentiated from collision kinetics. It is well established [17] that collision kinetics are driven by shear from which a collision frequency can be derived. This entails the number of events in a given time that could potentially lead to a collision between two solids. Hydrodynamic interactions between the solids will affect the outcome of such collision events and are usually represented by collision efficiency [18,19]. In the case of turbulence induced collision, kernels obtained by Saffman and Turner [20] yield very similar results to that derived by Smoluchowski with a shear value derived from the energy dissipation. It is, therefore, a common assumption that, as long as particles remain small when compared to the Kolmogorov turbulence length scale, their collision efficiency can be estimated from their interactions in a plane shear flow.

In this study, only spherical particles have been considered since non-metallic inclusions such as calcium aluminate found in the ladle simulated later are not completely solid but are rather highly viscous and have globular shapes. As illustrated in Figure 2, collision cross sections have been determined from simulations of particle pairs starting from various relative positions along the y z axes. Particles are initially far enough away from each other (along the x axis) and do not see the flow disturbance induced by each other. Then, the shear flow drives them closer and, depending on their initial relative positions, particles collide or avoid each other. The collision cross section is the set of all the initial positions in the upstream plane of one particle relatively to the other that lead to collision. It has been determined with an iterative process by identifying sets of relative positions leading to collision and avoidance. This procedure is illustrated in Figure 2 in which collision outcomes are represented for various upstream relative positions of two particles in which empty circles mark avoidance outcomes while black dots mark actual collisions. The upstream plane is iteratively subdivided in order to estimate the cross-section, which is represented by the gray surface.

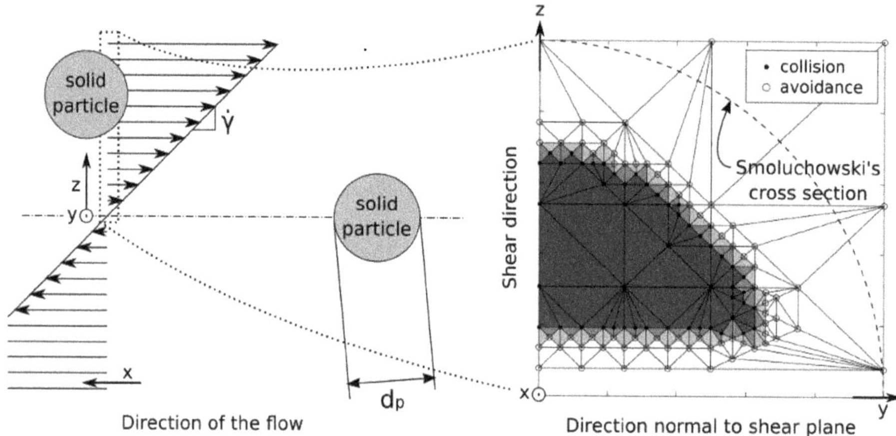

Figure 2. Simulation setup and example cross section derivation from 363 simulations of two identical spherical particles at $Re_{\gamma} = 0.213$.

The same procedure has been followed for several Reynolds numbers where the Reynolds number is defined by the equation below:

$$Re_{\dot{\gamma}} = \frac{\dot{\gamma} d_p^2}{\upsilon} \tag{1}$$

Cross-sections (surface $\mathbb{S}_{Col}$) have been calculated (see Figure 3) for different Reynolds numbers and diameter ratios between interacting particles. Collision efficiencies are derived from these cross-sections by comparing the particle rate going through them with the Smoluchowski's collision kernel.

$$\eta = \frac{\iint\limits_{S_{Col}} \dot{\gamma} z dy dz}{\iint\limits_{S_{Smo}} \dot{\gamma} z dy dz} \tag{2}$$

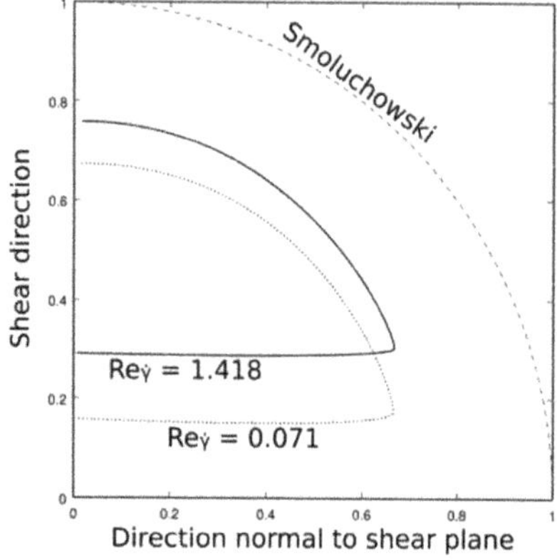

Figure 3. Evolution of the cross section with the shear-based Reynolds number for two spheres of the same size.

2.1.2. Aggregate Restructuring

A discrete element method (DEM) to account for non-hydrodynamic interactions between solids such as contact forces has also been used beside the LBM and IBM when studying deformable aggregate behaviors. Although this adds a new level of modeling in particle interactions, simulations remain deterministic and have to be repeated in order to extract statistical data. Therefore, a set of 10 aggregates with the same fractal dimension has been algorithmically created, exposed to various shear flows, and to various cohesive force intensities. The aggregate creation procedure [16] starts with a single particle and iteratively adds another one by picking the one location, which yields the fractal dimension closest to the target among a set of random possible locations in contact with other particles in the aggregate. Cohesive forces have been modeled by a normal force superposing a Van der Waals and born potentials [16] and by a tangential force in the form of a capped resistance to bending [21]. Aggregate morphological evolutions for different ratios of cohesive-to-shear forces have been investigated via numerical simulations.

2.2. *Macroscale Modeling of the Steel Ladle*

The approach adopted for reactor modeling is divided into two parts. The bubble plumes play an important role in metal bath mixing while, owing to their small weight fraction (<0.01%), the inclusions do not affect the flow. First, the two-phase turbulent flow of gas bubbles and liquid metal is simulated for the full 3D geometry of the ladle and a strong coupling is achieved between the liquid metal and the bubbles [7].

In a second step, the population balance equation is solved in each control volume by applying the cell-average technique proposed by Kumar [22], which is a variant of the fixed pivot method of Kumar and Ramkrishna [23]. This leads to a significant reduction in numerical diffusion. The solution to the system of resulting equations is based on the Fluent CFD V17.1 software in which a large number of user defined functions (UDF) has been incorporated.

2.2.1. Hydrodynamic Simulation

The modeling of liquid-gas flows applied to ladle treatment of molten steel is treated using a Euler-Euler method. Following this approach, the dispersed phase (gas) is governed by a set of transport equations (continuity and momentum) similar to the equations applied for the continuous phase (liquid). Because of the strong turbulent nature of the fluid flow, a multiphase k-ε model has been selected in order to assess the effective viscosity in both phases. In this model, called the "two-fluid model," the two sets of transport equations are coupled through the interaction force ($F_g = -F_l$) between the molten steel and the bubble swarm where F_g is expressed as the sum of four contributions F_{gD}, F_{gL}, F_{gVM}, and F_{kTD}, which are, respectively, drag, lift, virtual mass, and turbulent dispersion. A detailed description and expression of these contributions are reported in Reference [7].

2.2.2. Population Balance Modeling (PBM)

The population balance equation (PBE) is a generalized transport equation of the number density of a particle population. It accounts for various ways in which a population with a specific set of properties (here only the size of inclusions is considered) may appear or disappear [23]. The PBM treats both convective processes as well as the interaction at the individual particle scale within the particle population setting. The PBE may be solved using different numerical methods such as the Method of Moments (MOM) and the Class Method (CM). We have chosen the Class Method based on the discretization of the particle size distribution into M classes where N_i is the number of inclusions of class "i" per unit volume of liquid metal or particles with their diameter comprised between d_{pi} and d_{pi+1}. The discretized PBE is given in Equation (3) where the left-hand side represents the macroscopic transport of inclusions and the right-hand side models mesoscopic phenomena such as bubble-inclusion (flotation Z_{bi}) and inclusion-inclusion (aggregation $B_i - D_i$) interactions.

$$\frac{\partial \alpha_l N_i}{\partial t} + \text{div}(\alpha_l \mathbf{u}_l N_i) = \alpha_l (B_i - D_i) - \alpha_l Z_{bi} - S_i \tag{3}$$

The transient solution to this equation is obtained by separating the transport and collision operators [24]. In the first step, the transport equation of the quantity N_i is solved using a Finite Volume Method.

$$\frac{\partial \alpha_l N_i}{\partial t} + \text{div}(\alpha_l \mathbf{u}_l N_i) = 0 \tag{4}$$

In the second step, the population balance or Equation (5) is solved in each control volume applying the cell-average technique [22].

$$\frac{\partial \alpha_l N_i}{\partial t} = \alpha_l (B_i - D_i) - \alpha_l Z_{bi} - S_i \tag{5}$$

The flotation kernel Z_{bi} has been an issue of important studies in the literature and the reader will find details of the physical phenomena and applied model in References [25,26]. The aggregation term $(B_i - D_i)$ is readily obtained through the use of the well-known Smoluchowski in Equation [17] where the expression of the collision frequency between two inclusion classes i and j is given by the equation below.

$$Z_{ij} = \eta_{ij} \beta_{ij} N_i N_j \tag{6}$$

where β_{ij} is the Smoluchowski collision kernel and η_{ij} is the collision efficiency. The expression used for $(\eta_{ij}\beta_{ij})$ stems from the modeling work performed at the mesoscopic scale of the inclusion (see Section 3.1 and Equation (8)).

Concerning the separation of particles induced by gravity, following the decomposition of particle velocity into local fluid velocity and Stokes' velocity, the source term S_i for the transport equation is shown below.

$$S_i = \text{div}(\alpha_l \mathbf{u}_s N_i) \tag{7}$$

where u_s is the vertical Stokes' velocity in the case of small inclusions from whom $Re_p < 1$.

Particular attention has been paid to inclusion entrapment at the liquid metal/slag interface. It is modeled following the approach based on a turbulent deposition law developed initially by Wood [27] for aerosols and adapted to hydrosols by Xayasenh [28]. The entrapment of inclusions at the ladle walls is not considered yet but has been found to be negligible [29].

3. Results and Discussion

3.1. Calculation of the Shear Rate Collision Kernel

Aggregation kinetics has been quantified through a collision efficiency (η), which is defined in Equation (2). Some simulation results for colliding pairs of particles are presented in Table 1. In typical ladle conditions, for turbulence dissipation rates lying between 10^{-3} and 10 m^2/s^3, the shear rate ranges between 30 and 3000 s^{-1}, which yields Reynolds numbers between 10^{-2} and 1 for an inclusion around 10 µm in size.

The other parameter is the ratio between the diameters of the two particles. Particle density is kept close to the fluid density so that the Stokes number remains low. Particle inertia is negligible in these simulations. Moreover, particles undergo no other action than hydrodynamics (there are no distance forces between particles).

Table 1. Collision efficiencies (η) as a function of Reynolds number and the size ratio of the two inclusions.

Collision Efficiency		Re_{fl}			
		0.028	**0.071**	**0.213**	**0.319**
$d_{p,1}/d_{p,2}$	1	$0.299^{+0.070}_{-0.062}$	$0.293^{+0.024}_{-0.034}$	$0.296^{+0.021}_{-0.020}$	$0.309^{+0.076}_{-0.071}$
	2				$0.197^{+0.043}_{-0.037}$
	3				$0.094^{+0.027}_{-0.026}$

From Table 1, a conclusion can be drawn that there is little variation in the collision efficiency with the Reynolds number. Although cross-sections vary, which is illustrated in Figure 3, calculation of the efficiency results in similar values. Even at a Reynolds number much lower than 1, this efficiency remains quite significant ($\approx$0.3) while the theoretical behavior in Stokes flow is particles avoiding each other at their initial relative positions, which leads to zero efficiency. However, the theoretical behavior at a high Reynolds number should be a collision efficiency of 1 since an inviscid fluid would induce no hydrodynamic interactions between particles. Therefore, they would follow their trajectories similar to the Smoluchowski's model. Although counter-intuitive, this observation is consistent with collision efficiency observations from experimental studies in turbulent conditions [30].

This observation can be explained by flow recirculation patterns that demonstrate inertia in the liquid phase, which cannot be completely neglected when simulating particle interactions even at Reynolds numbers considered very low such as Re = 0.028. This is significant information since many numerical investigations in such conditions [19,31] use Stokesian Dynamics, which is unable to account for such effects. They, therefore, cannot be used for calculating inclusion aggregation kinetics.

While the collision efficiency is not sensitive to the shear rate, the particle size ratio has a significant impact for the conditions tested in this paper. This is primarily due to the low Stokes number of the considered particles since small particles are strongly deflected by the flow perturbation induced by larger ones, which favors avoidance when the size ratio increases.

Based on the above analysis and for the range of shear rates (i.e., turbulence dissipation rates) prevailing in steel ladle conditions, collision efficiency has been modeled with a very coarse relation that does not depend on Reynolds number but only on size ratio.

$$\eta_{ij} = \max(0.4 - 0.1\,\frac{d_{p,\max}}{d_{p,\min}}, 0) \tag{8}$$

Although rather simple, this relation captures the two significant characteristics that have been extracted from simulations at a mesoscopic scale. Collision efficiency hardly depends on the shear rate and particle size that are both captured in the shear Reynolds number. It stays at about 0.3 for all flow conditions that inclusions experience in a steel ladle. However, it strongly depends on particle size ratio and it can be estimated by a linear fit for low size ratios. When size ratios increase, this coarse fitting law is not expected to apply, but since collision efficiency is very low, such size ratios do not play an important role in the overall aggregation process and do not require a finer description.

3.2. *Restructuring of Aggregates in a Pure Shear Flow*

Ten aggregates made of 50 fully resolved primary spheres were simulated in plane shear flow conditions using the aforementioned LBM+IBM+DEM software. All aggregates had the same fractal dimension of 2.3 at the beginning of the simulations. Radii of gyration of the aggregates have been recorded over time and converted to fractal dimensions using an empirical correlation [16] since fractal dimension is a more common descriptor in aggregate studies. The shear Reynolds number for primary particles (Equation (1)) was kept small in all simulations (less than 10^{-2}). Aggregate restructuring over time has been studied for different aggregate strengths relative to a hydrodynamic action by varying the magnitude of the normal and tangential forces between particles in contact. The scaling of the hydrodynamic force is achieved using the equation below.

$$F_{drag} = 3\pi\mu\dot{\gamma}d_p^2 \quad (9)$$

Then the dimensionless normal and tangential forces are expressed as the ratio of the corresponding force to the drag force expression from Equation (9).

As can be expected, dimensionless normal forces below 1 always led to aggregate breakage. However, as soon as the normal force is strong enough to hold the aggregate together, it seems to have very little impact on the aggregate morphology. The tangential force ratio, however, has a significant impact on the final structure of aggregates. Simulations with resolved hydrodynamic interactions show that aggregates of fractal dimension 2.3 tend to contract under shear conditions, which means their fractal dimension increases. Generally, the lower their contact tangential forces are, the greater their contraction is. However, the very different results between simulations with and without resolved hydrodynamics show that this aggregate behavior depends in a non-straightforward way on tangential forces. Higher tangential forces lead to stronger inter-particle bonds. However, stronger bonds increase the overall aggregate resistance to deformation and, therefore, favor bond breakage, which leads to a potential creation of new bonds.

Figure 4 shows the evolution of the fractal dimension of all 10 aggregates with and without account for hydrodynamic interactions. In this example, a dimensionless normal force is 1 and a dimensionless tangential force is 0.1. When hydrodynamic interactions are not resolved, the LBM solver is disabled and the fluid action is modeled by a simple Stokes drag force and torque in the DEM using isolated particles in a plane shear. As expected, Figure 4 clearly shows that the hydrodynamics play a significant role in aggregate restructuring and they impact both kinetics and asymptotic trends.

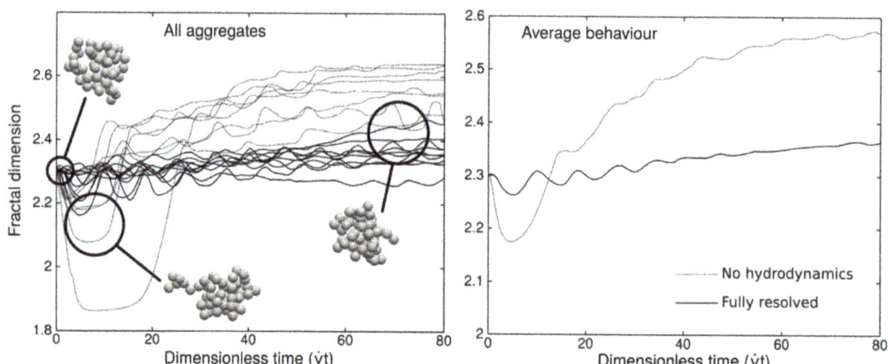

Figure 4. Evolution over time of the fractal dimensions of 50 particle aggregates. Dimensionless normal force is 1 and dimensionless tangential force is 0.1.

All simulation results especially when accounting for hydrodynamic interactions show that the impact of restructuring on aggregate morphology is not strong enough to play a significant role on inclusion size evolution at the process scale. This size evolution is dominated by aggregation phenomena. However, the complex relation between tangential contact forces and bond breakage suggests that including varying aggregate breakage rates in the population balance may be a future significant step towards better modeling of inclusion behavior in metallurgical processes.

3.3. Macroscale Simulation

The simulation at the macroscopic scale of a metallurgical reactor is applied to an industrial 60 t steel ladle. Argon is injected through two porous plugs located at the base of the ladle. Two successive hydrodynamic conditions have been simulated including a first step with a relatively high stirring intensity corresponding to 70 NL/min of argon for each plug over 10 min (which is 5 times the mixing time), which is followed by a second stir for 10 min at a smaller argon flow rate (less than 10 NL/min for each plug). The model assumes that liquid temperature is uniform and equal to 1600 °C.

The argon plume regions and the liquid steel velocity in a plane passing through the porous plugs of the ladle are shown in Figure 5a in the case of relatively weak aeration. One can observe the shapes of the two bubble plumes rising from the two porous plugs (the iso-surface of gas volume fraction equal to 1% is drawn). The turbulence, which mainly prevails in these regions, is characterized by a turbulent shear rate in the range of 150 to 1000 s^{-1}. The liquid metal flow is associated with two recirculation zones with one in each half of the plane of symmetry. The liquid steel velocities are consistent with the magnitude found in the literature for equivalent industrial configurations [32].

3.3.1. Hydrodynamic Validation

The hydrodynamic part of the numerical model was validated earlier using experimental data available in the literature. Physical modeling embodying aqueous as well as system with wood metal has been carried out to investigate the gas-stirred ladle system [33] used by many authors [34,35]. Comparisons between calculated and measured profiles of gas-retention α_g and liquid velocity u_l show a satisfactory prediction of the dispersion of the bubble swarm. Details of this first validation are available in Reference [7].

Another way to globally characterize the turbulent hydrodynamic behavior of the steel bath consists of introducing dissolved copper as a tracer and measuring the required time to full homogenization. Such a measurement of a characteristic mixing time can be easily compared to model predictions at an industrial scale, which involves completing the validation of the hydrodynamic simulations [36].

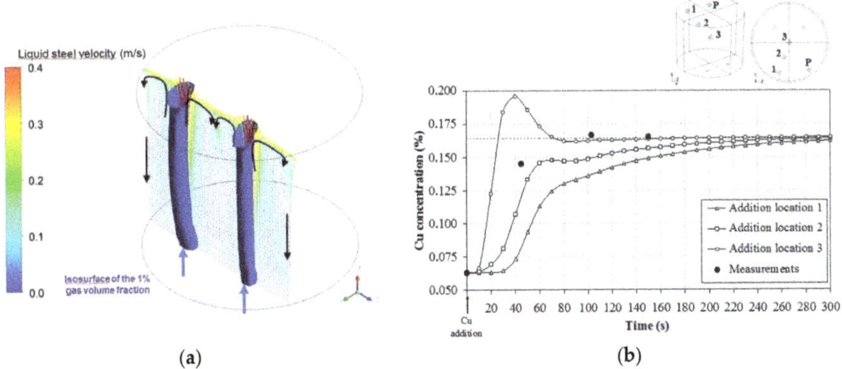

Figure 5. Hydrodynamic simulation of a 60 t steel ladle. (**a**) Predicted velocity of the liquid steel along with the argon plumes (isosurface of the 1% gas volume fraction) in a vertical plane passing through the porous plugs for a weak aeration. (**b**) Calculated Cu concentration profiles at the P sampling point compared with plant measurements for three platelet addition locations in the 60 t gas-stirred ladle.

Copper platelets were added into the 60 t steel bath and samples were taken at time intervals of about 50 s. Since the location of platelet immersion and dissolution is not accurate, several simulations were conducted with varying points of copper platelets (1–3) and a fixed sampling location (P).

Figure 5b shows a relatively strong dependence of mixing time with initial tracer locations especially when the addition is simulated on the symmetry axis of the ladle. This, consequently, results in low values for mixing time compared to other addition locations. Although numerical simulations seem to underestimate the bath mixing time, the order of magnitude (2 min) agrees with the measurements. This mixing time cannot be regarded as negligible in comparison with the characteristic time of aggregation. Therefore, non-homogeneity of the inclusion population in an industrial ladle requires the development of 3D modeling.

3.3.2. Evolution of the Inclusion Population in the Successive High and Low Stirring Steps

According to the literature and more specifically the review proposed by Zhang and Thomas [1], a log-normal distribution has been used as the initial PSD with a total mass content equal to 0.176 kg/m^3, which corresponds to 7.9 ppm of total oxygen content considering a population of calcium aluminate inclusions. This distribution matches relatively well with those analyzed in an industrial ladle for high steel quality since one kilogram of metal contains 5×10^6 inclusions smaller than 20 μm but only 6×10^4 larger than 20 μm. The initial distribution is plotted in red in Figure 6a where the inclusion population is rendered discretely into 20 different-sized classes.

After 10 min of high stirring, the obtained PSD (in blue on Figure 6a) noticeably evolves with the birth of inclusions larger than 20 μm. The turbulent stirring occurred mainly in the bubble swarms leads to aggregation. Therefore, the shift of PSD toward larger size classes and the regular increase of the Sauter diameter d_{p32} (see Figure 6b). The reduction of the total weight of inclusions in the whole ladle observed in Figure 6b clearly highlights the effect of the removal mechanisms, which are flotation, deposition, and settling. One-third of inclusion content is then removed during the 20 min of treatment. The combination of aggregation with removal mechanisms leads to a noticeable decrease in the number of inclusions smaller than 20 μm, which is shown in Figure 6a.

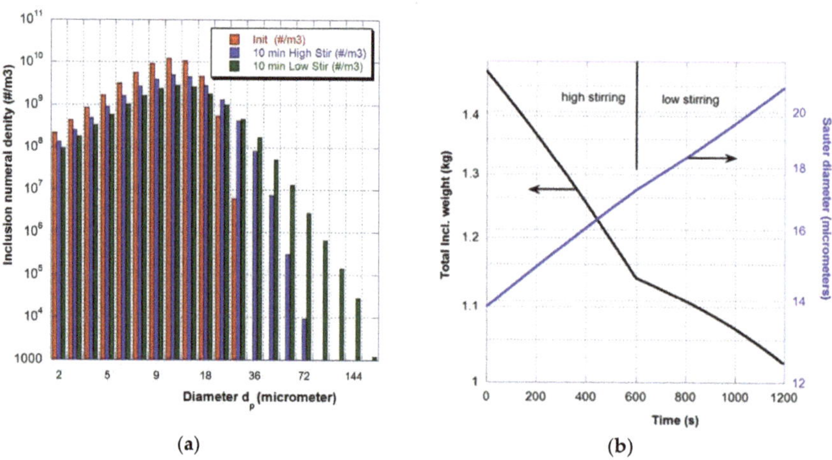

Figure 6. Average values of the non-metallic inclusion population in the ladle (**a**) PSD at three different times (initial in red, after 10 min of high stir in blue, and after 10 min low stir in green) (**b**) Time evolution of the total weight of non-metallic inclusions and of Sauter diameter.

The industrial practice usually consists in vigorous stirring to encourage particle collisions into larger ones followed by a "final stir" that slowly recirculates the steel to facilitate removal of large aggregates and to prevent any slag-free region of the steel surface and air re-oxidation. Although the developed turbulence is weaker, the 10 minutes of low stirring has a similar effect on the PSD (in green on Figure 6a) i.e., larger particles appear due to the aggregation of smaller particles leading to a fairly similar increase of the Sauter diameter d_{p32}.

Numerical simulations allow us to compare the relative importance of the different removal mechanisms on the inclusion population. The frequencies of the aggregation, flotation, settling, and deposition events have been reported in Figure 7a (at the beginning of the high stirring process i.e., at time t = 30 s). As an example, the aggregation frequency f_A is calculated by the equation below.

$$f_A = \int_{V_{ladle}} (B_i - D_i)\mathrm{d}V \tag{10}$$

In the specific case of aggregation, signs (−) or (+) (sign of f_A value) indicate that the numerical density of a given class decreases or increases.

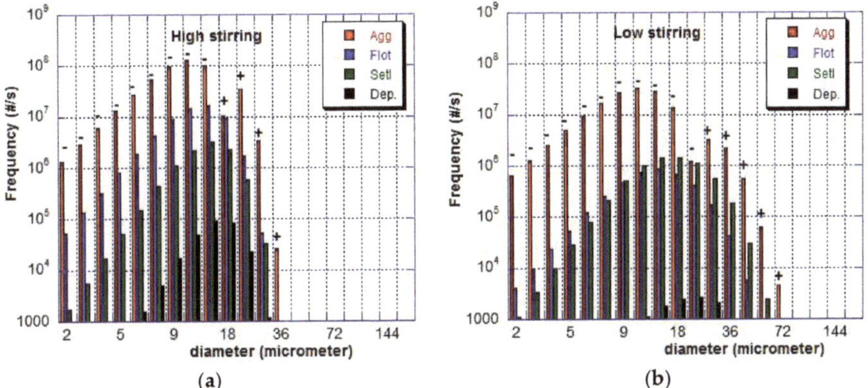

Figure 7. Frequency of mechanisms as a function of inclusion size (**a**) after 30 s of high stirring (**b**) after 30 s of low stirring. For aggregation, (−): reduced, (+): produced.

Figure 7 clearly highlights that the major role is played by aggregation (although aggregation efficiency is much lower than 1, see Equation (8)) since its frequency is two orders of magnitude higher than the frequencies of flotation and settling for most size classes. This is the main reason why the PSD continuously evolves over time. The number density of inclusions smaller than 20 μm continuously decreases over time because of the aggregation mechanism and because of the removal phenomena (flotation and settling). Another obvious conclusion is the negligible role of the turbulent deposition of inclusions on the top surface of the ladle.

4. Conclusions

The split scale approach, which is the coupling process scale simulations with local inclusion dynamics through statistical models, has proven effective for capturing inclusion cleanliness trends during metallurgical processing. Inclusion cleanliness is mainly ruled by aggregation, which is observed in Section 3.3.2. Aggregation has an event frequency roughly two order of magnitudes larger than the one of the removal mechanisms. All inclusion removal mechanisms are strongly dependent on the size of aggregates and, therefore, conditioned by the aggregation kinetics. Therefore, the need for accurate aggregation models relate operating conditions to process efficiency. In this study, this need has been addressed by mesoscopic scale simulations from which quantitative information on aggregation kinetics has been extracted and captured in a statistical model. These simulations have shown very little variations with the particle size and shear rate for a broad range of operating conditions, which yielded constant aggregation efficiency for all of the same-sized particle pairs.

However, this efficiency has been observed to drop drastically when increasing the particle size ratio. At the process scale, this means that the ruling aggregation kinetics are the ones between particles of the same size class.

Author Contributions: Conceptualization, J.-P.B. and J.-S.K.R.; Methodology, J.-P.B. and J.-S.K.R.; Software, J.-P.B., J.-S.K.R., A.S., M.G. and M.J.; Validation, J.-P.B., J.-S.K.R, A.S., M.G. and M.J.; Formal Analysis, J.-P.B., A.S., M.G. and M.J.; Investigation, J.-P.B., A.S., M.G. and M.J.; Resources, J.-P.B., A.J. and S.S.; Data Curation, J.-S.K.R.; Writing-Original Draft Preparation, J.-P.B. and J.-S.K.R.; Writing-Review & Editing, A.J. and S.S.; Visualization, J.-P.B., A.S., M.G. and M.J.; Supervision, J.-P.B., J.-S.K.R. and S.S.; Project Administration, J.-P.B., A.J. and S.S.; Funding Acquisition, J.-P.B., A.J. and S.S.

Funding: This research was funded by the French State through the program "Investment in the future" operated by the National Research Agency (ANR) and referenced by ANR-11-LABX-0008-01 (LabEx DAMAS). Moreover, the authors are grateful for two other program supports of the National Research Agency (ANR) (Grant No. ANR06 MATPRO 0005 and No. ANR-15-CE08-0040-01) and for the NSERC Industrial Research Chair in Pipeline Transport Processes (S. Sanders, Grant No. IRCPJ 363871-12).

Acknowledgments: Aubert & Duval is acknowledged for the supply of measurements of mixing time.

Conflicts of Interest: The authors declare no conflict of interest.

Nomenclature

Greek letters	
α	volume fraction
$\dot{\gamma}$	shear rate (s^{-1})
η	collision efficiency
μ	dynamic viscosity of the liquid metal (Pa·s)
v	kinematic viscosity of the liquid metal ($m^2 \cdot s^{-1}$)
π	Archimedes' constant
Latin letters	
B	population balance birth kernel (# $m^{-3} \cdot s^{-1}$)
D	population balance death kernel (# $m^{-3} \cdot s^{-1}$)
d	diameter (m)
F	force (N)
N	number of inclusions per unit volume of liquid metal (# m^{-3})
f	frequency (s^{-1})
S	gravity separation term in PBE (# $m^{-3} \cdot s^{-1}$)
$\mathbb{S}$	surface or surface area (m^2)
u	fluid velocity ($m \cdot s^{-1}$)
V	volume (m^3)
y	abscissa in the direction normal to the shear plane (m)
z	abscissa in the direction along the velocity gradient (m)
Z	collision frequency ($m^{-3} \cdot s^{-1}$)
Subscripts	
A	relative to the aggregation mechanism
b	relative to gas bubbles and flotation
Col	relative to the actual collision cross section
drag	relative to the action of the liquid phase on the solids
i, j	discrete size classes in the population balance equation
g, l	gas or liquid phase
ladle	relative to the steel ladle
p	particle property
Smo	relative to Smoluchowski's collision cross section
Compound	
$Re_{\dot{\gamma}}$	shear-based particle Reynolds number defined in Equation (1)

References

1. Zhang, L.; Thomas, B.G. State of the art in evaluation and control of steel cleanliness. *ISIJ Int.* **2003**, *43*, 271–291. [CrossRef]
2. Kato, T.; Shimasaki, S.; Taniguchi, S. Water model experiments for hydrodynamic forces acting on inclusion particles in molten metal under turbulent condition. In *Jim Evans Honorary Symposium: Proceedings of the Symposium Sponsored by the Light Metals Division of The Minerals, Metals and Materials Society (TMS)*; Wiley: Hoboken, NJ, USA, 2010.
3. Xie, Y.; Orsten, S.; Oeters, F. Behaviour of bubbles at gas blowing into liquid woods metal. *ISIJ Int.* **1992**, *32*, 66–75. [CrossRef]
4. Mendez, C.G.; Nigro, N.; Cardona, A. Drag and non-drag force influences in numerical simulations of metallurgical ladles. *J. Mater. Process. Technol.* **2005**, *160*, 296–305. [CrossRef]
5. Liu, H.; Qi, Z.; Xu, M. Numerical Simulation of Fluid Flow and Interfacial Behavior in Three-phase Argon-Stirred Ladles with One Plug and Dual Plugs. *Steel Res. Int.* **2011**, *82*, 440–458. [CrossRef]
6. Madan, M.; Satish, D.; Mazumdar, D. Modeling of mixing in ladies fitted with dual plugs. *ISIJ Int.* **2005**, *45*, 677–685. [CrossRef]
7. De Felice, V.; Daoud, I.L.A.; Dussoubs, B.; Jardy, A.; Bellot, J.P. Numerical modelling of inclusion behavior in a gas-stirred ladle. *ISIJ Int.* **2012**, *52*, 1274–1281. [CrossRef]
8. Li, L.; Liu, Z.; Li, B.; Matssra, H.; Tsukihashi, F. Water Model and CFD-PBM Coupled Model of Gas-Liquid-Slag Three-Phase Flow in Ladle Metallurgy. *ISIJ Int.* **2015**, *55*, 1337–1346. [CrossRef]
9. Bellot, J.P.; De Felice, V.; Dussoubs, B.; Jardy, A.; Hans, S. Coupling of a CFD and a PBE calculations to simulate the behavior of an inclusion population in a gas-stirred ladle. *Met. Trans. B* **2014**, *45*, 13–21. [CrossRef]
10. Kwon, Y.-J.; Zhang, J.; Lee, H.-G. A CFD-based Nucleation-growth-removal Model for Inclusion Behavior in a Gas-agitated Ladle during Molten Steel Deoxidation. *ISIJ Int.* **2008**, *48*, 891–900. [CrossRef]
11. Claudotte, L.; Rimbert, N.; Gardin, P.; Simonnet, M.; Lehmann, J.; Oesterle, B. A Multi-QMOM Framework to Describe Multi-Component Agglomerates in Liquid Steel. *AIChE J.* **2010**, *56*, 2347–2355. [CrossRef]
12. Lei, H.; Nakajima, K.; He, J.-C. Mathematical Model for Nucleation, Ostwald Ripening and Growth of Inclusion in Molten Steel. *ISIJ Int.* **2010**, *50*, 1735–1745. [CrossRef]
13. Eggels, J.G.M.; Somers, J.A. Numerical simulation of free convective flow using the lattice-Boltzmann scheme. *Int. J. Heat Fluid Flow* **1995**, *16*, 357–364. [CrossRef]
14. Sungkorn, R.; Derksen, J.J. Simulations of dilute sedimenting suspensions at finite-particle Reynolds numbers. *Phys. Fluids* **2012**, *24*, 123303. [CrossRef]
15. Niu, X.D.; Shu, C.; Chew, Y.T.; Peng, Y. A momentum exchange-based immersed boundary-lattice Boltzmann method for simulating incompressible viscous flows. *Phys. Lett. A* **2006**, *354*, 173–182. [CrossRef]
16. Saxena, A.; Kroll-Rabotin, J.-S.; Sanders, R.S. A numerical approach to model aggregate restructuring in shear flow using DEM in lattice-Boltzmann simulations. In Proceedings of the 12th International Conference on Computational Fluid Dynamics in the Oil & Gas, Metallurgical and Process Industries, Trondheim, Norway, 30 May–1 June 2017; pp. 761–772.
17. Smoluchowski, M. Drei Vorträge über Diffusion, Brownsche Molekularbewegung und Koagulation von Kolloidteilchen. *Z. Phys.* **1916**, *17*, 557–585.
18. Frungieri, G.; Vanni, M. Dynamics of a Shear-Induced Aggregation Process by a Combined Monte Carlo-Stokesian Dynamics approach. In Proceedings of the 9th International Conference on Multiphase Flow, Florence, Italy, 22–27 May 2016.
19. Frungieri, G.; Vanni, M. Shear-induced aggregation of colloidal particles: A comparison between two different approaches to the modelling of colloidal interactions. *Can. J. Chem. Eng.* **2017**, *95*, 1768–1780. [CrossRef]
20. Saffman, P.G.; Turner, J.S. On the collision of drops in turbulent clouds. *J. Fluid Mech.* **1956**, *1*, 16–30. [CrossRef]
21. Becker, V.; Briesen, H. Tangential-force model for interactions between bonded colloidal particles. *Phys. Rev. E* **2008**, *78*, 061404. [CrossRef] [PubMed]

22. Peglow, M.; Kumar, J.; Warnecke, G.; Heinrich, S.; Morl, L. A new technique to determine rate constants for growth and agglomeration with size- and time-dependent nuclei formation. *Chem. Eng. Sci.* **2006**, *61*, 282–292. [CrossRef]
23. Ramkrishna, D. *Population Balances: Theory and Applications to Particulate Systems in Engineering*; Academic Press: Cambridge, MA, USA, 2000; ISBN 9780125769709.
24. Toro, E.F. *Riemann Solver and Numerical Methods for Fluid Dynamics: A Pratical Introduction*; Springer: Berlin, Germany, 1999; ISBN 10: 3540659668/ISBN 13: 9783540659662.
25. Kostoglou, M.; Karapantsios, T.D.; Matis, K.A. Modeling local flotation frequency in a turbulent flow field. *Adv. Colloid Interface Sci.* **2006**, *122*, 79–91. [CrossRef] [PubMed]
26. Mirgaux, O.; Ablitzer, D.; Waz, E.; Bellot, J.P. Mathematical modelling and computer simulation of molten aluminium purification by flotation in stirred reactor. *Metall. Mater. Trans. B* **2009**, *40*, 363–375. [CrossRef]
27. Wood, N.B. A simple method for the calculation of turbulent deposition to smooth and rough surfaces. *J. Aerosol Sci.* **1981**, *12*, 275–290. [CrossRef]
28. Dupuy, M.; Xayasenh, A.; Waz, E.; Duval, H. Analysis of non-Brownian particle deposition from turbulent liquid-flow. *AIChE J.* **2015**, *62*, 891–904. [CrossRef]
29. Lou, W.; Zhu, M. Numerical Simulations of Inclusion Behavior in Gas-Stirred Ladles. *Metall. Mater. Trans. B* **2013**, *44*, 762–782. [CrossRef]
30. Higashitani, K.Y.; Kiyoyuki, K.; Matsuno, Y.; Hosokawe, G. Turbulent coagulation of particles dispersed in a viscous fluid. *J. Chem. Eng. Jpn.* **1983**, *16*, 299–304. [CrossRef]
31. Ren, Z.; Harshe, Y.M.; Lattuada, M. Influence of the Potential Well on the Breakage Rate of Colloidal Aggregates in Simple Shear and Uniaxial Extensional Flows. *Langmuir* **2015**, *31*, 5712–5721. [CrossRef] [PubMed]
32. Zhang, L. Transport Phenomena and CFD Application during Process Metallurgy. In *Advanced Processing of Metals and Materials (Sohn International Symposium), Volume 4, New, Improved and Existing Technologies: Non-ferrous Materials Extraction and Processing*; Wiley: Hoboken, NJ, USA, 2006; Volume 2.
33. Xie, Y.K.; Oesters, F. Experimental studies on the flow velocity of molten metals in a ladle model at centric gas blowing. *Steel Res.* **1992**, *63*, 93–104. [CrossRef]
34. Ling, H.; Li, F.; Zhang, L.; Conejo, A.N. Investigation on the Effect of Nozzle Number on the Recirculation Rate and Mixing Time in the RH Process Using VOF plus DPM Model. *Metall. Mater. Trans. B* **2016**, *47*, 1950–1961. [CrossRef]
35. Aoki, J.; Zhang, L.; Thomas, B.G. Modeling of Inclusion Removal in Ladle Refining. In Proceedings of the 3rd International Congress on Science & Technology of Steelmaking, Charlotte, NC, USA, 9–11 May 2005; AIST: Warrendale, PA, USA; pp. 319–332. [CrossRef]
36. Warzecha, M.; Jowsa, J.; Warzecha, P.; Pfeifer, H. Numerical and Experimental Investigations of Steel Mixing Time in a 130-t Ladle. *Steel Res. Int.* **2008**, *79*, 852–860. [CrossRef]

Article

Detailed Modeling of the Direct Reduction of Iron Ore in a Shaft Furnace

Hamzeh Hamadeh [1,2], Olivier Mirgaux [1,2] and Fabrice Patisson [1,2,*]

1 Institut Jean Lamour, CNRS, Université de Lorraine, 54011 Nancy, France; hamzeh.hamadeh@univ-lorraine.fr (H.H.); olivier.mirgaux@univ-lorraine.fr (O.M.)
2 Laboratory of Excellence on Design of Alloy Metals for Low-Mass Structures (DAMAS), Université de Lorraine, 57073 Metz, France
* Correspondence: fabrice.patisson@univ-lorraine.fr; Tel.: +33-372-742-670

Received: 6 September 2018; Accepted: 26 September 2018; Published: 1 October 2018

Abstract: This paper addresses the modeling of the iron ore direct reduction process, a process likely to reduce CO_2 emissions from the steel industry. The shaft furnace is divided into three sections (reduction, transition, and cooling), and the model is two-dimensional (cylindrical geometry for the upper sections and conical geometry for the lower one), to correctly describe the lateral gas feed and cooling gas outlet. This model relies on a detailed description of the main physical–chemical and thermal phenomena, using a multi-scale approach. The moving bed is assumed to be comprised of pellets of grains and crystallites. We also take into account eight heterogeneous and two homogeneous chemical reactions. The local mass, energy, and momentum balances are numerically solved, using the finite volume method. This model was successfully validated by simulating the shaft furnaces of two direct reduction plants of different capacities. The calculated results reveal the detailed interior behavior of the shaft furnace operation. Eight different zones can be distinguished, according to their predominant thermal and reaction characteristics. An important finding is the presence of a central zone of lesser temperature and conversion.

Keywords: ironmaking; direct reduction; iron ore; DRI; shaft furnace; mathematical model; heterogeneous kinetics; heat and mass transfer

---

1. Introduction

The direct reduction (DR) of iron ore, usually followed by electric arc steelmaking, is an alternative route to the standard, blast furnace, basic oxygen route for making steel. Annual DR iron production (86 Mt in 2017) remains small, compared to the production of 1180 Mt of blast furnace pig iron [1]. However, an attractive feature of DR, compared to blast furnace reduction, is its considerably lower CO_2 emissions, which are 40 to 60% lower for the DR-electric arc furnace route, compared to the blast furnace, basic oxygen route [2]. Among DR processes, shaft furnaces represent over 82% of the world's DR iron production, with the two main processes being MIDREX (65%), as shown in Figure 1, and HYL-ENERGIRON (17%) [3].

In a DR shaft furnace, a charge of pelletized or lump iron ore is loaded into the top of the furnace and is allowed to descend, by gravity, through a reducing gas. The reducing gas, comprised of hydrogen and carbon monoxide (syngas), and obtained by the catalytic reforming of natural gas, flows upwards, through the ore bed. Reduction of the iron oxides occurs in the upper section of the furnace, at temperatures up to 950 °C. A transition section is found below the reduction section; this section is of sufficient length to separate the reduction section from the cooling section, allowing an independent control of both sections. The solid product, called direct reduced iron (DRI) or reduced sponge iron, is cooled in the lower part of the furnace, down to approximately 50 °C, prior to being discharged.

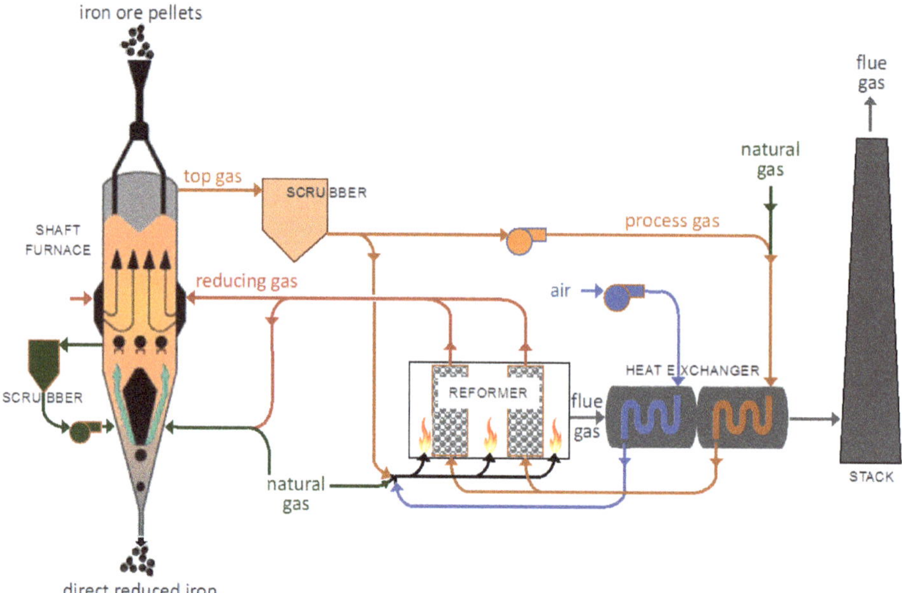

Figure 1. MIDREX process flowsheet.

The modeling of a shaft furnace, simulating the reduction of iron ore by syngas, is a powerful tool for defining optimal operating conditions. Use of such a model can lead to the maximization of conversion or the minimization of energy consumption, among other effects capable of reducing carbon dioxide emissions. As such, numerous iron ore shaft furnace models have been proposed in the literature. Initial studies addressed the reduction of a single pellet by H_2, CO, or H_2-CO mixtures [4–9]. Subsequent studies developed models that simulated the reduction zone of the shaft furnace in one dimension [10,11]. With the aim of correctly describing the lateral gas feed, some studies have introduced two-dimensional models [12–14]; however, these models did not consider the presence of methane, which is responsible for important reactions in the process. More recently, several authors introduced other reactions [15] and accounted for the cooling zone [16,17]. Some even developed plant models [18]; however, these works were limited to one-dimensional models.

In this work, we developed further the model of Ranzani Da Costa and Wagner, built to simulate the reduction section of DR shafts, operated with pure hydrogen [13,14,19]. We extended this model to consider CO-H_2-CH_4 reducing gas, and accounted for transition and cooling sections. The present model, named REDUCTOR, is 2-dimensional in the steady-state regime. The model includes a sophisticated, pellet sub-model. We consider eight heterogeneous and two homogeneous chemical reactions. These features represent a more advanced and detailed model, compared to previous studies. Moreover, the results were validated against two sets of plant data.

The present model, REDUCTOR, differs from the other model we recently reported [18] on the following points. REDUCTOR is a computational fluid dynamics (CFD)-type, two-dimensional model, which describes the shaft furnace alone. The other model is of the systemic type, is one-dimensional, and aims to simulate the whole DR plant. The shaft furnace description included in the plant model, though based on similar equations, was intentionally made simpler and faster to run, on process simulation software. Thus, REDUCTOR is more detailed and more precise, but, of course, requires longer computation times.

2. Mathematical Model

2.1. Principle

The reduction of hematite ore to iron occurs via two intermediate oxides, namely, magnetite and wüstite (considered as $Fe_{0.95}O$ [19]), and by two gaseous reactants, namely, H_2 and CO. The following six reduction reactions were therefore considered:

$$3Fe_2O_{3(s)} + H_{2(g)} \rightarrow 2Fe_3O_{4(s)} + H_2O_{(g)} \quad (1)$$

$$Fe_3O_{4(s)} + \frac{16}{19}H_{2(g)} \rightarrow \frac{60}{19}Fe_{0.95}O_{(s)} + \frac{16}{19}H_2O_{(g)} \quad (2)$$

$$Fe_{0.95}O_{(s)} + H_{2(g)} \rightarrow 0.95Fe_{(s)} + H_2O_{(g)} \quad (3)$$

$$3Fe_2O_{3(s)} + CO_{(g)} \rightarrow 2Fe_3O_{4(s)} + CO_{2(g)} \quad (4)$$

$$Fe_3O_{4(s)} + \frac{16}{19}CO_{(g)} \rightarrow \frac{60}{19}Fe_{0.95}O_{(s)} + \frac{16}{19}CO_{2(g)} \quad (5)$$

$$Fe_{0.95}O_{(s)} + CO_{(g)} \rightarrow 0.95Fe_{(s)} + CO_{2(g)} \quad (6)$$

Methane reforming and water gas shift reactions also occur in the gas phase, based on the composition of reduction gas and temperature, through the following reactions:

$$CH_{4(g)} + H_2O_{(g)} \rightleftharpoons CO_{(g)} + 3H_{2(g)} \quad (7)$$

$$CO_{(g)} + H_2O_{(g)} \rightleftharpoons CO_{2(g)} + H_{2(g)} \quad (8)$$

We also considered two other side reactions that could occur in the reactor, especially where an iron layer has formed:

- Methane decomposition reaction

$$CH_{4(g)} \rightleftharpoons C_{(s)} + 2H_{2(g)} \quad (9)$$

- Carbon monoxide disproportionation (inverse Boudouard reaction)

$$2CO_{(g)} \rightleftharpoons C_{(s)} + CO_{2(g)} \quad (10)$$

The model itself is two-dimensional, axisymmetrical, and steady-state. It is based on the numerical solution of local mass, energy, and momentum balances, using the finite volume method. The geometry in the reduction and transition sections is cylindrical, while conical in the cooling section. This corresponds to the geometry of the shaft furnaces and is necessary to describe correctly the lateral gas feed and outlet cooling gas, as shown in Figure 2. The reactor modeled is a shaft furnace of the MIDREX type.

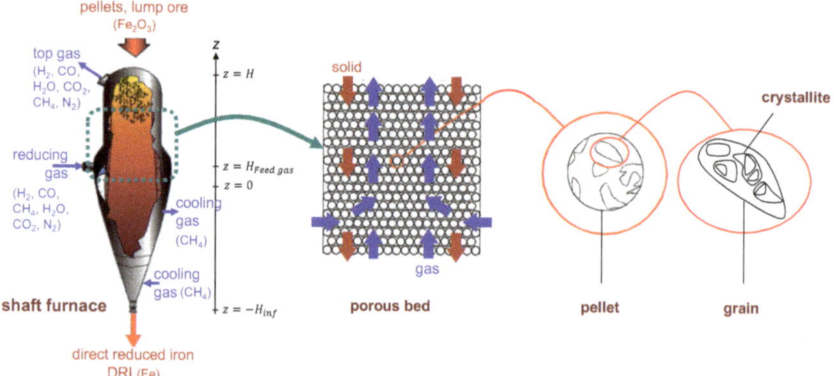

Figure 2. Schematic representation of the REDUCTOR model, from the reactor scale to the crystallite scale (see Appendix A for notations).

The solid load is fed from the top of the reactor (z = H) to form a moving bed of solid particles composed of spherical iron ore pellets that descend by gravity. The pellet diameter (d_p) is assumed to be unique and unchanging during the reduction reaction, and the initial pellet composition is known. The gas phase is composed of six species: H_2, CO, H_2O, CO_2, N_2, and CH_4. The reducing gas is injected from the sidewall, at a height of z = $H_{Feed,gas}$ which then moves upward, against the solid flow, before finally exiting the furnace at the top. The temperature and composition of this reducing gas are known. A secondary feed gas—the cooling gas—which is introduced from the bottom of the furnace (z = $-H_{inf}$), is also considered. This cooling gas exits the furnace, through the wall in the upper part of the conical section. The temperatures of the solid and gas are different and vary, according to their position (r, z) within the furnace. The solid temperature is assumed to be uniform in the interior of the pellets. Thus, this model is based on a faithful description of the physical-chemical and thermal phenomena, from the reactor scale to the crystallite scale, as shown in Figure 2. In the pellet sub-model, the pellet is assumed to be initially comprised of dense grains; these grains later fragment into smaller crystallites at the wüstite stage, in agreement with microscopic observations [19]. Thus, from the reactor to the crystallites, we have a 4-scale model.

2.2. Equations

2.2.1. Gas Phase

The descending solid pellets, through which the ascending gas flows, can be considered a porous medium, consisting of quasi-stationary solid spheres (the gas velocity is much greater than that of the solid). The Ergun equation (see Appendix A for nomenclature), combined with the continuity equation, thus gives

$$\frac{1}{r}\frac{\partial}{\partial r}\left(\frac{rc_t}{K}\frac{\partial p}{\partial r}\right)+\frac{\partial}{\partial z}\left(\frac{c_t}{K}\frac{\partial p}{\partial z}\right)=S_{mol,tot}=2v_7+v_9-v_{10} \tag{11}$$

where the terms are in units, mol $m^{-3}s^{-1}$; K is the permeability coefficient, calculated as

$$K=\frac{150(1-\varepsilon_b)^2}{\varepsilon_b^3 d_p^2}u_g+\frac{1.75(1-\varepsilon)}{\varepsilon_b^3 d_p}\rho_g u_g \tag{12}$$

and the source term $S_{mol,tot}$ corresponds to the net gas production by the non-equimolar reactions. Equation (11) is used to calculate the pressure field, and the gas velocity vector is calculated, using Equation (13):

$$\boldsymbol{u}_g = -\frac{1}{K}\nabla P \tag{13}$$

The mass balance for a gaseous species, i, considering axial and radial dispersion, in addition to convection, is written:

$$\frac{1}{r}\frac{\partial(rc_t x_i u_{g,r})}{\partial r} + \frac{\partial(c_t x_i u_{g,z})}{\partial z} = \frac{1}{r}\frac{\partial}{\partial r}\left(rc_t D_r \frac{\partial x_i}{\partial r}\right) + \frac{\partial}{\partial z}\left(c_t D_z \frac{\partial x_i}{\partial z}\right) + S_i \tag{14}$$

with the source term, S_i, given in Table 1.

Table 1. Source terms for the gas species mass balances.

Species i	S_i mol m^{-3} s^{-1}
H_2	$S_{H_2} = -v_1 - \frac{16}{19}v_2 - v_3 + 3v_7 + v_8 + 2v_9$
CO	$S_{CO} = -v_4 - \frac{16}{19}v_5 - v_6 + v_7 - v_8 - 2v_{10}$
H_2O	$S_{H_2O} = v_1 + \frac{16}{19}v_2 + v_3 - v_7 - v_8$
CO_2	$S_{CO_2} = v_4 + \frac{16}{19}v_5 + v_6 + v_8 + v_{10}$
CH_4	$S_{CH_4} = -v_7 - v_9$

The heat balance for the gas phase—considering convection and conduction—as well as the heat exchanged with the solid and heat brought by the gases evolving from the solid, gives:

$$\rho_g c_{pg}\left(u_{gr}\frac{\partial T_g}{\partial r} + u_{gz}\frac{\partial T_g}{\partial z}\right) = \frac{1}{r}\frac{\partial}{\partial r}\left(r\lambda_g \frac{\partial T_g}{\partial r}\right) + \frac{\partial}{\partial z}\left(\lambda_g \frac{\partial T_g}{\partial z}\right) + a_b h(T_s - T_g) + \sum_i S_i \int_{T_g}^{T_s} c_{pi} dT \tag{15}$$

2.2.2. Solid Phase

Regarding the grain flow, in the upper cylindrical section, it is considered that pellets descend vertically. In contrast, in the lower section of the conical shape, a radial component of the solid velocity must be introduced. A bibliographical study of granular flows led us to use the model of Mullins [20,21], in which the radial velocity is calculated as proportional to the radial gradient of the axial velocity:

$$u_{s,r} = -B\frac{\partial u_{s,z}}{\partial r} \tag{16}$$

where B is taken, as proposed in Reference [20]:

$$B = 2d_p \tag{17}$$

The mass balance for a gaseous species j gives:

$$-\frac{\partial(\rho_b u_{s,z} w_j)}{\partial z} + \frac{1}{r}\frac{\partial(r\rho_b u_{s,r} w_j)}{\partial r} = S_j \tag{18}$$

with the source term, S_j, given in Table 2.

Table 2. Source terms for the solid species mass balances.

Species j	S_j kg m^{-3} s^{-1}
Fe_2O_3	$-3M_{Fe_2O_3}(v_1 + v_4)$
Fe_3O_4	$M_{Fe_3O_4}(2v_1 - v_2 + 2v_4 - v_5)$
$Fe_{0.95}O$	$M_{Fe_{0.95}O}\left(\frac{60}{19}v_2 - v_3 + \frac{60}{19}v_5 - v_6\right)$
Fe	$0.95M_{Fe}(v_3 + v_6)$
C	$M_c(v_9 + v_{10})$

The heat balance for the solid phase takes into account axial and radial convection, conduction, and heat exchange with the gas phase. The heat of the reactions is attributable to the solid phase, considering that all the reactions occur either inside the pellets (heterogeneous reactions) or at their surfaces (homogeneous reactions, catalyzed by the solid); thus:

$$-\rho_b u_{s,z} c_{ps} \frac{\partial T_s}{\partial z} + \rho_b u_{s,r} c_{ps} \frac{\partial T_s}{\partial r} = \frac{1}{r}\frac{\partial}{\partial r}\left(r\lambda_{eff,r}\frac{\partial T_s}{\partial r}\right) + \frac{\partial}{\partial z}\left(\lambda_{eff,z}\frac{\partial T_s}{\partial z}\right) + a_b h(T_g - T_s) + \sum_{n=1}^{10}(-v_n \Delta_r H_n) \quad (19)$$

2.3. Transport Coefficients

The various transport coefficients, D_r, D_z, $\lambda_{eff,r}$, and $\lambda_{eff,z}$, as well as other parameters, like specific heats, are calculated as functions of temperature and composition. Details regarding the relationships were given in [22].

2.4. Reaction Rates

2.4.1. Iron Oxide Reduction

Unlike most of the previous approaches published which are based on the shrinking core model (with one or three fronts, separating the oxides in the pellet), we developed a specific, pellet sub-model. The sub-model was built, according to our experimental findings, to simulate the reduction of a single pellet by H_2-CO. The reaction rate was used as a function of the local reduction conditions (temperature and gas composition), inside the reactor. We used the law of additive reaction times [23], which considers the different resistances (chemical reaction, diffusion, external transfer) involved in series. Therefore, the time required to attain a certain conversion is approximately the sum of the characteristic times: τ_i [14,23]. This sub-model was initially developed for simulating reduction by H_2 only, as detailed previously [14]; we extended this model for reduction by CO. The characteristic times and the reaction rates are listed in Appendix B.

2.4.2. Methane Reforming and Water Gas Shift Reactions

Methane reforming and water gas shift reactions are known to be catalyzed by iron or iron oxides [24,25]; thus, their rates are functions of the composition of the reduction gas, temperature, and mass of the catalyst. The methane reforming rate equation considering the forward and reverse reactions is given by Equation (20):

$$v_7 = k_7(1 - \varepsilon_b)(1 - \varepsilon_{interg})\left(P_{CH_4}P_{H_2O} - \frac{P_{CO}P_{H_2}^3}{K_{eq,7}}\right) \quad (20)$$

The expression of the reaction rate constant, k_7, is given in Table 3. Because the reforming of CH_4 was hardly observed on the iron oxide catalysts, as reported in the literature [25], it was considered that such reforming only occurs with iron as a catalyst. We assumed that sufficient iron was formed on the outside of the pellet, when the reduction degree exceeded 50%.

Table 3. Kinetic constants.

Reactions		Reaction Rate Constants k_i	References
7		$k_7 = 392\exp\left(\frac{6770}{RT}\right)\left(\text{mol cm}^{-3}\text{s}^{-1}\right)$	[25]
8	Fe	$k_8 = 93.3\exp\left(-\frac{7320}{RT}\right)\left(\text{mol cm}^{-3}\text{s}^{-1}\right)$	[25]
	$Fe_{0.95}O$	$k_8 = 1.83\times10^{-5}\exp\left(\frac{7.84}{RT}\right)\left(\text{mol cm}^{-3}\text{s}^{-1}\right)$	[25]
	Fe_3O_4	$k_8' = 2.683372\times10^{5}\exp\left(-\frac{112000}{RT}\right)\left(\text{mol kg}_{\text{cat}}^{-1}\text{s}^{-1}\right)$	[24]
	Fe_2O_3	$k_8' = 4.56\times10^{3}\exp\left(-\frac{88000}{RT}\right)\left(\text{mol kg}_{\text{cat}}^{-1}\text{s}^{-1}\right)$	[24]
9		$k_9 = 16250\exp\left(-\frac{55000}{RT}\right)\left(\text{mol m}^{-3}\text{s}^{-1}\right)$	[16,26]
10		$k_{10} = 1.8\exp\left(-\frac{27200}{RT}\right)\left(\text{mol m}^{-3}\text{s}^{-1}\right)$ $k_{10}' = 2.2\exp\left(-\frac{8800}{RT}\right)\left(\text{mol m}^{-3}\text{s}^{-1}\right)$	[16,26]

Similarly, the rate expression for the water gas shift reaction is given by Equation (21):

$$v_8 = k_8(1-\varepsilon_b)(1-\varepsilon_{interg})\left(P_{CO}P_{H_2O} - \frac{P_{CO_2}P_{H_2}}{K_{eq,8}}\right) \tag{21}$$

when occurring on Fe or $Fe_{0.95}O$, and by Equation (22)

$$v_8 = k_8'\rho_c(1-\varepsilon_b)(1-\varepsilon_{interg})\left(P_{CO}P_{H_2O} - \frac{P_{CO_2}P_{H_2}}{K_{eq,8}}\right) \tag{22}$$

when occurring on Fe_2O_3 or Fe_3O_4. Here, besides iron, various iron oxides also catalyze the reaction. The corresponding expressions for k_8 and k_8' are given in Table 3, according the literature [24,25].

2.4.3. Carbonization Reactions

In the DR furnace, carbon can be formed, either from methane decomposition (Equation (9)) or from CO disproportionation (Equation (10)). Both reactions are reversible, and the reverse reactions are functions of the carbon activity. The carbon activity was calculated from Chipman's relationship [27]:

$$\log a_c = \frac{2300}{T} - 0.92 + \left(\frac{3860}{T}\right)C + \log\left(\frac{C}{1-C}\right) \tag{23}$$

where C is the ratio of atomic C to atomic Fe. For sake of simplicity, we did not distinguish between C and Fe_3C in the solid, with both being considered as C.

The rate equation of the methane decomposition reaction is given by Equation (24)

$$v_9 = \frac{k_9}{P_{H_2}^{0.5}}(1-\varepsilon_b)(1-\varepsilon_{interg})\left(P_{CH_4} - \frac{P_{H_2}^2 a_c}{K_{eq,9}}\right) \tag{24}$$

The expression of the reaction rate constant, k_9, included in Equation (24) was determined, as per the literature [16,26], as listed in Table 3.

The rate equation of the carbon monoxide disproportionation reaction is given by Equation (25)

$$v_{10} = \left(k_{10}P_{H_2}^{0.5} + k'_{10}\right)(1-\varepsilon_b)(1-\varepsilon_{interg})\left(P_{CO}^2 - \frac{P_{CO_2}a_c}{K_{eq,10}}\right) \quad (25)$$

and the values of the reaction rate constants, k_{10} and k'_{10}, are also provided in Table 3, from the same references.

2.5. Boundary Conditions

The balance equations need a set of associated boundary conditions to be solved. First, the temperature and composition of the solids and gases are assumed to be known (the operating conditions) at their respective inlets (at the top for solids, and at the bottom and sides for gases). In addition, because of axisymmetry and tight walls, one has:

$$\begin{array}{l} -\text{ Symmetry axis : zero fluxes} \frac{\partial T_s}{\partial r} = \frac{\partial T_g}{\partial r} = \frac{\partial x_i}{\partial r} = 0 \\ -\text{ Side wall (except gas inlet) : } \frac{\partial T_s}{\partial r} = \frac{\partial T_g}{\partial r} = \frac{\partial x_i}{\partial r} = 0 \end{array} \quad (26)$$

For the gas flow, a known pressure condition is also required at the exits. The top pressure was known but not the pressure of the cooling gas outlet, as shown in Figure 3, at Point 4. The latter was estimated to obtain approximately 90% of the inlet cooling gas expelled from this outlet and, approximately, 10% flowing upwards.

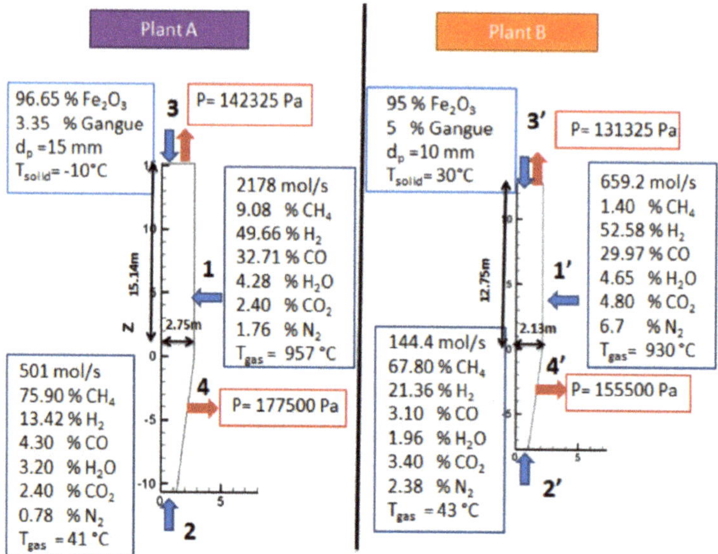

Figure 3. Operating conditions of plants A and B.

Figure 3 shows the values of the known boundary conditions for the two simulations conducted, corresponding to two different plants. Plant A is a North American, MIDREX plant, currently in operation, the main operating data of which were provided to us. Plant B was the first MIDREX plant operated in the USA, for which published data are available [10]. The production capacity of plant A is 4.5 times greater that of plant B.

2.6. Meshing and Numerical Solution

The system of partial derivative equations was discretized and solved, according to the finite volume method [28]. Meshing of the cylindrical reduction and transition sections is orthogonal, with cells made finer next to the top, as shown in Figure 4, on the left. For the conical section, a non-orthogonal grid was used, as shown in Figure 4, on the right. To easily connect the two sections, the number of radial cells was kept the same. The numerical code was written in the language of FORTRAN 1995.

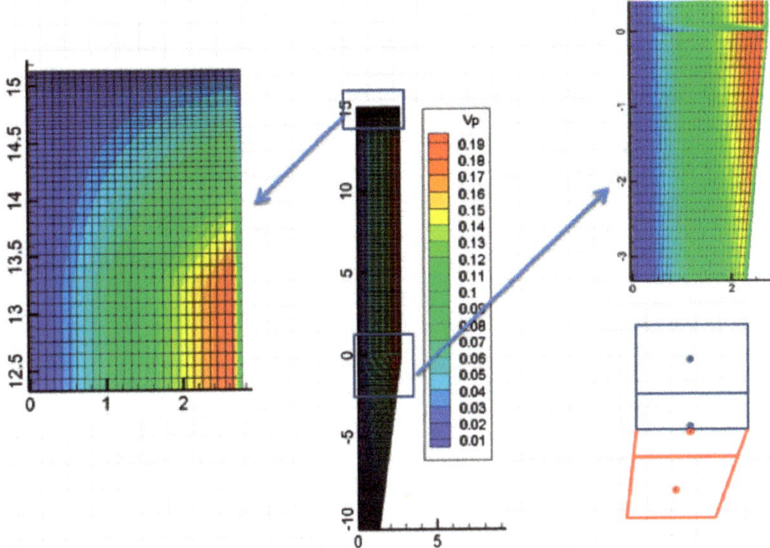

Figure 4. Meshing and volumes (m^3) of the cells.

3. Results and Discussion

In this section, the results of the Plant A simulation are first presented and discussed, then a comparison between the calculated and measured data for both plants is given. Results for the values of the different variables, throughout the reactor, are given in separate figures; however, all of these variables must be considered simultaneously for interpretation purposes.

3.1. Pressure Field, Velocity of Gas and Temperature Field

Figure 5a shows the pressure and velocity fields inside the bed, throughout the reactor. The color scale refers to the pressure, and the lines refer to the streamlines. The large arrows indicate the locations of the various gas and solid inlets and outlets. These locations are the same (and not repeated everywhere) in the following figures. The pressure decreases almost linearly, from bottom to top. The reducing gas, injected at the sidewall (z = 5.32 m), enters radially and then flows essentially vertically, except in the transition zone. The cooling gas first flows upwards, and then, most of it leaves the furnace radially, at the cooling gas outlet, except for a fraction that rises in the reduction section.

Figure 5b,c show the temperature distribution of the gas and solid phases in the reactor. First, it was found that the gas and solid temperatures were very close to each other. This similarity resulted from the high gas-to-solid heat transfer, as was described in a previous study [14]. Downwards from the solid inlet, the solid temperature rapidly increased to reach the gas temperature. Second, the temperatures were not axially or radially uniform, throughout the reactor. The hottest zone was near the reducing gas inlet, with gas introduced at 957 °C. Above this inlet, the temperature decreased, because of methane reforming (as shown later, in Figure 7), an endothermic reaction. Third, the cooling gas not only cooled the solid in the bottom section but also influenced the temperature field in the reduction section, with the gas rising from the cooling zone to the central part of the reduction zone. This maintained a lower temperature alongside the center of the shaft.

From these results, radial gradients of temperature were revealed to influence, together with the gas composition profiles, the reduction of the solids and the metallization degree achieved.

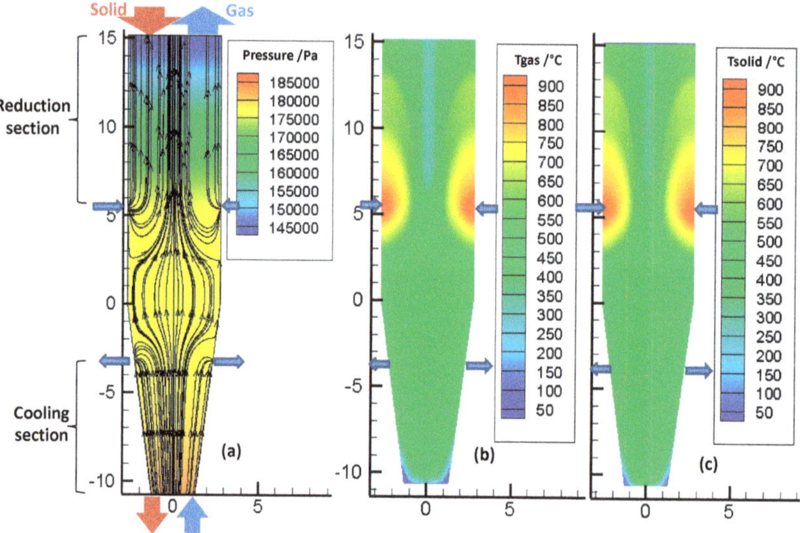

Figure 5. (**a**) Pressure field and velocity streamlines of gas flow inside the bed, (**b**) temperature distribution of the gas phase, and (**c**) temperature distribution of the solid phase.

3.2. Solid Mass Fractions

Figure 6 plots the evolution of solid mass fractions, throughout the reactor. Figure 6a shows that the hematite was fully converted to magnetite very rapidly in the upper part of the reactor. Subsequently, magnetite was reduced to wüstite, as shown in Figure 6b. Afterwards, wüstite slowly began to reduce to iron, as seen in Figure 6c,d. In the external two-thirds of the reduction section, above the reducing gas inlet—a zone where the gas was rich in H_2 and CO and the temperature high—the conversion to iron was completed, in approximately 7 m. In the central part of the reactor, where the temperature was lower and the gas, lower in H_2 and CO, the conversion was not completed and some wüstite remained in the cooling zone. Though the average metallization degree was approximately 94%, metallization was not uniform, with most pellets being completely reduced, whereas others were not.

Figure 6e shows the carbon mass fraction, throughout the reactor. We observed that the carbon was in the same location as Fe, in accordance with the catalytic effect of iron on carbon formation.

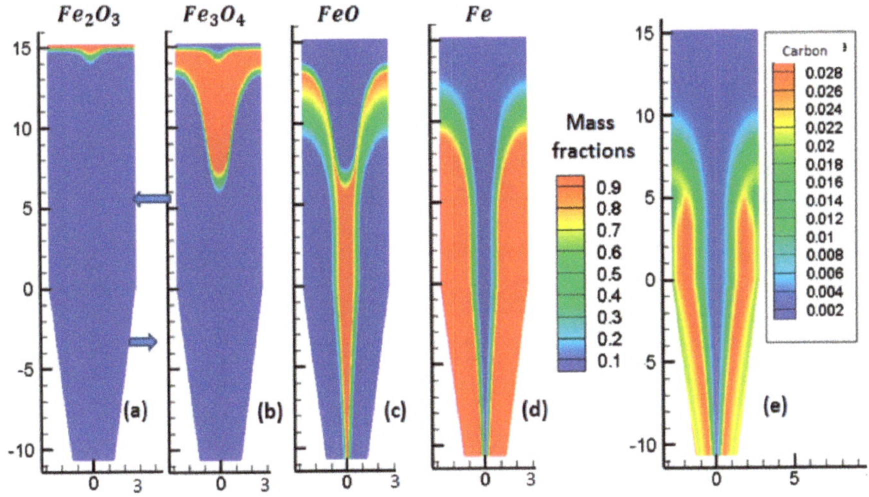

Figure 6. Mass fractions of the solid phases.

3.3. Gas Mole Fractions

As showcased by Figure 7, the situation here is more complex, due to the numerous reactions occurring. The main features of these reactions are as follows. Near the reducing gas inlet, the reforming of methane occurred, which increased the H_2 and CO contents. Above the gas inlet, the H_2 and CO contents decreased, while H_2O and CO_2 were formed, as a result of the reduction reactions. In the central zone, with less reduction, lower amounts of H_2O and CO_2 were formed, and part of the cooling gas, rich in CH_4, was present.

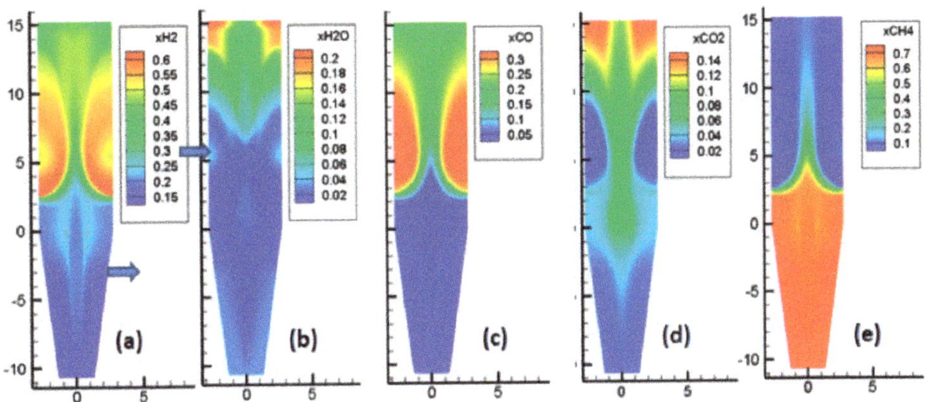

Figure 7. Mole fractions of the gas phase.

3.4. Overall Picture

Figure 8 is a summary diagram, based on the above results. The shaft furnace was divided into eight zones and distinguished according to the main chemical and thermal processes occurring. On the left part of the diagram are indicated the molar percentages of H_2 and CO, involved in each reaction, and the molar percentage of methane, reformed by H_2O or CO_2, or decomposed to carbon and H_2. This diagram is an illustration of how modeling work can help one to understand the detailed behavior of a reactor. Clearly, these results could not be obtained from other means.

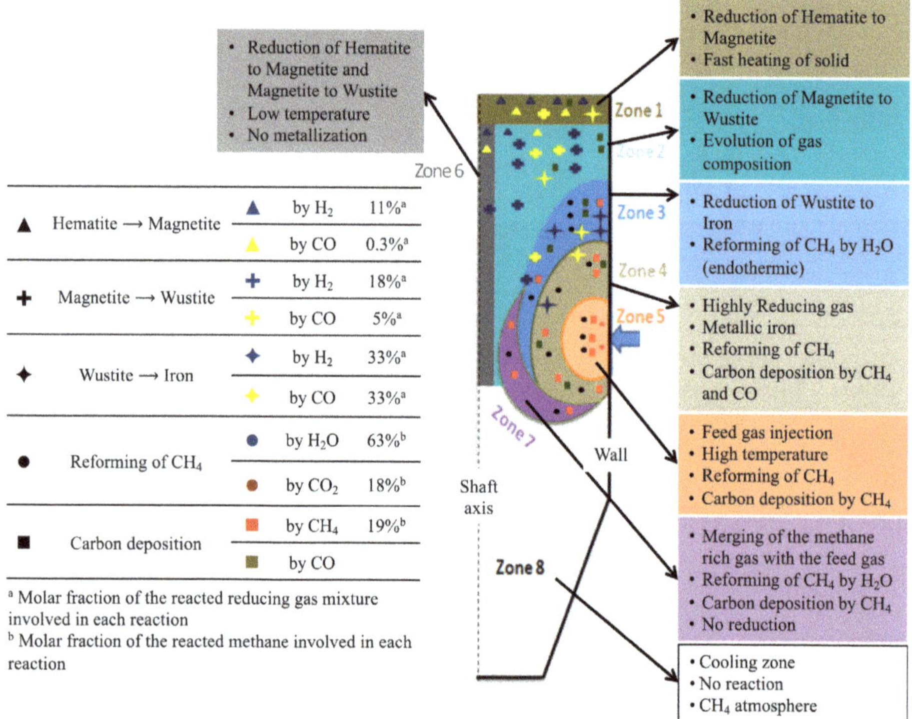

Figure 8. Diagram, illustrating the different zones in the shaft furnace.

3.5. Validation

Unfortunately, neither interior measurements of solid or gas temperatures, nor compositions, were available for comparison with the calculations. However, from some published data regarding Plant B, and from plant data measurements from Plant A, an overall validation of the model was possible.

Table 4 provides a comparison of the simulation results, with the available plant data. It can be seen that the model reproduced the outlet temperatures and compositions quite satisfactorily. From this strong agreement, obtained by simulations of two plants of differing capacities, the model can be considered validated.

Table 4. Comparison of the Plant A and Plant B outlet data with the REDUCTOR model calculations.

		Plant A		Plant B		Unit
		Plant Data	**Reductor Results**	**Plant Data**	**Reductor Results**	
Outlet solid	**Composition (%)**					
	Fe_2O_3	0	0	0	0	wt %
	Fe_3O_4	0	0	0	0	wt %
	FeO	7.47	7.1	n.a.	4.3	wt %
	Fe	85.72	85.9	n.a.	87.77	wt %
	C	2	2.2	2	0.91	wt %
	Gangue	4.71	4.8	6.3	7.02	wt %
	Production	119.2	119.8	26.4	27.33	t/h
	Metallization	93.8	94	93	95.3	%
Outlet gas	**Flow rate**	193	200	n.a.	54	kNm^3/h
	Composition (%)					
	H_2	40.28	40.41	37	37.72	vol %
	CO	19.58	19.89	18.9	20.87	vol %
	H_2O	19.03	19.52	21.2	20.61	vol %
	CO_2	17.09	14.69	14.3	13.13	vol %
	CH_4	2.95	3.91	}8.6	7.67	vol %
	N_2	1.02	1.55			vol %
	Temperature	285	284	n.a.	285	°C

n.a.: not available.

4. Conclusions

This article presented the modeling and simulation of an iron ore, direct reduction shaft furnace. We developed a new mathematical model, with the aim of introducing a more-detailed description of the chemical processes, compared to previous studies. The model presented is two-dimensional, describes three sections in the shaft, and accounts for eight heterogeneous and two homogeneous reactions. The model was validated against plant data from two MIDREX plants of notably different capacities. From the analysis of the calculated 2D maps of temperature and composition of the gas and solid phases, it was possible to gain new insights into the interior behavior of the shaft furnace and identify different zones, according to the chemical and thermal phenomena occurring. One significant result is the presence of a central zone of the shaft of lesser temperature and conversion.

Such a model can be helpful in: Investigating the influence of various parameters and operating conditions (including the reducing gas composition), comparing different furnace configurations, and suggesting improvements [29]. These investigations will be the subject of a future paper.

Author Contributions: Conceptualization, O.M. and F.P.; Investigation, H.H., O.M. and F.P.; Methodology, H.H., O.M. and F.P.; Project Administration, F.P.; Resources, F.P.; Software, H.H. and F.P.; Supervision, O.M. and F.P.; Writing—Original Draft, H.H.; Writing—Review and Editing, F.P.

Funding: This research was supported by the French State, through two programs: 'Investment in the future': (i) one, operated by the French Environment and Energy Management Agency (ADEME), 'Valorization of CO_2 in industry', 2014–18, VALORCO, No 1382C0245; the authors thank Mrs. Nathalie Thybaud and Aïcha El Khamlichi, and the coordinator, Eric de Coninck; and (ii) one operated by the National Research Agency (ANR) and referenced by ANR-11-LABX-0008-01 (LabEx DAMAS).

Acknowledgments: The authors also express their thanks to the staff of ArcelorMittal, at Maizières-lès-Metz, France, and Contrecœur, Canada, for their encouragement, information, and discussion, especially S. Bertucci, J. Borlée, T. Quatravaux, and J. Farley.

Conflicts of Interest: The authors declare no conflict of interest.

Appendix A. Notation

Latin	
a_b	specific area of the bed (m^2/m^3)
a_c	activity of carbon
c_t	total molar concentration of the gas (mol m^{-3})
c_{pg}	molar specific heat of the gas (J mol^{-1} K^{-1})
c_{ps}	mass specific heat of the solid (J kg^{-1} K^{-1})
d_p	pellet diameter (m)
D	diffusion or dispersion (D_a, D_r) coefficient (m^2/s)
h	heat transfer coefficient (W m^{-2} K^{-1})
H	height of the cylindrical section of the shaft (m)
$H_{feed\ gas}$	height of the reducing gas inlet (m)
H_{inf}	height of the conical section of the shaft (m)
K_{eq}	equilibrium constant
K	permeability coefficient (kg m^{-3} s^{-1})
k	mass transfer coefficient, or reaction rate constant
M	molar weight (kg mol^{-3})
p	gas pressure
P_i	partial pressure of component i (bar)
r	radius (m)
R	ideal gas constant (J mol^{-1} K^{-1})
S	source term
T	temperature (K)
u	velocity (m s^{-1})
v	reaction rate (mol m^{-3} s^{-1})
w_j	mass fraction of solid j
X	degree of conversion
x_i	molar fraction of i in the gas
z	height (m)
Greek	
$\Delta_r H$	heat of reaction (J mol^{-1})
ε	porosity
τ	characteristic time (s)
λ	thermal conductivity (W m^{-1} K^{-1})
μ_g	viscosity of the gas (Pa s)
ρ_g	mass density of the gas (kg m^{-3})
ρ_b	mass density of the bed (kg m_{bed}^{-3})
$\widetilde{\rho}_j$	molar density of species j in the bed (mol m_{bed}^{-3})
Subscripts	
b	bed
c	catalyst
cryst	crystallite
chem	chemical
diff	diffusional
interg	intergranular
ini	initial
intrac	intra-crystallite
interc	inter-crystallite
∞	in the bulk gas
eff	effective (for the bed)

eq	at equilibrium
g	gas
grain	grain
p	pellet
r	radial
s	solid
z	axial

Appendix B. Characteristic Times and Reaction Rates

Table A1. Kinetic sub-model of a single pellet. Expressions of the characteristic times. *i*: reaction number (see Section 2.1), *k*: H_2 or CO.

	Hematite → Magnetite	Magnetite → Wüstite	Wüstite → Iron
External transfer	$\tau_{ext,i} = \frac{\tilde{\rho}_{Fe_2O_3,ini} d_p}{18 k_g c_t (x_{k,\infty} - x_{k,eq(i)})}$	$\tau_{ext,i} = \frac{8 \tilde{\rho}_{Fe_3O_4,ini} d_p}{57 k_g c_t (x_{k,\infty} - x_{k,eq(i)})}$	$\tau_{ext,i} = \frac{\tilde{\rho}_{Fe_{0.95}O} d_p}{6 k_g c_t (x_{k,\infty} - x_{k,eq(i)})}$
Intergranular diffusion	$\tau_{diff,interg(i)} = \frac{\tilde{\rho}_{Fe_2O_3,ini} (d_p)^2}{72 (D_{k,eff})_{interg,i} c_t (x_{k,\infty} - x_{k,eq(i)})}$	$\tau_{diff,interg(i)} = \frac{2 \tilde{\rho}_{Fe_3O_4,ini} (d_p)^2}{57 (D_{k,eff})_{interg,i} c_t (x_{k,\infty} - x_{k,eq(i)})}$	/
Intragranular diffusion	/	$\tau_{diff,intrag(i)} = \frac{2 \tilde{\rho}_{Fe_3O_4,ini} (d_{grain,ini})^2}{57 (D_{k,eff})_{intrag,i} c_t (x_{k,\infty} - x_{k,eq(i)})}$	/
Inter-crystallite diffusion	/	/	$\tau_{diff,interc(i)} = \frac{\tilde{\rho}_{e0.95}O (d_p)^2}{24 (D_{k,eff})_{interc,i} c_t (x_{k,\infty} - x_{k,eq(i)})}$
Intra-crystallite diffusion (solid phase)	/	/	$\tau_{diff,intrac,(i)} = \frac{\tilde{\rho}_{Fe_{0.95}O} d^2_{cryst,ini}}{24 D_{sol} (c_{ox,eq} - c_{ox,\infty})}$
Chemical reaction	$\tau_{chem,i} = \frac{\tilde{\rho}_{Fe_2O_3} d_{grain,ini}}{6 k_i c_t (x_{k,\infty} - x_{k,eq(i)})}$	$\tau_{chem,i} = \frac{\tilde{\rho}_{Fe_3O_4} d_{grain,ini}}{2 k_i c_t (x_{k,\infty} - x_{k,eq(i)})}$	$\tau_{chem,i} = \frac{\tilde{\rho}_{Fe_{0.95}O} d_{cryst,ini}}{2 k_i c_t (x_{k,\infty} - x_{k,eq(i)})}$

Table A2. Kinetic sub-model of a single pellet. Expressions of the reaction rates. *i*: reaction number (see Section 2.1).

Reaction *i*	Reaction Rate mol m^{-3} s^{-1}
i = 1 and 4	$v_i = \frac{1}{3} \tilde{\rho}_{Fe_2O_3,ini} \left\{ \tau_{ext,i} + 2 \tau_{diff,interg(i)} \left[(1 - X_i)^{-\frac{1}{3}} - 1 \right] + \frac{\tau_{chem,i}}{3} (1 - X_i)^{-\frac{2}{3}} \right\}^{-1}$
i = 2 and 5	$v_i = \frac{1}{3} \tilde{\rho}_{Fe_2O_3,ini} \left\{ \tau_{ext,i} + 2 \left(\tau_{diff,interg(i)} + \tau_{diff,intrag(i)} \right) \left[(1 - X_i)^{-\frac{1}{3}} - 1 \right] + \frac{\tau_{chem,i}}{3} (1 - X_i)^{-\frac{2}{3}} \right\}^{-1}$
i = 3 and 6	$v_i = \frac{1}{3} \tilde{\rho}_{Fe_2O_3,ini} \left\{ \tau_{ext,i} + 2 \left(\tau_{diff,interc(i)} + \tau_{diff,intrac,(i)} \right) \left[(1 - X_i)^{-\frac{1}{3}} - 1 \right] + \frac{\tau_{chem,i}}{3} (1 - X_i)^{-\frac{2}{3}} \right\}^{-1}$

References

1. World Steel in Figures 2018. 2018, p. 10. Available online: https://www.worldsteel.org/en/dam/jcr:f9359dff-9546-4d6b-bed0-996201185b12/World+Steel+in+Figures+2018.pdf (accessed on 2 July 2018).
2. Duarte, P.E.; Becerra, J. Reducing greenhouse gas emissions with Energiron non-selective carbon-free emissions scheme. *Stahl und Eisen* **2011**, *131*, 85.
3. 2017 World Direct Reduction Statistics. 2018. Available online: https://www.midrex.com/assets/user/news/MidrexStatsBook2017.5_.24_.18_.pdf (accessed on 2 July 2018).
4. McKewan, W.M. Reduction kinetics of hematite in hydrogen–water vapor–nitrogen mixtures. *Trans. Metall. Soc. AIME* **1962**, *224*, 2–5.
5. Spitzer, R.H.; Manning, F.S. Mixed control reaction kinetics in the gaseous reduction of hematite. *Trans. Metall. Soc. AIME* **1963**, *23*, 6726–6742.
6. Tien, R.H.; Turkdogan, E.T. Gaseous reduction of iron oxides: Part IV. Mathematical analysis of partial internal reduction–diffusion control. *Metall. Trans.* **1972**, *3*, 2039–2048. [CrossRef]

7. Kam, E.K.T.; Hughes, R. A model for the direct reduction of iron ore by mixtures of hydrogen and carbon monoxide in a moving bed. *Trans. Inst. Chem. Eng.* **1981**, *59*, 196–206.
8. Negri, E.D.; Alfano, O.M.; Chiovetta, M.G. Direct reduction of hematite in a moving bed reactor. Analysis of the water gas shift reaction effects on the reactor behavior. *Ind. Eng. Chem. Res.* **1991**, *30*, 474–482. [CrossRef]
9. Valipour, M.S.; Hashemi, M.Y.M.; Saboohi, Y. Mathematical modeling of the reaction in an iron ore pellet using a mixture of hydrogen, water vapor, carbon monoxide and carbon dioxide: An isothermal study. *Adv. Powder Technol.* **2006**, *17*, 277–295. [CrossRef]
10. Parisi, D.R.; Laborde, M.A. Modeling of counter current moving bed gas-solid reactor used in direct reduction of iron ore. *Chem. Eng. J.* **2004**, *104*, 35–43. [CrossRef]
11. Arabi, S.; Hashemipour Rafsanjani, H. Modeling and Simulation of Noncatalytic Gas-Solid Reaction in a Moving Bed Reactor. *Chem. Prod. Process Model.* **2008**, *3*, 1–29. [CrossRef]
12. Valipour, M.S.; Saboohi, Y. Numerical investigation of nonisothermal reduction of haematite using syngas: The shaft scale study. *Model. Simul. Mater. Sci. Eng.* **2007**, *15*, 487–507. [CrossRef]
13. Ranzani da Costa, A.; Wagner, D.; Patisson, F.; Ablitzer, D. Modelling a DR shaft operated with pure hydrogen using a physical-chemical and CFD approach. *Rev. Métall.* **2009**, *10*, 434–439. Available online: https://arxiv.org/pdf/0911.4949.pdf (accessed on 2 July 2018). [CrossRef]
14. Ranzani da Costa, A.; Wagner, D.; Patisson, F. Modelling a new, low CO_2 emissions, hydrogen steelmaking process. *J. Clean. Prod.* **2013**, *46*, 27–35. [CrossRef]
15. Alamsari, B.; Torii, S.; Trianto, A.; Bindar, Y. Heat and Mass Transfer in Reduction Zone of Sponge Iron Reactor. *ISRN Mech. Eng.* **2011**, 1–12. [CrossRef]
16. Alhumaizi, K.; Ajbar, A.; Soliman, M. Modelling the complex interactions between reformer and reduction furnace in a midrex-based iron plant. *Can. J. Chem. Eng.* **2012**, *90*, 1120–1141. [CrossRef]
17. Shams, A.; Moazeni, F. Modeling and Simulation of the MIDREX Shaft Furnace: Reduction, Transition and Cooling Zones. *JOM* **2015**, *67*, 2681–2689. [CrossRef]
18. Béchara, R.; Hamadeh, H.; Mirgaux, O.; Patisson, F. Optimization of the iron ore direct reduction process through multiscale process modeling. *Materials* **2018**, *11*, 1094. [CrossRef] [PubMed]
19. Ranzani da Costa, A. La Réduction du Minerai de fer par L'Hydrogène: Étude Cinétique, Phénomène de Collage et Modélisation. Ph.D. Thesis, Institut National Polytechnique de Lorraine, Nancy, France, 2011.
20. Zhang, K.F.; Ooi, J.Y. A Kinematic model for solids flow in flat-bottomed silos. *Géotechnique* **1998**, *48*, 545–553. [CrossRef]
21. Choi, J.; Kudrolli, A.; Bazant, M.Z. Velocity profile of granular flows inside silos and hoppers. *J. Phys. Condens. Matter* **2005**, *17*, 2533–2548. [CrossRef]
22. Wagner, D. Etude Expérimentale et Modélisation de la Réduction du Minerai de Fer par L'Hydrogène. Ph.D. Thesis, Institut National Polytechnique de Lorraine, Nancy, France, 2008.
23. Sohn, H.Y. The law of additive reaction times in fluid-solid reactions. *Metall. Trans.* **1978**, *9B*, 89–96. [CrossRef]
24. Smith, R.J.B.; Loganathan, M.; Shantha, M.S. A Review of the Water Gas Shift Reaction Kinetics. *Int. J. Chem. React. Eng.* **2010**, *8*, 1–32. [CrossRef]
25. Takahata, M.; Kashiwaya, Y.; Ishii, K. Kinetics of Methane Hydrate Formation Catalyzed by Iron Oxide and Carbon under Intense Stirring Conditions. *Mater. Trans.* **2010**, *51*, 727–734. [CrossRef]
26. Grabke, H.J. Die Kinetik der Entkohlung und Aufkohlung von y-Eisen in Methan-Wasserstoff-Gemischen. *Berichte der Bunsengesellschaft* **1965**, *69*, 409–414. [CrossRef]
27. Chipman, J. Thermodynamics and phase diagram of the Fe-C system. *Metall. Trans.* **1972**, *3*, 55–64. [CrossRef]
28. Patankar, S.V. *Numerical Heat Transfer and Fluid Flow*; Hemisphere Publishing Corp: New York, NY, USA, 1980; ISBN 978-0-89116-522-4.
29. Hamadeh, H. Modélisation Mathématique Détaillée du Procédé de Réduction Directe du Minerai de Fer. Ph.D. Thesis, Université de Lorraine, Nancy, France, 2017. Available online: https://tel.archives-ouvertes.fr/tel-01740462/ (accessed on 2 July 2018).

Article

Optimization of the Iron Ore Direct Reduction Process through Multiscale Process Modeling

Rami Béchara [1,2], Hamzeh Hamadeh [1,2], Olivier Mirgaux [1,2] and Fabrice Patisson [1,2,*]

1 Institut Jean Lamour, CNRS, Université de Lorraine, 54011 Nancy, France; rami.bechara@univ-lorraine.fr (R.B.); hamzeh.hamadeh@univ-lorraine.fr (H.H.); olivier.mirgaux@univ-lorraine.fr (O.M.)

2 Laboratory of Excellence on Design of Alloy Metals for Low-Mass Structures (DAMAS), Université de Lorraine, 57073 Metz, France

* Correspondence: fabrice.patisson@univ-lorraine.fr; Tel.: +33-372-742-670

Received: 29 May 2018; Accepted: 20 June 2018; Published: 27 June 2018

Abstract: Iron ore direct reduction is an attractive alternative steelmaking process in the context of greenhouse gas mitigation. To simulate the process and explore possible optimization, we developed a systemic, multiscale process model. The reduction of the iron ore pellets is described using a specific grain model, reflecting the transformations from hematite to iron. The shaft furnace is modeled as a set of interconnected one-dimensional zones into which the principal chemical reactions (3-step reduction, methane reforming, Boudouard and water gas shift) are accounted for with their kinetics. The previous models are finally integrated in a global, plant-scale, model using the Aspen Plus software. The reformer, scrubber, and heat exchanger are included. Results at the shaft furnace scale enlighten the role of the different zones according to the physico-chemical phenomena occurring. At the plant scale, we demonstrate the capabilities of the model to investigate new operating conditions leading to lower CO_2 emissions.

Keywords: ironmaking; direct reduction; iron ore; DRI; shaft furnace; mathematical model; simulation; CO_2 emissions

---

1. Introduction

Steel is one of the key materials of today's industrial world. Moreover, its production is characterized by high energy consumption along with important carbon dioxide emissions. World steel production amounts to 6% of anthropogenic CO_2 emissions [1]. Beside the current reference steelmaking route (integrated plant with sinter and coke making, blast furnace, and basic oxygen furnace), less CO_2 emitting alternative routes are attracting greater interest. Direct reduction (DR) in a shaft furnace followed by electric arc steelmaking thus results in 40 to 60% lower CO_2 emissions compared with the reference route [2]. Direct reduction converts solid iron ore pellets into so-called direct reduced iron (DRI), using a mixture of CO and H_2 produced by reforming natural gas. Due to independence from coke and coal imports, sizeable units to meet demand, reduced investment costs, and reduced construction time [3], direct reduction units are becoming more numerous, particularly in countries where natural gas is cheap and abundant. These advantages have spurred increased research activity, namely in modeling and simulation, in order to correctly evaluate these processes and ultimately optimize them. The present work contributes to this effort with the key objective to provide an accurate simulation of the whole DR process, enabling us to propose new avenues for CO_2 emission mitigation. A distinctive feature of our approach is the multiscale nature of modeling, from, say, 1 μm (grain size) to 1 hm (plant size). We thus developed a single pellet model, a shaft furnace model, and a plant model, using different and complementary modeling strategies. These are presented in the next sections, together with references to previous works.

2. Single Pellet Model

Iron ore pellets (roughly spherical, typically 7–15 mm diameter) for DR are industrially produced from natural hematite grains (irregular, typically 20 μm). In a shaft furnace, these pellets are chemically reduced at 600–950 °C by both CO and H_2 in three steps, hematite $Fe_2O_3 \longrightarrow$ magnetite $Fe_3O_4 \longrightarrow$ wustite $Fe_{0.95}O \longrightarrow$ iron Fe, thus involving six gas-solid reactions at the grain scale. Numerous models for gas-solid reactions are available in the literature, from the simplest unreacted shrinking core model to pore models and grain models (Figure 1) [4]. The grain model matches well the grainy structure of the DR pellets; however, it was not used as such in the current study because it requires a numerical solution of the diffusion-reaction equations, which is incompatible for computation time reasons because of its integration in a multi-particle reactor model (next section). Instead, we used an alternate approach, the additive reaction times law, which gives an approximate analytical solution to calculate the reaction rates and can account for mixed regimes [5]. According to experimental observations, we also introduced an evolution of grain and pore structure with the reactions as follows: the hematite grains become covered with small pores after conversion to magnetite and wustite. At the wustite stage, the pores enlarge and the wustite grains break down into smaller particles termed "crystallites". These subsequently grow (typically from 1 to 10 μm) and join to form the molten-like structure of sponge iron (another name for DRI). These changes of course influence the kinetics of the transformation, mostly via the diffusion resistances. Details of the single pellet model used and of its results are given in [6,7].

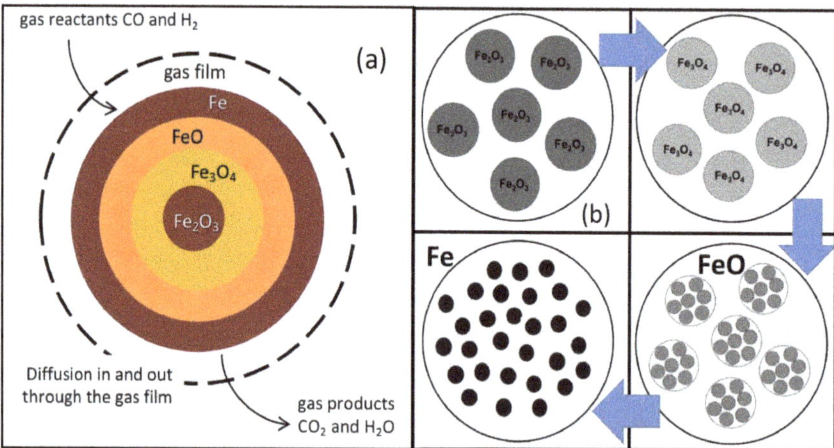

Figure 1. Evolution of pellet structure along with reaction: (**a**) Unreacted Shrinking Core Model; (**b**) Grain Model. The porous structure evolution (**b**) was determined from experimental observations [7].

3. Shaft Furnace Model

3.1. Previous Works

The shaft furnace (Figure 2) is the core of the DR process. Iron ore pellets are charged at the top, descend due to gravity, and encounter an upward counter-flow of gas. The reducing gas (CO and H_2, plus CH_4, CO_2, and H_2O, at about 950 °C) is injected peripherally at mid-height and exits at the top. In the lower section of the furnace, of conical shape, cold natural gas is injected to cool the iron pellets produced. The upper section (reducing zone and intermediate zone) is cylindrical (typically height 15 m; diameter 5 m). Two processes, MIDREX and HYL-ENERGIRON, share most of the DR market. Their shaft furnaces exhibit some differences (mostly in gas composition and pressure, size, and internal equipment details) but their basic characteristics are similar.

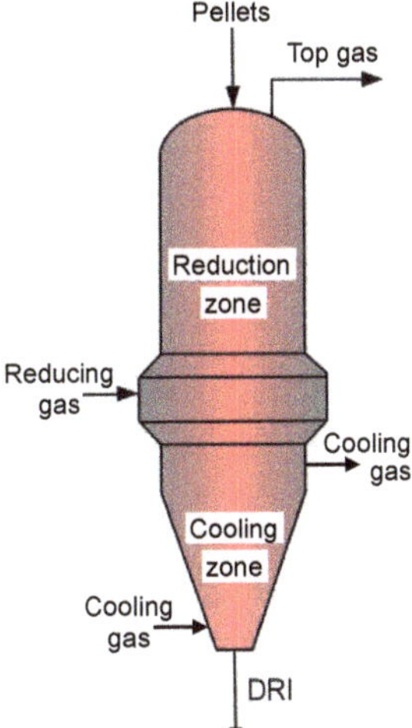

Figure 2. Schematic layout of a direct reduction shaft furnace.

Reflecting the great effort put in the simulation and development of the DR processes, numerous mathematical and numerical models of the shaft furnace were published, differing by the assumptions, the physico-chemical phenomena accounted for, the numerical scheme, the sections of the furnace considered, and the use of the model, etc. The most recent ones [6–15] are classified in Tables 1 and 2 and below. The basic description is that of an axisymmetric, porous moving bed reactor with counter-current flow between an ascending gas and a descending solid. The main differences are as follows:

- The number of reduction steps. This is related to the intermediary iron oxides taken into account. Works with one reduction step consider the direct transformation of hematite to iron, those with two steps consider further the presence of wustite, whereas those with three steps consider also magnetite.
- The nature of the inlet gas mixture. Most authors considered an inlet gas consisting of CO and H_2 only. Two models [13,14] considered the actual mixture, with CO_2, H_2O, N_2, and CH_4 as additional gas components.
- The number of dimensions included in the geometrical description. The standard is one-dimensional models. However, three works [6,14,15] considered two dimensions: along the height and the radius of the furnace.
- The description of the pellet transformation: shrinking core or grain models.
- The presence of supplementary reactions. The reduction reactions (in one, two or three steps) are always included. Additionally, methane decomposition and steam methane reforming were also taken into account in [10,12–14], together with the water gas shift and Boudouard reactions in [14].
- The type of heat transfer. All works included heat convection between solid and gas, as well as the heat of reaction. Heat transfer by conduction and radiation, through a contribution to effective conductivity, was also sometimes considered [7,8,11,14,15].
- Pressure drop was only included in [10,14].
- The presence of the cooling zone was only taken into account in three papers [10,13,14].
- Some models were validated against plant data, others not.

Results that are common to these works are as follows:

- The molar content of H_2 and CO decreases from the reduction zone bottom to its top whereas that of CO_2 and H_2O increases. Inversely, the content of iron oxides decreases from top to bottom with mainly iron exiting from the shaft bottom.
- The solid and gas temperatures are equal along the shaft except in a thin layer at the top near the solid inlet where a great temperature difference exists, the pellets are charged cold. Moreover, both temperatures increase from the shaft top to its bottom.
- Reaction enthalpy is of great importance, namely, the rather endothermic nature of H_2 reduction reactions vs. the exothermic nature of CO reduction reactions, as well as for the models that include it, the endothermic nature of steam methane reforming.

Key points that can be deduced when comparing differing results indicate that:

- the inclusion of three reduction steps and the grain model better represent the pellet transformation;
- all the components of the gas mixture (excepting N_2 but including CH_4) play a role in the transformations;
- two dimensions depict more accurately the reactor internal behavior and the output results; 2D results revealed the presence of two zones: one peripheral where the bustle gas is the reducing gas, and one central where the gas stems from the cooling and transition zones;
- taking supplementary reactions into consideration along with the cooling and transition zones better represents gas phase transformations and can account for carbon deposition.

The present shaft model has been built considering these results and on the basis of the most detailed (2D, 3 zones, 10 reactions–named REDUCTOR) shaft model [6,14]. However, the goal here is to simulate the whole DR plant, thus it was not practicable to incorporate such a comprehensive, 2D, CFD-type model in the plant simulation software. We selected Aspen Plus, a commercial simulation tool widely used in the chemical industry, as the software. Processes like the steelmaking blast furnace and the reverse chemical looping process are examples of processes somewhat close to the DR process that have been modeled with Aspen Plus. Thus, we had to build an Aspen Plus model of the DR shaft derived from REDUCTOR.

Table 1. Characteristics of DR shaft furnace mathematical models (1).

Reference	Reduction Steps	Gas Mixture	Dimen-sions	Geometric Model	Supplementary Reactions
Rahimi and Niksiar [8]	3	H_2-CO	1	Grain Model	no
Nouri et al. [9]	1	H_2-CO-H_2O-CO_2-N_2-CH_4	1	Grain Model	no
Shams and Moazeni [10]	2	H_2-CO-H_2O-CO_2-N_2-CH_4	1	Unreacted Shrinking Core Model	methane decomposition + reforming
Palacios et al. [11]	1	H_2-CO	1		no, but heat of reaction considered
Bayu and Alamsari [12] Alamsari et al. [13]	3	H_2-CO-H_2O-CO_2-N_2-CH_4	1		methane reforming and water gas shift
Ranzani da Costa et al. [7]	3	H_2	2	Grain Model (hematite & magnetite), Crystallite Model (Wustite)	no
Hamadeh et al. [6,14]	3	H_2-CO-H_2O-CO_2-N_2-CH_4	2		methane decomposition, reforming, water gas shift, Boudouard
Ghadi et al. [15]	3	H_2-H_2O	2	Unreacted Shrinking Core Model	no

Table 2. Characteristics of DR shaft furnace mathematical models (2).

Reference	Heat Contribution	Presence of Cooling & Transition Zones	Pressure Drop	Validation against Industrial Data
Rahimi and Niksiar [8]	reaction, convection, diffusion	no	no	no
Nouri et al. [9]	reaction, convection	no	no	yes
Shams and Moazeni [10]		yes	yes	yes
Palacios et al. [11]	reaction, convection, diffusion	no	no	no
Bayu and Alamsari [12] Alamsari et al. [13]	reaction, convection	yes	no	yes
Ranzani da Costa et al. [7]	reaction, convection, diffusion	no	no	no
Hamadeh et al. [6,14]		yes	yes	yes
Ghadi et al. [15]		no	no	yes

3.2. Aspen Plus Shaft Model

The results given in [14] show that, in addition to the initial 3-stage division, the reduction zone can be sub-divided into two zones. In the first one, Zone 1, the widest, peripheral zone, the oxides are almost completely reduced at the bottom of the zone. The inlet gas in this zone is the major fraction of the bustle reducing gas plus a part of the gas coming upward from the transition zone. The inlet solid is the major fraction of the oxide pellets charged. The second zone, Zone 2, is a narrow central zone in which only partial metallization occurs. Its inlet gas comes from the transition zone, whereas its inlet solid is a fraction of the oxide pellets. Figure 3 depicts this partitioning and also shows its translation into an Aspen Plus based configuration.

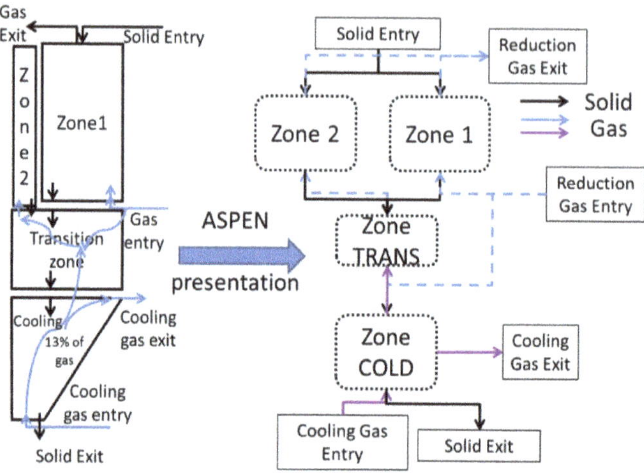

Figure 3. Sectioning of the shaft furnace for the Aspen Plus model.

Nevertheless, due to the countercurrent of the solid and gas flows, this description implies using several loops for the numerical solution; e.g., the gas coming from the transition zone influences the solid in the central zone, which in turn influences the transition zone gas. Likewise, the solid from the transition zone influences the gas from the cooling zone, which in turn influences the solid. These loops make it difficult and time-consuming to simulate the process. An effort was thus made to simplify the model. The flow rate of the solid in the central zone being considerably smaller than that in the peripheral zone, the former was not considered as an input to the transition zone, thus cancelling the first loop. Moreover, the transition zone had every little impact on the temperature of the descending solid, thus the link between the descending transition solid and the cooling zone was removed. In place, a fictitious solid stream equal to that leaving the peripheral zone was used as an input. The resulting temperature was thus imposed on the solid leaving the transition zone. This solid was considered as the ultimate output of the peripheral zone. Lastly, an additional cooling zone was added for the central zone. Attention was made to avoid any loops between the two sections. Considering this, Figure 4 highlights the restructuring of the Aspen Plus configuration to avoid the aforementioned loops. Herein, the cross signs emphasize deleted streams, which were responsible for loops, whereas the dotted lines indicate new streams. As a result of this restructuring, a new cooling zone (COLD2) was added and the resulting temperature from the cold section was returned to the exit of the transition zone. Of course, these changes, made for the sake of simplification, do not alter the overall mass and heat flowrates and balances.

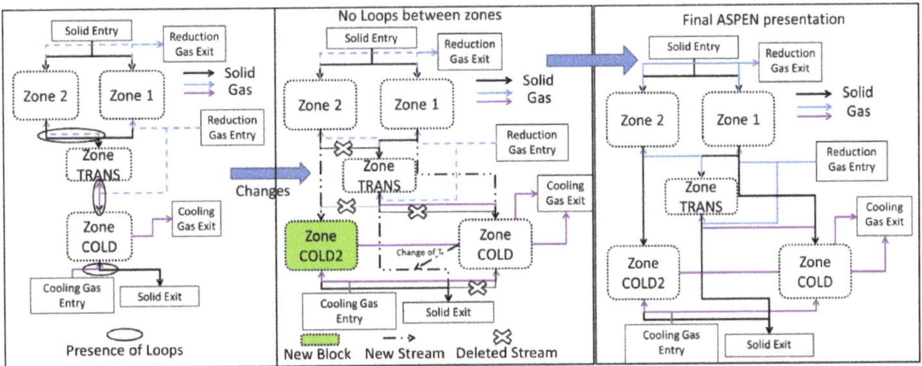

Figure 4. Restructuring of Aspen Plus configuration of the shaft furnace to avoid loops.

This representation however poses the problem of finding the adequate split for the streams between the various sections. We identified three critical splits. The first is at the reduction gas entry level, the bustle gas being split between the transition zone and Zone 1. The second is at the top solid inlet, the entering solid being split between Zone 1 and Zone 2. The third split is at the cooling gas stream that goes up to the transition zone. The first split was set to a predefined value. The second split was calculated considering that the ratios of the inlet solid flow to the input gas flow for each zone are equal. The third split was set to the constant value of 0.13 as given in [6].

This representation, set and done, is not yet sufficient to directly calculate the final conversion and temperatures. Especially for the transition and reduction zones, Aspen Plus does not offer any built-in model that allows their calculation in a way that could correctly depict the reaction kinetics and the variations of the different variables along the height of the shaft. An external calculator, written in Fortran and based on REDUCTOR's equations, was therefore used to compute these values, which were later rendered to the corresponding Aspen Plus blocks. Conversely, for the cooling section, a simple Aspen Plus heat exchanger could be used. The principle of the calculation, namely in the reduction and transition zones, and the list of reactions are given in Appendix A.

3.3. Results from the Aspen Plus Shaft Model

Two case studies were considered in this work, based on real industrial data. The first case corresponds to the Contrecoeur plant, located near Montreal, Canada, and currently operated by ArcelorMittal. It was constructed in the late 1970's and is a MIDREX series 750 module. It is characterized by a rather cold input solid and a high content of C in the input gas. The second case relates to the Gilmore plant, built near Portland, Oregon, USA. Now decommissioned, it was the first operating MIDREX plant. It is one of the most referenced plants in literature, a MiniMod module with a CDRI production of 26.4 tons/h. The input operating conditions for both plants are given in Appendix B. These are of quite different capacities, with different reducing gas compositions and different pellet diameters, and thus represent a good test for validating a model.

Table 3 compares the results obtained in the model for the outlet gas and solid streams with those provided in literature for the studied cases. The difference for each parameter was calculated as well as the total error. As can be seen, the absolute error for the Contrecoeur case is equal to 6%, whereas that for Gilmore is 4%. The biggest differences pertain to methane and nitrogen composition as well temperature. Other differences include hydrogen and water compositions in the Gilmore case, and the carbon dioxide composition in the Contrecoeur case. These differences can be related to the formulas chosen from the literature for the gas phase reaction rates. For example, methane decomposition and Boudouard reactions seem to be somewhat underestimated in Contrecoeur and overestimated in Gilmore. These values could only be further consolidated through the realization of up-to-date experiments to correctly characterize these reactions. Nonetheless, the results seem to be globally satisfactory, especially if they are compared with other model results [10,14]. The related differences are comparable although the present model is of a different type and simpler than the two CFD-type models.

Table 3. Simulation of a shaft furnace. Comparison of the results with industrial data and other model results.

	Parameter	Contrecoeur			Gilmore			
		Plant Data	Present Model	REDUCTOR Model [14]	Plant Data	Present Model	REDUCTOR Model [14]	Shams Model [10]
Top gas	CO (vol %)	19.58	18.88	19.89	18.90	19.00	20.87	19.44
	CO_2 (vol %)	17.09	15.60	14.69	14.30	14.60	13.13	17.54
	H_2O (vol %)	19.03	19.50	19.52	21.20	17.60	20.61	18.04
	H_2 (vol %)	40.28	40.20	40.41	37.00	42.20	37.70	37.96
	N_2 (vol %)	1.02	1.63	1.55	7.67	6.00	8.60	7.02
	CH_4 (vol %)	2.95	4.19	3.91	-	-	-	-
	Temperature (°C)	285	312	293	n.a.	417	285	474
Bottom solid	Fe_2O_3/Fe_{tot} (wt %)	0	0	0	0	0	0	0
	Fe_3O_4/Fe_{tot} (wt %)	0	0	0	0	0	0	0
	FeO/Fe_{tot} (wt %)	6.20	6.1	6.00	7.00	7.00	4.70	10.00
	Fe/Fe_{tot} (wt %)	93.80	93.90	94.00	93.00	93.00	95.30	90.00
	C (wt %)	2.00	2.30	2.20	2.00	1.80	0.91	1.42
	Average absolute difference with plant data (%)	-	6.0	4.4	-	4.0	6.9	6.5

Figure 5 shows the evolution of the solid component flow rates with shaft height in the first reduction zone for the Contrecoeur (a) and Gilmore (b) cases. Height zero corresponds to that of the bustle gas inlet. As can be seen, hematite disappears very rapidly near the solid inlet in both cases. Magnetite on the other hand has a different behavior; it disappears after 2.6 m in the Contrecoeur case, against 4.6 m in the Gilmore case, with more of this oxide formed in the latter case. Wustite on the other hand disappears rapidly in the Gilmore case, its presence window being only 2 m, against about 7 m in the Contrecoeur case. Finally, as expected, both cases give a 100% conversion of iron oxide to pure iron in this Zone 1. Figure 5 also shows the evolution of the carbon deposited in the pellets, which continuously increases from solid entrance before dropping near gas entrance. Exit solid has some carbon content in the Contrecoeur case, whereas the carbon content drops to zero in the Gilmore case. This is related to the inversion of the Boudouard equilibrium ($C + CO_2 \rightleftharpoons 2CO$), which is favorable to carbon deposition at lower temperatures and lower CO_2 content.

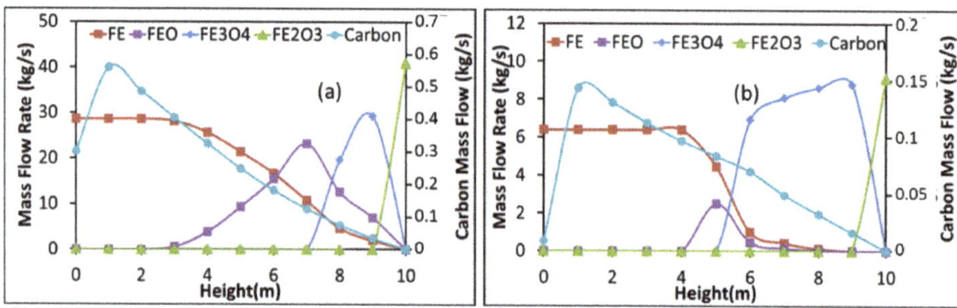

Figure 5. Evolution of the solid component flow rates with shaft height in Zone 1, (**a**) Contrecoeur; (**b**) Gilmore.

These profiles can be related to the gas phase reactions, which are presented for both cases in Figure 6. It can be seen that methane, water, and carbon dioxide flow rates decrease above the gas entrance, whereas those of hydrogen and carbon monoxide increase. This is the opposite in the upper half of the shaft, except for methane, which sees its flow rate reach equilibrium. This inversion can be explained by the preponderance of the iron oxide reduction reactions, as well as the inverse Boudouard reaction ($2CO \rightleftharpoons C + CO_2$) as evidenced by the carbon production in Figure 5. The decrease in methane can be related mainly to the steam reforming ($CH_4 + H_2O \rightleftharpoons CO + 3H_2$) with carbon deposition from methane ($CH_4 \rightleftharpoons C + 2H_2$) having a rather negligible impact. This little impact is emphasized by the decline in carbon flow rate near the gas inlet, which is due to the direct Boudouard reaction.

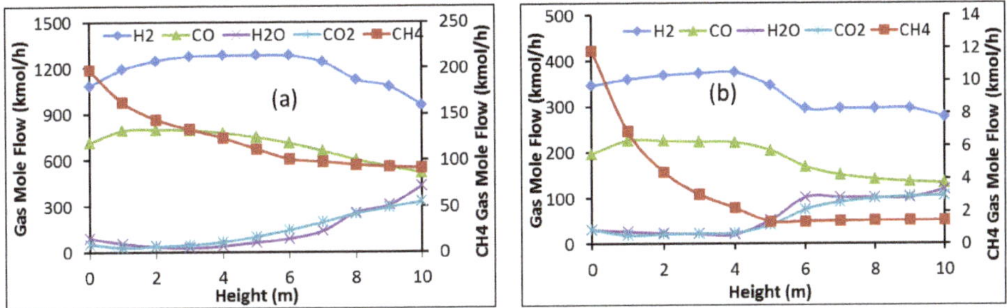

Figure 6. Evolution of the gas component flow rates with shaft height in Zone 1, (**a**) Contrecoeur; (**b**) Gilmore.

Among the other calculated results (temperature, reaction rates, and values for the other zones), we selected for presentation and comparison the evolution of the iron compounds flow rates with shaft height in Zone 2 (Figure 7). The situation differs from that of Zone 1. Whereas hematite is readily reduced into magnetite as previously reported, magnetite remains present over most of the height, being slowly reduced in the Contrecoeur case and almost not changing in the Gilmore case until the gas inlet. Conversion to wustite and iron only occurs at the zone bottom, with 60% wustite remaining in the solid at the exit. This situation results from a lower temperature and less CO and H_2 for the reduction in Zone 2 compared with Zone 1. The other reactions, involving methane, carbon monoxide and carbon dioxide, hardly take place in Zone 2. These results emphasize the impact this second zone has on the low exit gas temperature and on the average metallization degree.

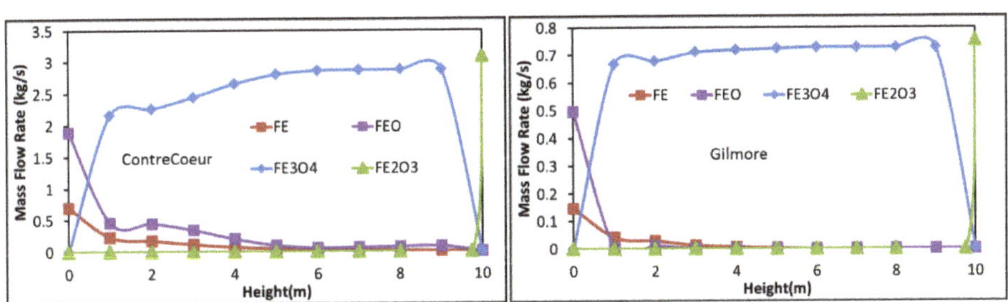

Figure 7. Evolution of the solid component flow rates with shaft height in Zone 2, (**a**) Contrecoeur; (**b**) Gilmore.

The importance of the transition zone should also be noted. The main reaction in this zone is methane decomposition ($CH_4 \rightarrow C + 2H_2$). Carbon is herein deposited on produced iron, leading to the final carbon content of the DRI. This carbon comes in addition to the carbon possibly deposited via the inverse Boudouard reaction in Zone 1 (Figure 5). At the rather low temperature of the transition zone, no iron oxide reduction by solid carbon occurs. The hydrogen produced by the methane decomposition reaction is sent to the second zone, contributing to the final metallization degree. Moreover, the gas-solid temperature difference is reversed in this zone, the descending solid being hotter than the ascending gas.

4. Plant Model

The next scale is that of the whole DR plant, a schematic diagram of which is given in Figure 8 (case of a MIDREX-type plant). The main units around the shaft furnace are: the reformer, the scrubber, and the heat exchanger. Downstream from the shaft furnace the top gas is scrubbed and divided into two streams. The first stream loops in the system. It is mixed with natural gas, reheated, and sent to the catalytic tubes of the reformer to produce the reducing gas. The second one (plus possibly some extra natural gas) is burnt with air to provide the reformer with heat. The hot flue gas is used in an exchanger to heat both gas streams entering the reformer. The MIDREX reformer is a kind of 'dry' reformer where the primary reaction is $CH_4 + CO_2 \longrightarrow 2H_2 + 2CO$. In other DR processes, methane is rather reformed by H_2O (HYL), the reformer is smaller or missing (HYL-ZR) or the reducing gas is produced from other fuels (ENERGIRON). This variety was a further incentive to develop the present DR plant model using Aspen Plus, since we expect to use it for process optimization, including modifying parts of the configuration. Contrary to the shaft furnace, the plant scale was scarcely investigated through modeling. Only two authors studied the interaction between the reformer and the shaft furnace, using mathematical models for each unit [6,16].

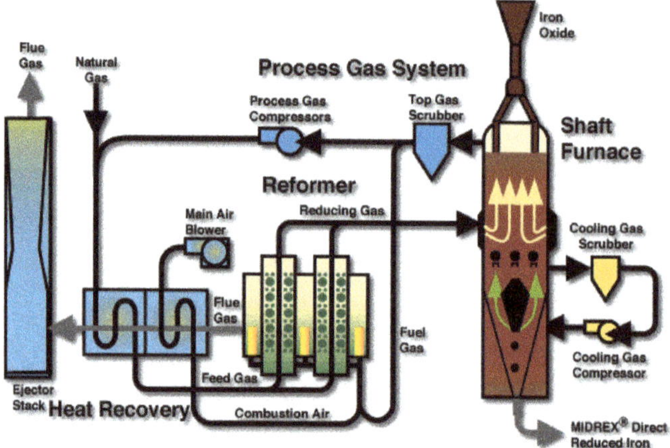

Figure 8. MIDREX direct reduction process [17].

4.1. Aspen Plus Plant Model

Figure 9 illustrates the corresponding Aspen Plus model. Starting from the shaft furnace top gas exit the scrubber is the first step, in which part of the water vapor is separated from the rest of the stream. The adopted Aspen sub-model is a splitter, in which the separated H_2O fraction is defined ($split_{H_2O}$). The remaining off-gas going is then split in two streams: one ($split_{gas}$) going to the reformer and the other to the burner. On the other hand, reformer natural gas is also split in two fractions, one sent prior to the reformer ($split_{CH_4}$) and the other after it. The reformer was modeled using the built-in Aspen Plus 'Gibbs reactor'. This reactor does not consider kinetics, but calculates the product composition minimizing its Gibbs energy; i.e., at equilibrium. This is a simplification often used in the literature [18,19]. The key design variable is T_{ref}, the reforming temperature. The gas stream exiting the reformer is later mixed with natural gas and air prior to being sent back to the shaft furnace. A prior step was added in the form of a Gibbs reactor where O_2 injection and burning is realized. Herein, the reactor temperature (T_{r,O_2}) is the key design variable. The combustion system on the other hand has as inlets the remaining off-gas, natural gas, and combustion air. These streams are heated to the combustion temperature, burnt in the burner producing hot flue gases later cooled prior to discharge. The burner was also modeled as a stoichiometric reactor, employing gas combustion reactions, with its temperature (T_{bur}) also as a key design variable. The burner's goal is to provide energy for the reforming system. This step is then followed by flue gas cooling in order to recover energy, namely for pre-heating purposes. The key flue gas cooling variable is (T_{flue}).

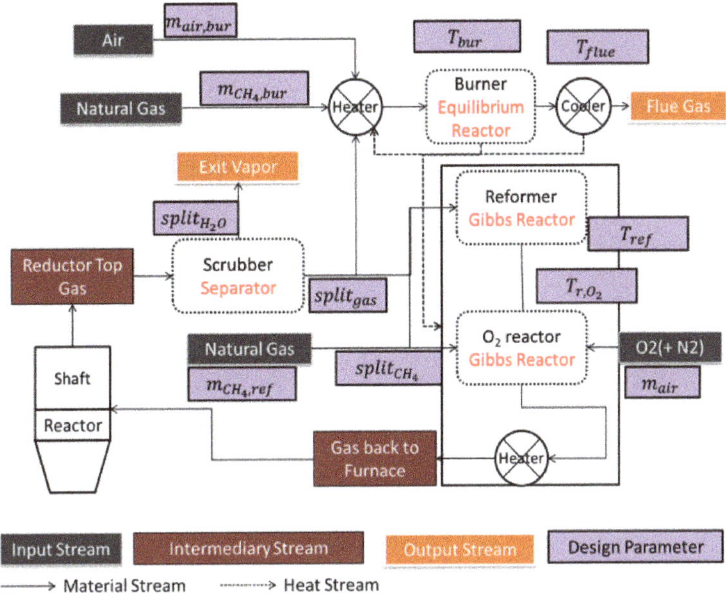

Figure 9. Whole plant Aspen Plus model, with gas recirculation and key design variables.

The principle used for modeling the plant in Aspen Plus was to make use of Aspen Plus Design Specs. A Design Spec is a condition that the Aspen Plus solver will make satisfied by acting on a 'controlled variable'. In this case, on the one hand, reformer variables were controlled so that the gas exiting the reformer is identical to the bustle gas, leading to a closed looped process. On the other hand, burner variables were controlled in order to guarantee an adequate heat balance in the process. The design specs refer to the equality of flow rates for hydrogen, water vapor, carbon monoxide, carbon dioxide, methane, and nitrogen between the bustle-gas and the exit gas respectively. The controlled variables are: $split_{gas}$, $split_{H_2O}$, T_{ref}, T_{r,O_2}, $n_{CH_4,ref}$, n_{air}, $split_{CH_4}$, $n_{CH_4,bur}$, T_{flue} and $n_{air,bur}$.

4.2. Results from the Aspen Plus Plant Model

Table 4 lists the various design specifications as well as the corresponding controlled variables, and the resulting values. The obtained values were compared to available data in the case of Contrecoeur; no such data were available for Gilmore. As it can be seen, the values do not differ in the first case except for $split_{CH_4}$. Indeed, this is not really a discrepancy given that the same amount of methane is reformed. According to plant data, 92% of the methane is sent to the reformer where 46% of it is reformed, the yield being limited by kinetics. Conversely, the reformer modeled as a Gibbs reactor gives an almost complete reforming and thus only 43% of the methane is needed in the reformer, the rest being by-passed. Another approach would have been to adopt a 1D model of the reformer with kinetics, like [6,16]. Downstream of the reformer, the gas temperature is adjusted using a small air injection.

Table 4. Design specifications, controlled variables and corresponding values.

Equality to Respect	Controlled Variable	Value (Contrecoeur)	Data (Contrecoeur)	Value (Gilmore)
$n_{H_2.ref_{out}} = n_{H_2.bustle}$	$split_{gas}$	0.64	0.65	0.68
$n_{H_2O.ref_{out}} = n_{H_2O.bustle}$	$split_{H_2O}$	0.62	0.60	0.44
$n_{CO.ref_{out}} = n_{CO.bustle}$	T_{ref}	837 °C	-	830 °C
$n_{CO_2.ref_{out}} = n_{CO_2.bustle}$	$T_{r.O_2}$	860 °C	-	670 °C
$n_{CH_4.ref_{out}} = n_{CH_4.bustle}$	$n_{CH_4.ref}$	338 mol/s	380 mol/s	71 mol/s
$n_{N_2.ref_{out}} = n_{N_2.bustle}$	$n_{air,ref}$	11.6 mol/s	-	14.9 mol/s
$n_{C.ref_{out}} = 0$	$split_{CH_4}$	0.434	0.92	0.92
$Q_{comb} = Q_{ref}$	$n_{CH_4.bur}$	0 mol/s	-	8 mol/s
$Q_{flue} = Q_{heat}$	T_{flue}	295 °C	-	345 °C
$n_{O_2.flue} = 0$	$n_{air.bur}$	871 mol/s	-	216 mol/s

Results in the Gilmore case resemble the Contrecoeur case for T_{ref} and $n_{CH_4.bur}$. The first is related to equilibrium whereas the second showcases the lack of need for excess fuel. Both processes however differ for the remaining variables. In the Gilmore case, $split_{gas}$ has a higher value leading to greater off-gas recycling. This may indicate the need for greater CO_2 reforming as well as the need to conserve unreacted CO-H_2 reductants present in the off-gas. The smaller $split_{H_2O}$ value can be understood in light of a greater H_2O reforming in the connected reformer. The smaller amount of natural gas sent to the reformer is related to the smaller gas flow rate entering the shaft furnace. The equivalent input air flow rate can be related to the greater amount of nitrogen present in the gas entering the furnace. Higher $split_{CH_4}$ values can be understood in light of the equilibrium that is reached in the reformer but also a smaller methane fraction in the gas stream entering the shaft furnace. The greater amount of input process air $n_{air,ref}$ required in the burner is related to the greater N_2 flow rate in the bustle gas.

4.3. Process Visualization

The adopted process model and simulation tool further enable us to create a mimic board of the plant. Figure 10 highlights the flow diagrams of this process in the Gilmore case. As can be seen, all process units are shown, the shaft reactor, scrubber, reformer, oxygen injection, burner, and air preheater. Moreover, all streams to and from these units are highlighted along with their composition. The figure puts a great emphasis on the gas loop, by providing temperature and composition changes along the operations; e.g., the reduction in H_2, CO and CH_4 content in the shaft reactor and their recovery in the reforming system. Moreover, it can be seen that the burner provides the reformer system with heat by burning part of the scrubbed off-gas, with little need for input methane. Certain key parameters are shown in yellow, like the mass flow rate of DRI, its metallization and carbon content, together with the ratio of equivalent carbon present in the flue gas to the quantity of DRI produced (C/DRI, here equal to 0.119) as well as the same ratio but for equivalent CO_2 (CO_2/DRI). In fine, this visualization makes it easier to compare simulations to plant data or to compare simulation runs between each other.

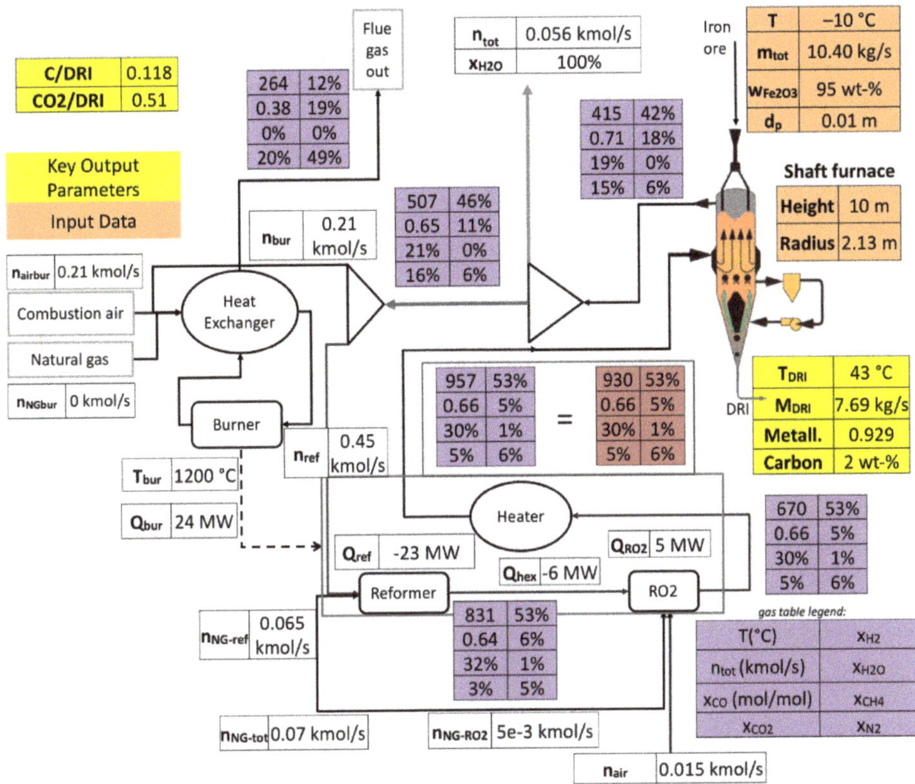

Figure 10. Simulated process flow diagram (Gilmore case).

4.4. Process Optimization

Having modeled the shaft reactor as well as the gas loop system, various process optimization schemes can be considered. Our aim in this paper was to address the reduction in CO_2 emissions. Keeping the same plant configuration (MIDREX-type), the chosen means was the modification of the inlet reducing gas composition. According to the literature, the two important ratios are H_2/CO and $(H_2 + CO)/(H_2O + CO_2)$; i.e., the hydrogen-to-carbon monoxide ratio and the reducing power of the reducing gas. The initial values for these parameters were respectively equal to 1.5 and 12 for Contrecoeur and 1.75 and 8.73 for Gilmore. Optimal ratios were recorded close to 1 and 12 for both parameters respectively [9,20]. Considering this, optimization runs are provided hereafter for Gilmore. The initial value for the carbon-to-DRI ratio was 0.119, a figure also in line with the literature, where the range of 0.105–0.120 was reported [2].

Table 5 highlights the optimization runs carried out by decreasing H_2/CO from 1.75 to 1.05 and increasing $(H_2 + CO)/(H_2O + CO_2)$ from 8 to 16. As can be seen, the carbon-to-DRI ratio decreases by as much as 12% between the original case and the optimal case. This case corresponded to a H_2/CO ratio of 1.23 and $(H_2 + CO)/(H_2O + CO_2)$ of 12. Attempts to further decrease H_2/CO led to model divergence, the reformer not being able to produce the desired bustle gas composition. Increasing the $(H_2 + CO)/(H_2O + CO_2)$ ratio over 12 did not further reduce the carbon-to-DRI ratio.

Table 5. Optimization runs in the Gilmore case.

Parameter	Original	Opti-1	Opti-2	Opti-3	Opti-4	Opti-5	Opti-6
$\frac{H_2}{CO}$	1.75	1.57	1.5	1.35	**1.23**	1.05	1.23
$\frac{H_2+CO}{H_2O+CO_2}$	8	12	12	12	**12**	12	16
$\frac{C}{DRI}$ (kg/kg)	0.119	0.119	0.113	0.109	**0.105**	Diverged	0.106
Metallization (%)	93%	93.3%	93.3%	93.3%	93.3%		93%
x_C(wt %)	1.8%	2%	2%	2%	2%		2%

These results illustrate the potential of the tool. The key differences between the optimized and the original operations pertain to greater $split_{H_2O}$ (0.72 vs. 0.44) and $split_{gas}$ ratios (0.75 vs. 0.64). More H_2O is thus withdrawn in the scrubber, whereas greater gas recycling occurs. This can be translated in a greater importance for CO_2 reforming and a greater conservation of the reducing gases. This conservation further leads to a smaller requirement of reformer natural gas (64 mol/s vs. 71 mol/s). Also, a smaller energy demand occurs in the reformer (22 MW vs. 24 MW). It should be further noted that these results were obtained without little (but positive) changes in metallization degree (93.3% for the optimal case vs. 93% for the original case) and carbon content (2% vs. 1.8%).

Calculations were also performed for the Contrecoeur case. The best carbon-to-DRI ratio was 0.099 kg/kg (a 17.5% reduction from the original case 0.12 kg/kg), obtained using H_2/CO and $(H_2 + CO)/(H_2O + CO_2)$ ratios of 1.39 and 12.5, respectively. These values are in line but somewhat different from those of the Gilmore case, due to the content in other gas components of the bustle gas.

5. Discussion

The obtained results were already given a first interpretation in the previous sections. Hereafter, we focus on the context and significance of the present work. The developed process model differs from previously published works by its multi-scale nature, from grains (μm) to the shaft furnace (hm), and ultimately to integrated plant simulation (Aspen Plus).

The core of the DR process is the reduction shaft furnace. The corresponding Aspen Plus model was made sufficiently sophisticated to reproduce the main results from the more complex, CFD-type models of this reactor, but sufficiently light enough to be integrated in a whole plant flowsheet model. Thus, based on the recent results of [14], which proposed a virtual division of the shaft into zones distinguished according to the physico-chemical and thermal phenomena occurring, the shaft furnace was modeled as a set of interconnected zones. The ones in the reduction and intermediate sections were discretized in horizontal slices. This description, intermediate between 1D and 2D, enables us to retain the advantages of describing the axial variations of the calculated variables and of a short computation time (typically 3 h on a standard PC). The results from this model agree well with both available industrial data and results from previous models. Moreover, the two cases simulated corresponding to plants of different operating conditions and of quite different throughput, this demonstrates the adaptability and the robustness of the model.

The overall Aspen Plus model of a DR plant includes models for the principal units: the shaft furnace, the natural gas reformer, the scrubber, and the main heat exchanger. To the best of our knowledge, this is the first published work on a systems model of the whole DR plant using process simulation software. The decisive interest of such an approach is that the recycling gas can be accounted for. In MIDREX-type DR plants, most of the top gas exiting the shaft furnace is recycled, being first sent to the reformer then fed as the reducing gas at the gas inlet of the shaft furnace. Using the overall model, it becomes possible to study the interactions between the respective performances of the reformer and of the shaft furnace.

As a test seeking to mitigate the CO_2 emissions from the plant, the reducing gas composition at the shaft gas inlet was varied. A series of simulations show that CO_2 emissions and natural gas consumption could be reduced tuning the ratios H_2/CO and $(H_2 + CO)/(H_2O + CO_2)$ at 1.23 and 12, respectively. The C-to-DRI ratio can be lowered from 0.119 to 0.105 kg/kg, a 12% reduction.

However, this optimization by hand is a first piece of work. The next stage would be to use a mathematical optimizer for the same purpose. The overall model could also be used to test different plant configurations. A variety of DR plants (MIDREX-type of different sizes, HYL-type of different sizes, HYL-ZR, ENERGIRON with different reducing gas sources) were designed and built. This means that further configurations could be designed to optimize criteria like the CO_2 emission, but also energy consumption, operating costs, etc. The present model can give useful answers.

6. Conclusions

In conclusion, this paper showcases a systemic, multiscale process model developed for the simulation, visualization, and further optimization of a gas-based DR iron reduction plant. A multi-scale approach was adopted, going from grain-size to model the iron pellet and its transformation, to reactor-size where reaction kinetics and heat and mass exchanges were simulated, and ultimately plant-scale where the gas recycling loop was modeled to obtain a closed process, and the subsequent carbon emissions were calculated for two cases. Simulation results were found in good agreement with industrial data. Moreover, hand optimization tests provided a 12% decrease in carbon emissions. Future works will address computer-based optimization along with the testing of novel process configurations.

Author Contributions: Conceptualization, F.P., H.H. and R.B.; methodology: all; Fortran software: H.H.; Aspen Plus software: R.B. and O.M.; investigation: H.H. and R.B.; writing: R.B. and F.P.; supervision: F.P. and O.M.; Project Administration and Funding Acquisition: F.P.

Funding: This research was supported by the French State through two programs 'Investment in the future': (i) one operated by French Environment and Energy Management Agency (ADEME), 'Valorization of CO_2 in industry', 2014-18, VALORCO, No. 1382C0245; the authors thank Nathalie Thybaud and Aïcha El Khamlichi, and the coordinator, Eric de Coninck; and (ii) one operated by the National Research Agency (ANR) and referenced by ANR-11-LABX-0008-01 (LabEx DAMAS).

Acknowledgments: The authors also express their thank to the staff of ArcelorMittal at Maizières-lès-Metz, France, and Contrecoeur, Canada, for encouragements, information. and discussion, especially S. Bertucci, J. Borlée, T. Quatravaux, and J. Farley.

Conflicts of Interest: The authors declare no conflict of interest.

Appendix A

The calculation scheme for the reduction and transition zones is explained hereafter. A 1D finite difference method was employed. The reactor was partitioned into (4000) horizontal slices of constant surface and equal height (Figure A1).

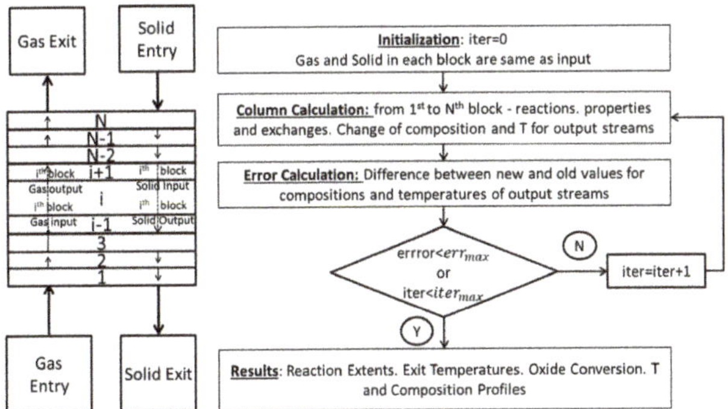

Figure A1. Resolution scheme for the reduction and transition zones.

The calculation loop can be broken down as follows. All gas and solid streams are initialized as equal to the input gas and solid streams respectively. Calculation then starts from the first slice up to the last. Reaction and heat and mass exchanges occur between the ascending gas (i) and the descending solid (i + 1) entering the slice. This exchange impacts the exiting ascending gas (i + 1) and descending solid (i). The difference between the newly calculated values and the previous values is then recorded for each stage. Once the calculation is over, the overall relative error is calculated as the sum of the difference between the old and new value divided by the sum of the total values for each of the following variables: molar flow rate for gas and solid components as well as the temperatures for gas and solid phases. The iteration then continues until either this error is lower than a certain tolerance (10^{-4}) or the number of iterations is superior to a maximum number of iterations (7000). Once the calculation is over, results are returned to the Aspen Simulation. These include: the reaction extents, the exit solid and gas temperatures, as well as the temperature and composition profiles for both gas and solid streams along the shaft.

Table A1 provides the list of all the reactions considered in the reduction and transition zones. The first set of equations relate to iron oxide reduction: $Fe_2O_3 \rightarrow Fe_3O_4 \rightarrow Fe_{0.95}O \rightarrow Fe$, with CO and H_2 as reduction agents. Their reaction rates are controlled by various resistances (inter-granular, intra-granular, inter-crystallite, intra-crystallite, and chemical) summed together courtesy of the additive reaction times law that considers that the different matter transportation steps occur in series. The rest of reactions are the methane steam reforming, water gas shift, methane decomposition and Boudouard. The kinetic rates are calculated based on stoichiometric factors with a possibility for reverse reactions. The expressions for the kinetic constants are based on literature [17,21–23], with the role of iron as a catalyst for methane reactions.

Table A1. Considered reactions.

Reaction Number	Name	Formula
Reaction 1	Hematite H_2 reduction	$3\,Fe_2O_3 + H_2 \rightarrow 2\,Fe_3O_4 + H_2O$
Reaction 2	Hematite CO reduction	$3\,Fe_2O_3 + CO \rightarrow 2\,Fe_3O_4 + CO_2$
Reaction 3	Magnetite H_2 reduction	$1.19\,Fe_3O_4 + H_2 \rightarrow 3.75\,Fe_{0.95}O + H_2O$
Reaction 4	Magnetite CO reduction	$1.19\,Fe_3O_4 + CO \rightarrow 3.75\,Fe_{0.95}O + CO_2$
Reaction 5	Wustite H_2 reduction	$Fe_{0.95}O + H_2 \rightarrow 0.95\,Fe + H_2O$
Reaction 6	Wustite CO reduction	$Fe_{0.95}O + CO \rightarrow 0.95\,Fe + CO_2$
Reaction 7	Steam reforming	$CH_4 + H_2O \rightarrow CO + 3\,H_2$
Reaction 8	Water gas shift	$CO + H_2O \leftrightarrow CO_2 + H_2$
Reaction 9	Methane decomposition	$CH_4 \rightarrow C + 2\,H_2$
Reaction 10	Boudouard (inverse)	$2\,CO \leftrightarrow C + CO_2$

Reaction heat is calculated as the sum of each equation's heat of reaction and is attributed to the solid phase. Moreover, solid and gas phases transport heat by convection and conduction, and exchange heat through convective transfer.

Appendix B

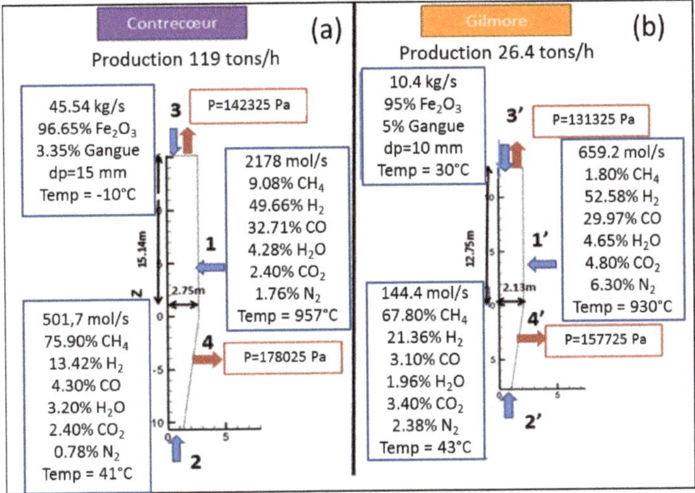

Figure A2. Input conditions of Contrecoeur (**a**) and Gilmore (**b**) plants.

References

1. IEAGHG. Iron and Steel CCS Study (Technoeconomics Integrated Steel Mill). IEAGHG Report. 4 July 2013. Available online: http://ieaghg.org/publications/technical-reports (accessed on 19 May 2018).
2. Duarte, P.E.; Becerra, J. Reducing greenhouse gas emissions with Energiron non-selective carbon-free emissions scheme. *Stahl Und Eisen* **2011**, *131*, 85.
3. Chatterjee, A. *Sponge Iron Production by Direct Reduction of Iron Oxide*; PHI Learning Pvt. Ltd.: Delhi, India, 2010; ISBN 978-81-203-3644-5.
4. Patisson, F.; Ablitzer, D. Modeling of gas-solid reactions: Kinetics, mass and heat transfer, and evolution of the pore structure. *Chem. Eng. Technol.* **2000**, *23*, 75–79. [CrossRef]
5. Sohn, H.Y. The law of additive reaction times in fluid-solid reactions. *Metall. Trans. B* **1978**, *9*, 89–96. [CrossRef]
6. Hamadeh, H. Modélisation Mathématique Détaillée du Procédé de Réduction Directe du Minerai de fer. Ph.D. Thesis, Université de Lorraine, Nancy, France, 2017. Available online: https://tel.archives-ouvertes.fr/tel-01740462 (accessed on 19 May 2018).
7. Ranzani da Costa, A.; Wagner, D.; Patisson, F. Modelling a new, low CO_2 emissions, hydrogen steelmaking process. *J. Clean. Prod.* **2013**, *46*, 27–35. [CrossRef]
8. Rahimi, A.; Niksiar, A. A general model for moving-bed reactors with multiple chemical reactions, part I: Model formulation. *Int. J. Min. Proc.* **2013**, *24*, 58–66. [CrossRef]
9. Nouri, S.M.M.; Ebrahim, H.A.; Jamshidi, E. Simulation of direct reduction reactor by the grain model. *Chem. Eng. J.* **2011**, *166*, 704–709. [CrossRef]
10. Shams, A.; Moazeni, F. Modeling and Simulation of the MIDREX Shaft Furnace: Reduction, Transition and Cooling Zones. *JOM* **2015**, *67*, 2681–2689. [CrossRef]
11. Palacios, P.; Toledo, M.; Cabrera, M. Iron ore reduction by methane partial oxidation in a porous media. *Int. J. Hydrogen Energy* **2015**, *40*, 9621–9633. [CrossRef]
12. Alamsari, B.; Torii, S.; Trianto, A.; Bindar, Y. Study of the Effect of Reduced Iron Temperature Rising on Total Carbon Formation in Iron Reactor Isobaric and Cooling Zone. *Adv. Mech. Eng.* **2010**, *2*, 192430. [CrossRef]
13. Alamsari, B.; Torii, S.; Trianto, A.; Bindar, Y. Heat and mass transfer in reduction zone of sponge iron reactor. *ISRN Mech. Eng.* **2011**, *2011*, 324659. [CrossRef]

14. Hamadeh, H.; Patisson, F. Reductor, a 2D Physicochemical Model of the Direct Iron Ore Reduction in a Shaft Furnace. In Proceedings of the AISTech 2017 Conference, Nashville, TN, USA, 8–11 May 2017; pp. 937–946. Available online: http://digital.library.aist.org/pages/PR-372-106.htm (accessed on 19 May 2018).
15. Ghadi, A.Z.; Valipour, M.S.; Biglari, M. CFD simulation of two-phase gas-particle flow in the Midrex shaft furnace: The effect of twin gas injection system on the performance of the reactor. *Int. J. Hydrogen Energy* **2017**, *42*, 103–118. [CrossRef]
16. Alhumaizi, K.; Ajbar, A.; Soliman, M. Modelling the complex interactions between reformer and reduction furnace in a midrex-based iron plant. *Can. J. Chem. Eng.* **2012**, *90*, 1120–1141. [CrossRef]
17. KOBELCO Site. Available online: http://www.kobelco.co.jp/english/products/dri/img/flow.gif (accessed on 19 May 2018).
18. Ye, G.; Xie, D.; Qiao, W.; Grace, J.R.; Lim, C.J. Modeling of fluidized bed membrane reactors for hydrogen production from steam methane reforming with Aspen Plus. *Int. J. Hydrogen Energy* **2011**, *34*, 4755–4762. [CrossRef]
19. Tripodi, A.; Compagnoni, M.; Ramis, G.; Rossetti, I. Process simulation of hydrogen production by steam reforming of diluted bioethanol solutions: Effect of operating parameters on electrical and thermal cogeneration by using fuel cells. *Int. J. Hydrogen Energy* **2017**, *42*, 23776–23783. [CrossRef]
20. Guo, X.; Wu, S.; Xu, J.; Du, K. Reducing gas composition optimization for COREX® pre-reduction shaft furnace based on CO-H_2 mixture. *Procedia Eng.* **2011**, *15*, 4702–4706. [CrossRef]
21. Takahashi, R.; Takahashi, Y.; Yagi, J.I.; Omori, Y. Operation and simulation of pressurized shaft furnace for direct reduction. *Trans. Iron Steel Inst. Jpn.* **1986**, *26*, 765–774. [CrossRef]
22. Byron Smith, R.J.; Loganathan, M.; SHANTHA, M.S. A review of the water gas shift reaction kinetics. *Int. J. Chem. React. Eng.* **2010**, *8*. [CrossRef]
23. Grabke, H.J. Die Kinetik der Entkohlung und Aufkohlung von γ-Eisen in Methan-Wasserstoff-Gemischen. *Ber. Bunsenges. Phys. Chem.* **1965**, *69*, 409–414. [CrossRef]

Article

Revealing Grain Boundary Sliding from Textures of a Deformed Nanocrystalline Pd–Au Alloy

Laszlo S. Toth [1,2,*], Werner Skrotzki [3], Yajun Zhao [1,2], Aurimas Pukenas [3], Christian Braun [4] and Rainer Birringer [4]

1. Laboratory of Excellence on Design of Alloy Metals for Low-Mass Structures (DAMAS), Université de Lorraine, 57073 Metz, France; yajun.zhao@univ-lorraine.fr
2. Laboratoire d'Etude des Microstructures et de Mécanique des Matériaux (LEM3), Université de Lorraine, 57073 Metz, France
3. Institut für Festkörper- und Materialphysik, Technische Universität Dresden, D-01062 Dresden, Germany; skrotzki@mail.zih.tu-dresden.de (W.S.); aurimas.pukenas@tu-dresden.de (A.P.)
4. Experimentalphysik, Universität des Saarlandes, D-66041 Saarbrücken, Germany; c.braun@nano.uni-saarland.de (C.B.); ph12rb@zimbra.uni-saarland.de (R.B.)

* Correspondence: laszlo.toth@univ-lorraine.fr

Received: 29 December 2017; Accepted: 23 January 2018; Published: 25 January 2018

Abstract: Employing a recent modeling scheme for grain boundary sliding [Zhao et al. *Adv. Eng. Mater.* **2017**, doi:10.1002/adem.201700212], crystallographic textures were simulated for nanocrystalline fcc metals deformed in shear compression. It is shown that, as grain boundary sliding increases, the texture strength decreases while the signature of the texture type remains the same. Grain boundary sliding affects the texture components differently with respect to intensity and angular position. A comparison of a simulation and an experiment on a Pd–10 atom % Au alloy with a 15 nm grain size reveals that, at room temperature, the predominant deformation mode is grain boundary sliding contributing to strain by about 60%.

Keywords: Pd–10Au alloy; shear compression; texture; grain boundary sliding

---

1. Introduction

As grain size decreases, the mechanical properties of polycrystalline materials become increasingly determined by their grain boundaries. This grain-boundary-mediated plasticity dominates at grain sizes below approximately 20 nm [1], leading to "grain size softening", also called "inverse Hall–Petch behavior" [2]. Moreover, texture formation is significantly reduced [1,3]. These characteristic features indicate a kind of superplastic behavior dominated by grain-boundary-mediated deformation. How this deformation process actually takes place is not clear so far. For nanocrystalline (nc) materials, grain-boundary-mediated deformation is considered a cooperative shearing process in which several grains are involved [4]. Another approach is to assign the grain boundaries to an amorphous structure in which shear shuffling or shear transformation takes place [5]. By applying general terminology, this process is simply called grain boundary sliding (GBS) here. It should also be noted that, besides GBS, grain boundary migration also takes place, leading to grain growth due to stress-induced shear-coupling [1,6].

Recently, a new GBS model was presented and applied successfully on the texture evolution of a Pd–10 atom % Au nc alloy [7]. That model is applied in the present work using experimental results published recently on the same alloy [5] for a different strain path. By comparing experimental and simulated texture results, the amount of GBS was estimated.

2. Experimental Details

2.1. Material, Deformation, and In-Situ X-ray Microdiffraction

The material system studied was an nc Pd–10 atom % Au alloy prepared by inert-gas condensation followed by powder compaction at a pressure of 2 GPa to disk-shaped samples with a diameter of 8 mm and a thickness in the range of 0.5–1.0 mm [8]. The grain size distribution was lognormal and the texture was almost random [1]. The area-averaged grain size determined by X-ray line profile analysis was about 15 nm [1]. Note that the Pd–10 atom % Au alloy is a solid solution with a relatively high stacking fault energy of 150 mJ·m^{-2} [3].

Deformation was carried out on a small-scale specimen, the specific geometry of which is shown in Figure 1. Such a specimen is termed shear compression specimen (SCS) [9] and was cut by spark erosion from the as-prepared disk. Two parallel and oblique slits recessed on opposing faces formed the gauge section. The dimensions of the SCS were 7 mm × 0.95 mm × 0.77 mm (H × W × T). The gauge section was thinned to h = 123 μm with a width of about s = 120 μm (Figure 1a). Applying a compression stress, the SCS was forced to shear along the gauge section oriented at 45° with respect to the compression axis (Figure 1b). Since plastic deformation is confined to the gauge section [10], the SCS is ideally suited for synchrotron radiation-based transmission experiments. Such experiments were done at the high energy micro-diffraction end-station of beamline ID15A at the European Synchrotron Radiation Facility (ESRF, Grenoble, France). Mechanical testing was performed by a tension/compression device from Kammrath–Weiss (Dortmund, Germany) made up of two load plungers moving symmetrically towards the center, thus keeping the gauge section in a fixed position relative to the incoming beam. The lower plunger was replaced by a roll bearing wagon to substantially reduce the friction coefficient of the horizontal sample movement that goes along with the enforced shear deformation (Figure 1b) [10]. The applied strain rate was about 10^{-3} s^{-1}.

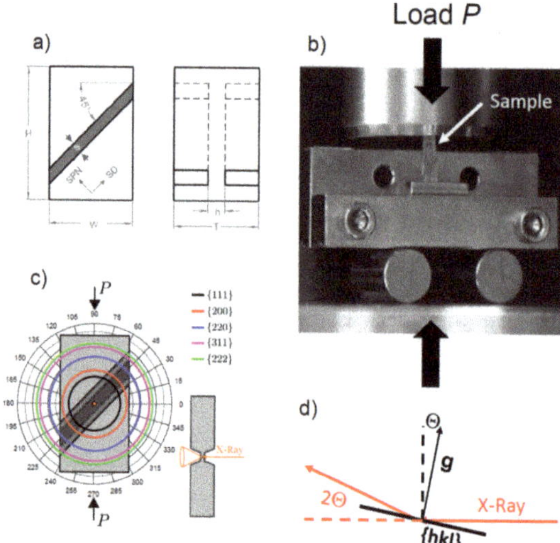

Figure 1. (**a**) Shear compression specimen in front and side view with dimensions marked; (**b**) experimental setup of shear compression; (**c**) sketch of a shear compression specimen (SCS) in front and side view serving as reference frame for obtaining polar plots. In all cases, the slit of the SCS is positioned 45° from the lower left to the upper right, and load is applied along the vertical direction. The smaller side view shows the trapezoid slit geometry. (**d**) Sketch explaining the deviation of the diffraction vector g from the position normal to synchrotron beam corresponding to an outer great circle in the pole figures.

The sample deformation was recorded by a CCD camera from Pixelink (Ottawa, ON, Canada) using an inline microscope lens for magnification and several light sources to illuminate the sample. The images were processed by digital image correlation for extracting true sample displacements to establish load–displacement curves. As the gauge section of the SCS deforms under dominant simple shear but also under an appreciable superimposed plane strain compression, there are no simple formulas to compute stress–strain curves from load–displacement data. Instead, finite element analysis was employed using Abaqus to generate load–displacement curves that best approximated the experimental curves. More details on SCS testing are given in [5,11,12].

The testing device with the installed SCS was mounted on top of a six-axis goniometer, which enables the exact placement of the sample with respect to the incoming synchrotron beam. The focused beam with a cross section of 8 µm × 20 µm was directed to the center of the thin gauge section, thus penetrating approximately 30 billion grains. The high energy X-ray radiation had a wavelength of λ = 0.0178 nm. To avoid shadowing of the scattering cones by the edges of the gauge slits, a trapezoid slit profile was used. The diffracted signal was recorded on a 2D-area detector (mar CCD), capable of recording one scan every seven seconds. The detector–sample distance was set to capture the first five Bragg reflections (Debye–Scherrer rings), so enabling a reasonable compromise between recording sufficiently high diffraction orders while still maintaining a good angular resolution. For data analysis, the 2D scans were averaged over 2°-wide polar segments to give 180 radial line scans, which were then fitted by split-Pearson-VII functions to obtain peak parameters such as the 2θ peak position, the integral intensity, and the integral peak width. For more details related to data reduction, see recent work by Lohmiller et al. [11]. Because of the short wave length of the high energy synchrotron beam, the only grains that are diffracting are those that have the diffracting vector g (lattice plane normal) almost normal to the incoming beam direction (deviation Θ = arcsin ($\lambda/2d_{(hkl)}$): (111): 2.27°, (200): 2.62°, (220): 3.64°, with the Bragg angle Θ and lattice plane spacing $d_{(hkl)}$), Figure 1c. It should also be mentioned that high monochromacy of the synchrotron radiation strongly reduces the fraction of grains in the Bragg condition. Thus, because of the high transmission, synchrotron radiation is an appropriate way to reach high grain statistics to a level reliable for measurements of weak textures, which is very important in the present texture analysis. Note that not all of the crystallographic texture was measured in the experiments because the sample was not rotated. Thus, since the specimen was not rotated, only low scattering angles of the diffraction circles were available, and this allowed for the determination of the peripheral parts of the pole figures only.

2.2. Texture Simulations

The texture development due to plastic strain was modeled using the Taylor viscoplastic polycrystal model. This uniform deformation polycrystal approach was justified in several previous studies of the present authors, which proved to be applicable to different nano-polycrystalline materials [1,6]. It was also found in those studies that the major operating slip mechanism was a <112> type partial dislocation slip on {111} planes. The usual {111}<110> slip systems were also considered in the simulations; the relative slip resistance of the two slip system families was kept at 1.5 during deformation ($\tau_0^{110}/\tau_0^{112} = 1.5$). Strain hardening was not considered because nc materials do not show significant strain hardening. The slip rates $\dot{\gamma}^{(slip)}$ were connected to the resolved shear stresses $\tau^{(slip)}$ using the strain rate sensitive slip law of Hutchinson [13]:

$$\dot{\gamma}^{(slip)} = \dot{\gamma}_0{}^{(slip)} \left(\frac{\tau^{(slip)}}{\tau_0^{(slip)}} \right)^{1/m}. \tag{1}$$

Here $\dot{\gamma}_0{}^{(slip)}$ is a reference rate and m is the strain rate sensitivity index for slip. A relatively high value was used for m (m = 1/6), which can be justified by (i) the rounding effect of the yield surface produced by viscoplastic slip [14] and (ii) the enhanced strain rate sensitivity of nano-polycrystalline materials [15].

Grain boundary sliding (GBS) was taken into account as an additional deformation mechanism of dislocation slip. For a quantitative simulation of the GBS process, the grain boundaries (GBS) were modeled by flat planes in 3D Cartesian space, by constructing a regular dodecahedron with 12 facets for each grain, with different initial positions (see more about the GBS modeling approach in [7]). The deformation generates changes in the orientation of the GBS. To account for this important effect, the directions of the GB normals were updated by the sum of the rigid body rotations of the macroscopic velocity gradient and the one corresponding to the GBS. The constitutive law for GBS was taken similar to that of the slip, with the difference that linear viscous slip was assumed for GBS:

$$\dot{\gamma}^{(GB)} = \dot{\gamma}_0{}^{(GB)} \left(\frac{\tau^{(GB)}}{\tau_0{}^{(GB)}} \right). \tag{2}$$

Although the GBS process is considered here linear viscous, considering that the crystallographic slip within the grains is non-linear (Equation (1)), the overall behavior is also non-linear. The resolved shear stress acting on the GBS was calculated from the macroscopic stress tensor $\underline{\underline{\sigma}}^{(Mac)}$:

$$\tau^{k(GB)} = \sigma_{ij}{}^{(Mac)} m_{ij}{}^{k(GB)}. \tag{3}$$

The $m_{ij}{}^{k(GB)}$ Schmid-type GB orientation tensor was constructed from the GB normal $\underline{n}^k$ and the GB-slip $\underline{g}^k$ unit vectors: $\underline{\underline{m}}^{k(GB)} = \underline{g}^k \otimes \underline{n}^k$. The latter was assumed to be in the direction of the maximum resolved shear stress acting in the GB plane.

From the Schmid orientation tensor $\underline{\underline{m}}^{k(GB)}$ and the sliding rate $\dot{\gamma}^{k(GB)}$ of each GBS system, the velocity gradient $L_{ij}^{(GBS)}$ corresponding to the GBS of a grain can be calculated by

$$L_{ij}^{(GBS)} = \sum_{k=1}^{12} m_{ij}{}^{k(GB)} \dot{\gamma}_i^{k(GBS)}. \tag{4}$$

The grain interior is subjected to the difference between the macroscopically imposed and GBS-produced gradient tensor:

$$L_{ij}^{(grain)} = L_{ij}^{(Mac.)} - L_{ij}^{(GBS)}. \tag{5}$$

This velocity gradient $L_{ij}^{(grain)}$ was used to solve the crystal plasticity problem in full constraints Taylor deformation mode to obtain the slips in each slip system and then the lattice orientation change, from which the texture evolution can be calculated. Fifty thousand grains were considered in the simulations with initially perfectly random orientation distribution. The maximum applied compressive strain was 0.27, ideally yielding a maximum shear strain of 0.54. The texture simulations were done in strain steps of 0.01.

Note that the experiments were planned for simple shear tests; however, the experimental setup led to deviation from the ideal simple shear conditions. Namely, during compression, the length of the gauge section was reduced leading to an axial compression strain component. Therefore, additional to the ideal simple shear velocity gradient $\underline{\underline{L}}^{(shear)}$, a plane strain type velocity gradient had to be considered:

$$\underline{\underline{L}}^{(Mac.)} = \underline{\underline{L}}^{(shear)} + \underline{\underline{L}}^{(comp.)}. \tag{6}$$

As the simulations were done in the experimentally defined reference system, the following macroscopic velocity gradient tensor was used:

$$L_{ij}^{(Mac.)} = |\dot{\gamma}| \begin{pmatrix} 0.5 & -0.5 & 0 \\ 0.5 & -0.5 & 0 \\ 0 & 0 & 0 \end{pmatrix} + |\dot{\varepsilon}| \begin{pmatrix} 0 & 0 & 0 \\ 0 & -1 & 0 \\ 0 & 0 & 1 \end{pmatrix}. \tag{7}$$

Here the first term on the right side is the simple shear part—operating at 45° with respect to the x–y–z reference system in the x–y plane—and the second part is the plane strain compression part, see Figure 1 for the definition of the reference system. The effect of the compressive strain is very significant on the texture evolution, so its relative proportion to the shear was varied by the parameter $C = |\dot{\varepsilon}|/|\dot{\gamma}|$. The following values were employed: C = 0, 0.15, 0.3, and 0.45. Finite element simulations from [16] suggested that the most probable value is C = 0.3. As shown below, our simulations indeed led to the best agreement with the experiments when C was 0.3.

3. Results and Discussion

Typical pole figures for two cases, simple shear and a combination of simple shear and compression (C = 0.3) plus an additional 60% of GBS, are shown in Figure 2 together with the key figure of the typical components developing during the simple shear of fcc metals (e.g., [1]). The simple shear texture at low strain is composed of all main components, in contrast to severe plastic deformation by high pressure torsion (HPT), where a dominant $B/\overline{B}$ component was found [1]. The contribution of the compressive strain leads to a tilted <110> fiber texture, while GBS leads to texture weakening. For a quantitative analysis, the intensities and positions of the maxima marked at the edge (outer great circle) of the (111) pole figure are plotted as a function of GBS and a fraction of compressive strain in Figure 3. The intensities of Peak P2 (Figure 3a) and Peak P3 (Figure 3b) evolve with increasing GBS approaching 1.0 mrd, i.e., the texture becomes randomized. At constant amount of GBS, the intensity is lowest for C = 0.3. As the absolute intensity depends on the angular Gaussian spread used (15° was imposed), the intensity ratio of Peaks P2 and P3 has been chosen instead (Figure 3c). This ratio is greater or smaller than 1.0 and changes with the amount of GBS and compressive strain. As Peaks P2 and P3 are related to the shear texture components C + A_1* and C + A_2* (see key figure of Figure 2), respectively, it is demonstrated that GBS affects the texture components differently.

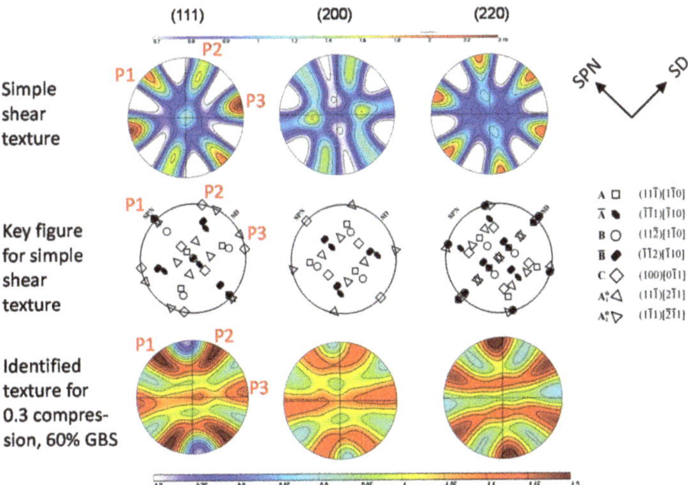

Figure 2. Pole figures simulated for simple shear and 60% grain boundary sliding (GBS) with C = 0.3. Maxima at the periphery evaluated are marked with P1, P2, and P3. The main texture components of simple shear deformed fcc metals (denoted by {hkl} parallel to the shear plane and <uvw> parallel to the shear direction) are given in the key figure.

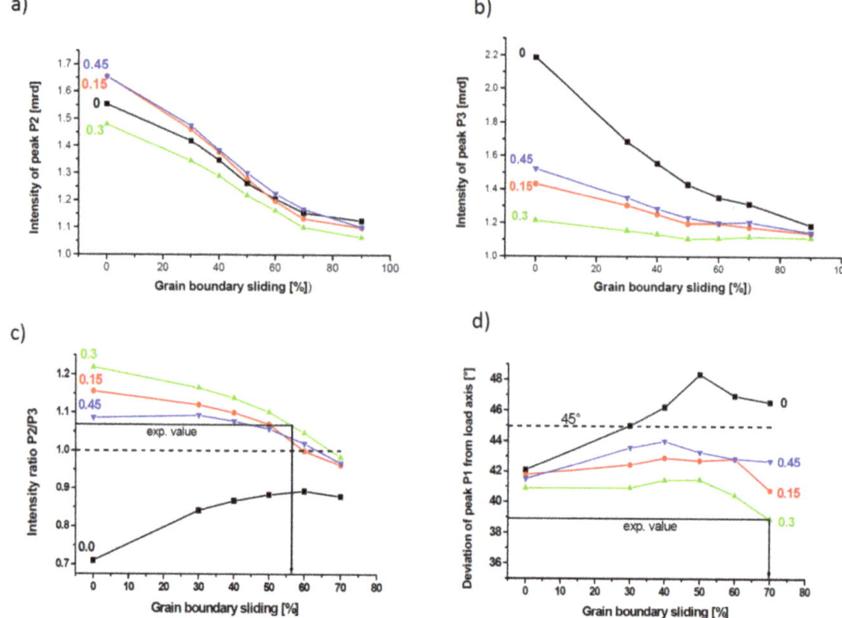

Figure 3. Dependence of the peak intensities P2 (**a**) and P3 (**b**), intensity ratio P2/P3 (**c**), and angular deviation of Peak P1 from the load axis (**d**) on the contribution of GBS to strain with the fraction of compressive strain as parameter. In (c,d), the amount of GBS is estimated from the experimentally measured parameters.

Another interesting result is that GBS changes the position of Peak P1. This peak is related to components of the so-called A-fiber, namely A_1*, A_2*, A, and $\overline{A}$, which are characterized by the alignment of the (111) plane parallel to the shear plane, i.e., its normal is positioned anticlockwise at 45° with respect to the loading direction (Figure 2). However, with an increasing amount of GBS for C = 0, an overshooting takes place (Figure 3d). Indeed, for the simple shear, the texture components tilt opposite to the shear direction with respect to their ideal positions if GBS is high [7].

The results of the simulations and those of the experiment are shown in Figure 4. This figure shows rose diagrams (Figure 4b–d) of the intensity along the Debye–Scherrer rings (Figure 4a) for a single fixed sample orientation as a function of the compressive strain applied to the gauge section in the loading direction, i.e., the intensity profile along the outer circle of the pole figures (with small deviation of Θ, see above and Figure 1), so a "partial texture" can be seen. Overlaid is the simulated best fitting intensity profile at maximum strain applied for 60% GBS and C = 0.3. There is a quite good fit for the (111) and (220) reflections, but less agreement for (200), where only the experimental minima are captured. However, this disagreement should not be overvalued, because the (200) reflection has the lowest experimental intensity due to the lowest plane multiplicity. A GBS amount of 55% and 70% can also be estimated based on data shown in Figures 3c and 3d, respectively. These values depend on the fraction of plane strain compression and are due to the semi-controlled boundary conditions of the simple shear applied to the sample. Good agreement was found for C = 0.3, which is also the value estimated from finite element simulations of the deformation process [16]. Previously, a smaller value (30%) was found in a texture simulation [1] with a larger grain size (>20 nm) due to stress-induced grain growth during large-strain HPT. This clearly shows the effect of grain size on the amount of GBS.

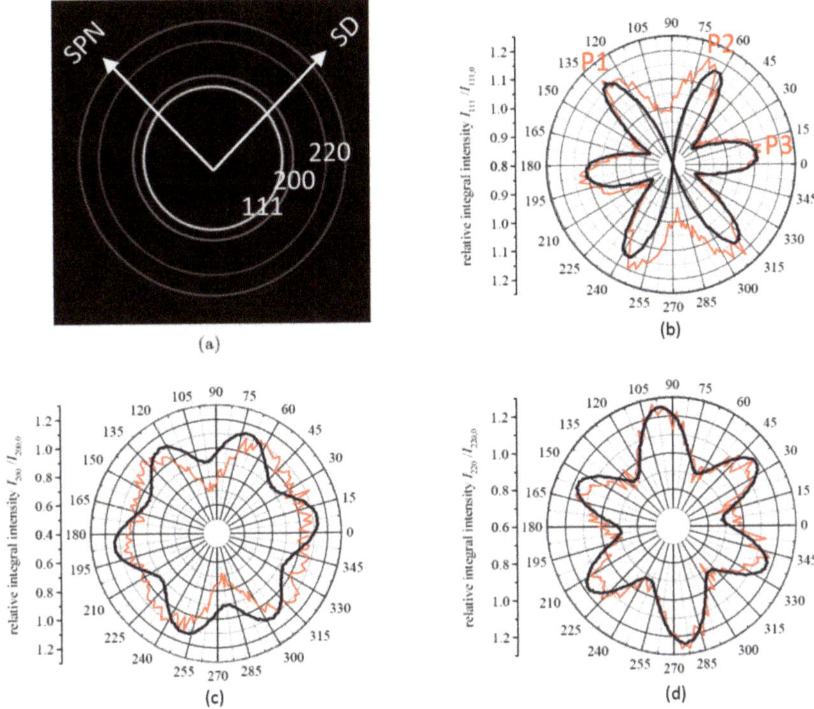

Figure 4. Experimentally measured intensity profile along the Debye–Scherrer rings (**a**) for the (111) (**b**), (200) (**c**), and (220) reflections (**d**) at a compressive strain of 0.27 applied to the gauge section in a load direction normalized with the intensity of the undeformed sample (taken from [5]). Overlaid are the simulated intensity profiles.

4. Conclusions

Based on texture simulations and a comparison with an experiment, the following conclusions can be drawn:

(i) GBS decreases the texture strength but keeps the signatures of the texture type.
(ii) GBS affects the texture components differently with respect to intensity and angular position.
(iii) The amount of GBS can be estimated from the texture evolution as a function of GBS.
(iv) In the investigated Pd–10 atom % Au alloy with a grain size of about 15 nm, GBS is the predominant deformation mode at room temperature, contributing to strain by about 60%.

Acknowledgments: The help received from Benoit BEAUSIR (Lorraine University) in the construction of the radar-type plotting is greatly appreciated. This work was supported by the French State through the program "Investment in the Future" operated by the National Research Agency (ANR) and referenced by ANR-11-LABX-0008-01 (LabEx DAMAS).

Author Contributions: For the present paper, Laszlo S. Toth performed the simulation work; Werner Skrotzki analyzed the experimental data and helped the simulation part; Yajun Zhao helped in constructing the computer code; Aurimas Pukenas, Christian Braun and Rainer Birringer performed the experimental work. The paper was written by Laszlo S. Toth and Werner Skrotzki.

Conflicts of Interest: The authors declare no conflict of interest.

References

1. Skrotzki, W.; Eschke, A.; Jóni, B.; Ungár, T.; Tóth, L.S.; Ivanisenko, Y.; Kurmanaeva, L. New experimental insight into the mechanisms of nanoplasticity. *Acta Mater.* **2013**, *61*, 7271–7284. [CrossRef]
2. Chokshi, A.; Rosen, A.; Karch, J.; Gleiter, H. On the validity of the Hall-Petch relationship in nanocrystalline materials. *Scr. Metall.* **1989**, *23*, 1679–1683. [CrossRef]
3. Ivanisenko, Y.; Skrotzki, W.; Chulist, R.; Lippmann, T.; Kurmanaeva, L. Texture development in a nanocrystalline Pd–Au alloy studied by synchrotron radiation. *Scr. Mater.* **2012**, *66*, 131–134. [CrossRef]
4. Hahn, H.; Mondal, P.; Padmanabhan, K.A. Plastic deformation of nanocrystalline materials. *Nanostruct. Mater.* **1997**, *9*, 603–606. [CrossRef]
5. Grewer, M.; Braun, C.; Deckarm, M.J.; Lohmiller, J.; Gruber, P.A.; Honkimäki, V.; Birringer, R. Anatomizing deformation mechanisms in nanocrystalline $Pd_{90}Au_{10}$. *Mech. Mater.* **2017**, *114*, 254–267. [CrossRef]
6. Li, L.; Ungár, T.; Toth, L.S.; Skrotzki, W.; Wang, Y.D.; Ren, Y.; Choo, H.; Fogarassy, Z.; Zhou, X.T.; Liaw, P.K. Shear-coupled grain growth and texture development in a nanocrystalline ni-fe alloy during cold rolling. *Met. Trans. A* **2016**, *47*, 6632–6644. [CrossRef]
7. Zhao, Y.; Toth, L.S.; Massion, R.; Skrotzki, W. Role of grain boundary sliding in texture evolution for nanoplasticity. *Adv. Eng. Mater.* **2017**, 2700212. [CrossRef]
8. Birringer, R. Nanocrystalline materials. *Mater. Sci. Eng. A* **1989**, *117*, 33–43. [CrossRef]
9. Rittel, D.; Lee, S.; Ravichandran, G. A shear-compression specimen for large strain testing. *Exp. Mech.* **2002**, *42*, 58–64. [CrossRef]
10. Ames, M.; Markmann, J.; Birringer, R. Mechanical testing via dominant shear deformation of small-sized specimen. *Mater. Sci. Eng. A* **2010**, *528*, 526–532. [CrossRef]
11. Lohmiller, J.; Grewer, M.; Braun, C.; Kobler, A.; Kübel, C.; Schüler, K.; Honkimäki, V.; Hahn, H.; Kraft, O.; Birringer, R.; et al. Untangling dislocation and grain-boundary-mediated plasticity in nanocrystalline nickel. *Acta Mater.* **2014**, *65*, 295–307. [CrossRef]
12. Ames, M.; Grewer, M.; Braun, C.; Birringer, R. Nanocrystalline metals go ductile under shear deformation. *Mater. Sci. Eng. A* **2012**, *546*, 248–257. [CrossRef]
13. Hutchinson, J.W. Bounds and self-consistent estimates for creep of polycrystalline materials. *Proc. R. Soc. A* **1976**, *348*, 101–127. [CrossRef]
14. Tóth, L.S.; Gilormini, P.; Jonas, J.J. Effect of rate sensitivity on the stability of torsion textures. *Acta Metall.* **1988**, *36*, 3077–3091. [CrossRef]
15. Meyers, M.A.; Mishra, A.; Benson, D.J. Mechanical properties of nanocrystalline materials. *Prog. Mater. Sci.* **2006**, *51*, 427–556. [CrossRef]
16. Braun, C. Plastizität von PdAu-Legierungen am unteren Ende der Nanoskala: Ein Übergang zu glasartigem Verhalten? Ph.D. Thesis, Universität des Saarlandes, Saarbrücken, Germany, April 2015. [CrossRef]

MDPI
St. Alban-Anlage 66
4052 Basel
Switzerland
Tel. +41 61 683 77 34
Fax +41 61 302 89 18
www.mdpi.com

Materials Editorial Office
E-mail: materials@mdpi.com
www.mdpi.com/journal/materials

www.ingramcontent.com/pod-product-compliance
Lightning Source LLC
LaVergne TN
LVHW071621170726
843515LV00009B/2451

* 9 7 8 3 0 3 9 3 6 1 5 8 8 *